ELECTRONIC PROPERTIES OF NOVEL NANOSTRUCTURES

Previous Proceedings in the Series of International Kirchberg Winterschools

	Year	Held in	Publisher	ISBN
XVIII	2004	Kirchberg, Austria	AIP Conf. Proceedings Vol. 723	0-7354-0204-3
XVII	2003	Kirchberg, Austria	AIP Conf. Proceedings Vol. 685	0-7354-0154-3
XVI	2002	Kirchberg, Austria	AIP Conf. Proceedings Vol. 633	0-7354-0088-1
XV	2001	Kirchberg, Austria	AIP Conf. Proceedings Vol. 591	0-7354-0033-4
XIV	2000	Kirchberg, Austria	AIP Conf. Proceedings Vol. 544	1-56396-973-4
XIII	1999	Kirchberg, Austria	AIP Conf. Proceedings Vol. 486	1-56396-900-9
XII	1998	Kirchberg, Austria	AIP Conf. Proceedings Vol. 442	1-56396-808-8

To learn more about these titles, or the AIP Conference Proceedings Series, please visit the webpage **http://proceedings.aip.org**

ELECTRONIC PROPERTIES OF NOVEL NANOSTRUCTURES

XIX International Winterschool/Euroconference on Electronic Properties of Novel Materials

Kirchberg, Tirol, Austria   12 –19 March 2005

EDITORS
Hans Kuzmany
Universität Wien, Austria

Jörg Fink
Institut für Festkörperphysik
Dresden, Germany

Michael Mehring
Universität Stuttgart, Germany

Siegmar Roth
Max-Planck-Institut für Festkörperforschung
Stuttgart, Germany

SPONSORING ORGANIZATIONS
Bundesministerium für Bildung, Wissenschaft und Kultur, Austria
Austrian Nano Initiative, FFG-Forschungförderungsgesellschaft GmbH, Austria
Verein zur Förderung der Internationalen Winterschulen in Kirchberg, Austria

Melville, New York, 2005
AIP CONFERENCE PROCEEDINGS ■ VOLUME 786

Sep/ae
phys

Editors:

Hans Kuzmany
Institut für Materialphysik
Universität Wien
Strudlhofgasse 4
A-1090 Wien
AUSTRIA
E-mail: kuzman@ap.univie.ac.at

Michael Mehring
2. Physikalisches Institut
Universität Stuttgart
Pfaffenwaldring 57
D-70550 Stuttgart
GERMANY
E-mail: m.mehring@physik.uni-stuttgart.de

Jörg Fink
IFW Dresden
Postfach 16
D-01171 Dresden
GERMANY
E-mail: j.fink@ifw-dresden.de

Siegmar Roth
Max-Planck-Institut für Festkörperforschung
Heisenbergstr. 1
D-70569 Stuttgart
GERMANY
E-mail: S.Roth@fkf.mpg.de

Authorization to photocopy items for internal or personal use, beyond the free copying permitted under the 1978 U.S. Copyright Law (see statement below), is granted by the American Institute of Physics for users registered with the Copyright Clearance Center (CCC) Transactional Reporting Service, provided that the base fee of $22.50 per copy is paid directly to CCC, 222 Rosewood Drive, Danvers, MA 01923, USA. For those organizations that have been granted a photocopy license by CCC, a separate system of payment has been arranged. The fee code for users of the Transactional Reporting Services is: 0-7354-0275-2/05/$22.50.

L.C. Catalog Card No. 2005932479
ISBN 0-7354-0275-2
ISSN 0094-243X
Printed in the United States of America

CONTENTS

FULLERENES, ENDOHEDRALS, AND FULLERIDES

CARBON NANOSTRUCTURE SYNTHESIS, PURIFICATION AND SEPARATION

CHARACTERIZATION AND PROPERTIES OF CARBON NANOTUBES AND
NANOSTRUCTURES

CHEMICAL TREATMENT OF CARBON NANOTUBES

NON-CARBONACEOUS NANOTUBES

THEORY OF NANOSTRUCTURES

ELECTRON TRANSPORT PROPERTIES

APPLICATIONS

WORKSHOP ON "APPLICATION OF MOLECULAR NANOSTRUCTURES"

PREFACE

The 19th International Winterschool on Molecular Nanostructures was held in Kirchberg/Tirol from the afternoon of March 12th until the morning of March 19th, 2005. The first Winterschool took place in March 1985 and this event represented the 20th anniversary in this series. Except for the first two, all Winterschools were arranged in the lecture rooms of the Hotel Sonnalp in Kirchberg.

The program of the Winterschool consisted of oral presentations, poster presentations, miniworkshops, and a special workshop on Applications of Molecular Nanostructures. Oral presentations of invited lecturers were scheduled for morning and evening sessions. The miniworkshops were arranged in the afternoon. Posters were presented in late evening sessions and covered various subjects in the field of molecular nanostructures.

In a special session on Monday, March 14th, the Landeshauptmann of Tirol, Dr. Dr. Herwig van Staa, presented a recognition of the 20th anniversary of the Winterschools in Kirchberg. In this session, awards were presented to the two senior participants of the IWEPNM2005, Prof. Robert Blinc and Prof. Fany Milia, as well as to Prof. Valy Vardeny. Prof. Vardeny was one of the invited speakers at the first Winterschool in 1985. This time he presented an honorary lecture on "Spin valves in organic semiconductors and single molecules."

On Tuesday, March 15th a special Workshop was arranged on the "Application of Molecular Nanostructures." This Workshop consisted of several invited lectures, a mini-workshop with contributions from industry, as well as a large number of posters dedicated to applications of molecular nanostructures.

The presentations at the IWEPNM2005 were concerned with the growth of carbon nanotubes, their optical and spectroscopic properties, quantum transport, and Luttinger liquid behavior. A large number of contributions were dedicated to applications such as sensors, nonlinear optics, or spintronics. Besides the contributions on carbon nanotubes, results were also presented for other tubes based on boron nitride, gold, TiO_2, or molybdenum. Additional topics covered functionalized fullerenes, organic semiconductors, and electronic structure of surface layers. In total, 4 tutorial lectures, 48 research lectures, and 152 posters were presented. A representative part of these presentations has been included in this conference proceedings volume.

The event benefited substantially from the support of the Universität Wien, the Austrian Nanoinitiative, and the Verein zur Förderung der Winterschulen in Kirchberg, as well as from numerous industrial sponsors. We greatly appreciate their financial contribution. Without the support from the sponsors and supporters all the engagement and enthusiasm could have been wasted.

We are very grateful to the manager of the Hotel Sonnalp, Frau Edith Mayer, and to her staff for local arrangements as well as for their patience with many special services during the meeting.

Finally, special acknowledgements go to all contributors to the IWEPNM2005 and to Viera Skákalová and Sabine Kessler for their efforts in compiling and editing the proceedings of the Winterschool 2005.

H. Kuzmany, J. Fink, M. Mehring, S. Roth
Wien, Dresden, Stuttgart 2005

Table of Previous Kirchberg Winterschools

Year	Title	Published By
2004	Electronic Properties of Synthetic Nanostructures	AlP Conference Proceedings 723 (2004)
2003	Structural and Electronic Properties of Novel Materials	AlP Conference Proceedings 685 (2003)
2002	Structural and Electronic Properties of Molecular Nanostructures	AlP Conference Proceedings 633 (2002)
2001	Electronic Properties of Molecular Nanostructures	AlP Conference Proceedings 591 (2001)
2000	Electronic Properties of Novel Materials - Molecular Nanostructures	AlP Conference Proceedings 544 (2000)
1999	Electronic Properties of Novel Materials - Science and Technology of Molecular Nanostructures	AlP Conference Proceedings 486 (1999)
1998	Electronic Properties of Novel Materials - Progress in Molecular Nanostructures	AlP Conference Proceedings 442 (1998)
1997	Molecular Nanostructures	World Scientific Publ. 1998
1996	Fullerenes and Fullerene Nanostructures	World Scientific Publ. 1996
1995	Physics and Chemistry of Fullerenes and Derivatives	World Scientific Publ. 1995
1994	Progress in Fullerene Research	World Scientific Publ. 1994
1993	Electronic Properties of Fullerenes	Springer Series in Solid State Sciences 117
1992	Electronic Properties of High-T_c Superconductors	Springer Series in Solid State Sciences 113
1991	Electronic Properties of Polymers - Orientation and Dimensionality of Conjugated Systems	Springer Series in Solid State Sciences 107
1990	Electronic Properties of High-T_c Superconductors and Related Compounds	Springer Series in Solid State Sciences 99
1989	Electronic Properties of Conjugated Polymers III - Basic Models and Applications	Springer Series in Solid State Sciences 91
1987	Electronic Properties of Conjugated Polymers	Springer Series in Solid State Sciences 76
1985	Electronic Properties of Polymers and Related Compounds	Springer Series in Solid State Sciences 63

ORGANIZER

Institut fuer Materialphysik Universitaet Wien, Austria

PATRONAGE

ELISABETH GEHRER
Bundesministerin fuer Bildung, Wissenschaft und Kultur, Austria

Magnifizenz
Univ. Prof. Dr. GEORG WINCKLER
Rektor der Universitaet Wien , Austria

EWALD HALLER
Bürgermeister von Kirchberg , Austria

SUPPORTERS

BUNDESMINISTERIUM FUR BILDUNG, WISSENSCHAFT UND KULTUR, Austria
AUSTRIAN NANO INITIATIVE, FFG-Forschungfoerderungsgesellschaft GmbH (Austrian
Research Promotion Agency), Canovagasse 7, A-1010 Vienna, Austria
VEREIN ZUR FOERDERUNG DER INTERNATIONALEN WINTERSCHULEN IN
KIRCHBERG, Austria

SPONSORS

BRUKER OPTIC GmbH, Rudolf-Planckstrasse 23, D-76275 Ettlingen, Germany
BUSSAN NANOTECH RESEARCH INSTITUTE, 8F Sumitomofudosan-Hamacho bldg., 3-42-3
Nihonbashi Hamacho, Tokyo, Japan
FUTURE CARBON, Gottlieb-Keim-Str. 60, D-95448 Bayreuth, Germany
HORIBA Jobin Yvin GmbH, Neuhofstrasse 9, D-64625 Bensheim, Germany
NANOCYL S.A., Rue de Seminaire 22, 5000 Namur, Belgium
n-TEC, PO Box 280, 1323 Høvik, Norway
PYROGRAF PRODUCTS Inc., Cedarville Ohio, USA
SAMSUNG ADVANCED INSTITUTE OF TECHNOLOGY; San 14-1, Nongseori, Giheung-eup,
Yongin-si, Gyeonggi-do, Korea
SCHAEFER TECHNOLOGIE GmbH, Moerfelder Landsstrasse 33, 63225 Langen, Germany

The financial assistance from the sponsors and the supporters is gratefully acknowledged.

FULLERENES, ENDOHEDRALS,
AND FULLERIDES

Band-like dispersion in the valence band photoemission spectra of K_6C_{60} (110) films

A.Goldoni, L. Petaccia, G. Zampieri and S. Lizzit

Sincrotrone Trieste, s.s. 14 km 163.5 in Area Science Park, 34012 Trieste, Italy

C. Cepek

Lab. TASC-INFM, s.s. 14 km 163.5 in Area Science Park, 34012 Trieste, Italy

E. Gayone, J. Wells and Ph. Hofmann

Institute for Storage Ring Facilities, University of Aarhus, 8000 Aarhus C, Denmark

ABSTRACT

Using angle resolved photoemission we show that band-like states in $K_6C_{60}(110)$ films at 40 K survive in spite of the strong electron-phonon coupling expected in fullerides that should give rise to polaron formation: the momentum dispersion of the HOMO and LUMO features agrees quite well with band structure calculations.

Introduction

The band structure in solid C_{60} compounds is far from being understood due to the experimental and theoretical challenges involved. Orientational disorder, temperature dependent phase transitions, very small Brillouin zone, weak molecular interaction and strong electron-phonon coupling have combined effects that are difficult (if not impossible) to disentangle and all together imply a level of complexity that, at best, has made difficult the application of theoretical models and the elucidation of experimental results. For example, several calculations [1-3] and recent experiments [4] have demonstrated that band dispersion and bandwidth in C_{60} compounds strongly depend on the molecular orientation: it is, therefore, not obvious that extended electronic states will survive in C_{60} crystals in presence of the inherent orientational disorder, which may disrupt the band formation or strongly renormalize the bandwidth compared to band structure calculations. In addition, as vibrational broadening plays a major role in determining the lineshape of gas phase photoemission spectral features [5], we should expect *a priori* a similar significant contribution also in the solid-state spectra. Indeed, the strong similarity between gas-phase and solid C_{60} photoemission spectra seems to support this hypothesis [6] and suggests that polaronic effects may be important in two-dimensional (2D) and three-dimensional (3D) fullerene structures. Brühwiler et al. [6] claimed that in C_{60} 2D-islands on graphite vibronic coupling accounts for most of the observed bandwidth, while the band dispersion is reduced to less than 70 meV by polaronic effects. These authors conclude that a similar phenomenon and band dispersion reassessment should persist for the 3D solid.

Conflicting pictures have been obtained for undoped C_{60} crystals and films. The first angle-resolved photoemission experiment on a cleaved C_{60} single crystal [7] showed very little or no band dispersion pointing, therefore, on the importance of the above vibronic effects on the band structure of the solid: the width and shape of the measured valence band photoemission features are dominated by phonon broadening and do not reflect the intrinsic bandwidth. Subsequent angle resolved photoemission experiments [8, 9] performed at lower photon energies on C_{60} thick films, were successful in observing intensity modulations and lineshape changes in the HOMO and HOMO-1 features, which were interpreted as band dispersion of both occupied and unoccupied (final states) bands.

CP786, *Electronic Properties of Novel Nanostructures*, edited by H. Kuzmany, J. Fink, M. Mehring, and S. Roth
© 2005 American Institute of Physics 0-7354-0275-2/05/$22.50

These low energy photoemission experiments were taken as evidence for the existence of band-like electronic states in *fcc* C_{60} at room temperature, despite the presence of fast orientational disorder, that mainly determine the width of the photoemission peak without the need of invoking vibrational broadening.

On the other hand, there is the indisputable evidence of important vibrational contributions to the gas-phase photoemission spectra and the strong similarity between the HOMO lineshape of these spectra and the spectra of 2D-C_{60} islands on graphite. Moreover, Brühwiler and co-workers [6] suggest that in 3D C_{60} solid the top most molecular layer contribution to the photoemission spectrum may be shifted with respect to the second molecular layer (and bulk) because of different screening effects, which could explain the width and lineshape of the HOMO and HOMO-1 peaks.

Angle resolved photoemission data on A_6C_{60} (A= K, Rb, Cs) compounds may help to shine some light on the electronic spectra of C_{60}. Several studies indicate that at saturation doping (6 electrons/molecule) the Rb, Cs and K fullerides form band insulators [10-12]. The crystal structure is *bcc*, the LUMO-derived bands are completely filled and there is no room for molecular motions. A_6C_{60} is therefore a very lucky case in which merohedral disorder and electron-electron interactions do not play a role in the photoemission spectra.

Experimental

Here we report on the angle resolved photoemission spectra of K_6C_{60}. An ordered K_6C_{60}(110) thick film was grown on Ag(100) in the ultra-high-vacuum experimental chamber of the SGM-3 beamline at ISA synchrotron facility in Aarhus (Denmark) [13].

In order to reduce as much as possible phonon effects, measurements of the K_6C_{60} band dispersion were performed at 40 K. The used photon energy was 25 eV, with energy and angular resolution of 25 meV and 0.5°, respectively. Angular scans were performed by moving the analyzer (step 0.25°) with the sample position fixed at 45° from the light beam. At the photon energy we used, considering an inelastic mean free path of 7 Å for the photoelectrons [14], the integration in $K_\perp$ is about 20% of the Γ-N direction.

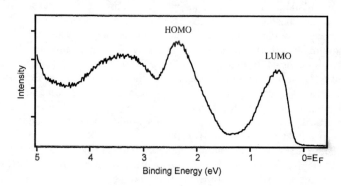

Fig. 1: K_6C_{60} photoemission spectrum.

Results and Discussion

In Fig. 1 is reported a typical photoemission spectrum of K_6C_{60} measured at normal emission. The 0 eV binding energy corresponds to the position of the Fermi level as measured on the clean Ag(100). Here, the nomenclature of the two main features visible in the spectra, Highest Occupied Molecular Orbital (HOMO) and Lowest Unoccupied

4

Molecular Orbital (LUMO), refers to the orbitals of the undoped C_{60} molecule. Obviously, in K_6C_{60} the HOMO- and LUMO-derived bands are both fully occupied and below the Fermi level (band insulator).

Looking at the available band calculations for K_6C_{60} [15, 16], we note that both LUMO-derived and HOMO-derived bands are quite grouped forming bundles (made of three bands and five bands, respectively) whose energy spread is small compared to the band dispersion. Under these conditions, there is the possibility to see at least a k-dependence of the measured band centroids. However, the coupling with on-ball vibrational modes should be similar to the case of pure C_{60} and, therefore, a similar vibrational broadening and possible polaronic renormalization of the band dispersion should be expected [6]. As recently pointed out by Wehrli et al. [17], strong coupling with K^+ optic modes, which is not screened in insulating compounds like K_6C_{60}, is mainly responsible of the broad width of the HOMO and LUMO peaks observed in photoemission, and the ground state of a hole created by photoemission in these bands should be polaronic.

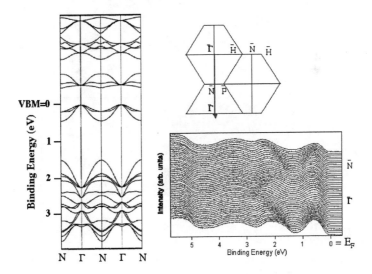

Fig. 2: Calculated bulk band structure repeated along the Γ-N direction (left panel) [15] and measured band dispersion along the $\bar{\Gamma}$ -$\bar{N}$ direction (right panel). The scanning direction on the surface Brillouin zone is shown on the top

The measured angular dependence along two main symmetry directions of the bcc(110) surface, $\bar{\Gamma}$ -$\bar{N}$ and $\bar{\Gamma}$ -$\bar{H}$ - $\bar{N}$, are reported Fig. 2 and Fig. 3 respectively. In both cases there is a clear dispersion of the LUMO-derived and HOMO-derived bands and the dispersion is different for the two symmetry directions. The LUMO dispersion is about 0.5 eV while the HOMO bands show a larger dispersion (~1 eV). These numbers are comparable to the estimation based on one-electron band calculations, although the calculations are made for bulk bands (the photoemission spectra should have a sizable surface contribution, for which a slightly reduced dispersion should be expected because of the reduced coordination number of the C_{60} molecules). In spite of the fact that $K_\perp$ is actually undetermined, by comparing the calculated bulk dispersion along the Γ-N with our measurements, we note that our choice of the photon energy for this bcc(110) surface is particularly fortunate since it appears that $\overline{\Gamma} \sim \Gamma$ and $\overline{N} \sim N$.

These data indicate that in K_6C_{60} at low temperature (40 K) there is no evidence of polaronic renormalization of the band dispersion in photoemission spectra: the measured angle-resolved photoemission spectra reflect the K_6C_{60} electronic structure and show a good agreement with the one-electron band calculation results. However, looking at the calculations, we can see that at $\Gamma \sim \overline{\Gamma}$ the bandwidth due to the presence of the three LUMO bands should be very sharp and, even considering our integration in $K_\perp$, in the worst case we must expect a width of the photoemission feature of 0.3 eV. The LUMO width observed in the photoemission spectra is a factor 1.5-2 larger than expected, which should be imputed to phonon broadening according to Wehrli et al. [17].

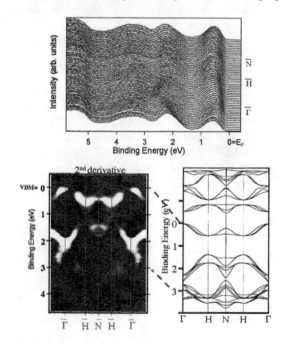

Fig. 3: Calculated bulk band structure [15] repeated along the Γ-H-N direction (bottom left panel) and measured band dispersion along the $\overline{\Gamma}$ $-\overline{H} - \overline{N}$ direction (top panel). The second derivative of the spectra symmetrized around $\overline{N}$ is also shown as intensity plot in gray scale (white=max, black=min).

The above observations may be considered in conflict, but they can be reconciled by a recent work of Rösch and Gunnassonn [18]. Using an adiabatic approximation, Rösch and Gunnassonn have developed a method for calculating the photoemission spectra from systems with strong coupling of doped carriers to phonons. This method is particularly simple for systems where the electron-phonon coupling can be neglected in the initial state, e.g. the undoped t-J model. According to the authors, their calculations naturally explains why the electron-phonon coupling just leads to a broadening of spectra calculated neglecting electron-phonon coupling, without changing the dispersion. This is in what we observed in our angle-resolved photoemission data on K_6C_{60}, supporting the interpretation in terms of band dispersion consistent with band calculations with the main influence of the electron-phonon interaction on the width of the photoemission features.

Conclusions

Low temperature angle-resolved photoemission spectra of $K_6C_{60}(110)$ films demonstrate that the HOMO and LUMO features show momentum dispersion comparable to band calculations, at variance with the case of undoped C_{60} and in spite of the strong electron-phonon coupling expected in fullerides which should favor a polaronic ground state for the photo-hole.

Acknowledgments

This work has been financed via EU contracts HPRI-CT-2001-00122 and RII3-CT-2004-506008

References

[1] O. Gunnarsson et al., Phys. Rev. Lett. 67, 3002 (1991); N. Laouini et al., Phys. Rev. B 51, 17446 (1995).
[2] M.P. Gelfand and J.P. Lu, Phys. Rev. Lett. 68, 1050 (1992).
[3] R.P. Gupta and M. Gupta, Phys. Rev. B 47, 11635 (1993).
[4] V. Brouet et al., Phys. Rev. Lett. 93, 197601 (2004).
[5] D.L. Lichtenberger et al., Chem. Phys. Lett. 176, 203 (1991).
[6] P.A. Brühwiler et al., Chem. Phys. Lett. 279, 85 (1997).
[7] J. Wu et al., Phisica C 197, 252 (1991).
[8] G. Gensterblum et al., Phys. Rev. B 48, 14756 (1993).
[9] P.J. Benning et al., Phys. Rev. B 50, 11239 (1994).
[10] R. Tycko et al., Science 253, 884 (1991).
[11] P.J. Benning et al., Phys, Rev. B 47, 13843 (1993).
[12] P.J. Benning et al., Phys, Rev. B 48, 9086 (1993).
[13] S.V. Hoffmann et al., Nucl. Instrum. Methods Phys. Res. A 523, 441 (2004).
[14] A. Goldoni et al., Phys. Rev. Lett. 87, 076401 (2001).
[15] S.C. Erwin and M.R. Pederson, Phys. Rev. Lett. 67, 1610 (1991)
[16] M. Marangolo et al., Phys. Rev. B 58, 7594 (1998).
[17] S. Wehrli et al., Phys. Rev. B 70, 233412 (2004).
[18] O. Rösch and O. Gunnarsson, Eur. Phys. J. B 43, 11 (2005).

Spatially Dependent Inelastic Tunneling in Gd@C$_{82}$

M. Grobis,[1] K. H. Khoo,[1] R. Yamachika,[1] Xinghua Lu,[1] K. Nagaoka,[1,*] Steven G. Louie,[1] H. Kato,[2] H. Shinohara,[2] and M. F. Crommie[1,†]

[1] Department of Physics, University of California at Berkeley, Berkeley California 94720-7300, USA and Materials Sciences Division, Lawrence Berkeley Laboratory, California 94720-7300, USA
[2] Department of Chemistry and Institute for Advanced research, Nagoya University, Nagoya 464-8602, Japan
* Current Address: Nanomaterials Laboratory, National Institute for Materials Science, Tsukuba, 305-0044, Japan
† Corresponding author. E-mail: crommie@socrates.berkeley.edu

Abstract. We have measured the elastic and inelastic tunneling properties of isolated Gd@C$_{82}$ molecules on the Ag(001) using cryogenic scanning tunneling microcopy and spectroscopy. We observe several inelastic excitations in our tunneling spectra, and the dominant inelastic channel is strongly spatially localized. Density functional theory calculations show that the observed localization in inelastic tunneling arises from localization in the electron-phonon coupling to a C$_{82}$ cage phonon mode.

Keywords: Endohedral Metallofullerenes, Scanning Tunneling Microscopy, Inelastic Tunneling
PACS: 73.63.-b, 63.22.+m, 68.37.Ef

INTRODUCTION

Endohedral fullerenes are exciting candidates for future nanotechnological devices due to their electronic and magnetic flexibility.[1] Transport properties in such devices are expected to be strongly influenced by the inelastic excitations of molecular systems.[2,3] Coupling between phonons and electrons, for example, can lead to creation of additional conductance channels in low-energy tunneling through single molecules.[2,4] Here we present a scanning tunneling spectroscopy study aimed at understanding the spatial dependence of inelastic tunneling in a single endohedral fullerene molecule. We probed isolated Gd@C$_{82}$ molecules residing on the Ag(001) surface and found several inelastic excitations. We found that tunneling through the dominant inelastic channel is spatially localized to a particular region on the molecule. Using density functional theory we show that the observed localization in inelastic tunneling arises from localization in the electron-phonon coupling to a C$_{82}$ cage phonon mode. We further find that the spatial localization of electron-phonon coupling shows no correlation with the spatial distribution of the constituent electronic wave function and phonon atomic displacements.

CP786, *Electronic Properties of Novel Nanostructures*, edited by H. Kuzmany, J. Fink, M. Mehring, and S. Roth
© 2005 American Institute of Physics 0-7354-0275-2/05/$22.50

EXPERIMENT

Our experiments were conducted using a homebuilt ultrahigh vacuum (UHV) STM with a PtIr tip. The single-crystal Ag(001) substrate was cleaned in UHV, cooled to ~80K, and then dosed with Gd@C$_{82}$ before being cooled to 7K in the STM stage. dI/dV spectra and images were measured through lock-in detection of the ac tunneling current driven by a 450Hz, 1-10mV (rms) signal added to the junction bias under open-loop conditions (bias voltage here is defined as the sample potential referenced to the tip). Over fifty Gd@C$_{82}$ molecules were examined in this study. We deposited an additional dilute coverage of C$_{60}$ as a reference molecule for determining the quality of the STM tip. All data were acquired at 7K.

The electronic structure of Gd@C$_{82}$ was probed by performing dI/dV spectroscopy at various points on the molecule. Figure 1(a) shows the wide bias Gd@C$_{82}$ spectrum (black curve) which has a series of high amplitude molecular resonances located between 0.5 V and 2 V. Spectra performed at smaller tip-sample separations (gray curve, $\Delta z = 3.5$Å) reveal two more resonance at -0.85 ± 0.2 V and 0.10 ± 0.02V.[5]

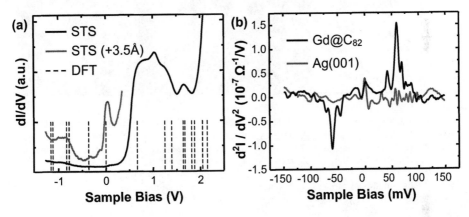

FIGURE 1. (a) Experimental STS spectra (solid lines) performed on an isolated Gd@C$_{82}$ molecule on the Ag(001) surface. The solid black curve is a spatial average of 30 spectra taken on a grid covering the molecule (tip stabilized at 2.0V, 0.3nA). The gray curve (marked +3.5Å) was obtained at the spatial maximum of the 0.1V resonance, after moving the tip 3.5Å closer to the molecule. The dashed lines show the calculated energies of free C$_{82}^{4-}$ orbitals. (b) Experimental d^{2}I/dV^2 spectra taken on (black) and off (gray) a Gd@C$_{82}$ molecule. The on-molecule spectrum was performed at the spatial maximum of the 60mV inelastic channel. The tip height was stabilized at 0.15V, 1.5nA in both spectra.

Vibrational excitations of individual Gd@C$_{82}$ molecules were examined by measuring d^{2}I/dV^2 spectra at low bias (dI/dV was numerically differentiated). Inelastic channels appear as pairs of anti-symmetric peaks in d^{2}I/dV^2, shown in Fig. 1(b). Gd@C$_{82}$ exhibits at least three well-resolved inelastic tunneling channels, which are not seen when tunneling into the bare Ag(001) surface. The dominant Gd@C$_{82}$ inelastic channel corresponds to an excitation of 60meV (±1 meV).

The Gd@C$_{82}$ inelastic signals show large spatial inhomogeneity and are detectable only in a localized region of the molecule. This inhomogeneity can be seen in spatial d^{2}I/dV^2 map of the dominant inelastic channel (60mV), shown in Fig. 2. The largest changes in differential conductance (corresponding to 25% jumps in dI/dV) are localized to a small region in the upper half of the molecule. Extraneous contributions to this image, such as cross-talk from electronic and topographic structure,[5] are minimal as evidenced by the nearly featureless d^2I/dV^2 map taken at 80mV, an electronically similar energy range that does not show any significant features in d^2I/dV^2.

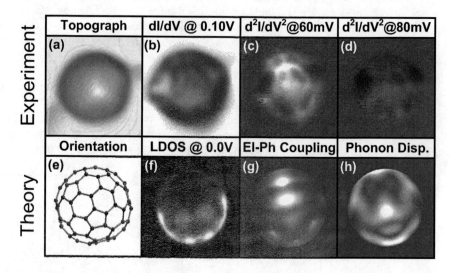

FIIGURE 2. Top Row: Experimental STM and STS images (35Å X 35Å) of an isolated Gd@C$_{82}$ molecule. (a) Topograph (V = 2.0, I = 0.3nA). (b) dI/dV map at 0.1V. (c) d^2I/dV^2 map at 60mV. (d) d^2I/dV^2 map at 80mV. **Bottom Row:** Theoretical modeling of isolated Gd@C$_{82}$. (e) Best fit molecular orientation of the Gd@C$_{82}$ molecule seen in (a)-(d). (f) Simulated dI/dV map of C$_{82}$ state at E$_F$. (g) Simulated spatial map of inelastic tunneling due to the dominant phonon mode at 52meV. (h) Spatial map of the amplitude of bare atomic displacements of the 52 meV phonon mode (broadened by 1.0Å). The images in (f) and (g) were calculated over a simulated tip trajectory, as in Ref. 5.

THEORY

In order to explain the localization of inelastic tunneling seen in our experiment, we first consider how an inelastic tunneling signal arises from the vibrational and electronic structure of Gd@C$_{82}$. Examination of the coupling between phonon modes and electronic states reveals that the spatially averaged inelastic tunneling signal can be expressed as,[6,7]

$$\Delta dI/dV\Big|_{eV=\hbar\omega} \sim \left| \sum_i \frac{dE_{E_F}}{dQ_i} \frac{1}{\sqrt{\omega}} \right|^2 \tag{1}$$

Here, Q_i is the displacement of the ith atom along the canonical phonon coordinate. If the electron is injected locally into the molecule, as in the case of STM tunneling, the spatial dependence of the inelastic tunneling signal varies as [6,7]

10

$$\Delta dI/dV\big|_{eV=\hbar\omega}(r_0) \sim \left| \sum_i \left\langle \frac{d\psi_{E_F}}{dQ_i} \bigg| r_0 \right\rangle \frac{1}{\sqrt{\omega}} \right|^2 \qquad (2)$$

We used *ab-initio* density functional theory computations to evaluate Eq. (1) and (2) for the case of Gd@C_{82}. The details of our calculation method are presented in Refs. 7 and 8. For simplicity we only considered tunneling into a bare C_{82} cage and incorporated the effects of the inner Gd atom and Ag(001) substrate by treating the C_{82} charge state as a free parameter. The best fit C_{82} charge state of -4|e| was determined by comparing experimental and theoretical spectra (shown in Fig. 1(a)), dI/dV maps, and electronic state symmetries. The computed C_{82} phonon modes are shown in Fig. 3, along with the simulated inelastic tunneling spectrum. The theoretical and experimental inelastic tunneling spectra are in reasonable agreement and indicate the cage mode at 52 meV is responsible for the dominant inelastic channel in our experiment.

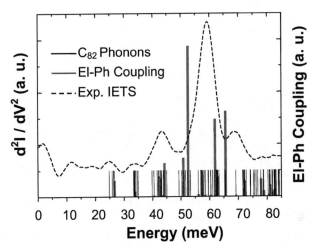

FIGURE 3. Theoretical vibrational spectrum for an isolated C_{82} molecule charged with four electrons is shown as black vertical lines. Gray vertical lines show the theoretical electron-phonon coupling between each phonon mode and the C_{82} electronic state at E_F. Dotted curve is the experimental inelastic tunneling signal.

The calculated spatial dependence of inelastic tunneling through the 52meV phonon mode is shown in Fig. 2(g). The molecular orientation used in calculating this image was deduced from comparing all experimental and theoretical dI/dV plots. In the best fit orientation (Fig. 2(e)), the Gd molecule resides close to the Ag interface. The theoretical inelastic tunneling plot shows a striking localization of the inelastic tunneling signal to the upper half of the molecule, in good agreement with the experimental plot (Fig. 2(c)). We note that some of the deviations between the theoretical experiment inelastic tunneling data are likely to originate from the inner Gd atom and underlying Ag(001) substrate,[9] which we ignored in our calculation. However, these effects are expected to have minimal impact on the qualitative features seen in our experiment since they should not strongly change the C_{82} phonon modes and electronic states responsible for the observed behavior.[10]

DISCUSSION

DFT calculation allows us to compare the spatial distribution of the inelastic tunneling signal with the spatial distribution of the electronic and vibrational states from which it originates. As seen in Figs. 2(f), (g), and (h), the spatial distribution of the electronic state and phonon amplitude show little, if any, correlation with the local electron-phonon coupling (i.e., the source of the inelastic tunneling signal). This behavior stems from the fact that inelastic tunneling is determined not by the amplitude of electronic states and phonon modes, but by the *rate* at which electronic wave functions are deformed by phonon displacements, as dictated by Eq. (3). Maxima in $d\Psi/dQ$ need not be correlated with maxima in Ψ or Q, and this interplay between electronic wave functions and phonon dynamics can produce surprising localization in the inelastic tunneling, as seen in our study.

ACKNOWLEDGMENTS

This work was supported in part by NSF Grant Nos. DMR04-39768 and EIA-0205641 and by the Director, Office of Energy Research, Office of Basic Energy Science, Division of Material Sciences and Engineering, U.S. Department of Energy under contract No. DE-AC03-76SF0098. Computational resources have been provided by DOE at the National Energy Research Scientific Computing Center.

REFERENCES

1. H. Shinohara, Reports on Progress in Physics **63** (6), 843 (2000); J. Twamley, Physical Review A **67** (5), 052318 (2003).
2. B. C. Stipe, M. A. Rezaei, and W. Ho, Science **280** (5370), 1732 (1998).
3. H. Park, J. Park, A. K. L. Lim et al., Nature **407** (6800), 57 (2000).
4. Paul K. Hansma, *Tunneling spectroscopy : capabilities, applications, and new techniques*. (Plenum Press, New York, 1982); J. I. Pascual, J. Gomez-Herrero, D. Sanchez-Portal et al., Journal of Chemical Physics **117** (21), 9531 (2002).
5. X. H. Lu, M. Grobis, K. H. Khoo et al., Physical Review Letters **90** (9), 096802 (2003).
6. N. Mingo and K. Makoshi, Physical Review Letters **84** (16), 3694 (2000); N. Lorente and M. Persson, Physical Review Letters **85** (14), 2997 (2000); N. Lorente, M. Persson, L. J. Lauhon et al., Physical Review Letters **86** (12), 2593 (2001).
7. M. Grobis, K. H. Khoo, R. Yamachika et al., Physical Review Letters **94**, 136802 (2005).
8. X. H. Lu, M. Grobis, K. H. Khoo et al., Physical Review B **70** (11), 115418 (2004).
9. B. Kessler, A. Bringer, S. Cramm et al., Physical Review Letters **79** (12), 2289 (1997); J. Kirtley and P. K. Hansma, Physical Review B **13** (7), 2910 (1976).
10. J. Lu, X. W. Zhang, X. G. Zhao et al., Chemical Physics Letters **332** (3-4), 219 (2000); L. Senapati, J. Schrier, and K. B. Whaley, Nano Letters **4** (11), 2073 (2004); M. Krause, P. Kuran, U. Kirbach et al., Carbon **37** (1), 113 (1999).

Preparation, surface characteristics and electrochemical properties of electrophoretically deposited C_{60} films

Wlodzimierz Kutner[a,b]*, Piotr Pieta[a], Robert Nowakowski[a],
Janusz W. Sobczak[a], Zbigniew Kaszkur[a],
Amy Lea McCarty[c], and Francis D'Souza[c,]*

[a] Institute of Physical Chemistry, Polish Academy of Sciences, 44/52 Kasprzaka, 01-224 Warsaw, Poland
[b] Cardinal Stefan Wyszynski University in Warsaw, Faculty of Mathematics and Natural Sciences,
School of Science, Dewajtis 5, 01-815 Warsaw, Poland
[c] Department of Chemistry, Wichita State University, 1845 Fairmount, Wichita, KS 67260-0051, USA

Abstract. Thin fullerene films of controlled roughness were electrophoretically deposited from C_{60} suspensions formed in mixed toluene-ethanol solutions. Mass of the deposited films, determined by piezoelectric microgravimetry (PM) with the use of an electrochemical quartz crystal microbalance, exponentially increased with time. Size of the AFM imaged C_{60} grains in the films depended both on time of C_{60} aggregation in bulk solution prior to deposition and strength of the electric field applied. In the accessible potential range, cyclic voltammetry (CV) curves for the films in 0.1 M (TBA)PF_6, in acetonitrile, featured four main cathodic peaks formed during the negative potential excursion. These peaks corresponded to four one-electron reductions. Simultaneously recorded PM and CV curves showed an overall mass decrease, corresponding to stepwise C_{60} electroreduction and the complete dissolution of the C_{60}^{3-} film. The CV, XPS and XRD analyses indicated the film swelling and reversible ingress of both TBA$^+$ counter- and PF_6^- co-ion into the C_{60}^- film.

INTRODUCTION

Size of C_{60} crystallites in thin fullerene films, film porosity, possible solvent and electrolyte entrapment as well as electrochemical properties depend upon procedure used for deposition of these films [1, 2]. The film topography vary from molecularly smooth to highly porous for the vapor [3] and electrophoretic [4-7] deposition, respectively. So far, topography of surface-developed fullerene films can be directly controlled only by electrophoretic deposition [4-7]. In this procedure a C_{60} film is deposited from suspension of C_{60} aggregates in mixed polar and non-polar solvent solutions. First, C_{60} nanoclusters are allowed to grow in bulk solution and, then, constant electric field is applied. In effect, fullerene aggregates gain partial negative charge and migrate to the surface of a target positive electrode depositing onto its surface. Apparently, the stronger electric field the lower is the film porosity [4].

Here we show that the time allowed for C_{60} aggregation in bulk solution prior to electrophoretic deposition can also be effectively used to control film topography.

CP786, *Electronic Properties of Novel Nanostructures*, edited by H. Kuzmany, J. Fink, M. Mehring, and S. Roth
© 2005 American Institute of Physics 0-7354-0275-2/05/$22.50

EXPERIMENTAL

Chemicals. C_{60} (99.5% purity) was from M.E.R. Corp. (Tucson AZ, USA). Toluene (pro analysis) of Chempur (Piekary Slaskie, Poland) and acetonitrile ($\geq$99.5% purity, $H_2O \leq 0.001$) of Fluka were deaerated before measurements by a solvent-saturated argon purge. Line ethanol (96.3% by weight in rectified ethyl alcohol and 3.7% by weight in diethyl ether) was from Linegal Chemicals Co., Ltd. (Warsaw, Poland). Tetra(n-butyl)ammonium hexafluorophosphate, $(TBA)PF_6$, (puriss, electrochemical grade, $\geq$99.0%) was from Fluka. Methylene chloride (pro analysis) was from Chempur (Piekary Slaskie, Poland). All chemicals were used as received.

Instrumentation and procedures. For electrophoretic deposition of C_{60} films, 1×1 cm^2 Pt tab and 5 mm diameter Pt-quartz film electrode (separated by 4 mm), served as the auxiliary (negative), and working (grounded) electrode, respectively. A sample of 0.1 mM C_{60} in toluene was added to line ethanol to reach a toluene-to-ethanol ratio of 1 : 10 ($v : v$). Electrophoretic deposition of C_{60} was monitored by piezoelectric microgravimetry with the use of an electrochemical quartz crystal microbalance EQCM 5710 of the Institute of Physical Chemistry (Warsaw, Poland). This instrument allows for simultaneous measurement of the current, resonant frequency, and dynamic resistance changes of a 10-MHz quartz resonator vs. potential or time. An AUTOLAB computerized electrochemistry system of Eco Chemie (Utrecht, The Netherlands) was used for CV measurements. A glass three-electrode conical cell was used with a Pt/quartz, Pt tab and Ag wire serving as the working, auxiliary, and pseudo-reference electrode, respectively.

The AFM imaging was performed on the TMX 2000 Discoverer microscope of TopoMetrix (Santa Clara CA, USA).

The XPS analysis was accomplished with an ESCALAB-210 spectrometer of VG Scientific (East Grinstead, UK) using Al K_α ($h\nu$ = 1486.6 eV) X-ray radiation.

The XRD patterns were recorded with a Siemens D5000 powder diffractometer of Bruker AXS GmbH (Karlsruhe, Germany) with a sealed-off Cu tube, which provided K_α X-ray radiation of λ = 0.154184 nm.

All experiments were performed at ambient temperature (22 $\pm$ 1)° C.

RESULTS AND DISCUSSION

First, electrophoretic deposition of C_{60} films was investigated by PM. Then, the resulting films were imaged by AFM and, next, their electrochemical properties were investigated by simultaneous CV and PM measurements. Finally, composition and structure of the films was determined by the XPS and XRD analysis, respectively.

After mixing a sample of C_{60} toluene solution and line ethanol, a C_{60} suspension was allowed to grow at an open circuit. Then, dc voltage was switched on and, after ca. 2 min, a C_{60} film started to deposit, as manifested by the measured resonant frequency decrease with time. This frequency decrease was recalculated into mass gain with the Sauerbrey equation (Fig. 1). The mass of the film increased exponentially with time.

The AFM imaging revealed presence of spherical C_{60} clusters in films (Fig. 2) of 200-250 nm diameter obtained under all deposition conditions. These clusters were welded to form larger aggregates. For 4-min deposition, the aggregate diameter was 600-700 and 700-1500 nm at 50 and 100 V dc, respectively. Films were more porous for 4- rather than 12-min aggregation time and 50 rather than 100 V applied.

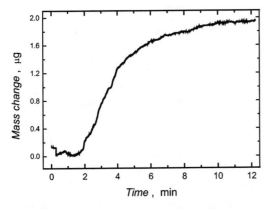

FIGURE 1. Change of mean mass with time during electrophoretic deposition of C_{60} films on 5-MHz Pt-quartz resonators from mixed solvent solutions at the toluene-to-ethanol ratio of 1 : 10 ($v : v$).

FIGURE 2. AFM image of an electrophoretically deposited for 2 min at 50 V C_{60} film on a HOPG electrode after 12-min aggregation from a mixed solvent solution of the toluene-to-ethanol ratio of 1 : 10 ($v : v$).

Highly reduced C_{60}^{n-} ($n \geq 3$) films readily dissolved, as manifested by abrupt frequency increase (Fig. 3). In order to unravel the electrochemical properties, a limited negative reversal potential was imposed such that only the first two cathodic and related two anodic CV peaks, corresponding to the C_{60}/C_{60}^{-} and C_{60}^{-}/C_{60}^{2-} redox couples, were developed. Electroreduction and subsequent electro-oxidation of the films under multi-scan CV conditions, accompanied by the counter ion dynamic equilibrium, caused changes in film swelling with the acetonitrile solvent. Simultaneous measurement of the current, frequency change and dynamic resistance under multi-scan CV conditions showed that these changes affected visco-elastic properties of the film (not shown).

From the slope of tangent of the raising portion of the frequency change vs. charge curve (Fig. 4), characteristic for the foot of the first cathodic peak, apparent electrochemical equivalent of the counter ion entering the film was determined (by invoking the Sauerbrey equation and Faraday law) as being ca. 588 g/mole. By considering the molecular weights of $TBA^{+} \approx 242$ and $PF_{6}^{-} \approx 145$, presumably, one TBA^{+} counter ion and, additionally, one TBA^{+}, PF_{6}^{-} ion pair enters the film per one C_{60}^{-} anion generated.

From the relative integrated intensities of the respective C 1s, N 1s, F 1s, and P 2p core level spectra a mole ratio of $C_{60} : TBA^{+} : PF_{6}^{-} \approx 1 : 2 : 1$ was determined by XPS for the C_{60}^{-} film indicating that the charge of each C_{60}^{-} anion was compensated by one TBA^{+} cation and, moreover, one pair of TBA^{+} and PF_{6}^{-} ions entered the film per one molecule of C_{60}. Furthermore, neither TBA^{+} nor PF_{6}^{-} was detected by XPS in the

15

electro-oxidized film. These results were confirmed by the powder XRD analysis of the initial C_{60} film, C_{60}^- film, and subsequently electro-oxidized C_{60} film. That is, crystals of (TBA)PF$_6$ were found in the C_{60}^- film and not in the electro-oxidized film.

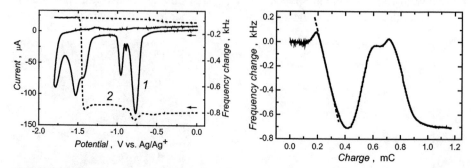

FIGURE 3. Simultaneously recorded curve of (*1*) CV and (*2*) frequency change vs. potential for a C_{60} film in 0.1 M (TBA)PF$_6$, in acetonitrile. After initial aggregation in bulk solution for 1 min, the film was electrophoretically deposited for 12 min at 50 V dc from of toluene:ethanol, 1 : 10 (*v* : *v*), solution on a 5 MHz Pt-quartz electrode. Potential sweep rate 0.1 V s^{-1}.

FIGURE 4. Frequency vs. charge curve for a C_{60} film corresponding to the first two cathodic peaks; 1 min aggregation, 3 min deposition at 50 V, 10 MHz Pt-quartz resonator, toluene : ethanol, 1 : 10 (*v* : *v*).

CONCLUSIONS

The C_{60} nanocrystalline films were prepared by bulk electrophoresis and their roughness was controlled by the time of aggregation and dc voltage of deposition. At the same voltage, roughness was higher for shorter rather than longer aggregation time. The higher dc voltage applied the larger was the aggregate diameter.

Exhaustive C_{60}^-/C_{60}^0 electroreduction was accompanied by both TBA$^+$ counter- and PF$_6^-$ co-ion ingress into the film. Subsequent back electro-oxidation of C_{60}^- to C_{60}^0 resulted in a complete removal of ions of the supporting electrolyte from the film.

ACKNOWLEDGMENTS

This work was financially support by the State Committee for Scientific Research of Poland, Project No. 4 T09A 160 23 (to W.K.), and ACS-PRF (to F.D.).

REFERENCES

1. J. Chlistunoff, D. Cliffel, A. J. Bard, *Thin Solid Films* **257** 166-184 (1995).
2. K.Winkler, D. A. Costa, A. L. Balch, *Polish J. Chem.* **74**, 1-37 (2000).
3. P. Janda, T. Krieg, L. Dunsch, *Adv. Mater.* **10**, 1434-1438 (1998).
4. S. Barazzouk, S. Hotchandani, P. V. Kamat, *Adv. Mater.* **13**, 1614-1617 (2001).
5. K. Vinodgopal, M. Haria, D. Meisel, P. V. Kamat, *Nano Lett.* **4**, 415-418 (2004)
6. S. Barazzouk, S. Hotchandani, P. V. Kamat, *J. Mater. Chem.* **12**, 2021-2025 (2002).
7. P. V. Kamat, S. Barazzouk, S. Hotchandani, K. G. Thomas, *Chem. Eur. J.* **6**, 3914-3921 (2000).

Nanosegregation in Na_2C_{60}

G. Klupp*, K. Kamarás*, N. M. Nemes†, P. Matus*, D. Quintavalle**, L. F. Kiss*, É. Kováts*, S. Pekker* and A. Jánossy**

*Research Institute for Solid State Physics and Optics, Hungarian Academy of Sciences, P. O. Box 49, H-1525 Budapest, Hungary
†Instituto de Ciencia de Materiales de Madrid, Cantoblanco, 28049 Madrid, Spain
**Department of Experimental Physics, Budapest University of Technology and Economics, Budafoki út 8, Budapest, Hungary H-1111

Abstract.
There is continuous interest in the nature of alkali metal fullerides containing C_{60}^{4-} and C_{60}^{2-}, because these compounds are believed to be nonmagnetic Mott–Jahn–Teller insulators. This idea could be verified in the case of A_4C_{60}, but Na_2C_{60} is more controversial. By comparing the results of infrared spectroscopy and X-ray diffraction, we found that Na_2C_{60} is segregated into 3-10 nm large regions. The two main phases of the material are insulating C_{60} and metallic Na_3C_{60}. We found by neutron scattering that the diffusion of sodium ions becomes faster on heating. Above 470 K Na_2C_{60} is homogeneous and we show IR spectroscopic evidence of a Jahn–Teller distorted C_{60}^{2-} anion.

INTRODUCTION

The proposed [1] Mott–Jahn–Teller ground state of A_4C_{60} (A = K, Rb, Cs) fullerides was experimentally verified by EELS [2], NMR [3] and infrared (IR) [4] spectroscopies. Theory predicts a similar effect in fullerides with C_{60}^{2-} ions, the electron-hole inverted analogue of C_{60}^{4-}. Na_2C_{60} is the only known alkali-metal fulleride that is believed to contain the C_{60}^{2-} dianion and there were several attempts to verify the Mott–Jahn–Teller insulating ground state in this compound as well. Experimental results were not unambiguous [3], e.g. a very weak metallic behavior was found in Na_2C_{60}, but it was concluded that K_4C_{60} and Na_2C_{60} have essentially the same Mott–Jahn–Teller ground state.

In contrast, our infrared, X-ray and neutron scattering experiments indicate that Na_2C_{60} is not a good model system to investigate the ground state of A_2C_{60} fullerides. Above 470 K Na_2C_{60} is homogeneous and at high temperatures we find indication for Jahn–Teller distorted C_{60}^{2-} ions. The IR spectroscopy and X-ray diffraction results at ambient temperatures show, however, a nanosegregation of Na_2C_{60} into two phases.

EXPERIMENTAL

Na_2C_{60} was obtained by the reaction of stoichiometric amounts of Na and C_{60} at high temperature in a stainless steel capsule. The typical annealing sequence was first 23 days

CP786, Electronic Properties of Novel Nanostructures, edited by H. Kuzmany, J. Fink, M. Mehring, and S. Roth
© 2005 American Institute of Physics 0-7354-0275-2/05/$22.50

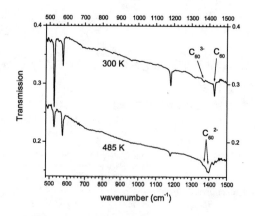

FIGURE 1. The IR spectra of Na_2C_{60} at 300 K and at 485 K. At high temperature the absorption of the $T_{1u}(4)$ mode could be fitted with two Lorentzians centered at the indicated positions, corresponding to C_{60}^{2-}, while at 300 K lines corresponding to C_{60} and C_{60}^{3-} appear.

at 350 °C and then 7 days at 450 °C. To homogenize the samples, we reground them about once every five days. Due to the air sensitivity of Na_2C_{60}, the reaction and the handling of the samples were carried out under inert atmosphere.

The sample was characterized by X-ray diffraction, which showed the material to be single phase $Pa\bar{3}$ (simple cubic) Na_2C_{60}. This result is identical to those in the literature [5].

Infrared measurements were performed on pressed KBr pellets in a Bruker IFS 28 FTIR instrument in a cryostat under dynamic vacuum. For neutron scattering measurements 1.2 g of Na_2C_{60} powder was placed in an annular aluminum sample holder. Temperature dependent elastic fixed-window scans were taken on the High Flux Backscattering Spectrometer of the NIST Center for Neutron Scattering [6]. In this measurement, the incident neutron energy was fixed at 2.08 meV and scattering processes were detected near the elastic line within the 1 μeV energy window of the resolution of the instrument[7]. Thus when a dynamic process became faster than the corresponding timescale of 0.8 ns, the measured intensity decreased.

RESULTS AND DISCUSSION

The charge state of C_{60} anions can be evaluated from the quasilinear relationship between the charge state and the line position of the $T_{1u}(4)$ IR mode [8]. We expect from this relation a line of C_{60}^{2-} at about 1380 cm^{-1}. Contrary to this expectation, there is no such line in the measured 300 K spectrum (Fig. 1) instead lines characteristic of neutral C_{60} and of C_{60}^{3-} appear. In contrast to the narrow line of C_{60}, that of C_{60}^{3-} is smeared into a broad line. This kind of line broadening is common for A_3C_{60} phases and is caused by metallic electrons. The metallic character of an aggregate as small as a few C_{60}^{3-} mole-

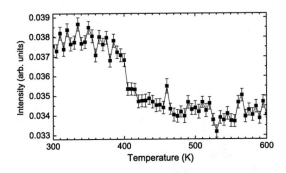

FIGURE 2. Elastic fixed window scan intensity of Na_2C_{60} at $Q = 1\mathring{A}^{-1}$.

cules is sufficient to broaden the IR line. Thus C_{60}^{3-} and C_{60} are not homogeneously distributed in the crystal lattice.

The observation of segregated Na_3C_{60} and C_{60} regions in the material by IR spectroscopy apparently contradicts the observation of a single phase by X-ray diffraction. However, if the size of segregated regions is smaller than about 10 nm, then X-ray diffraction cannot resolve the two phases but detects their average. Thus from the comparison of IR and X-ray diffraction measurements we conclude that Na_2C_{60} is nanosegregated at ambient temperatures.

The amount of neutral C_{60} in the material could be determined by its selective extraction with toluene. The concentration of the obtained C_{60} solution was measured with high-pressure liquid chromatography (HPLC). From this we could estimate the C_{60} content to be 26-33 % C_{60} in the nominally Na_2C_{60} material.

Heating the material to high temperatures proves that the stoichiometry of the sample is indeed Na_2C_{60} and that at high temperatures there is no neutral C_{60} left. The lines of neutral C_{60} disappear above 470 K, and a pair of lines appears at 1369 cm^{-1} and 1394 cm^{-1} (Fig. 1). This line pair is at about the expected frequency for the $T_{1u}(4)$ line of C_{60}^{2-}. Thus at room temperature the C_{60} content is not material left unreacted during the high temperature synthesis, but rather the product of the reaction $3C_{60}^{2-} \longrightarrow C_{60} + 2C_{60}^{3-}$ taking place on cooling after preparation. This reaction is reversible on heating and cooling, though a total retransformation at ambient temperature following treatment at high temperature is reached only after about two weeks.

The line pair of C_{60}^{2-} is the lower frequency analogue of the line pair of C_{60}^{4-} [4], indicating that C_{60}^{2-} is distorted to a D_{3d} or a D_{5d} geometry due to the molecular Jahn–Teller effect [4].

The synproportion reaction on heating ought to be accompanied by diffusion of Na ions in the lattice. This could be proven by neutron scattering. Figure 2 shows the temperature dependence of the elastic line at $Q = 1\mathring{A}^{-1}$. The intensity shows a Debye-Waller-type overall decrease, but has an unusual drop near 400 K. We interpret this with the increase of the jump diffusion of sodium ions between tetrahedral and off-centered octahedral sites. Above 400 K it becomes fast enough to be resolved by the instrument and the incoherent scattering contribution of the sodium ions is removed from the fixed-window intensity. The different transition temperatures found in the IR and neutron measurements may be due to the different timescales.

CONCLUSION

We propose that Na_2C_{60} is nanosegregated at room temperature. The two main phases of Na_2C_{60} are insulating C_{60} and metallic Na_3C_{60}. The size of the homogeneous regions is about 3-10 nm. The segregation disappears on heating when the jump diffusion of sodium ions becomes faster. A similar segregated phase, the "intermediate phase" is known for KC_{60} [9]. In Na_2C_{60} at high temperatures, the sodium distribution is homogeneous and a Jahn–Teller distortion of C_{60}^{2-} ions is observed.

ACKNOWLEDGMENTS

This work was supported by OTKA grants T 034198, T 049338, T 046700 and T 043255. This work utilized facilities supported in part by the National Science Foundation under Agreement No. DMR-0086210. We acknowledge the support of the National Institute of Standards and Technology, U. S. Department of Commerce, in providing the neutron research facilities used in this work.

REFERENCES

1. Fabrizio, M., and Tosatti, E., *Phys. Rev. B*, **55**, 13465 (1997).
2. Knupfer, M., and Fink, J., *Phys. Rev. Lett.*, **79**, 2714 (1997).
3. Brouet, V., Alloul, H., Garaj, S., and Forró, L., *Phys. Rev. B*, **66**, 155122 (2002).
4. Kamarás, K., Klupp, G., Tanner, D. B., Hebard, A. F., Nemes, N. M., and Fischer, J. E., *Phys. Rev. B*, **65**, 052103 (2002).
5. Yildirim, T., Hong, S., Harris, A. B., and Mele, E. J., *Phys. Rev. B*, **48**, 12262 (1993).
6. Meyer, A., Dimeo, R. M., Gehring, P. M., and Neumann, D. A., *Rev. Sci. Instrum.*, **74**, 2759 (2003).
7. Becker, T., and Smith, J. C., *Phys. Rev. E*, **67**, 021904 (2003).
8. Pichler, T., Winkler, R., and Kuzmany, H., *Phys. Rev. B*, **49**, 15879 (1994).
9. Faigel, G., Bortel, G., Tegze, M., Gránásy, L., Pekker, S., Oszlányi, G., Chauvet, O., Baumgartner, G., Forró, L., Stephens, P. W., Mihály, G., and Jánossy, A., *Phys. Rev. B*, **52**, 3199 (1995).

Study of Defects in Polymerized C_{60}: A Room-Temperature Ferromagnet

A. Zorko,[1] T. L. Makarova,[2] V. A. Davydov, A. V. Rakhmanina,[3] and D. Arčon[1, 4]

[1] *"Jožef Stefan" Institute, Jamova 39, SI-1000 Ljubljana, Slovenia*
[2] *Department of Experimental Physics, Umeå University, S-90187 Umeå, Sweden*
[3] *Institute of High Pressure Physics, 142092 Troisk, Russia*
[4] *Faculty of Mathematics and Physics, University of Ljubljana, SI-1000 Ljubljana, Slovenia*

Abstract. A X-band ESR study of a family of rhombohedral two-dimensional polymerized C_{60} samples is presented. The investigated polymerized fullerenes exhibit ferromagnetic features up to temperatures around 500 K. No ferromagnetic ESR modes are detected, which is probably due to extremely broad signals. However, a narrow ESR absorption line is systematically observed. Although this line does not originate directly from the system of ferromagnetic moments, temperature dependence of the line position speaks in favor of a coupling with the former system providing important insights.

Keywords: magnetic carbon, polymerized C_{60}, ferromagnetism, graphite.
PACS: 81.05.Tp, 75.50.Dd, 76.30.-v

INTRODUCTION

Ferromagnetism in nonmetallic molecular materials is a relatively rare phenomenon, which is usually observed only at rather low temperatures. An observation of ferromagnetic features in polymerized C_{60} at room temperature [1- 8] has opened a new field of prospectively very useful materials and introduced a challenge to the current understanding of ferromagnetism.

The molecules in solid C_{60}, which form van der Walls *fcc* crystalline solid at ambient conditions, become covalently bonded when exposed to UV light [9]. Likewise, polymerization is observed, if a high-pressure high-temperature (HPHT) treatment is applied to pristine C_{60}, which depending on conditions results in one-, two-, or three-dimensional polymers [10]. Ferromagnetism has been observed in two-dimensional rhombohedral (Rh) and tetragonal (T) C_{60} polymers [1-5]. UV irradiation can cause ferromagnetic behavior in one-dimensional orthorhombic (O) phase [6-8].

Since the experimentally observed transition temperatures into the magnetic phase were reported to be as high as 500 K [1, 4] or even 800 K [3, 6], these polymers have been stimulating a great deal of theoretical work in recent years [4, 11-13]. The ferromagnetism is believed to be intrinsic to the C_{60} polymer. Magnetic force microscopy measurements were able to image magnetic stripe domains in regions with negligible (i.e., less than 1 µg/g) magnetic impurity concentration [4]. On the other hand, it was shown experimentally [11] as well as theoretically [12] that ideal C_{60}

CP786, *Electronic Properties of Novel Nanostructures*, edited by H. Kuzmany, J. Fink, M. Mehring, and S. Roth
© 2005 American Institute of Physics 0-7354-0275-2/05/$22.50

polymers are not magnetic. Therefore, one should seek for a defect-related mechanism of magnetism.

The origin of the induced magnetic moments in the polymerized samples still has to be unambiguously identified. A credible explanation has been provided by Andriotis et al. [13] who argued that interplay between carbon vacancies and sp^2-sp^3 hybridization should be involved favoring ferromagnetism in the Rh-C_{60} polymeric phase. However, this mechanism can not account for the magnetic order, which requires long range magnetic coupling between local moments [5]. In order to elucidate the role of the defects in polymerized ferromagnetic samples, we initiated an electron spin resonance (ESR) study, since this is a powerful spectroscopic technique capable of detecting magnetic moments on a local scale.

RESULTS AND DISCUSSION

X-band ESR experiments were conducted on a family of 2D polymerized samples displaying weak ferromagnetism at room temperature. The samples were polymerized at a pressure of 8 GPa and at temperatures in the range between 873 K and 1073 K with an exception of the sample prepared at 923 K and 4 GPa. The ferromagnetic character was confirmed by SQUID magnetization measurements.

The initial aim of the ESR investigation was to observe broad ferromagnetic resonance modes and follow their temperature evolution in order to get valuable information about the properties of the magnetic moments participating in the ferromagnetic ordering. However, up to now no such modes have been unambiguously detected, which could be due to very broad absorption lines usually encountered in the ferromagnetic phase. On the contrary, narrow ESR spectra were systematically observed in all samples with their g-factor values close to the free electron value at room temperature. The lineshape is Lorentzian in all cases with an exception of the sample synthesized at 1023 K, which exhibits Dysonian absorption lines. The latter lineshape is typical of metallic samples where the dispersion spectrum is admixed into the absorption one due to skin effects.

Regarding the intensity of the ESR spectra, there seem to exist two groups of materials among our samples. The ESR susceptibility of the first group amounts to around $12 \cdot 10^{-9}$ emu/g at room temperature, while it is somewhat reduced in the second group, i.e., to $3.5 \cdot 10^{-9}$ emu/g. The temperature dependence of the ESR susceptibility exhibits Curie-like behavior in most of the samples and a Pauli-like behavior in the sample with the Dysonian ESR line (Figure 1). This is an indication that the observed signals do not originate from the ferromagnetically ordered moments. Moreover, the ESR susceptibility is drastically reduced compared to the SQUID value of $7.1 \cdot 10^{-7}$ emu/g. The portion of the ESR detectable moments can be then estimated to be around $c \approx \chi_{ESR} / (\chi_{SQUID} + \chi_{dia}) \approx 1\%$, where $\chi_{dia} = -3.4 \cdot 10^{-7}$ emu/g represents the diamagnetic contribution of the pristine material, which cannot be detected by ESR.

Similar conclusions can be drawn from the investigation of the line-width of the observed spectra, which is around 1.6 G at room temperature and shows only a slight temperature dependence. If the signal is assumed to originate from diluted charged C_{60} cages coupled by dipolar coupling, the ESR linewidth will be given by [14]

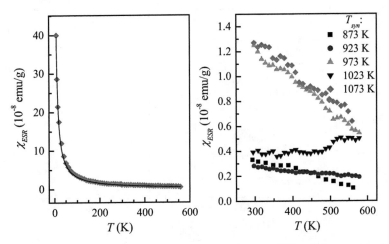

FIGURE 1. a) Curie-like temperature dependence of the ESR susceptibility of the sample polymerized at 1073 K (diamonds) and b) the high-T ESR susceptibility in different polymerized samples.

$$\Delta H_{pp} = c \cdot 5.3 \frac{2}{\sqrt{3}} \frac{\mu_0}{4\pi} \frac{g\mu_B}{d^3} \approx c \cdot 150\,\mathrm{G}, \tag{1}$$

where μ_B represents the Bohr magneton and d the distance between C_{60} cages. It should be stressed at this point that the Orbach relaxation mechanism dominant in many fullerides is expected to become ineffective in our samples because of the distorted symmetry of the defect C_{60} cages. The main contribution to the line-width should, therefore, come from the dipolar interaction. The above relation (1) again yields the impurity concentration of $c = 1\%$. It has to be emphasized that this evaluated value is several orders of magnitude above the estimated amount of magnetic impurities in the polymerized C_{60} samples [4].

Although the defects detected by ESR do not seem to be directly participating in ferromagnetic ordering, there is an indication that a coupling to the latter moments exists. Namely, the temperature dependence of the line position (g-factor) shows a rather strange and well pronounced behavior above approximately 450 K in all samples as shown in Figure 2a. This value is far beneath the depolymerization temperature for O-phase (560 K), Rh-phase (545 K), and T-phase (575/603 K) [15, 16]. Furthermore, subsequent ESR measurements showed the same room-temperature g-factor values. Therefore, we propose that the observed shift of the resonance field should be related to the development of an internal magnetic field in the ferromagnetically ordered phase. The internal field is given by $B_{int} = B_0 - N \cdot M(T)$, where B_0 represents the external field, N the demagnetization factor and $M(T)$ the magnetization of the sample. In fact, the temperature dependence of the effective "g-factor" presented in Figure 2a closely resembles the behavior of the magnetization of the polymerized samples in the same temperature range [1, 4]. For the above-mentioned reasons, we believe that the ESR detected defects directly sense the magnetization of the ferromagnetically ordered moments. Moreover, similar anomalous behavior is observed also in the change of the Dysonian character of the sample polymerized at 1023 K (Figure 2b).

23

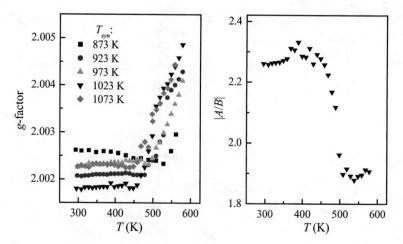

FIGURE 2. Temperature dependence of a) the g-factor in polymerized C_{60} samples and b) the Dysonian character of the sample polymerized at 1023 K (ratio of peaks of derivative ESR lines).

CONCLUSIONS

In summary, the observed ESR signal in polymerized C_{60} ferromagnetic samples has been assigned to intrinsic defects. There are at least two possibilities of their origin. These C_{60} defects could be placed at grain boundaries or there is a phase separation between ferromagnetic and nonferromagnetic domains. The ESR detected magnetic moments sense local magnetic fields originating from the ferromagnetically ordered moments as indicated by the peculiar temperature dependence of the line position. A comprehensive ESR study performed at different resonant frequencies is currently underway in order to rationalize the above-presented results.

REFERENCES

1. Makarova, T. L., et al., *Nature* **413**, 716-718 (2001).
2. Wood, R. A.., et al., *J. Phys.: Condens. Matter* **14**, L385-L391 (2002).
3. Narozhnyi, V. N., et al., *Physica B* **329**, 537-538 (2003).
4. Han, K.-H., et al., *Carbon* **41**, 785-795 (2003).
5. Chan, J. A., et al., *Phys. Rev. B* **70**, 041403(R) (2004).
6. Murakami, Y., and Suematsu, H., *Pure Appl. Chem.* **68**, 1463-1467 (1996).
7. Makarova, T. L., et al., *Carbon* **41**, 1575-1584 (2003).
8. Owens, F. J., Iqbal, Z., Belova, L., and Rao, K. V., *Phys. Rev. B* **69**, 033403 (2004).
9. Rao, A. M., et al., *Science* **259**, 955-957 (1993).
10. Nuñez-Requeiro, M., et al., *Phys. Rev. Lett.* **74**, 278-281 (1995).
11. Boukhvalov, D. W., et al., *Phys. Rev. B* **69**, 115425 (2004).
12. Okada, S., Oshiyama, A., *Phys. Rev. B* **68**, 235402 (2003).
13. Andriotis, A. N., et al., *Phys. Rev. Lett.* **90**, 026801 (2003).
14. Abragam, A., *The Principles of Nuclear Magnetism*, Oxford: Oxford University Press, 1961, pp. 125-128.
15. Korobov, M. V., et al., *Chem Phys. Lett.* **381**, 410-415 (2003).
16. Korobov, M. V., et al., *Carbon* **43**, 954-961 (2005).

Unusual two-dimensional polymer network in Li_4C_{60} – an ESR study

P. Cevc[1], D. Arčon[1,2], D. Pontiroli[3], and M. Ricco[3]

[1] *"Jožef Stefan" Institute, Jamova 39, SI-1000 Ljubljana, Slovenia*
[2] *Faculty of Mathematics and Physics, University of Ljubljana, SI-1000 Ljubljana, Slovenia*
[3] *Universita degli Studi di Parma, Dip. Di Fisica, Parma, Italy*

Abstract. At room temperature Li_4C_{60} forms a unique two-dimensional polymer network where the bonding encounters [2+2] cycloaddition along one direction and a single C-C bond along the perpendicular direction. This structure thus involves both so-far known bonding motifs and is unique among fullerene structures. The structural and electronic properties of this system were investigated by temperature dependent cw-ESR. The ESR signal of the polymer phase has an extremely narrow Lorentzian line with a linewidth of about 0.3 G. On heating, the transition from the low-temperature polymer phase to a metallic monomeric phase occurs above ~250 °C. The high-temperature ESR signal has a characteristic Dysonian lineshape.

Keywords: electron spin resonance, polymerization, fullerenes.
PACS: 61.48.+c, 61.72.Hh, 72.80.Le

INTRODUCTION

Ever since the discovery of the one-dimensional polymer phase in RbC_{60} [1], bonding of fullerenes attracted enormous attention. The prime interest in fullerene-bridged arrays comes from their reduced dimensionality and often accompanying electronic and magnetic instabilities. There are two ways to form a bond between two neighbouring C_{60} molecules. The predominant mechanism involves the [2+2] cycloaddition reaction where carbons atoms from different molecules form a four-member ring. Such a mechanism is present in photopolymerised C_{60} [2], in pressure induced polymer phases [3] as well as in most of the A_1C_{60} polymer structures (here A=K, Rb, Cs). Another structural motif has been discovered in Na_2RbC_{60} [4,5]. Here the C_{60}-C_{60} bond relies on the single C-C covalent bond.

Doping of C_{60} with Li does not necessary lead to same effects as doping with other alkali metals. The small ionic radius of the Li^+ ion makes a big difference from the structural point of view and very often one encounters even an incomplete charge transfer to C_{60}. To investigate these effects Li_xC_{60} and especially Li_4C_{60} phases were revisited. Initial reports [6,7] suggested that Li_4C_{60} forms a two-dimensional polymer phase with [2+2] cycloaddition bonds. However recent high-resolution synchrotron X-ray diffraction studies [8] lead to a surprising conclusion: Li_4C_{60} at room temperature indeed forms a two-dimensional polymer network, but the bonding encounters [2+2] cycloaddition along one direction and a single C-C bond along the perpendicular

CP786, *Electronic Properties of Novel Nanostructures*, edited by H. Kuzmany, J. Fink, M. Mehring, and S. Roth
© 2005 American Institute of Physics 0-7354-0275-2/05/$22.50

direction. This structure thus involves both known bonding motifs and is unique among fullerene structures.

Solution of the room-temperature structure of polymerized Li_4C_{60} opened many important questions like: the temperature and mechanism of depolymerisation and what are the electronic and magnetic properties of polymerised as well as monomeric phases of Li_4C_{60}. We have tried to address these questions by using electron spin resonance (ESR) technique.

RESULTS AND DISCUSSION

As prepared Li_4C_{60} samples were first checked by XRD, which proved that as-prepared samples are polymerized into a two-dimensional network as described in [8]. On these samples we measured at room temperature a very narrow ESR signal ($\Delta H_{pp} = 0.27(3)\,G$), which is well simulated by a Lorentzian lineshape. The g-factor of this line at room temperature is just slightly larger than 2 (more precisely 2.0009(3)) (Fig. 1). The calibrated intensity corresponds to $0.24 \cdot 10^{-4}$ emu/mol. The extreme narrowness of the signal could arise from the C_{60}^- defects in the polymer structure. On the other hand extremely narrow ESR signals were found in many polymerized doped C_{60} phases and were explained in terms of reduced dimensionality [1,9,10]. Additional experiments are needed to clarify this point.

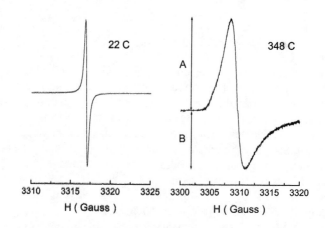

FIGURE 1. A comparison of the ESR lineshapes measured in Li_4C_{60} at 22 °C and 348 °C. Please note the difference in the linewidth and anisotropy of the line.

On heating dramatic changes could be observed in the ESR signal. The ESR lineshape becomes slightly anisotropic already at around 100 °C. On further heating to temperatures above 250 °C the anisotropy in the ESR signal becomes even more pronounced. In figure 1 we compare the lineshapes measured on as prepared samples at room temperature (22 °C) and after heating to 348 °C. If we define the asymmetry of the ESR line as a ratio between the height of the positive (A) and negative (B) peak

(Fig. 1), then we can follow the changes in the lineshape. At room temperature this ratio is $A/B=1$ in agreement with the Lorentzian lineshape. However on heating this ratio first increases to around 1.10(4) and remains almost temperature independent up to 250 °C. Above 250 °C ratio A/B starts to increase dramatically (Fig. 2) and reaches $A/B=1.6$ at 348 °C. Such anisotropy of the ESR spectra is characteristic for the metallic samples and the lineshape resembles the Dysonian lineshape.

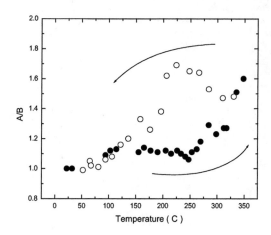

FIGURE 2. The temperature dependence of the lineshape asymmetry ratio A/B for heating (solid circles and for cooling (open circles).

Simultaneous X-ray diffraction studies [11] showed, that at these temperatures the polymer phase of Li_4C_{60} starts to decompose and that a monomeric phase of Li_4C_{60} is stable at temperatures higher than 300 °C. Our ESR results thus prove, that the high-temperature monomeric phase of Li_4C_{60} is in fact metallic.

Surprisingly we find that on cooling the anisotropy in the lineshape persists to much lower temperatures (Fig. 2). The ratio A/B remains approximately constant 1.6(1) down to 200 C. It thus seems that the high-temperature metallic phase starts to appear between 250 and 300 °C, while it persists down to 200 °C on cooling.

CONCLUSIONS

In conclusion, we have studied a new two-dimensional polymer Li_4C_{60} with a high-temperature electron spin resonance technique. A very narrow ESR signal observed in virgin sample originates either from the defect C_{60}^- sites or is a result of the reduced dimensionality in the polymer phase. On warming the concentration of defects increases due to a gradual thermal depolymerisation of a two-dimensional network. A monomeric Li_4C_{60} phase has been identified above 250-300 °C. This phase was found to be metallic and on cooling persists down to much lower temperatures.

27

REFERENCES

1. O. Chauvet et al., Phys. Rev. Lett. 72, 2721 (1994).
2. Rao, A. M., et al., *Science* **259**, 955-957 (1993).
3. Nuñez-Requeiro, M., Marquez, L., Hodeau, J.-L., Béthoux, O., and Perroux, M., *Phys. Rev. Lett.* **74**, 278-281 (1995).
4. K. Prassides et al., J. Am. Chem. Soc. 119, 834 (1997).
5. G.M. Bendele et al., Phys. Rev. Lett. 80, 736 (1998).
6. M. Yusukawa et al., Chem. Phys. Lett. 341, 467 (2001).
7. T. Wagberg et al., J. Phys. Chem. Solids 65, 317 (2004).
8. S. Margadonna et al., J. Am. Chem. Soc. 126, 15032 (2004).
9. G. Oszlanyi et al., Phys. Rev. B 58, 5 (1998).
10. D. Arčon et al., Phys. Rev. B 60, 3856 (1999).
11. D. Pontiroli et al., unpublished.

Magnetic Properties of TDAE–C$_{70}$

R. Blinc, P. Jeglič, T. Apih, P. Cevc, A. Omerzu, and D. Arčon

Jožef Stefan Institute, Jamova 39, 1000 Ljubljana, Slovenia

Abstract. The ^{1}H and ^{13}C NMR as well as the X-band and high-field EPR spectra of TDAE–C$_{70}$ are compared with those of TDAE–C$_{60}$. The observation of a weak temperature dependent high-field EPR line below 50 K may be connected with the occurrence of a ferromagnetic modification of TDAE–C$_{70}$ in addition to the well known paramagnetic one.

Keywords: Fullerenes, Organic ferromagnetism, NMR, EPR
PACS: 76.60.-k, 71.20.Tx, 75.50.Dd

INTRODUCTION

It is well known [1,2] that TDAE–C$_{60}$ (where TDAE stands for tetrakis-dimethylamino ethylene) exists at normal pressures in two different modifications: (i) The α-phase which exhibits a ferromagnetic transition at $T_c = 17$ K and is found in well annealed samples and crystals grown above room temperature; (ii) The α'-phase which stays paramagnetic at all temperatures and is found in fresh samples and crystals grown at lower temperature.

In contrast, only one phase is known in TDAE–C$_{70}$ [3,4] which stays paramagnetic at least down to 4.5 K. The question arises whether this paramagnetic phase is analogous to the α' phase of TDAE–C$_{60}$ and whether still another ferromagnetic α phase exists in TDAE–C$_{70}$.

For this reason, we decided to reinvestigate the proton and ^{13}C NMR spectra of TDAE–C$_{70}$ as well as the corresponding X-band and high field EPR spectra.

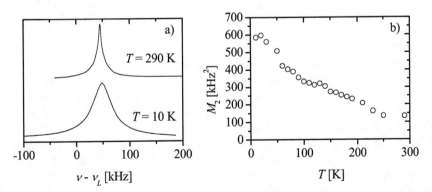

FIGURE 1. a) 270 MHz proton NMR of powdered TDAE–C$_{70}$ spectra at 290 K and 10 K. b) Temperature dependence of the proton second moment in powdered TDAE–C$_{70}$.

CP786, *Electronic Properties of Novel Nanostructures*, edited by H. Kuzmany, J. Fink, M. Mehring, and S. Roth
© 2005 American Institute of Physics 0-7354-0275-2/05/$22.50

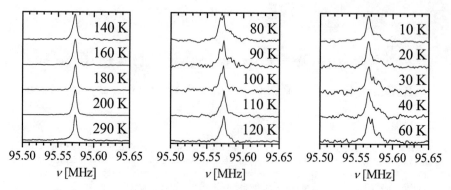

95.50 95.55 95.60 95.65   95.50 95.55 95.60 95.65   95.50 95.55 95.60 95.65
ν [MHz] ν [MHz] ν [MHz]

FIGURE 2. ^{13}C NMR spectra of TDAE–C$_{70}$ between 290 K and 10 K.

EXPERIMENTAL

TDAE–C$_{70}$ was prepared in the standard way [5] in analogy to TDAE–C$_{60}$. A certain amount of C$_{70}$ was dissolved in a degassed toluene solvent and excessive TDAE was added under nitrogen flow. Care was taken to purge oxygen from the solvent before the reaction. A black precipitate was obtained and the unreacted TDAE was removed. The reaction was performed in a quartz EPR tube to exclude the atmosphere. The tube was sealed under nitrogen flow. Several different samples were prepared. The proton and ^{13}C NMR spectra were measured at 9 T.

RESULTS AND DISCUSSION

Whereas the proton NMR spectra of normally prepared TDAE–C$_{60}$ show two proton lines – corresponding to the α and α' modifications – we observed only a single proton NMR line in powdered TDAE–C$_{70}$ (Fig. 1a). This seemed to indicate that this sample of TDAE–C$_{70}$ was homogeneous and composed of only one modification. The position of this proton line in TDAE–C$_{70}$ was practically temperature independent, whereas its width and second moment (Fig. 1b) continuously increase with decreasing temperature in analogy to the behaviour of the α'-line in TDAE–C$_{60}$. The magnitude of the second moment ($\approx$ 600 kHz^2 at 10 K) shows that the proton linewidth is dominated by paramagnetic electron-nuclear coupling. The absence of a shift of the resonance line shows that there is no contact hyperfine Fermi field, i.e., there is no unpaired spin density at the methyl proton sites. This contrasts sharply to the case of the α-modification of TDAE–C$_{60}$, in which the proton line shift amounts to nearly 1 MHz between room temperature and T_c = 17 K.

The above results show the absence of hyperexchange via the TDAE groups in TDAE–C$_{70}$ in contrast to the situation in the α-modification of TDAE-C$_{60}$. The ^{13}C NMR spectra of powdered TDAE–C$_{70}$ between room temperature and 10 K are shown in Fig. 2. The width $\Delta\nu$ of the ^{13}C spectrum at room temperature is $\approx$ 7.4 kHz whereas in TDAE–C$_{60}$ $\Delta\nu$ is about 2.1 kHz. The difference is due to the existence of five chemically non-equivalent carbon sites in C$_{70}$ as well as to more anisotropic motion connected with the "rugby ball" shape of the C_{70} molecule. At and below room temperature the lineshape is characteristic of fast uniaxial rotation around the long molecular axis leading to a logarithmic type singularity [6] at $\langle x \rangle = \frac{1}{3}$, where $x = (v - v_0)/\Delta$, $\Delta = v_L(\sigma_\perp - \sigma_\parallel)$ and $v_0 = v_L(1 - \sigma_\perp)$. Here v_L is the Larmor frequency, whereas $\sigma_\parallel$ and $\sigma_\perp$ are the two components of the chemical shift tensor which is assumed to be axially symmetric. The lineshape is not a δ-function, as the component of the shift tensor parallel to the rotational axis is not averaged out by uniaxial rotations. Around 110 K a shoulder starts to appear in the spectrum and at 90 K we have a two-peak spectrum. Below 60 K the peak at v_0 is already stronger then the one at $\langle x \rangle = \frac{1}{3}$. Below 40 K this peak gradually disappears and we see a static powder spectrum with a peak at $v_0 = v_L(1 - \overline{\sigma}_\perp)$ and a shoulder at $v = v_L(1 - \overline{\sigma}_\parallel)$. Here $\overline{\sigma}_\perp$ and $\overline{\sigma}_\parallel$ are weighted averages of the corresponding elements of the five chemically non-equivalent C_{70} ^{13}C shift tensors.

Such a powder spectrum has never been observed in the α-modification of TDAE–C_{60} where the spectral shape at low temperatures is determined by the non-uniform belt-like unpaired electron spin density distribution on the C_{60}^- ion. It should be noted that the width in TDAE–C_{60} is $\approx 10^4$ ppm below 10 K whereas in TDAE–C_{70} it is approximately $\overline{\sigma}_\parallel - \overline{\sigma}_\perp \approx 110$ ppm at 10 K.

The room temperature X-band EPR spectra of TDAE–C_{70} and TDAE–C_{60} are compared in Fig. 3a. The similarity of the lineshapes and intensities shows that we deal in both cases with a one electron transfer from the TDAE to C_{70} respectively C_{60} molecules. The g-value of TDAE–C_{60} 2.003 at room temperature is significantly larger than that of TDAE–C_{70} 2.0022 [7] which is close to the free electron value.

The X-band EPR susceptibility χ_{EPR} of TDAE–C_{70} follows a Curie law down to about 15 K where it starts to decrease with decreasing temperature (Fig. 3b). The product $\chi_{EPR}T$ is as well temperature independent down to 15 K, below which it starts to decrease with decreasing temperature indicating the formation of a singlet state as in the α'-modification of TDAE–C_{60} (Fig. 3c).

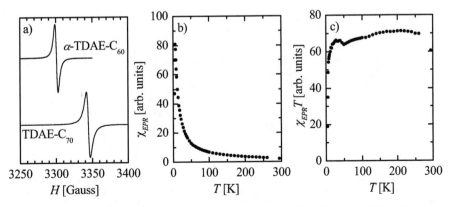

FIGURE 3. a) Comparison between the X-band EPR spectra of α-TDAE–C_{60} and TDAE–C_{70} at room temperature. Temperature dependence of b) χ_{EPR} and c) $\chi_{EPR}T$ in TDAE–C_{70}.

The high field EPR spectrum of TDAE–C_{70} at $\omega_L/2\pi$ = 245 GHz at room temperature is narrower ($\Delta v \approx$ 1.6 mT) than that of powdered TDAE–C_{60} ($\Delta v \approx$ 2.7 mT). The position of the line is temperature independent between 290 K and 50 K. Below 50 K a new line appears at the low field side of the main line (Fig. 4). It is strongly temperature-dependent and continues to move to lower fields down to 4 K. Between 50 K and 4 K the low field shift amounts to 4.2 mT.

The observation of this line might indicate that a transition from the paramagnetic to a weak ferromagnetic phase takes place in TDAE–C_{70} at low enough temperatures. It is not yet clear whether this new temperature-dependent "ferromagnetic-like" line is due to the presence of an α-like modification of TDAE–C_{70} in the sample used or whether it is magnetic field induced.

It should be also noted that the intensity of the new temperature-dependent high field EPR line is at most 20 % of that of the main line.

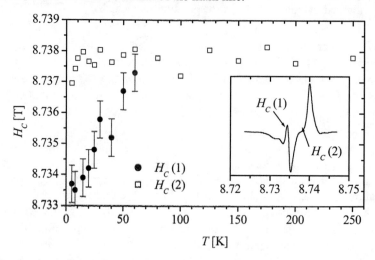

FIGURE 4. Temperature dependence of the position of high-field EPR spectra of TDAE–C_{70}. In the inset the EPR lineshape is shown.

REFERENCES

1. Allemand, P. M., Khemani, K. C., Koch, A., Wudl, F., Holczer, K., Donovan, S., Gruner, G., and Thompson, J. D., *Science* **253**, 301-303 (1991).
2. Arčon, D., Cevc, P., Omerzu, A., and Blinc, R., *Phys. Rev. Lett.* **80**, 1529-1532 (1998).
3. Tanaka, K., Zakhidov, A. A., Yoshizawa, K., Okahara, K., Yamabe, T., Yakushi, K., Kikuchi, K., Suzuki, S., Ikemoto, I., and Achiba, *Phys. Rev. B* **47**, 7554-7559 (1993).
4. Arčon, D., Blinc, R., Cevc, P., Chouteau, G., and Barra, A.-L., *Phys. Rev. B* **56**, 10786-10788 (1997).
5. Johnson, R. D., Meijer, G., Salem, J. R., and Bethune, D. S., *J. Am. Chem. Soc.* **113**, 3619-3621 (1991).
6. Blinc, R., Seliger, J., Dolinšek, J., and Arčon, D., *Phys. Rev. B* **49**, 4993-5002 (1994).
7. Tanaka, K., Zakhidov, A. A., Yoshizawa, K., Okahara, K., Yamabe, T., Yakushi, K., Kikuchi, K., Suzuki, S., Ikemoto, I., and Achiba, Y., *Phys. Lett. A* **164**, 221-226 (1992).

Photoluminescence and Relaxation Processes in the Optically Excited States of Fluorinated Fullerenes

*I. Akimoto, M. Mori, K. Kan'no

Department of Materials Science and Chemistry, Faculty of Systems Engineering,
Wakayama University, 930 Sakaedani, Wakayama 640-8510, Japan

Abstract. Origin of the photoluminescence and relaxation processes in the optically excited states of fluorinated fullerenes has been investigated by site selective excitation and time resolved spectroscopy. The intrinsic luminescence consists of at least two components with the different kind of decay kinetics. One of which is of single exponential with $\tau_d \sim 6$ ns and the other is of non-exponential but proportional to t^1. The latter component seems to be recombination luminescence due to an intermolecular polaron, in which an electron and a hole are separately localized with accompanying some structural relaxation.

Keywords: Excited states, Decay, Time resolved luminescence
PACS: 71.20.Tx, 33.50.Dq

INTRODUCTION

Fluorinated fullerenes $C_{60}F_x$ (x<=48) are novel wide-band gap insulating materials of C_{60} derivatives. The fundamental absorption edge of thin films is largely shifted at energies more than 4 eV from 1.8 eV of the starting C_{60} material [1, 2], in favor of higher transition energies of the residual C=C π-electron systems and the created C-F σ-bonds. Then, highly fluorinated fullerenes are promising transparent materials among fullerene family.

So far, there are only primitive reports about the optical properties of fluorinated fullerenes in solution, powder and spin-coated films [1, 3-5]. The luminescence spectrum exhibits large Stokes shift with relatively wide band width. The photo-induced desorption of CF and C_2F was also reported. These results imply that fluorinated fullerenes are strong electron-nuclear coupling systems. In present work, we report the optical properties of highly fluorinated fullerene in crystal with the ordinary and time-resolved spectroscopic techniques.

EXPERIMENTAL

The powder of fluorinated fullerenes $C_{60}F_x$ (TermUSA Co.) was prepared as a mixture of various fluorine compositions x (22≤x≤ 48) by reacting C_{60} powder with F_2 gas [3]. Poly-crystals of transparent pale yellow typically 1 x 1 x 0.5 mm^3 in size were

CP786, *Electronic Properties of Novel Nanostructures*, edited by H. Kuzmany, J. Fink, M. Mehring, and S. Roth
© 2005 American Institute of Physics 0-7354-0275-2/05/$22.50

grown in a vacuum by sublimation method from such mixed powder. A sublimated film specimen was prepared from a piece of poly-crystals on a LiF substrate. Absorption spectrum of the film was measured by VUV monochromator facilities (SHIMAZU, SGV-50). Luminescence spectrum and decay profiles of poly-crystals excited by nano-second pulses of OPO laser (SP, MOPO) were measured using a combination of a gated CCD (Roper Scientific, PI-MAX) and a photomultiplier (HAMAMATSU, R3896) attached to a monochromator (ACTON, 300i). Spectral distribution of the photoluminescence and its time evolution were observed with the minimum time gate of 10 ns under site-selective excitation into the tail region of 2.8 - 5.4 eV.

RESULTS

Figure 1 shows an absorption spectrum of a sublimated film of $C_{60}F_x$ at RT and a luminescence spectrum of a poly-crystal of $C_{60}F_x$ at 5 K. The absorption spectrum is peaked at 6.6 eV with FWHM ~ 2.0 eV and accompanied by shoulder structure at 4.0 eV and long tail below 4.0 eV. The luminescence spectrum consists of at least two bands peaked at 3.0 and 2.4 eV with FWHM of 1.3 and 0.5 eV, respectively. Such broadening feature of spectra is totally different from that of C_{60} single crystals [2], where well-resolved vibronic structure is observed as a typical character of molecular crystals.

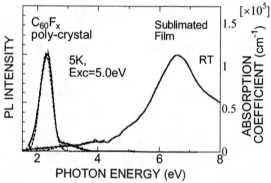

FIGURE 1. Absorption spectrum of a sublimated film of $C_{60}F_x$ at RT and luminescence spectrum of a poly-crystal of $C_{60}F_x$ at 5K excited at 5.0 eV. The luminescence spectrum was temporally fitted by two Gaussian functions (broken curves).

Resonant effects by selective excitation make the origin of broad and draged absorption structure more clearly. Figure 2 shows evolution of luminescence spectra at 5K obtained by selective excitation at the lower energy side of optical absorption into 2.8-5.4 eV. It seems reasonable to divide the region in two parts by 4.4 eV. At excitations into 4.4-5.4 eV, two broad luminescence bands were observed at fixed energies of 3.0 and 2.4 eV. On the other hand, at excitations into 2.8-4.3 eV, the band at higher energy side is more intensive and shifts according to the excitation energy. Therefore, we suppose that the appearance of two broad luminescence bands at 3.0

and 2.4 eV is an intrinsic nature of fluorinated fullerenes crystals, corresponding to the bulky absorption. The absorption in the low energy tail region reflects the existence of $C_{60}F_x$ molecules with relatively lower degree of fluorination because HOMO-LUMO gap becomes closer when the degree of fluorination x is lower [6].

We focused on the origin of the intrinsic luminescence bands. Figure 3 shows the decay profiles of luminescence monitored at 3.0 eV and 2.4 eV when excited by 5.0 eV pulses at 5 K. These two bands exhibit completely different kind of decay kinetics. The decay profile at 3.0 eV is of single exponential with a time constant $\tau_d \sim 6$ ns, while one at 2.4 eV is of non-exponential but proportional to t^{-1}. The decay profiles of the two bands were more clarified by time resolved measurement. Figure 4 shows normalized time resolved luminescence spectra at delay time, 0ns, 6 ns, 13ns, 15ns, 24ns, 100ns and 500 us at 5 K excited at 4.96 eV (lower panel). Each spectrum was decomposed into three Gaussian distribution functions peaked at 3.3, 2.9 and 2.4 eV (broken lines).

We confirmed that the long-lived component is not observed in a dilute solution sample of $C_{60}F_x$ in C_6F_6. Then, the non-exponential long-lived component is proper to the crystal. The Stokes shift of 2.4 eV band is large as, at least, 2.0 eV. It is attributed to be recombination luminescence due to an intermolecular polaron, in which an electron and a hole are separately localized with accompanying some structural relaxation as reported in other systems [7]. On the other hand, the single

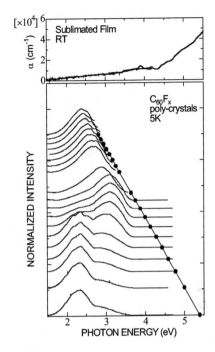

FIGURE 2. Evolution of luminescence spectra at 5K obtained by selectively excited at the low energy side of optical absorption. The positions of excitation were denoted by solid circle dots. The absorption spectrum was redrawn in the upper panel.

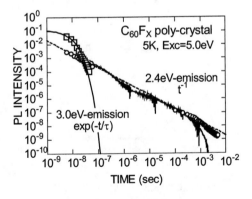

FIGURE 3. Decay profiles of luminescence at 3.0 eV (open squares) and 2.3 eV (solid lines and open circles) at 5 K in logarithmic scales. A single exponential decay curve of a time constant 6 ns was denoted by a solid line and a t^{-1} dependence was denoted by a broken line.

exponential component around 3.0 eV is tentatively attributed to the HOMO-LUMO transition of a molecule.

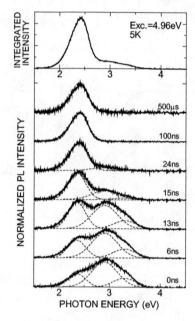

FIGURE 4. Normalized time resolved luminescence spectra at delay time, 0ns, 6 ns, 13ns, 15ns, 24ns, 100ns and 500 us at 5 K excited at 4.96 eV (lower panel). Each spectrum was reconstructed by three Gaussian functions peaked at 3.3, 2.9 and 2.4 eV (broken lines). For comparison, an integrated spectrum was redrawn in the upper panel.

ACKNOWLEDGMENT

This work was supported in part by the Grant-in-Aid for Science Research from the Ministry of Education, Culture, Sports, Science and Technology (No. 15340098, 15740214).

REFERENCES

1. K.Kan'no; Materials Science Forum, **239-241** (1997), 197-202.
2. I.Akimoto, K.Kan'no; J. Jpn. Phys. Soc., **71** (2002), 630-643.
3. A.A.Tuinman, P.Mukherjee, J.L.Adcock, R.L.Hettich and R.N.Compton; J.Phys.Chem.**96** (1992), 7584.
4. I.Akimoto, K. Kan'no, M. Shirai, F. Okino, H. Touhara, M. Kamada, V.G.Stankevitch; J. Lumin. **72-74** (1997), 503-504.
5. A.Kolmakov, V.Stankevitch, V.Bezmelnitsin, A.Rizkov, V.Sokolov, N.Svechnikov, I.Akimoto, T.Matsumoto, K.Kan'no, S.Hirose and M.Kamada: UVSOR Activity Report 1994 (1995) 88.
6. V.Stankevitch, in private communication.
7. C.J.Delbecq, Y.Toyozawa and P.H.Yuster; Phys. Rev. **B9** (1974),4497.

Model Description of Aggregation in Fullerene Solutions

V.L. Aksenov[1,2], M.V. Avdeev[1], T.V. Tropin[1], V.B. Priezzhev[1], J.W.P. Schmelzer[1,3]

[1]Joint Institute for Nuclear Research, Dubna, Moscow Reg., 141980, Russia
[2]Physics Department, Lomonosov Moscow State University, Moscow, Russia
[3]Physics Department, University of Rostock, Rostock, Germany

Abstract. A kinetic theory of aggregate formation and growth in fullerene solutions is developed. Two basic models for aggregates evolution are analyzed and qualitatively compared with experimental data. For supersaturated solutions it is shown, that the liquid drop model cannot appropriately describe the observed kinetics of aggregate growth. The model of limited growth is an adequate model and can qualitatively describe the experimental data.

INTRODUCTION

In recent time, interest to fullerene solutions has considerably grown mainly due to their much promising medical and biological applications. Investigating any fullerene solution one encounters the problem of describing aggregate formation and growth. There exist a lot of experiments which show different aspects of this problem (for brief review see, e.g. [1]). Current theoretical description restricts itself to chemical thermodynamics [2] or simplest phenomenological models [3]. In this work, we have made an attempt of presenting a stage-by-stage description of the kinetics of aggregate growth in fullerene solutions. For this purpose, kinetic equations of the nucleation theory [4] were used. The kinetics of cluster growth is considered on the base of two regimes: diffusion-limited aggregation (DLA) and kinetic-limited aggregation (KLA). Two simple models of aggregate evolution – the liquid drop model and the limited growth model for unsaturated and supersaturated solutions are discussed.

INITIAL EQUATIONS AND APPROXIMATIONS

Our aim is to obtain the evolution of the cluster size distribution function, $f(n,t)$ with time. Function $f(n,t)$ represents the bulk concentration of fullerene aggregates consisting of n monomers (further – «aggregate of size n») at the moment t. We suppose that fullerenes inside the aggregate are densely packed and aggregate form is quasispherical, independent of its size (the aggregation number n). It is also assumed that the formation and growth of aggregates is only possible by means of aggregation or emission of monomers. The assumption is based on the fact that for monomers the diffusion coefficient is considerably higher than that for aggregates.

CP786, *Electronic Properties of Novel Nanostructures*, edited by H. Kuzmany, J. Fink, M. Mehring, and S. Roth
© 2005 American Institute of Physics 0-7354-0275-2/05/$22.50

Based on these assumptions, we may write the following equation [4]:

$$\frac{\partial f(n,t)}{\partial t} = w_{n-1,n}^{(+)} f(n-1,t) + w_{n+1,n}^{(-)} f(n+1,t) - w_{n,n+1}^{(+)} f(n,t) - w_{n,n-1}^{(-)} f(n,t). \quad (1)$$

Here $w_{n-1,n}^{(+)}$ is the probability a monomer aggregation to a cluster of the size $(n-1)$, and $w_{n,n-1}^{(-)}$ is the probability of monomer emission per unit of time. The initial and boundary conditions are determined by the initial concentration of free monomers, c_0, in the solution:

$$f(n,t=0) = \begin{cases} 0, & n > 1, \\ c_0, & n = 1. \end{cases} \qquad \sum_{n=1}^{\infty} nf(n,t) = c_0. \quad (2)$$

Changes in the free energy of the system ΔG due to the formation of an n or $(n-1)$ size cluster are determined by the monomer aggregation to monomer emission probability ratio:

$$\frac{w_{n-1,n}^{(+)}}{w_{n,n-1}^{(-)}} = \exp\left\{-\frac{\Delta G(n) - \Delta G(n-1)}{k_B T}\right\}. \quad (3)$$

The probability $w_{n-1,n}^{(+)}$ depends on the kinetics of aggregate growth. As a rule, two regimes of growth are considered: diffusion- and kinetic-limited aggregation. In DLA-regime, the time of monomer aggregation to the surface of the aggregate is negligibly small compared to the particle drift time towards the aggregate and $w_{n-1,n}^{(+)}$ is then:

$$w_{n,n+1}^{(+)} = 4\pi Dc\left(\frac{3w_s}{4\pi}\right)^{1/3} n^{1/3}, \quad (4)$$

where D is the diffusion coefficient, c is the concentration of free monomers in solution at the given moment of time, w_s is the excluded volume of the C_{60} molecule.

In the case of KLA-regime, when the time of incorporation of a monomer to the aggregate is larger that the time of particle diffusion towards the aggregate surface, the following equation can be obtained:

$$w_{n,n+1}^{(+)} = 4\pi D' ca_m \left(\frac{w_s}{w_m}\right)^{2/3} n^{2/3}, \quad (5)$$

where D' is the diffusion coefficient characterizing the aggregation rate of particles to the aggregate. Index m stands for the parameters of the aggregate.

In general, both processes of transport of the segregating particles to the aggregate as well as the rate of incorporation may be of importance for the rate of aggregate formation and growth. As the result, we obtain an intermediate regime (DLA+KLA), and the probability $w_{n-1,n}^{(+)}$ is described by the following equation:

$$w_{n,n+1}^{(+)} = 4\pi D' c\left(\frac{3w_s}{4\pi}\right)^{1/3} n^{1/3} \left\{ \frac{\left(a_s/a_m\right)n^{1/3}}{1 + \left[\left(D'/D\right)\left(a_s/a_m\right)\right]n^{1/3}} \right\}. \quad (6)$$

Substituting (3) and one of the Eqs. (4-6) into (1) we obtain the system of kinetic equations that describe the system considered. For more details, consider [4,5].

AGGREGATION REGIME

Before obtaining numerical solutions of kinetic equations for various parameters for different fullerene solutions, we must determine, which kind of aggregation regime properly describes our system – DLA, KLA or DLA+KLA regime.

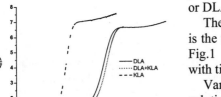

The main difference between these regimes is the time-scale of aggregation processes. On Fig.1 the evolution of mean aggregate size with time is present, for different regimes.

Various experiments on different fullerene solutions show that the time of aggregate formation and growth in these systems may be of the order of days, weeks and even months. From that point of view, DLA and DLA+KLA regimes must be used in order to obtain agreement with experiments. In this work, we choose DLA+KLA regime as a more adequate model to describe our system.

FIGURE 1. Liquid drop model - <n> evolution.

AGGREGATION MODELS

If the liquid drop model is used, the aggregate formation work takes the form:

$$\Delta G(n) = -n\Delta\mu + \alpha_2 n^{2/3}, \quad \alpha_2 = 4\pi\sigma\left(\frac{3w_s}{4\pi}\right)^{2/3}, \quad \Delta\mu = k_B T \ln\left|\frac{c}{c_{eq}^{(\infty)}}\right|, \quad (7)$$

where σ is the surface tension on the aggregate-surrounding phase interface, $\Delta\mu$ is the difference between the chemical potential of the monomer in a free state in the solution and the chemical potential of the monomer in the aggregate, $c_{eq}^{(\infty)}$ is the concentration of segregating particles in the solution necessary for equilibrium coexistence of the solution with a solid phase at a planar interface.

In unsaturated solution ($c_0/c_{eq}^{\infty} < 1$) a stable cluster size distribution establishes in $\Delta t \sim 10^{-6}$ sec [5]. The major part of the C_{60} concentration is in the form of monomers.

A supersaturated solution ($c_0/c_{eq}^{\infty} > 1$) is metastable. The evolution of system goes in several stages, the final stage of the system is one large aggregate in equilibrium with monomers, dimers, etc. From Fig.1 it is seen that at a certain stage (stage of independent growth), the mean aggregate size in the solution remains constant. If this stage lasts a sufficiently long time, it may be the very size and concentration of aggregates that is observed in different experiments.

For various fullerene solutions, the

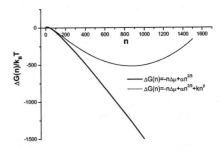

FIGURE 2. The $\Delta G(n)$ curves for the liquid drop model and the limited growth model.

calculated [5] time length of the stage of independent growth ($\sim10^{-3}$ sec.) is too short in comparison with experimental time. Therefore, the liquid drop model cannot appropriately describe aggregate formation in fullerene solutions.

It is clear then, that the function $\Delta G(n)$ should be modified so that some finite size would become energetically beneficial. This, in particular, can be achieved with the help of an additional term of the type kn^{β} in the equation for $\Delta G(n)$:

$$\frac{\Delta G(n)}{k_B T} = -n \ln \frac{c_0(t')}{c_{eq}^{\infty}} + \frac{\alpha_2}{k_B T} n^{2/3} + \frac{k}{k_B T} n^{\beta}. \tag{8}$$

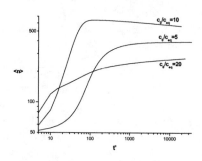

FIGURE 3. Evolution of <n> for limited growth model.

Such a modification and its comparison with $\Delta G(n)$ for the liquid drop model is qualitatively illustrated in Fig. 2.

For this model, the unsaturated solution behaves similar to liquid drop model. In the case of metastable solutions ($c_0 / c_{eq}^{\infty} > 1$), the behavior of the $f(n,t')$ evolution changes qualitatively.

Fig. 3 shows the evolution of <n> in supersaturated solution for the limited growth model. This behavior qualitatively explains observed aggregate growth in studied fullerene solutions[5].

CONCLUSIONS

In the present work in the frame of nucleation theory the kinetics of fullerene behavior in solutions have been studied. It is shown that for kinetic study of the aggregation processes in the solution DLA+KLA model is more suitable. For description of cluster (aggregates) evolution two models – the liquid drop model and the model of limited growth – were used. As the result of the numerical solutions of kinetic equations the evolution of mean aggregate size <n> was obtained.

For supersaturated solutions, we show that liquid drop model is inappropriate for description of cluster state in fullerene solutions, while the limited growth model qualitatively describes observed kinetics of C_{60} dissolution. However, the mechanism of stabilization of aggregate growth stays unclear and requires further investigations.

REFERENCES

1. Aksenov, V.L., Avdeev, M.V., et al., "Study of Fullerene Aggregates in Pyridine/water Solutions" in *Electronic Properties of Molecular Nanostructures-2001*, edited by H. Kuzmany et al., AIP Conference Proceedings **591**, New York: American Institute of Physics, 2001, pp. 66-69.
2. Smith, A.L., et al., *J. Phys. Chem.,* **100** 6775 (1996).
3. Bezmelnitsyn, V.N., Eletskii, A.B., Okun, M.B., *Usp. Fiz. Nauk,* **168** 1195 (1998) (in Russian).
4. Schmelzer J.W.P., et al., Chapters 2-3 in *Nucleation Theory and Applications*, edited by J.W.P. Schmelzer, G. Röpke, V.B. Priezzhev, JINR, Dubna, 1999, pp. 6-130.
5. Aksenov, V.L., Tropin, T.V., et al., "Kinetics of Cluster Growth in Fullerene Molecular Solutions" in *Physics of Particles and Nuclei*, 2005 (to be published).

Electron delocalization and dimerization in solid $C_{59}N$ doped C_{60} fullerene

A. Rockenbauer[*], Gábor Csányi[†], F. Fülöp[**], S. Garaj[‡], L. Korecz[§],
R. Lukács[**], F. Simon[**], L. Forró[‡], S. Pekker[¶] and A. Jánossy[**]

[*]*Chemical Research Center, Institute of Structural Chemistry, P.O. Box 17 H-1525 Budapest,
Hungary*
[†]*Theory of Condensed Matter group, Cavendish Laboratory, University of Cambridge, Cambridge
CB3 0HE, United Kingdom*
[**]*Budapest University of Technology and Economics, Institute of Physics, and Solids in Magnetic
Fields Research Group of the Hungarian Academy of Sciences, P.O. Box 91, H-1521 Budapest,
Hungary*
[‡]*Institute of Physics of Complex Matter, École Polytechnique Fédérale de Lausanne, CH-1015
Lausanne-EPFL, Switzerland*
[§]*Chemical Research Center, Institute of Chemistry, P.O. Box 17 H-1525 Budapest, Hungary*
[¶]*Research Institute for Solid State Physics and Optics, P.O. Box 49, H-1525 Budapest, Hungary*

Abstract. Electron spin resonance and ab initio electronic structure calculations show an intricate relation between molecular rotation and chemical bonding in the dilute solid solution $C_{59}N$:C_{60}. The unpaired electron of $C_{59}N$ is delocalized over several C_{60} molecules above 700 K, while at lower temperatures it remains localized within short range. The data suggest that below 350 K rigid $C_{59}N$–C_{60} heterodimers are formed in thermodynamic equilibrium with dissociated rotating molecules. The structural fluctuations between heterodimers and dissociated molecules are accompanied by simultaneous electron spin transfer between C_{60} and $C_{59}N$ molecules. The calculation confirms that in the $C_{59}N$–C_{60} heterodimer the spin density resides mostly on the C_{60} moiety, while it is almost entirely on $C_{59}N$ in the dissociated case.

Keywords: electron spin resonance, fullerene, molecular semiconductor, density functional theory
PACS: 71.20.Tx, 72.80.Rj, 76.30.Rn

INTRODUCTION

Pure C_{60} fullerene is an insulating solid with an energy gap of about 1.5 eV between the valence and conduction bands[1, 2, 3, 4, 5, 6]. It is natural to expect that suitably doped fullerenes maybe semiconductors. However, introducing donors or acceptors in a controllable way into C_{60} is very difficult.

In the dilute solid solution of $C_{59}N$ in C_{60} the $C_{59}N$ molecules substitute for C_{60}. There are two possible low energy configurations of the system. In one case, the $C_{59}N$ molecule remains isolated, in the other case a covalent bond forms with one of the neighbouring C_{60} molecules. First principles calculations show two distinct energy minima with a barrier between them. In the ESR experiments, we observe below 350 K a temperature dependent dynamic fluctuation between these two states. The ground state is the covalently bound heterodimer, in which the rotation of the molecules is not possible or is limited to rotations around the covalent bond joining the C_{60} and $C_{59}N$ molecules. At higher temperatures, the free energy associated to the large rotational

CP786, *Electronic Properties of Novel Nanostructures*, edited by H. Kuzmany, J. Fink, M. Mehring, and S. Roth
© 2005 American Institute of Physics 0-7354-0275-2/05/$22.50

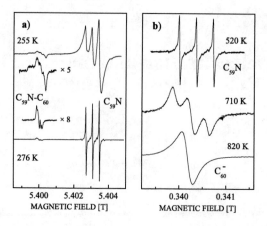

FIGURE 1. ESR spectra in dilute $C_{59}N:C_{60}$ solid solution. a) Spectra at 150 GHz near the $T_c = 261$ K phase transition of solid C_{60}. The two ESR lines are assigned to non-rotating $C_{59}N–C_{60}$ heterodimers and neutral, rotating $C_{59}N$. We propose that structural fluctuations lead to a simultaneous observation of the two species. The $C_{59}N–C_{60}$ signals are also shown magnified. b.) 9.4 GHz ESR spectra at high temperatures. Above 600 K the $C_{59}N$ ESR is gradually replaced by the ESR of C_{60}^- anions. This is attributed to delocalization of electrons over many C_{60} sites.

entropy overcomes the binding energy of the heterodimer. Between 350 K and 600 K all $C_{59}N$ molecules are in the unbound state. Above 600 K, the configurational entropy of the unpaired electron overcomes the coulomb attraction, and the electron is observed to delocalize in the crystal[7].

RESULTS AND DISCUSSION

Figure 1 shows the ESR spectra in the various states of $C_{59}N$ in the C_{60} matrix. Below 350 K, two ESR active species appear in $C_{59}N:C_{60}$ (Figure 1a). Both species are embedded in the C_{60} matrix since their ESR spectra change abruptly at the $T_c = 261$ K phase transition of solid C_{60}. We assign one component of the ESR spectrum to $C_{59}N–C_{60}$ heterodimers formed from $C_{59}N$ and one of its 12 C_{60} neighbors. The other component arises from $C_{59}N$ free radicals and was investigated in detail in an earlier work[6].

Between 120 and 600 K the molecular dynamics of $C_{59}N$ measured by ESR resembles the dynamics of neutral C_{60} in pure solid C_{60} measured by NMR[8]. At T_c, the *sc* phase with a well defined order of C_{60} molecular orientations changes to the *fcc* phase where molecules rotate quasi independently. The activation energies of rotational correlation times of C_{60} and $C_{59}N$ molecules are similar both above and below T_c. Below 200 K, the large g factor and hyperfine anisotropies of $C_{59}N$ result in a broad ESR line with a complicated structure in the powder samples. Between 220 and 261 K, the frequency of $C_{59}N$ rotations between energetically similar positions becomes faster than the frequency spread of the anisotropy, and the ESR spectrum gradually narrows. Above T_c, the $C_{59}N$ molecules rotate nearly freely and the ESR spectrum consists of extremely narrow lines of a ^{14}N triplet and a series of weak satellites from ^{13}C doublets.

We find a dramatic change in the ESR spectrum between 600 and 820 K (Figure 1b). The ^{14}N triplet with an isotropic g factor of 2.00137 transforms gradually into a single line shifted to lower fields at $g = 2.0024$ and the $C_{59}N$ spectrum disappears. We explain these observations as a delocalization of electrons, first over a cluster of $C_{59}N$ and 12 first neighbor C_{60} molecules and then at higher temperatures to larger distances. The decrease of free energy associated with this delocalization overcomes the Coulomb attraction of the $C_{59}N^+$ ion left behind.

We now turn to the assignment of the second ESR active species as $C_{59}N$–C_{60} heterodimers that form below 340 K. At 261 K the extra line changes abruptly (Figure 1a), the g factor anisotropy narrows but does not disappear. This g factor anisotropy is due to a molecular distortion; the g factor of an undistorted C_{60}^- ion would be isotropic. Thus the second species does not rotate freely since in that case the g factor distribution would be motionally narrowed. We suggest that this complex ESR spectrum reflects a thermal distribution of two molecular structures: a ground state $C_{59}N$–C_{60} heterodimer with a covalent bond between the component molecules and a higher energy state of dissociated neutral molecules. The heterodimer appears as a distorted C_{60} molecule in the ESR spectrum as the unpaired electron spin resides mostly on the C_{60} moiety while in the dissociated state the spin is on the $C_{59}N$ molecule. This spin transfer is plausible since the covalent bond takes one electron from the $C_{59}N$ and the other from the C_{60} moiety.

The following observations support this assignment. The extra line is intrinsic to $C_{59}N$:C_{60} since its intensity relative to the $C_{59}N$ line is sample independent, in a wide range of $C_{59}N$ concentrations. Yet, for all samples, the concentration ratio of the species of the extra line to monomeric $C_{59}N$ is the same at any given temperature. This is well explained if the unpaired electron spin hops with a low frequency between two states: in one it is localized to $C_{59}N$ and in the other mainly to a neighboring C_{60}. Since the observed C_{60} like radical has a static distortion, it is natural to assign it to C_{60} covalently bound to the neighboring $C_{59}N$ molecule.

The *ab initio* calculation of the electronic structure with all atomic positions relaxed shows a $C_{59}N$–C_{60} heterodimer with a $d_0 = 0.164$ nm long C–C covalent bond between the C1 first neighbor to the nitrogen atom of $C_{59}N$ (in the notation of [6]) and a C atom of C_{60} (Figure 2a). It has a mirror plane that includes the N atom and the C atoms of the intermolecular covalent bond. Twisting the C_{60} and $C_{59}N$ molecules around the bond confirms that the minimum energy configuration is the *trans* conformation with a mirror plane, shown in Figure 2a. This type of bonding is expected since C1 is the most reactive atom, as it has the largest spin density in the isolated $C_{59}N$ molecule. Figure 2a also shows the extra spin density, $\rho(z)$, along the molecular bond direction z, integrated over (x,y) planes perpendicular to z. Remarkably, $\rho(z)$ is almost entirely on the C_{60} moiety and is not at all spherically distributed. These characteristics are in agreement with the ESR assignment of a distorted C_{60} like molecule. We calculated a very small ^{14}N hyperfine constant of $A = 0.016$ mT from the 10% spin density remaining on the $C_{59}N$ molecule. We observed, however, no hyperfine splitting with a precision of 0.01 mT above and 0.05 mT below T_c, showing again that the spin transfer to the C_{60} moiety is nearly complete.

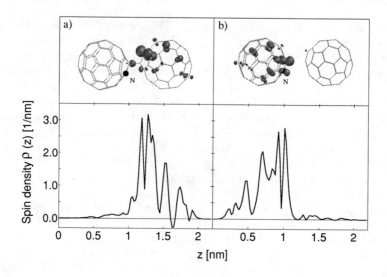

FIGURE 2. *Ab initio* calculation of molecular structures and the spin densities, $\rho(z)$, along the molecular axis z integrated in the plane perpendicular to z. a) the ground state heterodimer and b) the dissociated pair with a fixed intermolecular C–C separation $d = 0.30$ nm. Note the almost complete transfer of an electron spin from C_{60} to $C_{59}N$ with the dissociation of the heterodimer.

We calculated the way the heterodimer is torn apart by fixing larger and larger nearest intermolecular C–C distances, d, while relaxing all other atomic positions. The total energy of the molecule as a function of d is shown on Figure 3 and Figure 2b shows the structure of the dissociated molecule for $d = 0.30$ nm, which is about the nearest neighbor distance in solid C_{60}. Note that the lack of an energy minimum in the curve for the dissociated species reflects the fact the density functional theory does not include attracive dispersion forces. In contrast to the heterodimer, in the dissociated case most of the spin density resides on the $C_{59}N$ molecule. The energy difference between the heterodimer and the dissociated pair of $C_{59}N$ and C_{60} molecules is small. We find a barrier between the heterodimer and the dissociated pair with an energy maximum of 0.5 eV at $d = 0.21$ nm. The energy difference between the dimer molecule and the dissociated species (both indicated by arrows on Figure 3) is 0.18 eV, which is in excellent agreement with the experimental binding energy of dimer (2400 K corresponds to about 0.2 eV.)

CONCLUSION

Thus, the calculation is in quantitative agreement with the picture derived from the experiment. At finite temperatures the configuration fluctuates between the covalently bonded heterodimer and dissociated molecules since these configurations have similar energies and the barrier between them is small. Dissociation of the heterodimer is accompanied by transfer of an electron spin. In the experiments at intermediate tem-

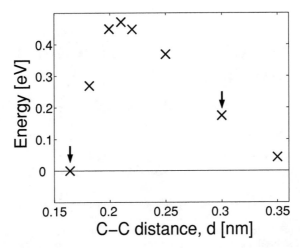

FIGURE 3. *Ab initio* calculation of the dissociation curve of $C_{59}N-C_{60}$, showing the total energy as a function of the length of the C–C bond being broken. Black arrows indicate the bond lengths corresponding to the bound heterodimer and the equilibrium intermolecular distance in the solid.

peratures we detect simultaneously the distorted C_{60} moiety of the heterodimer and the neutral $C_{59}N$ of the dissociated pair. he extra electron is localized to $C_{59}N$ from 350 K to about 700 K and only above this high temperature is $C_{59}N:C_{60}$ a doped semiconductor.

ACKNOWLEDGEMENTS

We are indebted to A. Hirsch for supplying pure $(C_{59}N)_2$, to K.–P Dinse for discussion and ESR measurements at an early stage of the study. We acknowledge Hungarian State grants OTKA 043255 (A. J.), TS040878 (F. S), T 032613 (S. P.), Bolyai Hungarian Fellowship (F. S.) and the Swiss National Science Foundation (L. F. and S. G.). Computational work was carried out at the CCHPCF, University of Cambridge and supported by grant EC HPRN-CT-2000-00154.

REFERENCES

1. A. Oshiyama, S. Saito, N. Hamada, Y. Miyamoto J. Phys. Chem. Solids **53**, 1457 (1992).
2. W. Andreoni, F. Gygi, M. Parrinello Chem. Phys. Lett. **190**, 159 (1992).
3. T. Guo, C. Jin, R. E. Smalley J. Phys. Chem. **95**, 4948 (1991).
4. J.C. Hummelen, B. Knight, J. Pavlovich, R. González, F. Wudl Science **269**, 1554 (1995).
5. B. Nuber, A. Hirsch, J. Chem. Soc. Chem. Commun 1421 (1996).
6. F. Fülöp, A. Rockenbauer, F. Simon, S. Pekker, L. Korecz, S. Garaj, A. Jánossy, Chem. Phys. Lett. **334**, 233 (2001).
7. A. Rockenbauer, G. Csányi, F. Flöp, S. Garaj, L. Korecz, R. Lukács, F. Simon, L. Forró, S. Pekker, A. Jánossy, Phys. Rev. Letts. **94** 066603 (2005).
8. K. Mizoguchi, Y. Maniwa, K. Kume Mater. Sci. Eng. B **19** 146 (1993).

Photoconductivity Associated with Thermal Dissociation of Frenkel Excitons in Pristine C_{60} Crystals

K. Kan'no, R.Tanaka, T. Koyanagi and I. Akimoto

Department of Materials Science and Chemistry, Faculty of Systems Engineering,
Wakayama University, 930 Sakaedani, Wakayama 640-8510, Japan

Abstract. Temperature dependence of the optical absorption and photo-conductance spectra of pristine C_{60} single crystals have been investigated in the range of 4-300 K. At low temperatures below 150 K, two different contributions, one from Frenkel excitons and the other from localized excitons, were clearly resolved above and below the fundamental absorption edge around 1.81 eV, whereas no photo-conductivity was stimulated in the whole spectral range. At high temperature above 200 K, however, photoconductivity is stimulated with an appreciable red-shift of the threshold energy, and it can be attributed to mobile electrons generated via the thermal dissociation of the bulk Frenkel excitons. On the basis of these results, an origin of the "anomalous" photoconductivity peak, which was first reported by Matsuura *et al.* (Phys. Rev. **B51** (1995) 10217), has been reconsidered.

Keyword: fullerene; photoconductivity; exciton; thermal dissociation; relaxation dynamics
PACS: 71.35.Aa , 72.40.+w

INTRODUCTION

In previous studies of photoluminescent processes and near-band-edge electronic structures in pristine C_{60} single crystals, we clarified a characteristic role of unexpected transitions found at 1.69 eV just below the fundamental absorption edge [1,2]. As a result, the origin of the well-known photoluminescence spectrum and its relaxation kinetics is fully understood in terms of the inhomogeneously broadened localized exciton states. These states are presumably due to perturbed Frenkel excitons stabilized with symmetry lowering by adjacent intrinsic defects or some lattice disorder, so-called *"X-traps"* [3]. On the other hand, an anomalous peak of photoconductivity was reported by Matsuura *et al.* [4] evolving around 1.65 eV upon raising temperature above 200 K. They assigned this peak to the transition to the t_{1u} molecular orbital, being distinctive from photo-conductance excited at a energies higher than 1.8 eV, which was ascribed to autoionization of excitons involving higher molecular orbitals. Their assignment, however, seems not to be in line with our current understanding of photoluminescence studies [2]. Therefore, in the present study the origin of this "anomalous" photoconductivity peak has been reconsidered on the basis of its temperature dependence of the near-edge optical absorption spectrum and that of the photoconductance spectrum, both of which were measured in one and the same C_{60} crystal.

CP786, *Electronic Properties of Novel Nanostructures*, edited by H. Kuzmany, J. Fink, M. Mehring, and S. Roth
© 2005 American Institute of Physics 0-7354-0275-2/05/$22.50

EXPERIMENTAL

Crystals of C_{60} were grown in vacuum by sublimation starting from a 99.98% pure powder material (Term Co.). For measuring photoconductivity, two electrodes were formed by depositing gold on the same (111) surface of the fcc crystal platelet with 1mm spacing. Considering the effect of non-Ohmicity between sample surface and electrodes, carrier generation processes were also reconsidered on the basis of transient photoconductivity obtained by time-of-flight measurements using blocking electrodes.

RESULTS AND DISCUSSION

Origin of the "Anomalous" Photoconductivity Peak

Figure 1 shows the log (I_0/I) spectra measured just below the fundamental absorption edge at various temperatures 4.2 - 300 K, where I_0 and I are intensities of incident and transmitted light, including effects of optical loss due to reflectivity and scattering. As typically seen in the spectrum at 4.2 K (bold line 1), a steep increase of absorption coefficient is readily recognized at 1.81 eV. This corresponds to the onset of the lowest absorption band associated with the $h_u \rightarrow t_{1u}$ transition of the bulk Frenkel exciton. Below this edge, additional absorption bands appear, originating from the localized exciton (so-called "deep X trap"), with a prominent peak at 1.68 eV and three sub-peaks at 1.72, 1.75 and 1.78 eV. These spectral features, which agree well with those of our previous study, are attributed to the inhomogeneously broadened vibronic progressions of "gerade"-symmetry [2].

When raising the temperature, the steep edge absorption from bulk excitons gradually shifts towards lower energy, while the absorption bands of localized excitons hardly change except broadening. Thus, the localized exciton absorption peaks are finally obscured by the red-shift of the fundamental absorption edge, and can hardly be recognized above 150 K.

Figure 2 shows excitation spectra for the photoconductance in the pristine C_{60} single crystal obtained at temperatures between 200 - 300 K. At temperatures less than 150 K, the photoconductivity was hardly observed. Upon raising the temperature, the photoconductance band emerges above 1.8 eV, the threshold of the $h_u \rightarrow t_{1u}$ transition. Higher energy spectral features agree qualitatively well with earlier reported results obtained using thin films [5]. Also, they exhibit anti-correlation with the absorption spectra of thin film; that is, the dip structures at 2.7, 3.6, and 4.9 eV

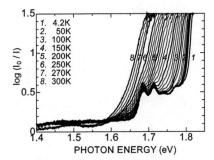

FIGURE 1. Absorption spectra obtained for a pristine C_{60} single crystal at various temperatures 4.2 K $\sim$ 300 K with regular intervals of 10 K.

47

correspond to characteristic absorption peaks of C_{60}, i. e., $h_u \rightarrow t_{1g}$, $t_{1u} \rightarrow h_g$, $h_g + g_g \rightarrow t_{1u}$, and $h_u \rightarrow h_g$, respectively [6].

The evolution of photo-conductivity most likely results from thermal decomposition of excitons, as indicated by correlation with thermal quenching of photoluminescence (called type A emission in ref. [2]). It is noteworthy that the photoconductivity spectra shown in Fig. 2 are essentially the same as the excitation spectrum for the photoluminescence obtained at low temperatures [7]. This fact suggests that both the photoluminescence at low temperatures and the photoconductivity at high temperatures originate from a common initial relaxation process.

Here we focus our attention on the spectral range around 1.65 eV, in which Matsuura et al. [4] found an "anomalous" photoconductance peak. Figure 3 shows the photoconductivity spectra for 1.4 - 1.9 eV with expanded scales. In agreement with their report, above 200 K one can see the evolution of a separated peak structure around 1.65 eV. It should be remarked, however, that the dip around 1.7 eV, and not the growing peak, corresponds to the position of the lowest peak of localized exciton absorption (Fig.1). In contrast with the previous interpretation given in ref. [4], we suggest that evolution of the "anomalous" peak around 1.65 eV, as well as growth of photoconductivity above 1.8 eV, should be attributed to the thermal dissociation of the bulk Frenkel excitons. The dip structure arises from the fact that the photo-excitation into the bulk exciton absorption results finally in photo-conductivity, but the direct excitation into the localized exciton absorption does not at all contribute to the generation of carriers.

According to this new interpretation, the temperature dependence of photoconductivity spectra is computer-simulated by assuming an invariant absorption

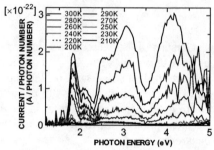

FIGURE 2. Photoconductivity spectra obtained for a pristine C_{60} single crystal at various temperatures 200～300K.

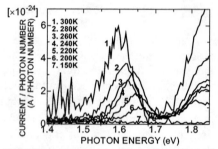

FIGURE 3. Photoconductivity spectra obtained for a pristine C_{60} single crystal at various temperatures 150K～300K.

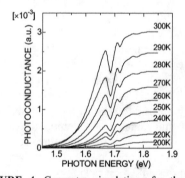

FIGURE 4. Computer simulations for the photoconductance spectra of C_{60} single crystal at various temperatures 200K – 300K.

48

profile of the localized band, a red-shift of the exponential tail of the Frenkel exciton absorption, and thermal dissociation of Frenkel excitons with activation energy of 0.15 eV.[4] As seen in Fig. 4, the main features of the experimental spectra can well be reproduced.

Carrier Generation in C_{60} Crystals

Figure 5 shows an Arrhenius plot of the photocurrent intensity, measured by a time-of-flight set-up using blocking electrodes. Comparing with the temperature dependence of the photoluminescence, which has been attributed to the existence of localized excitons, it is found that the thermally activated behavior of the photoconductance exhibits anti-correlation to photoluminescence quenching. Furthermore, the activation energy of 95 meV is similar. This behavior is independent of excitation energy in the range 1.85-3.65 eV, indicating that carrier generation in C_{60} crystals originates from a common initial state.

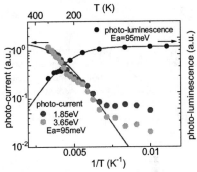

FIGURE 5. the Arrhenius plot of photocurrent and photoluminescence.

SUMMARY

It is concluded that the "anomalous" photoconductivity peak is attributed to the existence of localized states which are not involved in photoconduction. We suppose that mobile Frenkel excitons thermally decompose into free carriers before being trapped into the localized exciton state, under condition of sufficiently elevated temperatures.

This work was supported in part by the Grant-in-Aid for Science Research from the Ministry of Education, Culture, Sports, Science and Technology (No. 15340098, 15740214).

REFERENCES

[1] I. Akimoto, M. Ashida, K. Kan'no, Chem. Phys. Lett. **292** (1998) 561.

[2] I. Akimoto, K. Kan'no, J. Phys. Soc. Jpn **71** (2002) 630.

[3] D.J. van den Heuvel, G.J.B. van den Berg, E.J.J. Groenen, J. Schmidt, I. Holleman, G. Meijer, J. Phys. Chem. **99** (1995) 11644.

[4] S. Matsuura, T. Ishiguro, K. Kikuchi, Y. Achiba, Phys. Rev. **B51** (1995) 10217.

[5] C.H. Lee, G. Yu, B. Kraabel, D. Moses, Phys. Rev. **B49** (1994) 10572.

[6] M.S. Dresselhaus, G. Dresselhaus, P.C. Eklund, Science of Fullerenes and Carbon Nanotubes, Academic Press, 1996, p.500.

[7] I. Akimoto, Bussei_Kenkyu **67-1** (1996) 57. (in Japanese)

First Chemical Synthesis of H2@C60

Koichi Komatsu

Institute for Chemical Research, Kyoto University
Uji, Kyoto 611-0011, Japan

Abstract. First production of fullerene C_{60} encapsulating molecular hydrogen, H2@C_{60}, has been achieved by organic synthesis with the technique of molecular surgery. The method involves opening a hole on the surface of C_{60}, enlargement of the hole, insertion of a hydrogen molecule, and closure of the hole without loss of any hydrogen molecule. Interaction between the inside hydrogen and outside pi-conjugated system of the C_{60} cage is quite small as shown by NMR and electrochemical measurements.

INTRODUCTION

Fullerene, C_{60}, is a typical nanometer-sized material, and, because of its characteristic electronic properties originating from its low-lying LUMOs and highly symmetric structure, it has great possibility for application.[1,2] Thus, in the area of organic chemistry, a wide variety of derivatives have been synthesized.[3] But these are of course the products of the exterior modification of the fullerene cage. In contrast, the internal structural modification, which is concerned with the science of so-called endohedral fullerenes appears to be not so well developed as compared with the external modification. This is ascribed to the severe limitation for the formation of endohedral fullerenes, which has relied on the rather hardly controllable physical method for their production, that is, for example, arc discharge of the graphite rods containing metal oxides[4,5] or high-temperature/high-pressure treatment of C_{60} with noble gase.[6-8] Here we describe our recent success in production of a new endohedral fullerene encapsulating hydrogen, H_2@C_{60}, by the use of a molecular surgical method, that involves opening a hole on the surface of C_{60}, insertion of molecular hydrogen through the hole, and complete closure of the hole to regenerate the C_{60} structure.[9]

OPEN-CAGE FULLERENES

As an approach to the molecular surgery of C_{60}, Rubin is the first to show that it is possible to open a 14-membered-ring hole (compound **1**)[10] on C_{60} and introduce a small molecule and atom into the fullerene cage.[11] Under high temperature and high

CP786, *Electronic Properties of Novel Nanostructures*, edited by H. Kuzmany, J. Fink, M. Mehring, and S. Roth
© 2005 American Institute of Physics 0-7354-0275-2/05/$22.50

pressure, the amount introduced was 5% for H_2 (400 °C, 100 atm) and 1.5% (300 °C, 475 atm),[11] but these values are much higher than 0.1% reported previously.

On the other hand, in a study independent of Rubin's work, we found that a solution-phase thermal reaction of C_{60} with a 1,2,4-triazine derivative is suitable for obtaining an open-cage fullerene **2**, which has an eight-membered-ring hole, in good yield (85% yield based on consumed C_{60}; recovery of C_{60} being 41%) (Fig. 2(a)).

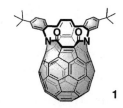

FIGURE 1.   Rubin's open-cage fullerene with a 14-membered-ring hole, **1**.

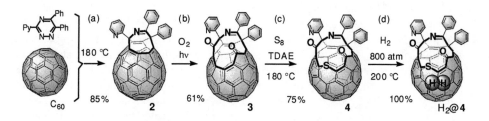

FIGURE 2.   Opening a hole on C_{60} and its enlargement to a 13-membered-ring hole.

Then, as shown in Fig. 2(b), the eight-membered-ring hole was enlarged by photochemical oxidation to give a C_{60} derivative **3** and its isomer, both with a 12-membered-ring opening.   Further enlargement of the hole of the major isomer **3** to a 13-membered ring was conducted by insertion of a sulfur atom on the rim of the opening by activation of compound **3** by the addition of a strong electron donor, tetrakis(dimethylamino)ethylene (TDAE) (Fig. 2(c)). [12]

The structure of the open-cage fullerene with a 13-membered-ring **4** was determined as shown in Fig. 3 by X-ray crystallography.   The opening appears to be fairly large, but the space-filling model indicates that it is just only a good size for a small molecule such as hydrogen.

51

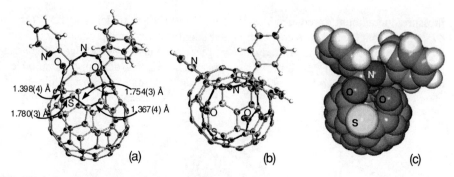

FIGURE 3. Molecular structure of an open-cage fullerene **4** determined by X-ray crystallography. (a) side view, (b) top view, (c) top view of a space-filling model.

INSERTION OF MOLECULAR HYDROGEN

Insertion of molecular hydrogen into open-cage fullerene **4** was conducted upon treating the powdery crystals of **4** in an autoclave under pressurized hydrogen at 200 °C for 8 hours (Fig. 2(d)). The rate of incorporation of hydrogen was 51% when the pressure was 180 atmosphere, 90% under 560 atmosphere, and finally 100% under 800 atmosphere.[13] The inserted hydrogen was confirmed by the ^1H NMR signal, which was shifted to unusually high field (δ −7.25 ppm) due to the strong shielding effect of the fullerene cage. The hydrogen incorporating compound $H_2@4$ is quite stable at the temperature below 160 °C, but when heated above 160 °C, it slowly released the encapsulated hydrogen at the rate according to the first-order kinetics. The activation parameters for this process have been determined as shown in Table 1. Thus, the following procedures for the closure of the opening have to be conducted at the temperature below 160 °C in order to avoid the release of hydrogen.

TABLE 1. Activation parameters for release of hydrogen from $H_2@4$.

E_{act} kcal mol^{-1}	$\Delta G^{\ddagger\,a)}$ kcal mol^{-1}	$\Delta H^{\ddagger\,a)}$ kcal mol^{-1}	$\Delta S^{\ddagger\,a)}$ cal K^{-1}mol^{-1}
34.3	35.5	33.4	−7

a) At 25 °C.

CLOSURE OF THE OPENING

The size of the opening of $H_2@4$ was reduced in a stepwise fashion as shown in Fig. 4. First, the sulfur atom was eliminated by oxidation with m-chloroperbenzoic acid (MCPBA) to SO, which was followed by removal of the SO unit by photo-irradiation to give $H_2@3$ with a 12-membered-ring opening (Fig. 4(a) and (b)). Now the MALDI TOF MS spectrum of $H_2@3$ exhibited a base peak not for the molecular-ion peak but for a m/z with 722, which corresponds to $H_2@C_{60}$ as shown in Fig. 5. This means that the energy of laser irradiation on $H_2@3$ can cause rearrangement involving the elimination of addends such as benzene and pyridine rings on **3** to yield $H_2@C_{60}$ under vacuum to be detected by MS spectrometry.

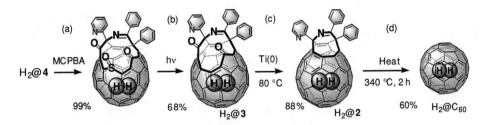

FIGURE 4. Chemical reactions to close the hole of hydrogen-containing open-cage fullerene $H_2@4$.

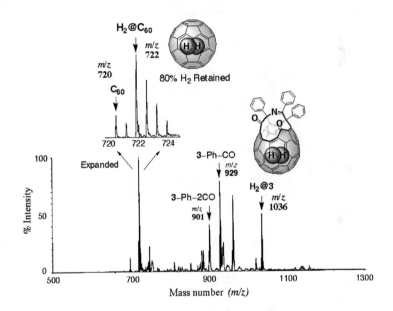

FIGURE 5. MALDI TOF MS spectrum of hydrogen incorporating open-cage fullerene $H_2@3$. Matrix, dithranol: negative-ion mode.

However in the MS spectrum, there is also observed a peak for empty C_{60} in an intensity of about 20%, indicating that a part of encapsulated hydrogen is released upon laser irradiation. Therefore, it is necessary to further reduce the opening to prevent the release of inside hydrogen before applying the hole-closure reaction. Thus, as shown in Fig. 4(c), the two C=O groups at the rim of the opening was coupled by the use of activated Ti(0) following the McMurry's method to reduce the 12-membered-ring opening into an eight-membered ring as in H_2@**2**.

Finally, the complete closure of the opening was achieved simply by the thermal reaction of the solid of H_2@**2** under vacuum. Application of 340 °C for 2 hours resulted in the formation of desired H_2@C_{60} in 61% contaminated by 9% of empty C_{60}. The purification of H_2@C_{60} was performed with recycling high-pressure liquid chromatography (HPLC) over Buckyprep column eluted with toluene. The 100% pure H_2@C_{60} was obtained as a peak with the retention time of 399 minutes while empty C_{60} was separated as a peak at 395 minutes.

PROPERTIES OF H_2@C_{60}

The ^{1}H NMR signal for the hydrogen in C_{60} was observed at δ –1.44 ppm, which is 5.98 ppm upfield-shifted relative to the signal of free hydrogen dissolved in the solvent (*o*-dichlorobenzene-d_4). This extent of upfield shift is comparable to that observed for ^{3}He for C_{60} encapsulating 0.1% ^{3}He in ^{3}He NMR,[6] indicating that the magnetic effect of the C_{60} cage is almost constant indifferent to the paramagnetic species encapsulated inside the cage. The ^{13}C NMR signal for the C_{60} cage of ^{3}He@C_{60} appeared at δ 142.844 ppm, which is only 0.078 ppm downfield-shifted from that of empty C_{60}. This is considered as the result of weak electronic and/or van der Waals interaction of inside hydrogen molecule and the outside C_{60} cage. Reflecting such a small interaction caused by the inside hydrogen, the UV-vis and IR spectra of purified H_2@C_{60} were found to be essentially the same as those of empty C_{60}. Even electrochemical reduction (up to –2.5 V vs Fc/Fc$^+$) or oxidation (up to +2 V) of the outer fullerene cage does not affect the behavior of encapsulated hydrogen molecule.

The purified H_2@C_{60} was thermally stable since no decomposition nor release of hydrogen was observed by heating at 500 °C.

This work is regarded as just a first step toward the controlled synthesis of various endohedral fullerenes. It is quite possible that the present method could be applied to the synthesis of endohedral fullerenes such as D_2@C_{60} as well as the homologous series with C_{70}. The road is long and steep, but we hope that this road would eventually lead to the actualization of macroscopic synthesis of endohedral metalofullerenes in future.

REFERENCES

1. S. Kobayashi, S. Mori, S. Iida, H. Ando, T. Takenobu, Y. Taguchi, A. Fujiwara, A. Taninaka, H. Shinohara, and Y. Iwasa, *J. Am. Chem. Soc.* **125**, 8116-8117 (2003).
2. H. Kato, Y. Kanazawa, M. Okumura, A. Taninaka, T. Yokawa, and H. Shinohara, *J. Am. Chem. Soc.* **125**, 4391-4397 (2003).
3. A. Hirsch and M. Brettreich, *Fullerenes - Chemistry and Reactions*, Wiley-VCH Verlag, Weinheim, 2005.
4. H. Shinohara, in *Fullerenes: Chemistry, Physics and Technology*, edited by K. M. Kadish and R. S. Ruoff, Wiley, New York, 2000, pp. 357-393.
5. H. Shinohara, *Rep. Prog. Phys.* **63**, 843-892 (2000).
6. M. Saunders, R. J. Cross, H. A. Jiménez-Vázquez, R. Shimshi, and A. Khong, *Science* **271**, 1693-1697 (1996).

7. M. Saunders, H. A. Jiménez-Vázquez, R. J. Cross, S. Mroczkowski, D. I. Freedberg, and F. A. L. Anet, *Nature* **367**, 256-258 (1994).

8. M. S. Syamala, R. J. Cross, and M. Saunders, *J. Am. Chem. Soc.* **124**, 6216-6219 (2002).

9. K. Komatsu, M. Murata, and Y. Murata, *Science*, **307**, 238-240 (2005).

10. G. Schick, T. Jarrosson, and Y. Rubin, *Angew. Chem. Int. Ed.* **38**, 2360-2363 (1999).

11. Y. Rubin, T. Jarrosson, G-W. Wang, M. D. Bartberger, K. N. Houk, G. Schick, M. Saunders, and R. J. Cross, *Angew. Chem. Int. Ed.* **40**, 1543-1546 (2001).

12. Y. Murata, M. Murata, and K. Komatsu, *Chem. Eur. J.,* **9**, 1600-1609 (2003).

13 Y. Murata, M. Murata, and K. Komatsu, *J. Am. Chem. Soc.* **125**, 7152-7153 (2003).

QUANTIZED ROTATIONAL STATES IN ENDOHEDRAL FULLERENES

M. Hulman[1], M. Krause[1,2], O. Dubay[1], G. Kresse[1], K. Vietze[3], G.Seifert[3],
C. Wang[4], H. Shinohara[4] and H. Kuzmany[1]

[1] Institut für Materialphysik, Universität Wien, Austria
[2] Institut für Festkörper- und Werkstofforschung, Dresden, Germany
[3] Institut für Physikalische Chemie, Technische Universität Dresden, Dresden, Germany
[4] Department of Chemistry and Institute for Advanced Research, Nagoya University, Japan

Abstract. Results of Raman measurements on endohedral fullerene $Sc_2C_2@C_{84}$ will be presented. We concentrated preferentially on the low-energy part of the spectrum where signatures of the rotational state of the C_2 molecule are expected. The positions of the Raman lines are consistent with those for the plane rotor. However, deviations appear in the very low-energy part of the spectrum. The intensity of Raman lines anticipates both the plane and free rotor, depending on the temperature. This puzzling behaviour is resolved by introducing a barrier the molecule encounters during rotation. Consequently, the low-lying rotational lines split and the line intensities redistribute in agreement with the experiment. The broadening of the line widths seen at higher temperatures likely points to thermally activated processes in $Sc_2C_2@C_{84}$.

Keywords: fullerenes, endohedral, Raman, rotation

PACS: 78.30.Na,33.15.Dj,33.15.Hp,33.15.Mt

INTRODUCTION

The idea that the interior of fullerene cages may provide an ideal space for growing unique material appeared shortly after the fullerenes had been discovered. Various atoms were encaged so far. In some cases, like in $Sc_2@C_{84}$, the encapsulated species are bound to the fullerene cage that restricts their dynamics allowing them to vibrate only [1]. In other cases, like for trimetallnitride cluster molecules in C_{80}, the weak interaction between the cluster and the cage allows for more complicated dynamics [2]. In agreement with theoretical models, a diffuse motion governs this dynamics.

From this point of view, the molecule $Sc_2C_2@C_{84}$ represents a new system. The two scandium atoms are still bound to the C_{84} cage, but the C_2 dimer is relative distant from both the Sc atoms and the cage [3]. Since there is no direct bonding, the C_2 dynamics is governed by the fields created by the surroundings Sc and carbon atoms. At certain circumstances, this might lead to a quite different dynamics and even a quasi-free rotation of the C_2 molecule can be anticipated.

The rotational spectra are well known for gases. On the other hand, the resolved rotational patterns are really rare in the solid state. The vast majority of examples deal

CP786, *Electronic Properties of Novel Nanostructures*, edited by H. Kuzmany, J. Fink, M. Mehring, and S. Roth
© 2005 American Institute of Physics 0-7354-0275-2/05/$22.50

with hydrogen because of its large rotational constant. Rotational lines of short-living C_2 molecules were observed only in the gas phase as structures accompanying electronic transitions in a hot carbon plasma [4]. The C_{84} cage stabilizes the molecule and provides a possibility to study its rotational properties for the first time by Raman scattering.

EXPERIMENTAL

The $Sc_2C_2@C_{84}$ powder was dissolved in toluene and drop coated on gold covered Si substrates. The latter were subsequently annealed at 500 K in high vacuum for several hours. Raman measurements were carried out in the backscattering geometry for red (647.1 nm) and green (514.5 nm) excitations with a 1.5 cm^{-1} resolution. The temperature of the sample varied between 25 and 200 K. Line positions and widths were determined by fitting the spectral lines by Voigtian profiles after the stray-light background had been subtracted.

RESULTS AND DISCUSSIONS

An example of the Raman spectrum taken at low temperatures is shown in Fig.1. Two lines at 100 cm^{-1} and 126 cm^{-1} represent the vibrations of the Sc atoms against the carbon cage. The series of almost equidistantly spaced lines below 100 cm^{-1} is a unique feature of $Sc_2C_2@C_{84}$. The regular spacing is a characteristic of rotational spectra of diatomic molecules.

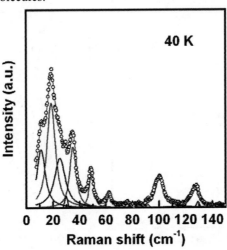

FIGURE 1. Low-energy Raman spectrum of $Sc_2C_2@C_{84}$ measured at 40 K with the red laser, λ=647.1 nm. The circles represent the as measured spectrum; the solid line connecting the circles is a sum of Voigtian profiles shown.

For the rotational lines with the shift larger than ~ 30 cm^{-1}, it turned out that the positions agree well with those predicted for a diatomic molecule rotating in a plane. From that, the rotational constant of 1.73 cm^{-1} and the internuclear distance of 0.127 nm were determined for the C_2 plane rotor [5]. However, the position of the line at 18 cm^{-1} is downshifted by about 2 cm^{-1} as compared to theory. Moreover, there are additional lines at 11 and 25 cm^{-1} that are not present in a simple energy–level scheme for a free unperturbed plane rotor. In fact, *ab-initio* density functionals theory (DFT) calculations yielded a small periodic potential of ~30 cm^{-1} the C_2 molecule encounters during the rotation. The height of the barrier is comparable or even higher than the energy of the lowest rotational states of the plane rotor. Therefore, the strong perturbation of such states may lead to the shift of the energy levels and to a lifting of degeneracy. This is fully justified by solving the Schrödinger equation including the additional potential. In order to obtain the best possible agreement between the as calculated and as observed line positions (<1 cm^{-1} in the whole spectrum), the rotational barrier height was set to ~13 cm^{-1}. This is less than one-half of the value from *ab-initio* calculations. On the other hand, it is a tiny fraction from the absolute binding energy of the C_2 unit in the $Sc_2@C_{84}$ cage (several eV) and can be accounted for by inaccuracies of density functionals [5].

The intensity profiles are another important characteristic of the rotational spectra. In Fig.2, the spectra measured at two different temperatures are displayed. For 40 K, the intensity decreases rapidly following approximately an exponential decay, as expected for the plane rotor. At 120 K, the spectral weight is shifted to higher rotational transitions, a feature typical for three dimensional (3D) rotational spectra.

The apparent discrepancy between the 3D-like intensity profile and the 2D rotation of C_2 molecules can be resolved when the exact rotational wavefunctions and the Raman polarizability tensor of the problem are carefully examined. The matrix

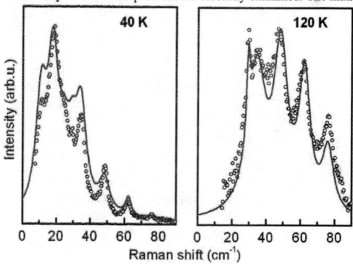

FIGURE 2. Experimental (circles) and as calculated (solid lines) rotational spectra of C_2 molecule for two different temperatures. The intensities were normalized to the maximum intensity of the experimental spectrum.

elements of the tensor must be calculated for each Raman transition. At the lowest temperatures, the rapidly decaying Boltzmann factor suppresses the influence of the matrix elements and governs basically the intensity profile. On the other hand, the matrix elements of the transitions start to play a significant role at higher temperatures where the Boltzmann factor flattens. The agreement between the theoretical intensity profile and the experiment is very good as shown in Fig.2.

The intensity is not the only quantity that changes with temperature. Fig.3. demonstrates the temperature dependence of the linewidths of the rotational lines as compared to the dependence for the Sc-cage mode at 100 cm^{-1}.

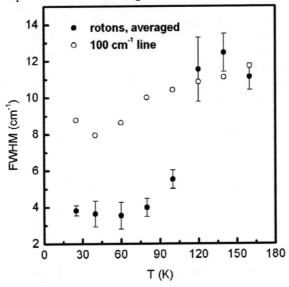

FIGURE 3. Full width at half maximum (FWHM) for the Sc-cage vibrational mode at 100 cm^{-1} (open circles) and the average for all the observed rotational lines (solid circles) as a function of temperature.

The two dependencies are essentially different. The FWHM of the Sc-cage vibration is temperature independent at low temperatures and then increases linearly. The line width increases by about 30 % between 25 and 160 K. The rotational lines are much narrower than the Sc-cage line at low temperatures and their FWHM is also temperature independent up to ~ 80 K. After that, it increases by a factor of three between 80 and 120 K. At even higher temperatures, the linewidth becomes comparable to the separation between the lines. Eventually, the rotational lines form a broad continuum and cannot be resolved in the spectra for temperatures above 160 K.

The linewidth dependence for the Sc-cage mode can be explained by taking into account a simple model of a phonon-phonon interaction. In such a case, the linewidth is proportional to the number of phonons, n_{ph}. At low temperatures where $\hbar\omega_{ph} \gg k_B T$ (k_B is Boltzmann constant), the phonon number is weakly temperature dependent. On the other hand, the phonon number $n_{ph} \sim T$ for $\hbar\omega_{ph} < k_B T$.

For the rotational lines, the steep temperature dependence of FWHM between 80 and 120 K can be fitted with the exponential function. This points out to a thermally activated mechanism. A similar type of behavior was found for the linewidth of the cage modes in $Sc_2@C_{84}$. However, the onset of the increase was observed at much higher temperatures around 160 K. In that case, a coupling between the cage vibrations and thermally activated cage librations and/or rotations was proposed to explain the temperature dependence.

For $Sc_2C_2@C_{84}$, one can also expect the coupling of the C_2 molecule rotations to the C_{84} cage degrees of freedom. The axis of rotation of the C_2 molecule is parallel to the C_2 axis of the C_{84} cage. The $Sc_2C_2@C_{84}$ molecule resembles a macroscopic gyroscope when the cage is static. The situation changes significantly when the cage starts to librate or rotate. The angular momentum of the C_2 unit may couple to the angular momentum of the cage librating/rotating around the axis. Unlikely the macroscopic gyroscope, the C_2 molecule angular momentum can be strongly disturbed by the cage. The C_2 rotational energy levels (depending on the angular momentum) become very broad and the Raman transitions disappear from the spectrum. Likely, the C_2 rotational levels couple to the librations of the C_{84} cage above 80 K, which eventually turn out to the cage rotation at 160 K.

ACKNOWLEDGMENTS

M.H. acknowledges support from the EU Project NANOTEMP (HPRN-CT-2002-00192).

REFERENCES

1. Krause, M., Hulman, M., Kuzmany, H., Dennis, T.J.S., Inakuma, M., Shinohara, H., *J. Chem. Phys.* **111**, 7976–7984 (1999)
2. Stevenson, S., Rice, G., Glass, T., Harich, K., Cromer, F., Jordan, M.R., Craft, J., Hadju, E., Bible, R., Olmstead, M.M., Maitra, K., Fisher, A.J., Balch, A.L., Dorn, H.C., *Nature (London)* **401**, 55 -57 (1999).
3. Wang, C.R., Kai, T., Tomiyama, T., Yoshida, T., Kobayashi, Y., Nishibori, E., Takata, M., Sakata, M., Shinohara, H., *Angew. Chem.* **40**, 397 – 399 (2001)
4. Weltner, W., van Zee, R.J., *Chem. Rev.* **89**, 1713-1747 (1989)
5. Krause, M., Hulman, M., Kuzmany, H., Dubay, O., Kresse, G., Vietze, K., Seifert, G., Wang, C., Shinohara, H., *Phys. Rev. Lett.* **93**, 137403 (2004)

Medium-energy proton irradiation of fullerene films: polymerization, damage and magnetism

K. – H. Han[a], T. L. Makarova[a,b], A. L. Shelankov[a,b], I. B. Zakharova[c], V. I. Sakharov[b] and I. T. Serenkov[b]

[a] Umea University, 90187 Umea, Sweden
[b] Ioffe Physico-Technical Institute of the RAS, St. Petersburg, 194021, Russia
[c] State Polytechnic University, St. Petersburg, 19525, Russia

Abstract. We study magnetism of C_{60} films prepared by the discrete evaporation in a quasi-closed volume and irradiated by protons within the energy range 20 – 95 keV. Structural changes are studied by Raman spectroscopy and atomic force microscopy (AFM). Whereas the proton bombardment does not produce noticeable changes in the surface AFM-morphology, the irradiated regions show pronounced magnetic structure as seen by means of magnetic force microscopy (MFM). The intrinsic nature of the MFM-images follows from the fact that the MFM-images are sensitive to the scanning height, the type of magnetic tip, reversal of magnetization of the tip, and are modified after application of magnetic field.

INTRODUCTION

The recent experiments of Esquinazi et al. have shown that the irradiation of ultra-pure graphite with a beam of protons at energy of 2.25 MeV creates structural defects and transforms the diamagnetic graphite into a ferromagnetic state [1]. These experiments offer a method for controlled manufacturing of a micrometer-size ferromagnetic pattern on otherwise nonmagnetic materials [2]. Defect-induced ferromagnetic behaviour has also been reported for the fullerene structures polymerized by different methods [3]. In the case of fullerene solids, the data may be interpreted   in different ways: either to regard fullerenes as an auxiliary unit that transforms into graphite-like structure or to look for a universal mechanism which accounts for magnetism in structures of buckyballs with some amounts of defects.

In this work we present the study of ferromagnetic spots on fullerene films produced by medium-energy protons.   The polymerized fullerene films with the stoichiometry $C_{60}H_{0.5}$ were thus created, which showed distinct images in magnetic force microscopy.

Images of a magnetic structure of the films as well as the film topography were obtained simultaneously with a Nanoscope IV, a scanning probe microscope from Digital Instruments. The microscope was operated in a "tapping / lift" scanning mode. The most common application of MFM is to characterize a sample's magnetic domain structure. The "tapping/lift" mode of operation can efficiently separate topography contrast and magnetic interaction. In this mode, the image is scanned twice: the to-

CP786, *Electronic Properties of Novel Nanostructures*, edited by H. Kuzmany, J. Fink, M. Mehring, and S. Roth
© 2005 American Institute of Physics 0-7354-0275-2/05/$22.50

pography is obtained in the tapping-mode, whereas the magnetic contrast is subsequently obtained in the lift-mode scan where the cantilever's phase shift is recorded in the process of rescanning the previously measured topography at a user-defined height. Another useful application of MFM is imaging a remanence after the application of the external magnetic field. Imaging at remanence eliminates the combined effect of tip stray field and external magnetic fields.

EXPERIMENTAL

In this work, the fullerene films were grown from the gas-dynamic vapor flow by the method of condensation in a quasi-closed volume. To ensure film-growing conditions close to the equilibrium ones we have developed a method to considerably decrease the vapor oversaturation [4]. Unlike thermal evaporation, our gas-dynamic flow method allows us to obtain films with large crystallites of the order of ~ 1 μm in size.

Fullerene films 800 nm in thickness were doped with hydrogen by irradiating the film protons within a circle of radius 1 mm. Sequential irradiation with the protons of the energy 20, 36, 55, 75, and 95 keV resulted in a quasi-homogeneous proton (*i.e.* hydrogen) distribution within the film depth. The evaluation of the proton distribution using the SRIM program [5] showed the peaks of hydrogen distribution at the depth 0.25, 0.4, 0.55, 0.7 and 0.8 μm; the peaks were strongly overlapping so that the hydrogen distribution in the film is almost homogenous. The total dose $5 \cdot 10^{16}$ cm^{-2} corresponds to the average stoichiometry $C_{60} H_{0.5}$ in the doped layer.

The structure of the films was studied by Raman technique using a Renishaw 1000 grating spectrometer with a notch filter to remove the Rayleigh line and a Peltier cooled charge-coupled device (CCD detector). As a probing laser we used an He-Ne laser operating at 632.8 nm. The power density of the laser on the sample was less than 0.5 W/cm^2. This wavelength, as well as low power density was used to ensure that no photoinduced changes occurred in the samples and to avoid sample heating.

RESULTS AND DISCUSSION

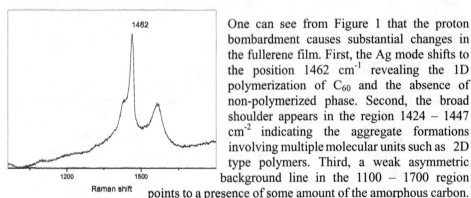

One can see from Figure 1 that the proton bombardment causes substantial changes in the fullerene film. First, the Ag mode shifts to the position 1462 cm^{-1} revealing the 1D polymerization of C_{60} and the absence of non-polymerized phase. Second, the broad shoulder appears in the region 1424 – 1447 cm^{-2} indicating the aggregate formations involving multiple molecular units such as 2D type polymers. Third, a weak asymmetric background line in the 1100 – 1700 region points to a presence of some amount of the amorphous carbon.

FIGURE 1. Raman spectrum of proton bombarded fullerene film

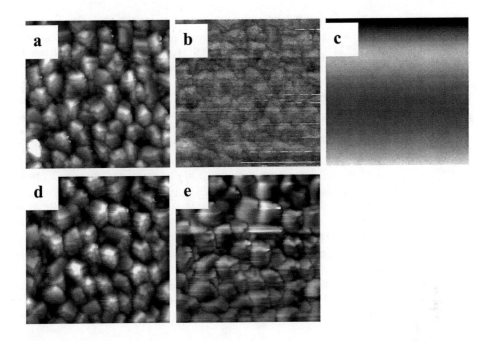

FIGURE 2. Topography (a, d) and magnetic force gradient (b, c, e) images on proton irradiated area of $C_{60}H_{0.5}$ film. The scan size is 3 μm × 3 μm and scan height is 50 nm for MFM.

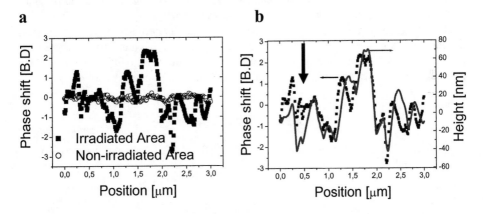

FIGURE 3. (a) The change of magnetic signal (phase shift) in proton irradiated and non-irradiated areas of the $C_{60}H_{0.5}$ thin film. (b) Line scanning of topography (right axis) and magnetic force gradient (left axis) obtained on the proton irradiated fullerene film. Note that the MFM images do not follow the topography.

Figure 2 shows topography (a) and magnetic force gradient images (b, c) in non-irradiated area. Topographic clusters with the average size of S_q equal to 26 nm are found. The magnetic force gradient images are similar to topography (Fig. 2 a and b) and magnitude of phase shift very small, below 0.5°. Fig. 2, c shows that in non-irradiated area, MFM images are insensitive to the reverse of the tip magnetization. Thus, the non-irradiated areas of the C_{60} film can be regarded as non-magnetic.

Figure 2 shows topography (d) and magnetic force gradient images (e) in the irradiated area of the film. Topographic clusters with the average size of S_q equal to 28 nm are found in all scanned irradiated areas of the surface. Figure 3, a shows the change of the strength of magnetic signal in the irradiated and non-irradiated areas of the C_{60} film. For all areas, enhancement of magnetic signal in the irradiated areas is recorded. As shown here, enhanced magnetic signal is observed on proton irradiated areas and maximum change of phase shift ($\Delta\Phi$) is ~ 5° and in non-irradiated change of $\Delta\Phi$ is below 0.5°. Moreover, difference between topography and magnetic image is found in some areas (see the arrow in Figure 3, b). Measurements of the phase shift at different heights reveal that magnetic signal decreases as scan height increases from 50 nm to 300 nm. Also, the phase shift show significant changes after the change of the MFM tip magnetization, from +z direction to –z direction. In both cases no changes in topographic images i.e. average surface roughness is same, however clear changes of MFM images in proton irradiated area have been found. Additionally, magnetic signal (magnitude of magnetic phase shift) changes with MFM tips. It reveals that origin of magnetic signal does not come from the contact of tip with samples surface but an intrinsic effect.

In conclusion, we have observed that medium-energy proton bombardment of fullerenes produces magnetic effects similar to those produced by the high-energy proton bombardment of graphite [1, 2]. The present experiments, however, cannot give an answer to a question whether the induced magnetic ordering is due to the polymerized fullerene structures or due to partially destroyed fullerenes, as both are observed in the Raman spectrum. Polymerization processes frequently compete with the damage of C_{60} molecules. Moderate doses of ion irradiation promote polymerization, whereas the increase of irradiation dose destroys both polymerization and fullerenes itself. Experiments with lower and higher doses of protons of various energies are planned.

ACKNOWLEDGMENTS

Supported by the Swedish Research Council, RFBR 05-02-17779 and the FP6 program "Ferrocarbon".

REFERENCES

1. P. Esquinazi, D. Spemann, R. Höhne, A. Setzer, K.-H. Han, and T. Butz, *Phys. Rev. Lett.* **91**, 227201 (2003)
2. K.-H. Han, D. Spemann, P. Esquinazi, R. Höhne, V. Riede, and T. Butz, *Advanced Materials* **15**, 1719 (2003)
3. T. L. Makarova. *J. Magn. Magn. Mater.* **272-276**, E1263-E1268 (2004)
4. T. L. Makarova, A. Ya. Vul', I. B. Zakharova and T. I. Zubkova *Phys. Sol. State* **41** 319-323 (1999).
5. Ziegler JF, SRIM 2003, Version 20, 2004. Obtained from http://www.srim.org

ISOLATION OF CARBON NANOSTRUCTURES

Martin Kalbáč[1,2], Ladislav Kavan[1,2], Hana Pelouchová[1], Pavel Janda[1],
Markéta Zukalová[1] and Lothar Dunsch[2]

[1]*J. Heyrovský Institute of Physical Chemistry, Academy of Sciences of the Czech Republic,*
Dolejškova 3, CZ-182 23 Prague 8
[2]*Leibniz Institute of Solid State and Materials Research, Helmholtzstr. 20, D - 01069 Dresden*

Abstract. Carbon nanostructures such a single wall carbon nanotubes (SWCNT), double wall carbon nanotubes (DWCNT) and fullerene peapods (e.g. C_{70}@SWCNT) usually occur in the form of bundles. Here, we present application of a novel simple and versatile method for deposition of small isolated nanoribbons of carbon nanotubes on annealed gold surface. The nanoribbons were characterized by Raman spectroscopy and exhibit characteristic features of individual carbon nanostructures. The resonance condition allowed the observation of a distinct spectrum of one inner tube in the nanoribbon from DWCNT. The signal of inner tubes of isolated DWCNT nanoribbons was found to be up to 50 times stronger than the sum of signals of the corresponding tubes in buckypaper sample. This dramatic enhancement is assigned to SERS (surface enhanced resonant Raman scattering) effect.

INTRODUCTION

Isolated SWCNT were obtained by sonication of SWCNT in the solution of surfactants (such as sodium dodecyl sulphate, SDS) [1,2]. The surfactant strongly interacts with SWCNT causing a disintegration of the bundle, but this is paid by the presence of the surfactant shell on each individual SWCNT. Sonication may also cut the tubes, creating defects in SWCNT[3] and trigger irreversible chemical reactions with the solvent [4]. An alternative method for fabrication of isolated SWCNT consists in their direct growth on Si substrate with deliberately patterned catalyst islands [5-7]. This method produces cleaner tubes than the surfactant-aided debundling, but only limited selection of structures is accessible, depending on the catalyst and the growth conditions. The growth of peapods and DWCNT would not be easy by this method. We have shown already that it is possible to prepare clean isolated nanoribbons from SWCNT, DWCNT and C_{60}@SWCNT by pressing of the particular buckypaper toward the annealed gold substrate [8]. Here we present new data for nanoribbon samples.

CP786, *Electronic Properties of Novel Nanostructures*, edited by H. Kuzmany, J. Fink, M. Mehring, and S. Roth
© 2005 American Institute of Physics 0-7354-0275-2/05/$22.50

EXPERIMENTAL SECTION

HiPco nanotubes (SWCNT) and peapods C_{70}@SWCNT were available from our earlier works. Double wall carbon nanotubes were synthesized by vacuum pyrolysis of C_{70}@SWCNT at 1200°C during 12 hours. The nanoribbon samples were prepared by pressing of the particular buckypaper toward the annealed gold substrate with simultaneous slow sweeping in horizontal direction [8].

The Raman scattering was excited by Ar^+ laser at 2.41 eV and by Kr^+ laser at 1.83 (Innova 300 series, Coherent). Spectra were recorded on a T-64000 spectrometer (Instruments SA) interfaced to an Olympus BH2 microscope (objective 100x). The laser power impinging on the sample was between 1 and 5 mW. The spectrometer was calibrated by using the F_{1g} mode of Si at 520.2 cm^{-1}.

RESULTS AND DISCUSSION

The Au-supported carbon nanostructures (SWCNT, DWCNT, C_{70}@SWCNT) prepared by contact deposition method occur almost exclusively in isolated two-dimensional arrays (nanoribbons). The vertical size of nanoribbon is approximately 1.2 nm, which is in good agreement with the diameter of one isolated SWCNT [8].

Fig 1 represents two examples of Raman spectra of gold-supported nanoribbons of HiPco nanotubes. The bulk material (not shown) exhibits the expected polydisperse mixture of tubes (with RBM wavenumbers from 150 to 400 cm^{-1}) that resonate with the used laser [9]. The Au-supported nanoribbons usually showed ca. 1-3 main peaks in the same spectral region, while the other peaks were efficiently suppressed.

Fig 2 shows the comparison of the signal of buckypaper sample of DWCNT and the DWCNT nanoribbon on annealed gold surface. Apparently the signal of nanoribbon is strongly enhanced. However the strong SERS effect was observed mostly for inner tubes. The "chemical" SERS effect requires a direct contact between the molecule and metal. This is not the case for inner tubes, therefore only electromagnetic field enhancement mechanism of SERS should be considered.

The RBM band is at 306.5 cm^{-1}. The calculated diameter of inner tube is 0.83 nm. The diameter of a possible outer tube should be ca. 1.5 nm, assuming the inter tube distance of 0.34 nm. The corresponding outer tube RBM band position should be close to 165 cm^{-1}. As no band was observed in this region, we assume that the particular outer tube is out of resonance at the used conditions.

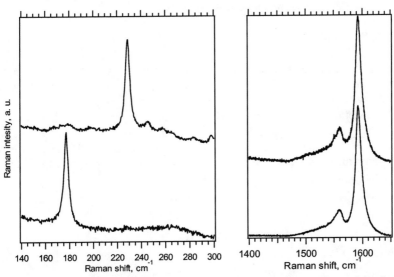

FIGURE 1. Raman spectra (excited at 1.83 eV) of HiPco nanotubes on Au surface. Sample was prepared by contact-deposition of nanotubes from a buckypaper. Two spectra represents two different spots on the surface. Spectra are offset for clarity, but the intensity scale is identical for all spectra in the respective window.

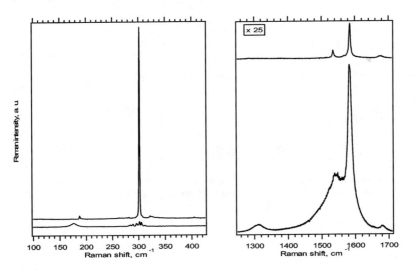

FIGURE 2. Raman spectra (excited at 1.83 eV) of DWCNT. Top: nanoribbons on Au surface obtained by contact-deposition of nanotubes from a buckypaper; Bottom: The same sample in form of bucky paper. Spectra are offset for clarity, but the intensity scale is identical for all spectra in the respective window.

Similarly to isolated SWCNT, two components of TG mode were observed also for DWCNT. The more intensive one (ω_G^+) was found at 1585 cm^{-1}, while the less intensive one (ω_G^-) at 1535 cm^{-1}. The difference ($\omega_G^+ - \omega_G^-$) scales inversely with the square of tube diameter as shown by Jorio et al.[10]:

$$\omega_G^- = \omega_G^+ - 47.7/d^2 \tag{1}$$

However the difference found: ($\omega_G^+ - \omega_G^-$) ≈ 50 cm^{-1} does not match the equation (1) neither for outer tube nor for inner tube. As the RBM mode of outer tube is not in resonance, we assume that the observed TG mode corresponds to inner tube. A more detailed inspection reveals a tiny band at around 1565 cm^{-1}. The diameter of the corresponding outer tube calculated using equation (1) is ca. 1.3 nm, which corresponds to the outer tube diameter. Therefore, the band at 1541 cm^{-1} is attributed to ω_G^- of inner tube and the strong curvature of inner tube is responsible for differences in the dispersion of ω_G^-.

ACKNOWLEDGEMENTS

This work was supported by IFW Dresden, by the Academy of Sciences of the Czech Republic (contract No. A4040306) and by the Czech Ministry of Education (contract No. LC-510).

REFERENCES

1. Matarredona O., Rhoads H., Li Z. R., Harwell J. H., Balzano L. and Resasco D. E. *J.Phys.Chem.B* **107**, 13357-13367 (2003).
2. O'Connell M. J., Bachilo S. M., Huffman C. B., et al. *Science* **297**, 593-596 (2002).
3. Lu K. L., Lago R. M., Chen Y. K., Green M. L. H., Harris P. J. F. and Tsang S. C. *Carbon* **34**, 814-816 (1996).
4. Niyogi S., Hamon M. A., Perea D. E., et al. *J.Phys.Chem.B* **107**, 8799-8804 (2003).
5. Jorio A., Saito R., Hafner J. H., et al. *Phys.Rev.Lett.* **86**, 1118-1121 (2001).
6. Jorio A., Fantini C., Dantas M. S. S., et al. *Phys.Rev.B* **66**, art-115411(2002).
7. Brar V. W., Samsonidze G. G., Dresselhaus M. S., et al. *Phys.Rev.B* **66**, art-155418(2002).
8. Kalbac M, Kavan L., Zukalova M, Pelouchova H., Janda P. and Dunsch L. *ChemPhysChem* **6**, 426-430 (2005).
9. Kavan L., Dunsch L. and Kataura H. *Chem.Phys.Lett.* **361**, 79-85 (2002).
10. Jorio A., Souza A. G., Dresselhaus G., et al. *Phys.Rev.B* **65**, 155412 (2002).

Mercator maps of orientations of a C_{60} molecule in single-walled nanotubes with distinct radii

K.H. Michel*, B. Verberck* and A.V. Nikolaev**

*Department of Physics, University of Antwerp, Universiteitsplein 1, 2610 Antwerp, Belgium
**Institute of Physical Chemistry of the Russian Academy of Sciences, Leninskii prospekt 31, 117915, Moscow, Russia

Abstract. We study the confinement of a C_{60} molecule encapsulated in a cylindrical nanotube as a function of the tube radius. Drawing the Mercator maps of the potential, we find two distinct molecular orientations; for tubes with small radii, $R_T \lesssim 7$ Å, a fivefold axis of the molecule coincides with the tube long axis, for larger radii, $R_T \gtrsim 8$ Å, a threefold axis of the molecule coincides with the tube long axis. These different orientations are caused by the relative importance of the repulsive and the attractive parts of the van der Waals potentials of the molecule with the tube wall for small and large tubes respectively. Experimental evidence is provided by the apparent splitting of A_g modes of the C_{60} molecule in resonant Raman scattering.

The synthesis [1] of self-assembled chains of C_{60} molecules inside single-walled carbon nanotubes (SWCNT), the so-called peapods $(C_{60})_n$SWCNT, has opened the road for the study of a unique class of nanoscopic hybrid materials with unusual electronic [2] and structural properties. High-resolution transmission electron microscopy observations on sparsely filled tubes [3] demonstrate the motion of molecules along the tube axis and imply that the interaction between C_{60} and the surrounding tube is of van der Waals type. This interaction comprises a short-range repulsive and a long-range attractive part. By now it is possible to prepare fullerene encapsulating tubules with high filling rate of C_{60} or other fullerene molecules, thereby obtaining one-dimensional (1D) molecular crystals [4] inside the tube. In the following we will show that these systems are unique in as far as the confinement of the C_{60} molecules by the tube walls does not only cause the 1D arrangement of the molecular centers-of-mass along the cylindrical tube axis but implies also an orientational confinement of the molecules. The most probable molecular orientation depends on the radius of the tube.

We start from a model where one C_{60} molecule is located with its center-of-mass on the axis of an infinitely long tube. The tube is taken as a smooth cylinder with radius R_T. The axis of the cylinder (symmetry $D_{\infty d}$) coincides with the Z axis of a space fixed Cartesian system (X,Y,Z). We consider a molecular fixed system of axes (x,y,z) such that these axes coincide with twofold axes of the molecule. When both systems coincide, the molecule is in standard orientation [5]. The molecule (symmetry I_h) is taken as a rigid body, an arbitrary orientation of the molecule with respect to the space fixed system is described by Euler angles (α, β, γ). The molecule is characterized by a skeleton of interaction centers located on a sphere with radius $d = 3.55$ Å. The interaction centers are of three types: 60 atomic -, 30 double bond - and 60 single bond centers interacting with the tube wall by repulsive Born–Mayer and attractive van der Waals forces. Similar

CP786, *Electronic Properties of Novel Nanostructures*, edited by H. Kuzmany, J. Fink, M. Mehring, and S. Roth
© 2005 American Institute of Physics 0-7354-0275-2/05/$22.50

potentials have been used for the description of interactions between C_{60} molecules [6].

For an arbitrary orientation the total interaction potential of the molecule with the cylinder walls is conveniently written in terms of molecular and site symmetry-adapted rotator functions. The result reads

$$V(R_T; \beta, \gamma) = \sum_{l=0,6,10,\ldots} v_l(R_T) \mathcal{U}_l(\beta, \gamma), \tag{1}$$

with rotator functions

$$\mathcal{U}_l(\beta, \gamma) = \sum_{n=-l}^{l} \alpha_l^{n(I_h)} \mathcal{D}_{n,0}^l(\beta, \gamma) \tag{2}$$

and potential coefficients

$$v_l(R_T) = \sum_{i=1}^{3} v^i(R_T) g_l^i. \tag{3}$$

The rotator functions $\mathcal{U}_l$ are linear combinations of Wigner's $\mathcal{D}_{n,m}^l(\alpha, \beta, \gamma)$ functions. In Eq. (2) the cylindrical symmetry implies that the angle α is absent and the index m has value 0. The Euler angles β and γ are the polar and azimuthal angles that specify the direction of the (molecular) z axis in the space fixed system. The molecular symmetry is accounted for by the structure factors $\alpha_l^{n(I_h)}$, $l = 0, 6, 10, \ldots$, n even $\leq l$. The index i labels the three types of interaction centers, g_l^i are molecular shape factors that account for the distribution of interaction centers [6].

For a fixed set of interaction potential parameters, we have plotted the potential $V(R_T; \beta, \gamma)$, for different tube radii, in form of Mercator [7] maps as a function of the angles $0 \leq \beta \leq \pi$ and $0 \leq \gamma \leq 2\pi$. Figs. 1(a) and (b) refer to $R_T = 7$ Å and $R_T = 8$ Å, respectively. In the former case twelve equivalent minima of the potential — which determine the most probable orientations of the molecule — are found. The values ($\beta \approx 58°, \gamma = 0$) correspond to the situation where the molecule has been rotated counterclockwise around the Y axis, away from the standard orientation by an angle $\beta = \arccos 2/(10 + 2\sqrt{5})^{\frac{1}{2}}$. Then, the nanotube's long axis intersects the centers of two opposing pentagons.

The minima in Fig. 1(a) correspond to the twelve pentagons on the C_{60} molecule. The twenty maxima appearing in Fig. 1(a) correspond to the energetically unfavorable orientation when the tube axis intersects two opposing hexagon faces (a threefold axis of C_{60} coincides with the tube axis). On the other hand, comparing Fig. 1(b) with Fig. 1(a), we find that for $R_T = 8$ Å, the positions of potential maxima and minima are reversed. Potential minima now correspond to orientations of the molecule where a threefold axis of the molecule coincides with the tube axis. The distinction of two preferred orientations as a function of the tube radius reflects the competition between the short-range repulsive and the long-range attractive parts of the van der Waals interactions. Considering the smaller nanotube with $R_T = 7$ Å but neglecting the repulsive interactions, we obtain a Mercator map similar to Fig. 1(b). We conclude that in small tubes repulsive forces are

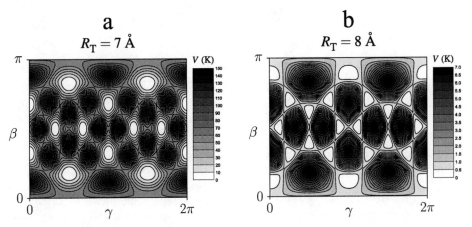

FIGURE 1. Crystal field $V(R_T; \beta, \gamma)$, units K: (a) $R_T = 7$ Å, (b) $R_T = 8$ Å. The lowest occurring value has been subtracted to make the minimal potential energies lie at zero.

responsible for the orientation of a fivefold axis of the molecule along the long cylinder axis, while in large tubes, the attractive forces are dominant and favor the orientation of a threefold axis of the molecule along the long cylinder axis. Distinct orientations of the molecule with respect to the tube axis reflect distinct "crystal fields" acting on the molecule.

Resonance Raman scattering [8] measurements of pristine single-walled carbon nanotubes filled with C_{60} molecules exhibit an unexpected splitting of the totally symmetric modes of the C_{60} molecule. In particular, the pentagonal pinch mode $A_g(2)$ "splits" into a doublet $A_g(2)'$ and $A_g(2)''$ below room temperature. This mode, located at 1469 cm^{-1} at room temperature in pristine C_{60}, can be used as an analytical probe for structural and electronic properties. As we have seen, for C_{60} in SWCNT there are essentially two distinct orientations at sufficiently low temperature. Due to the crystal fields there will be two distinct symmetry breakings of the C_{60} molecule, resulting in a shift of the $A_g(2)$ mode to higher ($A_g(2)''$) or lower ($A_g(2)'$) frequencies for smaller or larger radii of the encapsulating tubes respectively. Since even for the best available materials experiments are carried out on C_{60}@SWCNT with a dispersion of tube radii, these shifts just appear as a splitting of the $A_g(2)$ modes. This resolves the paradox that there should be no splitting for the non-degenerate A_g modes. The present explanation is corroborated by the experimental fact that thinner tubes tend to yield stronger $A_g(2)''$ components and that this component seems to increase relative to $A_g(2)'$ with decreasing temperature.

The results on orientational confinement of the C_{60} molecule by the encapsulating tube have also consequences for the formulation of theoretical models of 1D crystals of C_{60} molecules in nanotubes. Beside correlations between the center-of-mass positions of the molecules, also orientational correlations have to be taken into account. We have derived the rotational interaction potential V^{RR} of a chain of N C_{60} molecules inside a nanotube as a O_2-symmetry rotor model on a linear chain:

$$V^{RR} = \sum_{n=1}^{N} J^s \vec{S}_s(n) \cdot \vec{S}_s(n-1), \tag{4}$$

where the rotational coordinate reads

$$\vec{S}_s(n) = \Big(\cos[s\psi(n)], \sin[s\psi(n)] \Big), \tag{5}$$

$\psi(n)$ being the rotation angle of the nth molecule about the tube axis (Z) measured away from the X axis. Here $s = 5$ or 3 refers to the s-fold molecular axis which is parallel to the long tube axis for small or large tubes, respectively. We find that in both cases $J^s < 0$, hence the rotational interaction between neighboring molecules is maximum and attractive if both molecules have the same orientation angle ψ and an s-fold axis parallel to the tube axis. Since the C_{60} molecule has a center of inversion symmetry, the same orientation of neighbouring molecules implies that neighboring faces are in staggered orientation.

We expect that rotational interactions as described by V^{RR} are relevant for the theoretical explanation of inelastic neutron scattering experiments [9] on the dynamics of C_{60} molecules inside SWCNTs.

Although we have restricted ourselves here to C_{60} in carbon nanotubes, we believe that the present analytical methods and results can be extended to the study of C_{60} encapsulated in boron-nitride nanotubes [10].

The present work has been supported by the Bijzonder Onderzoeksfonds, Universiteit Antwerpen (BOF – NOI). B.V. is a research assistant of the Fonds voor Wetenschappelijk Onderzoek – Vlaanderen.

REFERENCES

1. B.W. Smith, M. Monthioux, and D.E. Luzzi, Nature (London) **396**, 323-324 (1998).
2. D.J. Hornbaker *et al.*, Science **295**, 829-831 (2002).
3. B.W. Smith, M. Monthioux, and D.E. Luzzi, Chem. Phys. Lett. **315**, 31-36 (1999).
4. K. Hirahara *et al.*, Phys. Rev. B **64**, 115420 - 115425 (2001).
5. W.I.F. David *et al.*, Nature (London) **353**, 147 - 149 (1991).
6. J.R.D. Copley and K.H. Michel, J. Phys.: Condens. Matter **5**, 4353 (1993).
7. Gerardus Mercator (1512 - 1594), Flemish cartographer, inventor of the cylindrical projection.
8. R. Pfeiffer *et al.*, Phys. Rev. B **69**, 035404 — 1 - 7 (2004).
9. J. Cambedouzou *et al.*, Phys. Rev. B **71**, 041403(R) (2005).
10. W. Mickelson *et al.*, Science **300**, 467 (2003).

Isolation and spectroscopic characterization of two isomers of the metallofullerene Nd@C$_{82}$

Kyriakos Porfyrakis[*a], Mito Kanai[b], Gavin W. Morley[c], Arzhang Ardavan[d], T. John S. Dennis[b] and G. Andrew D. Briggs[a]

[a] Department of Materials, University of Oxford, Parks Road, Oxford, OX1 3PH, U.K. Fax: +44 1865 273789; Tel: +44 1865 273724; E-mail: kyriakos.porfyrakis@materials.ox.ac.uk
[b] Centre for Materials Research, Queen Mary, University of London, Mile End Road, London, E1 4NS, U.K.
[c] Centre for interdisciplinary Magnetic Resonance, National High Magnetic Field Laboratory, Florida State University, Tallahassee, FL 32310, U.S.A.
[d] Clarendon Laboratory, Parks Road, Oxford, OX1 3PU, U.K.

Abstract. For the first time, two types of the metallofullerene Nd@C$_{82}$ have been isolated and characterized. HPLC was used to isolate Nd@C$_{82}$(I, II). The two isomers were characterized by mass spectrometry and UV-Vis-NIR absorption spectroscopy. Nd@C$_{82}$(I) was found to be similar in structure to the main isomer of other lanthanofullerenes such as La@C$_{82}$, as was previously reported. We assign Nd@C$_{82}$(I) to have a C$_{2v}$ cage symmetry. Nd@C$_{82}$(II) showed a markedly different UV-Vis-NIR absorption spectrum to Nd@C$_{82}$(I). Its spectrum is in good agreement with that of the minor isomer of metallofullerenes such as Pr@C$_{82}$. We therefore assign Nd@C$_{82}$(II) to have a C$_s$ cage symmetry. In contrast to other metallofullerenes, both isomers appear to be equally abundant.

INTRODUCTION

Endohedral metallofullerenes have been the target of widespread research due to their promising electronic properties. Obtaining pure isomers of metallofullerenes remains challenging, but it is necessary for studying their properties in detail. It has been a matter of some dispute whether the different isomers have different cage structures or the same cage structure with different positions of the metal atom(s) inside the cage. In this paper we present the first isolation and characterization of two isomers of Nd@C$_{82}$.

EXPERIMENTAL

Nd@C$_{82}$ was produced by the DC arc discharge method[1]. A Nd-doped graphite rod (Nd content of the rod is 0.8%) was kept a few mm apart from a graphite block in a vacuum chamber. The rod and block were connected to an external power supply and high current was passed through them (300-500 A). The electric power was dissipated

CP786, *Electronic Properties of Novel Nanostructures*, edited by H. Kuzmany, J. Fink, M. Mehring, and S. Roth
© 2005 American Institute of Physics 0-7354-0275-2/05/$22.50

in an arc and the rod begun to evaporate. The vaporisation took place in high-purity He atmosphere (50-100 mbar). The resulting soot contained 10-20% fullerenes. The fullerenes were extracted from the soot in a soxhlet extraction apparatus during several hours of operation, using boiling N,N-Dimethylformamide (DMF) as solvent.

RESULTS AND DISCUSSION

A two-stage High-performance liquid chromatography (HPLC) method was employed to isolate individual fullerene species. In the first stage, the DMF extract was dissolved in toluene and the solution was passed through a 5-PYE (pyrenyl ethyl) column (20 x 250 mm) with pure toluene eluent (flow rate: 18 ml/min). Figure 1 shows the HPLC stages for the purification of Nd@C_{82}.

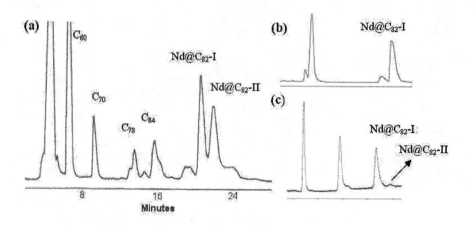

FIGURE 1. (a) Fist stage of purification of Nd@C_{82}. The two peaks eluting between 21 and 23 mins are Nd@C_{82} isomers. (b) Nd@C_{82}-I was passed through HPLC in recycling mode in order to further purify it. (c) Further purification of the Nd@C_{82}-I isomer through a Byckyprep-M column showed that it contained a small amount (~5%) of Nd@C_{82}-II.

During this stage, C_{60} that has the highest mobility was separated first. C_{70} eluted second. The peaks that eluted after C_{70} correspond to higher fullerenes (C_{78}, C_{84}, etc.) and Nd-containing endohedral metallofullerenes. Mass spectrometry has shown that the two peaks eluting between 21 and 23 mins correspond to Nd@C_{82}.

A second HPLC stage was used to separate the Nd@C_{82} isomers. Each of the Nd@C_{82} fractions was re-injected in recycling mode. After a few cycles, Nd@C_{82}-I (which elutes first) and Nd@C_{82}-II were individually isolated. Figure 1(b) shows the HPLC chromatogram of Nd@C_{82}-I. Mass spectrometry confirmed that both isomers are Nd@C_{82}. Their mass spectra are in good agreement with the theoretically predicted isotopic distribution.

Nd@C_{82}-I was then passed through a Buckyprep-M column (toluene eluent, flow rate 18 ml/min) in recycling mode. After a few cycles it became evident that the peak corresponding to NdC$_{82}$-I, also contains a small quantity (~ 5%) of Nd@C$_{82}$-II (figure 1(c)).

In order to further elucidate the structure of Nd@C$_{82}$-I and Nd@C$_{82}$-II, UV-Vis-NIR spectroscopy was used. Figure 2 shows the UV-Vis-NIR absorption spectra of the two Nd@C$_{82}$ isomers in CS$_2$ solution. The spectrum for Nd@C$_{82}$-I shows absorption peaks at 389, 638, 1025 nm and a broad peak at 1412 nm. Nd@C$_{82}$-II shows absorption peaks at 383, 710, 1090 nm and a broad peak at 1832 nm. The MALDI mass spectrum of Nd@C$_{82}$ is shown in the inset.

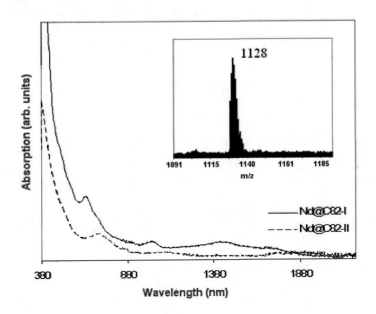

FIGURE 2. UV-Vis-NIR absorption spectra of Nd@C$_{82}$-I and Nd@C$_{82}$-II in CS$_2$ solution. The MALDI mass spectrum of Nd@C$_{82}$ is shown in the inset.

The spectrum of Nd@C$_{82}$-I looks remarkably similar with the major isomer of other metallofullerenes such as La@C$_{82}$ and Pr@C$_{82}$ etc. The spectrum of Nd@C$_{82}$-II is also very similar with the minor isomer of La@C$_{82}$ and Pr@C$_{82}$ [2]. This close similarity is evidence that the electronic structure of the C$_{82}$ cage is the same in all cases[2, 3]. It was previously shown that the major and minor isomers of La@C$_{82}$ and Pr@C$_{82}$ have C$_{2v}$ and C$_s$ cage symmetry, respectively[4, 5]. We therefore conclude that Nd@C$_{82}$-I has C$_{2v}$ symmetry while Nd@C$_{82}$-II has C$_s$ symmetry. Using X-ray photoelectron spectroscopy (XPS), it was previously shown[3, 6] that the oxidation state of the metal encapsulated in the C$_{82}$ cage is 3+. Hence, in the case of Nd@C$_{82}$ as well as in all the metallofullerenes mentioned above, 3 valence electrons are transferred from the metal atom to the fullerene cage and the molecule acquires the following electronic structure: M^{3+}C$_{82}$$^{3-}$ (where M represents the metal atom).

In most metallofullerenes the C$_{2v}$ isomer is the dominant one. Indeed an earlier study had shown that only one isomer (C$_{2v}$) of Nd@C$_{82}$ could be isolated[6]. On the contrary, we have shown in this work that two structural isomers of Nd@C_{82} can be produced and isolated in almost equal abundance.

ACKNOWLEDGMENTS

This research is part of the QIP IRC (GR/S82176/01) and is supported through the Foresight LINK award (Nanoelectronics at the quantum edge) by EPSRC (GR/R66029/01) and Hitachi Europe Ltd. GADB thanks EPSRC for a Professional Research Fellowship (GR/S15808/01). AA is supported by the Royal Society. We would like to thank Prof. H. Shinohara (Nagoya University, Japan) for his valuable assistance during the production of Nd@C_{82}.

REFERENCES

1. H. Shinohara, *Rep. Prog. Phys.*, **63**, (2000), 843-892.
2. Akasaka, T., Okubo, S., Kondo, M., Maeda, Y., Wakahara, T., Kato, T., Suzuki, T., Yamamoto, K., Kobayashi K. and Nagase, S., *Chem. Phys. Lett.*, **319**, (2000), 153-156.
3. Ding, J. and Yang, S., *J. Phys. Chem. Solids*, **58**, 11, (1997), 1661-1667.
4. Akasaka, T., Wakahara, T., Nagase, S., Kobayashi K., Waelchli, Yamamoto, K., Kondo, M., Shirakura, S., Maeda, Kato, T., Kako M., Nakadaira, Y., Gao, X., Van Caemelbecke, E and Kadish, K.M., *J. Phys. Chem. B*, **105**, (2001), 2971-2974.
5. Hosokawa, T., Fujiki, S., Kuwahara, E., Kubozono, Y., Kitagawa, H., Fujiwara, A., Takenobu, T. and Iwasa, Y., *Chem. Phys. Lett.*, **395**, (2004), 78-81.
6. Ding, J., Lin, N., Weng, L., Cue, N. and Yang, S., *Chem. Phys. Lett.*, **261**, (1996), 92-97.

Metal Oxides and Low Temperature SWCNT Synthesis via Laser Evaporation

M. Rümmeli[1], O. Jost[2], T. Gemming[1], M. Knupfer[1], E. Borowiak-Palen[3],
T. Pichler[1], S. Ravi P. Silva[3], B. Büchner[1]

*1.Leibniz Institute for Solid State and Materials Research Dresden, IFW Dresden, P.O. Box 27016
D-01171 Dresden, Germany.
2. Dresden University of Technology, D-01062 Dresden, Germany.
3. Advanced Technology Institute, University of Surrey, Surrey, GU2 7XH, United Kingdom.*

Abstract. Studies using metal oxides show they are active as catalysts for single wall carbon nanotubes synthesis in laser ablation, even at room temperature. In addition when combined with Ni and Co, large diameter SWCNT can be synthesized at temperatures below those required when only using Ni and Co as a binary catalyst. The results suggest a nucleation mechanism previously not identified.

Keywords: Laser ablation, synthesis, nanotubes, microscopy, raman and optical absorption spectrocopy

PACS: 79.20.Ds 82.80.Dx 68.37.EF,Hk,Lp,Nq 81.07.De

INTRODUCTION

The available catalysts for laser ablated SWCNT production, until now, have been limited to the requirement of at least one of a select few transition metals[1] (Co, Ni, Rh & Pt) being present in the reaction. These require a relatively high synthesis temperature. In addition, a recent number of investigations into nanotubes nucleation and growth mechanisms point to a dominant role played by fullerenes limiting the lower temperature synthesis limits. An external supply of fullerenes lowers this limit[2] to temperatures below 600°C. So called low temperature SWCNT have also been formed using continuous wave CO_2 lasers (e.g. refs 3) or solar furnaces (e.g. ref 4) since no additional heating is required as is usual in pulsed laser ablation systems. However, in these cases the required heat for SWCNT synthesis and fullerene production is provided by the laser or the sun and so differ from low temperature SWCNT reactions using a standard pulsed laser ablation system, as we report here. Low temperature SWCNT synthesis is attractive, as it would allow their synthesis to be compatible with microelectronics (below 400-500°C) or biomolecular electronics (60-100°C). The select metals available for laser evaporation SWCNT synthesis can be combined with each other (e.g. ref 5) or other metal based compounds (e.g. ref. 6) to alter the mean diameter, diameter distribution and yield. Indeed, our initial studies

CP786, *Electronic Properties of Novel Nanostructures*, edited by H. Kuzmany, J. Fink, M. Mehring, and S. Roth
© 2005 American Institute of Physics 0-7354-0275-2/05/$22.50

began by admixing metal oxides with a more standard Co-Ni binary catalyst. The results showed unique processes, that on closer inspection suggested two growth mechanisms were present.

EXPERIMENTAL

The SWCNT were synthesized using a furnace based pulsed laser evaporation method. The setup is very similar to that described in[1], with the only difference being that in this case the outer and inner (to restrict the reaction volume) tubes are made from alumina since the oven can reach temperatures as high as 1600°C. In the present study, temperatures between 1200°C and room temperature were used for the synthesis of the SWCNTs. A Q-switched high power Nd:YAG laser (2.5 GW per pulse, pulse width = 8 ns) was used to evaporate catalyst containing graphite targets using very high purity materials. The evaporated products were then swept away by the carrier gas (Nitrogen) to a water-cooled copper cold finger behind the target, which provides a well-defined reaction point. Pressures of 1 bar and a gas flow rate of 0.4 l/min. were used for all experimental conditions discussed here. Optical absorption spectroscopy (OAS) measurements were conducted with the product dispersed in acetone in an ultrasonic bath and dropped onto a KBr single crystal giving a thin homogeneous film. A Bruker IFS 113V/88 spectrometer was used to obtain the optical absorption spectrum in the energy range 0.35 eV to 2.35 eV, with a spectral resolution of 0.25 meV. For transmission electron microscopy these films were floated off the KBr crystal, in water, and collected on standard Cu TEM grids (TEM, FEI Tecnai F30). Raman measurements were performed on a Bruker Fourier Transform Raman spectrometer with a resolution of 2 cm^{-1}. In addition, Raman measurements were recorded using a T-64000 spectrometer with a Kr$^+$ laser at 1.91 eV (Innova 300 series, Coherent).

RESULTS AND DISCUSSION

We found that adding a metal oxide (In$_2$O$_3$ or Fe$_2$O$_3$) to a standard binary catalyst of Ni/Co yielded different mean diameters and diameter distributions when compared to SWCNT produced by an ablation target containing only the binary catalyst of Ni/Co under the same conditions. Depending on the target to cold finger distance even bi-modal diameter distributions could be obtained. Further, reducing the temperature we also synthesized SWCNT at 600°C, where normally, with only the binary Ni/Co catalysts, no SWCNT are formed since, it is argued, no fullerenes are produced at this temperature. The results show, for the first time, that the addition of a metal oxide to a standard catalyst (Co, Ni, Rh or Pt) can yield SWCNT at lower temperatures than required without the presence of a metal oxide. The data also points an additional or altered growth mechanism as suggested by the results at higher temperatures. Furthermore, it strongly suggests a SWCNT growth mode independent of fullerenes.

To better understand the growth mechanism for the larger diameter SWCNT we examine first the effect of oxygen when synthesizing SWCNT using the reference target (only Ni/Co) at 1200°C. This was done by feeding small quantities of oxygen gas mixed with the carrier gas during ablation (up to 5%). To our surprise, despite oxygen levels close to that found in air, we still successfully synthesized SWCNT. The corresponding OAS and RBM modes from Raman data showed negligible changes in the mean diameter of the produced SWCNT. Further, an ablation target was prepared with only In added to Ni/Co which was then ablated at 1200°C in oxygen free conditions. The resultant soot showed no SWCNT when analyzed in OAS or Raman spectroscopy and TEM studies showed only multiwalled carbon nanotubes (MWCNT) present. A reasonable question then arises as to whether SWCNT growth can occur without Ni/Co being included in the target with a metal oxide. This we tried firstly at 1200°C and found very low yields of SWCNT and found these SWCNT had very narrow diameters. This can be explained by nucleation via fullerenes. The presence of only narrow diameter SWCNT would suggest that an alternate mechanism nucleates the larger diameter SWCNT. TEM studies showed that the catalyst particles from the combined metal oxide and Ni/Co ablated samples were far too large (3 to 50 nm) to explain the SWCNT growth via circumference growth as can be found in CVD[17] and indicates that the presence of Ni/Co plays a role in the large diameter SWCNT formation when including metal oxides at a temperature of 1200°C. Energy dispersive X-ray (EDX) analysis shows that the catalyst particles are primarily composed of a single metal with only trace quantities of other metal species, oxygen/impurities. This would suggest that O_2 is reduced in the reaction, probably by C. In addition, this suggests that the growth temperature is closer to the carbon - metal eutectic than the carbon - metal-oxide eutectic. Assuming the metal-carbon eutectic is connected to the first stable melt and therefore the upper SWCNT formation limit[7] we tried using temperatures so low that a first possible melt should (to the best of our knowledge) be present at somewhat higher temperatures. Thus, we attempted synthesizing SWCNT using only MgO at 500 °C (Mg melts at 650°C). The resultant sample showed the presence of SWCNT as confirmed by Raman studies, OAS and TEM. The OAS spectrum showed a mean diameter of 1.4 nm. TEM studies showed the presence of SWCNT, amorphous carbon species and catalysts particles along with the presence of some MWCNT. We also tried synthesizing SWCNT using only PbO_2 and In_2O_3 at room temperature (Pb and In melt at 327°C and 156°C respectively). OAS and Raman studies showed the presence of SWCNT for both samples and TEM studies confirmed the presence of SWCNT. In the case of PbO_2, so-called bamboo-MWCNT were also observed whilst with the In_2O_3 sample, no MWCNT were observed. Despite the lower SWCNT yield when using metal-oxide catalysts as compared to standard catalysts, the results do show, for the first time, that metal-oxides can be used as catalysts for laser ablated SWCNT synthesis and at temperatures well below those normally associated with laser ablation SWCNT synthesis.

The results show the growth of the SWCNT unlikely depends on fullerenes and that oxygen plays a key role. We propose a growth mechanism based on mix of molten and solidified particles where the solidified particles have precipitated carbon and formed a carbon shell. Oxygen, which is highly reactive, can then etch away part of the graphitic shell surrounding the solidified particle to a point where a stable oxide ring forms (forming the embryonic stage of the SWCNT) and the solidified particle can merge into the molten metal particle. The formation of a stable oxide ring is not unreasonable and is considered to be the reason why preferential oxidation of SWCNT caps occurs when heated in air. Mazzoni et al.[8] showed that preferential opening of caps as opposed to the walls is due to the release of its strain energy. In addition they showed that an oxidized rim is stable.

Following the formation of a cap with a stable oxidized ring, carbon can substitute the oxygen and thus SWCNT growth takes place. Fig 1. outlines the growth mechanism.

To summarize, we have shown for the first time that metal oxides can be used as catalysts for SWCNT growth and that by selecting a metal oxide (where the metal has a low melting temperature) SWCNT can be synthesized even at room temperature without the need for fullerenes. The results make the controlled growth (diameter and chirality) of SWCNT applications in microelectronics and biomolecular electronics more accessible. The presence of oxygen is key to preventing, at least some, metals from precipitating carbon freely and so forming MWCNT. In addition the role of oxygen is vital to etch open the nucleating carbon shell and in providing a stable oxidized rim forming an embryonic cap from which subsequent SWCNT growth takes place. The results strongly suggest a growth mechanism previously not identified which we term nucleation via etched carbon shells (NECS).

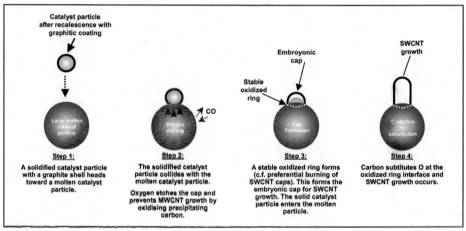

Figure 1. Illustration of nucleation via encapsulated carbon shell (NECS) and subsequent SWCNT growth.

REFERENCES

1. O. Jost, A. Gorbunov, X. Liu, W. Pompe, J. Fink, *J. Nanosci. Nanotech.*, **2004**, 4, 433-440.
2. Y. Zhang, S. Iijima, *Appl. Phys. Lett.* **1999**, 75, 3087-3089.
3. W.K. Maser, E. Munoz, A.M. Benito, et al., *Chem. Phys.Lett.* **1998**, 292 587-593.
4. Laplaze D, Bernier P, Maser WK, Flamant G, Guillard T,Loiseau, *Carbon* **1998**, 36, 685–689.
5. T. Guo, P. Nikolaev, A. Thess, D.T. Colbert, R. E. Smalley, *Chem. Phys. Lett.*, **1995**, 243 49-54
6. S. Lebedkin, P. Schweiss, B. Renker, S. Malik, F. Hennrich, M. Neumaier, C. Stoermer, M. M. Kappes, *Carbon* **2002**, 40, 417-423.
7. H. Kataura, Y. Kumazawa, Y. Maniwa, Y. Ohtsuka, R. Sen, S. Suzuki, Y. Achiba, *Carbon* **2000**, 38, 1691-1697.
8. M.S.C Mazzoni, H. Chacham, P. Ordejon, D. Sanchez-Portal, J. M. Soler, E. Artacho, *Phys. Rev. B* **1999**, 60, 2208 – 2211.

CARBON NANOSTRUCTURE SYNTHESIS, PURIFICATION AND SEPARATION

Large-scale Synthesis of Carbon Nanotubes by Catalytic Chemical Vapor Deposition Method and Their Applications

Morinobu Endo

Faculty of Engineering, Shinshu Univeristy, 4-17-1 Wakasato, Nagano-shi 380^8553, Japan

Abstract. Carbon nanotubes consisting of rolled graphene layer built from sp^2-units have attracted the imagination of scientists as one-dimensional macromolecules. Their unusual physical and chemical properties make them useful in the fabrication of nanocomposite, nano-electronic device and sensor etc. In this study, the recent hot topics "highly pure and crystalline double walled carbon nanotubes" will be described because it is expected that these tubes are thermally and structurally stable, and also contain small-sized tubes (below 2 nm). Among the recent applications of carbon nanotubes, micro-catheter fabricated from high purity carbon nanotubes as filler and nylon as matrix exhibited quite low blood coagulation and also reduced thrombogenity. It is envisaged that carbon nanotubes will play an important role in the development of nano-technology in the near-future.

Keywords: Carbon Nanotube; Catalytic chemical vapor deposition (CVD); structure; thermal treatment; micro-catheter
PACS: 61.48.+c

INTRODUCTION

Extraordinary chemical and physical properties of carbon nanotubes and also the success of large-scale production by a catalytic chemical vapor deposition method [1], particularly with the use of a floating reactant technique [2, 3], make them applicable in the fabrication of adsorbent, electrochemical electrode, field emitter and functional filler in composite at a possible low cost [4]. Through judicious selection of transient metal, support materials and synthetic conditions, it is possible to produce different types of carbon nanotubes such as multi-walled carbon nanotubes (MWNTs), double-walled carbon nanotubes (DWNTs) and single-walled carbon nanotubes (SWNTs) selectively. In this study, we will describe the selective synthesis of various carbon nanotubes including highly pure and crystalline DWNTs [5], and structural modification by post-treatments will be discussed in terms of nano-structure, and finally their practical applications of these carbon nanotubes, such as micro-catheter, will be described from the industrial point of view.

CP786, *Electronic Properties of Novel Nanostructures*, edited by H. Kuzmany, J. Fink, M. Mehring, and S. Roth
© 2005 American Institute of Physics 0-7354-0275-2/05/$22.50

FABRICATION OF DOUBLE WALLED CARBON NANOTUBES

DWNTs, which consist of two concentric graphene cylinders, have attracted the attention of numerous scientists because their intrinsic coaxial structures may exhibit intriguing electronic and mechanical properties that have not been reported hitherto. Here we report the fabrication of a new type of a paper-like material consisting of high-purity DWNTs in high yields using a catalytic chemical vapor deposition (CVD) method in conjunction with an optimized purification treatment. These tubes are hexagonally packed in bundles and show a narrow diameter distribution. Obtained a dark and stable paper-like sheet is very flexible and mechanically stable (tough) (Fig. 1 (a)). Careful HRTEM observations revealed an extremely high-yield of DWNTs (more than 95%) arranged in bundles (Fig. 1 (b)). The DWNT paper contains nanotubes with a narrow diameter distribution, exhibiting a hexagonal packing (Fig. 1 (b)). From Raman studies (Fig. 1 (c)), it is possible to define two pairs of DWNTs; (inner diameter: outer diameter) = (0.77nm:1.43nm) and (0.9nm:1.60nm), respectively. With the availability of this pure material available, researchers will now be able to answer: (1) Do DWNTs behave as quantum wires?; (2) Is there a chirality relationship between the concentric tubes during growth?; (3) Is there any significant effect of the concentric tubes on the electronic conductance?; (4) What are the adsorption properties of co-axial nanotube ropes?, etc. In addition to fundamental research, we envisage this material to be useful in the fabrication of novel sensors, nano-composites, field emission sources, nanotube bi-cables and electronic devices.

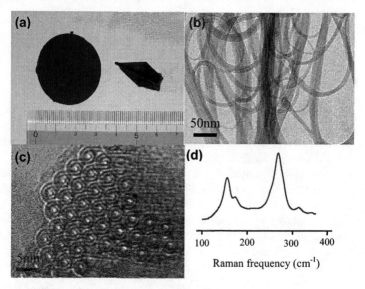

FIGURE 1. (a) Photographs of DWNTs sheet, (b) low resolution TEM image, (c) cross-sectional high resolution TEM image and (d) low frequency Raman spectrum.

STRUCTURAL MODIFICATION OF DOUBLE WALLED CARBON NANOTUBES

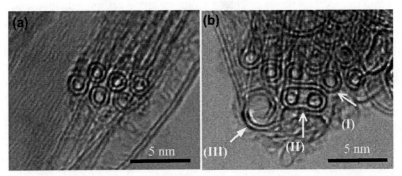

FIGURE 2. (a) A typical high resolution transmission electron microscopy (HRTEM) image of DWNTs in a bundle state, (b) HRTEM image of DWNTs at 2100°C. This image exhibits a sequential reconstruction process of a DWNT: (I) two outer tubes start to merge, through a zipping mechanism, (II) two outer tubes are completely combined into a single large outer tube with an oval shape containing two SWNTs, and (III) two inner SWNTs in a confined space might decompose along the inner wall of an outer shell, to form one inner single shell, like the formation of a DWNT derived from a peapod

The thermal stability and structural changes suffered by DWNTs as a function of heat treatment temperature were evaluated using Raman spectroscopy and HRTEM. We demonstrated that DWNTs are much more stable than SWNTs, and are able to stand temperatures as high as 2000°C without experiencing considerable morphological changes. Due to this thermal stability, we envisage DWNTs being used in the fabrication of field emission devices operating at high current and high temperatures. In addition, DWNTs could be used as fillers in nanotube composite materials. In particular, we noted that DWNT bundles are stable and start coalescing above 2000°C. Subsequent annealing resulted in the formation of shortened large-sized DWNTs, MWNTs and the flake-like sp^2 carbons. Our results indicate that the removal of structural defects and metal particles present within DWNTs could occur by thermal annealing without altering significantly the structure and diameter of the tubules. Therefore, in the near future, it may be possible that DWNTs replace SWNTs in specific carbon nanotube devices used today. On top of that, by coalescing double-walled carbon nanotubes through thermal treatment above 2100oC, a novel and stable structure consisting of flattened tubules containing two single-walled tubes (SWNTs) is created [6]. The process occurs due to the coalescence and reconstruction of the outer shells of DWNTs, leaving the inner cylinders almost intact, the latter being encapsulated inside the large diameter coalesced tube (bi-cable) (see Figure 2). We propose that the coalescence process is due to: (1) the thermal activity of the outer shells, which is driven by a surface energy minimization process, and (2) a zipping mechanism followed by atom reconstruction, in which two outer shells interact and anneal.

HIGH PERFORMANCE OF MICRO-CATHETER

We demonstrated that novel catheter by incorporating carbon nanotube into nylon-12 is very promising for medical applications due to their improved mechanical property, less thrombogenicity and less coagulability. On top of that, increased mechanical strength derived from intrinsic nature of carbon nanotube resulted in high resistance to fracture and highly improved handleability in operation. It is noteworthy that no carbon nanotube exposed on the outer surface of catheter indicates the low possibility of direct reaction of carbon nanotube with blood as well as the intraluminal surface of the new catheter.

CONCLUSIONS

Even though practical applications of carbon nanotubes have been realized step by step, some challenges have to be solved in order to realize the potential uses of CNTs. The first one is the large-scale synthesis producing defect-free CNTs at low cost. Secondly, it is important to control the diameter and chirality of CNTs. The final and most important issue is how to manipulate these tiny molecules in order to fabricate novel nano-electronic devices and materials. We envisage that in less than ten years various nanotube-based devices will be playing a significant role in emerging technologies. We foresee that carbon nanotubes will be utilized in versatile applications and will take an important place in the development of emerging technologies in the near future.

ACKNOWLEDGMENTS

This work was supported by the CLUSTER of Ministry of Education, Culture, Sports, Science and Technology.

REFERENCES

1. Dresselhaus MS, Dresselhaus G, Eklund P. Science of fullerene and carbon nanotubes. Academic Press, 1995.
2. Oberlin A, Endo M, Koyama T. Filamentous Growth of Carbon Through Benzene Decomposition, Journal of Crystal Growth 1976;32:335-349.
3. Endo M. Grow Carbon Fibers in the Vapor Phase. Chemtech 1988;568-576.
4. Endo M, Kim YA, Hayashi T, Nishimura K, Matsushita T, Miyashita K, et al. Vapor-grown carbon fibers (VGCFs): Basic properties and battery application. Carbon 2001;39: 1287-1297.
5. Endo M, Koyama S, Matsuda Y, Hayashi T, Kim YA. Thrombogenicity and blood coagulation of a micro-catheter prepared from carbon nanotube-nylon based composite. Nanoletters 2005;5(1):101-106.
6. Endo M, Hayashi T, Muramatsu H, Kim YA, Terrones H, Terrones M, Dresselhaus MS. Coalescence of double-walled carbon nanotubes: formation of novel carbon bicables. Nanoletters 2004;4(8):1451-1454.Brown, M. P., and Austin, K., *The New Physique*, Publisher City: Publisher Name, 1997, pp. 25-30.

The Influence of Sulfur Promoter on the Production of SWCNTs by the Arc-Discharge Process

M. Haluška[1], V. Skákalová[1], D. Carroll[2], S. Roth[1]

[1]Max Planck Institute für Festkörperforschung, D-70569 Stuttgart, D
[2] Center for Nanotechnology, Wake Forest University, Winston-Salem, NC, USA

Abstract. The influence of sulfur addition on the production of SWCNTs by the arc-discharge method with Fe/Y catalysts was investigated. It was found that the yield of SWCNTs is strongly influenced by the presence of sulfur. The dependence of the SWCNTs yield on the concentration of sulfur in the anode reaches its maximum around 1.5 at%. The yield of SWCNTs was estimated by thermal gravimetry and by optical absorption.

INTRODUCTION

The arc-discharged method has been one of the most commonly used methods to produce single wall carbon nanotubes (SWCNTs) since 1993, when Iijima (1) and Bethune et. al (2) produced SWNTs using the modified Krätschmer generator (3). SWCNTs were first grown using a single transient metal catalyst (Fe, Co or Ni) added to the carbon anode. The higher yields of SWCNTs were obtained for bimetal compositions such as Fe/Ni (4). The highest yield and the best purity were found for the combination Ni and Y (5). Because the promoting role of sulfur in catalytically produced vapor-grown carbon fibers was well known (6), sulfur was added to the transient metal catalyst for SWCNT production (7). The authors in (7) used 2 and 4 at% of sulfur and 4 at % of Co mixture with graphite. They found that addition of S at least doubled the tube abundance and broadened the tube diameter distribution that extends out to diameters of about 6 nm. Some double walled NTs (DWCNTs) were found as well. In later work, sulfur was used in the anode carbon mixture together with Ni/Fe/Co (8, 9), Ni/Co (10), Ni/Y/Fe or Ni/Ce/Fe (11) catalysts. Sulfur was used as promoter in other nanotube production methods as well. It was used in the floating catalyst method with Fe (in form of ferrocene) (12), in the solar energy evaporation method with Ni/Co (13), and in the laser vaporization method with Ni/Co/Fe (14). It was found that sulfur alone without metal catalysts does not catalyze the SWCNTs growth process (7, 13). Typically, for all methods mentioned and all catalysts used, the addition of a small amount of sulfur increases the yield of nanotubes and broadens the tube diameter distribution. In spite of the experience with the use of sulfur for both carbon fibers and carbon nanotubes, its role is not well known. This work is focused on the effect of S added to Fe/Y catalysts on the yield of nanotubes, with the aim to contribute to the clarification of the role of sulfur in the production of SWCNTs.

CP786, *Electronic Properties of Novel Nanostructures*, edited by H. Kuzmany, J. Fink, M. Mehring, and S. Roth
© 2005 American Institute of Physics 0-7354-0275-2/05/$22.50

EXPERIMENTAL

SWCNTs production experiments were performed in an horizontally oriented water-cooled cylindrical stainless steel chamber. The chamber was equipped with a water-cooled static cathode and a mobile anode, respectively, and with a window. A constant current mode of the DC source was used. Typical values were $I = 77 - 86$ A, $U_{dc} = 33 - 38$ V. The diameter and length of the cathode and anode graphite rods were 1.3 cm and 9 cm and 0.63 cm and 10 cm, respectively. The graphite rods used as anodes were first drilled to make a 4.4 cm deep hole with 2.7 mm in diameter. The holes were filled with mixtures of graphite, Fe, FeS_2 and Y powders. The concentration of Fe and Y in the anode was approximately 6 and 1 at%, respectively. The chamber was filled with helium. Obtained arc-products were weighted, investigated by Raman spectroscopy, thermal gravimetry (TG) and optical spectroscopy. Thermal gravimetry (TG) was measured using Pyris 1 TGA Perkin Elmer. Approximately 7 mg of material was inserted into a TG ceramic crucible. The samples were heated in a 40 ml/min flow of air under a 10°C/min temperature increase rate, up to 950°C. Raman spectra were measured using microscope laser Raman spectroscopy with a Jobin Yvon-LabRam spectrometer. The laser excitation wavelengths were 632 nm and 785 nm. Optical absorption was measured by a Perkin-Elmer Lambda spectrometer in DMF solution.

RESULTS

In this work we focused on the web material because it contains less impuritiest than cathode collar and wall deposit. Each sample was first checked by Raman spectroscopy to confirm the presence of SWCNTs. The yield of the web was calculated as the ratio of web amount to the amount of evaporated anode material. In Fig. 1 the dependence of the web yield on the concentration of sulfur is shown for four various initial He pressures. None or a very small amount of web was found if no sulfur was present and for a concentration above 2.7 at%. Yields for various He

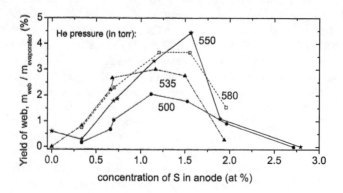

FIGURE 1. Dependence of he web yield on sulfur concentration in the anode. Numbers correspond to the initial He pressure in the reactor.

Pressures reach their maximum for slightly different sulfur concentrations. The highest yield was repeatedly obtained for 1.56 at % of S at 550 torr of helium. The analysis was performed to know which of the materials (metals, SWCNTs or carbon impurities) cause the increase of the web yield. The content of metals was estimated by TG. No TG was performed for the experiments with the lowest (0) and highest (2.7 at%) concentration of sulfur, since the amounts of web were very small. Fig. 2a shows TG weight loss curves of web prepared at $p_i(He)= 550$ torr with different sulfur concentrations. All curves have a similar shape. There is a visible small mass gain

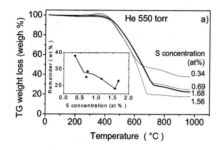 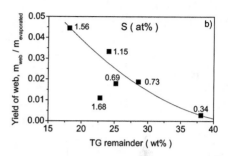

FIGURE 2. a) TG weight loss curves of some samples prepared with various sulfur concentrations in the anode. The inset shows remainder dependence on sulfur concentration. b) Relation between metal content and yield for samples prepared with various sulfur concentrations. This picture is made using data from Fig. 1 and Fig. 2a). Numbers give sulfur concentration in the anode.

around 350°C. It is followed by mass loss as the temperature increases. The non-burned remainder was red and contains metals and their oxides. The inset in Fig. 2a shows the dependence of the remainder's amount on sulfur concentration in the anode. Fig 2b shows the relation between the TG remainder and the yield of web for samples prepared with various sulfur concentrations. Numbers give the sulfur concentration in the anode. The content of metals decreases with increasing yields of web. The smallest remainder was found for the sample with the highest yield. It means that carbon materials rather than metals contribute to the yield increase shown in Fig. 1. The ratio of SWCNTs to carbon impurities can be estimated by deconvolution of the first derivative of the TG mass loss curves (DTG) (15) and by optical absorption (16). DTG is a common method for quantitative analysis, the content of each species corresponding to its peak area in DTG. The DTG curves were fitted by 4 gaussian peaks in the part of highest mass losses. One can expect that the first of these peaks corresponds to burning of amorphous carbon, the next peaks to SWCNTs, cage like carbon nanostructures and turbostratic-graphite, respectively. Fig. 3a shows the dependence of the area of the DTG SWCNTs gaussian peak on sulfur concentration. The analysis of optical absorption spectra according to (16) gives the SWCNT's relative purity (relative content of SWCNTs) in ratio to other carbon materials. The dependence of the relative purity on sulfur concentration is shown in Fig. 3b. Both analyses for the ratio of SWCNTs to other carbon materials give similar dependences on sulfur concentration. Figs. 3c and 3d show the relation between relative content of SWCNTs and yield of webs for samples prepared with various sulfur concentration obtained by analyzing DTG curves and optical adsorption, respectively. Both figures

89

show similar dependences and indicate that the highest ratio of SWCNTs to other carbon materials exists for samples prepared with 0.7 at % of sulfur. The smallest content of SWCNTs is found for small (0, 0.34 at %) and highest tested (1.68 at%) sulfur concentrations.

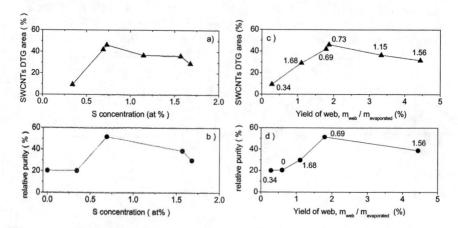

FIGURE 3. a) Dependence of the area of the SWCNT gaussian peak of DTG curves on sulfur concentration. b) Dependence of the relative SWCNT purity estimated from optical absorption according to (16) on sulfur concentration. Relation between SWCNTs content and yield for samples prepared with various sulfur concentrations estimated c) from DTG and d) from optical absorption. Numbers give sulfur concentration in the anode.

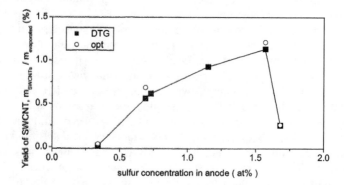

FIGURE 4. a) Dependence of the yield of SWCNTs on sulfur concentration in anode. Circles and squares represent data obtained by combination of TG with DTG and optical absorption methods, respectively.

The concentration of SWCNTs in web material now can be estimated, because we estimated the content of metals by TG and estimated the relative purity/concentration by optical absorption/DTG. Fig. 4 shows the estimated yield dependence of SWCNTs (in weight percent of evaporated anode) on sulfur concentration in the anode. Both methods (optical absorption and DTG) used for estimation of the ratio of SWCNTs to other carbon materials give very similar results. The yield of SWCNTs is not shown for 0 and 2.8 at % of sulfur because metal content was not determined (not enough web for TG).

CONCLUSION

The addition of sulfur to Fe/Y catalysts in the production of SWCNTs by the arc-discharge method plays a very important role. Metal-sulfur interactions change the surface tension and melting point of small droplets of metal. This can support the formation of SWCNTs from metals, which catalyze badly the formation of SWCNTs in their pure form. The overcritical concentration of S has a poisoning effect on SWCNTs growth, similar to that of many other catalytically controlled processes. The yield of SWCNTs was estimated by TG and optical adsorption. Both methods give similar results. The highest yield of SWCNTs was obtained for a carbon anode containing approximately 6, 1 and 1.5 at % of Fe, Y and S, respectively.

ACKNOWLEDGMENTS

This work was supported by EU projects SPANG, CARDECOM and CANAPE and U.S. Air Force project. M.H. thanks to A. Lachgar and his students for helping with TG.

REFERENCES

1. Iijima S., Ichihashi T., Nature **363**, 603-605 (1993).
2. Bethune D. S., Kiang C. –H, de Vries M. S, et al.,Nature **363**, 605-607 (1993).
3. Krätschmer W., Lamb L.D., Fostiropoulos K., Huffman D.R., Nature **347**, 354 (1990).
4. Saito Y., Koyama T., Kawabata K., Z. Phys. D **40**, 421-424 (1997).
5. Journet C., Maser W. K., Bernier P., et al., Nature **388**, 756-758 (1997).
6. Tibbettts G. G., Bernardo C. A., et al, Carbon **32**, 569-576 (1994).
7. Kiang C.-H., Goddart W. A., Beyers R., at al. S., J. Phys. Chem. **98,** 6612-6618 (1994).
8. Liu C., Cong H. T., Li F., et al., Carbon **37**, 1865-1868 (1999).
9. Park Y. S., Kim K. S., Jeong H. J., et al., Synthetic Metals **126**, 245-251 (2002).
10. Huang H., Kajiura H., Tsutsui S., et al., Chem. Phys. Lett., **343**, 7-14 (2001).
11. Zhu H-W., Jiang B., Xu C-L., et al., J. Phys. Chem., **107**, 6514-6518 (2003).
12. Ci L., Wei J., et al., Carbon **39**, 329-335 (2001).
13. Alvarez L., Guillard T., Sauvajol J.L., et al., Chem. Phys. Lett., **342,** 7-14 (2001).
14. Lebedkin S., Schweiss P., Renker B., et al., Carbon **40,** 417-423 (2002).
15. Shi Z., Lian Y., et al., Solid State Com., **112**, 35-37 (1999).
16. Itkis M., Haddon R. C., et al. Nano Lett. **3**, 309-314 (2003).

Influence of Fe-doped Graphite Electrode Characteristics on Ar-H$_2$ Carbon Arc Plasma and SWCNT Formation

A. Huczko[*1], H. Lange[1], M. Bystrzejewski[1], Y. Ando[2], X. Zhao[2] and S. Inoue[2]

[1]Department of Chemistry, Warsaw University, 1 Pasteur , 02-093 Warsaw, Poland
[2]Meijo University, Shiogamaguchi 1-501, Tempaku-ku, Nagoya 468-8502, Japan

Abstract. Two Fe-doped (ca. 1 at.%) homogeneous graphite electrodes (different graphite microcrystals, degree of graphitization and, thereby, electrical conductivities) electrodes were used in the process of production of single-walled carbon nanotubes in Ar-H$_2$ arc plasma under pressure 26 kPa. The C$_2$ content (namely carbon vapor pressure) and temperature distributions in the arc plasma were determined using optical emission spectroscopy. The mechanism of CNT formation based on carbon dimers as the building blocks seems to be at least questionable.

Keywords: Carbon nanotubes; arc discharge; electron microscopy; Raman spectroscopy.
PACS: 52.80.Mg; 78.30.-j; 81.07.De; 87.64.Ee.

INTRODUCTION

The proper choice of the catalyst in the case of single-walled carbon nanotubes (SWCNTs) formation is considered to be a crucial factor affecting the process yield. Ando et al. [1] used Fe-doped homogeneous graphite to produce SWCNTs in carbon arc under argon-hydrogen atmosphere. To learn more about the process we applied the emission spectroscopy [2] for the diagnostics of the plasma zone of this process. Surprisingly, we found that SWCNTs formation yield differed heavily depending on the characteristics of the starting graphite from different batches (despite the same Fe content).

EXPERIMENTAL

Two different (A and B – Table 1) Fe-doped (ca. 1 at. %) homogeneous graphite electrodes were used in the process of SWCNT synthesis in Ar-H$_2$ in the system described elsewhere [3]. The electrodes differed with the size of graphite microcrystals, degree of graphitization and, thereby with electrical conductivities. It has already been found that specifically selected graphite electrode with the low conductivity can lead to a very efficient SWCNT formation [1]. In this work, we have focused our attention to the effect of the electrode material morphology on plasma

CP786, *Electronic Properties of Novel Nanostructures*, edited by H. Kuzmany, J. Fink, M. Mehring, and S. Roth
© 2005 American Institute of Physics 0-7354-0275-2/05/$22.50

parameters and final products. The morphology and characteristics of the SWCNT produced were investigated by using SEM, HRTEM and Raman spectroscopy.

TABLE 1. Starting graphite characteristics and experimental conditions.

Density [g cm^{-3}]	Resistivity [$\mu\Omega$ m]	Fe contents [at. %]	Arc discharge conditions
Electrode A			P_o=26 kPa
1,38	62	0,96	40% H$_2$; 60% Ar
Electrode B			I=50 A; U=23÷33 V
1,61	32,7	0,99	Anode erosion rate 1÷1.5 mg s^{-1}

RESULTS AND DISCUSSION

The characteristic feature of the XRD patterns of the material A is much broader graphite peak at 2θ equal to 26° (Fig. 1). It indicates to a lower graphitization degree and/or smaller size of graphite crystallites comparing to the material B. There is also a diversity in Fe peaks. Surprisingly, in the case of the electrode A, Fe peaks correspond to the fcc and bcc structures. In the case of the anode A within the seconds after arc ignition, the formation of a dense and fine web filling the whole reactor volume was observed. TEM and HR TEM images of the nanotubes are shown in Figs. 2 and 3, respectively. In the case of the material B, only soot (with very few nanotubes) condensed on the reactor wall. One should mention here that the reproducibility of the experiments was high. The high content of SWCNTs formed while using A graphite is confirmed by Raman spectra of the products (Fig. 4).

Surprisingly, despite of the drastic changes in the product morphology, any significant differences in the carbon arc plasma characteristics have been observed (Fig. 5). The content of 'small' carbon species calculated from the LTE suggests that there are also other important bigger starting carbon 'blocks' related to the nanocarbon formation since there has not been apparently any correlation between C$_2$ radical concentration in the arc plasma zone and SWCNT content in the product.

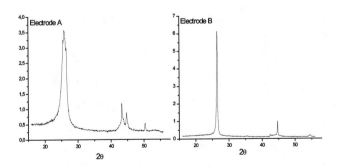

FIGURE 1. XRD patterns of electrode A and B.

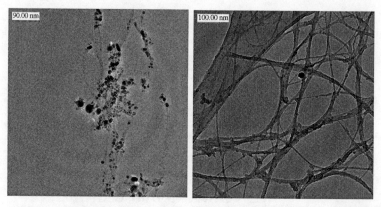

FIGURE 2. TEM images of the web (graphite A) before and after purification, left and right, respectively.

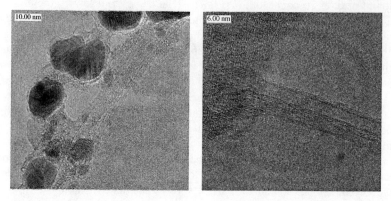

FIGURE 3. HRTEM images of the web (graphite A) before and after purification, left and right, respectively.

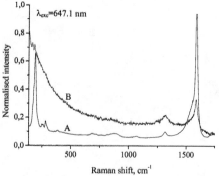

FIGURE 4. Raman spectra of products obtained during arcing of two different graphites, A and B

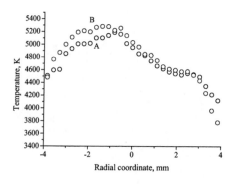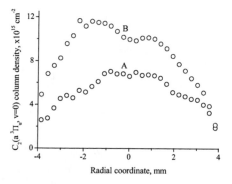

FIGURE 5. Average temperature and C_2 column density distributions across the arc dischrage for both, A and B, electrodes, left and righht, respectively.

CONCLUSIONS

The web produced from the less-graphitized carbon (material A) evidently contains high amount of SWCNTs. In the case of electrode B, very few CNTs were observed. The product contains however, a high concentration of iron encapsulates. Regardless of the starting carbon characteristics, there is no significant difference in plasma temperatures. Higher C_2 content (and, consequently, total pressure of atomic and molecular species) was found, however, in the case of well-crystallized carbon electrode (material B). These results shed new light on the mechanism of CNT formation in the carbon arc plasma. It seems that not only C_2 radicals (as it has been commonly assumed) play an important role in CNT growth. It is very likely that some other bigger specific carbon nanostructures are chemically activated and transported by hydrogen away from the sublimation to coalescence zone. The process seems to be more efficient in the case of a loose carbon structure (material A). However, much work remains to be done to elucidate the role of these factors on CNT formation.

ACKNOWLEDGMENTS

This work was supported by the Comittee for Scientific Research (KBN) through the Department of Chemistry, Warsaw University, under Grant No 4 T08D 021 23.

REFERENCES

1. X. Zhao, S. Inoue, M. Jinno, T. Suzuki, Y. Ando, *Chem. Phys. Lett.* **373**, 266-270 (2003).
2. H. Lange, K. Saidane, M. Razafinimanana, A. Gleizes, *J. Phys. D: Appl. Phys.* **32**, 1024-1030 (1999).
3. H. Lange, P. Baranowski, A. Huczko, P. Byszewski, *Rev. Sci. Instrum.* **68**, 3723-3727 (1997).

Magnetic Carbon Cluster Formation Process: Optical Spectroscopy of Laser-Ablated Carbon Plume

A. V. Rode[1], N. R. Madsen[1], A. G. Christy[2], J. Hermann[3], E. G. Gamaly[1], and B. Luther-Davies[1],

[1]*Laser Physics Centre, Research School of Physical Sciences and Engineering,*
[2]*Department of Earth and Marine Science, Faculty of Science,*
The Australian National University, Canberra, ACT 0200 Australia

[3]*Laboratoire Lasers, Plasmas et Procédés Photoniques,*
Université Aix - Marseille II, Marseille Cedex 9 France

Abstract. Cluster-assembled carbon nanofoam produced by high-repetition-rate laser ablation of graphite in Ar exhibits numerous unique properties including extremely low bulk density, large surface area and DC conductivity similar to that of amorphous diamond-like carbon films. In addition it exhibits para- and ferromagnetic behavior measurable up to 90 K. We present preliminary in-situ spectroscopic studies of the laser ablated plume during creation of the carbon nanofoam material.

INTRODUCTION

A new form of carbon, a cluster-assembled carbon nanofoam, has been recently synthesized by high-repetition-rate laser ablation of graphite in an ambient atmosphere of argon at ~1-100 Torr [1-5]. The material is an exceptionally low-density aerogel with a hierarchical nanostructure. Our structural studies revealed the presence of hyperbolic "schwarzite" layers [6,1] inside the clusters. Schwarzites are anticlastic (saddle-shaped), warped graphite-like sheets, in contrast to the synclastic (ellipsoidally curved) sheets of fullerenes [7]. The schwarzite layers form spheroidal clusters of 6 nm typical diameter, which in turn are connected into dendritic forms.

It has been shown [5] that the carbon vapour temperature in the formation zone of the laser plume can be in the range 1-10 eV, hence the formation process takes place in a partly ionized carbon plasma. The formation process involves periodic cycling between vapour heating and cluster formation stages, with the time period dependent upon the initial Ar density, the evaporation rate, and reaction rate, which in turn is a function of the temperature and density of the atomic carbon vapour [4,5]. Ultimately it is desirable to investigate the effect of changes in these in-situ parameters upon the properties of the deposited nanofoam. To achieve this it is vital that the mechanisms of formation be understood through the implementation of in-situ spectroscopic (and imaging) diagnostic techniques. Such measurements enable one to deduce information regarding the gas kinetic temperature, electron temperature and density, plume constituent density and other information regarding the plume expansion dynamics.

OPTICAL SPECTROSCOPY OF LASER-CREATED PLUME

In the experiments presented Ar was introduced into a vacuum chamber at a flow rate of 300 sccm while maintaining a constant pressure of 2 Torr. The graphite target was irradiated with an in-house designed and built Nd:YVO laser [8] operating at second harmonic 532nm, repetition rate 1.5 MHz, pulse duration 12 ps and focal spot size 30 microns giving an incident intensity on the target surface of 2×10^{11} W/cm^2. The laser beam was scanned over the target

CP786, *Electronic Properties of Novel Nanostructures*, edited by H. Kuzmany, J. Fink, M. Mehring, and S. Roth
© 2005 American Institute of Physics 0-7354-0275-2/05/$22.50

surface in a Lissajous pattern of 1cm² at a frequency of about 50 Hz. Time averaged spectra were obtained by focussing upon the aperture of an optic fibre. The fibre was then coupled into an OceanOptics HR2000 spectrometer for measurement of spectra in the range 200 to 840 nm at resolution of approximately 0.2 nm.

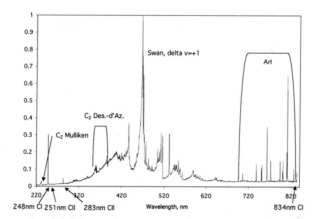

Figure 1. Sample spectrum for laser ablated carbon plume at a distance of 3 mm from the target. Strong line at 532 nm is scattered light from the ablation laser.

The spectrum contains numerous lines from neutral argon in the region 700-840 nm in addition to a number of lines from neutral (CI) and singly ionised carbon (CII). Examples of line radiation from atomic carbon species are the 248 nm CI emission and the 283 nm CII ion emission. The remaining spectral features are attributable to C_2 molecular species which sit atop a broad continuum emission. In the work presented here, we use the most intense emission bands of C_2 Swan emission due to the electronic transition between the upper $d\ ^3\Pi_g$ and the lower $a\ ^3\Pi_u$ molecular states [9]. The C_2 Swan system is situated in the visible spectral range from 400 to 520 nm (vibrational sequences $\Delta\upsilon= 0, +1, +2$) to fit to theoretically predicted spectra and thus obtain rotational and vibrational temperatures for the species.

One can generate a theoretical prediction for the C_2 Swan system with rotational and vibrational temperatures used as fitting parameters. Each vibrational band of a diatomic molecule is composed of three branches R, P and Q. The wavenumber of each rotational transition is determined by computing the total energies of the corresponding upper and lower levels [9]

$$v = T'_e - T''_e + G' + G'' + F' - F'' \tag{1}$$

where T_e, G and F are the energies of electronic state, vibrational and rotational levels, respectively. The indexes (') and (") stand for the upper and lower level, respectively. The vibrational and rotational energies have been computed using the term formula for triplet transitions given by Herzberg [10] with the molecular constants given by Prasad and Bernath [11]. The relative intensities of the rotational transitions have been computed using

$$I_{v',v'',J',J'} = C^{st} \frac{q_{v',v''}S_{J',J'}}{\lambda^4 Q_{rot}(v')} \exp\left\{-F'(J')\frac{hc}{kT_{rot}}\right\} \exp\left\{-G'(v')\frac{hc}{kT_{vib}}\right\} \tag{2}$$

where C^{st} is a numerical constant and $q_{\upsilon'\upsilon''}$ are the Franck-Condon factors from [12]. The line strengths $S_{J'J''}$ were computed using the expressions for triplet transitions and $\Delta\Lambda=0$ given by

Kovacs [13]. The wavelength of the rotational transition is $\lambda = 1/\nu$ while Q_{rot} is the partition function of rotational levels in the upper vibrational state υ'. To compute the spectrum in the range of interest, each rotational transition was approximated by a gaussian profile with a full width at half maximum equal to the apparatus width. The shape of the spectrum was obtained by superposition of all rotational transitions with $J \leq 200$.

A Boltzmann equilibrium distribution of the population densities of rotational levels is easily established by collisions of the molecules with other plasma species (atoms, ions, molecules). For that reason, the analysis of the intensity distribution of rotational levels is often used to determine the gas kinetic temperature (which is equal to the rotational temperature deduced from the spectra analyses). Equilibrium of vibrational levels with the collisional partners (atoms, ions, molecules) is more difficult to establish.

If $T_{rot} = T_{vib}$, a collisional equilibrium is established of both the rotational and vibrational levels with the collisional partners, and the deduced temperature is equal to the gas kinetic temperature. If T_{vib} differs from T_{rot}, the vibrational levels are not in equilibrium with the collisional partners. The distribution of the population densities of vibrational levels is thus due to collisional and radiative processes (recombination, etc...) and hence the vibrational temperature has no useful meaning in terms of the temperature of the plasma. Instead it is only a parameter that describes approximately the vibrational population density distribution.

Here the values deduced from theoretical predictions of the C_2 emission spectra for T_{rot} and T_{vib} are approximately 6000K and 50 000K respectively. It is clear that very high vibrational temperature deduced from the spectra (and the significant difference between T_{rot} and T_{vib}) indicates there is no collisional equilibrium between C_2 molecules and other plasma species. There are most likely strong recombination rates leading to the formation of diatomic and polyatomic carbon molecules. In light of the above considerations we can suggest that since the rotational levels are in equilibrium with the gas, the rotational temperature can be equated with the gas kinetic temperature that is about 6000K in this experiment.

The same experimental conditions discussed above were used to investigate the behavior of the plume as gas pressure and flow rate are increased. In these experiments a higher pressure was obtained via an increase in the flow rate. Ideally the flow rate should be kept constant, but in these preliminary experiments we are still able to obtain an indication of the plume's appearance as confinement is changed.

Figure 2 plots the spectrum of the plume as a function of the gas pressure, in the vicinity 0 to 0.5 mm from the target surface. The inset shows the change in relative intensity of ion and atomic carbon spectral features with gas pressure. It is important to note the fact that while most plume species increase in population near the target surface with increasing confinement, the concentration of CII atoms decreases. This can be understood in a view of recombination of CII with electrons to from CI. As pressure increases so does the recombination rate through greater collision frequency, and as a result ionised carbon is converted to neutral carbon closer to the target surface compared to lower pressure conditions.

In conclusion, theoretical fitting of the C_2 Swan features of the carbon plume spectrum indicats a lack of collisional equilibrium between C_2 molecules and other plasma species. Hence clusters are formed in highly non-equilibrium conditions. Such non-equilibrium conditions seems to be more favourable for formation of metastable carbon structures like carbon nanoitubes, fullerens, and also carbon schwarzites with unusual magnetic properties [1,2].

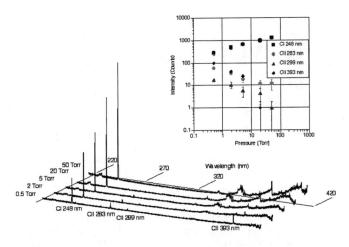

Figure 2. UV spectrum of plume near the target as a function of gas pressure. Inset: Relative intensities of identified atomic carbon species as a function of gas pressure.

ACKNOWLEDGEMENTS

Alessio Perrone is acknowledged for fruitful discussions on molecular spectra simulations.

REFERENCES

1. A. V. Rode, E. G. Gamaly, A. G. Christy, J. D. Fitz Gerald, S. T. Hyde, R. G. Elliman, B. Luther-Davies, A. I. Veinger, J. Androulakis, J. Giapintzakis, *Phys. Rev. B* **70**, 054407 (2004).
2. A. V. Rode, R. G. Elliman, E. G. Gamaly, A. I. Veinger, A. G. Christy, S. T. Hyde, B. Luther-Davies, *Appl. Surf. Sci.* **197-198**, 644-649 (2002).
3. E. G. Gamaly, A. V. Rode, and B. Luther-Davies, *Laser and Particle Beams*, **18**, 245-254 (2000).
4. E. G. Gamaly, A. V. Rode, *Nanostructures created by lasers*, in: *Encyclopaedia of Nanoscience and Nanotechnology*, Ed. H. S. Nalwa, (American Scientific Publishers, Stevenson Range, 2004), v. **7**, 783-809.
5. A. V. Rode, E. G. Gamaly, and B. Luther-Davies, *Appl. Phys. A* **70**, 135-144 (2000).
6. A. V. Rode, S. T. Hyde, E. G. Gamaly, R. G. Elliman, D. R. McKenzie, S. Bulcock, *Appl. Phys. A* **69**, S755-S758 (1999).
7. D. Vanderbilt and J. Tersoff, *Phys Rev Lett.* **68**, 511 (1991).
8. B. Luther-Davies, V. Z. Kolev, M. J. Lederer, N. R. Madsen, A. V. Rode, J. Giesekus, K.-M. Du, M. Duering, *Appl. Phys. A* **79**, 1051-1055 (2004).
9. B. Rosen, *Données Spectroscopiques relatives aux Molécules Diatomiques* (Oxford: Pergamon, 1970).
10. idem 2, p. 235.
11. C.V.V. Prasad and P.F. Bernath, *Astrophysical Journal* **426** (2), 812(1994).
12. Cooper and Nicholls, *Spec. Lett*, **9** (3), 149 (1976).
13. I. Kovaks, *Rotational Structure in the Spectra of Diatomic Molecules* (Budapest: Akadémiai Kiadó, 1969) p.131.

ACCVD Growth, Raman and Photoluminescence Spectroscopy of Isotopically Modified Single-Walled Carbon Nanotubes

Shigeo Maruyama and Yuhei Miyauchi

Department of Mechanical Engineering, The University of Tokyo
7-3-1 Hongo, Bunkyo-ku, Tokyo 113-8656, Japan

Abstract. Using alcohol catalytic CVD (ACCVD) technique optimized for the efficient production of SWNTs from very small amount of ethanol, SWNTs consisting of carbon-13 isotope ($SW^{13}CNTs$) were synthesized. Raman scatterings from $SW^{13}CNTs$ show no change from $SW^{12}CNTs$ in spectrum shape except for the Raman shift frequency down-shifted as much as square-root of mass ratio 12/13. On the other hand, SWNTs with mixed 12/13 isotopes from $^{13}CH_3\text{-}CH_2\text{-}OH$ and $CH_3\text{-}^{13}CH_2\text{-}OH$ show different amount of down-shift in Raman spectra, suggesting that smaller amount of site-1 carbon atom (next to OH) is incorporated into SWNTs. Based on this result, the initial decomposition reaction of ethanol on metal catalyst is discussed. Furthermore, near infrared luminescence of D_2O-surfactant dispersions of both $SW^{13}CNTs$ and $SW^{12}CNT$ were mapped. By comparing these maps, phonon sidebands of excitonic excitation were clearly identified.

Keywords: Single-Walled Carbon Nanotubes, Alcohol CVD, Raman, Photoluminescence.
PACS: 78.67.Ch, 81.07.De

INTRODUCTION

For the production technique of single-walled carbon nanotubes (SWNTs), various CVD approaches are being developed in stead of laser-furnace and arc-discharge methods. At present, CVD approaches using the high-pressure CO (HiPco) [1] is dominant for the mass production of SWNTs. Recently, we have proposed the use of alcohol for the carbon feedstock [2,3] for the generation of high-purity SWNTs at low temperatures. Furthermore, it was demonstrated that vertically aligned high quality SWNTs mat could be synthesized directly on silicon and quartz substrates [4]. This versatile nature of this technique made it possible to use rather expensive isotopically modified ethanol for the production of SWNTs. In this report, ACCVD method is modified for the efficient production of SWNTs from very small amount of ethanol with the similar technique used for the SWNTs generation from fullerene [5]. The SWNTs with various amount of ^{13}C abundance were generated by ACCVD technique from isotopically modified ethanol: $^{13}CH_3\text{-}^{13}CH_2\text{-}OH$, $^{13}CH_3\text{-}CH_2\text{-}OH$, and $CH_3\text{-}^{13}CH_2\text{-}OH$, in addition to the normal ethanol. The resonant Raman scattering of those SWNTs shows a simple shift of Raman scatting frequency for ^{13}C SWNTs. On the other hand, SWNTs with mixed 12/13 isotopes from $^{13}CH_3\text{-}CH_2\text{-}OH$ and $CH_3\text{-}^{13}CH_2\text{-}$

CP786, *Electronic Properties of Novel Nanostructures*, edited by H. Kuzmany, J. Fink, M. Mehring, and S. Roth
© 2005 American Institute of Physics 0-7354-0275-2/05/$22.50

OH show different amount of down-shift in Raman spectra, suggesting that smaller amount of site-1 carbon atom (next to OH) is incorporated into SWNTs. Based on this result, the initial decomposition reaction of ethanol on metal catalyst is discussed.

Photoluminescence (PL) has been intensively studied for the characterization of SWNTs [6, 7]. By plotting PL emission intensities as a function of emission and excitation photon energy, each peak in the PL-map is assigned to each chirality (n, m) [7]. As far as semiconductor SWNTs, PL-map is the most promising approaches for a quick determination of the structure distribution on a bulk SWNT sample [8], Hence, photoluminescence spectroscopy is a powerful tool not only for investigations of electronic properties of SWNTs but challenges to the (n, m)-controlled synthesis of SWNTs. However, in a PL map, we can find some PL peaks whose origins have not been elucidated other than bright PL peaks already assigned to particular nanotube (n, m) [7, 8]. Since these unassigned features may overlap with other PL peaks if a measured sample is an assemblage of various (n, m) structures, it is very important to understand the origins of all the features in a PL map for the accurate measurement of relative PL intensities of each (n, m) nanotube. Phonon assisted excitonic recombination [9] or phonon sideband due to strong phonon-exciton interaction [10] are proposed for some of extra excitation features. By using isotopically modified SWNTs with different phonon energies, PL peaks originates from exciton-phonon (exciton-phonon) interactions can be clearly distinguished from PL peaks without exciton-phonon interaction.

EXPERIMETAL PROCEDURES

Alcohol CVD Growth of Isotopically Modified SWNTs

We synthesized $SW^{13}CNTs$ from 0.5 gram of isotope-modified ethanol by alcohol catalytic chemical vapor deposition (ACCVD) method [2, 3] optimized for the efficient production of SWNTs from very small amount of ethanol, which is similar to the technique used for the SWNT synthesis from fullerene [5]. The detailed preparation of metal supporting zeolite powder was described in our previous reports [2, 3]. We prepared a catalytic powder by impregnating iron acetate $(CH_3CO_2)_2Fe$ and cobalt acetate $(CH_3CO_2)_2Co-4H_2O$ onto USY-zeolite powder (HSZ-390HUA over 99 % SiO_2). The weight concentration of Fe and Co was chosen to be 2.5 wt% each over the catalytic powder. A slightly modified apparatus from our standard ACCVD technique was employed as shown in Fig. 1. The catalyst was placed on a quartz boat and the boat was set in the center of a quartz tube (i.d. = 26 mm, length = 1 m). Small amount of ethanol was placed in an end of a smaller test-tube guiding the carbon source flow to the catalyst. One end of the quartz tube was connected to a rotary pump by two different paths, one 25 mm and the other 6 mm diameter tubes to select the pumping efficiency. The central 30 cm of the quartz tube was surrounded with an electric furnace. While the furnace was heated up from room temperature, about 200 sccm of Ar was flowed so that the inside of the quartz tube was maintained at 1 atm. After the electric furnace reached desired temperature, 800 °C, Ar flow was stopped and the larger evacuation path was opened to bring the inside of the quartz tube vacuum. Subsequently, ethanol vapor was injected to catalyst for 5 min from the end of the smaller test-tube ethanol reservoir. The synthesized SWNTs were characterized by micro Raman scattering measurements using CHROMEX 501is and ANDOR DV401-FI for the spectrometer and CCD system, respectively, with an optical system of SEKI TECHNOTRON STR250.

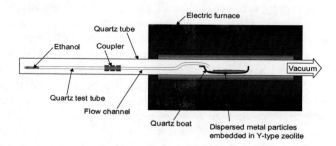

FIGURE 1. Schematic diagram of alcohol CVD apparatus for efficient generation of SWNTs

Dispersion and Photoluminescence Spectroscopy

In order to measure PL spectra from individual SWNTs in surfactant suspension, 'as-grown' materials were dispersed in D_2O with 0.5 wt % sodium dodecylbenzene sulfonate (NaDDBS) by heavy sonication with an ultrasonic processor (Hielscher GmbH, UP-400S with H3/Micro Tip 3) for 1 h at a power flux level of 460 W/cm^2. These suspensions were then centrifuged (Hitachi Koki himac CS120GX with S100AT6 angle rotor) for 1 h at 386 000 g and the supernatants, rich in isolated SWNTs, were used in the PL measurements.

Near infrared emission from the samples were recorded while the excitation wavelength was scanned from VIS to NIR range. The measured spectral data were corrected for wavelength-dependent variations in excitation intensity and detection sensitivity. The excitation and emission spectral slit widths were 10 nm (15~30 meV for excitation and ~10 meV for emission in the measuring range), and scan steps were 5 nm on both axes. In addition to PL maps of wide energy range of emission, we scanned PLE spectra of (7,5) nanotubes (emission at 1026.5 nm (1.208 eV)) with narrower excitation spectral slit width (5 nm : 8~15 meV in the measuring range) and scan steps (2nm) to obtain spectra with higher resolution. The photoluminescence spectra were measured with a HORIBA SPEX Fluorolog-3-11 spectrofluorometer with a liquid-nitrogen-cooled InGaAs near IR detector.

RAMAN SPECTRA AND GROWTH MECHANISM

Raman Spectra of Isotopically Modified SWNTs

In addition to normal ethanol, 3 isotopically modified ethanol were employed for the production of isotopically modified SWNTs. Three isotopically modified ethanol were $^{13}CH_3$-$^{13}CH_2$-OH (1,2-$^{13}C_2$, 99%), $^{13}CH_3$-CH_2-OH (2-^{13}C, 99%), CH_3-$^{13}CH_2$-OH (1-

^{13}C, 98%), supplied from Cambridge Isotope Laboratories, Inc. The carbon atoms in normal ethanol CH_3-CH_2-OH are composed of isotope mixtures of natural abundance, ie. 98.892% ^{12}C and 1.108% ^{13}C. Hence, by using normal ethanol, generated SWNTs should be composed of carbon atoms with the same abundance ratio. On the other hand, by using ethanol ($^{13}CH_3$-$^{13}CH_2$-OH), SWNTs principally made of ^{13}C can be generated. By using other 2 isotopically labeled ethanol ($^{13}CH_3$-CH_2-OH or CH_3-$^{13}CH_2$-OH), the abundance of ^{13}C isotopes in generated SWNTs should depend on the reaction process.

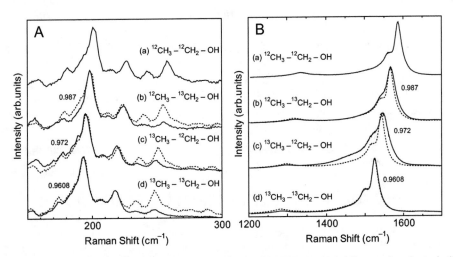

FIGURE 2. Comparison of RBM (A) and G-band (B) of SWNTs generated from various isotopically modified ethanol. (a): SWNTs from $^{12}CH_3$-$^{12}CH_2$-OH. (b): SWNTs from $^{12}CH_3$-$^{13}CH_2$-OH. (c): SWNTs from $^{13}CH_3$-$^{12}CH_2$-OH. (d): $^{13}CH_3$-$^{13}CH_2$-OH. Dotted lines are shifted spectra of ^{12}C SWNTs by multiplying the following factor to the frequency: (b) 0.987, (c) 0.972, (d) $\sqrt{12/13}$.

Fig. 2 compares Raman scattering for 4 kinds of SWNTs from isotopically modified ethanol excited with 488 nm laser spectra. Spread in G-band peak at 1590 cm^{-1} for normal SWNTs and very small D-band signal at 1350 cm^{-1} suggest that high quality SWNTs were generated by these experiments from tiny amount of ethanol. The strong radial breathing mode (RBM) peaks at around 150-300cm^{-1} confirms this observation. By multiplying the mass-ratio factor $\sqrt{12/13}$ =0.9608 to frequency of SW^{12}CNT (Fig. 2(a)) spectrum, SW^{13}CNTs (Fig. 2(d)) spectrum is almost completely reproduced, where dotted line is the shifted spectrum from SW^{12}CNT.

The Decomposition of Ethanol on Catalyst

The interpretation of 2 mixed isotope cases in Figs. 2 (b, c) is quite important. In addition to slight broadening of spectra, the different shifts of the frequency was observed for 2 mixed cases. The difference in the vibrational frequency between (b) from $^{12}CH_3$-$^{13}CH_2$-OH and (c) from $^{13}CH_3$-$^{12}CH_2$-OH means that two carbon atoms in an ethanol molecule are not equally used for the SWNT formation. Apparently, the carbon atom at site 2, further from OH, is more likely to be incorporated into an SWNT. The scaling factor on frequency of the dotted lines for (b) and (c) were 0.972 and 0.987, respectively. Assuming that this factor is the frequency change due to the average mass of carbon atoms, the factor should equal to $\sqrt{12/m_{ave}}$. Then, the average mass for (b) and (c) are 12.32 amu (Site 1: Site 2 = 32:68) and 12.70 amu (Site 1: Site 2 = 30:70), respectively. Both (b), (c) results conclude that only about 30 % carbon atoms in an SWNT are from the site 1 carbon atom. This result gives an important key for the analysis of the reaction mechanism of nanotube formation. On the metal catalyst surface, ethanol is expected to decompose. After losing hydrogen atoms, we believe that the oxygen atom has an important role of cleaning the carbon atoms with dangling bonds [2].

Let's assume that the ratio of original site 1 carbon atom in an SWNT is α, and that of site 2 is $(1-\alpha)$. If we further assume that the probability of complete break of C-O bond is P_{free}, and the free oxygen atom can randomly find its partner to become gas phase CO. Then, the incorporation of site 1 carbon atom can be possible by the C-O breaking and the choice of originally site 2 carbon atom by this free oxygen atom. Hence the probability is described as $\alpha = P_{free} \times (1-\alpha)$. With α = 30 %, P_{free} can be calculated as 45 %. So, we can conclude that about half of C-O bond of an original ethanol is completely broken in the typical ACCVD process.

PHOTOLUMINESCENCE FROM ^{13}C NANOTUBES

Figure 3(a,b) shows PL maps for SW^{13}CNTs and normal SWNTs and Fig. 3c compares PLE spectra corresponding to (7, 5) nanotubes (emission at 1.208 eV) for both SW^{13}CNTs and SW^{12}CNTs. Peak positions of major PL peaks of SW^{13}CNT sample were in good agreement with those of normal SWNT sample, indicating electronic properties and average environment around nanotubes are almost equivalent in each sample. Hence, only phonon energies are the principal difference between SW^{13}CNTs and normal SWNTs. Small PL peaks around the main peaks were observed as shown in Fig.3. For example, we find three small peaks above and below the main PL peaks corresponding to E_{11} (1.208 eV, out of measuring range) and E_{22} (1.923 eV) transition energies of (7,5) nanotubes in Fig. 3. Since the emission energies of these peaks were almost identical with the emission energy of the main peak at E_{22} excitation energy of (7, 5) nanotubes, these peaks are also attributed to photon emission of (7, 5) nanotubes. In this paper, we focus on these unassigned PL peaks of (7, 5) nanotubes. Hereafter, we refer to these unassigned peaks as peak A, B and C as shown in Fig. 3.

In the case of peak A and C, energy differences from the main peaks were clearly changed depending on whether the sample is SW^{13}CNTs or SW^{12}CNTs, while positions of the main peaks were almost identical. If a certain PL peak is a phonon sideband corresponding to the main PL peak, the amount of the change of the energy difference between the main peak and its sideband should be consistent with the value estimated from the difference of phonon energies confirmed by Raman spectroscopy shown in Fig.2. Isotope shift for peak A and C confirmed that these peaks are phonon side-band from E_{11} and E_{22}, respectively [11]. On the other hand, we observed almost no peak shift of peak B as shown in Fig.3c, indicating that peak B is due to 'pure electronic' transition without electron-phonon interaction. Further study shows that this peak is due to transverse excitation of nanotubes [12].

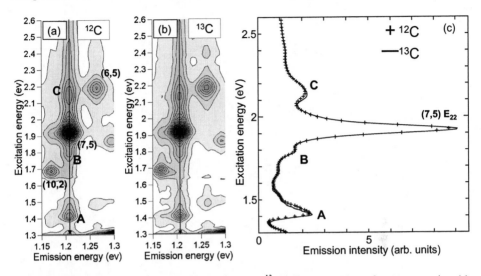

FIGURE 3. PL maps of (a) normal SWNTs and (b) SW^{13}CNTs dispersed in surfactant suspension. (c) : Comparison of PLE spectra of SW^{13}CNTs and normal SWNTs at the emission energy of 1.208 eV (corresponding to E11 energy of (7,5) nanotubes).

REFERENCES

1. P. Nikolaev, M.J. Bronikowski, R.K. Bradley, F. Rohmund, D.T. Colbert, K.A. Smith, R.E. Smalley, Chem. Phys. Lett., 313, 91-97 (1999).
2. S. Maruyama, R. Kojima, Y. Miyauchi, S. Chiashi, M. Kohno, Chem. Phys. Lett. 360, 229 (2002).
3. Y. Murakami, Y. Miyauchi, S. Chiashi, S. Maruyama, Chem. Phys. Lett. 374 (2003) 53.
4. Y. Murakami, S. Chiashi, Y. Miyauchi, M. Hu, M. Ogura, T. Okubo, S. Maruyama, Chem. Phys. Lett., 385, 298-303 (2004).
5. S. Maruyama, Y. Miyauchi, T. Edamura, Y. Igarashi, S. Chiashi, Y. Murakami, Chem. Phys. Lett. 375, 553 (2003).
6. M.J. O'Connell, S.M. Bachilo, C.B. Huffman, V.C. Moore, M.S. Strano, E.H. Haroz, K.L. Rialon, P.J. Boul, W.H. Noon, C. Kittrell, J. Ma, R.H. Hauge, R.B. Weisman, R.E. Smalley, Science 297, 593 (2002).
7. S.M. Bachilo, M.S. Strano, C. Kittrell, R.H. Hauge, R.E. Smalley, R.B. Weisman, Science 298, 2361 (2002).
8. Y. Miyauchi, S. Chiashi, Y. Murakami, Y. Hayashida, S. Maruyama, Chem. Phys. Lett. 387, 198 (2004).
9. S. G. Chou, F. Plentz, J. Jiang, R. Saito, D. Nezich, H. B. Ribeiro, A. Jorio, M. A. Pimenta, Ge. G. Samsonidze, A. P. Santos, M. Zheng, G. B. Onoa, E. D. Semke, G. Dresselhaus, M. S. Dresselhaus, Phys. Rev. Lett. 94, 127402 (2005).
10. V. Perebeinos, J. Tersoff, Ph. Avouris, Phys. Rev. Lett. 94, 027402 (2005).
11. Y. Miyauchi, S. Chiashi, S. Maruyama, Phys. Rev. Lett., to be submitted.
12. Y. Miyauchi, S. Maruyama, Phys. Rev. Lett., to be submitted.

CHARACTERIZATION AND PROPERTIES OF CARBON NANOTUBES AND NANOSTRUCTURES

HR-TEM study of atomic defects in carbon nanostructures

K. Urita[1], Y. Sato[1], K. Suenaga[1] and S. Iijima[1]

[1]*Research Center for Advanced Carbon Materials, National Institute of Advanced Industrial Science and Technology (AIST), Tsukuba, 305-8565, Japan*

Abstract. Direct observation of the defect formation in carbon nanostructures during electron irradiation is carried out by high-resolution transmission electron microscopy (HR-TEM) on the double-wall carbon nanotubes (DWNT) and carbon nanopeapods. The irradiation damage due to the knock-on collision with high energy electron beam are found to be dependent on the curvature of the constituent graphen layers. The formation of knock-on defects is more abundant at an inner tube than an outer tube in DWNT, and at a fullerene than a carbon nanotube in carbon nanopeapods. These results definitely indicate that the effects of electron irradiation in nanocarbon materials are quite more prominent on higher surface curvature, such as fullerene.

Keywords: Electron irradiation, Point defect, HR-TEM
PACS: 62.71.ji, 61.72.Ff, 61.80.Lj, 81.07.De

INTRODUCTION

Since their discovery, the carbon nanomaterials, such as fullerenes [1] and carbon nanotubes [2] have attracted a great deal of attention in the fields of physics and chemistry, because their properties and structures are interestingly different from the graphite and the possibility of new physical properties have been expected [3,4]. They exhibit a wide range of surface curvatures, which modify both the bond length and angle between adjacent carbon atoms, and a lot of theoretical and experimental studies have focused on the curvature effects of their properties [5-7]. A large-diameter tube (with a small curvature) is energetically stable from theoretical viewpoints. It is therefore expected that any atomic defect can be more likely to be induced in carbon nanostructure with a higher curvature, however, any direct evidence for the curvature dependence on the defect formation rate has never been experimentally provided at atomic-level. Recently, the atomic defects in carbon nanostructures have been identified by a HR-TEM study [8-10]. Here, we carry out a systematic study about the knock-on defect formation rate in various carbon nanostructures with different curvatures in order to clarify their curvature dependence. The defect formation rates in a DWNT and a carbon nanopeapod will be compared.

CP786, *Electronic Properties of Novel Nanostructures*, edited by H. Kuzmany, J. Fink, M. Mehring, and S. Roth
© 2005 American Institute of Physics 0-7354-0275-2/05/$22.50

EXPERIMENTAL

To visualize the formation of knock-on defects in DWNTs and fullerene peapods (C_{92}@SWNT), we have operated a field emission high-resolution transmission electron microscope (JEOL-2010F). The accelerating voltage has been chosen as 120kV, which is just close to the threshold of the knock-on effect of carbon [11,12]. The specimens were dispersed in *n*-hexane and then fixed on holey C-coated copper grids for electron microscopy. The typical electron dose is estimated as 60,000 electrons per nm^2 for one-second exposure. The specimen stages for cryo-observation (Gatan 626 DH) were used to detect an obvious effect of the surface curvature on DWNTs at low temperature. Furthermore, carbon nanopeapods encapsulating Tb dimetallofullerene (Tb_2@C_{92}) and codoped with monometallofullerenes and dimetallofullerenes (Gd@C_{82} and Gd_2@C_{92}) were used to indicate a new dynamical behavior thorough fullerene coalescences triggered by the knocked-on carbon atoms in fullerene. Detailed procedures for these specimen syntheses can be found in our previous reports [13,14].

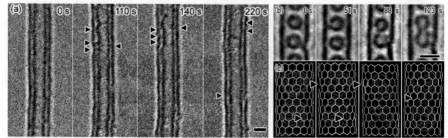

FIGURE 1. (a) Sequential HR-TEM images of the knock-on defects on electron-irradiated DWNT recorded at 300K. The black arrows indicate possible interlayer defects. The inner nanotube more easily accomodates structural defects than the outer nanotube. (b) A series of HR-TEM images for the irradiation-induced coalescence of the fullerene cage (C_{92}) aligned in a SWNT. The formation of interlayer couplings is indicated by black arrow in the schematic presentation (c). The knock-on damage is clearly seen at the fullerene cage. Scale bar = 1nm.

RESULTS AND DISCUSSION

Figure 1 shows two sequential HR-TEM images of the DWNTs and fullerene peapods recorded at 300K. In the first sequential images (Figure 1(a)), each layer of DWNT is clearly visible at *t* = 0 s. Subsequently, several bridges in dark contrast (indicated black arrows) unambiguously appear and then disappear between an inner and an outer layer in DWNT (*t* = 110 to 220 s). This is a Frenkel type defects involving a vacancy and interstitial atoms. The outer nanotube keeps its shape during observation while the inner tube exhibits a considerable deformation. From this viewpoint, the inner tube always suffers damage faster than the outer tube, and the outer tube seems more resistive to the electron irradiation. This is because the inner tube exhibits a large curvature (a smaller diameter) and more easily suffers knock-on damage as expected. Therefore the atomic vacancies are more likely to be induced in the inner tube. Figure 1(b) shows a series of HR-TEM images of atomic defects on C_{92}@SWNT peapod, which contains empty C_{92} fullerenes aligned in a SWNT in order to investigate the electron irradiation effects in a fullerene, which exhibits larger curvature than a carbon nanotube. Reasonable atomic models as the observed defective structures are also given in Figure 1(c). Because the exact positions for carbon atoms cannot be derived from the HR-TEM images, a most stable atomic configuration based on a semiempirical potential calculation has been employed for each model to achieve a better fit with the HR-TEM image simulations. At the beginning of the HR-TEM observation ($t = 0$ s and 51 s), the interlayer coupling between the nanotube and defected fullerene, forming a Frenkel pair, has been frequently observed. Such induced atomic vacancy on fullerene, which gives a passage for the imprisoned atoms from inner space to the outer, would act as an atomic pathway. Afterward, the two adjacent defected fullerenes (probably turn around to face its once defected position to each other and then) start to coalesce at $t = 123$ s. Most interestingly, the observed HR-TEM image undoubtedly indicates that an atomic path has been clearly induced between the two adjacent fullerene and a bridge has been formed to interconnect the two inner spaces, however, the structure of the outer SWNT is maintained during the observation. It should be mentioned that the presence of many pentagons, corresponding to a higher curvature, of the fullerene cage is one of the crucial reasons why the defect is more likely induced at the fullerene site rather than at the nanotube.

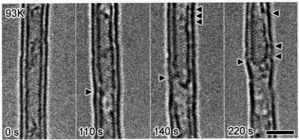

FIGURE 2. Sequential HR-TEM images of the knock-on defects on electron-irradiated DWNT recorded at 93K. The electron irradiation damage at the inner layer is more remarkable. Scale bar = 2nm.

The electron irradiation effect inducing the structural defects is more prominent at lower temperature. Figure 2 shows a series of the sequential HR-TEM images of DWNTs recorded at 93K. The black arrows indicate possible interlayer defects. In the initial state of HR-TEM observation ($t = 0$ s), two layers of DWNT with less defects are clearly visible. Afterward, a large number of defects are seen as electron irradiation time increases, and the inner layer in DWNT is heavily damaged ($t = 140$ s). At $t = 220$ s, no more tubular structure is kept for the inner tube while the outer layer still maintains the tubular structure. This heavier damage of the inner layer in DWNT is due to a possible higher knock-on rate of adjacent carbon atoms nearby the defect created in the previous knock-on event, which could hardly recover. As seen in the example of the HR-TEM observation of knock-on damage on DWNT at low temperature, we can reasonably assume that the vacancy is more likely to be generated in the inner nanotube.

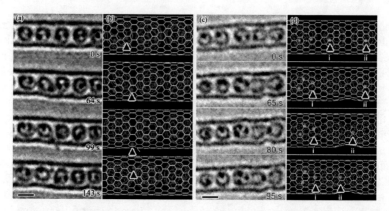

FIGURE 3. (a) Sequential HR-TEM images showing the break out of the imprisoned Tb atoms from $Tb_2@C_{92}$ peapod. One of the Tb atoms goes out through an atomic path induced between tube wall and fullerene cage (indicated by black arrows (b)). (c) A series of HR-TEM images showing the migrating Gd atom from cage to cage at ($Gd@C_{82}$ and $Gd_2@C_{92}$) peapod. (d) The migrating Gd atoms are indicated by black arrows in the reasonable atomic models. Scale bar = 1nm.

We next show two phenomena connected to the knock-on defects in fullerene proven by our *in-situ* HR-TEM observations of the carbon nanopeapod; (i) breakout of one of the imprisoned Tb atoms through such induced atomic pathway on the encaging fullerene molecule, and (ii) migration of the imprisoned Gd atoms from one cage to the next thorough another pathway interconnecting two adjacent fullerenes. Figure 3(a) shows a series of HR-TEM images of the breakout of the imprisoned Tb atoms from $Tb_2@C_{92}$ peapod. Each fullerene encapsulating two Tb atoms is clearly visible at $t = 0$ s. After electron irradiations ($t = 64$ and 99 s), HR-TEM images successfully captured the dynamic behavior of a Tb atom breaking out from the metallofullerene indicated by an arrow in their schematic presentation (Figure 3(b)). The Tb atom breakout clearly proves that the atomic pathway has been definitively induced on the fullerene cage as a consequent of the knock-on damage. The breakout of the Tb atom is attributed to three different interactions; (i) a strong Coulombic repulsive force between the two imprisoned Tb atoms in each cage, (ii) ionic attractive force between the Tb atoms and fullerene (($Tb^{+3})_2@C_{92}{}^{6-}$)), and (iii) a strong interaction exerted between encapsulating Tb atom and the defective area of the fullerene cage.

Additionally, to obtain clear proof of transferring the imprisoned atoms from cage to cage, an irradiated heteropeapod, which contains $Gd@C_{82}$ and $Gd_2@C_{92}$ encapsulated within a SWNT, was also observed by HR-TEM. Figure 3(c) shows that Gd atoms migrate inside fused fullerene cages by electron irradiations. The migrating Gd atoms are indicated by black arrows in the schematic presentation (Figure 3(d)). In the initial state of HR-TEM observation ($t = 0$ s), the numbers of Gd atoms in each cage are 1, 1, 2, 1, 1 (from left). After electron irradiation, three fullerene cages coalesce, and then Gd atoms go through atomic pathways ($t = 65$ s). Finally, a peanut-like fullerene, which engulfs three Gd atoms at one chamber, has been generated. It is clear evident that the atomic path and the atomic transfer from cage to cage are induced by electron irradiation.

Our results provide the first experimental evidence that the electron irradiation defects in nanocarbon materials are more likely to be generated on higher surface curvature. Then, the atomic path and the atomic transfer connected to the knock-on defects in fullerene suggest the possibility to create a new material from bottom-up process.

ACKNOWLEDGMENTS

We thank prof. Hisanori Shinohara in Nagoya University for providing us specimens investigated in this study. The work in electron microscopy was supported by the NEDO Nano-carbon Technology project. We are grateful to Professor Morita for the valuable discussion of defect formation on DWNTs.

REFERENCES

1. H. W. Kroto, J. R. Heath, S. C. O'Brien, R. F. Curl, and R. E. Smally, *Nature*, **318**, 162 (1985).
2. S. Iijima, *Nature*, **354**, 56 (1991).
3. A. Yu. Kasumov, R. Dblock, M. Kociak, B. Reulet, H. Bouchiat, I. I. Khodos, Yu. B. Gorbatov, V. T. Volkov, C. Journet, and M. Burghard, *Science*, **284**, 1508 (1999).
4. T. Ohba, H. Kanoh, and K. Kaneko, *J.Am.Chem.Soc.*, **126**, 1560 (2004).
5. R. F. Eggerton, M. Takeuchi, *Appl. Phys. Lett.*, **75**, 1884 (1999).
6. O. Gülseren, T. Yildrim, and S. Ciraci, *Phys. Rev.B*, **65**, 153405 (2002).
7. J. W. Ding, X. H. Yan, J. X. Cao, D. L. Wang, Y. Tang and Q. B. Yang, *J. Phys.:Condens. Matter*, **15**, L439 (2003).
8. A. Hashimoto, K. Suenaga, A. Gloter, K. Urita, and S. Iijima, *Nature*, **430**, 870 (2004).
9. K. Urita Y. Sato, K. Suenaga, A. Gloter, A. Hashimoto, M. Ishida, T. Shimada, H. Shinohara, and S. Iijima, *Nano Lettt.*, **4(12)**, 2451 (2004).
10. K. Urita, K. Suenaga, T. Sugai, H. Shinohara, and S. Iijima, *Phys. Rev. Lett.*, **94**, 155502 (2005).
11. F. Banhart, *Rep. Prog. Phys.*, **62**, 1181 (1999).
12. B. W. Smith, M. Monthioux, and D. E. Luzzi, *Nature*, **396**, 323 (1998).
13. T. Iwata and T. Nihira, *J. Phys. Soc. Jpn.*, **31**, 1761 (1971).
14. M. Takeuchi, S. Muto, T. Tanabe, S. Arai, and T. Kuroyanagi, *Phi. Mag. A*, **76**, 691 (1997).
15. T. Sugai, H. Yoshida, T. Shimada, T. Okazaki, S. Bandow, and H. Shinohara, *Nano Lett.*, **3**, 769 (2003).
16. H. Shinohara, *Rep. Prog. Phys.*, **63**, 843 (2000).

Polarized Raman spectroscopy study of SWCNT orientational order in an aligning liquid crystalline matrix

Giusy Scalia [(1)(2)], Miro Haluska[(3)], Ursula Dettlaff-Weglikowska[(3)], Frank Giesselmann[(1)] and Siegmar Roth[(3)]

[(1)]University of Stuttgart, Institute for Physical Chemistry, Pfaffenwaldring 55, Stuttgart, Germany; [(2)] ENEA, SS7 Appia km 713.7, Brindisi, Italy; [(3)] Max-Planck Institute for Solid State Research, Heisenbergstrasse 1, Stuttgart, Germany

Abstract. Liquid crystals are self-organizing anisotropic fluids. In this work the transfer of order from liquid crystals to single-wall carbon nanotubes embedded in them is exploited within the geometry of standard cells used for large-scale alignment of liquid crystals. Using Polarized Raman Spectroscopy the alignment of the SWCNTs along the common liquid crystal direction is unambiguously demonstrated through the analysis of the variation of the peak intensity of the nanotube Raman modes with respect to the polarization direction of the light.

Keywords: Single-wall carbon nanotubes, liquid crystal, alignment, polarized Raman spectroscopy
PACS: 78.30.Na; 61.30.-v; 61.46.+w

INTRODUCTION

Liquid crystals (LCs) are partially ordered fluids. Their family is formed by a variety of molecules and mixtures that exhibit an imperfect but long-range orientational order, and in some cases also translational order. The least ordered phase is the nematic where the molecules, commonly of rod-like shape, tend to align towards a common direction, *i. e.* are orientationally ordered. The common average direction can be easily identified by means of a polarizing optical microscope since the nematic is strongly birefringent. As self-organizing ordered fluids they are highly interesting to use as templates transferring their order onto objects dispersed in them. In this work the alignment imposed on single-wall carbon nanotubes (SWCNTs) is studied. Carbon nanotubes are anisotropic objects just like liquid crystalline molecules, but their aspect ratio is enormously larger! Nevertheless the transfer of order is possible because the carbon nanotubes have to deal not with single liquid crystal molecules but with a phase where every deformation from the common alignment direction is counteracted by elastic forces.

Indications of alignment of CNTs induced by LCs have been reported in a few earlier works [1] but so far with no direct prove of the alignment of CNTs by simple LC self-organization. In the present work we apply Polarized Raman spectroscopy to analyze the CNT-LC composites as this technique can provide direct information on

CP786, *Electronic Properties of Novel Nanostructures*, edited by H. Kuzmany, J. Fink, M. Mehring, and S. Roth
© 2005 American Institute of Physics 0-7354-0275-2/05/$22.50

the orientation of single SWCNTs [2] or in ropes [3]. The intensity of the nanotube Raman modes is strongly orientation-dependent, having a maximum for light polarized parallel to the SWCNT axes. In our investigations we monitor the polarized Raman signal from nanotubes dispersed at very small concentration in the LC host, filled into a standard LC-cell. The alignment direction of the liquid crystal molecules was fixed by the cell and with respect to this alignment direction the orientation of the nanotube axes was investigated. The analysis results in an unambiguous demonstration of the alignment of SWCNTs imposed by LCs. Polarized Raman spectroscopy thus appears to be a powerful tool not just for LCs [4] and CNTs, but also for their combination.

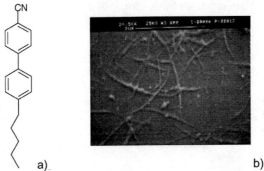

a)   b)

FIGURE 1. Materials used in this work: a) nematic liquid crystal 5CB b) SEM image of the shortened laser-ablation SWCNTs.

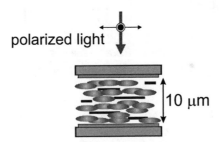

FIGURE 2. Standard LC-cell filled with LC-SWCNT. The drawing is a schematic where the shape of liquid crystal molecules is "cigar-like", and their length does not respect the real scale with CNTs. The direction of incidence/reflection (perpendicular to the glass plates) of the light for the Raman spectroscopy is also indicated, with the two chosen polarizations (parallel and perpendicular to the alignment of LC respectively).

EXPERIMENTAL

As liquid crystalline matrix the single-component nematic liquid crystal 5CB (Merck, Darmstadt) was chosen. This liquid crystal is well studied as it was the first commercial room temperature liquid crystal synthesized and the assignment of the peaks in its Raman spectrum with specific vibrational modes is well known. SWCNTs from different sources were used, either produced by laser ablation technique (afterwards shortened and then purified) or by high pressure CO conversion (HiPCO) used as purchased. For both types of nanotube a similar behavior in the 5CB matrix was observed. In Figure 1 the molecular structure of 5CB is shown as well an SEM image of the laser-ablation SWCNTs. Liquid crystal and carbon nanotubes were mixed through sonication using a very low percentage of nanotubes (less than 0.1 % w/w).

In order to obtain a macroscopic alignment of the LC matrix a standard LC-cell was used, constituted by two glass plates glued together at a distance fixed by spacers (in our case 10 micrometer thick) as shown in Figure 2. Each glass plate was previously spin-coated with a thin layer of polymer that was subsequently unidirectionally rubbed, to promote a uniform macroscopic alignment. The obtained LC-SWCNT mixture was placed at the entrance of an LC cell into which it filled by capillary action, forming a 10 micrometer thick film with the LC aligned uniformly along the rubbing direction of the substrate plates.

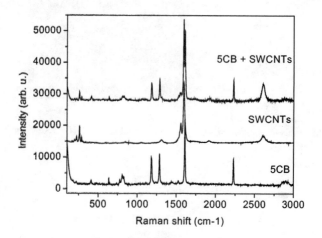

FIGURE 3. Raman spectrum of the pure liquid crystal 5CB (lower trace), SWCNTs (middle trace) and the mixture of LC-SWCNT (upper trace). The signal intensity of the middle and top Raman spectra has been shifted for clearer discriminating the different features.

The Raman study was performed with a micro-Raman setup using a He-Ne laser source (632.8 nm) in the backscattering geometry. Incident and scattered light were polarized and two polarizations were analyzed: one parallel to the direction of alignment of the liquid crystal and the other perpendicular to it. This geometry is schematically depicted in Figure 2.

RESULTS AND CONCLUSIONS

Raman spectroscopy was performed on pure 5CB, pure SWCNTs and their mixture as shown in Figure 3. The Raman spectrum of the mixture allows to discriminate clearly the modes coming from the two components either for a drop of the mixture observed on a substrate or for the mixture as a film, within the glass cell. The C-C stretching mode of aromatic rings in 5CB at 1606 cm^{-1} is very close to, yet not overlapping with, the G band peaks from SWCNTs, shifted towards lower wavenumbers. For both materials these were the most intense peaks, always more clearly distinguishable especially in the glass cell. Thus we focused our analysis on these peaks to detect the alignment of SWCNTs in the LC matrix.

In Figure 4 two extracts from the Raman spectra from the mixture LC-SWCNTs in the cell are shown. These correspond to the incident and scattered polarization directions being parallel to the LC alignment direction (pp) and orthogonal to it (oo) respectively. The decrease of intensity for the nanotube peaks as well as the LC peaks (as expected) in the orthogonal direction demonstrates that both components of the mixture are orientationally ordered along the same direction. Several points of the sample were investigated with similar results as well as other Raman modes. By fitting suitable functions to the peaks quantitative measurements of the orientational order of the SWCNTs, transferred from the LC matrix can be obtained. These results will be presented elsewhere.

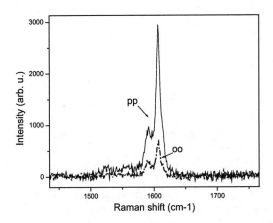

FIGURE 4. Extracts from the Raman spectra of SWCNT-5CB in a 10 micron thick cell for input and output light polarized along the direction of alignment of liquid crystal (pp) and orthogonal to it (oo) respectively).

ACKNOWLEDGMENTS

Dr. Piero Morales from ENEA, C.R. Casaccia, Rome, Italy, is kindly thanked for providing the shortened carbon nanotubes as well their SEM image. Giusy Scalia gratefully acknowledges financial supports from Vetenskapsrådet. We thank for financial support the EU-project SPANG, contract number 505483-1, in the Sixth Framework Programe, Priority 3-NMP, FP6-2002-NMP-1

REFERENCES

1. M. D. Lynch and D. L. Patrick, *Nano lett.*, 2, 1197 (2002); I. Dierking, G. Scalia, P. Morales and D. LeClere, *Adv. Mat.* **16**, 11, 865-869 (2004).
2. G. S. Duesberg, I. Loa, M. Burghard, K. Syassen ans S. Roth, *Phys. Rev. Lett.* **85**, 5436-5439 (2000); R. Saito, T. Takeya, T. Kimura, G. Dresselhaus and M. S. Dresselhaus, *Phys. Rev. B* **57**, 7, 4145-4153 (1988).
3   E. Anglaret, A. Righi, J. L. Sauvajol, P. Bernier, B. Vigolo and P. Poulin, *Phys. Rev. B* **65**, 165426 (2002
4. S. Jen, N. A. Clark, P. S. Pershan, E. B. Priestly, *J. Chem. Phys.* **66**, 10, 4635-4661 (1997).

Extraction and Local Probing of Individual Carbon Nanotubes

L. de Knoop*, K. Svensson*, H. Pettersson* and E. Olsson*

*Department of Applied Physics, Chalmers University of Technology, SE– 412 96 Göteborg, Sweden

Abstract. A method to extract individual carbon nanotubes with a good electrical contact for investigation by transmission electron microscopy (TEM) and scanning tunneling microscopy (STM) has been developed and evaluated. The extraction method includes the use of a combined focused ion beam workstation and a scanning electron microscope (SEM) with an *in situ* manipulator. The electrical conductance of individual tubes was subsequently probed in a TEM using a dedicated sample holder that joins TEM and STM (TEM-STM). Our study shows that we can extract individual CNTs and make good electrical contacts. However, great care has to be taken in order to minimize contamination from the background pressure inside the SEM.

Keywords: carbon nanotube, transmission electron microscopy, scanning tunneling microscopy
PACS: 68.37.Lp; 73.63.Fg; 81.07.De; 87.64.Dz; 87.64.Ee

INTRODUCTION

Carbon nanotubes (CNTs) display unique mechanical and electrical properties with potential applications e.g. as field emitters and in micro- and nano-electromechanical devices (MEMS and NEMS). The properties are determined by the structure of the nanotubes where atomic-scale structure, size and chemistry are important factors. It is therefore of significant interest to directly correlate the properties to the microstructure of individual carbon nanotubes using *in situ* experiments. The combination of a scanning tunneling microscope (STM) and a transmission electron microscope (TEM) allows simultaneous imaging and characterization of the transport properties [1]. There are other examples where individual CNTs have been studied using *in situ* experiments in the TEM [2, 3, 4]. However, a common denominator for all experiments is that they have been carried out on ensambles of CNTs, and which individual nanotube that is within reach is random. Further information about the properties of CNTs can be gained by extracting individual tubes. They should be suspended and have good and well defined electrical contacts at fixed ends.

Here we present a method to extract CNTs from bundles and to attach individual tubes, rigidly and with a good electrical contact onto a support. The nanotubes were manipulated and extracted using an *in situ* manipulator in a combined scanning electron microscope and focused ion beam workstation (SEM-FIB). The contacts were made by deposition of platinum on the nanotube and support.

CP786, *Electronic Properties of Novel Nanostructures*, edited by H. Kuzmany, J. Fink, M. Mehring, and S. Roth
© 2005 American Institute of Physics 0-7354-0275-2/05/$22.50

EXPERIMENTAL

The CNTs were made by pyrolysing ferrocene powder at 800 °C in a mixed air and Ar atmosphere at a pressure <100 Torr[1]. Soot flakes containing bundles of iron filled CNTs were obtained. The extraction of the tubes and the deposition of platinum was carried out using a FEI Strata 235 DualBeam SEM-FIB instrument. The CNTs were put on a conducting carbon tape on a stub holder for SEM investigations. The holder was modified with four additional holes, allowing the mounting of four gold wires serving as supports for subsequent TEM-STM investigations. The platinum contacts were made using the FEI gas injection system with the $(CH_3)_3Pt(C_pCH_3)$ gas. An *in situ* Omniprobe was used for the manipulation of the CNTs in the SEM-FIB. The TEM-STM investigations were performed using a Philips CM200 supertwin TEM, with a field emission gun, operated at 120 kV in order to reduce electron beam induced damages of the CNTs during observation [5, 6]. The TEM-STM holder is based on a recently developed instrument [7] where the STM is controlled by software and electronics from Nanofactory Instruments AB[2].

RESULTS AND DISCUSSION

Extraction of an individual CNT

We have tried several different ways to extract individual CNTs from bundles of nanotubes on the carbon tape. Entangled tubes were in most cases too strongly bound to the bundles and could not easily be withdrawn exploiting only van der Waals forces between the movable Omniprobe and the CNTs. Even securing, i.e. welding, a nanotube, using platinum deposition, to the probe was not sufficient in order to simply pull out the tube from a bundle. Using the ion beam one could possibly cut off the CNT but this would damage the tube and would also lead to an undesired implantation of gallium from the ion source into the nanotube which could obscure the native electron transport properties of the nanotubes. Instead we used brute force in order to separate the entangled nanotubes in a bundle, and a few tubes, or a small bundle, could then be transferred to the Omniprobe using only van der Waals forces. From this smaller bundle an individual nanotube was selected and welded onto the gold wire that would form the sample for TEM-STM investigations. The nanotube could then be extracted by moving the Omniprobe away from the sample. In Fig. 1 a bundle of CNTs is attached to the manipulator, to the left, and a CNT has been welded onto the gold wire to the right. This is in the middle of the process of separating the manipulator and the gold wire support, and the CNT is just about to detach from the bundle on the manipulator. In this case we also welded the bundle to the manipulator in order to avoid a detachment of the whole bundle.

---

[1] The nanotubes were made by H. Pettersson, M. Terrones and N. Grobert at the Sussex Nanoscience and Nanotechnology Centre at the University of Sussex, UK.

[2] See www.nanofactory.com.

FIGURE 1. An SEM image showing the manipulator to the left and the gold wire to the right. A small bundle of CNTs has been attached to the manipulator, and a single CNT has been welded to the gold wire.

Welding the CNT to the gold wire

In order to ensure a well-defined and good electrical contact between the CNT and support, we deposited a small platinum electrode onto the nanotube end and the gold wire. This firm connection to the support would also aid the extraction of the tube from the small bundle on the Omniprobe. In order to ensure a good electrical contact with the gold wire, the contact area was first cleaned by sputtering away a 5 nm thin layer using the ion beam. The nanotube was then placed with one end on the cleaned area and a platinum electrode was deposited on top, using the gas injection needle and the electron beam to decompose the platinum precursor. The electron beam was preferred over the ion beam here, again in order to avoid structural damages and implantation of gallium into the nanotube structure. Figure 2(a) shows an SEM image of a nanotube that has been secured to the support by depositing a 400×400 nm^2 large electrode to the tip (the height of the electrode was 1 μm). The Omniprobe was retracted, leaving a single nanotube attached to the gold wire support. This procedure was repeated for other nanotubes, giving a gold wire with three nanotubes secured at suitable locations for subsequent TEM-STM characterizations.

Platinum deposits were also found in regions extending out from the intended electrode (as can be seen on the nanotube in Fig. 2(a) near the electrode). This platinum deposition comes from an unwanted effect originating in the Gaussian form of the electron beam. If, for example, a narrow line is being made, the Gaussian form of the beam is transferred to the pattern and the line will have a Gaussian cross section. So when our rectangular electrodes were made, the result was electrodes with rounded edges and

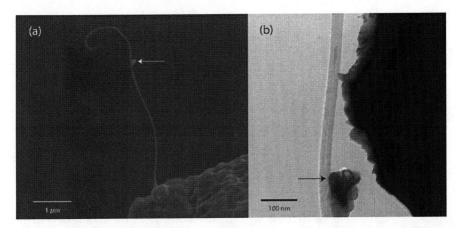

FIGURE 2. (a) SEM image of a CNT welded onto the tip of a gold wire. The roughness of the gold surface comes from the ion beam sputtering which enhances the grain structure of the material. (b) TEM image of the gold tip in contact with the CNT. The arrows in (a) and (b) points at the same carbon debris, used as a reference mark.

long "tails". By reducing the amount of deposited material, i.e. the thickness of the electrodes, the unwanted layer thickness was reduced proportionally, although it could never be completely avoided.

The background pressure in the SEM stayed at all times below 2×10^{-5} torr and most of the time it was around 5×10^{-6} torr. Even so, there was a risk to form a contamination layer on the CNTs. This is caused by hydrocarbons in the background gas being decomposed by incident electrons, and thus forming a contamination layer wherever the electron beam has scanned. Wilms *et al.* have reported about this as well [8]. Some have even used this effect in order to make electrodes out of the hydrocarbons [9]. The rather poor electrical conductivity of such layers have been investigated by Kim *et al.* in [9]. Quite contrary to this Avouris *et al.* have seen an increase in conductance by irradiating the contact area [10]. According to Banhart [11] hydrocarbons can transform to amorphous and graphitic carbon under an electron beam with a resulting adequate conductivity. For our study such contacts were not considered to be adequate and instead we took great care in order to minimize the amount of electron illuminations during the experiments.

TEM-STM characterizations

The gold wire with secured individual tubes was transferred to the TEM-STM holder which enables local probing of electrical properties while imaging using the TEM. A sharp gold tip was used as an electrical probe and contacts at various positions on the nanotubes were made. Figure 2(b) shows the gold tip in contact with the CNT in Fig. 2(a). For different contacts along the tube we measured the current-voltage characteristics (IV-curve) and obtained the overall conductance, including both contacts

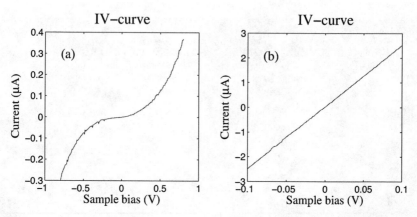

FIGURE 3. Typical IV-curves (a) before, and (b) after a heating current has removed the hydrocarbon contamination layer.

to the nanotube. Since the nanotubes had been exposed to the atmosphere there seemed to be a very thin layer of hydrocarbons that was not visible using the TEM, but would give rise to a rather poor electrical contact (see Fig. 3(a)). This was circumvented by first driving a high electrical current through the tube in order to desorb the contamination layer. Subsequent IV-curves then displayed a higher conductance and by repeating the heating at increasingly higher currents we could conclude that a current of about 50 μA was needed in order to sufficiently get rid of the hydrocarbon layer (see Fig. 3(b)) [5]. It should be noted that the resistivity obtained in Fig. 3(b) is comparable to previously reported resistivities for these iron filled CNTs [1], which also are in accordance with four point measurements on supported CNTs [12].

CONCLUSIONS

A method where individual CNTs can be extracted and welded onto the tip of a gold wire, for later TEM-STM analysis has been developed. The CNTs were extracted using an *in situ* manipulator in a combined SEM-FIB workstation. Electrical contacts were deposited on-top of the nanotube and support. This method can be used on any other CNT or similar structure that is of interest to investigate on a nanometer scale.

ACKNOWLEDGMENTS

Financial support from the Swedish Natural Science Research Council (NFR) and the Swedish Research Council (VR) are gratefully acknowledged.

REFERENCES

1. K. Svensson, H. Olin, and E. Olsson, *Physical Review Letters*, **93**, 145901 (2004).
2. P. Poncharal, Z. L. Wang, D. Ugarte, and W. A. de Heer, *Science*, **283**, 1513 (1999).
3. Z. L. Wang, R. P. Gao, W. A. de Heer, and P. Poncharal, *Applied Physics Letters*, **80**, 856 (2002).
4. P. Poncharal, C. Berger, Y. Yi, Z. L. Wang, and W. A. de Heer, *Journal of Physical Chemistry B*, **106**, 12104 (2002).
5. L. de Knoop, *Master Thesis: Investigation of Iron Filled Multiwalled Carbon Nanotubes*, Chalmers University of Technology, Göteborg, 2005.
6. R. F. Egerton, *Philosophical Magazine*, **35**, 1425 (1977).
7. K. Svensson, Y. Jompol, H. Olin, and E. Olsson, *Review of Scientific Instruments*, **74**, 4945 (2003).
8. M. Wilms, J. Conrad, K. Vasilev, M. Kreiter, and G. Wegner, *Applied Surface Science*, **238**, 490 (2004).
9. K. S. Kim, S. C. Lim, I. B. Lee, K. H. An, D. J. Bae, S. Choi, J.-E. Yoo, and Y. H. Lee, *Review of Scientific Instruments*, **74**, 4021 (2003).
10. P. Avouris, T. Hertel, R. Martel, T. Schmidt, H. R. Shea, and R. E. Walkup, *Applied Surface Science*, **141**, 201 (1999).
11. F. Banhart, *Nano Letters*, **1**, 329 (2001).
12. C. Schönenberger, A. Bachtold, C. Strunk, J.-P. Salvetat, and L. Forro, *Applied Physics A: Materials Science and Processing*, **69**, 283 (1999).

Interband Transition Energy Shifts in Photoluminescence of Single-Walled Carbon Nanotubes under Hydrostatic Pressure

Sergei Lebedkin [a], Katharina Arnold [a,b], Oliver Kiowski [a,b], Frank Hennrich [a] and Manfred M. Kappes [a,b]

[a] Forschungszentrum Karlsruhe, Institut für Nanotechnologie, D-76021 Karlsruhe, Germany
[b] Institut für Physikalische Chemie, Universität Karlsruhe, D-76128 Karlsruhe, Germany

Abstract. Photoluminescence (PL) spectroscopy was applied to determine shifts of electronic interband transition energies, E_{11} and E_{22}, of semiconducting single-walled carbon nanotubes (SWNTs) under hydrostatic pressure of several kbar in a diamond anvil cell. Our results show that the energy shifts depend not only on the (n,m) structure (helicity) of nanotubes, but also on nanotube aggregation and interaction with the surrounding medium. The latter factors are also affected by application of pressure. This can explain a discrepancy between the experimental data and theoretical predictions for 'ideal' (non-interacting) SWNTs under hydrostatic pressure deformation.

Keywords: carbon nanotubes, photoluminescence, hydrostatic pressure.
PACS:

INTRODUCTION

Excellent mechanical strength and flexibility are among the most attractive properties of SWNTs for potential applications [1]. Importantly, according to theoretical studies, mechanical stress also has a significant effect on the electronic properties of nanotubes. It can result in shifts, splittings and coalescence of van Hove singularities in the electronic density of states of SWNTs and can cause metal-semiconductor transitions, depending on the kind of deformation and the specific (n,m) structure (helicity) of nanotubes [2]. One particular feature of this complex behavior is that at relatively small uniaxial strains SWNTs are predicted to show characteristic shifts of interband transition energies E_{ii} (i=1, 2..), with the signs and magnitudes of these shifts determined by the $(n-m)$ $mod3$ parameters and helicity angles, respectively [2]. Good agreement with these predictions was recently demonstrated by us and other researchers: photoluminescence (PL) measurements of semiconducting SWNTs axially compressed up to ~50 kbar in an ice matrix at low temperatures allowed determination of the corresponding E_{11} and E_{22} shifts [3, 4]. The band gap (E_{11}) vs. strain dependence has also been verified by electron transport measurements on SWNTs stretched with an AFM tip [5].

CP786, *Electronic Properties of Novel Nanostructures*, edited by H. Kuzmany, J. Fink, M. Mehring, and S. Roth
© 2005 American Institute of Physics 0-7354-0275-2/05/$22.50

Hydrostatic pressure deformation is closely related to uniaxial deformation, therefore a similar $(n-m)$ mod 3 family behavior is expected for semiconducting tubes $((n-m)mod3 = 1, 2)$ [6]. However, the recent absorption spectroscopy study of water-surfactant dispersions of SWNTs under hydrostatic pressure in a diamond anvil cell (DAC) revealed no sign alternation of the energy shifts according to the $(n-m)$ mod 3 rule, but rather uniform and relatively large down-shifts of E_{11} and smaller down-shifts of E_{22} energies [7].

In contrast to the absorption spectra, where bands corresponding to different (n,m) nanotubes are overlapped with each other and with the plasmon absorption background, PL provides pairs of E_{11} (emission) and E_{22} (excitation) energies which can be reliably assigned to specific (n,m) structures [8]. Furthermore, PL differentiates individual nanotubes (and likely small nanotube aggregates) from large nanotube bundles since only the former luminesce. Here we present the first PL experiments on water-surfactant dispersions of SWNTs under hydrostatic pressure of several kbar in a DAC.

EXPERIMENTAL

HiPco nanotubes from Rice University and PLV nanotubes produced in our lab by pulsed laser vaporization of C:Co:Ni targets were used in this work. Preparation of D_2O/ surfactant dispersions of isolated (debundled) SWNTs by powerful sonication and ultracentrifugation has been described elsewhere [9]. For comparison, several anionic surfactants were tested including sodium dodecyl sulfate (SDS), sodium dodecylbenzene sulfonate (SDBS) and sodium cholate.

Pressure experiments were performed using a conventional lever-arm driven DAC (Diacell) and 0.25 mm thin stainless steel gaskets with a 0.5 mm central hole. Because of the rather low pressures applied, the standard tedious procedure of gasket preindentation was not necessary. Instead, ready-to-use, laser-machined gaskets were positioned on a diamond anvil with the aid of a miniature X-Y-Z stage, filled with a sample dispersion and sealed with another anvil. This simple procedure required only a few minutes to assemble the DAC for a measurement. Ruby microcrystals were added to the dispersion for the standard pressure calibration using the ruby emission.

Photoluminescence in the DAC was measured with a home-built near-infrared laser luminescence microscope. In this apparatus, a CW Ti-Sapphire laser (Spectra Physics) could automatically step-scan (typically 2 nm step size) an excitation range of 708-850 nm at a controlled output power (~1-5 mW on a sample). At each excitation wavelength a PL spectrum was acquired using a liquid nitrogen cooled InGaAs photodiode array (~800-1600 nm) attached to a near-infrared spectrograph (Roper Scientific). The PL spectra obtained were combined into a PL map like that shown in Fig. 2. The ruby emission was excited with a HeNe laser at 633 nm and analysed using a separate spectroscopic setup coupled to the microscope via an optical fiber.

RESULTS AND DISCUSSION

We found the effects of hydrostatic pressure on photoluminescence (energy shifts) of SWNT dispersions to be much more complex than for SWNTs thermomechanically compressed in an ice matrix [3]. For example, Fig. 1 shows large and nearly uniform PL shifts observed for a dispersion of nanotubes in D_2O/ SDS. The large emission (E_{11}) shifts and apparent absence of *(n-m) mod 3* family behavior are consistent with the results of Wu et al. for a similar SDS dispersion of HiPco tubes [7]. However, we also found a pronounced time dependence of the PL under pressure (on time scales of hours to days), irreversible broadening and red shifts of the PL peaks after pressure release, and poor reproducibility of these effects for dispersions from different preparations and with different storage times. Fig. 2 shows an example of relatively fast and dramatic pressure-induced changes in the PL of a one year old SWNT dispersion in D_2O/SDBS. We tentatively attribute such effects to aggregation of nanotubes. Hydrostatic pressure may strongly accelerate this process because of the higher effective density of nanotube aggregates as compared to individual tubes [8]. Pressure can also destroy 'isolating' micelle structures around nanotubes and in this way drive them to aggregate. In contrast to non-emissive large SWNT bundles, still-luminescent aggregates presumably consist of a few nanotubes with 'loose' binding contacts. Their emission is expected to be weaker, broader and on average red-shifted as compared to individual tubes [3]. It is obvious that the proposed pressure-driven aggregation will depend on the surfactant, dispersion preparation conditions, and, finally, on the specific structures of nanotube-surfactant assemblies which are difficult to control.

Nevertheless, reproducible pressure effects having only a weak dependence on the sample history and duration of pressure treatment (for pressures up to several kbar) were obtained for relatively fresh dispersions with SDBS and, in particular, for sodium cholate dispersions. During our experiments the latter dispersions were stored for up to three-four months. Similar PL shift patterns obtained for HiPco and PLV tubes clearly showed *(n-m) mod3* family behavior as illustrated in Fig. 3.

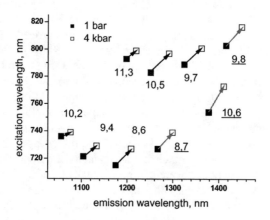

FIGURE 1. Shifts of the photoluminescence maxima of *(n,m)* HiPco nanotubes in a D_2O/ SDS dispersion under hydrostatic pressure of 4 kbar. Underlined are *(n,m)* indices of nanotubes with *(n-m) mod 3 = 1*, other indices indicate nanotubes with *(n-m) mod 3 = 2*.

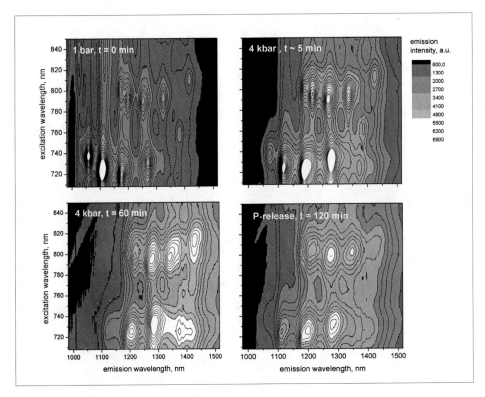

FIGURE 2. Fast and irreversible changes in the photoluminescence of an 'old' (stored for more than one year) dispersion of HiPco nanotubes in D_2O/ SDBS after application and release of 4 kbar pressure. The average PL intensity decreased with the time (the PL emission intensity levels are not on the same scale for the four maps).

The observed PL shifts can be qualitatively separated into two components (Fig. 3). One is a uniform, approximately (n,m)-independent down-shift of E_{11} and E_{22} (red shift of the emission and excitation maxima). The direction of the second component is determined by the $(n-m)$ *mod* 3 rule. The tubes of the $(n-m)$ *mod* 3 = 1 family show an increase of E_{11} and a decrease of E_{22} energies, whereas the tubes with $(n-m)$ *mod* 3 = 2 show the opposite trends. The magnitude of this shift component apparently increases with increasing $(n-m)$, *i.e.* with decreasing helicity angle. All these features are consistent with theoretical predictions and can be attributed to intrinsic electronic energy shifts upon hydrostatic pressure deformation of SWNTs.

The first uniform shift component can be assigned to the pressure modulated interaction between nanotubes and the surrounding medium (D_2O plus surfactant). It is known that E_{11} and E_{22} energies of dispersed SWNTs depend slightly on the surfactant used and shift, for example, upon addtion of polyvinyl pyrrolidone which can associate with nanotubes [3]. Such interactions are expected to be influenced by pressure. To study the pressure effects in more detail, we plan to compare PL shifts of SWNTs in a dispersion with those of SWNTs deposited onto a substrate and pressurized with various fluid media.

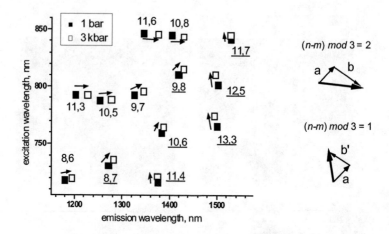

FIGURE 3. Shifts of the PL maxima of (n,m) PLV nanotubes in a $D_2O/$ SDBS dispersion under hydrostatic pressure of 3 kbar. Underlined are (n,m) indices of nanotubes with $(n-m)$ *mod* $3 = 1$, other indices indicate nanotubes with $(n-m)$ *mod* $3 = 2$. The right-hand pictures show schematically two contributions to the observed PL shifts (thick arrows): pressure effects on the SWNT interaction with the surroundings *a* and strain-induced E_{11} and E_{22} energy shifts *b, b'* (for details see text).

ACKNOWLEDGMENTS

The support of this work by the Deutsche Forschungsgemeinschaft and by the BMBF is gratefully acknowledged. The authors thank Prof. Smalley for a sample of HiPco nanotubes.

REFERENCES

1. Saito, R., Dresselhaus, G., Dresselhaus, M. S. *Physical Properties of Carbon Nanotubes*; Imperial College Press: London, 1998.
2. see, e.g., Yang, L., and Han, J., *Phys. Rev. Lett.* **85**, 154 (2000).
3. Arnold, K., Lebedkin, S., Kiowski, O, Hennrich, F., and Kappes, M. M., *Nano Lett.* **4**, 2349 (2004).
4. Li, L. J., Nicholas, R. J., Deacon, R. S., et al., *Phys. Rev. Lett.* **93**, 156401 (2004).
5. Minot, E. D., Yaish, Y., Sazonova, V., Park, J.-Y., Brink, M., and McEuen, P. L., *Phys. Rev. Lett.* **90**, 156401 (2003).
6. Capaz, R.B., Spataru, C. D., Tangney, P., Cohen, M. L., and Louie, S.G., *Phys. Stat. Sol.(b)* **241**, 3352 (2004).
7. Wu, J.; Walukiewicz, W.; Shan, W., et al, *Phys. Rev. Lett.* **93**, 017404 (2004).
8. O'Connell, M. J., Bachilo, S. M., Huffman, C. B., et al., *Science* **297**, 593 (2002); Bachilo, S. M., Strano, M. S., Kittrell, C., et al., *Science* **298**, 2361 (2002).
9. Arnold, K., Lebedkin, S., Hennrich, F, Krupke, R., Renker, B., and Kappes, M. M., *New J. Phys.* **5**, 140 (2003).

Purity Evaluation of Bulk Single Wall Carbon Nanotube Materials

U. Dettlaff-Weglikowska[a], J. Wang[a,b], J. Liang[a,b], B. Hornbostel[a], J. Cech[a] and S. Roth[a]

[a]Max Planck Institute for Solid State Research, 70569 Stuttgart, Germany,

[b]Yangtze Nanomaterials Co. Ltd, Shanghai, China

Corresponding author; E-mail: u.dettlaff@fkf.mpg.de

Abstract. We report on our experience using a preliminary protocol for quality control of bulk single wall carbon nanotube (SWNT) materials produced by the electric arc-discharge and laser ablation method. The first step in the characterization of the bulk material is mechanical homogenization. Quantitative evaluation of purity has been performed using a previously reported procedure based on solution phase near-infrared spectroscopy [1]. Our results confirm that this method is reliable in determining the nanotube content in the arc-discharge sample containing carbonaceous impurities (amorphous carbon and graphitic particles). However, the application of this method to laser ablation samples gives a relative purity value over 100 %. The possible reason for that might be different extinction coefficient meaning different oscillator strength of the laser ablation tubes. At the present time, a 100 % pure reference sample of laser ablation SWNT is not available, so we chose to adopt the sample showing the highest purity as a new reference sample for a quantitative purity evaluation of laser ablation materials. The graphitic part of the carbonaceous impurities has been estimated using X-ray diffraction of 1:1 mixture of nanotube material and C_{60} as an internal reference. To evaluate the metallic impurities in the as prepared and homogenized carbon nanotube soot inductive coupled plasma (ICP) has been used.

INTRODUCTION

On 10th of December 2004 news@nature.com published online a report criticizing the quality of carbon nanotubes products supplied by various companies. "Batches of carbon nanotubes are often contaminated", says Matthew Nordan of Lux Research in New York, leader of a consulting firm. According to his survey a semiconducting company found that one third of a purchased sample of carbon nanotubes was highly contaminated by iron residues from the synthesis process. Due to permanently upgrading of equipment most nanotube suppliers are not able to provide comparable SWNT batches. The current situation in quality assurance of carbon nanotube materials is unlikely to change until a standardized purity evaluation methods are available.

In this paper we propose a preliminary protocol for quality control of bulk SWNT materials which is one of the deliverable for the EU project SPANG. The purpose of this protocol is to provide standard analytical instructions for purity evaluation of SWNT samples. It is hoped that by applying this protocol reproducible results will be obtained at different laboratories. The proposed procedure is applicable for a 10-50 g batch. Three analytical methods are utilized: solution phase near-infrared spectroscopy for the quantitative evaluation of SWNTs purity, X-ray diffraction for the quantitative evaluation of graphitic impurities, and inductive coupled plasma analysis for the

CP786, Electronic Properties of Novel Nanostructures, edited by H. Kuzmany, J. Fink, M. Mehring, and S. Roth
© 2005 American Institute of Physics 0-7354-0275-2/05/$22.50

quantitative evaluation of metallic impurities. We tested the protocol using as prepared arc-discharge materials produced at Yangtze Nanomaterials (Shanghai), Nanoledge (Montpellier) and laser ablation nanotubes purchased from University of Karlsruhe and from National Research Laboratory in Ottawa.

EXPERIMENTAL, RESULTS AND DISCUSSION

Our protocol begins with a homogenization process, which is required to obtain spectral data that represent the average purity of large batches of nanotubes. This step is of outmost importance, since the typical as prepared SWNT soot contains a significant amount of impurities, and is usually very inhomogeneous. Characterization experiments have to be carried out on the same batch, and a part of the batch has to be stored away for future verifications and for blind experiments.

Batches of 10-50 g as prepared SWNT soot are mechanically homogenized in a simple kitchen mixer until a fine powder is obtained (typically 1-5 min). This step is followed by shaking in a tumbler for 24 hours (container filed to ½ at maximum, usually this corresponds to 20-30 g).

Quantitative evaluation of the relative nanotube purity

Near-infrared (NIR) spectroscopy has been established as an important technique for the characterization of the electronic structure of SWNTs. Itkis et al. [1] have shown that solution phase NIR can be used for the quantitative evaluation of the sample purity. To this end a 50 mg of initially homogenized soot has been dispersed in 100 ml of dimethylformamide (DMF) by ultasonication and mechanical stirring for 5 min. Then, a few (5-8) drops of the homogeneous slurry were dissolved in 10 ml of DMF to obtain a faintly colored non-scattering liquid after 2 min of ultrasonication. Absorption spectra were recorded and the spectral range of the S_{22} interband transition of 7750-11750 cm^{-1} was chosen for the purity evaluation of arc-discharge samples. The integration of the spectra gives the total area under the absorption curve and the area under the curve after the baseline subtraction. The ratio of the areas is compared to the spectra of a reference sample (set to 100 % purity) [1]. The purity P of the unknown sample against the reference is calculated from the expression

$$P(sample) = \frac{A(peak, sample)/A(total, sample)}{A(peak, reference)/A(total, reference)} \times 100\%$$

where the ratio of areas A(peak,reference)/A(total,reference) is equal to 0.141 as derived from the standard used in Ref. [1].

Fig.1 illustrates the NIR spectroscopic analysis of two as prepared arc-discharge samples provided by different suppliers: left, obtained from Yangtze Nanomaterials, Shanghai, right, obtained from Nanoledge, Montpellier. The baseline separates the peak area from the background. The ratio of the areas under the curves is equal to 0.0486 and 0.0336, respectively. Thus, the purity of the Yangtze sample is estimated to be 34.5 % (0.0486/0.141) against the reference, and the purity of the Nanoledge sample is 23.9 % (0.0337/0.141) against the same reference. The light hatched part of the background corresponds to the non-peak contribution of the nanotubes. What is left over (dark hatched area is attributed to non-tubular carbonaceous by-products (amorphous carbon and graphitic particles).

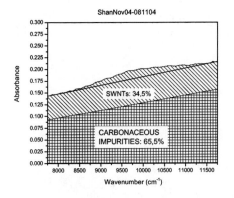 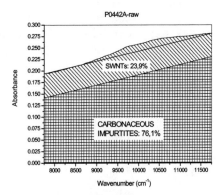

FIGURE 1. left: Purity of an as prepared arc-discharge sample from Yangtze Nanomaterials, Shanghai; right: as prepared sample from Nanoledge, Montpellier.

To test the reliability of this purity evaluation procedure, we carried out independent purity evaluations on five samples of the same homogenized 20 g batch of as prepared Yangtze material. The results presented in Table 1 demonstrate the reproducibility of this method in evaluating of bulk nanotube material within 3 % of standard deviation (SD).

Table 1. Statistical analysis of the purity of an as prepared homogenized arc-discharge nanotube material supplied by Yangtze Nanomaterials, Shanghai.

Sample	1	2	3	4	5	mean value (SD)
Relative purity, % [1]	36.7	34.2	34.5	39.4	37.8	36.5 (± 3)

Application of the method to one of our laser ablation samples gave a purity of 150 %. A possible reason for this unrealistically high value might be a different extinction coefficient (different oscillator strength) of the laser ablation tubes. Due to the different spectral features originating from a different distribution of SWNT diameters in laser ablation material, we changed the spectral range for integration of the spectra and the baseline correction from 7750-11750 cm^{-1} for arc-discharge to 8000-12750 cm^{-1} for laser ablation material. There is no 100 % pure laser ablation sample available up to now, so we used a sample showing the highest purity (Fig. 2 top right with the ratio of areas equal to 0.211) as a reference. This is a nanotube sample provided by M.

Kappes at the University of Karlsruhe and purified by controlled thermal oxidation and acid treatment at Max Planck Institute, Stuttgart. Fig. 2 (bottom) shows the determination of the areas under the absorption curves used for purity evaluation of two as prepared laser ablation samples (left: purchased from the same lab at the University of Karlsruhe, right: from the National Research Laboratory, Ottawa). Thus, the purity of the nanotube sample from Karlsruhe is calculated to be 28.9 % (0.061/0.211) and the purity of the nanotube sample from Ottawa is 22.7 % (0.048/0.211) with respect to the reference with the ratio of areas $A(S_{22})/A(T) = 0.211$.

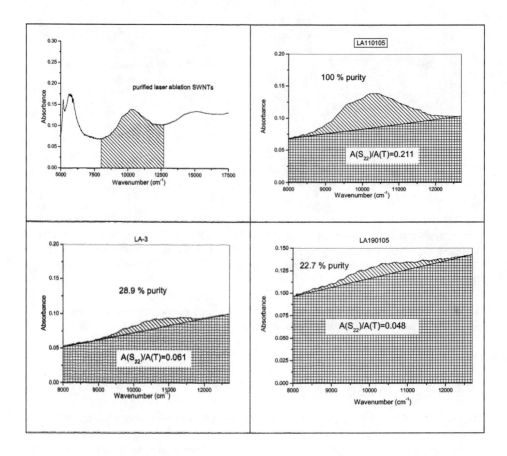

FIGURE 2. top left: Determination of the spectral range for integration of the spectra and the baseline correction for laser ablation samples; top right: Estimation of the ratio of the areas $A(S_{22})/A(T)=0.211$ for the laser ablation sample used as a reference; bottom left: The purity of laser ablation sample from Karlsruhe 0.061/0.211 is equal to 28.9 %; bottom right: The purity of laser ablation sample from Ottawa 0.048/0.211 is equal to 22.7 %.

Quantitative evaluation of the graphitic impurities

In order to evaluate the graphitic impurities of bulk SWNT materials we applied X-ray powder analysis using C_{60} as an internal reference. Fig. 3 shows the diffraction pattern of an "artificial" sample of C_{60} and graphite powder in a weight ratio of 1:1 recorded on a flat bed sample to avoid de-mixing. The areas of the graphite peak located at 26° and the second fullerene peak located at 16° are about equal. To estimate the graphitic impurities in a real material the nanotube sample is mixed in a ratio of 1:1 with C_{60}. The area of the graphite peak at 26° is compared with that of the second fullerene peak at 16 ° (which is used as a marker). In Fig. 3 (right hand panel) the pattern of a sample with unknown graphite concentration is shown. Now the peak area between graphite and fullerene marker compare like 1:5, indicating that the sample contains about 20 % of graphite contamination. In addition to the crystalline graphite (peak at 26°) there is probably also an important contamination with turbostratic graphite leading to the broad hump between 15° and 35°.

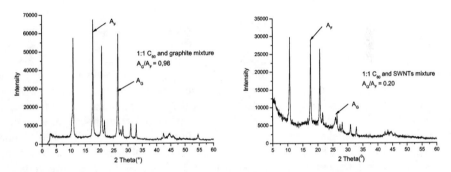

FIGURE 3. left: X-ray pattern of a 1:1 graphite and C_{60} mixture. The ratio of the peak areas of graphite and fullerene A_G/A_F is equal to 0.98; right: the X-ray pattern of a 1:1 mixture of C_{60} and nanotube sample. The A_G/A_F ratio reveals about 20 % of graphitic impurities.

Quantitative evaluation of the metallic impurities

Inductive coupled plasma analysis has been successfully used to analyze the amount of metallic impurities left over in the as prepared SWNT material from transition-metal catalysts. For sample preparation about 5 mg of SWNT soot have been put into a teflon tube with 3 ml conc. HCl and 1 ml conc. HNO_3. The teflon tube has been placed in a closed stainless steel container and heated for 18 h at 200° C. After cooling down to room temperature, the carbonaceous solid has been filtered off and the remaining solution has been diluted up to 50 ml. The intensity of the emission lines attributed to Ni, Y and Co have been measured in the analyzed solution and compared with that of corresponding standard solutions with the known concentrations of the metals. We carried out the ICP analysis on several arc-discharge and laser ablation nanotube samples. The results are presented in Table 2. Depending on the production method, batch and supplier, the average amount of the metallic impurities varied from about 2 wt % in laser ablation nanotubes to about 24 wt % in as prepared arc-discharge tubes.

Table 2. Metal content (wt %) in as-delivered arc-discharge and laser ablation nanotube samples determined by ICP.

Arc-disch	Ni	Y
NL-1	20.7	2.2
Clem-1	19.0	4.7
Clem-2	15.8	3.3
NL-2	12.3	5.2
NL-3	4.0	2.4
Yang-1	10.4	1.4

Laser-abl	Ni	Co
K-4	10.4	9.9
CNI-1	3.5	3.5
CNI-2	1.2	3.0
K-5	1.6	1.6
K-10	0.8	0.9

CONCLUSIONS

This report provides analytical instructions for purity evaluation of bulk SWNT materials. Combined analytical methods are used for quantitative evaluation of purity: solution phase near-infrared spectroscopy for the concentration of nanotubes, X-ray diffraction for probing graphitic particles and ICP for determination of metallic impurities. It is pointed out that so far the nanotube concentrations are only relative, with respect to a standard sample, and that a reliable reference sample (100 % pure nanotubes, or at least with a definitely known nanotube concentration) is still missing. We stress the importance of working with large homogenized batches (presently, we use 100 g batches whenever possible). SEM and Raman spectroscopy are assigned the role of presenting first evidences of the presence of nanotubes in a sample, but they are classified as not appropriate for quantitative analysis [2].

ACKNOLEDMENTS

This work is supported by the European Commission's Sixth Framework Programme (SPANG Project Contract No. NMP4-CT-2003-505483). Regular updates of a (preliminary) quality control protocol can be obtained from the authors. We thank M. Kappes (University of Karlsruhe) for providing laser ablation SWNT material for comparative study.

REFERENCES

1. M. E. Itkis, D. E. Perea, S. Niyogi, S. M. Rickard, M. A. Hamon, H. Hu, B. Zhao, and R. C. Haddon, *Nano Lett.* **2003**, *3*, 309.
2. M. E. Itkis, D. E. Perea, R. Jung, S. Niyogi, and R. C. Haddon, *J. Am. Chem. Soc.* **2005**, (in press).

Characterization of Carbon Nanotubes on Insulating Substrates using Electrostatic Force Microscopy

T.S. Jespersen*, P.E. Lindelof* and J. Nygård*

*Niels Bohr Institute, NanoScience Center, University of Copenhagen, Denmark

Abstract. We report on the use of electrostatic force microscopy (EFM) for the characterization of carbon nanotubes (CNT) grown on insulating substrates. In contrast to traditional atomic force microscopy (AFM) which relies on the van der Waals forces, EFM measures the long range Coulomb interaction between a conducting cantilever and the nanotube. This makes large area scans possible for rapid assessment of CNT samples and we present a statistical method of extracting the CNT density.

The samples used for most carbon nanotube based electrical devices and for many other studies of isolated CNT's have the same structure as starting point[1]: A wafer of doped silicon (Si) capped with a layer of insulating silicon dioxide (SiO_2) onto which a sub-monolayer of CNT's is either deposited from a suspension of CNT's in an organic solvent[2] or grown directly on the substrate by chemical vapor deposition (CVD)[3]. Therefore, techniques for characterizing such samples are often required. Most often atomic force microscopy (AFM) or scanning electron microscopy (SEM) is employed but both methods have their drawbacks. The SEM is relatively fast but the resolution is often limited by charge build-up in the insulating substrate and the sample may change during imaging due to defects induced by the electron beam or deposition of hydro-carbons thus debilitating device performance. The AFM does not change the sample but is relatively slow and limited to imaging only a small area (about $30\mu m \times 30\mu m$) if single walled CNT's with diameters of 1-3nm are to be clearly resolved.

The slow imaging of an AFM is due to the short range of the van der Waals forces exploited in imaging the topography of a surface. Thus high resolution and slow movement of the AFM tip is needed to not "miss" the CNT's. Electrostatic force microscopy (EFM), however, is a scanning probe technique similar to AFM, but relying on the long range Coulomb interaction between the scanning tip and the sample. Here we emphasize the usefulness of EFM for the characterization of the standard CNT samples described above.

The EFM operation of a scanning probe microscope (Figure 1a) is a dual scan technique where the topography of each scan line is first obtained by standard tapping mode AFM. In the second scan the topographic data is used to retrace the first line with a constant tip-sample separation h. In the second scan the cantilever is oscillated at its free resonant frequency ω_0 and the EFM signal is the phase difference between the driving force and the actual oscillation of the tip. In the presence of a force F between the tip and the

CP786, *Electronic Properties of Novel Nanostructures*, edited by H. Kuzmany, J. Fink, M. Mehring, and S. Roth
© 2005 American Institute of Physics 0-7354-0275-2/05/$22.50

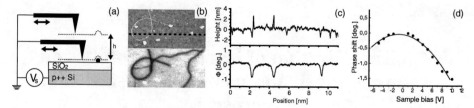

FIGURE 1. (a) Schematic illustration of the EFM measurement. (b) Topography (top) and EFM (bottom) of CVD-grown CNT's. (c) Height profile (top) and EFM phase shift along the line in (b). (d) The nanotube EFM signal Φ_{NT} as a function of V_s. The solid line is a fit to a quadratic form $\Phi_{NT} = aV_s^2 + b$.

sample the phase difference (in radians) is given by[4, 5]

$$\varphi = \tan^{-1}\left(\frac{k}{QF'}\right) \approx \frac{\pi}{2} + \frac{Q}{k}F' \tag{1}$$

where $F'(z) = \partial F(z)/\partial z$ is the force gradient. The standard convention of EFM is to use $\Phi = \varphi - \pi/2$ as the EFM signal. If a voltage V_s is applied between tip and sample and a capacitive coupling between them is assumed then

$$\Phi = \frac{Q}{2k}C''V_s^2, \tag{2}$$

where C'' is the second derivative of the tip-sample capacitance[5]. Usually an off-set is chosen such that $\Phi = 0$ at the naked substrate, and by considering the tip-sample capacitances with and without a nanotube on the substrate, it can be shown that CNT's will always appear with at negative phase shift Φ_{NT}[4]. Furthermore both metallic and semiconducting CNT's have relatively high conductances and both types appear in EFM images [5].

All CNT samples discussed here are single walled CNT's grown by CVD on highly doped silicon substrates with 400nm oxide (for details on the growth, see ref. [6]). The EFM measurements were performed using a Digital Instruments Dimension 3100 operated in air at room temperature and using conducting PtIr-coated cantilevers[7] with resonant frequencies $\omega_0 \sim 60$kHz, spring constants $k \sim 2.8$N/m, and quality factors $Q \sim 225$.

Figure 1b shows topography and EFM of a typical sample (EFM parameters: $V_s = -5V, h = 60$nm). With the chosen color scale the negative phase shift of the CNT's appear as dark lines in the EFM panel. The height profile and EFM along the dotted line of panel b is shown in panel c. The diameter of the CNT's are $1 - 3$nm but since EFM measures the long range Coulomb interaction the CNT's appear with an effective width of about 0.5μm. This makes it possible to observe CNT's in fast large area scans where they do not show up in standard topography AFM. It is our experience that with a resolution of a 512×1024 (lines $\times$ points per line) a standard topography AFM image is limited to an area of approximately $10 \times 10\mu$m if individual CNT's are to be clearly identified. With this resolution CNT's can be clearly resolved in EFM scans of 100μm $\times 100\mu$m. Thus, for these samples the EFM technique is at least 100 times

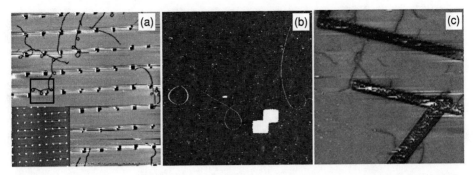

FIGURE 2. (a) 100μm × 100μm EFM scan of CVD grown CNT's and metal alignment marks. The resolution of the corresponding topography (inset) is too low to resolve the CNT's. (b) 12μm × 12μm topography scan of the region marked in (a). (c) EFM image of CNT's between metal electrodes. Electrode distance, 5μm

more time efficient than standard AFM in assessing the overall properties. Figure 1d shows Φ_{NT} as a function of V_s and the quadratic dependence expected for a capacitive coupling (eq 2) is clearly observed. Below we show two cases where the properties of EFM is used for: (i) Identification of special CNT structures with respect to predefined alignment marks (in this case CNT loops for electrical devices) and (ii) for fast density characterization of as-grown CNT samples using a statistical analysis of the EFM data. In the process of making electrical devices we identify CNT's with respect to predefined alignment marks. For low density samples, or devices which require rare CNT structures (e.g. crossing CNT's or CNT loops), a large area may have to be scanned in order to find an adequate structure, and this process can be very time consuming using standard AFM. Figure 2a shows an example where EFM is used to identify CVD grown CNT-loops in a 100μm × 100μm area with a grid of metal alignment marks made by e-beam lithography. The inset shows the corresponding topographic image in which the tubes are not resolved. However, for information about the diameter of the tubes we still rely on standard AFM and Figure 2b shows a high resolution topography image of the area marked in 2a. The final devices may also be examined by EFM if the electrode gap is not to small. Figure 2c shows such an example.

The possibility of large area scans makes EFM ideal for rapid characterization of CNT samples. Figure 3a,b,c show three 90μm × 90μm EFM images with a resolution of 512×1024 points of as-grown CNT samples. The samples have different tube densities due to differences in the growth conditions. The histogram in Figure 3d shows the measured phase shifts from Figure 3a and the inset shows the data along the dashed line. For the specific experimental conditions the nanotubes appear with a phase shift of $\Phi_{NT} \sim -1.2°$ with respect to the substrate. In order to estimate the tube density we find the percentage of pixels with phase shifts below a cut-off value $\Phi_0 = \frac{1}{3}\Phi_{NT} \approx -0.4°$ chosen to ensure a safe CNT/substrate distinction. Correlating with the apparent width of the CNT's at this value (0.6μm) the entire length of tubes in the image is found. In the case of Figure 3a 1.02% of the pixels have phase shifts below Φ_0 resulting in a total length of tubes of 180μm very close to the actual length (190μm) found by measuring

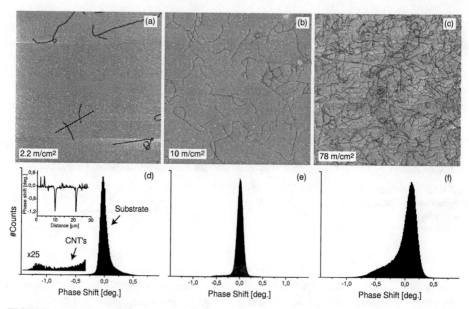

FIGURE 3. (a),(b),(c) $90\mu m \times 90\mu m$ EFM images of CVD grown SWNT samples with varying tube density. Images were measured with $V_s = -5V$ and $h = 60nm$. (d),(e),(f) Histograms of measured phase shift values from (a),(b),(c), respectively. Inset to (d) shows the EFM phase along the dashed line in (a).

directly the length of each NT in the image. This gives a density of $2.2 m/cm^2$ on sample a. On the higher density samples shown in Figure 3b,c the direct measurement is very time consuming but the statistical approach above directly gives a rough idea of the amount of tubes: $830\mu m$ and $6300\mu m$ for Figure 3b and c, respectively giving densities of $10 m/cm^2$ and $78 m/cm^2$.

In summary we have shown that EFM is a powerful technique for rapid characterization of CNT's on insulating substrates, however, the technique does have limitations: Unlike AFM it cannot be used on conducting surfaces and it does not give information about the tube diameter which is often desired and, to some extent, can be used to distinguish between isolated tubes and ropes.

REFERENCES

1. S. Reich, et. al., *Carbon Nanotubes*, Wiley-Vch (2004).
2. See, e.g., J. Nygård, et. al., *Appl. Phys. A*, **69**, 297 (1999).
3. J.H. Hafner, et. al., *J. Phys. Chem. B*, **105**, 743 (2001).
4. C. Staii, et. al., *Nano Lett.*, **4**, 859 (2001).
5. M. Bockrath, et. al., *Nano Lett.*, **2**, 187 (2002).
6. T.S. Jespersen, et. al., (submitted)
7. SCM-PIT cantilevers from Veeco Instruments, www.veeco.com.

Detection Of Single Carbon Nanotubes in Aqueous Dispersion *via* Photoluminescence

Oliver Kiowski [a,b], Katharina Arnold [a,b], Sergei Lebedkin [a], Frank Hennrich [a] and Manfred Kappes [a,b]

[a] *Forschungszentrum Karlsruhe, Institut für Nanotechnoligie, D-76021 Karlsruhe, Germany*
[b] *Institut für Physikalische Chemie, Universität Karlsruhe, D-76128 Karlsruhe, Germany*

Abstract. We report on the 'on-line' photoluminescence (PL) detection of individual single-walled carbon nanotubes (SWNTs) in water-surfactant dispersions. This was achieved by using a near-infrared confocal laser luminescence microscope specially optimized for the PL spectroscopy of SWNTs . A detection of single nanotubes was possible by using dispersions with a relatively low concentration of nanotubes and reducing the PL acquisition time so that a 'switch-on-switch-off' behavior of emission peaks was observed, which reflected a random diffusion of different nanotubes through the laser focus volume. These results demonstrate the analytical potential of PL spectroscopy using SWNTs as luminescent markers.

Keywords: carbon nanotubes, photoluminescence, confocal laser microscopy
PACS: 78.67.Ch ; 42.62.Fi

INTRODUCTION

The near-infrared photoluminescence of isolated (debundled) semiconducting SWNTs is now a well established method to study electronic properties of nanotubes as a function of their (n,m)-structure or helicity [1,2]. Particularly interesting is the possibility to detect PL from a single nanotube. PL microscopy of this kind is complicated by the fact that SWNTs usually emit light in the near-infrared spectral range (available photodetectors are less sensitive than for the visible light) with a quantum efficiency determined from ensemble measurements of only $\sim 10^{-3}$. However, these two negative factors are greatly compensated by the very large absorption cross-sections, α, associated with such big 'molecules' as nanotubes. According to our estimate, SWNTs with a length of ~ 0.5 μm (comparable to a diffraction-limited laser excitation spot) show $\alpha \sim 10^9$ $M^{-1}cm^{-1}$ compared to $\sim 10^5$ $M^{-1}cm^{-1}$ for typical organic fluorophors. Furthermore, SWNTs show an outstanding photostability which allows a high excitation power. Indeed, several groups have reported PL measurements on spatially fixed single nanotubes which were either catalytically grown on a substrate [3] or deposited by spin-coating from a water-surfactant dispersion [4,5].

Here we demonstrate the possibility of PL detection and spectroscopy of single semiconducting nanotubes freely drifting in an aqueous dispersion. This method requires short measurement times (~ 1 s or less) at a good signal-to-noise ratio and could be realized by using optimized laser microscopy techniques [6]. The

CP786, *Electronic Properties of Novel Nanostructures*, edited by H. Kuzmany, J. Fink, M. Mehring, and S. Roth
© 2005 American Institute of Physics 0-7354-0275-2/05/$22.50

demonstrated sensitivity of the PL detection provides support for proposed analytical applications of SWNTs, for example, as near-infrared luminescent markers in biological systems [7]. Furthermore, such 'single nanotube' spectroscopy can give information about intrinsic properties (e.g., PL lineshape) of nanotubes in a fluid dispersion, free from ensemble averaging. Diffusion of nanotubes in a dispersion and a dependence of the relative PL quantum efficiencies on the (n,m)-structure of nanotubes can be estimated from these data as well [8].

EXPERIMENTAL

Raw (bundled) nanotubes were produced in our lab by pulsed laser vaporization of C:Co:Ni targets (PLV nanotube material). Preparation of D_2O/ surfactant dispersions of isolated (debundled) SWNTs by powerful sonication and ultracentrifugation has been described elsewhere [9]. The concentration of nanotubes in the final dispersion could be adjusted (reduced) by increasing the ultracentrifugation time. PLV nanotubes have typical diameters of ~1.0-1.4 nm, an emission wavelength range of ~1200-1700 nm, and an average length after the dispersion procedure of ~0.8-1 μm according to AFM analysis of spin-coated substrates.

Photoluminescence of SWNTs was measured with a home-built near-infrared confocal laser luminescence microscope and equipped with a Leica HCX PL APO oil immersion objective (NA = 1.4) [6]. Although this objective is designed for the visible spectral range, we found that it also performs quite well in the near-infrared range. In a typical experiment, a ~10 μm layer of a dispersion of SWNTs was confined between a base quartz plate and a microscopic cover glass. The laser luminescence microscope was focused in the middle of the dispersion layer. PL could be excited in the range of ~710-950 nm with a CW Ti-Sapphire laser. A typical excitation intensity was ~1 mW. Emission was detected with a liquid nitrogen cooled InGaAs photodiode array attached to a near-infrared spectrograph (Roper Scientific). The latter was coupled to the microscope through an optical fiber which also acted in this setup as a confocal pinhole.

RESULTS AND DISCUSSION

For high concentrations of SWNTs in a water-surfactant dispersion and long PL acquisition times, emission spectra obtained with the laser microscope were time-independent and similar to those obtained for the same bulk dispersion by the FTIR-luminescence technique (operating with a ~0.1 ml excitation and emission collection volume) [9]. The PL spectra were apparently contributed by numerous different (n,m)-nanotubes in both measurements. Dramatic changes were observed with the laser microscope when the nanotube concentration corresponded to an optical density of less than ~0.2 at 700-1000 nm and the PL acquisition time was reduced to 0.5-1 s. Emission peaks corresponding to different (n,m) nanotubes started to appear and disappear in a random fashion on the time scale of seconds as illustrated in Fig. 1. For some short time intervals there was practically no PL observed. This 'switch-on-switch-off' behavior apparently reflects a transition from ensemble to single nanotube measurement mode and a random drift (diffusion) of nanotubes through the excitation and emission collection volume of the laser microscope.

We estimate this volume for typical SWNT excitation and emission wavelengths in our confocal microscope setup as ~0.2 μm^3. One tube per this volume would correspond to a molar concentration of ~10^{-8} M. Taking into account the average length of dispersed PLV nanotubes (~1 μm), we roughly estimate that a pronounced 'switch-on-switch-off'

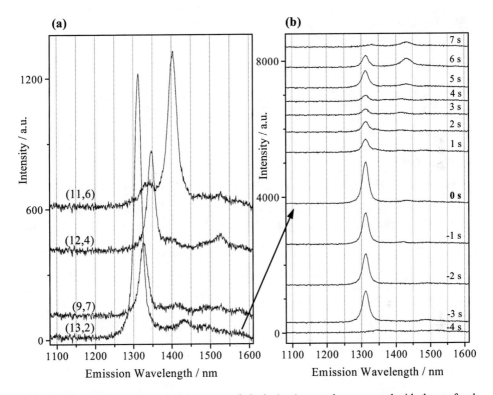

FIGURE 1. (a) Typical PL emission spectra of single (n,m) nanotubes measured with the confocal laser microscope. Nanotubes were dispersed in D_2O/ 1 wt.% sodium dodecylbenzene sulfonate and excited at 843 nm. (n,m) were determined by comparison with an ensemble spectrum (b) Time evolution of the (13,2) nanotube emission peak shown in (a) 4 seconds before, and 7 seconds after reaching the maximum intensity. Each spectrum was acquired within 1 s. The spectra are vertically shifted for convenience.

PL behavior was encountered at nanotube concentrations of less than ~10^{-9} M. Strong variations of the PL peak intensities during the measurement can be explained by different (random) orientations of nanotubes relative to the laser beam axis. This results in varying PL signals due to anisotropy of nanotube excitation and emission.

We found rather symmetric lineshapes for the majority of observed emission peaks with linewidths between 15 and 20 meV. These properties are similar to those found for the photoluminescence of surface-deposited nanotubes at ambient temperature [4]. The PL lineshape data as well as PL peak statistics for different (n,m)-nanotubes will be presented and discussed in detail elsewhere [8].

ACKNOWLEDGMENTS

The support of this work by the Deutsche Forschungsgemeinschaft and by the BMBF is gratefully acknowledged.

REFERENCES

1. O'Connell, M. J., Bachilo, S. M., Huffman, C. B., et al., *Science* **297**, 593 (2002).
2. Bachilo, S. M., Strano, M. S., Kittrell, C., et al., *Science* **298**, 2361 (2002).
3. Lefebvre, J., Fraser, J. M. , Finnie, P., and Homma, Y., *Phys. Rev. B* **69**, 075403 (2004).
4. Hartschuh, A., Pedrosa, H. N., Novotny, L., and Krauss, T. D., *Science* **301**, 1354 (2003).
5. Htoon, H., O'Connell, M. J., Cox, P. J., Doorn, S. K., and Klimov, V. I., *Phys. Rev. Lett.* **93**, 027401 (2004).
6. Kiowski, O., Diploma Thesis, University of Karlsruhe, 2004.
7. Cherukuri, P., Bachilo, S. M., Litovsky, S. H., and Weisman R. B., *J. Am. Chem. Soc.* **126**, 15638 (2004).
8. Kiowski, O., Lebedkin, S., and Kappes, M. M., in preparation.
9. Arnold, K., Lebedkin, S., Hennrich, F, Krupke, R., Renker, B., and Kappes, M. M., *New J. Phys.* **5**, 140 (2003).

Diameter Dependence of the Elastic Modulus of CVD-Grown Carbon Nanotubes

Kyumin Lee, Branimir Lukic, Arnaud Magrez, Jin Won Seo, Andrzej Kulik, László Forró

Institute of Physics of Complex Matter, Ecole Polytechnique Fédérale de Lausanne (EPFL)
CH-1015 Lausanne, Switzerland

Abstract. Due to their exceptional physical strength, carbon nanotubes are promising candidates for reinforcement material of super-strong composites. While multi-walled carbon nanotubes grown by arc-discharge or single-walled carbon nanotubes grown by laser ablation methods can easily reach the ideal physical strength of 1 TPa elastic modulus, the multi-walled carbon nanotubes grown by chemical vapor deposition (CVD) are significantly weaker, having elastic moduli below 100 GPa. The CVD is preferable over other methods due to its high yield, but the growth mechanism is still not fully understood. A systematic study by our group revealed that the Young's modulus of a CVD-grown multi-walled nanotube decreases exponentially with its diameter. This relationship has implications on the growth mechanism of the CVD-grown multi-walled nanotubes.

INTRODUCTION

Our lab has measured the mechanical strength (namely the elastic or Young's modulus) of various batches of carbon nanotubes, using the method developed by Jean-Paul Salvetat [1-3]. The multi-walled carbon nanotubes grown by catalytic chemical vapor deposition (CVD) were found to have disappointingly low Young's modulus values around or below 100 GPa [4], in contrast with the 1 TPa values for MWCNTs or SWCNTs grown by arc-discharge. The collected data suggested a relationship between the diameter and the Young's modulus of a CVD-grown MWCNT, calling for a further investigation on this subject.

EXPERIMENTAL

An improved method [5] was developed to increase the throughput of the experiment. While the original method by Salvetat relied on chance to find nanotubes suspended over a pore in polished alumina surface, AC electrophoresis [6] was used in the new method to deposit nanotubes at the electrodes formed on a GaAs structure with grill-like patterns. Fig. 1a is an AFM image of nanotubes prepared by this method. The nanotubes lie more or less perpendicular to the electrode border, which, with the right sample orientation, allows AFM force-displacement curve technique to be used for data acquisition. The slopes of the force-displacement curves on a flat

CP786, *Electronic Properties of Novel Nanostructures*, edited by H. Kuzmany, J. Fink, M. Mehring, and S. Roth
© 2005 American Institute of Physics 0-7354-0275-2/05/$22.50

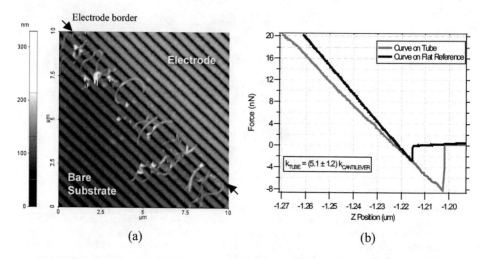

(a)   (b)

FIGURE 1. a) MWCNTs deposited at an electrode border by AC electrophoresis. The cantilever axis was actually perpendicular to the grill-like trenches in the GaAs structure, and the image was taken with a 45-degree rotation in the scan direction. The opposing electrode is not visible on this image. b) The black line is the reference force-displacement curve on the flat part of the substrate. The gray line is the curve acquired on the center of the suspended portion of a nanotube. Its slope after the jump-to-contact gives the effective spring constant of the tip-nanotube system.

reference surface and on the center of the suspended nanotube are compared to calculate the spring constant of the suspended nanotube structure (see Fig. 1b). This method is much faster than the original successive imaging technique.

The diameter and the suspended length of a nanotube are obtained from the AFM images. We calculate the Young's modulus using the acquired values, modeling the suspended nanotube as a uniformly filled cylindrical beam clamped at both ends.

RESULTS AND DISCUSSION

Fig. 2 shows the Young's modulus values of CVD-grown MWCNTs, obtained over the last 2-year period. The values for small-diameter SWCNT bundles [2], MWCNTs grown by arc-discharge [3], and CVD-grown double-walled carbon nanotubes [7] are also included in the figure, to serve as references. Of the CVD-grown MWCNT data, the data for batches #2 and #3 were obtained using the original Salvetat method, and the data for batch #1 were acquired using the new improved method.

The graph shows an exponential dependence between the tube diameter and the Young's modulus. The authors do not believe that the diameter dependence is due to inter-shell sliding of perfect shells, since the MWCNTs grown by arc-discharge exhibit ideal 1 TPa Young's moduli at comparable diameters. The phenomenon is also not a shortcoming of the simple uniform-cylinder model that neglects the inner tube diameter. The simple model does induce false diameter dependence, but only as a 3rd order polynomial function of the diameter, which cannot explain the data. Thus the discovered exponential dependence is not an artifact of our simple model.

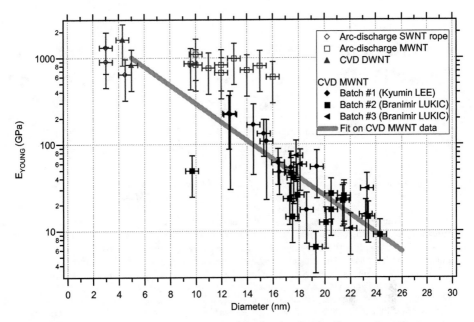

FIGURE 2. Young's modulus versus tube diameter plot showing the diameter dependence.

The authors believe that the diameter dependence is related to the mechanism and kinetics of carbon nanotube growth in CVD. The inner and outer shells probably grow simultaneously, restricting the mobility of carbon atoms or dimers that dissolve or diffuse out of the catalyst particle. Then any lattice sites that were "skipped" during the growth will remain as vacancies or stabilize into other forms of defects. As larger vacancy "islands" can occur in shells with larger diameters, the large-diameter shells may be less crystalline than the small-diameter shells, leading to the diameter dependence of the Young's modulus of the whole tube.

Future work will involve correlating the AFM data with HRTEM images to verify the relationship between the tube diameter and the defect ratio.

REFERENCES

1. Salvetat J. P., J. M. Bonard, N. H. Thomson et al., *Appl. Phys. A-Mater. Sci. Process.* **69** (3), 255-260 (1999).
2. Salvetat J. P., G. A. D. Briggs, J. M. Bonard et al., *Phys. Rev. Lett.* **82** (5), 944-947 (1999).
3. Salvetat J. P., A. J. Kulik, J. M. Bonard et al., *Adv. Mater.* **11** (2), 161-165 (1999).
4. Lukic B., J. W. Seo, E. Couteau et al., *Appl. Phys. A-Mater. Sci. Process.* **80** (4), 695-700 (2005).
5. Lee Kyumin, *to be published*.
6. Yamamoto K., S. Akita, and Y. Nakayama, *J. Phys. D-Appl. Phys.* **31** (8), L34-L36 (1998).
7. Lukic Branimir, *to be published*.

Femtosecond Transient Absorption Spectra and Relaxation Dynamics of SWNT in SDS Micellar Solutions

V.A. Nadtochenko[1], A.S. Lobach[1*], F.E. Gostev[2], D.O. Tcherbinin[2], A. Sobennikov[2], O.M. Sarkisov[2]

[1]*Institute of Problems of Chemical Physics RAS, 142432 Chernogolovka, Moscow Region, Russia*
[2]*Institute of Chemical Physics RAS, 4 Kosygina St., 119991 Moscow, Russia*

Abstract. Transient absorption spectra and relaxation dynamics of excited SWNT were studied by femtosecond absorption spectroscopy as a function of: the energy of excitation quanta ($\hbar\omega$ = 2 eV, 2.5 eV, 4 eV); the density of the excitation energy; polarizations of the pump and probe pulses. The transient absorption spectra were monitored by white supercontinuum light pulse in the spectral region of $\sim 1.2 \div 3.6$ eV. The induced transient absorption spectra of SWNT are considered as filling of the size-quantized energy bands with nonequilibrium carriers; renormalization of the one-dimensional energy bands at high density of the induced plasma; quantum confined Stark effect and screening of excitons. The anisotropic relaxation rate is observed.

INTRODUCTION

Theoretical and experimental evidences exist that unusual electron and electron-transport properties of SWNT correspond to the effects of electron–electron (e–e) interactions and peculiarities of relaxation dynamics [1, 2]. Femtosecond spectroscopy is a unique and efficient method to study the dynamics of excited states and e–e interactions. It allows us to study the dynamics of elementary excitations (excitons, electron–hole (e–h) pairs, plasmons, etc. [3]). A distinctive feature of present experiments compared to previous ones [4–9] is the use of probe pulses of the broadband femtosecond white continuum as well as the excitation of a sample by laser pulses with different energies of photons of 2, 2.5, and 4 eV. Preliminary results of this work published in [10,11] are in a good agreement with independent measurements of transient spectra in the recent work [12].

EXPERIMENTAL

SDS water micellar solutions of SWNT. Pristine SWNT HiPco were used to prepare suspension in a 1-wt % water-micellar solution of SDS. Details of SWNT water-micellar solutions preparations are published in [10,11].

Femtosecond laser pump-probe measurements. Pump-probe femtosecond experiments were carried out with two laser setups: 1) CPM dye laser of λ = 616 nm

CP786, *Electronic Properties of Novel Nanostructures*, edited by H. Kuzmany, J. Fink, M. Mehring, and S. Roth
© 2005 American Institute of Physics 0-7354-0275-2/05/$22.50

was used with the 50 fs pulse after amplification and the energy up to 1 mJ. This part of the work was fulfilled at the Institute of Chemical Physics RAS, Moscow; 2) a Clark MX 100 femtosecond laser with a subsequent nonlinear optical transformation of light into a 50-fs pulse with a wavelength of 485 nm (Humboldt University, Berlin).

RESULTS AND DISCUSSION

Transient absorption spectra (TAS). Fig. 1 demonstrates TAS (Δ Absorbance = Abs.(t) - Abs.(t<0)) of SWNT in SDS/H_2O solution after the excitation by 45 fs, $\hbar\omega$ = 2.5 eV laser pulse for two different densities of excitation energy. For a comparison the absorbance spectra of the same SWNT solution are shown. The main features of the TAS are: 1) the strong wide bleach band is superimposed with the bleach peaks; 2) the intensity and the width of the wide bleach band depend on concentration of the excited states in SWNT; 3) the bleach peaks correspond to the absorbance peaks in the spectrum of nonexcited SWNT, which relates to van Hove singularities; 4) some shifts of transient bleach peaks relative to the peaks of nonexcited SWNT are detected. The shift value depends on the delay time, the excitation energy density; 5) the absorption peaks appear after delay. The intensity and the delay time of absorption peaks appearance depend on the concentration of excitations. The common features of TAS are similar after excitation pulses with the different energy of quanta of 2 eV, 2.5 eV and 4 eV. In all cases, it was observed superposition of the wide bleach band and bleach peaks with the similar spectral structure, which relax to the system of absorption and bleaching peaks at longer delays.

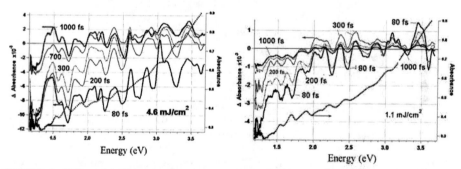

FIGURE 1. Differential TAS of suspension of HiPco SWNT in 1% SDS/H_2O solution after the excitation by 45 fs, $\hbar\omega$ = 2.5 eV laser pulse (left axis). In the left subplot, the energy density of excitation was 4.6 mJ/cm^2. In the right subplot the energy density was 1.1 mJ/cm^2. Time delays are shown as tags. In both subplots, for a comparison, the absorbance of HiPco SWNT in 1% SDS/H_2O spectrum is shown by dotted line (right axis). Time delays are shown in subplots.

Transient decay. Fig. 2 demonstrates decays of bleaching signal at different probe wavelengths. The relaxation curves are well fitted by the exponent. Fig. 2 depicts that decay time τ of the bleach curves depends on the probe wavelength. The further relaxation of transient absorption is observed at the picosecond time scale. The decay of transient absorption component depends on the wavelength too.

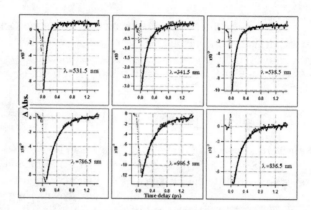

FIGURE 2. The decay of bleach component of SWNT transient spectra at different wavelength. The solid lines are exp fit of experimental data (circles). The excitation was $\hbar\omega = 2.5$ eV.

Fig. 3 shows the dependence of the decay time τ of bleaching curves on the probe photon energy. Fig. 3 depicts the correspondence between bleach peaks of transient spectra and peaks in the dependence $1/\tau$ vs. energy.

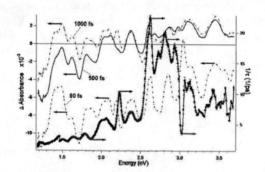

FIGURE 3. The dependence of the bleaching relaxation time τ as a function of the probe energy (left axis). For comparison TAS at time delays of 80 fs, 500 fs, and 100 fs are shown. The density of excitation energy was 4.6 mJ/cm^2. The excitation was $\hbar\omega = 2.5$ eV.

Fig. 4 shows that time decay depend not only on the energy of probe photon but on the density of excitation energy too. The decrease of the density of excitation energy and due to this the decrease of concentration of elementary excitations in SWNTS leads to the increase of the decay rate of the bleaching components in transient spectra.

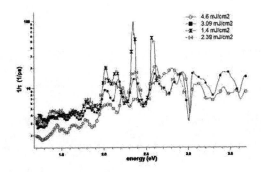

FIGURE 4. The dependence of relaxation time τ of the bleaching component as a function of the probe energy with different densities of excitation energy. The excitation was at $\hbar\omega = 2.5$ eV.

Transient spectra manifest the changes of absorption coefficient $\Delta\alpha$. The induced TAS of SWNT can be considered as filling of the size-quantized energy bands with nonequilibrium carriers; renormalization of the one-dimensional energy bands at high density of the induced plasma; quantum confined Stark effect and screening of excitons. The anisotropic relaxation rate depicted in Figs. 2-4 can be explained by two reasons: 1) an inhomogeneous nature of SWNT HiPco samples; 2) an anisotropic relaxation rate in individual nanotubes, because the conservation laws for momentum and energy in e–e, and electron-phonon scattering in quantum one-dimensional system. The shape of curves $1/\tau$ vs. probe energy and the increase of the decay rate with the decrease of excitations concentration in Figs. 3,4 suggest that the effect of the anisotropic relaxation rate is meaningful for individual nanotube.

ACKNOWLEDGMENTS

Authors thank to Dr. S. Kovalenko for the great assistance in experiments and RFBR (grants 03-03-32727, 03-03-32668, and 04-02-17618) for financial support.

REFERENCES

1. T. Hertel, R. Fasel, and G. Moos, Appl. Phys. A: Mater. Sci. Process. **75**, 449 (2002).
2. C. L. Kane and E. J. Mele, Phys. Rev. Lett. **90**, 207401 (2003).
3. I. E. Perakis and T. V. Shahbazyan, Surface Sci. Repts. **40**, 1 (2000).
4. J.-S. Lauret, C. Voisin, G. Cassabois, *et al.* Phys. Rev. Lett. **90**, 057404 (2003).
5. O. J. Korovyanko, C.-X. Sheng, Z. V. Vardeny, *et al.* Phys. Rev. Lett. **92**, 17403 (2004).
6. Y.-Z. Ma, J. Stenger, J. Zimmerman, *et al.* J. Chem. Phys. **120**, 3368 (2004).
7. J. Kono, G. N. Ostojic, S. Zaric, *et al.* Appl. Phys. A: Mater. Sci. Process. **78**, 1093 (2004).
8. G. N. Ostojic, S. Zaric, J. Kono, *et al.* Phys. Rev. Lett. **92**, 117402 (2004).
9. H. Hippler, A.-N. Unterreiner, J.-P. Yang, *et al.* Phys.Chem. Chem. Phys. **6**, 2387 (2004).
10. V. A. Nadtochenko, A.S. Lobach, **F. E. Gostev,** *et al. JETP Letters, 80, 176 (2004).*
11. V. A. Nadtochenko, A.S. Lobach, **F. E. Gostev,** *et al., Doklady Physics, 50, 12 (2005).*
12. I.V.Rubtsov, R.M. Russo, T. Albers, *et al.,* Appl. Phys. A 79, 1747 (2004)

X-ray Spectroscopy Characterization of Carbon Nanotube Film Texture

A.V. Okotrub[1], V.V. Belavin[1], L.G. Bulusheva[1], A.G. Kudashov[1],
D.V. Vyalikh[2], S.L. Molodtsov[2]

[1]Nikolaev Institute of Inorganic Chemistry SB RAS, av.Ak.Lavrentieva 3, Novosibirsk 630090, Russia
[2]BESSY II, Albert-Einstein-Strasse15, D-12489 Berlin, Germany

Abstract. X-ray absorption spectra measured for film of the aligned multiwall carbon nanotubes at different grazing angles of incident beam showed a variation in π^*/σ^* peak ratio. An average deviation of carbon nanotubes from vertical orientation was estimated from comparison of the experimental data with the results quantum-chemical calculation of (6,6) carbon tube crystal. Total imperfectness of tubes ordering in the film and tube layers packing was found to be about 60°.

Keywords: X-ray Spectroscopy, Carbon Nanotubes.
PACS: 78.70. Dm 61.46.+w

INTRODUCTION

Anisotropy of chemical bonds in carbon nanotube gives an opportunity to divide its electronic structure on σ- and π-subsystems. The π-orbitals are constructed from C2p-atomic orbitals and oriented perpendicular to a cylindrical surface of a nanotube, while the σ-orbitals are tangentially directed to its surface and may have contribution of C2s-atomic orbitals also. Depending on orientation of E vector of incident X-ray radiation, the π- and σ-subsystems will have different contribution to an absorption spectrum. Therefore, the angular dependence of X-ray absorption spectrum can be used for restoration of the form of π-system in the graphite-like materials (see for example [1]) and for definition of sample perfection. Anisotropy of electronic structure of single-wall carbon nanotube film has been revealed by electron energy loss spectroscopy [2]. Recently [3] it was proposed to use the angular dependence of X-ray absorption spectrum of a film of aligned carbon nanotubes for definition of nanotubes ordering. Measurement of X-ray absorption spectra from individual nanotube is not obviously possible, so far, therefore we have used quantum–chemical calculations for separation of density of π- and σ-electrons.

In the present work we have numerically estimated a degree of ordering of multiwall carbon nanotubes in the film grown on a substrate, from angular dependence of X-ray absorption spectra and results of quantum–chemical calculation on carbon tube crystal.

CP786, *Electronic Properties of Novel Nanostructures*, edited by H. Kuzmany, J. Fink, M. Mehring, and S. Roth
© 2005 American Institute of Physics 0-7354-0275-2/05/$22.50

EXPERIMENTAL AND CALCULATIONS

The film of vertically aligned carbon nanotubes on silicon substrate was prepared by thermal catalytic chemical vapour decomposition (CCVD) of $C_{60}/Fe(C_5H_5)_2$ mixture like described in [4]. The CCVD apparatus consists of a stainless steel gas flow reactor of 1 m length and 3.4 cm diameter and a tubular furnace with a heating length of 30 cm. A ceramic boat with a mixture of fullerene C_{60} and ferrocene $Fe(C_5H_5)_2$ taken in the ratio of 1:1 was placed inside the quartz tube under the silicon plate of size 10×10 mm. The pyrolysis was performed at 950°C and atmospheric pressure in an argon flow (3 l/min). The micrographs of the product were obtained with a JSM-T200 scanning electron microscope (Fig. 1(a, b)).

Carbon K-edge X-ray absorption spectra of the film were measured using the Berlin synchrotron radiation facility at the Russian-German laboratory in BESSY-II. The data were acquired in the total yield of electrons mode and normalized to the primary photon current from a gold-covered grid, which was recorded simultaneously. The energy resolution of incident radiation was about 0.06 eV. The sample was rotated around vertical axis, while a synchrotron radiation beam was polarized in horizontal plane (Fig. 1(c)). Absorption spectra were recorded for three different grazing angles of incident X-ray radiation $\Theta = 0°, 30°, 50°, 80°$.

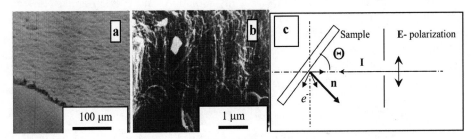

FIGURE 1. a, b – Scanning electron microscopy images of aligned carbon nanotubes on silicon substrate (a, b). c – Scheme of sample location relative to the incident X-ray beam (I).

The modeling of X-ray absorption spectra was based on results of quantum–chemical calculation of C2p partial electronic density of states for crystal of (6,6) carbon tubes. The quantum–chemical calculation was performed in a framework of density functional theory in GGA approximation with Perdew-Burke-Ernzerhof 96 exchange–correlation functional [5], included into Wien2k package [6]. Cross-sections of (6,6) carbon tube in a crystal (space group 175 P6/m) formed a hexagonal lattice with parameters: $a = b = 11.49$ Å, $c = 2.461$ Å. Diameter of nanotubes had a value 8.14 Å. Geometry was not relaxed. Atomic sphere radii for construction of pseudo-potentials for C2s и C2p atomic states have a value 1.3 Å. Decomposition of the valence states functions on a basis of plain waves was performed till the energy -50 Ry. Electronic density of states were calculated by integration on the irreducible wedge of the Brillouine zone with 54 k- points and then were smoothed by 0.02-eV Gaussian.

RESULTS AND DISCUSSION

The angular dependence of X-ray absorption spectra results from the angular dependence of transfer matrix element of dipole moment and a symmetry of the nearest environment of radiating atom. Prior to model the spectra we have assumed that carbon nanotubes in a film have cylindrical form, close values of diameter and lengths. Besides we shall suppose that angular distribution of nanotubes in a sample is characterized by Gaussian-like shape $\rho(\theta, w) = N \cdot \exp\left(-(\theta / w)^2 \cdot \ln(2)\right)$, where N is the normalizing coefficient considering a constancy of nanotubes quantity in a film. The value $\rho(\theta, w)$, at given half-width w, defines a quantity of nanotubes in a film which axes make an angle θ with a normal to a substrate plane. On the basis of geometrical considerations the following expressions for angular dependence of π- and σ-contributions to absorption spectrum have been received:

$$I_\pi^{abs}(\Theta,\theta) = I_{\pi^*}^0 \cdot \left(0.75 - 0.25 \cdot \cos^2(\theta) + 0.75 \cdot \cos^2(\theta) \cdot \cos^2(\Theta) - 0.25 \cdot \cos^2(\Theta)\right),$$

$$I_\sigma^{abs}(\Theta,\theta) = I_{\sigma^*}^0 \cdot \left(1.25 + 0.25 \cdot \cos^2(\theta) - 0.75 \cdot \cos^2(\theta) \cdot \cos^2(\Theta) + 0.25 \cdot \cos^2(\Theta)\right).$$

Here Θ is a grazing angle of incident X-ray radiation, and θ is an angle between nanotube axis and normal (n) to a substrate plane (see Fig. 1(c)). The angular dependence of X-ray absorption spectra for carbon nanotubes film can be presented in the form:

$$I(\Theta, w) = \int_0^{\pi/2} (I_\pi^{abs}(\Theta,\theta) + I_\sigma^{abs}(\Theta,\theta)) \cdot \rho(\theta, w) \cdot \sin(\theta) d\theta,$$

where an appearance of $\sin(\theta)$ is caused by cylindrical symmetry of the task. Figure 2 presents the experimental X-ray absorption spectra taken for the aligned nanotubes film (left top picture) and theoretical spectra, calculated for three different angular distributions of the tubes (with the width of $w = 10°$, $30°$ and $50°$) in a film. Theoretical and experimental spectra were normalized on the intensity of the σ^*-peak.

Relative energy position of the A and C maxima in the theoretical spectra agrees well with the corresponding π^* и σ^* peaks in the experimental spectra. An appearance of maximum B is obviously caused by a presence of impurities and imperfections on a film surface and could not be explained by a structure of carbon nanotube itself. By comparing the results of theoretical modeling with experimental spectra we have determined texture of the carbon nanotube film. The determined width of nanotubes angular distribution (w) is about 60°. It should be noted, that shells of multiwall carbon nanotubes can have a cone-like arrangement. In this case, the obtained angular distribution would be a result of convolution of carbon nanotubes ordering in the film and graphitic layers packing in the tubes. Fourier analysis of the microscopic images of the aligned carbon nanotubes produced using the same synthetic technique revealed the angular distribution of nanotubes in film has a width $w_1 \approx 45°$ [7]. One can see this value is less than that obtained from the X-ray absorption data.

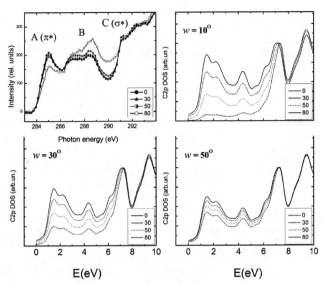

FIGURE 2. Experimental X-ray absorption spectra of the aligned carbon nanotubes film measured at $\Theta = 0°, 30°, 50°, 80°$ (lop left picture) and theoretical spectra simulated for angular tube distribution with the width of $10°, 30°,$ and $50°$.

The total width w_1 includes a component characterized disordering of carbon nanotube layers. Taking into account the values obtained from spectral and spectroscopic experiments, an average width of this component estimated by $w_2 = \sqrt{w^2 - w_1^2}$ is equal to $40°$.

ACKNOWLEDGMENTS

We thank Dr V.S. Danilovich for SEM characterization of the sample. The work was financially supported by the RFBR (grant 03-03-32336).

REFERENCES

1. Skytt, P., Glans, P., Mancini, D.C., Guo, J.-H., Wassdahl, N., Nordgren, J., and Ma, Y., *Phys. Rev. B*, **50**, 10457 (1994).
2. Liu, X., Pishler, T., Knupfer, M., Golden, M.S., Fink, J., Walters, D.A., Cassavant, M.J., Schmidt, J., and Smalley, R.E., *Synth. Met.*, **121**, 1183 (2001).
3. Shiessling, J., Kjeldgaard, I., Rohmund, F., Falk, L.K.L, Campbell E.E.B., Nordgren, J., and Bruhwiler, P.A., *J. Phys.: Condens. Matter.*, **15**, 6563 (2003).
4. Grobert, N., Hsu, W.K., Zhu, Y.Q., Hare, J.P., Kroto, H.W., Walton, D.R.M., Terrones, M., Terrones, H., Redlich, P., Ruhle, M., Escudero, R., Morales, F., *Appl. Phys. Lett.*, **75**, 3363 (1999).
5. Perdew J.P., Burke K., and Ernzerhof M., *Phys. Rev. Lett.*, **77**, 3865 (1996).
6. Schwarz, K., Blaha, P., and Madsen, G.K., *Comp. Phys. Comm.*, **147**, 71 (2002).
7. Okotrub, A.V., Dabagov, S.B., Kudashov, A.G., Gusel'nikov, A.V., Kinloch, I., Windle, A.H., Chuvilin, A.L., and Bulusheva, L.G., *JETP Lett.*, **81**, 34 (2005).

STM Images of Atomic-Scale Carbon Nanotube Defects Produced by Ar$^+$ Irradiation

Z. Osváth, G. Vértesy, L. Tapasztó, F. Wéber, Z. E. Horváth, J. Gyulai, and L. P. Biró

Research Institute for Technical Physics and Materials Science,
P.O. Box 49, H-1525 Budapest, Hungary

Abstract. Multi-wall carbon nanotubes (MWCNTs) dispersed on graphite (HOPG) substrate were irradiated with Ar$^+$ ions of 30 keV, using a low-dose of D = 5×10^{11} ions/cm^2. The irradiated samples were investigated by scanning tunneling microscopy (STM) under ambient conditions. Atomic resolution STM images reveal individual nanotube defects, which appear as hillocks of 1-2 angstroms in height, due to the locally changed electronic structure. After annealing at 450 °C in nitrogen atmosphere, the irradiated MWCNTs were investigated again by STM. The effect of the heat treatment on the irradiation-induced nanotube defects is also discussed.

Keywords: Carbon Nanotube, STM, Electronic Structure, Ion Irradiation.
PACS: 07.79.Cz, 73.22.-f

INTRODUCTION

The carbon nanotubes (CNTs) usually are not perfect tubes, vacancies or non-hexagonal carbon rings are often present in their structure. Defects of this type are responsible for the formation of complex nanostructures [1] and their presence also influences the transport properties [2]. Nanotube defects can form during the synthesis or can be introduced for example by chemical treatment or irradiation. Experiments show that both electron and heavy ion irradiation can modify the structure and dimensions of CNTs [3,4]. Zhu *et al.* [5] pointed out that energetic Ar$^+$ ions produce dangling bonds (vacancies) on the surface of nanotubes. Irradiation with heavy ions may induce interesting effects like welding, cross-linking or coalescence of the nanotubes. A brief overview on the irradiation-induced effects in carbon nanotubes has been given recently [6].

The STM signatures of topological CNT defects were also simulated [7]. Krasheninnikov and co-workers calculated and predicted the STM images of individual vacancies in single wall CNTs [8]. In the present work we report experimental atomic resolution STM images of individual CNT defects, created by irradiation with Ar$^+$ ions of 30 keV using a low dose of D = 5×10^{11} ions/cm^2 (15 minutes of irradiation time at normal incidence). Before irradiation the MWCNTs were sonicated in toluene and dispersed on HOPG substrate.

CP786, *Electronic Properties of Novel Nanostructures*, edited by H. Kuzmany, J. Fink, M. Mehring, and S. Roth
© 2005 American Institute of Physics 0-7354-0275-2/05/$22.50

RESULTS AND DISCUSSION

The STM observation of irradiated MWCNTs revealed the signatures of the defects induced by the Ar^+ ions. These defects appear as protrusions (tunneling-current maxima) in the STM images, which is in agreement with theoretical predictions [8]. The observed protrusions are similar to the hillocks observed earlier on irradiated HOPG surfaces [9,10]. Figure 1(a) presents an atomic resolution STM image of an individual CNT defect. The cross-sectional line in Fig. 1(b) shows that the apparent height of the hillock is around 2.4 angstroms.

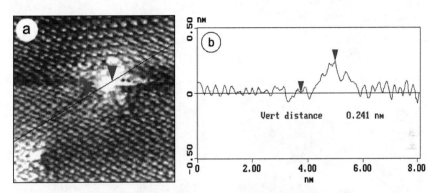

FIGURE 1. Atomic resolution STM image of a MWCNT defect produced by irradiation (a) and a cross-sectional line showing its apparent height (b).

One can observe that atomic resolution is lost at the hillock, and the type of the defect cannot be determined directly. The defects induced by irradiation act as local perturbations on the electronic structure and induce enhancements in the tunneling current due to the presence of the defect-induced electronic states near the Fermi energy [8].

Carbon nanotube defects can be removed by annealing the samples at high temperatures. In order to monitor with STM the effect of a heat treatment, we annealed the irradiated nanotube sample at 450 °C for 1.5 hours under nitrogen atmosphere of 5 bars. After annealing, we investigated the sample again and found that the defect sites could be still observed with STM, although their apparent heights were lower. Figure 2(a) shows such a portion of a MWCNT with a defect after the heat treatment. The cross-sectional line in Fig. 2(b) shows that the hillock corresponding to this defect site has a height of around 1 angstrom, which is much lower than in the case of the defect measured before heating the sample. This effect can be also observed in Fig. 3. A CNT portion containing approximately 30 defects is presented in Fig. 3(a). The hillocks corresponding to the defect sites are clearly visible here. On the other hand Fig. 3(b) shows a CNT portion after the heat treatment. One can see that – using the same tunneling conditions – the irradiation-induced hillocks were not very well observed after annealing. However, when scanning smaller areas ($\sim$10x10 nm^2), the defect sites could be well distinguished again (see Fig. 2, 4). These results show how CNT defects tend to heal already at moderate temperatures.

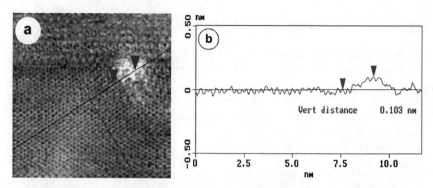

FIGURE 2. STM image of a CNT defect after the heat treatment (a) and a cross-sectional line showing its apparent height (b).

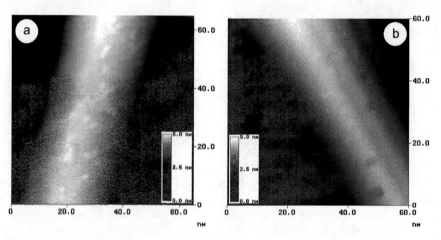

FIGURE 3. STM images of (a) an irradiated MWCNT: the protrusions correspond to defect sites; (b) an irradiated MWCNT after annealing: the protrusions have smaller apparent heights.

The results also show that the defects did not disappear completely, but changed. One possible mechanism is that the vacancy-type defects transformed into non-hexagonal ring-type defects by dangling bond saturation, predicted by molecular dynamical simulations [11]. We also performed spectroscopic measurements at the defect sites (the results are not shown here) and we found [12] that additional electronic states appear above the Fermi level after annealing. Such kind of states can appear assuming that nitrogen adsorbs at the defect-sites [13].

Another atomic resolution STM image is presented in Fig. 4 (image recorded after annealing), showing three defect sites separated from each other by 10-20 angstroms. In the close vicinity of the defect in the left-center of Fig. 4(a) we observed periodic oscillations with a period of about 4.06 angstroms (Fig. 4(b)), which is larger than the period indicated by the atomic structure (~2.46 angstroms). The amplitude of these oscillations decreases with increasing distance, and it vanishes within a distance of about 3 nm from the defect.

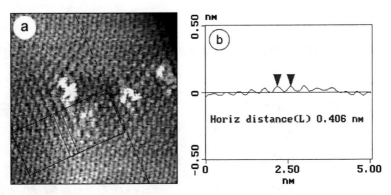

FIGURE 4. Electronic superstructure observed in the vicinity of a defect site (a). The cross-sectional line (b) shows the period of the superstructure.

These superstructures closely resemble the "$\sqrt{3} \times \sqrt{3} \, R$" type superstructures observed earlier on HOPG surfaces [9,10] and predicted theoretically also for single wall CNTs with vacancies [8] which appear due to the interference between normal and scattered electron waves.

In conclusion, we have investigated the CNT defects produced by irradiation with atomic resolution STM. The defects appear as protrusions in the STM images, in agreement with theory. We showed that annealing at moderate temperature did not heal the defects completely, the nature of the defects changed during the heating. We observed electronic superstructures in the vicinity of some CNT defects, similar to the superstructures observed earlier on HOPG.

ACKNOWLEDGMENTS

This work has been done in the framework of the GDRE n°2756 "Science and applications of the nanotubes - NANO-E". The authors acknowledge the support of OTKA grants T 43685 and T 43704 in Hungary.

REFERENCES

1. M. Menon, D. Srivastava, Phys. Rev. Lett. **79**, 4453 (1997).
2. J. W. Park, J. Kim, J.-O. Lee, K.C. Kang, and J.-J. Kim, K.-H. Yoo, App. Phys. Lett. **80**, 133 (2002).
3. P. M. Ajayan, V. Ravikumar, J.-C. Charlier, Phys. Rev. Lett. **81**, 1437 (1998).
4. M. S. Raghuveer, P. G. Ganesan, J. D'Arcy-Gall, G. Ramanath, Appl. Phys. Lett. **84**, 4484 (2004).
5. Y. Zhu, T. Yi, B. Zheng, L. Cao, Appl. Surf. Sci. **137**, 83 (1999).
6. A. V. Krasheninnikov, K. Nordlund, Nucl. Instrum. Meth. B **216**, 355 (2004).
7. V. Meunier, Ph. Lambin, Carbon **38**, 1729 (2000).
8. A. V. Krasheninnikov, Phys. Low-Dim. Struct. **11/12**, 1 (2000).
9. L. Porte, C. H. de Villeneuve, and M. Phaner, J. Vac. Sci. Technol. B **9**, 1064 (1991).
10. J. R. Hahn and H. Kang, Phys. Rev. B **60**, 6007 (1999).
11. A. V. Krasheninnikov, K. Nordlund, Sov. Phys. Solid State **44**, 470 (2002).
12. Z. Osváth *et al.*, Phys. Rev. B, submitted.
13. A. H. Nevidomskyy, G. Csányi, and M. C. Payne, Phys. Rev. Lett **91**, 105502 (2003).

Microdiffraction Study Of Iron-Containing Nanotube Carpets

Pichot V. [1], Launois P. [1], Pinault M. [2], Mayne-L'Hermite M. [2], Reynaud C. [2], Burghammer M. [3], Riekel C. [3]

[1] Laboratoire de Physique des Solides (UMR CNRS 8502), bât. 510, Université Paris Sud, 91405 Orsay Cedex, France
[2] Laboratoire Francis Perrin (URA CNRS 2453), CEA Saclay, DRECAM-SPAM, bât. 522, 91191 Gif sur Yvette, France
[3] The European Synchrotron Radiation Facility (ESRF), 6 rue Jules Horowitz, 38043 Grenoble, France

Abstract. X-ray microdiffraction can be a powerful technique to study nanotube-based materials, as shown here for carpets of aligned multiwall carbon nanotubes synthesized by aerosol-assisted Catalytic Chemical Vapor Deposition. Catalytic particles are found to be iron oxide. Moreover, local analysis of nanotube alignment allows one to understand the effects of the synthesis procedure.

Keywords: Nanotubes, X-ray diffraction, aerosol-assisted Chemical Vapor Deposition
PACS: 61.46.+w, 61.10.Nz, 81.15.Gh

INTRODUCTION

The synthesis of macroscopic nanotube (NT) carpets of several cm^2, composed of mm long nanotubes aligned perpendicularly to the surface of a substrate is a challenge for many applications, from high density magnetic storage [1] to chemical separation [2]. Special interest is focused on Catalytic Chemical Vapor Deposition (CCVD) methods for low cost and large scale production [3]. However, recent experiments [4,5] showed that the structure of carpets could vary from their basis to their top, which is of course a crucial point for future applications. Therefore local analyses are needed in order to characterize in details carpet structure and subsequently to better control the synthesis methods. In this context, we performed X-ray microdiffraction experiments. To our knowledge, it is the first time that X-ray microdiffraction is used for the study of a nanotube-based material.

EXPERIMENTS

Carpets of aligned multi-wall carbon nanotubes (MWNT) were synthesized by CCVD of liquid aerosol obtained from toluene/ferrocene (5 wt. %) solution [6]. The experimental set-up is composed of an aerosol generator, a quartz reactor placed in a

CP786, *Electronic Properties of Novel Nanostructures*, edited by H. Kuzmany, J. Fink, M. Mehring, and S. Roth
© 2005 American Institute of Physics 0-7354-0275-2/05/$22.50

furnace and traps for the gases coming out. Synthesis is performed on thin silicon wafers (covered by a SiO_2 native layer) placed in the reactor heated at 850°C and fed during 15 min with the aerosol carried by argon gas. The quartz reactor is then slowly cooled down to room temperature. Two different experimental set-ups were used [7]. Our initial experimental set-up was composed of a single argon inlet. When stopping the aerosol generation and starting cooling, the argon gas flushes the remaining aerosol to the reactor. Aerosol characteristics (density, ferrocene concentration,...) are no more controlled. The experimental device has been improved by changing the argon distribution through the addition of a second argon inlet : aerosol feeding can be stopped instantaneously by sending a high argon flow through the additional inlet. Carpet 1 (C1) was obtained with the initial set-up and carpet 2 (C2) with the modified one; their thicknesses are about 600μm and 500μm, respectively.

X-ray microdiffraction experiments, with maximum beam size of 2μm x 2μm, were performed using high flux synchrotron radiation at ESRF (beamline ID13, wavelength 0.9755Å). The small beam size allowed us to analyze the carpet from its bottom to its top; fig. 1 inset details the geometry of the experiment. Diffraction patterns are recorded on a two-dimensional CCD detector. They exhibit scattering rings whose positions allow one to determine the structure of the scattering objects and whose modulations characterize preferential orientations if any.

RESULTS AND DISCUSSION

Analysis of diffraction patterns recorded from the bottom of carpets C1 and C2 (next to the wafer) to their top shows the occurrence of an iron oxide phase -magnetite Fe_3O_4 or isostructural maghemite γ-Fe_2O_3- mainly located just above the wafer. This is illustrated in fig. 1 where the intensity of the most intense diffraction peak of magnetite (or maghemite) is plotted as a function of the beam position with respect to the wafer. These microdiffraction results, together with X-Ray Photoelectron Spectroscopy and Transmission Electron Microscopy/Energy Dispersive X-ray results discussed in ref. [7], demonstrate that catalytic particles are iron oxide. Nanotube formation occurs through a base growth mechanism from these particles [6,7]. To our knowledge, only one result has been published in the literature for carpets obtained by CCVD of prevaporised hydrocarbon (xylene) / ferrocene solutions, which reports the occurrence of γ-Fe [8]. Note also that few papers give experimental evidence for the growth of nanotube or filamentous carbon from iron oxide by CCVD of hydrocarbon gas [9-11]. In most cases, catalytic particles are iron or iron carbide. For discussion of our result, beyond the scope of this article, the reader should refer to ref. [7]. Our aim here is to show that microdiffraction is a powerful tool to obtain local information about iron-based phases in NT carpets (catalytic particles, as shown here, but also nanowires inside NTs).

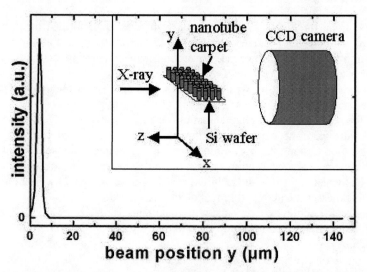

FIGURE 1. Intensity of the (3,1,1) peak of cubic magnetite (or of isostructural maghemite) at 2.5Å^{-1}, for carpet C1, as a function of the beam position y with respect to the silicon wafer (the wafer surface corresponds to y≈0, results shown here are for the 150 μm above the wafer). Inset: the normal to the carpet basis is placed perpendicular to the X-ray beam, its width Δz along the X-ray beam being chosen to be rather small (~0.2mm) to preserve good resolution. The structural information obtained for beam position (x,y) concerns the parallelepiped defined by the beam on the carpet (2μm x 2μm x Δz). Intensity as a function of y is obtained by translating the carpet perpendicularly to the beam at a constant x position.

Information about the nanotube orientation around the normal to the carpet basis is obtained from the angular modulation of the (002) MWNT diffraction peak at 1.83Å^{-1}. As explained in ref. [4], it is characterized by its Half-Width at Half Maximum (HWHM), which is displayed in fig.2. Figure 2 shows that nanotube orientation is roughly the same (+/- 15° with respect to the normal of the carpet base) in the core of carpets C1 and C2, sufficiently far away from the substrate. However, such alignment degree is strongly disturbed over 50 μm above the wafer for carpet C1 (+/- 30° alignment value is measured close to the wafer). The different alignment degrees are related to the synthesis procedures used. Considering that growth occurs from nanotube basis, the low alignment degree at the basis of carpet C1 is attributed to the changes in ferrocene concentration and in aerosol density during the stopping procedure. On the contrary, the alignment degree at the basis of carpet C2 is not affected since the aerosol is suddenly stopped. Such results illustrate the crucial role of synthesis procedure on carpet structure.

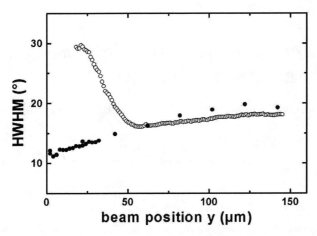

FIGURE 2. HWHM of the (002) NT peak modulation for carpets C1 (open circles) and C2 (filled circles) over the first 150 μm above the silicon wafer (wafer's surface corresponds to y≈0); note that due to absorption effects, data analysis could not be performed very close to the wafer for carpet C1.

ACKNOWLEDGMENTS

This work has been done thanks to collaborations established in the framework of the CNRS GDR n° 1752 'Nanotubes mono- et multi-éléments' and of the GDRE n°2756 'Science and applications of the nanotubes - NANO-E'. FIT2D program provided by A.P. Hammersley and the ESRF was used for the data analysis.

REFERENCES

1. Grobert N., Hsu W.K., Zhu Y.Q., Hare J.P., Kroto H.W., Walton D.R.M., Terrones M., Terrones H., Redlich Ph., Rulhe M., Escudero R., Morales F., *Appl. Phys. Lett.* **75**, 3363-3365 (1999)
2. Hinds B.J., Chopra N., Rantell T., Andrews R., Gavalas V., Bachas L.G., *Science* **303**, 62-65 (2004)
3. Singh C., Shaffer M.S.P., Koziol K.K.K., Kinloch I.A., Windle A.H., *Chem. Phys. Lett.* **372**, 860-865 (2003)
4. Pichot V., Launois P., Pinault M., Mayne-L'Hermite M., Reynaud C., *Appl. Phys. Lett.* **85**, 473-475 (2004)
5. Ruskov T., Asenov S., Spirov I., Garcia C., Mönch I., Graff A., Kozhuharova R., Leonhardt A., Mühl T., Ritschel M., Schneider C.M., Groudeva- Zotova S., *J. Appl. Phys.* **96**, 7514-7518 (2004)
6. Pinault M., Mayne-L'Hermite M., Reynaud C., Beyssac O., Rouzaud J.N., Clinard C., *Diamond and related materials* **13**, 1266-1269 (2004)
7. Pinault M., Mayne-L'Hermite M., Reynaud C., Pichot V., Launois P., Ballutaud D., submitted
8. Jung Y.J., Wei B., Vajtai R., Ajayan P.M., Homma Y., Prabhakaran K. and Ogino T., *Nano Lett.* **3**, 561-564 (2003)
9. Fan S., Chapline M.G., Franklin N.R., Tombler T.W., Cassell A.M., Dai H., *Science* **283**, 512-514 (1999)
10. Baker R.T.K., Alonzo J.R., Dumesic J.A., Yates D.J.C., *J. of Catalysis* **77**, 74-84 (1982)
11. Mauron Ph., Emmenegger C., Züttel A., Nützenadel Ch., Sudan P., Schlapbach L., *Carbon* **40**, 1339-1344 (2002)

Chirality dependence of the high-energy Raman modes in carbon nanotubes

H. Telg*, J. Maultzsch*, S. Reich† and C. Thomsen*

*Institut für Festkörperphysik, Technische Universität Berlin, Hardenbergstr. 36, 10623 Berlin, Germany
†Department of Engineering, University of Cambridge, Cambridge CB2 1PZ, United Kingdom

Abstract. By resonant Raman scattering we found an (n_1, n_2)-assignment of single-walled carbon nanotubes. Based on this assignment we analyzed the high-energy modes as a function of chiral angle. We discuss the differences between spectra from metallic and semiconducting tubes. We show that the high-energy mode can be used as an indicator for metallic tubes even if the tubes are separated and have small diameters.

The separation of different nanotube structures is still an unsolved problem. But for the mass production of electronic devices based on nanotubes the separation is crucial since the electronic properties strongly depend on the nanotube structure [1]. The first approach has been undertaken by separating semiconducting from metallic tubes [2, 3, 4]. A common way of monitoring the ratio of metallic and semiconducting nanotubes in a sample is to perform Raman spectroscopy. Especially the lineshape of the high-energy mode (HEM) is used to distinguish between semiconducting and metallic tubes [5, 6]. The assignment of the different features in the HEM to metallic and semiconducting nanotubes was proposed by Pimenta *et al.* and is based on a simple tight-binding description of the nanotube band structure [5]. For the assignment they performed measurements on bundled tubes. In contrast, most processes which physically separate metallic from semiconducting tubes use debundled or isolated tubes. Results from calculations by Kempa *et al.* predict a decrease of the metallic feature in the Raman spectrum with decreasing bundle size [7, 8]. This gives rise to the question if the high-energy mode can still be used to observe metallic tubes.

In this paper we present a revisited assignment of the features in the high-energy mode to metallic and semiconducting nanotubes. It is based on a comparison of the high-energy mode to the radial breathing modes which were recently assigned to chiral indices (n_1, n_2) [9]. We show that the line shape characteristic for metallic tubes, eventhough it is indeed less pronounced, can still be observed for isolated tubes. A number of small peaks, which appear on the low-energy side of the HEM could be assigned to TO-like phonons of semiconducting tubes.

In Fig. 1(a) we show high-energy Raman spectra of carbon nanotubes excited between 1.93 and 2.14 eV. The nanotubes were produced by the HiPCO method and dispersed in D_2O containing a surfactant (sodium dodecyl sulfate) [10]. We conducted the Raman measurements on a Dilor XY 800 spectrometer in backscattering geometry. For excitation we used a dye laser running with Rhodamin 6G. The spectra show the characteristic features, a relatively sharp peak at 1590 cm^{-1} and a broader peak at ≈ 1560 cm^{-1}

CP786, *Electronic Properties of Novel Nanostructures*, edited by H. Kuzmany, J. Fink, M. Mehring, and S. Roth
© 2005 American Institute of Physics 0-7354-0275-2/05/$22.50

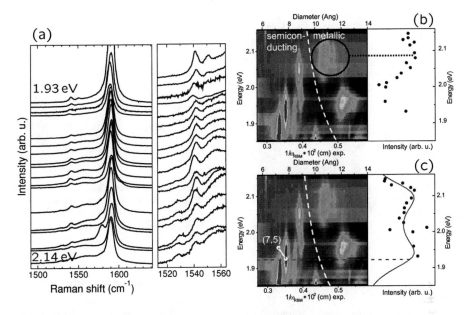

FIGURE 1. (a) Spectra of the high-energy Raman mode. The excitation energy was varied between 1.93 and 2.14 eV. On the right of (a) an enlargement of the small peaks around $1540\,\mathrm{cm}^{-1}$ is shown. (b) and (c) compare the Raman intensities of the RBMs (false-color plot on the left) and the HEM (diagram on the right) as a function of excitation energy. The right of (b) and (c) shows the intensity of the broad peak at $1560\,\mathrm{cm}^{-1}$ and of the small peak at $1541\,\mathrm{cm}^{-1}$, respectively. The solid line in (c) is a resonance profile fitted to the data. The transition energy correspond to the incoming resonance which is close to the maximum intensity at lower energy, marked by the horizontal dashed line.

with a full width at half maximum of 9 and $30\,\mathrm{cm}^{-1}$, respectively. Additionally a number of small peaks appear on the low-energy side of the HEM at 1526, 1541 and $1550\,\mathrm{cm}^{-1}$ [enlarged on the right in Fig. 1 (a)]. Each of these small peaks is only seen for a limited range of excitation energies.

As it was shown recently, each RBM can be assigned to a particular nanotube structure [9]. By comparing the intensities of the HEM peaks with the RBM intensities it is possible to assign the HEM peaks to a particular nanotube chiral index or group of tubes with similar properties, say from the same branch in a Kataura plot [11, 12]. The left part of Fig. 1 (b) and (c) shows a false-color plot of the Raman signal as a function of the excitation energy and the inverse RBM frequency. On the right the intensity of the HEM is plotted versus the excitation energy. The vertical axis is in both the left and right part the energy axis with identical scaling. The intensity of the broad peak at $1560\,\mathrm{cm}^{-1}$ clearly shows a maximum at 2.08 eV. At the same excitation energy a number of RBMs are at maximum highlighted by an ellipse. The assignment from Ref. [9] shows that the RBMs of metallic and semiconducting nanotubes are well separated in the false-color plot [dashed line in Fig. 1 (b) and (c)]. Thus the RBMs inside the ellipse are all related to metallic tubes. Since the maximum of these peaks match the maximum intensity of the

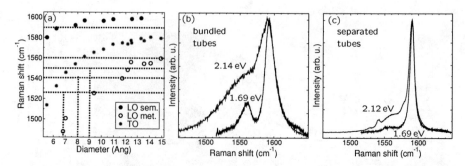

FIGURE 2. (a) Phonon frequencies as a function of tube diameter. LO-like phonons are given by circles. Filled circles correspond to semiconducting, open circles to metallic nanotubes obtained from *ab-initio* calculations from Ref. [13]. Experimental frequencies and the diameter of the assigned tubes are marked by dotted lines. Spectra in (b) and (c) show the different line shape of the HEM with and without metallic nanotubes in resonance, excited at 2.12eV (2.14eV) and 1.69 eV, respectively. In (b) tubes are bundled in bulk material. In (c) tubes are separated in solution.

broad peak, we can assign the broad peak at $\approx 1560\,\mathrm{cm}^{-1}$ to the same group of metallic tubes [(13,1),(12,3),(11,5),(10,7) and (9,9)].

We performed the same comparison with the remaining HEM peaks. In contrast to the broad peak, the intense peak at $1590\,\mathrm{cm}^{-1}$ decreases with increasing excitation energy. This corresponds to a decrease of the number of semiconducting RBMs in resonance with increasing excitation energy. Therfore this peak is definitely associated with semiconducting nanotubes. As mentioned above, the small peaks on the low-energy side of the HEM are only observed for a limited excitation-energy range. Since the width of this range is similar to the energy of the corresponding phonon it can be assumed that each of the small peaks originates from a particular type of tube. In Fig. 1 (b) we show the intensity of the small peak at $1541\,\mathrm{cm}^{-1}$ as a function of excitation energy compared to the RBM intensities. We fit a resonance profile to the data (solid line) and find a transition energy of the corresponding nanotube of 1.92 eV, marked by the dashed horizontal line. The false-color plot shows that this energy matches exactly the semiconducting (7,5) tube. In case of the peaks at 1526 and $1550\,\mathrm{cm}^{-1}$ the energy range of our measurements did not cover the whole resonance profiles. To find the correct assignment we modeled resonance profiles on the basis of the transition energies from Ref. [9]. The experimental resonance profile of the $1526\,\mathrm{cm}^{-1}$ peak is best fit by the profile of the (6,4) tube. The peak at $1550\,\mathrm{cm}^{-1}$ could not be assigned to a particular tube but to a group of tubes. This group, also referred to as branch [9, 11, 12], contains the (11,0), (10,2), (9,4) and (8,6) tube.

Additionally we compared our results with theoretical predictions. Fig. 2 (a) shows the Γ-point frequencies as a function of tube diameter from *ab-initio* calculations by Dubay *et al.* [13]. Stars and circles are the quasi TO and LO phonons, respectively. The LO phonon frequencies strongly differ between semiconducting (filled circles) and metallic (open circles) tubes due to the strong electron-phonon coupling in metallic tubes. The dotted lines mark the peak positions of each peak observed in our measurements. The large peak at $1590\,\mathrm{cm}^{-1}$ can clearly be assigned to the LO phonons of semiconducting tubes. The frequencies of the small peaks on the low-energy side match the calculated TO frequencies for small-diameter tubes, in good agreement with our assignment outlined above. The strong electron-phonon coupling which causes the softening of the phonon frequency also results in a broadening of the Raman peak. Therefore, our assignment of the broad peak at $1560\,\mathrm{cm}^{-1}$ to the metallic tubes of the $(13,1)$ branch fits very well with the calculation in Fig. 2 (a).

In Fig. 2 (b) we show the spectra of bundled HiPCO nanotubes excited at 1.69 eV (lower curve) and 2.12 eV (upper curve). The spectra show the characteristic line shapes for metallic tubes which are in resonance at 2.12 eV and off resonance at 1.69 eV [6, 5]. Figure 2 (c) contains the data from dispersed tubes at similar excitation energies as for the bundled tubes. The spectra still have the characteristic lineshape although the broadening and intensity due to metallic tubes is much weaker. This effect was explained by an electron-plasmon interaction in metallic tubes which gives rise to a strong Breit-Wigner-Fano line only in bundled tubes [7].

In conclusion, we assigned the components of the high-energy Raman mode of carbon nanotubes to particular tubes and groups of tubes with similar properties. The assignment is based on the comparison of the low-energy and the high-energy part of the nanotube Raman spectrum and supported by *ab-initio* calculations [13]. We confirmed independently the assignment of the strong peak at $1590\,\mathrm{cm}^{-1}$ to semiconducting and the broad peak at $1560\,\mathrm{cm}^{-1}$ to metallic nanotubes. Our results show that the line shape of the HEM can be used to test for metallic tubes in a sample even if the tubes are debundled. Several smaller peaks on the low-energy side of the HEM could be assigned to TO-like phonons of semiconducting tubes with small diameters.

We thank F. Hennrich for providing us with the samples. S.R. acknowledges support from Oppenheimer Fund and Newnham College.

REFERENCES

1. S. Reich, C. Thomsen, and J. Maultzsch, *Carbon Nanotubes: Basic Concepts and Physical Properties*, Wiley-VCH, Berlin, 2004.
2. R. Krupke, F. Hennrich, H. v. Löhneysen, and M. M. Kappes, *Science*, **301**, 344 (2003).
3. D. Chattopadhyay, I. Galeska, and F. Papadimitrakopoulos, *J. Am. Chem. Soc.*, **125**, 3370–3375 (2003).
4. M. Zheng, A. Jagota, E. Semke, B. Diner, and R. McLean, *Nature Materials*, **2**, 338–342 (2003).
5. M. A. Pimenta, A. Marucci, S. A. Empedocles, M. G. Bawendi, E. B. Hanlon, A. M. Rao, P. C. Eklund, R. E. Smalley, G. Dresselhaus, and M. S. Dresselhaus, *Phys. Rev. B*, **58**, R16 016 (1998).
6. P. M. Rafailov, H. Jantoljak, and C. Thomsen, *Phys. Rev. B*, **61**, 16 179 (2000).
7. K. Kempa, *Phys. Rev. B*, **66**, 195406 (2002).
8. C. Jiang, K. Kempa, J. Zhao, U. Schlecht, U. Kolb, T. Basché, M. Burghard, and A. Mews, *Phys. Rev. B*, **66**, 161404(R) (2002).
9. H. Telg, J. Maultzsch, S. Reich, F. Hennrich, and C. Thomsen, *Phys. Rev. Lett.*, **93**, 177401 (2004).
10. S. Lebedkin, F. Hennrich, T. Skipa, and M. M. Kappes, *J. Phys. Chem. B*, **107**, 1949 (2003).
11. J. Maultzsch, H. Telg, S. Reich, and C. Thomsen (2005), this volume.
12. J. Maultzsch, H. Telg, S. Reich, and C. Thomsen (2005), to be published.
13. O. Dubay, G. Kresse, and H. Kuzmany, *Phys. Rev. Lett.*, **88**, 235506 (2002).

Raman Scattering as a Probe of the Electronic Structure of Single-Wall Carbon Nanotubes Under High Pressure

S.V. Terekhov[1], E.D. Obraztsova[1], H.D. Hochheimer[2], P.Teredesai[3], J.L. Yarger[3], A.V. Osadchy[1]

[1] Natural Sciences Center of A.M. Prokhorov General Physics Institute, RAS, 119991, Moscow, Russia
[2] Department of Physics, Colorado State University, Fort Collins, CO 80523, USA
[3] Department of Chemistry and Biochemistry, Arizona State University, Tempe, AZ 85287-1604, USA

Abstract. Single-wall carbon nanotubes (SWNTs) under *in-situ* pressures of 0-0.5 GPa have been studied using Raman scattering in a diamond anvil cell (DAC). The pressure-induced changes in electronic structure of SWNT were related to changes observed in resonant Raman spectra. In particular, the radial breathing Raman modes show a strong pressure dependence. At ambient conditions the band at 191 cm^{-1}, corresponding to SWNTs with a diameter of 1.29 nm was a dominant spectral feature, whereas at pressures above 0.3 GPa, the band at 211 cm^{-1}, corresponding to nanotubes with a smaller diameter (1.16 nm), became dominant. The switching of resonance enhancement in selective diameter SWNTs is interpreted as a pressure-induced narrowing of the corresponding electronic gap. Upon decompression these effects are reversed. Our estimation of the band gap narrowing with pressure is in agreement with the direct measurement from optical absorption experiments on SWNT surfactant liquid suspension pressurized in a DAC.

INTRODUCTION

Raman scattering has been used as a common diagnostics of the electronic properties in SWNTs (single-wall carbon nanotubes), primarily due to the photo-selective resonance response for the radial breathing modes (RBM) [1]. The SWNTs exist in ropes, where each nanotube possesses an individual 1D electronic density of states (DOS). It is dominated by a set of van Hove singularities situated symmetrically relatively to the Fermi level (Figure 2) [2]. Optical transitions are allowed between the symmetrical maxima, and the energy separation ("pseudo"-gap) is almost inversely proportional to the tube diameter. Hence, laser Raman spectroscopy provides a diameter size selective resonance excitation when the laser photon energy matches one of the allowed optical transitions for a specific diameter SWNT. The most common way to achieve selective Raman excitation of different diameter carbon nanotubes in the rope is through tuning of the laser wavelength to match the resonance optical gap. However, another way to achieve selectivity is either through pressure or temperature tuning of the electronic structure in SWNTs. In our previous work, elevated temperature was used to narrow the SWNT pseudo-gap and providing a selective excitation of Raman scattering in narrow

CP786, *Electronic Properties of Novel Nanostructures*, edited by H. Kuzmany, J. Fink, M. Mehring, and S. Roth
© 2005 American Institute of Physics 0-7354-0275-2/05/$22.50

diameter tubes [3,4]. The aim of this work was to explore the effect of pressure-tuning in SWNTs using resonant Raman scattering. In this letter, we explore the influence of *hydrostatic pressure* on the electronic structure of bundled SWNTs. The experiments described here are compared to recent experimental (optical absorption) [5] and theoretical [6] data on pressure-induced gap narrowing of individual SWNTs suspended in a surfactant-water solution or in vacuum.

EXPERIMENTAL

SWNT samples have been synthesized by the HiPco (high pressure CO decomposition) method. According to Raman and HRTEM (high resolution transmission electron microscopy) data [7] the material consisted of bundles of nanotubes with diameters between 0.7- 1.4 nm. The raw material has been heated up to 450° C to remove the smallest tubes with diameter in the range of 0.7-1.0 nm [7].

The high-pressure Raman studies of HiPco nanotubes were performed at room temperature in a Merrill-Bassett-type diamond anvil cell (DAC). Potassium bromide was used as a pressure-transmitting medium. The pressure inside the DAC was determined using the ruby fluorescence method. The DAC in-situ high pressure Raman scattering was excited with radiation from an Ar^+-ion laser (λ=514.5 nm, power ~20 mW). The back-scattering geometry was used for the experiment. A computer-controlled Action 300i spectrometer with a liquid nitrogen-cooled CCD detector was used to collect the Raman spectra. The spectral resolution was 1 cm^{-1}.

RESULTS AND DISCUSSION

The Raman scattering from optical and acoustical phonons in SWNT consists of numerous modes over a wide spectral region (100 – 3500 cm^{-1}). In this study, we focus on the spectral range between 150–300 cm^{-1}, which is the typical region for the Raman radial "breathing" modes (RBM). The RBM represents a collective radial vibration of all nanotube atoms perpendicularly to the tube axis. The frequency of this mode depends strongly on the diameter of the SWNT. In a typical Raman spectrum there are several RBM modes corresponding to nanotubes with different diameters. The mode corresponding to the nanotubes having an optical transition energy coinciding with the laser photon energy is resonantly enhanced and becomes the dominant spectral feature. For comparison, we also recorded Raman spectra in the 1200-1600 cm^{-1} region. This corresponds to the tangential Raman modes, where the carbon atoms vibrations are parallel to the tube surface. A pressure-induced shift to higher frequencies was observed for the tangential mode, which is consistent with previous studies [8-10] and can therefore be used as an additional pressure calibrant.

The Raman spectrum of the annealed HiPco SWNT measured at ambient pressure is dominated by RBM with a frequency position at 191 cm^{-1}, corresponding to tubes with a diameter 1.29 nm [11]. Another smaller Raman peak is observed at 211 cm^{-1} and corresponds to smaller nanotubes with a diameter of 1.16 nm. The weaker peak intensity of the 211 cm^{-1} mode is primarily due to the off-resonant nature of the excitation and does not reflect the quantitative distribution of diameters of the SWNTs. Weak peaks at 250 cm^{-1} and 265 cm^{-1} correspond to tubes with diameters of 0.98 nm and 0.92 nm, respectively, and are suppressed due to a negligible number of such nanotubes after annealing of the pristine HiPco material and to non-resonant excitation conditions.

Increasing the pressure to 0.3 GPa and 0.5 GPa leads to the intensity redistribution between RBM peaks. The 211 cm^{-1} mode becomes dominant in the Raman spectrum. Upon decompression, the original ambient condition features of the Raman spectrum is recovered for both the RBM and tangential modes. This confirms the reversible nature of the pressure-induced transformations in the low-pressure regime (<1 GPa). Data to be discussed in an upcoming publication show that increasing pressure up to 3 GPa leads to the reversible shift toward high frequencies and broadening of RMB modes, causing the modes to disappear completely [8-10].

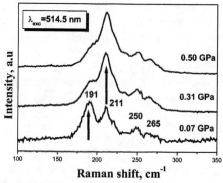

Figure1. Pressure-induced changes of the Raman spectra of HipCO single wall carbon nanotubes in the frequency range of the radial breathing modes.

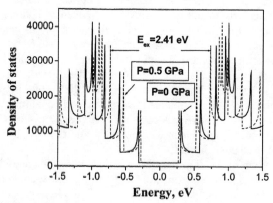

Figure.2 An illustration of the pressure induced change of the density of states (DOS): *full line*- DOS at ambient pressure, *dashed line*- DOS at a pressure of 0.5 GPa

The change of the spectra with pressure, shown in Figure 1, confirms indirectly the narrowing of the pseudo-gap for the nanotube with diameter 1.16 nm. At ambient conditions the laser photon energy (2.41eV) was too small to excite selectively the RBM for this nanotube (Figure 2). The photon energy matches only a gap of the bigger nanotube (with diameter of 1.29 nm). However, as pressure is increased the gaps narrow for all tubes in the rope, including the tube with diameter 1.16 nm. Hence, the pressure can be used to tune the optical energy gap. A similar effect of resonance switching, or tuning, from bigger to smaller nanotubes was observed using temperature [3,4] or through tuning of the excitation laser wavelength [7]. In the latter case, to achieve the same RBM resonance spectrum, as shown in Figure 1, the excitation laser energy had to be tuned from 2.41 eV to 2.47 eV. Therefore, we can consider this energy difference (0.06 eV) as a rough estimation of the gap narrowing under the pressure of 0.5 GPa.

CONCLUSION

A pressure-induced narrowing of the gap in HiPco single-wall carbon nanotubes has been observed indirectly - through a change of the resonance conditions for a selective excitation of the radial breathing Raman modes, corresponding to two tubes of close diameters. At ambient conditions the band at 191 cm^{-1}, corresponding to tubes with a diameter of 1.29 nm was a dominant spectral feature, whereas at pressures above 0.3 GPa, the band at 211 cm^{-1}, corresponding to nanotubes with a smaller diameter (1.16 nm), became dominant. This effect is due to a pressure-induced narrowing of the optical gaps for all nanotubes. A comparison of the change of the Raman spectra with pressure and the one obtained by tuning the excitation laser energy provided a rough estimate of the gap narrowing when high pressure is applied.

ACKNOWLEDGEMENT

The work is supported by *the NATO Collaborative Linkage Grant SA(PST.CLG.978929)6993/FP*, and by *RFBR project 04-02-17618*. J.L. Yarger would like to acknowledge support from *the US National Science Foundation* and *the Department of Defense Army Research Office*.

REFERENCES

1. A. M. Rao, E. Richter, S. Bandow et. al., *Science, 275, 187 (1997)*.
2. J. W. Mintmire, C. T. White, *Phys Rev. Lett. 81 (1998) 2506*.
3. E.D. Obraztsova, J.-M. Bonard, V.L. Kuznetsov et al., *Nanostructured Materials 12 (1999) 567*.
4. S.N. Bokova, V.I. Konov, E.D. Obraztsova et al., *Quantum Electronics 33 (2003) 645*.
5. J.Wu, W. Walukiewicz vet al., *Phys. Rev. Lett. 93 (2004) 017404*.
6. S.G.Louie, *Book of abstracts of XIX IWEPNM, 2005, p. 15*.
7. E.D. Obraztsova, S.N. Bokova, V.L. Kuznetsov et al., *AIP Proceedings v. 685 (2003) 215*.
8. U. D.Venkateswaran, *et al. Phys. Stat. Sol. B 235 (2003)364*.
9. E.D. Obraztsova, H.Th. Lotz, J.A. Schouten et al., *AIP Proceedings v.486 (1999) 333*.
10. P.V. Teredesai, A.K. Sood, D.V.S. Muthu et al., *Chem. Phys. Lett. 319 (2000) 296*.
11. H. Kuzmany, W. Plank, H. Hulman, Ch. Kramberger et al., *Eur. Phys. J. B 22 (2000) 307*.

Integrated Study of Ion Irradiated Singlewall and Multiwall Carbon Nanotubes by Spectroscopic Methods

Mariya M. Brzhezinskaya [*], Eugen M. Baitinger [*], Vladimir V. Shnitov [†], and Aleksey B. Smirnov [†]

[*] *Department of Physics, Chelyabinsk State Pedagogical University, Chelyabinsk 454080, Russia*
[†] *Ioffe Physico-Technical Institute of the Russian Academy of Sciences, St. Petersburg 194021, Russia*

Abstract. The results of experimental study of SWNTs and MWNTs are presented. They were obtained by XPS, REELS and AES. The samples of SWNTs and MWNTs were periodically irradiated by argon ions (Ar^+). The Ar^+ energy was 1 keV. The maximum dose (Q) of Ar^+ irradiation was 360 $\mu C/cm^2$. The process of Ar absorption by SWNTs and MWNTs reveals nonlinear character. The dependence of the $\pi+\sigma$-plasmon energy and the full width at the half of the maximum on Q were determined. Possible causes of the observed effects are discussed. The microscopic model is proposed.

INTRODUCTION

Carbon nanotubes (CNTs) are interesting and promising objects for research and nanotechnology. Unique properties of CNTs make it possible to produce functional elements for nanoelectronics (for example, diodes, transistors) out of CNT fragments [1] and to fabricate nanolasers.

Ion irradiation is relatively simple technique, which provides good opportunity to controllably modify CNT structure and to change CNT physical properties [2]. The irradiation by moderate-energy (0.5 – 10 keV) ions is a useful method for nanoelectronics because it can weld adjacent CNTs due to formation of covalent bonds between driven-out carbon atoms and CNT walls [3].

Nondestructive methods of rapid structure and properties of the CNTs during irradiation and after it are very desirable. The study of collective longitudinal oscillation of electrons (plasmons) is such effective method which gives a great opportunity to monitor important details of nanoscale processes that take place during ion (electron, photon, etc.) irradiation of CNTs.

In this paper the effect of dependence of plasmon spectra on CNT properties and specific experimental parameters was used to study interaction of Ar ions with single-walled (SWNTs) and multi-walled carbon nanotubes (MWNTs). The primary steps of destruction of SWNTs and MWNTs under Ar ion irradiation were studied by analyzing of π- and $\pi+\sigma$-plasmon behavior. The Ar^+ energy was chosen 1 keV. The results were obtained using reflection electron energy loss spectroscopy (REELS) and X-ray photoelectron spectroscopy (XPS). The energy of the primary electrons was 1 keV. Auger electron spectroscopy AES) was used to control condition of CNT surface and to determine the concentration of argon absorbed in the near surface area of the sample.

CP786, *Electronic Properties of Novel Nanostructures*, edited by H. Kuzmany, J. Fink, M. Mehring, and S. Roth
© 2005 American Institute of Physics 0-7354-0275-2/05/$22.50

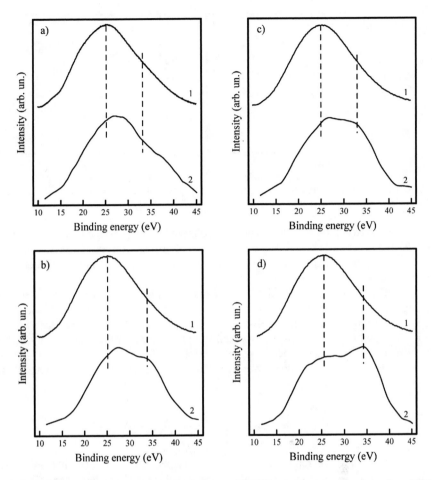

FIGURE 1. The section of original EEL and XPS spectra of SWNT sample. $\pi+\sigma$-plasmon section of EEL (curve 1) and XPS (curve 2) spectra of SWNTs after the different Ar^+ irradiation doses Q: (a) $Q = 0$, (b) $Q = 90 \ \mu C/cm^2$, (c) $Q = 126 \ \mu C/cm^2$, (d) $Q = 360 \ \mu C/cm^2$.

EXPERIMENT

The samples of both SWNTs and MWNTs were used in these experiments. MWNTs were prepared by the arc-discharge evaporation and produced by the company «Astrin», Russia. MWNTs were ~ 10 nm in diameter and ~ 10 μm in length. SWNTs were prepared by the electric-arc-discharge synthesis [4]. The SWNTs were 1.2 - 1.6 nm in diameter and 1 - 10 μm in length.

The samples under test were periodically irradiated by flux of Ar^+ with a current density of $j = 0.75 \ \mu C/cm^2 \cdot s$ and energy of 1 keV. The total irradiation dose Q was determined as $Q = j \cdot t$, where t was the irradiation time. Exposure to Ar^+ was carried out in ultra-high vacuum (UHV) better than 10^{-9} Torr.

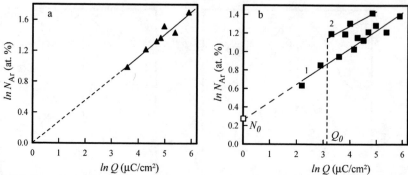

FIGURE 2. Dependence of argon concentration on the surface of samples on the dose of ion irradiation Q for MWNTs (a) and SWNTs (b).

Each irradiation of the sample was followed by Auger spectra measurement ($\Delta E = 0.6$ eV) in order to determine the concentration of argon absorbed in the near surface area of the sample. The primary electron beam energy was 2.5 keV for Auger spectra measurement.

REELS was used: absolute energy resolution ($\Delta E = Const$) was 0.1 - 0.2 eV, the energy of the primary electrons was 1 keV. The measurements were carried out using UHV electron spectrometer PHI-5500, manufactured by Perkin-Ermler (in case of SWNTs), and UHV electron spectrometer with the multichannel energy analyzer (in case of MWNTs). Experimental details were described in [5].

The XPS measurements were carried out using electron spectrometer PHI-5500 using Al Kα line ($\Delta E = 0.2$ eV).

RESULTS AND DISCUSSION

Previously we reported that π-plasmon spectrum of CNTs changed drastically under Ar$^+$ irradiation. However, π-plasmons were observed and studied only in the case of REELS. It was found that π-plasmon energy E_π value decreases and plasma peak broadens with increasing dose of Ar$^+$ irradiation for both MWNTs and SWNTs [5]. It was shown that broadening of π-plasmon peak and simultaneous increase of π-plasmon energy E_π value were interdependent. The same effect was observed for $\pi+\sigma$-plasmon also; however, the changes of spectrum of this plasmon are not so significant. Complex and asymmetrical shape of this plasmon is a possible cause of such less changes. Figure 1 shows the section of original EEL and XPS spectra of SWNT sample. In this figure spectra of $\pi+\sigma$-plasmon from EELS measurements are compared with spectra of $\pi+\sigma$-plasmon from XPS measurements for several doses of Ar$^+$ irradiation. The $\pi+\sigma$-plasmon spectra of CNTs have doublet shape. The vertical dash lines in Figure 1 show positions of two components on which $\pi+\sigma$-plasmon could be decomposed, using Gauss function. Figure 1 illustrates how contribution of these components in the $\pi+\sigma$-plasmon peak changes during ion irradiation. The contribution of the second component increases with the increased dose of Ar$^+$ irradiation. The value of this increase is about 1.2, when the dose of absorbed Ar reached 360 μC/cm^2. The energy of this second component changes from 32.9 eV to 35 eV. Contrary to XPS, this peak was not observed in the EEL spectra

172

(see Figure 1). These values are close to the energy of $\pi+\sigma$-plasmon in diamond which is three-dimensional allotropic modification of carbon [6].

Figure 2 shows the experimental dependence of the concentration of absorbed argon on the dose of ion irradiation Q for SWNTs and MWNTs. Figure 2 illustrates, that dependence of N_{Ar} on Q can be approximated by power function. Therefore, experimental data could be approximated by two straight lines (1 and 2). The line 1 is universal and corresponds to argon absorption at both, small and big doses of irradiation. Extrapolation of line 1 to the zero value $Q = 0$ gives non-zero value of the concentration $N_0 \approx 1.25\%$. Fig. 2 demonstrates that there is the second way of Ar absorption, when $Q > Q_0 \approx 40 - 50$ $\mu C/cm^2$ (line 2). Therefore, N_{Ar} increases faster at the big doses of Ar^+ irradiation $Q > Q_0$.

Influence of ion irradiation on the energy of C $1s$ peak and on its FWHM was determined. The C $1s$ peak energy decreases non-monotone and the peak broadens slightly from 1.4 to 1.8 eV with the increase of the dose Q of Ar^+ irradiation.

CONCLUSIONS

This study has shown complicated and multi-stage process, which takes place in carbon nanotubes during ion irradiation. Data of $\pi+\sigma$-plasmon study by REELS were augmented successfully by XPS data. Possible causes of observed effects are discussed. The difference in electronic selection rules could be one of the causes of distinction in $\pi+\sigma$-plasmon spectra from REELS and XPS measurements.

According microscopic model there are two interrelated ways of interaction of plasmons with defects induced by ion irradiation. First, this is direct resonance scattering of plasmons on the defects when plasmon "characteristic length" is equal to distance between the defects. It is possible that the type of atomic orbital hybridization changes in the places of defect formation. Second, the defects cause deformation of the CNT atomic structure. Consequently, CNT band structure also changes, changing the energies of interband transitions and symmetry of wave functions of valence electrons.

ACKNOWLEDGMENTS

This work was supported by the Russian Ministry of Education under grant No.PD02-1.2-170. We thank Dr. Lobach A.S. (Institute of Problems of Chemical Physics, the Russian Academy of Sciences, Chernogolovka, Russia) for samples of SWNTs.

REFERENCES

1. Avouris, P.; Appenzeller, J; Martel, R.; and Wind, S.J. *Proceedings of the IEEE*. **91**, 1772-1784 (2003).
2. Krasheninnikov, A.V.; Nordlund, K; and Keinonen, J. *Phys. Rev. B*. **65**, 165423 (2002).
3. Klusek, Z.; Datta, S.; Biszewski, P.; and Kowalcsuk, P. *Surf. Sci*. **507**, 577-581 (2002).
4. Lobach, A.S.; Spitsina, N.G.; Terekhov, S.V.; and Obraztsova, E.D. *Physics of the Solid State*. **44**, 475-477 (2002).
5. Brzhezinskaya, M.M.; Baitinger, E.M.; and Shnitov, V.V. *Phisica B*. **348**, 95-100 (2004).
6. Raether, H. *Excitation of plasmons and interband transitions by electrons,* Springer-Verlag, Berlin, 1980, p. 192.

Experimental Symmetry Assignment of the D Band: Evidence from the Raman Spectra of Soluble "Molecular Graphite"

C. Castiglioni[a], M. Tommasini[a], M. Zamboni[a], L. Brambilla[a], G. Zerbi[a], K. Müllen[b]

a. *Dipartimento di Chimica, Materiali e Ingegneria Chimica, Politecnico di Milano e INSTM UdR Milano, Italy*
b. *Max Planck Institut für Polymerforshung, Mainz, Germany*

Abstract. The experimental and predicted (through Density Functional Theory) Raman spectrum of a soluble polycyclic aromatic hydrocarbon (HBC-C12) is presented. The proof of the totally symmetric nature of the normal modes associated to the so-called D band is given.

Keywords: Raman Spectroscopy, Polycyclic Aromatic Hydrocarbons, Graphite
PACS: 33.20.Fb

INTRODUCTION

The vibrational assignment of the so called D band, arising in the Raman spectra of defective graphite, has been matter of scientific debate since a long time [1]. The defect induced resonant activation of phonons near the **K** point in the first BZ of graphite is the basic concept introduced by the solid state approach proposed by Thompsen et al. [2]. On the other hand, a "molecular approach" [3] has been proposed which starts from the interpretation of the Raman spectra of polycyclic aromatic hydrocarbons (PAHs) and allows to reach the following conclusions: (i) the modes involved in the D band are ascribed to collective breathing vibrations of condensed rings and show vibrational displacements strictly correlated with those of the totally symmetric A_1' phonon at **K** (in graphite) [4]; (ii) the very responsible of the strong intensity of the D band (enhanced by resonance) is the fact that any structural defect in graphite (as for instance the presence of edges in microcrystalline graphite) leads to a long range relaxation of the geometrical structure [5].

The above conclusions have their basis on the comparison between the experimental Raman spectra of PAHs and the results obtained from Density Functional Theory calculations of the vibrational spectra and of the equilibrium structures of the molecular models [6].

One of the crucial points of the molecular approach is the validation of the correctness of the vibrational assignment of the D band, as predicted by Quantum Chemistry. In particular, an independent proof of the reliability of the symmetry assignment reached through DFT calculations is highly desirable. Indeed, it would

CP786, *Electronic Properties of Novel Nanostructures*, edited by H. Kuzmany, J. Fink, M. Mehring, and S. Roth
© 2005 American Institute of Physics 0-7354-0275-2/05/$22.50

give a definitive indication about the choice of the right phonon branch of graphite required by the model proposed in ref. [2]. In particular, the assignment to a totally symmetric phonon implies that the phonon branch involved coincides with the one showing a marked Kohn anomaly at the **K** point [7].

In the case of molecules, the experimental symmetry assignment of vibrational normal modes can be straightforwardly inferred by measuring the depolarization ratios of the Raman lines of a sample in solution. In what follows we present a study performed on a soluble PAH molecule (namely hexakisdodecyl-hexaperi-hexabenzocoronene [8], HBC-C12, see the sketch in Figure 1) which allows to prove that the modes associated to the D line are totally symmetric. High level DFT calculations on HBC-C12 will also be presented. These predictions allow to obtain the accurate description of the molecular normal modes and to extrapolate the observed behavior to graphitic domains of any shape and size.

HBC-C12: RAMAN SPECTRUM AND DEPOLARISATION RATIOS

The functionalisation of hexabenzocoronene with six long alkyl chains (chemical formula of each chain ($-CH_2)_{11}CH_3$) makes this molecule soluble in common organic solvents like carbon tetrachloride. A solution of HBC-C12 in CCl_4 has been prepared and FT- Raman spectra have been recorded in back-scattering geometry (excitation wavelength of 1064 nm). The two complementary experiments with polarized light have been carried out, namely parallel/parallel and parallel/perpendicular polarization of the exciting and of the scattered beams (see Figure 1). In this way the experimental depolarization ratios can be determined (see Table 1). The main spectral features observed in Figure 1 can be described as two doublets in the frequency region of the G and the D bands of graphite, respectively. Notice that the presence of more than one transition in the G and D region is due to the coupling of bulk vibrations (whose vibrational displacements are related to graphite phonons) with vibrations involving bending of the peripheral CH bonds [6] and/or the alkyl chains.

From group theory it can be immediately established that transitions associated to non totally symmetric modes must have a fixed depolarization ratio ($\rho = I^{\parallel,\perp} / I^{\parallel,\parallel} = $ ¾) that is they are "un-polarized" bands. On the opposite, totally symmetric vibrations give rise to "polarized" bands with $0 \leq \rho \leq$ ¾ [9]. A look to the experimental ρ values reported in Table 1 shows that the G bands of HBC-C12 can be classified (within the experimental error) as "un-polarized" bands, while the *D bands are "polarized"*. *Therefore the Raman experiments unambiguously prove that the D band is due to totally symmetric vibrations.* It is important to mention that, due to the conformational disorder of the alkyl chains at room temperature, a molecule of HBC-C12 in solution belongs to the point group C_1. As a consequence, all the vibrational modes are, in principle, of A symmetry. However, as proven by our DFT calculations, the polarizability changes associated to the Raman intensities are mainly due to the contribution from the "graphene-like" HBC core and not from the alkyl chains. For this reason the normal modes in the G and D band regions can be adequately classified according to the local D_{6h} point symmetry characteristic of the HBC moiety.

175

Accordingly, Raman active totally symmetric (A_{1g}) and non totally symmetric modes (E_{2g}) are expected to show up in the Raman spectrum with markedly different depolarization ratios, as indeed it happens in the case here studied.

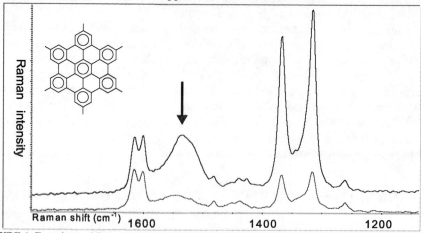

FIGURE 1. Experimental Raman spectra of HBC-C12 in CCl4 solution ($\lambda_{exc} = 1064$ nm). Broken line: parallel parallel polarization. Full line: parallel/perpendicular polarization. The arrow indicate a solvent band.

TABLE 1. Comparison between predicted (DFT-BLYP/6-31G**) and measured Raman parameters of HBC-C12 for the strongest transitions in the D and G band regions.

Spectral region	Experimental frequency [cm^{-1}]	Experimental ρ	Calculated frequency [cm^{-1}]	Calculated intensity [A^4/amu]	Calculated ρ
D band	1316	0.20	1289	2865	0.11
			1298	559	0.26
			1299	3396	0.14
	1368		1336	2343	0.13
G band	1601	0.85	1557	2135	0.75
	1615		1592	350	0.33
			1593	1345	0.72

In Figure 2 and Table 1 the results from DFT simulations are reported. In order to reduce the demand of computational resources while keeping high enough the quality of the calculations, we have considered a model molecule where the six alkyl chains of HBC-C12 have been replaced with shorter -CH_2CH_3 units. The real mass effect of the peripheral chains is obtained with a calculation of vibrational frequencies where the mass of the six -CH_3 units has been artificially increased (*i.e.* the mass of a $(CH_2)_{10}CH_3$ chain has been attributed to each methyl group). The simulation of the Raman spectrum shows a good agreement with the experimental one (see Figure 2). Moreover, the DFT calculation give concordant indication regarding the symmetry assignment, as it can be directly inferred looking at the computed eigenvectors (Figure 2). The fact that the DFT symmetry assignment has been experimentally checked in the case of a soluble PAH is a strong support for the use of DFT calculations for symmetry diagnosis. On this basis we can conclude that the symmetry of the normal vibration associated to the D line is indeed totally symmetric, as previously inferred from Quantum Chemical calculations carried out on many PAHs of increasing size.

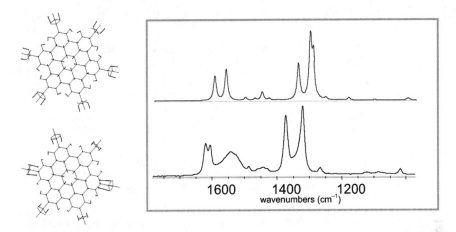

FIGURE 2. Comparison between non polarized experimental (lower spectrum) and DFT Raman spectrum (upper spectrum) of HBC-C12. On the right hand side: eigenvectors associated to the stronger transitions responsible of the D bands.

ACKNOWLEDGMENTS

This work was supported by a grant from MURST (Italy) (FIRB project "Carbon based micro and nano structures", RBNE019NKS).

REFERENCES

1. for a state of the art, see for instance: Ferrari, A.C., and Robertson J.,eds. , *Theme issue on: Raman Spectroscopy in Carbons: from Nanotubes to Diamond, Phil. Trans. R. Soc. Lond. A.,* **362** (2004)
2. Thomsen, C., and Reich S, *Phys. Rev. Letters,* **85**, 5214-5217 (2000); Reich, S., and Thomsen, C., *Phil. Trans. R. Soc. Lond. A.,* **362,** 2271-2288 (2004)
3. Castiglioni, C., Tommasini, M., and Zerbi,, G. *Phil. Trans. R. Soc. Lond. A.,* **362**, 2425-2459 (2004)
4. Castiglioni, C., Negri, F., Rigolio, M., and Zerbi G., *J. Chem. Phys.* **115**, 3769-3778 (2001)
5. Tommasini, M., Di Donato, E., Castiglioni, C., Zerbi, G., Severin, N., Böhme, T., and Rabe, J., Proceedings of IWEPNM2004 "Electronic Properties of Novel Materials", edited by Kuzmany, H., Fink, J., Mehring, M., Roth, S., AIP Conference Proceedings 723, American Institute of Physics (AIP), 2004
6. Negri, F., Castiglioni, C., Tommasini, M., and Zerbi, G. , J. Phys. Chem. A, **106**, 3306-3317 (2002)
7. Piscanec, S., Lazzeri, M., Mauri, F., Ferrari, A.C. and Robertson, J., Phys. Rev. Letters, 93, 185503 (2004)
8. Watson, M.D., Fechtenkotter, A., and Müllen, K., Chem. Rev. 101,1267-1300 (2001); Van de Craats, A.M., Stutzmann, N., Bunk, O., Nielsen, M.M., Watson, M., Mullen, K., Chanzy, H.D., Sirringhaus, H., Friend, R.H., Adv. Mat. **15**, 495-499 (2003).
9. Wilson, E. B., Decius, J.C., and Cross, P.C., Molecular Vibrations: the Theory of Infrared and Raman Vibrational Spectra, New York: Mc.Graw-Hill, 1955

Intermediate Frequency Raman Modes in Metallic and Semiconducting Carbon Nanotubes

C. Fantini*, A. Jorio*, M. Souza*, R. Saito†, Ge. G. Samsonidze**,
M. S. Dresselhaus** and M. A. Pimenta*

*Departamento de Física, Universidade Federal de Minas Gerais, Belo Horizonte, MG,
30123-970, Brazil.
†Department of Physics, Tohoku University and CREST JST, Aoba Sendai 980-8578, Japan.
**Massachusetts Institute of Technology, Cambridge, MA 02139-4307, USA.

Abstract. Resonance Raman spectra of bundled and isolated single-wall carbon nanotubes (SWNTs) have been investigated in the spectral range between 600 and 1100 cm^{-1}, associated with intermediate frequency modes (IFM). The IFMs have been poorly studied before despite the rich set of spectral features and their relation to defects and finite size effects in SWNTs. Different kinds of samples and many laser excitation energies (E_{laser}) have been used in the experiments. The differences in the IFM spectra for different SWNT samples are discussed, and from a line-shape analysis of the spectra we get an (n,m) assignment based on IFM features.

Keywords: Raman, nanotubes, intermediate frequency modes
PACS: 78.30.Na, 78.20.Bh, 78.66.Tr, 63.22.+m, 36.20Kd, 36.20.Ng

INTRODUCTION

Many important results about the electronic and phonon structure of single-wall carbon nanotubes (SWNTs) have been obtained by resonant Raman spectroscopy from analysis of the first-order Raman features; the radial breathing mode ($100–400\,cm^{-1}$) and the G-band ($1500–1600\,cm^{-1}$) [1]. Moreover, some information about defects in the nanotube structure have been obtained from the analysis of the D-band feature ($1200–1400\,cm^{-1}$), which is explained by a double-resonance mechanism [2, 3]. Besides these well studied features, there are also several Raman features with low intensity in the intermediate frequency region (between $600–1100\,cm^{-1}$) that show a dependence on the nanotube diameter and moreover are related to defects and finite-size effects [4, 5] in SWNTs. A rich set of peaks is observed, some of them independent of the excitation laser energy and some exhibiting a step-like dispersive behavior, characterized by peaks at constant frequency increasing and then decreasing in intensity when E_{laser} is changed [6]. The origin of the step-like dispersive IFM peaks was explained based on a second-order Raman process involving optical absorption/emission in different electronic states (E_{ii}), and internal resonance scattering mediated by two phonons, one optic and one acoustic-like. In the present work we report the differences of IFM spectra in different kinds of samples, showing a diameter dependence and associating the observed spectra with metallic or semiconducting SWNTs. From the model proposed to explain the origin of step-like dispersive IFM peaks [6], we report an (n,m) assignment based on the IFM analysis, and this (n,m) assignment is compared with a result for an isolated nanotube.

CP786, Electronic Properties of Novel Nanostructures, edited by H. Kuzmany, J. Fink, M. Mehring, and S. Roth
© 2005 American Institute of Physics 0-7354-0275-2/05/$22.50

EXPERIMENTAL RESULTS

Figure 1 shows the Raman spectra of the intermediate frequency modes (IFM) for four different bundled SWNT samples, using the same laser excitation energy $E_{\text{Laser}} = 2.41$ eV. We can see clearly in this figure that the IFM spectra for each sample are completely different from each other, because the IFM spectra are related to the nanotube diameter and the presence of defects in the sample. In both cases, the presence of a central peak is observed at $\omega_{\text{oTO}} \sim 860$ (845) cm^{-1} for electric-arc and fiber (CoMoCat and HiPco) samples, originating from the oTO phonon mode in graphite. The step-like dispersive effect is observed for IFM peaks below and above ω_{oTO}, with negative and positive dispersions, respectively, when we change the laser excitation energy [6, 7]. In the case of the electric arc and fiber samples, two phonon, with E_3 symmetry, connect the electronic states E_{33}^S and E_{44}^S to each other, and the observed IFM spectra are related to semiconducting nanotubes. For HiPco and CoMoCat samples the observed IFM spectra are related to metallic nanotubes and the two phonon with E_2 symmetry connect the electronic states E_{11}^{M-} and E_{11}^{M+} originating from the splitting on the electronic state E_{11}^M caused by the trigonal warping effect.

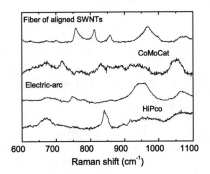

FIGURE 1. Raman spectra of IFM obtained in different kinds of nanotube samples with $E_{\text{laser}} = 2.41$ eV.

The IFM frequencies can be fitted by the relation $\omega_{IFM}^{\pm} = \omega_\mathcal{O} \pm \omega_\mathcal{A}$, where $\omega_\mathcal{O}$ and $\omega_\mathcal{A} = v_\mathcal{A} q$ are the frequencies of the optic and acoustic phonons, respectively. This model can be applied to both semiconducting and metallic nanotubes, but, considering different selection rules for the phonons symmetries. In the case of semiconducting nanotubes, the phonon wavevector is $q = 6/d_t$, because the phonon has E_3 symmetry, and in the case of metallic nanotubes the phonon wavevector is $q = 4/d_t$ because the phonon symmetry is now E_2.

Next, we will carefully examine the line-shape of the IFM features. Figure 2(a) shows the spectrum obtained from the electric arc sample with $E_{\text{laser}} = 2.17$ eV originating from electronic transitions between E_{33}^S and E_{44}^S. In this spectrum we can see that the two well-resolved features at 780 and 810 cm^{-1} exhibit an asymmetric profile on the high energy side. The profiles are asymmetric because these features are not related to

179

only one SWNT, but rather reflect the contribution of different SWNTs, as discussed below.

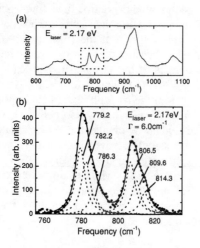

FIGURE 2. (a) Raman spectra of IFMs measured with $E_{laser} = 2.17$ eV. (b) Lorentzian fit of two IFM features observed at $E_{laser} = 2.17$ eV inside the dotted rectangle in (a). The line-widths of all the Lorentzians in (b) are taken to be $\Gamma = 6$ cm^{-1}.

Figure 2(b) shows the fit of two well resolved IFM features, with frequencies at 780 and 810 cm^{-1}, observed for $E_{laser} = 2.17$ eV. Each IFM feature in Fig. 2(b) belongs to a different $(2n + m)$ SWNT family. In each case, the highest intensity Lorentzian comes from the SWNT with the lowest chiral angle, where the electronic states E_{33}^S and E_{44}^S are closest in energy for S1 nanotubes ($2n + m$ mod 3 = 1). The other Lorentzians come from the other SWNTs within the same $(2n + m)$ family, the intensity decreasing as the chiral angle increases. All the Lorentzians used in the fit have the same line-width ($\Gamma = 6.0$ cm^{-1}). From this fit and using the relation $\omega_{IFM}^- = \omega_0 - v_{s4} 6/d_t$, we can obtain the diameter associated with each IFM frequency. The diameters thus obtained are in agreement with the ideal diameter for the SWNTs belonging to the two $(2n + m) = 40$ and 43 families. The analysis of the results are shown in Table 1, where the ideal diameter is calculated by the equation $d_t^{theor} = a_{C-C} \sqrt{3}(n^2 + nm + m^2)^{1/2}/\pi$, where $a_{C-C} = 0.142$ nm is the distance between two carbon atoms.

TABLE 1. IFM frequency values obtained by the Lorentzian fit, the diameter (d_t) obtained from each IFM frequency, and the ideal diameters (d_t^{thero}). The left and right sides correspond to families $2n + m = 40$ and 43, respectively.

(n,m)	ω_{IFM}	d_t	d_t^{theor}	(n,m)	ω_{IFM}	d_t	d_t^{theor}
(20,0)	779.2	1.580	1.585	(21,1)	806.5	1.699	1.705
(19,2)	782.2	1.598	1.591	(20,3)	809.6	1.713	1.716
(18,4)	786.3	1.615	1.609	(19,5)	814.3	1.734	1.738

180

Figure 3 shows a Raman spectrum of an isolated single-wall carbon nanotube, grown across opened slits of a Si_3N_4 membrane, obtained with $E_{laser} = 1.96$ eV. A very intense radial breathing mode peak is observed at 143.5 cm^{-1}. This frequency gives a diameter $d_t = 1.71$ nm, according to the relation between ω_{RBM} and d_t proposed in Ref. [8]. We observe in the IFM spectrum of the isolated SWNT the presence of a intense feature at 874 cm^{-1} and the presence of a low intensity feature at 810 cm^{-1}. By the nanotube diameter and the E_{laser} value, we obtain the (n,m) assignment for this isolated nanotube as a (20,3) and the resonance process here is the transition between $E_3 - E_4$. As we can see in table 1, this nanotube shows a low intensity IFM feature at ~ 810 cm^{-1}, in agreement with the feature observed in the Fig. 3. The intense feature at 874 cm^{-1} may be related to the non-dispersive ω_{oTO} peak, but with higher frequency due the large nanotube diameter. A more detailed investigation of IFM in isolated nanotubes needs be performed for a better understanding about the dependence of IFMs with (n,m) parameters.

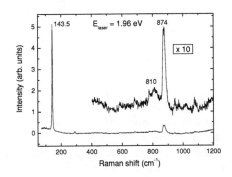

FIGURE 3. Raman spectrum of an isolated SWNT obtained with $E_{laser} = 1.96$ eV where we can see both the RBM and IFM features.

ACKNOWLEDGMENTS

The Brazilian authors acknowledge financial support from the Instituto de Nanociências - CNPq, and CAPES. R.S. acknowledges a Grant-in-Aid (No. 13440091) from the Ministry of Education, Japan. MIT authors acknowledge support under NSF 04-05538.

REFERENCES

1. A. Jorio et al., Phil. Trans. R. Soc. Lond. A, **362**, 2311 (2004).
2. C. Thomsen and S. Reich, Phys. Rev. Lett. **85**, 5214 (2000).
3. R. Saito et al., New Journal of Physics, **5**, 157 (2003).
4. R. Saito et al., Phys. Rev. B **59**, 2388 (1999).
5. A. Rahmani et al., Phys. Rev. B **66**, 125404 (2002).
6. C. Fantini et al., Phys. Rev. Lett. **93** 087401 (2004).
7. C. Fantini et al., (umpublished) (2005).
8. A. jorio et al., Phys. Rev. B (2004) **71**, 075401 (2005).

Raman Spectroelectrochemistry - A Way of Switching the Peierls-like Transition in Metallic Single-walled Carbon Nanotubes

P. M. Rafailov*, J. Maultzsch* and C. Thomsen*

*Institut für Festkörperphysik, Technische Universität Berlin
Hardenbergstr. 36, 10623 Berlin, Germany

Abstract. The high-energy vibrational modes of metallic nanotubes are believed to be softened compared to the semiconducting ones by a Peierls-like transition. We examined the frequency shifts and the intensity of the peaks of the high-energy band in SWNT Raman spectra in dependence on the doping level, as excited with a red laser to enhance the metallic tubes. The metallic modes were indeed found to be exceptionally sensitive to electrochemical doping, exhibiting large frequency shifts and intensity fluctuations. Our data may be interpreted as controlling the Peierls-like instability in metallic tubes with the applied potential.

INTRODUCTION

Single-wall carbon nanotubes (SWNT) possess remarkable properties that are important for both fundamental science and device applications. Plenty of these applications are related to doping of SWNTs as their electronic and mechanical properties are very sensitive to charge transfer. This sensitivity, on the other hand, makes Raman spectroscopy a powerful and favored tool to examine doping-induced phenomena.

Especially, electrochemical doping, *i.e.*, charge induction by varying the double-layer potential at their contact interface with an electrolytic solution[1], allows a fine tuning of the added charge and hence the doping level. This is of key importance for the SWNT application as actuators, because SWNT expansion or contraction can be driven by the electrochemically induced charge, i.e., a fine tuning of the strain can be achieved.

The response of the high-energy SWNT mode (HEM) to doping in an electrolytic solution has been intensively studied[2, 3]. An interesting behavior was found for excitation energies at which metallic SWNTs are resonant: these lie typically in the $1.7 - 2.2$ eV region for nanotubes with diameters $1.1 - 1.5$ nm. In this "metallic resonance window" doping successively restores the HEM shape characteristic for semiconducting SWNTs, causing the broad "metallic" band at the low frequency side of the HEM to vanish[4, 5]. This effect was attributed to a depletion of charge carriers from the valence/conduction bands that participate in the first resonant optical transition in metallic SWNTs[5]. Such an explanation, however, assumes that electrochemical doping yields doping levels comparable to those of intercalation (up to ≈ 0.1 holes/C-atom) which are able to deplete the first "metallic" transition at ≈ 1.9 eV.

The HEM phonon softening in metallic nanotubes was explained recently by a Peierls-like mechanism[6], which should be effective when the valence-conduction band cross-

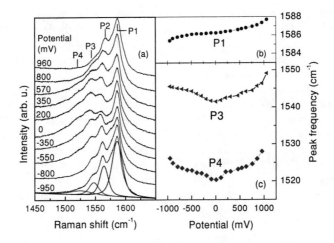

FIGURE 1. *(a):* Raman spectra at 1.95 eV excitation energy of the high-energy mode (HEM) at several potentials applied to the SWNT electrode. *(b):* The frequency of the main HEM peak P1 as a function of the applied potential. *(c):* Same as *(b)* for the peaks P3 (triangles) and P4 (diamonds).

ing occurs at the Fermi wave-vector k_F. This would imply that even a small external influence shifting the crossing point in the hexagonal plane away from k_F, should be able to weaken the softening mechanism. Here we report a resonant Raman investigation of SWNTs exposed to both *p-* and *n-*type electrochemical doping and establish a strong hardening of the HEM phonons of metallic SWNTs upon charge injection of either sign at very low doping levels, which is consistent with a possible removal of a Peierls-like instability[7].

EXPERIMENTAL

A stripe of SWNT "buckypaper" with a nanotube diameter distribution ranging from 1.3 nm to 1.5 nm was prepared as a working electrode in a three-electrode cell. A platinum wire and Ag/AgCl/3 M KCl served as auxiliary and reference electrode, respectively, and the electrolyte used was NH_4Cl.

RESULTS AND DISCUSSION

The Raman spectra for several constant potentials at an excitation energy of 1.95 eV are displayed in Fig. 1(a). All spectra could be fitted well by four peaks: P1, P2, P3 and P4 at 1587, 1560, 1542 and 1520 cm^{-1}, respectively. P3 and P4 consist of bond-stretching modes of metallic SWNTs that are known to be resonant at red excitation.

The fitted peak frequencies are plotted as a function of the applied potential in Figs. 1(b) for P1 and 1(c) for P3 and P4, respectively. Within the interval from -900 mV to 900 mV the frequency of P1 shifts linearly with doping with slightly different slopes for positive and negative voltages of about $1 - 1.5$ cm^{-1}/V. In contrast, P3 and P4 exhibit a strong hardening upon doping of either sign: about 6 cm^{-1} and 8 cm^{-1} in going to -1 V and 1 V, respectively. An important feature of this hardening is its non-linearity and an apparent dramatic loss of intensity with increasing the doping level. Similar effects were already reported in literature[4, 5], the maximum observed shift of the "metallic" modes being even 11 cm^{-1}. Some authors[5] explain the shift and the intensity decrease of these modes assuming that the doping level for applied potentials above 0.6 V becomes sufficiently high for the Fermi level E_F to pass through one of the mirror-imaged van Hove singularity governing the first optical transition of metallic SWNTs at ≈ 1.9 eV. However, E_F shifts of the order 1 eV are only achievable upon heavy doping as, e. g., chemical intercalation provides.

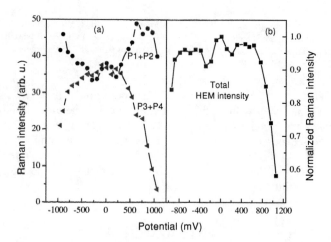

FIGURE 2. *(a)*: Integrated Raman intensities of the peaks P3+P4 of metallic SWNTs (triangles) and P1+P2 (circles) of the high-energy band as a function of the applied potential. *(b)*: Total HEM intensity (P1+P2+P3+P4) as a function of the applied potential.

On the other hand, from both theoretical and experimental studies[8, 9] we estimate a Fermi level shift not higher than 0.6 eV at $|f| = 0.005$. Especially, the first metallic van Hove singularity (≈ 0.9 eV) should not be reached and depleted at $U = 1$ V for most of the metallic SWNTs in our sample. Next we discuss the peak intensities in the high-energy mode. In Fig. 2(a) we plotted the Raman intensity of the "metallic" modes, and the intensity sum of P1 and P2 that comprise what is known as the "semiconducting" HEM shape outside the metallic resonance region, as a function of the applied potential. As is nicely seen from Fig. 2(a), besides the expected diminishing of the P3+P4 intensity, there is actually a gain in the P1+P2 intensity as doping progresses. What appears to be a mere intensity loss of metallic modes can thus be viewed as a redistribution of Raman intensity within the HEM, the modes of metallic SWNTs moving under the envelope of the semiconducting HEM. Combining the behavior of peak frequencies and Raman intensity of the metallic modes, it is important to note that both are strong and non-linear effects that take place upon a relatively small doping ($0 \leq |f| \leq 0.005$). For comparison, the P1 shift, which reflects the mere phonon softening due to doping-induced change in the C-C bond stiffness, is linear and almost 10 times smaller than the shift of the metallic modes. The metallic modes thus behave like in a position of unstable equilibrium. That is why we consider this as an experimental confirmation for the theory of Dubay et al.[6] that the softening of metallic modes is due to a Peierls-like mechanism, which leads to an opening of a temporary band gap with a size oscillating in unison with the displacement pattern of the A_1(LO) modes. The formation of this gap lowers the energy of the filled states thus reducing the energy required to distort a metallic nanotube[6].

Electrochemical doping of either sign shifts the Fermi level away from the valence-conduction band crossing point which should weaken the coupling of the metallic A_1(LO) phonons and restore the frequencies they would have in absence of this coupling.

CONCLUSIONS

We performed a Raman investigation of the high-energy modes of metallic SWNTs upon electrochemical doping. We found these modes to be extremely sensitive to quite low doping levels which is manifested in large frequency shifts and a dramatic intensity redistribution within the high-energy Raman band. We argue that these phenomena occur due to a removal by electrochemical doping of the Peierls-like instability that governs the softening of the high-energy modes in metallic carbon nanotubes. Our results thus support the explanation of the phonon softening in metallic tubes by a Peierls-like mechanism.

ACKNOWLEDGEMENTS

We gratefully acknowledge H. Kataura for providing us with the SWNTs samples. W. Frenzel is acknowledged for his help with the electrochemical measurements. We thank M. Stoll for supplying us with some of the Raman spectra used in this study.

REFERENCES

1. R. H. Baughman et al. *Science* **284**, 1340 (1999).
2. L. Kavan et al., *J. Phys. Chem. B* **105**, 10764 (2001).
3. M. Stoll, P. M. Rafailov, W. Frenzel and C. Thomsen, Chem. Phys. Lett. **375**, 625 (2003).
4. C. P. An et al., *Synthetic Metals* **116**, 411 (2001).
5. P. Corio et al., *Chem. Phys. Lett.* **370**, 675 (2003).
6. O. Dubay, G. Kresse and H. Kuzmany, *Phys. Rev. Lett.* **88**, 235506 (2004).
7. P. M. Rafailov, J. Maultzsch and C. Thomsen, *Phys. Rev. B*, submitted.
8. S. Kazaoui, N. Minami and R. Jacquemin, *Phys. Rev. B* **60**, 13339 (1999).
9. C. Jo, C. Kim, and Y. H. Lee, *Phys. Rev. B* **65**, 035420 (2002).

Raman Scattering And Electronic Properties Of The Cyclic Anthracene Tetramer (Picotube)

Ch. Schaman*, F. Simon*, H. Kuzmany*, D. Ajami#, K. Hess#, R. Herges#, O. Dubay* and G. Kresse*

*) Institut für Materialphysik, Universität Wien, Strudlhofgasse 4, A-1090 Wien, Austria
#) Institut für org. Chemie, Christian-Albrechts-Universität zu Kiel, Otto-Hahn-Platz 3, D-24098 Kiel, Germany

Abstract. We present a study on the electronic and vibrational properties of the cyclic anthracene tetramer. FTRaman and Raman spectroscopy in the visible were used to investigate the vibrational properties of the molecule. Gaussian'98 and Gaussian'03 were employed to calculate the Raman response; the electronic properties were modeled by VASP, resulting in a gap energy of $E_g = 2.24$ eV. A strong correlation of the calculated modes to the measured Raman frequencies provides evidence for the D_{2d} symmetry.

Keywords: Picotube, Raman scattering, VASP.

PACS: 78.30.Na

INTRODUCTION

We present recent spectroscopic results from the cyclic anthracene tetramer (picotube). This compound was first prepared by R. Herges *et al.* [1,2]. It consists of four anthracene units arranged in a tube-like configuration, thus resembling a (4,4) nanotube. However, unlike the usually considered infinitely long tube, this system has molecular character, as the picotube only represents 3 unit cells of the analogous (4,4) tube, with a length of 8.2 A. It was previously not possible to grow tubes with such a small diameter of only 5.4 A. The exception to this is found in the well known lower limit for zeolite grown nanotubes, which are unstable outside the zeolite channels. The system also provides us with a new approach to nanotube production by chemical methods. Here we focus on the vibrational and electronic properties of this novel system.

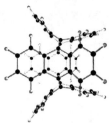

FIGURE 1: The chemical shape of the picotube. The four anthracene units are saturated with hydrogens, and connected by only one double bond to form a tube-like configuration. The image represents the optimized structure with D_{2d}-symmetry obtained by Gaussian'98.

Figure 1 depicts the chemical shape of the picotube. Due to the tilting of the upper and lower anthracene unit the systems symmetry is reduced from D_{4h} to D_{2d}. This is in contrast to NMR- and X-ray diffraction experiments [3], which show the higher symmetry D_{4h}. The discrepancy between the calculation and the experiments originates from a thermal averaging of the tilting of the anthracene planes.

CP786, *Electronic Properties of Novel Nanostructures*, edited by H. Kuzmany, J. Fink, M. Mehring, and S. Roth
© 2005 American Institute of Physics 0-7354-0275-2/05/$22.50

EXPERIMENTAL, VIBRATIONAL ANALYSIS AND CALCULATIONS

FTRaman spectra were recorded from a Bruker FRA106/s IR/FTRaman spectrometer, with an excitation wavelength of 1064 nm in 180° backscattering geometry. Alternatively, Raman spectra were also recorded in the visible spectral region from a Dilor xy triple spectrometer with blue laser excitation. All spectra were recorded in normal resolution mode, which corresponds to a resolution of 3 cm^{-1} for blue lasers. The FTRaman was recorded with a spectral resolution of 2 cm^{-1}. The calculation of the Raman active modes was performed for both geometries by Gaussian'03. To get the accurate line positions Voigtian lines were fit to the spectral response.

A vibrational analysis of the molecule in its D_{2d} symmetry yields a total of representation of the modes as

$$\Gamma^{(tot)} = 34 \cdot A_1 (Raman) + 34 \cdot A_2 (silent) + 34 \cdot B_1 (Raman) + 34 \cdot B_2 (Raman, IR) \quad (1)$$
$$+ 64 \cdot E(Raman, IR)$$

230 of these modes are Raman-active, as indicated.

The analysis of the D_{4h}-symmetric geometry yields

$$\Gamma^{(tot)} = 16 \cdot A_{1u} (silent) + 17 \cdot A_{1g} (Raman) + 15 \cdot A_{2u} (IR) + 16 \cdot A_{2g} (silent) + 16 \cdot B_{1u} (silent) + \quad (2)$$
$$17 \cdot B_{1g} (Raman) + 16 \cdot B_{2u} (silent) + 17 \cdot B_{2g} (Raman) + 33 \cdot E_u (IR) + 31 \cdot E_g (Raman)$$

which would result in a total of 82 Raman-active modes.

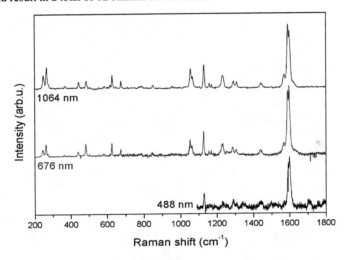

FIGURE 2: Recorded Raman spectra for excitation wavelengths as indicated. (*) supplied by the Bundestechnische Versuchsanstalt Braunschweig.

RESULTS

Raman spectra are depicted in Figure 2 as excited with various lasers. Notice the absence of any dispersive modes, as expected for the molecule. Also, best visible in the FTRaman spectrum if the spectral scale is blown up, the strongest component of the spectrum at 1600 cm^{-1} contains at least 7 Voigtian features, clearly indicating the proposed D_{2d} symmetry.

As we used different excitation wavelengths, we could tune towards the region of resonance which for this molecule would be in the UV spectral region at 319 nm. However, so far measurements from this spectral range did not yield proper results due to sample degradation by the high energy UV light. The spectra excited with blue lasers were rather noisy due to a strong luminescence background. Figure 3 shows the excellent agreement of measurement and calculation if the D_{2d} symmetry is assumed, while the D_{4h} symmetry reproduces the measured spectra to a lesser degree.

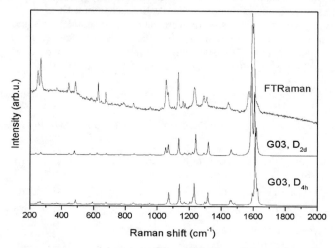

FIGURE 3: Comparison of the measured Raman response and calculations by Gaussian'03 for the D_{2d} and the D_{4h} symmetry. The calculated spectra were downscaled by 2 %.

A scaling factor of 2 % was applied to the calculated spectra in order to match the strongest line to the strongest component in the measured spectrum. The request for downscaling is well known for the Gaussian codes[4]. As far as the frequencies are concerned, excellent agreement is obtained. Interestingly, relative Raman intensities are also well reproduced from the calculation except for a systematic suppresion of intensities towards lower frequencies. Still, the excellent agreement allows us to correlate the individual calculated vibrational modes to the respective features in the measured spectrum.

TABLE 1: Correlation of the 14 strongest Raman modes of the experiment to calculated results. Given are calculated and measured frequencies, the relative intensities and the relative intensity ratios.

Calculated Frequency (cm-1)	I/I_{max}	FTR Frequency (cm-1)	I/I_{max}	$\Delta\nu$ (cm^{-1})	$(I/I_{(max)})_{exp}$ / $(I / I_{(max)})_{calc}$
236.18	0.02	248	0.13	11.8	6.5
267.54	0.02	266	0.25	-1.5	12.5
479.22	0.04	486	0.08	6.8	2
620.34	0.02	626	0.09	5.7	4.5
679.14	0.01	677	0.22	-2.1	22
1068.20	0.12	1066	0.22	-2.2	1.8
1132.88	0.20	1130	0.26	-2.9	1.3
1238.72	0.15	1234	0.09	-4.7	0.6
1291.64	0.03	1291	0.13	-0.6	4.3
1317.12	0.12	1309	0.15	-8.1	1.3
1461.18	0.06	1449	0.02	-12.2	0.3
1586.62	1	1573	0.16	-4.4	0.16
1588.58	0.91	1591	1	-2.4	1.1
1604.26	0.72	1602	0.71	2.3	1

The correlation for the blue excitation was more difficult due to strong luminescence. Not all modes were visible in this region. This is indicated by the ratio of measured versus calculated intensities, after the application of a normalisation to the strongest component of the spectra. Only the modes bigger than 10 % of the strongest component were taken into account.

The electronic states of the picotube were calculated by VASP using PAW[5,6]. Figure 4 presents results. Part (a) depicts the density of states for the D_{2d} symmetry.

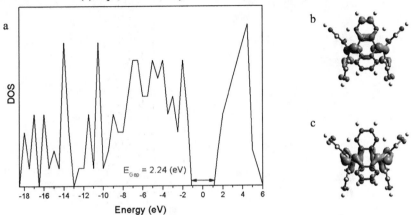

FIGURE 4: The calculated DOS, plotted per 0.5 eV (a); the isosurface for the HOMO state (b), and the isosurface for the LUMO state (c).

It shows several interesting features. The width of the gap is as low as 2.24 eV. As depicted in parts (b) and (c) of the figure, in the HOMO and LUMO states 70 % of the electrons are concentrated to the ring of double bonds connecting the anthracene units. Preliminary results from the calculations indicate that the anthracene units close up if the picotube is inserted into a SWCNT.

CONCLUSIONS

In conclusion, we have presented an experiemental and theoretical study of the electronic and vibrational properties of the picotube. This enabled us to give an assignment of the individual features of the measured spectra to the normal coordinates. Further prospects indicate a prossibility to close the picotube up to form a real piece of a (4,4) armchair tube.

ACKNOWLEDGEMENTS

We gratefully acknowledge the support of the Austrian Foundation of Science, project 17345.

REFERENCES

1. R. Herges, P. G. Jones and S. Kammermeier, *Angew. Chem. Int. Ed. Engl.* **35**, 2669-2671 (1996)
2. R. Herges and S. Kammermeier, *Angew. Chem. Int. Ed. Engl.* **35**, 417-419 (1996)
3. N. Treitel, M. Deichmann, T. Steinfeld *et al.*, *Angew. Chem. Int. Ed.* **42**, 1172-1176 (2003)
4. J. A. Pople, H. B. Schlegel, R. Krishnan *et al.*, Int. J. Quantum Chem. Symp. **15**, 269 (1981)
5. G. Kresse and J. Furthmüller, Phys. Rev. B **54**, 11169 (1996)
6. G. Kresse, and J. Joubert, Phys. Rev. B **59**, 1758 (1999)

CHEMICAL TREATMENT
OF CARBON NANOTUBES

Modulating Single Walled Carbon Nanotube Fluorescence in Response to Specific Molecular Adsorption

Paul W. Barone, Seunghyun Baik, Daniel A. Heller, and Michael S. Strano

Department of Chemical and Biomolecular Engineering
University of Illinois, Urbana - Champaign
Urbana, Illinois 61801 USA

Abstract. We report the synthesis and successful testing of solution phase, near-infrared sensors, using single walled carbon nanotubes that modulate their emission in response to the adsorption of specific biomolecules. Non-covalent functionalization using electron withdrawing molecules are shown to provide sites for transferring electrons in and out of the nanotube, and can be used as chemical mediators in the detection of β-D-glucose. We also demonstrate that two distinct mechanisms, fluorescence quenching and charge transfer, are responsible for the observed fluorescence modulation. The results demonstrate new opportunities for nanoparticle optical sensors that operate in strongly absorbing media of relevance to medicine or biology.

INTRODUCTION

Carbon nanotubes are particularly advantage as sensors because their 1-D electronic structure renders electron transport more sensitive to scattering from adsorbates[1]. Additionally, semiconducting single walled carbon nanotubes fluoresce in a region of the near-infrared (nIR) [2] where human tissue and biological fluids are particularly transparent to their emission[3, 4]. However, few organic molecules that absorb or emit in this region are photostable[5].

Methods of functionalizing the nanotube surface for ligand specificity[6], while simultaneously preserving optical properties and colloidal stability, is a central challenge for nanotube optical sensor development. Covalent functionalization necessarily destroys the delocalized 1-D electronic structure and desired optical properties[7]. Non-covalent modification using electro-active species, although difficult to control[8], provides a means of both preserving the carbon nanotube electronic structure (since no bonds are broken) and providing sites for selective binding.

To date, individual, fluorescent carbon nanotubes have been suspended using charged surfactants, non-ionic polymers[2], and certain DNA sequences[9]. These interfaces necessarily interfere with the adsorption of charged reagents[10] either via columbic interactions, or steric repulsion. We show in this work that it is possible to modify the nanotube coating such that the surface can be functionalized with electroactive species that can then act as chemical mediators. Reaction of the chemical

CP786, *Electronic Properties of Novel Nanostructures*, edited by H. Kuzmany, J. Fink, M. Mehring, and S. Roth
© 2005 American Institute of Physics 0-7354-0275-2/05/$22.50

mediator, in response to specific molecular adsorption, results in the quantifiable modulation of nanotube fluorescence.

EXPERIMENTAL

Single walled carbon nanotubes, produced via the HiPco process, were suspended according to a recently developed protocol[2] except that 2 wt. % sodium cholate surfactant in TRIS (pH 7.4) buffered solution is used. The solution is dialyzed against surfactant-free buffer for 20 hours in the presence of glucose oxidase (1:200 ratio with carbon atoms), replacing the surfactant with an immobilized, porous layer of protein at the surface. Nanotube fluorescence was obtained using 785 nm photodiode laser excitation, with emitted or light being collected 180° through a notch filter onto a thermoelectrically cooled CCD camera.

Protein Assembly on the Nanotube Surface

Charged and non-ionic surfactants that have been studied extensively appear to inhibit surface adsorption events due to electrostatic and entropic barriers, respectively. In this work, we bypass this problem by using stable monolayers of certain proteins or enzymes at nanotube surfaces. Macromolecules can be assembled on the nanotube surface using a dialysis method, with the resulting coating allowing molecular access to the surface while maintaining colloidal stability. We find that van der Waals forces alone can immobilize a wide range of proteins and enzymes as a monolayer on the nanotube surface, including glucose oxidase (GOx), a glucose specific enzyme. Above a threshold surface coverage, the dialysis process renders these systems individually isolated[2] in the absence of the surfactant phase, as evidenced by detection of nanotube fluorescence. Assembly of protein on the nanotube surface causes a fluorescence reduction by a factor of 2.2, which is constant above a starting concentration of 66 mg enzyme/mg SWNT but increases rapidly below this threshold. Additionally, the assembly in the case of glucose oxidase produces a shift in the emission maximum by 10 meV indicating that the tightly packed cholate adsorbed phase has been replaced by a more porous enzyme layer[10].

Fluorescence Attenuation upon Potassium Ferricyanide Adsorption

When single walled carbon nanotubes are suspended with macromolecules as reported above, electroactive species such as potassium ferricyanide, $K_3Fe(CN)_6$, can irreversibly adsorb on the surface and attenuate the nanotube fluorescence by shifting the Fermi levels into the valence bands, or by quenching the emission after photo-excitation. Figure 1 shows the fluorescence decrease as increasing amounts of ferricyanide are added to the buffered (pH 7.4) GOx suspended nanotube solution, with the effect saturating with 83.3% fluorescence attenuation at 225 mM ferricyanide. However, ferrocyanide, the ferricyanide reduction product, attenuates emission of the (6,5) nanotube only 27.4% under identical conditions. We find that this functionalization is irreversible, as removing ferricyanide from the bulk does not result in complete fluorescence restoration.

Figure 1: SWNT fluorescence as a function of analyte concentration. Potassium ferricyanide can be seen to attenuate nanotube emission to a greater degree than the redox partner potassium ferrocyanide. Experiments were carried out at 37 °C and pH 7.4.

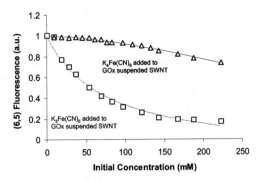

Mechanism of Fluorescence Attenuation

We explore the mechanism of fluorescence attenuation as ferricyanide adsorbs to the surface of the nanotube. This is done using photoabsorption and fluorescence measurements carried out on the same sample. For relatively large diameter species, a change in photo-absorption (Fig 2, right) is matched by a change in emission for the same nanotube upon ferricyanide functionalization (not shown). This indicates a partial electron transfer that localizes electrons and bleaches the optical transition. There is selectivity in the effect, as smaller band gap species show a greater response. However, for 6 of the smallest diameter nanotube species, there is minimal change in the absorption spectrum, while there is a ~ 60% decrease in fluorescence (Fig 2t). For these tubes the attenuation mechanism is one of fluorescence quenching. We postulate that the charged functional group at the nanotube surface behaves as a charge trap to stabilize a mobile exciton and lower the exciton energy below that necessary for radiative recombination.

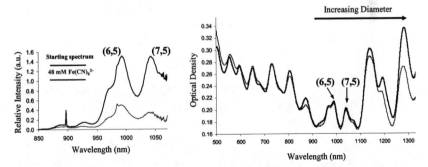

Figure 2: *(left)* Fluorescence spectrum of GOx suspended SWNT before (blue) and after (red) addition of 48 mM $Fe(CN)_6^{3-}$. *(right)* Absorption spectrum of the same sample before and after ferricyanide addition.

195

β-D-Glucose Sensing

The utility of the non-covalent functionalization is that the adsorbed electroactive species can now react selectively with a target analyte to modulate the fluorescence of the nanotube. For example, the $FeCN_6^{-3}$ functionality can be partially reduced by H_2O_2, creating a useful sensing application. Glucose oxidase catalyzes the reaction of β-D-glucose to the d-glucono-1,5-lactone with a H_2O_2 co-product. We show that partial reduction of the $FeCN_6^{-3}$ surface moiety due to the H_2O_2 at 37 °C and pH 7.4 can reversibly couple the carbon nanotube near infrared fluorescence to the glucose concentration in the nanotube analog of a flux-based sensor. The fluorescent emission of the (6,5) nanotube ($\lambda_{max} = 994$ nm) is shown to respond to the local glucose concentration after an 80 s transient (Fig 3, left). The response function maps a type I adsorption isotherm (Fig 3, right) with sensitivity in the general range of blood glucose regulation in diabetic patients (1 to 8 mM), and has a detection limit of 34.7 μM.

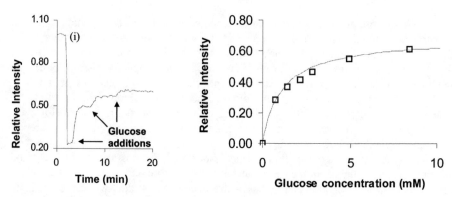

Figure 3: *(left)* Fluorescence response of GOx suspended SWNT after addition of 62.5 mM $Fe(CN)_6^{3-}$ (i) and subsequent glucose additions of 1.4, 2.4, and 4.2 mM. *(right)* The glucose response function follows a Type I Langmuir adsorption isotherm.

ACKNOWLEDGEMENT

This research has been supported by the NSF (CTS-0330350), the School of Chemical Sciences at UIUC, and a grant from the Molecular Electronics group, Dupont Co.

REFERENCES

1.   Saito, R., G. Dresselhaus, and M.S. Dresselhaus, *Physical Properties of Carbon Nanotubes.* 1998, London: Imperial College Press.
2.   O'Connell, M.J., et al., *Band gap fluorescence from individual single-walled carbon nanotubes.* Science, 2002. **297**(5581): p. 593-596.

3.	Kim, S., et al., *Near-infrared fluorescent type II quantum dots for sentinel lymph node mapping.* Nature Biotechnology, 2004. **22**(1): p. 93-97.
4.	McCartney, L.J., et al., *Near-infrared fluorescence lifetime assay for serum glucose based on allophycocyanin-labeled concanavalin A.* Analytical Biochemistry, 2001. **292**(2): p. 216-221.
5.	Frangioni, J.V., *In vivo near-infrared fluorescence imaging.* Current Opinion in Chemical Biology, 2003. **7**(5): p. 626-634.
6.	Chen, R.J., et al., *Noncovalent functionalization of carbon nanotubes for highly specific electronic biosensors.* Proceedings of the National Academy of Sciences of the United States of America, 2003. **100**(9): p. 4984-4989.
7.	Strano, M.S., et al., *Electronic structure control of single-walled carbon nanotube functionalization.* Science, 2003. **301**(5639): p. 1519-1522.
8.	Strano, M.S., et al., *Reversible, band-gap-selective protonation of single-walled carbon nanotubes in solution.* Journal of Physical Chemistry B, 2003. **107**(29): p. 6979-6985.
9.	Zheng, M., et al., *DNA-assisted dispersion and separation of carbon nanotubes.* Nature Materials, 2003. **2**: p. 338-342.
10.	Strano, M.S., et al., *The role of surfactant adsorption during ultrasonication in the dispersion of single-walled carbon nanotubes.* Journal of Nanoscience and Nanotechnology, 2003. **3**(1-2): p. 81-86.

Modification of the Electronic Properties of Carbon Nanotubes: Bundling and B – and N – doping

T. Skipa[a], P.Schweiss[a], F. Hennrich[b], S. Lebedkin[b]

Forschungszentrum Karlsruhe, [a]Institut für Festkörperphysik and Physikalisches Institut, Universität Karlsruhe,[b]Institut für Nanotechnologie, D-76021 Karlsruhe, Germany

Abstract. We report on a substantial modification of the electronic properties of single-walled carbon nanotubes (CNTs) via tube-tube interactions and incorporation of structural defects (boron and nitrogen atoms). Using Raman and photoluminescence spectroscopy, we observed a decrease of the electronic interband transition energies E_{ii} in tube bundles compared to individual tubes of pure CNTs. Weak (off-resonance) Raman signals from $B_xN_yC_z$ nanotubes at relatively high B and N concentrations can be attributed to an increase of the interband energies.

Keywords: Carbon nanotubes, boron and nitrogen-doped nanotubes, Photoluminescence spectroscopy, Raman spectroscopy.
PACS: 73.22.-f

INTRODUCTION

Raman and photoluminescence (PL) spectroscopy can now provide a reliable assignment of structural (n,m) indices of CNTs [1, 2]. However, the established assignment relations are ideally applicable to isolated (individual) tubes in a standard environment (water-surfactant dispersions). Therefore, there is an interest to the factors, which can affect the electronic structure of CNTs and thus interfere with the (n, m) assignment.

The tube-tube interaction has been shown theoretically [3] and experimentally [4] to influence strongly the band gap values E_g. But systematic studies of the intertube interaction are still missing because of the difficulty to control the size of nanotube bundles.

Another factor could be structural and substitutional defects in nanotubes. They might be modelled by incorporation of boron and nitrogen atoms into carbon nanotubes [5]. However, there is still no clear understanding of how the different B and N concentrations influence the electronic structure and vibrational properties of $B_xN_yC_z$ nanotubes.

Using Raman and near-infrared photoluminescence spectroscopy we observe substantial changes in the electronic properties of CNTs due to intertube interaction and B- and N- doping.

CP786, *Electronic Properties of Novel Nanostructures*, edited by H. Kuzmany, J. Fink, M. Mehring, and S. Roth
© 2005 American Institute of Physics 0-7354-0275-2/05/$22.50

EXPERIMENTAL

Materials and Methods

SWCNT material used in this study was obtained by laser vaporisation of graphite: Co:Ni targets. $B_xN_yC_z$ – CNTs were synthesized by the substitution reaction between SWCNTs and boric acid H_3BO_3 in NH_3 atmosphere. For the study of bundling effects we dispersed SWCNT material containing the large bundles of carbon nanotubes in D_2O/ sodium cholate by using an ultrasonic dispergator. In the next step, the dispersion was centrifuged at 100.000g in order to remove large bundles and heavy metal catalyst particles. On-line Raman measurements of sonicated dispersions were performed with a process Raman spectrometer from Kaiser Optical Systems (785 nm excitation). Micro-Raman spectra of $B_xN_yC_z$ nanotubes were obtained with a X-Y Jobin-Yvon spectrometer (532 nm excitation). Photoluminescence emission spectra were measured in the range of ~ 900–1600 nm using an InGaAs photodiode array coupled to a near-infrared spectrograph for detection and a CW Ti-Sapphire laser for excitation (λ_{exc} = 708-850 nm). High resolution transmission electron microscopy (HRTEM) analysis of $B_xN_yC_z$ nanotubes was performed on a Philips Tecnai F20 S-Twin TEM with a field emission gun operating at 200 kV.

RESULTS AND DISCUSSION

Bundling Effects in CNTs

Figure 1 shows the radial breathing mode (RBM) region of the Raman spectra of CNT dispersions in D_2O/ sodium cholate. The upper curve is for the starting material (before sonication), containing CNT bundles with different sizes. Here we see a peak from semiconducting tube (~204 cm^{-1}) and two peaks from metallic tubes (161 cm^{-1}) and (170 cm^{-1}).

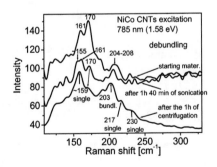

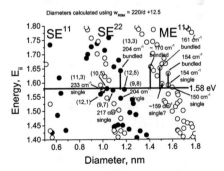

FIGURE 1. RBM pattern changes during the sonication process and after centrifugation of the CNT dispersions in D_2O/ sodium cholate.

FIGURE 2. Electronic interband transitions (E_{ii}) for tubes with diameters in the range of 0.5 - 1.8 nm, open symbols are for the tight binding calculations, closed are for the semiempirical formulas [1].

The electronic interband transitions E_{22} of these individual tubes can be estimated from the RBM frequencie: 1.64, 1.65, and 1.66 eV, respectively, which is far from the resonance with λ_{exc}= 785 nm (1.58 eV) (see Figure 2).

At the same time, we do not observe tubes, having interband transitions close to E_{exc} = 1.58 eV and appearing in the spectra after the sonication/ centrifugation procedure (for example, RBM of metallic tubes $\sim$ 154 and 159 cm^{-1}, and RBM of semiconducting tubes $\sim$ 217, and $\sim$ 233 cm^{-1}). This suggests a decrease of the interband energies E_{22} in bundled nanotubes. As a result, we see in the RBM patterns only contributions of the tubes having E_{22} above the E_{exc} = 1.58 eV.

During the sonication we break the big CNTs bundles to have more and more individual tubes in the solution. Applying centrifugation we get rid of the heavy bundles left to obtain a dispersion containing a large fraction of individual micelle-isolated tubes (the bottom curve in Fig.1). The curves in Fig. 1 represent how the bundled tubes of one (n,m) type leave the resonance and how the individual tubes of another (n,m) type come into the resonance during the debundling process.

Another evidence for the red shifts in energies of CNTs due to bundling came from the PL measurements of dispersed nanotubes (Fig. 3).

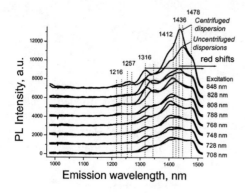

FIGURE 3. PL spectra of CNTs dispersed in D_2O/ NaCholate before and after centrifugation (excitation wavelengths 708 – 850 nm)

PL infrared emission peaks correspond to the first interband transitions E_{11} (band gap energies) of semiconducting CNTs. We see a decrease of E_{11} (red shifts of the emission wavelengths) for bundled tubes compared to individual tubes. In accordance with the Raman results, red shifts were also observed for the PL excitation resonances which correspond to the second interband transitions E_{22} (data not shown here).

B- and N- doping of CNTs

Boron and nitrogen doping was predicted to change the electronic properties of CNTs from metallic to insulating depending on the B and N concentration. Here we present Raman measurements on powders of $B_xN_yC_z$ – NTs. Our samples contain a mixture of the $B_xN_yC_z$ – NTs with different elemental compositions and a wide distribution of tube diameters (Fig. 4).

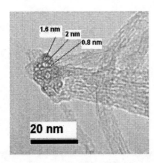 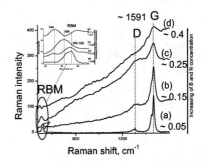

FIGURE 4. $B_xN_yC_z$ –NT bundle produced from as-prepared CNTs at 1200 ^{0}C (mean values for the bundle: $x = 0.12$, $y = 0.04$, $z = 0.84$).

FIGURE 5. Raman spectra from of $B_xN_yC_z$ –NT bundles with different mean B and N concentration $(x + y)$: (a) ~ 0.05; (b) ~ 0.15; (c) ~ 0.25; (d) ~ 0.4; λ_{exc}= 532 nm.

Fig. 5 shows Raman spectra obtained for solid samples having an inhomogeneous doping of the tubes. A mean composition of the tubes inside the areas probed by Raman beam (~ 30μm^2) was studied with energy dispersive X-ray analysis. The mean values of $(x + y)$ were < 0.05, ~ 0.15, ~ 0.25, ~ 0.4 for the curves (a), (b), (c), and (d), respectively. We observe the broadening of the G – band, increasing of the D-band intensity and the RBM frequencies lowering with the increasing of $(x + y)$.

Such a picture is consistent with the theoretical calculations in [6]: the sharp 1D van Hove singularities in the electronic density of states of $B_xN_yC_z$ – NTs are predicted to get more and more structureless and broad with increasing B and N concentrations. We estimate that the resonance Raman effect can be observed up to ~ 0.2 of $(x + y)$. $B_xN_yC_z$ – NTs with higher concentrations of B and N are expected to show much weaker Raman signals with visual excitation wavelength due to a large band gap (> 4 eV for $B_xN_yC_z$ – NTs with $(x + y)$ ~ 0.6 [6]).

ACKNOWLEDGMENTS

The authors gratefully acknowledge R. Fischer (group of Prof. M. Kappes, University of Karlsruhe) for the preparation of SWCNT materials used in this work.

REFERENCES

1. Bachilo S.M., Strano M.S., Kittrell C., Hauge R.H., Smalley R.E., Weisman RB., *Science* **298**, 2361-23662 (2002).
2. Fantini C., Jorio A., Souza M., Strano M.S., Dresselhaus M.S., Pimenta M.A., *PRL* **93**, 147406 (2004).
3. Reich S., Thomsen C., Ordejon P., *PRB* **65**, 155411 (2002).
4. O'Connell M.J., Sivaram S., Doorn K., *PRB* **69**, 235415 (2004).
5. Golberg D, Bando Y, Bourgeois L, Kurashima K, Sato T., *Carbon* **38**, 2017-2027 (2000).
6. Yoshioka T, Suzuura H, Ando T. *Journal of Phys Society of Japan* (available online).

^{7}Li NMR on Li intercalated carbon nanotubes

M. Schmid[1,2,3], C. Goze-Bac[1], M. Mehring[2],
S. Roth[3]

[1]LCVN, Univ. Montpellier II, F-34095 Montpellier cedex5, France
[2]2. Physikal. Inst., Universität Stuttgart, Pfaffenwaldring 57, D-70550 Stuttgart, Germany
[3]Max-Planck-Institut für Festkörperforschung, Heisenbergstr. 1, D-70569 Stuttgart, Germany

Abstract. Solid state ^{7}Li Nuclear Magnetic Resonance (NMR) measurements were performed on lithium intercalated single wall carbon nanotubes (SWNT). The temperature dependence of the static spectra, as well as the spin-lattice relaxation behavior of the intercalated Li nuclei reveal two coexisting kinds of Li intercalation sites. Our results can be interpreted in terms of a staging phenomenon similar to graphite intercalation compounds (GIC).

INTRODUCTION

Since several decades, ^{7}Li NMR on various lithium-carbon solid state compounds has been performed. Starting with graphite-Li compounds, this field of research was extended to many other kinds of carbonaceous Li hosts like disordered carbons, amorphous carbons, porous carbons and finally fullerenes. After the discovery of carbon nanotubes by Iijima in 1991 [1], the investigations were extended to Li intercalated SWNT [2]. The purpose of all these studies is to understand the intercalation chemistry of these carbon materials, as well as to find new materials for potential technical applications like new anode materials for secondary Li-ion batteries [3, 4]. We report here on the intercalation behavior of Li intercalated SWNT using temperature dependent ^{7}Li NMR.

SAMPLE PREPARATION

As carbon host, pristine SWNT grown by the electric arc method were Li intercalated in a solution with aromatic hydrocarbons of a given redox potential and tetrahydrofuran (THF, C_4H_8O). As described in detail in [5], the stoichiometries of LiC_6, LiC_7 and LiC_{10} were obtained using naphthalene, benzophenone and fluorenone, respectively. As this chemical Li intercalation process is similar to Li intercalation in graphite, we assume THF molecules being cointercalated with Li in the carbon nanotube host and a ternary compound with a stoichiometry of $Li(THF)_yC_x$ is expected [6]. A detailed ^{13}C NMR analysis of the Li intercalated SWNT can be found in Ref. [7].

CP786, *Electronic Properties of Novel Nanostructures*, edited by H. Kuzmany, J. Fink, M. Mehring, and S. Roth
© 2005 American Institute of Physics 0-7354-0275-2/05/$22.50

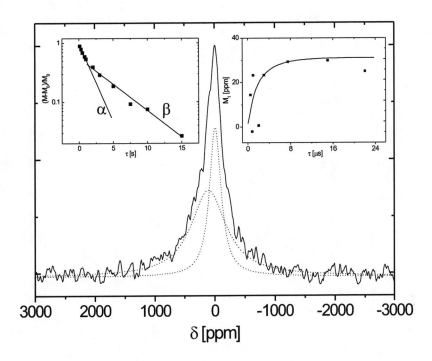

FIGURE 1. Static ^{7}Li NMR spectrum of LiC$_{10}$ at a temperature of 50 K and a magnetic field of 4.2 T. The dashed lines are fits using two Lorentzian lines. Left inset: Magnetization recovery after saturation for LiC$_{10}$ at a temperature of 50 K showing the two magnetization recovery components. Right inset: Evolution of the first moment of the ^{7}Li NMR line during magnetization recovery after saturation at 50 K. The drawn line is a fit assuming 2 Lorentzian lines recovering from saturation under the conditions extracted from a magnetization recovery fit at 50 K.

EXPERIMENTAL

The NMR experiments were performed using a home built pulsed NMR spectrometer working at a magnetic field of 4.2 T and 6.0 T. For ^{7}Li NMR, the Larmor frequency corresponded to 70.2 MHz and 100 MHz, respectively. The ^{7}Li NMR line shifts were referred to 1M LiCl as an external standard. All spectra were recorded using Hahn echo pulse sequence and Fourier transformation. Spin-lattice relaxation rates T_1^{-1} were obtained using a saturation recovery pulse technique and a Hahn echo as a detection pulse.

Fig. 1 shows the static ^{7}Li NMR spectrum of LiC$_{10}$ at a temperature of 50 K. No Li^0 metal was observed which is expected to be paramagnetic shifted up to $\delta_{metal} =$

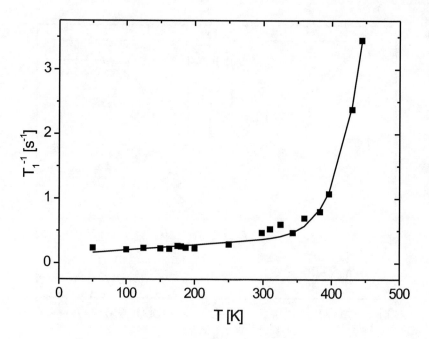

FIGURE 2. ^{7}Li NMR spin-lattice relaxation rate $1/T_1$ vs. temperature at a magnetic field of 6.0 T for LiC$_{10}$.

262 *ppm*. The spectrum in Fig. 1 consists of a Lorentzian line located around 0 ppm and a second Lorentzian line which is paramagnetic shifted to several tens of ppm. This suggests the presence of two inequivalent α and β sites for the *Li* with different environments in the system. The existence of the two different kinds of intercalation sites at 50 K is confirmed by ^{7}Li NMR spin-lattice (T_1) relaxation measurements, as the magnetization recovery after saturation clearly follows a biexponential behavior (left inset Fig. 1). As shown in Fig. 1 (right inset), this biexponential magnetization recovery originates from the evolution of two Lorentzian lines, one centered around 0 ppm and one paramagnetic shifted to around 65 ppm.

Whilst at 50 K, the fraction of the α-line is around 40%, this fraction decreases monotonically with increasing temperature, finally disappearing at 300 K. At the same time, with increasing temperature, the β-line position is diamagnetic shifted, finally appearing at 300 K as a quadrupolar broadened line at the former α-line position around 0 ppm. This reversible process can be directly followed in the temperature dependence of the first moment of the NMR line (not shown here).

The low temperature ^{7}Li NMR α-line position of LiC$_{10}$ centered around $\delta_\alpha \approx 0$ *ppm* suggests the presence of Li_α^+ cations with an ionicity close to +1. The paramagnetic shift of the β-line at $\delta_\beta \approx +65\ ppm$ can be interpreted as a Knight shift by an ionicity < 1, which is typical for a limited electronic charge transfer of the Li $2s$ electron to the SWNT. This partial charge transfer is in agreement with ^{13}C NMR measurements on the same sample [8].

The origin for the two inequivalent α and β sites is unknown, but it is related to the complex structure of bundles of carbon nanotubes which offers different locations and absorption sites for the alkali [9, 10].

Remarkable is the similarity of our [7]Li spectra with spectra obtained from Li graphite intercalation compounds (GIC) [2]. Depending on the stoichiometry, two coexisting lines located around +17 ppm and +45 ppm are observed for stage II and stage III Li-GIC. At higher stoichiometry of LiC_6 (stage I), only a quadrupolar shaped 45 ppm -line is observable. This suggests that the electronic states of our α- and β-Li are comparable to the ones in Li-GIC. The observed temperature dependence of the two lines in the present samples is difficult to model. One scenario is a thermal instability of the low temperature α- and β-Li structural order. An exchange process between the two adsorption sites could cause a rearrangement of the Li ions. Thereby, the high temperature Li configuration is similar to the 0 ppm-type Li in low intercalated Li-GIC. Of particular importance is the fact, that such a temperature dependence of the α and β-line position is only observed in the lowest intercalated sample LiC_{10}. For the higher intercalated samples LiC_7 and LiC_6, spin-lattice relaxation and first moment spectra analysis indicate a temperature independent ratio of the two lines of about 50:50 with their positions being unchanged in temperature. This indicates a much more stable Li configuration in these samples comparable to well defined Li intercalation in GIC.

In order to get further information on the dynamical behavior of the intercalated Li in LiC_{10}, we performed ^{7}Li NMR spin-lattice relaxation measurements by saturation-recovery experiments in the temperature range from 50 K to 450 K. As the α relaxation component disappears around room temperature, we present only the $\beta - T_1^{-1}$ relaxation rate for LiC_{10} in Fig. 2.

Below a temperature of ~ 300 K, the relaxation rate $1/T_1$ follows a linear behavior. Above ~ 350 K, a rapid exponential increase is observed. The linear relaxation regime is typical for hyperfine coupled nuclei to conduction electrons in metals. Therefore we interpret this linear behavior as a hyperfine coupling of the Li nuclei to the conduction electrons. The deviation from Korringa's law at higher temperatures can be explained by a diffusion or motion modulated quadrupolar interaction. A thermally activated diffusion path of the Li_β^+ nuclei is expected along the channels of the carbon nanotube bundles. By fitting our data using a BPP-type expression for motion modulated quadrupolar interaction [11], an activation energy of about $\Delta E \approx 0.4$ eV can be extracted.

SUMMARY

We recorded static ^{7}Li NMR spectra of Li intercalated SWNT with different stoichiometries. We were able to identify two coexisting Li α and β sites, which are chemically stable up to a temperature of 450 K in LiC_6 and LiC_7 and show no significant temperature dependence of their ratio of the two α and β lines. However, in LiC_{10}, with increasing temperature an exchange process between the two adsorption sites seems to cause a rearrangement of the Li ions. For the paramagnetic shifted β-type Li, we found a hyperfine coupling of the nuclei to the carbon nanotube host by spin-lattice relaxation measurements. Above a temperature of ~ 300 K, thermally activated diffusion with an activation energy of $\Delta E \approx 0.4$ eV starts to dominate the T_1 relaxation.

ACKNOWLEDGMENTS

We thank the CANAPE (No. NMP4-CT-2004-500096) European Project for funding and support. Furthermore we thank P. Petit and C. Mathis for providing the Li intercalated samples.

REFERENCES

1. Iijima, S., *Nature*, **354**, 56 (1991).
2. Conard, J., and Lauginie, P., "Lithium NMR in lithium-carbon solid state compounds," in *New Trends in Intercalation Compounds for Energy Storage*, edited by C. Julien, J. Pereira-Ramos, and A. Momchilov, Kluwer Academic Publishers, 2002, pp. 77–93.
3. Endo, M., Kim, C., Nishimura, K., Fujino, T., and Miyashita, K., *Carbon*, **38**, 183–197 (2000).
4. Wu, Y. P., Rahm, E., Holze, and R., *J. Power Sources*, **114**, 228–236 (2003).
5. Petit, P., Jouguelet, E., and Mathis, C., *Chem. Phys. Lett.*, **318**, 561–564 (2000).
6. Facchini, L., Quinton, M., and Legrand, A., *Physica B*, **99**, 525–530 (1980).
7. Schmid, M., Goze-Bac, C., Mehring, M., Roth, S., and Bernier, P., "13C NMR investigations of the metallic state of Li intercalated carbon nanotubes," in *Electronic Properties of Novel Materials – Molecular Nanostructures*, edited by H. Kuzmany, J. Fink, M. Mehring, and S. Roth, American Institute of Physics, Kirchberg, Austria, 2003, vol. 685, pp. 131–134.
8. Schmid, M., Goze-Bac, C., Kraemer, S., Mehring, M., Roth, S., Mathis, C., and Petit, P., *to be published*.
9. Lu, J., Nagase, S., Zhang, S., and Peng, L., *Physical Review B*, **69**, 205304–205308 (2004).
10. Liu, H. J., and Chan, C. T., *Solid State Commun.*, **125**, 77–82 (2003).
11. Mehring, M., and WeberruSS, A., *Object-Oriented Magnetic Resonance*, Academic Press, San Diego, San Francisco, New York, Boston, London, Sydney, Tokyo, 2001.

Facile Solubilization Of Single Walled Carbon Nanotubes Using A Urea Melt Process

Adrian Jung[a], William Ford[b], Ralf Graupner[c], Jurina Wessels[b],
Akio Yasuda[b], Lothar Ley[c] and Andreas Hirsch[a]

(a) Institute of Organic Chemistry, Henkestraße 42, Erlangen, Germany
(b) Materials Science Laboratory, Stuttgart Technology Center,
Sony Deutschland GmbH, Hedelfinger Straße 61, Stuttgart, Germany
(c) Institute of Technical Physics, Erwin-Rommel-Straße 1, Erlangen, Germany

Abstract. In this work we present a simple and effective process to overcome the water insolubility of mildly oxidized Single Walled Carbon Nanotubes (SWCNTs). As this insolubility mainly arises from strong van-der Waals and $\pi-\pi$ stacking interactions of the nanotubes, we applied a urea-melt process [1] to break up the bundles and simultaneously convert the carboxylic functional groups into primary amides. After intense washing and precipitation steps the urea functionalized nanotubes (U-SWCNTs) show a remarkably increased solubility in water and some alcohols which, makes them accessible for further functionalization and applications in these solvents.

Keywords: carbon nanotubes, solubility, urea, melt process

INTRODUCTION

Up to this day, most of the common production processes of Single Walled Carbon Nanotubes (SWCNTs), e.g. arc discharge, laser ablation or the HiPCO process, lead to nanotube material that is more or less contaminated by various amounts of amorphous carbon, fullerenes and the remains of the catalyst metals. In order to make this pristine material usable for modern applications, the SWCNTs have to be purified to remove most or all of the contaminants. Various purification processes have already been invented in the past, mostly relying on the oxidation of the SWCNT material with mineral acids or oxygen at elevated temperatures [2]. Unfortunately these processes are known to also affect the nanotubes by introducing oxidative defect groups [3]. Under harsh oxidation conditions the nanotubes are rapidly shortened which decreases the yield of purified material dramatically. In order to prevent nanotube loss through oxidation, we applied only a mild oxidative treatment followed by melting the oxidized SWCNTs in excess urea. This process leads to purified nanotube material (U-SWCNTs) that is to some degree functionalized by primary amides making the SWCNTs highly soluble in water.

CP786, *Electronic Properties of Novel Nanostructures*, edited by H. Kuzmany, J. Fink, M. Mehring, and S. Roth
© 2005 American Institute of Physics 0-7354-0275-2/05/$22.50

EXPERIMENTAL

Mildly oxidized SWCNTs (Carbon Solutions Inc., acid-purified) were mixed with a 40- fold mass of urea (Fluka) by applying a 30 second ball-milling. The resulting gray powder was then heated to 150°C for 7 minutes (Fig.1).

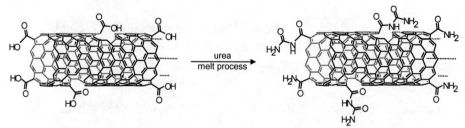

FIGURE 1. Functionalization of oxidized SWCNTs using the urea melt process.

At the melting point of urea (133°C) the gray powder instantly became a black liquid in which even larger SWCNT bundlles were rapidly dissolved. After the reaction mixture was cooled to room temperature, the SWCNTs were extracted with water by subsequent ultrasonication and centrifugation steps. The SWCNTs were precipitated from the aqueous extracts by adding an appropriate amount of 1M $NaClO_4$ followed by centrifugation. The U-SWCNTs were investigated by UV/Vis, Raman, NMR, XPS, AFM and TEM.

RESULTS AND DISCUSSION

UV/Vis measurements of the different aqueous extracts of U-SWCNTs revealed that, due to the centrifugation of the solutions, the first extract contained only amorphous carbon and catalyst remains. The purified SWCNTs were found in the second and third extracts in different amounts (Fig. 2a). Using a method described by Haddon et. al. [4], the nanotube concentration of the second extract was estimated to be 0.43 mg/mL based on the absorption of the solution at 800 nm. The Raman spectra (I_{ex}=514nm) showed dramatic decrease of the D-band and shifts in the RBM region in agreement with the removal of amorphous carbon (Fig. 2b). To prove the existence of the amides, the U-SWCNTs were investigated with proton NMR (Fig. 3a). Multiple peaks could be found in the amidic proton region around 6-9 ppm corresponding to (poly-) amides with different chemical surroundings and/or chain lengths. This result was in good agreement with the XPS measurements (Fig. 3b), which showed an amidic nitrogen content of 4% for U-SWCNTs at 400.2 eV. (The nitrogen content of the pristine SWCNTs can completely be assigned to $-NO_2$ groups at 406.0 eV.)

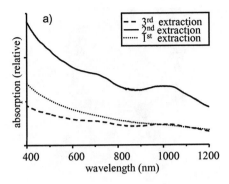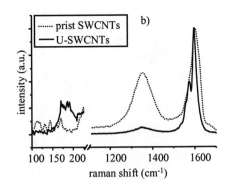

FIGURE 2. (a) UV/Vis spectra of three different aqueous extracts of U-SWCNTs. Only the second and the third extract contain nanotubes. (b) Raman spectra of pristine and urea-functionalized SWCNTs. U-SWCNTs show a dramatic decrease in D-band intensity as well as shifts in the RBM region.

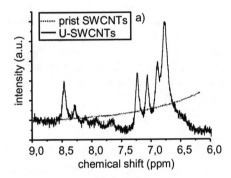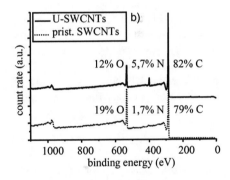

FIGURE 3. (a) Proton NMR of pristine and urea-functionalized SWCNTs showing the peaks resulting from introduced amides. (b) XPS spectra of pristine and urea-functionalized SWCNTs.

For a deeper insight in the modified structure of U-SWCNTs, AFM (not shown) and TEM images were taken of the functionalized material. The AFM images revealed that significant amounts of the nanotubes were somehow connected with ball-like structures that may consist of polyurea. Even though this obviously creates some high molecular weight nanotube networks, the solubility of the U-SWCNTs seems not to be affected too much by this process. The TEM images show that the amorphous carbon and most of the catalyst particles have been removed in U-SWCNTs (Fig. 4, left image). At higher magnification (Fig. 4, right image) it is possible to find some SWCNTs that were covered by urea in a high degree. This is mainly due to the fact that the melting process is to some degree inhomogeneous and therefore difficult to control.

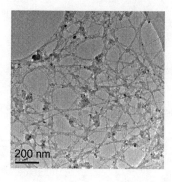 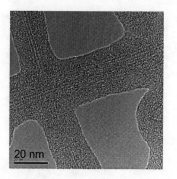

FIGURE 4 TEM images of U-SWCNTs at different magnifications.

CONCLUSIONS

We were able to show that the simple application of the urea-melt process leads to an improved solubility of the SWCNT material in alcohols and especially in water. The functional groups at the nanotube sidewalls consist mainly of primary amides which facilitate the development of hydrogen bonding with the solvent. The urea-melt process can easily be scaled-up to larger amounts of SWCNT-material thus allowing a one step purification / functionalization of the nanotubes.

ACKNOWLEDGMENTS

This work was supported by SONY Deutschland GmbH, Materials Science Laboratory.

REFERENCES

1. W. E. Ford, J. Wessels, A. Yasuda, *Eur. Pat. Appl.* (2004)
2. S. Nagasawa, M. Yudasaka, K. Hirahara, T. Ichihashi, S. Iijima. *Chem. Phys. Lett.*, **328**, 374-380 (2000).
3. M. A. Hamon, H. Hu, P. Bhowmik, S. Niyogi, B. Zhao, M. E. Itkis, R. C. Haddon. *Chem. Phys. Lett.*, **347**, 8-12 (2001).
4. M. E. Itkis, D. E. Perea, S. Niyogi, S. M. Rickard, M. A. Hamon, H. Hu, B. Zhao, and R. C. Haddon. *Nano Lett.*, **3**, 309-314 (2003).

Purification and Functionalisation of Nitrogen-Doped Single-Walled Carbon Nanotubes

Michael Holzinger[a,b], Johannes Steinmetz[c], Siegmar Roth[b], Marianne Glerup[d] and Ralf Graupner[e]

[a]LCVN, Université de Montpellier II, CC26, 34095 Montpellier Cedex 05, France; [b]Max Planck Institute of Solid State Research, 70569 Stuttgart, Germany; [c]Seoul National University, Seoul 151-747, Korea; [d]Department of Chemistry, University of Oslo, 0315 Oslo, Norway; [e]Institute of technical physics, Erwin-Rommel-Straße 1, Germany; Corresponding author; E-mail: holzinger@lcvn.univ-montp2.fr

Abstract. Nitrogen doped single-walled carbon nanotubes has been prepared using the arc discharge method. The material contains in average more impurities like pristine (pure) carbon nanotube material produced with the same method. We have tested several purification methods like the oxidation in hot air. First tests have shown that nitrogen-doped nanotubes exhibit a higher reactivity and are therefore less inert towards oxidative purification methods used for carbon nanotubes. We want to present a new purification method. It is based on the oxidation of the carbon impurities, homogeneously dispersed in boiling tetrachloroethan under airflow. The purity of the sample was investigated using scanning electron microscopy and transmission electron microscopy. Another part will present the characterization of chemically modified nitrogen-doped nanotubes after purification and first steps of the functionalisation of the hetero-nanotube material. Because of already existing model systems with hetero fullerenes, we have chosen a 'Mannich'-like reaction for the modification of nitrogen-doped nanotubes.

INTRODUCTION

Since nitrogen-doped single-walled carbon nanotubes are accessible in larger amounts [1], detailed investigation of this material is possible. The nitrogen in the nanotubes can be seen as regular defects which change the chemical behaviour of the tubes. The reactivity of these N-doped nanotubes is likely to be rather comparable to the reactivity of hetero-fullerenes. Theoretical calculations predict a localization of the unpaired electrons around the nitrogen-defect in the semiconducting hetero nanotubes [2]. In this case, more functionalization methods can be applied for this new type of doped SWCNT material than for pure single-walled carbon nanotubes.

EXPERIMENTAL, RESULTS AND DISCUSSION

Similar to pure carbon nanotubes, the main impurities for pristine nitrogen-doped single-walled nanotube samples are graphite, amorphous carbon, catalyst particles and other carbon species. The SEM and TEM images are representative for pristine, as grown N-doped SWCNT sample. Compared to pristine (pure) single-walled carbon

CP786, *Electronic Properties of Novel Nanostructures*, edited by H. Kuzmany, J. Fink, M. Mehring, and S. Roth
© 2005 American Institute of Physics 0-7354-0275-2/05/$22.50

nanotube samples, the N-doped material has more impurities. The nanotubes are well graphitisized and ordered in bundles.

FIGURE 1. representative SEM and TEM image of an as produced n-doped carbon nanotube sample.

Standard purification methods for carbon nanotubes like the oxidation in air or the oxidation in nitric acid are, in our opinion, too aggressive for nitrogen doped nanotubes. The incorporated nitrogen in the carbon lattice increases the reactivity of the nanotubes and decreases therefore the inertness towards strong oxidation methods. All 'standard' carbon nanotube purification methods have shown that the N-doped nanotubes are destroyed by these procedures.

For the purification of nitrogen doped single walled nanotubes we propose another, smoother oxidation method. N-doped nanotubes disperse very easily in tetrachloroethane (TCE). The dispersions are heated to reflux temperature (150°C) under air flow. Carbon containing impurities should be oxidized and dissolved. The impurities are better accessible in a homogeneous dispersion than as solid. Additionally, tetrachloroethane eliminates HCl when refluxed under airflow. This hydrochloric acid form salts with the catalyst metals.

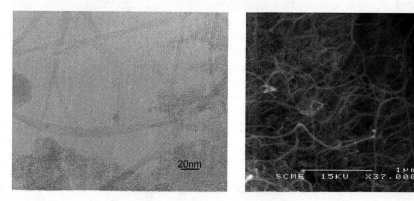

FIGURE 2. TEM and SEM image of a N-doped SWCNT sample after reflux in TCE under air-flow.

After this air-reflux step, the sample was filtered and washed with tetrachloroethane and THF to wash out the oxidized carbon species. A final washing step with water should also eliminate the metal salts. After drying, the nanotube sample was characterised with SEM, TEM, XPS and IR-spectroscopy.

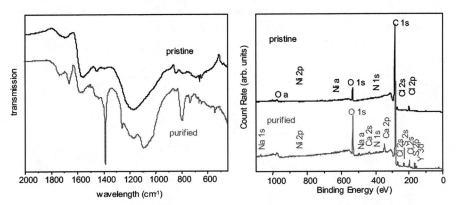

FIGURE 3. IR and XPS spectra of the purified N-doped nanotube samples compared with the pristine sample.

The IR and XPS spectra of the n-doped nanotube samples before and after the oxidation show clear differences. The IR spectrum of purified sample shows features of different organic compounds. The broad band with a minimum at 1084 cm^{-1} can be assigned to carbon-oxygen bonds. The formation of carbon oxides can also be seen in the XPS spectrum. The calculated quantity of atomic percent oxygen increases from 6.3% to 16%.

The results obtained from the purification experiments have shown that nitrogen-doped nanotubes have been chemically modyfied during this procedure. They are therefore very interesting for organic functionalisation.

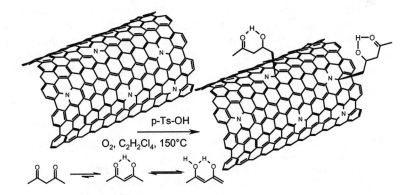

FIGURE 4. Reaction scheme for the functionalisation of N-doped carbon nanotubes during the purification process.

213

The pristine n-doped nanotubes were dispersed in tetrachloroethan together with p-toluene sulphonic acid and acetylaceton. Else, we used the same conditions as we used for the purification method. Under these conditions, the carbon atoms beside the nitrogen atoms in the nanotube lattice are oxidized by oxygen. Acetylaceton adds to these carbo-cations in a 'Mannich'-like reaction [3] and avoids further oxidation.

The IR and XPS spectra show clear changes between the functionalised samples and the pristine material. The formation of carbon-oxygen bonds can also be observed but not as drastic as in the spectra after purification.

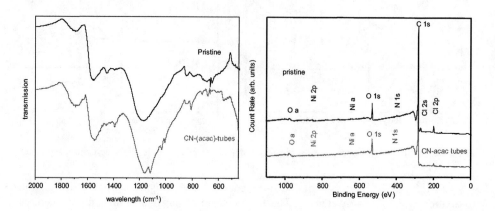

FIGURE 5. IR and XPS spectra of the functionalised N-doped SWCNTs compared with the spectra of the pristine smple.

CONCLUSION

This report represents the first investigation of nitrogen-doped single-walled cabon nanotubes and their chemical modification after purification. Purification methods used for pure carbon nanotubes are not recommended for the N-doped material because of the higher reactivity of this material. Reflux under air-flow in TCE is a effitient method for the purification of hetero-nanotubes and can still be optimised. Functionalisation with acetylacon durig this purification has shown that these attached acetylacetonate groupes serve as protection groups and avoid further oxidation of the N-doped nanotubes.

REFERENCES

[1]   M. Glerup, J. Steinmetz, D. Samaille, O. Stéphane, S. Enouz, A. Loiseau, S. Roth, P. Bernier, *Chem. Phys. Lett.* **2004**, *387*, 193.
[2]   A. H. Nevidomskyy, G. Csányi, M. C. Payne, *Phys. Rev. Lett.* **2003**, *91*, 105502.
[3]   F. Hauke, A. Hirsch, *J. Chem. Soc.; Chem. Commun.* **1999**, *21*, 2199.

Defect Analysis of Carbon Nanotubes

James L. Dewald[1], Jamal Talla[1], Tanja Pietrass[2] and Seamus A. Curran[1]

1 Department of Physics, New Mexico State University, Las Cruces, New Mexico, USA
2 Department of Chemistry, New Mexico Tech, Socorro, New Mexico, USA

Abstract. Raman spectroscopic analysis of plasma treated MWNTs was used characterize defects in MWNTs. Raman spectroscopy was also used to characterize defects induced via acid treatment. The Raman-active disorder modes are used to fingerprint PSF attachment to MWNTs via defect states.

Keywords: Raman, Nanotube, Defects

INTRODUCTION

Functionalization at defect sites using traditional chemical treatments such as wet oxidation in concentrated HNO_3/H_2SO_4 are still in use to create groups such as hydroxyl (-OH), carboxyl (-COOH) and carbonyl (>C=O). These generate significant alterations of the electronic nature of the nanotubes. The significance in changing the graphitic nature of CNT's by introducing defects along the nanotube body has been examined extensively. When we form defects along the nanotube backbone using acid treatment, we introduce dislocations, dangling bonds and other defects. We use the technique of Raman spectroscopy to understand the fundamental change in the nanotubes. Of particular importance, is the significance that outer wall alterations in MWNTs may have in defect formation.

EXPERIMENTAL

Vertically aligned carbon nanotubes were grown on SiO_2/Si substrates which were plasma etched top down at room temperature by a glow discharge chamber (Harrick Scientific) at 0.6 torr pressure. MWNTs synthesized by the arc-discharge method were sonicated in conc. H_2SO_4/HNO_3 (70/30 v/v) in an ultrasonic bath (600 W) for 6 h. Formation of Phenosafranin (3,7-diamino-5-phenyl-phenazine or PSF) functionalized MWNTs was achieved by sonicating 3 mg of carboxylated acid treated nanotubes in 5 ml of de-ionized water containing 0.1% v/v PSF for 5 minutes. Raman spectroscopy of the plasma treated MWNTs was performed using a Renishaw Raman microscope with the 514.5nm line from an Ar^+ laser.

CP786, *Electronic Properties of Novel Nanostructures*, edited by H. Kuzmany, J. Fink, M. Mehring, and S. Roth
© 2005 American Institute of Physics 0-7354-0275-2/05/$22.50

RESULTS AND DISCUSSION

Plasma Treatment

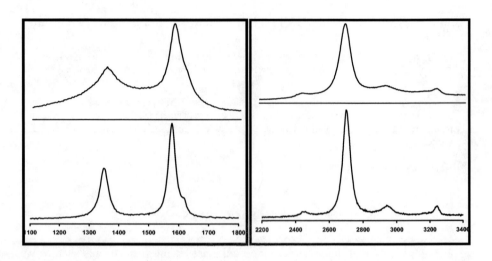

(A)   (B)

FIGURE 1. (A) First order Raman spectra of pristine MWNTs (top) and plasma treated MWNTs (bottom) and (B) Second order Raman spectra of pristine MWNTs (top) and plasma treated MWNTs (bottom).

Figure 1.a shows the first order Raman spectra of pristine MWNT's (top) and plasma treated MWNTs (bottom).[1] The spectrum of the pristine sample shows a very broad disorder induded peak (D band) centered around 1350 cm^{-1}. This band is due to the fact that the CVD grown aligned nanotubes have lattice defects due to the relatively low temperature in which the graphitization takes place. Although this peak is related to different defect structures and finite length effects, its width is related to the presence of amorphous carbon on the nanotubes surface. The plasma treatment has removed the amorphous constituents, as indicated by the reduced line width of the D mode from ~80 cm^{-1} to ~30 cm^{-1}. The plasma treatment has also caused the D band to shift from about 1349 cm^{-1} to 1354 cm^{-1}.

Also noticeable is the strong appearance of the D' mode at 1617 cm^{-1} induced by plasma treatment. This defect-induced state at 1617 cm^{-1} can be linked to the high density of phonon states in the density of states (DOS) of MWNT's and also in graphite.[2] This would indicate that plasma treatment has caused defects along the tube body, thus altering the DOS. This, in turn, indicates that the perfect 2D graphitization has been disordered, inducing strain in the C=C bond vibrations.

Figure 1.b shows the second order Raman spectra of pristine MWNT's (top) and plasma treated MWNTs (bottom).[1] The second order disorder mode, or D*, (at

2700 cm^{-1})[3] has increased in intensity. Of course this would be expected, since it is known that this mode is susceptible to increase in defect states[4], although it is not as susceptible as the first order mode. The E_{2g}+D peak (~2940 cm-1) has also narrowed, shifted, and become more intense upon plasma etching—which is to be expected given the changes seen in the first-order D mode. The mode at 3240 cm^{-1}, which is the second order of the defect state at 1617 cm^{-1}, has also increased upon plasma treatment, similar to the first-order D' peak. The mode at 2450 cm^{-1} has also been enhanced, and is linked to a disorder mode at 1230 cm^{-1} (2nd order).

Acid Treatment

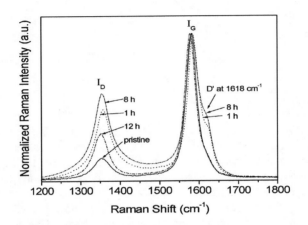

FIGURE 2. Raman spectra of 1200-1800 cm-1 region, showing the D and G-band for pristine, 1 h, 8 h, and 12 h acid treated MWNTs.

Figure 2 shows the normalized Raman scattering spectra of the pristine and 1 h, 8 h and 12 h acid digestions.[5] The D-band or defect mode appears at 1354 cm^{-1} and increases in intensity with increasing acid treatement. There is the appearance of a strong D' band, centered at 1618 cm^{-1}, after 1 hour of acid treatment. This is accompanied by an increase in the D peak. However, after 8 hours we see the strongest response to acid treatment and the D' peak is at a maximum, but the disorder created is simply by having multiple carboxyl groups on the surface. This mode has been reported elsewhere to be the defect-induced peak and has been assigned to the high density of phonon states in the density of states (DOS) of MWNTs, and to a lesser degree in graphite.[1]

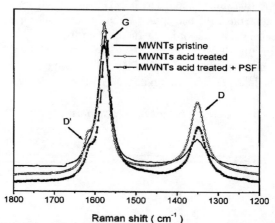

FIGURE 3. Raman spectra of – – pristine MWNTs, -o- MWNTs after acid treatment, and -□-MWNT after acid treatment and PSF dye.

Figure 3 shows the Raman spectra of pristine MWNTs, MWNTs after acid treatment, and MWNTs after acid treatment and PSF dye attachment[6]. The D-mode is the disorder induced mode, the G mode is representative of the E2g mode and the D' mode is another disorder induced mode, shown to be directly affected by defects on the nanotube body. Introducing defects along the nanotube body in a controlled manner using acid treatment, and using Raman to show the vibrational response allows us to understand changes along the nanotube body. Dye addition results in a decrease in the D-mode (1350 cm-1) intensity with a corresponding increase in the D' mode (1620 cm-1) corresponding to the PSF dye attaching to the nanotubes body. There is a shift in the G-mode (1580 cm-1, E_{2g}) by 5 cm^{-1}. As a consequence of 'removing' the defects from the MWNTs using PSF attachment, we introduce further defects along the MWNT body. This is seen as disorder as we no longer have a perfectly graphitized structure on the surface of the MWNTs, but one possessing dye attachments at positions of tubular defects.

CONCLUSION

In summary, we have introduced defects into MWNT's and looked at the corresponding changes that occurred to the different signature vibrational modes. Introducing defects, whether by plasma treatment or by acid treatment induces a significant change in the D' mode. When the nanotubes are treated with dye molecules we see that this mode correspondingly goes down.

ACKNOWLEDGEMENTS

This research is based upon work supported by the National Science Foundation under Grant No. 0132632

REFERENCES

1. Chakrapani, N., Curran, S., Wei, B. Q., Ajayan, P. M., Carrillo, A., and Kane, K., *J. Mater. Res.* **18**, 2515-2521 (2003).
2. Nemanich, R. J., and Solin, S. A., *Phys. Rev. B* **20,** 392-401 (1979).
3. Kaster, J., Pichler, T., Kuzmany, H., Curran, S., Blau, W., Weldon, D. N., Delamesiere, M., Draper, S. and Zandbergen, H., *Chem. Phys. Lett.* **221,** 53 (1994).
4. Jishi, R. A., and Dresselhaus, G., Phys. Rev. B 26, 4514-4522 (1982).
5. Pietraß, T., Dewald, J. L., Clewett, C. F. M., Tierney, D., Ellis, A. V., Dias, S. ,Alvarado, A., Sandoval, L., Tai, S., and Curran, S. A., submitted to *J. Nanoscience and Nanotechnology* (2005).
6. Curran, S. A., Ellis, A. V., Vijayaraghavan, A., and Ajayan, P.M., *J. Chem. Phys.* **20**, 4886-4889 (2004).

Tethering Carbon Nanotubes

Madhuvanthi A. Kandadai[1,2], Donghui Zhang[1], James Dewald[1], and Aditya Avadhanula[1,2] and Seamus A. Curran[1]

1. New Mexico State University, Department of Physics, PO Box 30001, Las Cruces, NM, USA.
2. New Mexico State University, Department of Chemistry and Biochemistry, PO Box 30001, Las Cruces, NM, USA.

Abstract. Using Phosphorous Pentasulphide a new single step electrophilic route has been achieved to link Multi-Walled Carbon Nanotubes covalently through a dithioester linkage. Raman spectra of the tethered tubes was collected using the In Via Renishaw Raman Spectrometer. FESEM images and AFM height analysis were also obtained for the linked tubes.

INTRODUCTION

Recent advancement in the direct functionalization of carbon nanotubes has been achieved through the reliance on chemical treatments such as wet oxidation in concentrated H2SO4/HNO3 [1,2] to produce hydroxyl (--OH) ,carboxyl (--COOH) and carbonyl (>C=O) groups [3,4]. These groups can then be readily derivatised. One such functional group of great interest is thiols, which have important applications in biosensors whereby the biological element can be integrated with an optical or electrochemical transducer[5]. Alkanethiol terminated single-walled carbon nanotubes (SWCNTs) have been created by Liu *et al.*[6] through the conversion of carboxyl groups into acid chlorides with subsequent amide linking with aminoalkanethiols. More recently carboxyl groups on SWCNTs have been derivatized by thionyl chloride to produce thiols on the ends of SWCNTs [7]. Although, nanotubes can now be successfully functionalized, direct covalent attachment of nanotube to nanotube to form"superstructures" has until now been unsuccessful. Here we report, for the first time, a unique approach to direct thiation of acid treated oxidized carbon nanotubes to dithioester functionalized nanotubes using a mild versatile catalyst, phosphorus pentasulfide (P4S10). Phosphorus pentasulfide is well known to provide an electrophilic single-step pathway for preparation of thiocarboxylic esters and dithioesters when carboxyl and carbonyl groups are reacted with thiols or alcohols in varying mole ratios.[8,9]

EXPERIMENTAL

Purified multi-walled carbon nanotubes (MWCNTs), synthesized by the arc-discharge method, were ultra-sonicated in concentrated H2SO4/HNO3. The resulting suspension was filtered under vacuum and thoroughly washed with deionized water. Synthesis of thiolated MWCNTs was achieved by adding dried acid treated MWCNTs to anhydrous toluene and phosphorus pentasulfide. The solution was refluxed at 140 °C and then allowed to cool before being filtered under vacuum. For NSOM imaging, dry samples of the thiol treated nanotubes were placed on a cover slip, which was then placed on the

CP786, *Electronic Properties of Novel Nanostructures*, edited by H. Kuzmany, J. Fink, M. Mehring, and S. Roth
© 2005 American Institute of Physics 0-7354-0275-2/05/$22.50

scanning stage of a Veeco Aurora3 Near-field Scanning Optical Microscope (NSOM). This system is capable of acquiring both topgraphic (Atomic Force Microscopy, AFM) and near-field optical images, but for this experiment we simply used the AFM option which operates in a non-contact mode. Normalized Raman spectra were obtained using a Renishaw InVia Raman spectrometer equipped with a Raman Leica RE02 microscope. 514.7-nm radiation was produced from a 20-mW air-cooled Ar+ Laser-Physics laser, operating at laser powers from 1-100 %. The Raman band of silicon (520 nm) was used to calibrate the spectrometer, with a resolution better than 1.5 cm-1.

RESULTS AND DISCUSSION

Covalent linkage of Acid Treated Multi Walled Carbon Nanotubes (MWCNTs) via a di-thio-ester linkage is taking place as shown in fig.1. The source of the di-thio-ester linkage is Phosphorous Pentasulphide, used as a reagent and catalyst for the reaction.

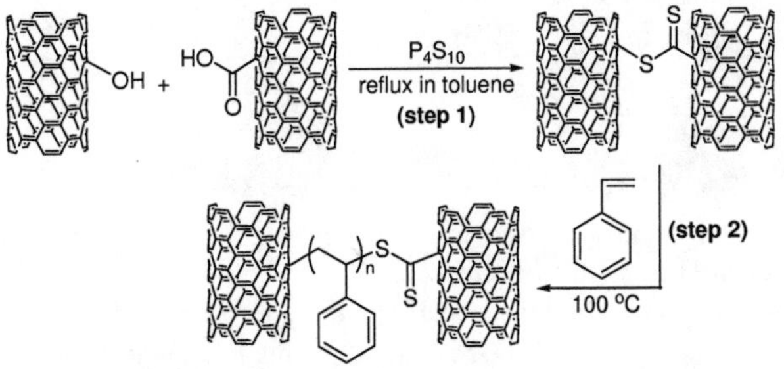

FIGURE 1. Proposed mechanism of substituted dithioester formation through the activation of carboxyl and hydroxyl groups by P_4S_{10}, resulting in covalently linked nanotubes.

The AFM image was obtained using a Veeco Aurora3 Near-Field Scanning Optical Microscope which utilizes an aluminum coated optical fiber tip, mounted on a piezoelectric tuning fork. This configuration uses a shear force feedback method of imaging topography. At the bottom of the image two separate MWNT's can be seen, then in the middle of the image they come together in a junction where a large blob type structure appears. Then on the other side of the blob, (upper part of the image) they continue on side by side. The most important data contained in the image are the height profile line analyses. At the bottom of the Figure 2C, where the two MWNT's are separate, line analysis measurements reveal that the tube on the left is approximately 4.0nm high and the tube on the right is approximately 5.3nm high. The line analysis measurement for the top of the image in figure 2A reveals two structures, side by side, with the left most structure being approximately 3.3nm high and the rightmost structure being approximately 5.1nm high. Although the measured heights of the tubes have some variation between the bottom of the image and the top of the image, these differences are well within the error tolerance of the instrument and could easily be due to a small amount of thermally induced drift. Thus, the line analysis measurements verify that we have two separate MWNT's coming together in a junction and then continuing on side by side. Also of interest is the blob-like junction itself. Although it has some topographical variation of its own, the line analysis reveals that it has an approximate height of about 6.2nm (Figure 2B). It is believed that this is an aggregation of linker molecules.

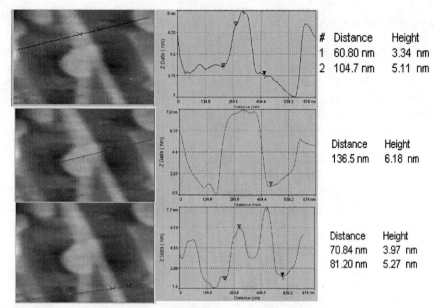

#	Distance	Height
1	60.80 nm	3.34 nm
2	104.7 nm	5.11 nm

Distance	Height
136.5 nm	6.18 nm

Distance	Height
70.84 nm	3.97 nm
81.20 nm	5.27 nm

FIGURE 2A-C. Height Profile Analysis of two Multi Walled Carbon Nanotubes joined together near the center by a linker molecule.

In terms of Raman analysis, the most notable is the appearance of a new peak centred at $v=$ 495 cm-1 in the thiolated nanotubes (see Fig. 3C). This is a combination of 3 vibrations $v =$ 475 cm^{-1}, 495 cm^{-1}, and 503 cm^{-1} representing the different S=C-S stretching and bending modes associated with a dithioester linkage[10]. The emergence of the peak at $v = 650-700$ cm^{-1} indicates the presence of a (C-C-S) stretching vibration in the dithioester linkage,[10] in particular, associated with a gauche conformation[11].

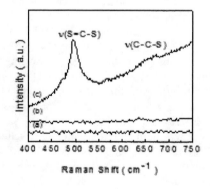

FIGURE 3A-C. Raman scattering spectra of (A) pristine MWCNTs; (B) acid treated MWCNTs; and (C)thiolated MWCNTs.

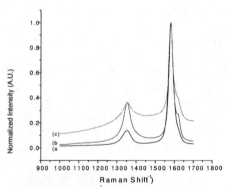

FIGURE 4A-C. High Frequency regions of A. Pristine Multi walled carbon nanotubes, B. Acid Treated Multi walled carbon nanotubes and C. Thiolated Multi walled carbon nanotubes

Figure 4A-C represents the high frequency region of the Raman spectra, showing the first order G-band (or in-plane C=C $E2g$ zone-center) centered at 1582 cm^{-1} and the D'-band representative of the disordered-induced at 1618 cm^{-1} which is shown to be directly affected by disorder on the nanotube body and associated with extrema in the density of states (DOS) of MWNTs[12, 13]. The D'-band is barely present in the pristine tubes (see Fig. 4A) but is prominent after acid treatment showing an increase in defects along the nanotube body (Fig 4B).This continual shift indicates the further presence of bonds other than $sp2$ carbon [14]. In addition, a second peak is observed at 1613 cm^{-1}, arising from the spitting of the $E2g$ mode and similar to acceptor states observed in graphite intercalation[14]. In this case the peak arises from the intercalated covalently bound dithioester linker between the nanotubes. The S-H stretching band, typically present at 2575 cm^{-1}, is missing in the spectra indicating covalent bonding of the thiols [11].

CONCLUSION

In summary we have provided strong evidence that the MWCNT surfaces have been thiolated by a novel technique involving an electrophilic catalyst, Phosphorous Pentasulphide, to produce dithioester linked MWCNTs.

ACKNOWLEDGEMENTS

This research is based upon work supported by the National Science Foundation under Grant No. 0132632

REFERENCES

1. S. A. Curran, A. V. Ellis, K. Vijayaraghavan and P. M Ajayan, *J. Chem. Phys.,* 2004, **120**, 4886.
2. A. V. Ellis, K. Vijayamohanan, R. Goswami, N. Chakrapani, L. S. Ramanathan, P. M. Ajayan and G.Ramanath, *Nano lett.*, 2003, **3**, 279.
3. H. Hu, P. Bhowmik, B. Zhao, M. A. Hamon, M. E. Itkis and R. C. Haddon, *Chem. Phys. Lett.*, 2001, **345**, 25.
4. J. Liu, A. G. Rinzler, H. Dai, J. H. Hafner, R. K. Bradley, P. J. Boul, A. Lu, T. Iverson, K. Shelimov, C. B. Huffman, F. Rodiguez-Macias, Y. -S. Shon, T. R. Lee, D.T.Colbert and R. E. Smalley, *Science,* 1998, **280**, 1253.
5. M.P. Byfield and R. A. Abuknesha, *Biosens. Bioelectron.*, 1994, **9**, 373.
6. J. Liu, A. G. Rinzler, H. Dai, J. H. Hafner, R. K. Bradley, P. J. Boul, A. Lu, T. Iverson, K. Shelimov, C. B. Huffman, F. Rodiguez-Macias, Y. -S. Shon, T. R. Lee, D. T. Colbert and R. E. Smalley, *Science,* 1998, **280**, 1253.
7. L. K. Lim, W. S. Yun, M. –H. Yoon, S. K. Lee, C. H. Kim, K. Kim and S. K. Kim, *Synthetic Met.,*2003, **139**, 521.
8. A. Sudalai, S. Kanagasabapathy and B. C. Benicewicz, *Org. Lett.*, 2000, **2**, 3213.
9. A. Kameyama, Y. Murakami and T. Nishikubo, *Macromolecules,* 1999, **32**, 1407.
10. J. Dong, L. Luo, P. –H. Liang, D. Dunaway-Mariano and P. R. Carey, *J. Raman Spectrosc.*, 2000, **31**, 365.
11. A. Kudelski, *J. Raman Spectrosc.*,2003, **34**, 853.
12. M. S. Dresselhaus, G. Dresselhaus, A. Jorio, A. G. Souza Filho and R. Saito, *Carbon*, 2002, **40**, 2043.
13. J. Kastner, T. Pichler, H. Kuzmany, S. Curran, W. Blau, D. N. Weldon, M. Delamesiere, S. Draper and H. Zandbergen, *Chem. Phys. Lett.*, 1994, **221**, 53.
14. R. J. Nemanich and S. A. Solin, *Phys. Rev. B,* 1979, **20**, 392.

New Polymer Nanotube Design from Graft Polymerization

Aditya Avadhanula, Wudyalew Wondmagegn, Madhuvanthi Kandadai,
Donghui Zhang and Seamus A. Curran

Physics Department, New Mexico State University, Las Cruses NM 88003-8001

Abstract. We have prepared a composite from polystyrene and multi-walled carbon nanotubes (MWCNT) and unlike traditional techniques of composite formations, we chose to polymerize styrene from the surface of dithiocarboxylic ester-functionalized MWCNTs to fabricate a unique nanocomposite material, a new technique of "GRAFT" polymerization.

Keywords: Nanotubes, Composites, polystyrene

INTRODUCTION

A nanocomposite system raises significant challenges, especially in terms of developing innovative routes to obtain control of the individual components at the nanoscale. To maximize the properties of nanocomposites, fundamental control of their morphology and interfacial chemical interactions is critical. While chemically doped polymers or organics containing conductive fillers such as carbon black (5-30% wt/wt loading) have been studied, these have several drawbacks which including high dopant or filler loadings, brittle thin films and opacity. Composite formation between carbon nanotubes and polymers has been investigated in the past [1] [2]. We believe that the GRAFT polymerization technique leads to covalent bonding of the polymer to nanotubes. We see an enhancement in the conductivity of the films as when compared to other means of composite preparation. There is also an increase in the thermal stability of the composite prepared by the RAFT technique as when compared to a blend of polystyrene and nanotubes.

METHODS

The first stage of film fabrication is the formation of covalently linked carbon nanotubes via an intercalated dithiocarboxylic ester functional group [3].Once we have the nanotubes linked through a dithiocarboxylic ester functional group, we use this functional group for cross-linking nanotubes with polystyrene in which it acts as a chain transfer agent (CTA).This controlled free radical reaction is called RAFT polymerization [4]. The thermal polymerization of Styrene occurs at 100^0C. The mechanism is shown in figure 1.

CP786, *Electronic Properties of Novel Nanostructures*, edited by H. Kuzmany, J. Fink, M. Mehring, and S. Roth
© 2005 American Institute of Physics 0-7354-0275-2/05/$22.50

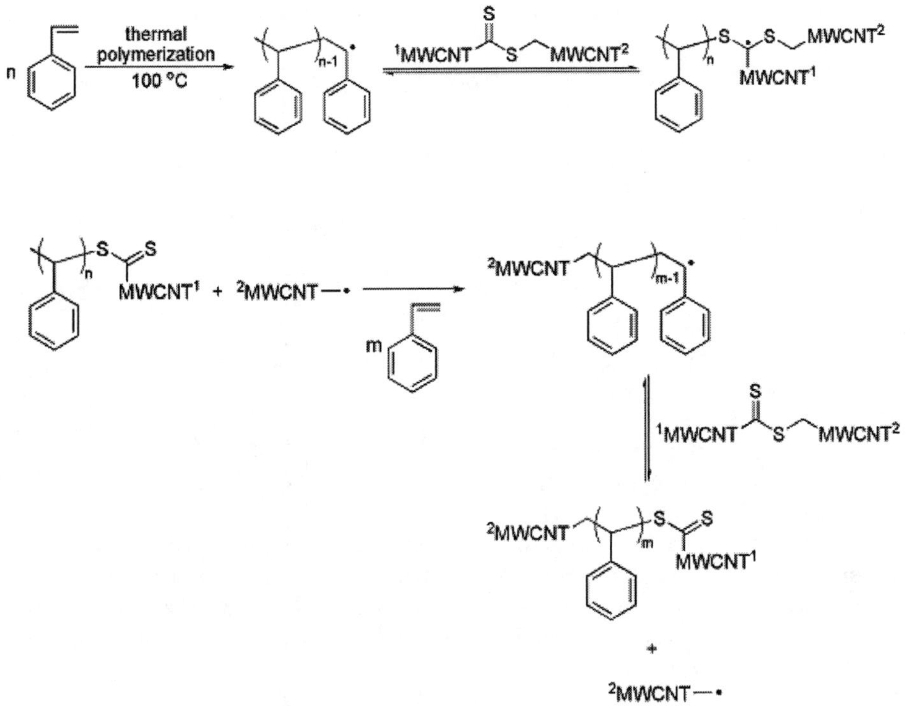

FIGURE 1. Chemical reaction schematic

RESULTS

In order to investigate the effect of covalently linked MWCNTs on the thermal stability of the polymer matrix, thermal gravimetric analysis (TGA) was performed on covalently linked MWCNT (0.33 wt %) - polystyrene composite, MWCNT (0.33 wt%)- polystyrene blend and pristine polystyrene sample. The TGA experiments were carried out under a nitrogen atmosphere to minimize the mass loss due to nanotube oxidation, while allowing the polystyrene to decompose completely. Comparison of the resulting thermograms between 0 and 700°C revealed that the thermal degradation of all three samples proceeded by a one step process with a maximum decomposition temperature. However the temperature, at the peak weight loss rate of covalently linked MWCNT-polystyrene composite was approximately 15°C higher than that observed for MWCNT-polystyrene blend with the same composition, although an earlier onset of polymer decomposition for the covalently linked composite was observed as compared with the blend. The increase in thermal stability observed for the composite can be attributed to an improved dispersion of the MWCNTs in the polymer matrix as well as the covalent bonding between the tubes and the polymer.

CONDUCTIVITY STUDY

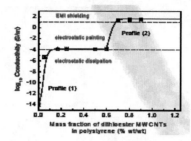

Figure 2. A semi logarithmic plot of covalently linked MWCNT-polystyrene composite conductivity as a function of mass fraction of MWCNT in polystyrene.

The solid line represents two curve fits using Eq.1. For profile 1 the parameters best fit are: p<0.6 wt% wt/wt; pc1 = 0.0425 wt% wt/wt; b= 60 and for profile (2) p>=0.6 wt% wt/wt; pc2 = 0.62 wt% wt/wt; b = 40.

In the past a percolative description has been used to explain the connectivity of the conductivity to the random two-phase system when examining nanotube/polymer composites. Based on the Fermi-Dirac distribution, Fourier et al. proposed an analytical model (Eq 1) to describe the critical insulator to conductor transition[5].

$$\log(\sigma_c) = \log(\sigma_n) + \frac{\log(\sigma_p) - \log(\sigma_n)}{1 + e^{[b(p - p_c)]}}$$

(1)

Where σ_c, σ_n and σ_p are the conductivities of the composite, nanotubes and pristine polymers respectively, p is the mass fraction of the nanotubes, and b is an empirical parameter that leads to the change in conductivity at the percolation threshold (or ballistic threshold) pc. The Eq. I has been fitted to Profile (1) and (2), individually (Figure 2), Profile (1) gives a percolation threshold (pc1 = 0.0425 wt%) for lower MWCNT loadings, due to the homogeneity of the MWCNTs in the polystyrene matrix. This ensures a low microscopic phase transition between polymer and nanotubes that allows charge transfer through a hopping transport regime. This results in an increase in conductivity from 10^{-14} sm^{-1} in pure polystyrene to 10^{-4} s.m^{-1} at MWCNT loadings of 0.02-0.06 wt%. At these loadings and conductivities, the composites are optically transparent (<500 μm). In profile (2) the polymerization has reached its limit due to high loadings of the chain transfer agent (dithioester MWCNTs) and we see a percolation threshold (pc2 =0.62wt%) with conductivities between 10^{-4} - 33 s.m-1 for nanotube loadings 0.06-0.09 wt%.

In normal circumstances the addition of nanotubes to an insulating matrix results in a singular percolative phenomenon where the random distribution of the conductive matrix and hopping mechanism between the insulating barriers would determine the bulk conductivity of the system[6]. The second regime is a result of the proximity of the nanotubes throughout the samples, rather than regions of isolated nanotube distribution that would normally conclude in hopping transport [7]. The mechanism of charge hopping has now changed to fractional quantum conductance, caused by nanotube proximity, as a function of chemical modification and increased length scales of individual nanotubes [8]. Sample conductance is now dominated by interwall conductance channels and redistribution of the current over the distorted outer graphitic planes.

CONCLUSIONS

We have demonstrated that MWCNT can be readily functionalized with thioester functional group to form an oriented and bundled structure. The functionalized MWCNTs can be covalently linked with polystyrene to from well-dispersed nanocomposite through a GRAFT technique. Since nanotubes dispersion is no longer hindered by aggregation, the cumulative effects have produced a smooth nanocomposite with enhanced electrical conductivity up to 33 sm-1 S/m with 0.90 wt % nanotube loadings and improved thermal stability linked MWCNT – polystyrene blend. Our approach shows a great potential for a practical route to creating transparent antistatic coatings, electromagnetic insulating (EMI) shields and reinforced polymeric membranes in fuel cells. These aspects are currently under study.

REFERENCES

1. S.A.Curran, P.M.Ajayan, W.J.Blau, D.L.Carroll, J.N.Coleman, A.B.Dalton, A.P.Drury, B.McCarthy, S.Maier and A.Strevens, *Adv.Mater*, 1998,10,1091.
2. G.Viswanathan, N.Chakrapani, H.Yang, B.Wei, K.Cho, C.Y.Ryu and P.M.Ajayan, *J.Am.Chem.Soc*, 2003, 125, 9258.
3. A.Sudalai, S.Kanagasabapathy and B.S.Beniecewicz, *Org. Lett.*, 2000, 2, 3213.
4. S. C. Farmer and T. E. Patter, *J. Polym. Sci. Pol. Chem.*, **2002**, *40*, 555
5. J. Fourier, G. Boiteax, G. Seytre and G. Marichy, *Synth. Met.*, **1997**, *84*, 839.
6. J. N. Coleman, S. A. Curran, A. B. Dalton, A. P. Davey, B. McCarthy, W. J. Blau and R. C. Barklie, *Phys. Rev. B*, **1998**, *58*, 7492.
7. F. Carmona and C. Mouney, *J. Mat. Sci.*, **1992**, *27*, 1322.
8. S. Sanvito, Y.-K. Kwon, D. Tomanek and C. J. Lambert, *Phys. Rev. Lett.*, **2000**, *84*, 1974.

Inertness of Near-Armchair Carbon Nanotubes Towards Fluorination

L.G. Bulusheva[1], P.N. Gevko[1], A.V. Okotrub[1], N.F. Yudanov[1], I.V. Yushina[1], E. Flahaut[2], U. Dettlaff-Weglikowska[3], and S. Roth[3]

[1]Nikolaev Institute of Inorganic Chemistry SB RAS, pr. Ak Lavrentieva 3, Novosibirsk 630090, Russia
[2]CIRIMAT/LCMIE, UMR CNRS 5085, 31062 Toulouse Cedex 4, France
[3]Max-Plank-Institute for Solid State Research, Heisenbergstr. 1, 70569 Stuttgart, Germany

Abstract. Reactivity of single-wall carbon nanotubes (SWNTs), obtained by HiPco method, and double-wall carbon nanotubes (DWNTs), grown during a chemical vapor deposition process, towards fluorinating agent BrF_3 has been examined using optical absorption spectra measurements. The spectra of the fluorinated samples exhibited sets of distinctly resolved peaks with energies having close values for DWNTs and HiPco nanotubes. It was found that the peaks can be associated with energy transitions in (n,n-1)-type carbon nanotubes, which structure remained unchanged with fluorination.

Keywords: Carbon nanotubes; fluorination; optical absorption.
PACS: 78.67.Ch; 61.46.+w.

INTRODUCTION

Fluorination is an effective way for attachment of large amounts of reagent to the carbon nanotube surface. A limiting stoichiometry for which the fluorinated tube still maintains its tube-like structure is close to C_2F [1]. Fluorinated carbon nanotubes are solvated in alcohols [2] that allow their further derivatization with nucleophiles [3]. Furthermore, computational investigations predict the fluorination of carbon nanotubes can yield insulating, semiconducting, or metallic nanotube derivatives, which might be used as building blocks for electronic devices [4].

Raman spectroscopy revealed the fluorination of single-wall carbon nanotubes (SWNTs) reduces intensity of the radial breathing mode (RBM) components corresponding to the thinner tubes [5-7]. By the result of quantum-chemical calculations, more susceptible fluorination of small diameter tubes is connected with an energetic gain due to the reduced pyramidalization of the π-system [8]. Raman spectrum excited with determined wavelength reveals only certain carbon nanotubes. Observation of complete set of carbon nanotubes contained in the sample can be achieved with optical absorption spectroscopy.

In present work, the optical absorption spectroscopy is used for defining of structural changes of carbon nanotubes samples with fluorination. Two types of samples were examined: SWNTs synthesized by HipCO method and double-wall carbon nanotubes (DWNTs) prepared using chemical vapor deposition (CVD)

CP786, *Electronic Properties of Novel Nanostructures*, edited by H. Kuzmany, J. Fink, M. Mehring, and S. Roth
© 2005 American Institute of Physics 0-7354-0275-2/05/$22.50

process. Both these samples contain variety of carbon nanotubes with different diameter and chirality.

EXPERIMENTAL DETAILS

SWNTs were synthesized by high pressure CO disproportionation over Fe catalyst [9] (HiPco technique) at CNI, Houston, USA. Diameter of tubes is varied for 0.7 nm to 1.4 nm. DWNTs were produced by decomposition of CH_4 over $Mg_{1-x}Co_xO$ solid solution with small addition of molybdenum [10]. Metal oxides were removed by treatment of the sample with concentrated aqueous HCl solution. High-resolution transmission electron microscopy showed the sample contains 77% DWNTs, 18% SWNTs, and 5% triple-wall nanotubes. The diameter distribution of the DWNTs ranged from 0.53 to 2.53 nm for inner shells and from 1.2 to 3.23 nm for outer shells.

Fluorination of carbon nanotubes samples was carried out following a procedure previously applied to the arc-produced multiwall carbon nanotubes [11]. A sample placed in a teflon flask was held in a vapor over a solution of BrF_3 in Br_2 for 7 days. Thereafter the flask content was dried by a flow of N_2 up to the termination of Br_2 evolution.

For absorption measurement the carbon nanotubes were deposited as thin films on sapphire substrates using an airbrush. For use with the airbrush the pristine and fluorinated nanotube-containing powders were suspended respectively in ethanol and heptane and then sonicated for about 1 hour. Evaporation of the solvent was aided by heating the substrate during airbrushing at about 70°C. The width and homogeneity of film obtained were controlled visually. The absorption spectra were recorded using Shimadzu UV 3101 PC instrument in a wavelength range of 190÷3200 nm with a resolution of 3 nm.

RESULTS AND DISCUSSION

The optical absorption spectra of pristine HiPco material and that after purification are compared in Fig. 1(a). The features in the energy region below ~1.32 eV correspond to the optical transitions across the band gap between the first van Hove singularities (VHS) in semiconducting nanotubes, which are denoted as E_{11}. The transitions between the second pair of VHS E_{22} in the semiconducting nanotubes and the E_{11} transitions in metallic nanotubes form the features between 1.32 and ~2.3 eV. The broad low-intensity features above 2.3 eV can be attributed to the E_{11} transitions in metallic nanotubes having the small diameters and the transitions between the third pair of VHS in the semiconducting nanotubes. The wide diameter distribution of SWNTs produced by HiPco method results in energetic overlapping of the features attributed to optical transitions between different pairs of VHS; the broadening of features is due to the tubes are combined into ropes [12]. The optical absorption spectrum of the fluorinated HiPco nanotubes exhibits a set of separated lines, which intensity is strongly reduced relative to that in the spectrum of pristine sample.

The optical absorption spectrum of the pristine DWNT-contained sample has broad and not clearly resolved structure especially in the region above 1.4 eV (Fig. 1(b)).

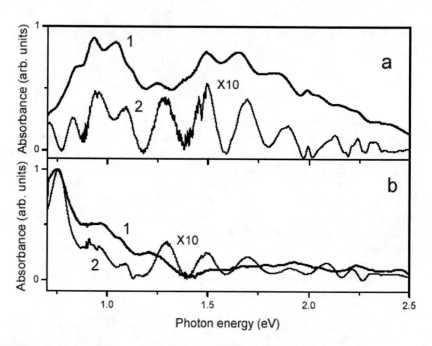

FIGURE 1. Background corrected optical absorption spectra of SWNTs (a) and DWNTs (b) samples before (1) and after (2) fluorination. The spectral intensity for the fluorinated nanotubes were increased by ten times.

Intense low-energy feature indicates a presence of large-diameter carbon nanotubes in the sample. Fluorination of DWNTs makes some spectral features more pronounced. The changes observed in the optical absorption spectrum of the DWNTs sample with fluorination are generally similar to those occurred in the HiPco spectrum. The fluorination results in increase of the transition energies by about 0.1–0.4 eV, that could be due to enlarging of intertube distances.

The energies of the optical absorption features for the fluorinated samples are collected in table. Five features are appeared in the spectrum of the fluorinated HiPco material below ~1.32 eV. According to the absorption peaks assignment made for debundled SWNTs [13] these features can be related to the E_{11} transitions in (n,n-1) tube family (see table). Diameter of carbon nanotubes occurring in the HiPco sample and belonging to this family varies from 7.5 Å to 12.9 Å. The E_{22} transitions in the selected tubes form features in the region of 1.38–2.15 eV. Although the value corresponding to the E_{22} transition in the thickest (10,9) tube (about 1.39 eV [14]) is absent in the table, this transition could be attributed to low-energy shoulder of the peak centered at 1.49 eV (Fig. 1). The features at 2.0, 2.11, and 2.31 eV have low intensity and are likely assigned to the E_{11} transitions in metallic carbon nanotubes [15]. Surprising, the energies of features exhibiting in optical absorption spectrum of the fluorinated DWNTs are close to those in spectrum of the fluorinated HiPco nanotubes. The E_{11} transition in (9,8) tube with energy about 0.83 eV could form a shoulder at high-energy side of the intense spectral feature.

TABLE 1. Energy (eV) and Assignment of Absorption Features in Spectra of Fluorinated Samples.

F-HiPco	F-DWNT	Assignment	F-HiPco	F-DWNT	Assignment
0.71	0.75	E_{11} (10,9)	1.69	1.70	E_{22} (8,7)
0.83	–	E_{11} (9,8)	1.89	1.91	E_{22} (7,6)
0.95	0.93	E_{11} (8,7)	2.00	–	E_{11} (9,9)
1.09	1.09	E_{11} (7,6)	2.11	2.09	E_{11} (11,5)
1.28	1.29	E_{11} (6,5)	2.23	2.21	E_{22} (6,5)
1.49	1.49	E_{22} (9,8)	2.31	2.31	E_{11} (10,1)

Appearance of optical transitions for the fluorinated samples indicates some carbon nanotubes remain intact under fluorination conditions. The most of these tubes are characterized by large chiral angle. It has been reported that methods using the oxygen-contained compounds for SWNTs growth produce mainly the near-armchair carbon nanotubes [14, 16]. Probably such tubes are stable towards oxidizing agents that cause their intertness to the fluorination.

ACKNOWLEDGMENTS

The work was supported by the Russian Foundation for Basic Research (project 03-03-32286).

REFERENCES

1. Mickelson, E.T., Huffman, C.B., Rinzler, A.G., Smalley, R.E., Hauge, R.H., and Margrave, J.L., *Chem. Phys. Lett.* **296**, 188-194 (1998).
2. Mickelson, E.T., Chiang, I.W., Zimmerman, J.L., Boul, P.J., Lozano, J., Liu, J., Smalley, R.E., Hauge, R.H., and Margrave, J.L., *J. Phys. Chem. B* **103**, 4318-4322 (1999).
3. Khabashesku, V.N., Billups, W.E., and Margrave, J.L., *Acc. Chem. Res.* **35**, 1087-1095 (2002).
4. Seifert, G., Kohler, T., and Frauenheim, T., *Appl. Phys. Lett.* **77**, 1313-1315 (2000).
5. Marcoux, P.R., Schreiber, J., Batail, P., Lefrant, S., Renouard, J., Jacob, G., Albertini, D., and Mevellec, J.-Y., *Phys. Chem. Chem. Phys.* **4**, 2278-2285 (2002).
6. Pehrsson, P.E., Zhao, W., Baldwin, J.W., Song, C., Liu, J., Kooi, S., and Zheng, B., *J. Phys. Chem. B* **107**, 5690-5695 (2003).
7. Bulusheva, L.G., Okotrub, A.V., Duda, T.A., Obraztsova, E.D., Chuvilin, A.L., Pazhetnov, E.M., Boronin, A.I., and Dettlaff-Weglikowska, U., in *Nanoengineered Nanofibrous Materials*, edited by S.I. Guceri et al., NATO Science Series 169, Kluver Academic Book Publishers, 2004, pp. 143-149.
8. Bettinger, H.F., Kudin, K.N., and Scuseria, G.E., *J. Am. Chem. Soc.* **123**, 12849-12856 (2001).
9. Nikolaev, P., Bronikowski, M.J., Bradley, R.K., Rohmund, F., Colbert, D.T., Smith, K.A., and Smalley, R.E., *Chem. Phys. Lett.* **313**, 91-97 (1999).
10. Flahaut, E., Bacsa, R., Peigney, A., and Laurent Ch., *Chem. Commun.* 1442-1443 (2003).
11. Yudanov, N. F., Okotrub, A. V., Shubin, Yu. V., Yudanova, L. I., Bulusheva, L. G., Chuvilin, A. L., and Bonard, J.-M., *Chem. Mater.* **14**, 1472-1476 (2002).
12. Hagen, A., Noos, G., Talalaev, V., and Hertel, T., *Appl. Phys. A* **78**, 1137-1145 (2004).
13. Wu, J., Walukiewicz, W., Shan, W., Bourret-Courchesne, E., Ager III, J.W., Yu, K.M., Haller, E.E., Kissell, K., Bachilo, S.M., Weisman, R.B., and Smalley, R.E., *Phys. Rev. Lett.* **93**, 017404 (2004).
14. Bachilo, S.M., Strano, M.S., Kittrell, C., Hauge, R.H., Smalley, R.E., and Weisman, R.B., *Science* **298**, 2361-2366 (2002).
15. Tel, H., Maultzsch, J., Reich, S., Hennrich, F., and Thomsen, C., *Phys. Rev. Lett.* **93**, 177401 (2004).
16. Jorio, A., Fantini, C., Pimenta, M.A., Capaz, R.B., Samsonidze, Ge.G., Dresselhaus, G., Dresselhaus, M.S., Jiang, J., Kobayashi, N., Gruneis, A., Saito, R., *Phys. Rev. B* **71**, 075401 (2005).

Effect of Solvents and Dispersants on the Bundle Dissociation of Single-walled Carbon Nanotube

Silvia Giordani [*], Shane D Bergin, Anna Drury, Éimhín Ní Mhuircheartaigh, Jonathan N Coleman & Werner J Blau

Molecular Electronics and Nanotechnology Group, Physics Department, Trinity College, Dublin 2, Ireland

Abstract. Single-wall carbon nanotubes (SWCNTs) are severely restricted in their applications, as they exist in rope-like bundles. Recently, J. Coleman et al. demonstrated a spectroscopic method to monitor bundle dissociation in low concentration polymer-nanotube dispersions.[1] The method relies on the measurement of the ratio of free-polymer to the nanotube-bound polymer in the SWCNT-polymer solutions via luminescent spectroscopy. A theory has been developed to transform this data into the bundle surface area, which is of course related to the bundle size. This method clearly shows that individual, isolated SWCNT are stable in low concentration dispersions. In an effort to broaden the understanding of the physical processes governing the NT de-bundling a short-chain molecule has been examined. We found a strong dependence of the concentration at which individual NTs become stable with the nature of the solvent.

Keywords: Single-wall carbon nanotubes, dispersion, de-bundling, fluorescence, microscopy.

PACS: 81.07.De

INTRODUCTION

Single-walled carbon nanotubes (SWCNTs) have generated much interest in recent years due to their unique properties and huge potential for applications. One of the fundamental problems that remain to be addressed is the fact that SWNTs are almost always present in the form of bundles. These bundles can be quite large with diameters of 10-200 nanometers. It would be advantageous to have access to samples of dispersed individual nanotubes or, at the very least, to have some control over the bundle size. This is the case not only for fundamental studies where individual SWCNTs are essential but for more applied studies such as composite research where uniform dispersions are required. While it is possible to obtain individual nanotubes by dispersion of SWCNTs in surfactants followed by ultra-centrifugation, this is a very inefficient method with more than 99% of SWCNTs being lost from the dispersion[2]. In this work we focus on the dispersing capability of a short chain analogue. Using the simple model based on adsorption/desorption equilibrium developed in-house, by Coleman et al.[1], we can calculate the concentration at which individual SWNTs are observed for each solvent system.

EXPERIMENTAL PROCEDURES

The purified nanotubes used in this work were prepared by the HiPco process[3], supplied by Carbon Nanotechnologies Inc and used without further treatment. The solvents Chloroform, N-methyl-2-pyrrolidone (NMP) and N,N-dimethylformamide (DMF) were purchased from commercial sources. The molecule used was 2,5-

CP786, *Electronic Properties of Novel Nanostructures*, edited by H. Kuzmany, J. Fink, M. Mehring, and S. Roth
© 2005 American Institute of Physics 0-7354-0275-2/05/$22.50

dioctyloxy-1,4-distyrylbenzene (pDSB), a short-chain analogues of the well known PPV, and it was synthesised in house. It is strongly fluorescent, displays minimal self-aggregation and disperses SWCNTs in the organic solvents. Stock solutions of pDSB and of SWCNT were made in three different solvents, and were subject to 2 mins sonication using a high-power sonic tip (200 Watts at 20%). Two identical solutions of the molecule in each solvent were made. Pure HiPCO SWCNTs were added to one solution, such that the pDSB-SWCNT mass ratio was 2:1. Each solution was then diluted, serially, on the order of twenty times to give a broad range of CNT concentrations (10^{-2} kg/m^3– 10^{-8} kg/m^3). Each solution was sonicated for 1 minute using a high power sonic tip and then allowed to stand for 24 hours to come to equilibrium. No sedimentation was observed over this period. The absorbance spectrum was recorded using a UV-Vis spectrometer and the emission of each solution was recorded using LS-55 Perkin Elmer luminescence spectrometer.

RESULTS AND DISCUSSION

Photoluminescence (PL) measurements were performed on all solutions. Figure 1 shows the PL spectra of pDSB in chloroform at three different concentrations before and after the addition of HiPco SWCNTs (excitation at 388 nm). The addition of the SWCNTs to pDSB results in a visible quenching of the emission, indicating interaction between the SWCNTs and the pDSB molecules. It should be noted that no shift occurred in the PL maxima. The molecule exists in two forms, free and bound to the SWNT. The PL efficiency for the bound form is expected to be extremely low as any photo-generated singlet excitons preferentially decay non-radiatively through the fast vibrational manifold of the nanotubes. Thus any observed PL from the composite solutions is due to the free form only. Therefore the ratio of PL intensity for a pDSB/SWNT solution to that of pDSB solution of equivalent concentration is a measure of the fraction of free molecule in the composite solution at that concentration. For all concentrations, the fraction of free molecule in each solvent was calculated from the intensity ratio and is plotted in Figure 2. The solvents investigated were Chloroform, N-methyl-2-pyrrolidone (NMP) and N,N-dimethylformamide (DMF). Due to constant adsorption/desorption from the nanotubes, this fraction of free molecule is not expected to be constant over the concentration range studied. Indeed this plot is highly non-linear with respect to concentration, but approaches 1 at very low concentrations indicating that most molecules are unbound at low concentration. The solid black lines represent the response of an individual-SWCNT system as deduced from Coleman's model, fitted for the results obtained in chloroform and in NMP. In the case of the chlorinated solvent the experimental data deviates from this curve at

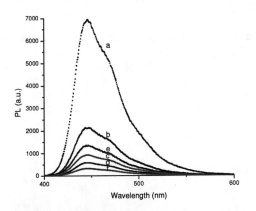

FIGURE 1. Photoluminescence spectra of pDSB in chloroform before (a, b, c) and after (d, e, f), the addition of SWCNT respectively. (a, d = 3.1 x 10^{-6}M, b, e = 7.7 x 10^{-7} M and c, f = 3.2 x 10^{-7} M).

SWCNT concentrations higher than 1×10^{-7} kg/m^3 indicating that bundles occur above this limit. In the case of the amide solvents the response does not deviate from the solid curve, indicating that individual SWCNTs or small bundles of SWCNTs are present up to the limit of the experiment.

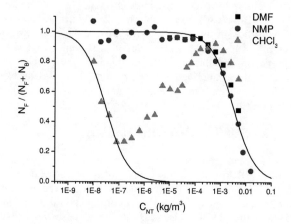

FIGURE 2. Graph of the fraction of free molecule as a function of concentration for pDSB in the different solvents.

Preliminary TEM results agreed with the spectroscopic results. Low-resolution images of solutions of DSB: SWCNTs on holey carbon grids were taken at concentrations of 0.0025 kg/m^3 in both CHCl$_3$ and NMP (Figure 3). The diameters distributions of the SWCNTs present in both solvents are presented in Figure 4 (CHCl$_3$ top, NMP bottom). There is a clear shift in the bundle size distribution, to smaller bundles, in the amide solvent versus the chlorinated solvent. The dominant feature is the presence of small SWCNT bundles at relatively high SWCNT concentrations in the case of the amide solvent.

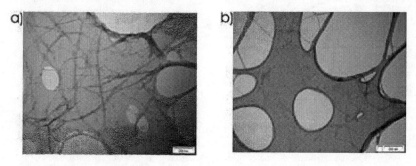

FIGURE 3. TEM images of pDSB : SWNT in Chloroform (a) and NMP (b).

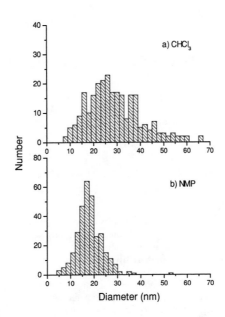

FIGURE 4. Diameter distribution in Chloroform (a) and NMP (b).

CONCLUSIONS

Stable dispersions of SWCNTs have been obtained and characterised via the use of non-covalent functionalisation with a luminescent short molecule. Isolated SWNT are stable at low concentrations. The small molecules have been shown to solubilise and de-bundle SWNT in chloroform at lower concentration compared to the polymers reported previously by Coleman et al. However, using amide solvents, the concentration at which isolated nanotubes are stable was found to increases by about five orders of magnitude.

ACKNOWLEDGMENTS

The authors wish to acknowledge financial support from the European Union, the Irish Higher Education Authority, Enterprise Ireland, and Intel.

REFERENCES

1. J.N. Coleman, S. Maier, A. Fleming, S. O'Flaherty, A. Minett, M.S. Ferreira, S. Hutzler, W.J. Blau, *J. Phy. Chem. B.*, 108, 3446, 2004.
2. M. J. O'Connell, S. M. Bachilo, C. B. Huffman, V. C. Moore, M. S. Strano, E. H. Haroz, K. L. Rialon, P. J. Boul, W. H. Noon, C. Kittrell, J. P. Ma, R. H. Hauge, R. B. Weisman, R. E. Smalley, *Science* **2002**, *297*, **297**, 593-596.
3. Nikolaev, P.; Bronikowski, M. J.; Bradley, R. K.; Rohmund, F.; Colbert, D. T.; Smith, K. A.; Smalley, R. E. *Chem. Phys. Lett.* **1999**, *313*, 91.

Silver intercalated carbon nanotubes

E. Borowiak-Palen*, M. H. Ruemmeli[¶], E. Mendoza*, S. J. Henley*, D. C. Cox*, C. H. P. Poa*, V. Stolojan*, T. Gemming[¶], T. Pichler[¶], and S. R. P. Silva*

* *Nano-Electronics Centre, Advanced Technology Institute, University of Surrey, Guildford, GU2 7XH, UK.*
[¶]*Leibniz Institute for Solid State and Materials Research Dresden, P.O. Box 270016 D-01171 Dresden, Germany.*

Abstract. The intercalation of metals within carbon nanotube structures has extended the potential applications of these materials to possible quantum memory elements as well as high density magnetic storage media. In our study we use methodologies based on wet chemistry and solid state physical (excimer laser) processes to incorporate silver nanoparticles in single and multiwall carbon nanotubes. We show high resolution TEM as evidence for the formation of very long (~ 100-150 nm) silver quantum wires within SWCNT, and their properties are probed using various analytical techniques. The variation of the silver intercalated nanotube percentage and yield are compared for the SW- and MW-CNTs, when using wet chemistry versus physical processes.

Introduction

It is well known that one of the easiest ways to modify the properties of carbon nanotubes in order to broaden their potential applications is using them as a template by which directing of the growth of metal modified carbon nanotubes can take place [1-4]. Among the many options of altering the properties of carbon nanotube is filling their hollow interior or decorating them with the metals. The filling of metals within carbon nanotube structures has extended the potential applications of these materials to possible quantum memory elements, high density magnetic storage media, semiconducting devices, field emitters, and scanning probe microscopes [5-10]. With wide variants of metals being used to fill within carbon nanotubes or deposited on to their surface, silver has received much attention as the metal with the highest electrical and thermal conductivity. The intercalation of silver within carbon nanotubes can be achieved using capillary action when liquid metals [11, 12] or solubilised silver salt is pulled into the cavities of CNTs [13-17].

In this paper, we optimise wet chemistry processes in order to produce Ag filled SWCNT. Very high filling ratios with respect to the individual nanotube, as well as to the bulk sample are achieved with our method. In our case, silver nanowires are up to 250 nm long with the entire length of the nanotube being filled (mean filling ratio ~ 80% for the individual tube when compared with the bulk sample obtained). On the other hand, the decoration of vertically aligned multiwall carbon nanotubes with silver using solid state physical (excimer laser) processes was also performed and compared with wet chemistry processes.

CP786, *Electronic Properties of Novel Nanostructures*, edited by H. Kuzmany, J. Fink, M. Mehring, and S. Roth
© 2005 American Institute of Physics 0-7354-0275-2/05/$22.50

Experimental

The pristine SWCNT material was produced by the laser ablation technique [18] with 40 % SWCNT yield with a mean diameter of 1.22 nm. The as produced material still contains amorphous carbon and catalyst particles (Ni, Co). In order to purify and open the tubes, acid treatment was performed. The soot was placed in 2M HNO_3 and refluxed for 30 h at about 400 K. After filtration and washing with distilled water, the opened and purified nanotubes were transferred to a beaker containing an over-saturated solution of silver nitrate ($AgNO_3$). The as-prepared solution was stirred for several hours at room temperature. In order to have pure Ag filled SWCNT, the washing out of $AgNO_3$ which is not incorporated into the tubes but remains on the walls or in the space between the bundles was crucial. This was accomplished by multiple centrifugations with distilled water, after each centrifugation cycle the water was removed and replaced. As the last step, the sample was heated in air, at 573 K, to reduce the silver nitrate to silver. The morphology and chemical composition of the sample was studied using high resolution transmission electron microscopy (HRTEM) and its energy dispersive X-ray (EDX) spectroscopy mode.

The decoration of MWCNT was investigated using laser ablation. In this technique, a target is vapourised using a high energy laser pulse. A target made of melted silver chloride was ablated with a laser (6 J/cm^2) operating at a wavelength of 248 nm. SEM and TEM were used in order to study the morphology of the resulting product.

Results

The acid treatment of SWCNT opens their ends which enables their filling with foreign elements. TEM micrograph in Figure 1 (a) shows the morphology of acid-treated SWCNT and SWCNT with open ends. As it can be seen in the figure, acid treatment results in purified and open-ended nanotubes.

High-angle annular dark field (HAADF) images show that SWCNT are forming a composite with $AgNO_3$ being distributed all over the carbon nanotube bundles and a significant number of the nanotubes being filled (Figure 1 b). The next step to obtaining pure samples was washing out the $AgNO_3$ crystals, which were on the outside of the carbon nanotubes. This then leaves the silver nitrate nanowires impregnated in the hollow interiors of the SWCNT, by multiple centrifugation of the sample in distilled water. Afterwards, the sample was heated at 572 K for an hour to reduce the salt to silver. The dark field image shows very thin nanowires with lengths from 20 up to 250 nm (Figure 1 c).

The detection of the silver filling using standard TEM micrographs was impossible as the diameter of the metal wires is so small that one can not differentiate between individual single nanowires and nanotube walls in the bundle. In the inset of the Figure 1 (c), the well defined Ag nanowires are showing in bundles which are only partially filled, whereas highly filled bundles of SWCNT are difficult to obtain because the nanowires do not all lie in the focal plane entirely and are twisted in three dimensions. This effect is seen in Figure 1 (d) EDX spectrum which shows only carbon, silver and copper (from the TEM grid) peaks.

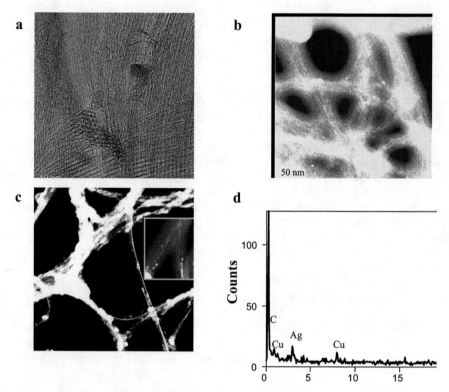

Figure 1. Bright field image of acid treated SWCNT (a), dark field image of SWCNT after AgNO₃ treatment (b), silver filled SWCNT (c), and EDX from bulk Ag filled SWCNT (d).

The ablated AgCl vapourized and decomposed into metallic silver and gaseous chlorine and the vapours of metal were deposited on the outer walls of the carbon nanotubes used as a template. Figure 2 (a) shows a SEM image of a CNT array which has been covered with Ag. Figure 2 (b) presents a magnification of the tubes that shows the Ag particles coating the walls extremely uniformly. Currently, the number of laser shots is being increased to achieve a homogeneous coverage of all the surface of the CNT. Figure 2 (b) shows a TEM image with a region subject to high magnification of the wall of the CNT. The metallic particles have well pronounced crystal structure with the lattice distance typical of a face centered cell (fcc) expected for silver. The size of the particles appears to be dependent of the number of laser shots. This effect is currently under investigation and the complete deposition of a silver layer expected.

Conclusions

We have optimised the wet chemistry technique to fill SWCNT with silver. It is important to note the significance of the diameter of carbon nanotubes, which has to be optimised. In the case of 1.22 nm carbon nanotubes, almost two days of stirring of the opened nanotubes resulted in an 80 % yield of filled nanotubes estimated using microscopic techniques.

CNT arrays with homogeneously decorated Ag particles have been produced. AgCl used as a target was decomposed into silver and chlorine. Therefore only silver was deposited on the surface of carbon nanotubes. This is a first step towards coating of the vertically aligned MWCNT with metals and using this for further functionalisation.

 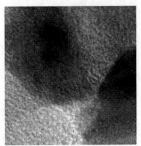

Figure 2. SEM image of a MWCNT array decorated by silver (a) and a TEM picture of the silver particles partially embedded in the body of the MWCNT (b).

Acknowledgement

The authors are grateful to the EPSRC for funding through the Portfolio Partnership and the Engineering Functional Materials Programme . The work has been funded also by the FP6 EU contract Sensation. The authors acknowledge O. Jost for delivering of SWCNT. E.M. acknowledges the Generalitat de Catalunya for sponsorship via the Nanotec Postdoctoral Fellowship Program.

References

[1] D. Golberg, Y. Bando, W. Han, K. Kurashima and T. Sato, *Chem. Phys. Lett.* **308**, 337 (1999).

[2] W. Han, K. Kurashima and T. Sato, *Appl. Phys. Lett.* **323**, 185 (1998).

[3] M. Monthioux, *Carbon* **40**, 1809 (2002).

[4] N. Keller, C. Pham-Huu, T. Shiga, C. EstournesJ.-M. Greneche and M.J. Ledoux, *J. Cryst. Growth* **265**, 184 (2004).

[5] J. Liu, A.G. Rinzler, H. Dai, J.H. Hafner, R.K. Bradley, P.J. Boul, A. Lu, T. Liverson, K. Shelimov, C. B. Huffman, F. Rodriguez-Macias, Y.-S. Shon, T.R. Lee, D.T. Colbert and R.E. Smalley, *Science* **280**, 1253 (1998).

[6] E.W. Wong, P.E. Sheehan, and C.M. Lieber, *Science* **277**, 1971 (1997).

[7] S. Saito, *Science* **278**, 77 (1997).

[8] S. J. Tans, M.H. Devoret, H. Dai, A. Thess, R.E. Smalley, L.J. Geerligs and C. Deeker, *Nature* **386**, 474 (1997).

[9] J.T. Hu, O.Y. Min, P.D. Yang and C.M. Lieber, *Nature* **399**, 48 (1999).

[10] S.J. Tans, A.R.M. Vershueren and C. Deeker, *Nature* **393**, 49 (1998).

[11] P.M. Ajayan and S. Iijima, *Nature* **361**, 333 (1993).

[12] U. Ugarte, A. Chatelain and A. de Heer, *Science* **274**, 1897 (1996).

[13] E. Dujardin, T. W. Ebbesen, H. Hiura and K. Tanigaki, *Science* **265**, 1850 (1994).

[14] A. Chu, J. Cook, R.J.R. Heeson, J.L. Hutchinson, M.L.H. Green and J. Sloan, *Chem. Mater.* **8**, 2751 (1996).

[15] J. Sloan, J. Hammer, M. Zwiefka-Sibley and M.L.H. Green, *Chem. Comm.* **347** (1998).

[16] Z.L. Zhang, B. Li, Z.J. Shi, Z.N. Gu, Z.Q. Xue and L.-M. Peng, *J. Mater. Res.* **15**, 2658 (2000).

[17] K. Matsui, B.K. Pradhan, T. Kyotani and A. Tamita, *J. Phys. Chem. B* **105**, 5682 (2001).

[18] O. Jost, A.A. Gorbunov, J. Moeller, W. Pompe, A. Graff, R. Friedlein, X. Liu, M.S. Golden and J. Fink, *Chem. Phys. Lett.* **339**, 297 (2001).

Characterisation of Single-walled Carbon Nanotube Bundle Dissociation in Amide Solvents

Shane D Bergin[*], Silvia Giordani, Donal Mac Kernan, Andrew Minett, Jonathan N Coleman & Werner J Blau

Molecular Electronics and Nanotechnology Group, Physics Department, Trinity College, Dublin 2, Ireland
*berginsd@tcd.ie

Abstract. The implementation of single-walled carbon nanotubes (SWNTs) into a wide array of applications is hindered by the formation of bundles. Many methods have been suggested to de-bundle the SWNTs, including both covalent and non-covalent funtionalisation with surfactants, polymers and macromolecules. These methods have their advantages but the ideal situation must be to dissolve and de-bundle the SWNTs in an appropriate solvent at concentrations that are useful for their implementation in applications. In this work we have concentrated on the amide solvent N-methyl-2-pyrrolidone (NMP). This solvent has been shown to be a good dispersant of SWNTs and we have analytically characterised the dispersing and de-bundling process. Optical absorbance based studies allowed the determination of the concentration at which these dispersions deviate from the Lambert Beer law. This concentration is thought to correspond to that where individual SWNTs are held in solution. Atomic force microscopy was carried out on all solutions confirming excellent dispersion and significant debundling even at high SWNT concentration.

INTRODUCTION

The large-scale implementation of single walled carbon nanotubes (SWCNTs) into a wide variety of applications is severely restricted by the formation of SWCNT bundles [1]. The existence of the ropes is responsible for the lack of solubility of SWCNTs in solvents. SWCNTs exist in bundles, as it is, thermodynamically, the optimum situation for that system. Thus, the thermodynamics of the system must be altered so that de-bundled SWCNTs result. Several approaches have been employed to dissolve the SWCNT bundles and to disperse the bundles to the point at which individual SWCNTs are stable in solution [2]. These methods include the use of amide solvents [3,4,5] surfactants [6,7], polymers [8,9], strong acids [10], and surface functionalisation [11]. However, the ideal situation must be to dissolve and de-bundle the SWCNTs in an appropriate solvent at concentrations that are useful for their implementation in applications. This would maintain the structural integrity of the SWCNTs, thus preserving their unique physical and electronic properties. This work focuses on the amide solvent N-methyl-2-pyrrolidone (NMP). Amide solvents are, to date, the best solvents reported for generating SWCNT dispersions [12].

Initial investigations into the dispersive effects of amide solvents show the combination of a low hydrogen bond parameter and a high electron pair donicity can lead to excellent dispersions. Landi et al. [13] carried out optical absorption analysis, in their work on the effects of alkyl amide solvents on the dispersion of SWCNTs, over a range of concentrations. Deviations from the linearity of the Lambert-Beer law were used to probe the extinction coefficient of the system. Whilst the initial research into the dispersive effects of amide solvents on SWCNTs has been novel, there remains a lack of understanding with regard to the physical processes that govern the dispersion. Our work aims to probe and characterise the dispersion and de-bundling of the SWCNTs in the

CP786, *Electronic Properties of Novel Nanostructures*, edited by H. Kuzmany, J. Fink, M. Mehring, and S. Roth
© 2005 American Institute of Physics 0-7354-0275-2/05/$22.50

amide solvent NMP, in an attempt to garner a basic grasp of the physics that lies behind the process.

EXPERIMENTAL DETAILS & RESULTS

The ability of NMP to disperse SWCNTs in solution was quantified by carrying out a sedimentation study of SWCNTs in NMP, over a range of concentrations. This was accomplished by sonicating (using a high power sonic tip, 200 Watts at 20%) the SWCNTs, in the solvent, for 1 minute and then monitoring the optical transmission of the samples.

Van Hove singularities, a spectroscopic footprint of SWCNTs, caused by the quasi 1-D nature of SWCNTs, were investigated. Absorbance Spectroscopy was carried out on dispersions of SWCNTs, in NMP, over a wide concentration range (from 1 mg / ml to 3×10^{-7} mg / ml). As the concentration dropped, the Van Hove peaks sharpened, indicating the de-bundling of the SWCNTs in the solvent. Figure 1 shows two spectra, at high (0.11mg / ml) and medium (0.0078 mg / ml) concentrations. Sharp Van Hove singularities are evident for the medium and low concentrations, clearly indicating the presence of small SWCNT bundles, or indeed individual SWCNTs.

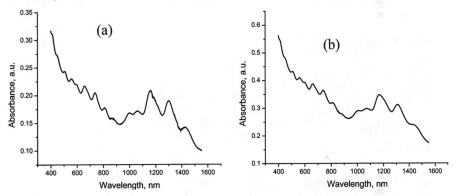

FIGURE. 1 Absorbance Spectra for HiPCO SWCNTs, in NMP, at two concentrations, representing a medium (a) and high (b) value, from the concentration range studied.

Absorbance, A, is governed by the Lambert-Beer law, as given by equation 2.

$$A = \alpha c l \qquad (2)$$

Plotting absorbance versus concentration, for a homogenous system should yield a linear response, as α and l are constants. However, as described by Landi et al. [13], α only remains constant for a homogenous system. Thus, deviations from the linearity of the Lambert-Beer law, in our system, are thought to be caused by the de-bundling of SWCNTs. Figure 2 shows a plot of α as a function of concentration. The data was extracted from the 1170nm Van Hove peak, in the absorbance spectra. At high SWCNT concentrations α remains constant (Fig. 2-A) as the SWCNT concentration is lowered. A stable α value is seen to represent a stable, unchanging, system. At lower SWCNT concentrations (Fig. 2-B), the rise in the α value, as the SWCNT concentration lowers further, is thus thought to represent the de-bundling of the SWCNTs. We expect that at concentrations below 1 x 10^{-6} mg / ml the α value will plateau again, representing a system comprised entirely of SWCNTs.

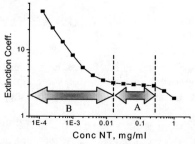

FIGURE 2. Extinction Coefficient, α, plotted as a function of SWCNT concentration. Data collected from the 1170nm Van Hove Peak.

Preliminary AFM results are seen to agree with the spectroscopic results. The image shown in Fig. 3 shows SWCNT, in NMP, at a concentration of 0.00156 mg / ml. The lengths and diameters of the SWCNTs were characterised. The dominant feature of the image, however, is the presence of small SWCNT bundles at relatively high SWCNT concentrations.

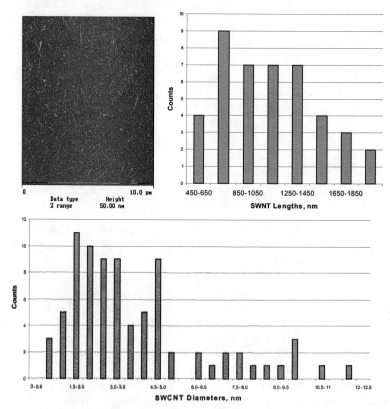

FIGURE 3 AFM image of SWCNTs in NMP. Concentration of SWCNTs : 0.0156 mg / ml

242

A wide variety of techniques have been utilised to assess and describe the dispersive capabilities of NMP. Preliminary computational simulations, carried out within our group [14], show NMP to be a relatively ordered solvent. The introduction of a SWCNT into that system introduces a local-disorder in the vicinity of the SWCNT, as the NMP molecules re-arrange themselves around the SWCNT in order to minimise the total energy of the system. This injection of local-disorder will cause the entropic term of the free-energy, of the total system (SWCNT in NMP), to increase, relative to the initial NMP-only system. This in turn will, thermodynamically, favour the de-bundling of the SWCNTs in NMP, where the activation energy for this process to occur is provided by sonication.

CONCLUSIONS

In this work we have illustrated the excellent dispersion of SWCNTs in the amide solvent NMP. This compares favourably with common organic solvents used. The de-bundling of SWCNTs has been characterised and monitored showing stable dispersions of SWCNTs at relatively high SWCNT concentrations.

ACKNOWLEDGEMENTS

The authors wish to acknowledge the financial support from the European Union, the Irish Education Authority, Enterprise Ireland, and Intel Ireland.

REFERENCES

1. Jian Zhao, 'Dispersion of Carbon Nanotubes and their Polymer Composites' Literature Review, University of Cincinnati, Oct 2001
2. Andreas Hirsch, Angew. Chem. Int. Ed. **2002**, 41, No. 11
3. C.A. Furtado, J.A.C.S. **2004**, 126, 6095 - 6105
4. K.D. Ausman, J. Phys. Chem. B, **2000**, 104, No. 38, 8911 – 8915
5. B.J Landi, J. Phys. Chem. B **2004**, 108, 17089 - 17095
6. M.J. O'Connell, Science **2002**, 297, 593 – 596
7. S.M. Bachilo, Science, **2002**, 298, 2361 – 2366
8. A. Star, Angew. Chem. Int. Ed. **2001**, 40, No.9
9. D.W.Steuerman, J. Phys. Chem B **2002**, 106, 3124 – 3130
10. A. Penicaud, J.A.C.S. **2005**, 127, 8 - 9
11. A. Koshio, Nano Lett., **2001**, 1 (7), 361 – 363
12. K.D. Ausman, J. Phys. Chem. B, **2000**, 104, 38
13. B. Landi, J. Phys. Chem. B, **2004**, 108, 17089 – 17095
14. Mac Kernan et al. in preparation.

Polymer-doped carbon nanotubes: Electronic properties and transport suppression

D. Grimm[*,†], A. Latgé[†], R. B. Muniz[†] and M. S. Ferreira[**]

[*]d.grimm@ifw-dresden.de; Leibniz Inst. for Solid State & Material Research Dresden, Germany
[†]Instituto de Física, Universidade Federal Fluminense, 24210-340 Niterói-RJ, Brazil
[**]Physics Department, Trinity College Dublin, Dublin, Ireland

Abstract. The field of conducting polymers and molecular carbon nanotubes have gained a lot of attention in the recent years mainly due to the large possibility of technological applications such as light emitting diodes and basic elements for molecular electronics. In this contribution we present an study of the influence of the interaction strengths between single-walled carbon nanotubes and the attached polymers on the electronic properties of these newly formed hybrid structures. By adopting a simple tight-binding model, we investigate the conductance of the functionalized CNs in the presence of a varied number of polymers attached on their walls. Random distribution for the set of grafted polymers are considered. The quantum conductance is calculated via the Kubo-Landauer formula, expressed in terms of single-particle Green functions, and we investigate the correlation between the added molecules concentration and the suppression of the electronic transport.

Keywords: Functionalized Carbon Nanotubes, Transport, Conductance, Green Function
PACS: 73.22.2f, 73.23.2b, 73.63.2b

The use of individual molecules, like carbon nanotubes (CN), as functional electronic devices is currently in focus due to the miniaturization strategies pursued by the electronic and computer industries. The construction of more complex structures based on CNs is crucial to expand their applicability[1]. There is growing interest in the possibility of controlling the nature, size and concentration of foreign objects that can interact with nanotubes[2]. Chemically altered CNs may change its mechanical and electronic properties, rising diverse potential applications as nanosensors and -devices[3, 4]. With such a motivation, here we investigate how the transport properties of a CN is affected by the absorption of a foreign molecule to its walls[5, 6]. Figure 1(a) shows an atomic representation of the simple systems considered here, recently published in Ref. [7].

Theory and Model

The quantum conductance Γ is calculated by the Kubo-Landauer formula expressed in terms of single-particle Green functions (GF), which are the key quantities in our formalism[8]. In a metallic CN with ideal electrical contacts one finds perfect transmission characterized by two units of quantum conductance $\Gamma = 2\Gamma_o = 4e^2/h$[9, 10]. Although high-precision *ab-initio* techniques have been used to study how the electronic structure of a CN is affected by the introduction of foreign atoms[11], they are limited to a few attachments and are too computer-time consuming to treat highly disordered environments such as the ones we want to investigate. Simple models for the electronic structure are then quite appropriate for the present investigation.

CP786, *Electronic Properties of Novel Nanostructures*, edited by H. Kuzmany, J. Fink, M. Mehring, and S. Roth
© 2005 American Institute of Physics 0-7354-0275-2/05/$22.50

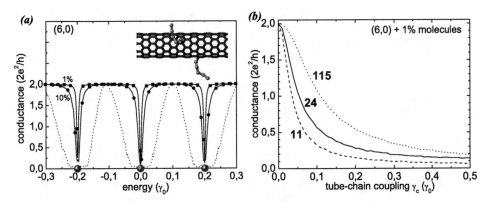

FIGURE 1. *(a)* Atomic configuration scheme of the composed system of a carbon nanotube and adhered molecules shown in inset. Conductance of this modelled system comprising a (6,0) CN with 1%(squares) and 10%(circles) of 24-atoms linear chain molecules attached ($\gamma_c = 0.3$, dotted curve 0.9), as a function of Fermi energy. *(b)* Conductance for a fixed molecular concentration (1%) as a function of the coupling energy γ_c between the CN and the chain, calculated at the central eigenvalue of the attached chains.

The electronic structure is known to be well described by a single-band tight-binding model[12], at least for energies close to the Fermi level. We represent the molecule by a finite-sized linear chain of atoms using the same model as for the CNs. For each system, the GFs are described by linking the local GF ($\hat{G}^0$) via the hopping matrices ($\hat{T}$), resulting in the matricial Dyson's equation $\hat{G}_{ij} = \hat{G}_{ij}^0 \delta_{ij} + \sum_k \hat{G}_{ii}^0 \hat{T}_{ik} \hat{G}_{kj}$ with the indices describing rings (for CN) or atoms (for the molecule)[13]. In what follows, the diagonal self-energies (in $\hat{G}$) and off-diagonal hopping energies (in $\hat{T}$) of the isolated parts are taken as ε_o and γ_o for the CN, and ε_m and γ_m for the molecule. As the energy of the 2p electrons represent just a shift in the electronic properties, we fix our energy origin at $\varepsilon_o = 0$. All energy parameters are given in terms of γ_o. Using real-space renormalization techniques, the surface GF of the tube $\hat{G}_{0,0}$ and the grafted polymer $\hat{G}_{1,1}$ can be calculated[14]. We connect the end atom of the molecule to the tube by switching on an electronic hopping (γ_c) between the parts, resulting in the arbitrary GF matrix element

$$\hat{G}_{i,j}(\varepsilon) = \hat{G}_{i,j}^0 + \hat{G}_{i,0}\,\gamma_c(1 - \hat{G}_{1,1}\,\gamma_c\,\hat{G}_{0,0}\,\gamma_c)^{-1}\,\hat{G}_{1,1}\,\gamma_c\,\hat{G}_{0,j}\,, \tag{1}$$

where i and j are ring sites belonging to the tube. In this way, the GF of the connected structure is written in terms of the GFs of the disconnected systems. The Kubo conductance may be directly expressed in terms of the calculated GFs

$$\Gamma = -\frac{2e^2}{h} \sum_{m \neq n} \sum_{k \neq l} \hat{T}_{mn} \mathrm{Im}\hat{G}_{nk}^+(\varepsilon)\hat{T}_{kl}\mathrm{Im}\hat{G}_{lm}^+(\varepsilon)\,, \tag{2}$$

with one conductance channel $\Gamma_0 = \frac{2e^2}{h}$. One immediate consequence of Eq.(1) & (2) is that the effect of the molecular attachment on the CN is explicitly displayed through the surface Green function of the molecular structure ($\hat{G}_{1,1}$). It is clear that small values of $\hat{G}_{1,1}$ lead to a regime of molecular transparency, in which the electron propagation is exclusively determined by the unperturbed GF of the isolated CN. On the other hand, at the eigenstates of the isolated molecule, $\hat{G}_{1,1}$ also carries information about the resonant states of the molecule leading to a divergence of the second term of Eq.(1).

245

Results and Discussion

The modelled structure consists of a certain number of molecules adhered randomly to the CN walls. One of the advantages of the present model is its ability to easily incorporate such kind of disorder within the Green function formalism in real space. We consider a large section of the CN containing 100 carbon rings to which molecules may be randomly docked. By summing over a large number of configurations (2000) we are able to calculate the average conductance for a given concentration of attached molecules. Figure 1(a) displays the conductance of a (6,0) CN with percentages equal to $x = 1\%$ (squares) and 10% (circles) of tube sites with connected molecules for a moderately low value of the coupling $\gamma_c = 0.3$. The simulated linear chains consist of 24 atoms and are polymer-like with $\varepsilon_m = 0.1$ and $\gamma_m = 0.8$. The eigenenergies obtained from the corresponding surface GFs of the isolated polymers are marked by full circles at the energy axis. It is clear that the conductance is suppressed exactly at those energies that coincide with these isolated chain eigenvalues. However, contrarily to the pure one-dimensional counterpart, the conductance is not totally blocked due to the possibility of the electrons of travelling alternative paths avoiding the polymerized sites[15]. Moreover, transparent regimes with no influence on the transport properties are also present. For higher concentrations x of grafted sites, the conductance reductions occur in extended regions around the isolated molecule eigenvalues. As expected, the effect is more pronounced for higher tube-molecule coupling values, as shown in fig. 1(a) by the dotted line for $\gamma_c = 0.9$ and a percentage equal to 10%.

The results discussed in this paper are focused on molecules that are weakly coupled to the nanotube, for instance, by means of a van der Waals type of interaction. Covalently-bounded molecules, which are much stronger coupled to the tube wall, would require calculations based on first principles and the inclusion of more proximate atom sites[6]. Nevertheless, it is instructive to investigate the effect of the coupling strength on the conductance by varying γ_c, as well as the role played by the molecule size. The transport is calculated at one of the central chain eigenvalues. Conductance results, as functions of γ_c, are shown in Figure 1(b) for different molecule sizes (consisting of 11, 24, and 115 atoms). As expected, the conductance suppression is more pronounced for stronger couplings. Likewise, the conductance decays exponentially with γ_c, and the decaying rate is inversely proportional to the size of the attached chain. As the amount of the resonant modes increases with the chain atom number, the weight of the corresponding eigenvalues on the transport decreases.

The dependence of the conductance suppression on the concentration of polymer chains with different lengths is shown in Fig. 2(a), calculated at one of the eigenvalues of the polymer. A similar conductance suppression effect as for the coupling dependence [fig. 1(b)] is found, i.e. the transport reduction for higher x values is also inversely proportional to the length of the attached molecule. We identify two distinct conductance regimes, separated by a transition region, indicating a passage from ballistic to insulating behavior. Γ decays exponentially for very low concentrations ($x < 1\%$), and decreases almost linearly for $x > 50\%$. For a better visualization the $\hat{x}$-axis for low concentrations is blown up. The inset highlights the exponential decaying regime by plotting $log(\Gamma - \Gamma_{plat})$, where Γ_{plat} is the average value of the nearly constant conductance plateau of the first regime between $1\% < x < 15\%$, shown in the left part of fig. 2(a). The interchange of the curves for the two different molecule lengths reach from the distinct plateau value, i.e. $\Gamma_{plat} = 0.255$ and 0.592 for the 24 and 115 atomic chains, respectively.

The evolution of the conductance suppression for energies around one of the chain's eigenvalues is depicted in Figure 2(b) with a gray-scale 2-dimensional plot. The diagram shows the conductance as a function of both the molecular concentration and the Fermi energy (9,0) CNs. A well-defined conductance gap centered at one of the molecule eigenvalue, indicated by the

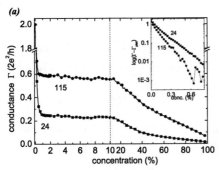 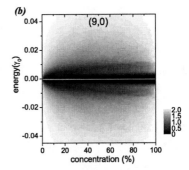

FIGURE 2. *(a)* Conductance decay at one of the eigenvalues of the chains, as a function of the concentration of molecules attached to a (9,0) CN. Inset shows the exponential decay for very low concentrations. *(b)* Gray-scale maps of the conductance as functions of the coverage percentage of attached 24-atomic molecules and the Fermi level. The white line mark the eigenvalue of the isolated molecule ($\varepsilon_m = 0.1$, $\gamma_m = 0.8$, and $\gamma_c = 0.3$).

white line, is clearly visible for concentrations $x \gtrsim 40\ \%$ (dark regions). For very high vestured tubes ($x \gtrsim 70\ \%$), the electron cannot travel through the structure without hitting a grafted site. The system appears quite ordered without major changes upon further attaching of molecules.

Summarizing, we have calculated the effects of polymer-like structures grafted to the walls of CNs on the transport properties. The adopted simple tight-binding model enable us to treat appropriately the disordered systems defined by the randomly distribution of the attached chains. By calculating the conductance we determined a transport suppression at particular energy values corresponding to the isolated chain resonant states. Different electronic transport regimes are found depending essentially on the concentration of attached molecules. This theoretical study should be helpful, for instance, on the analysis of the nature of the foreign structures which may conveniently modulate transport responses.

Acknowledgments Support by the Brazilian Agencies CNPq, CAPES, IM Nanociências, PRONEX; Science Foundation and Enterpr. Ireland. We thank A.T. Costa for useful discussions.

REFERENCES

1. A. Bachtold, P. Hadley, T. Nakanishi, and C. Dekker, Science **294**, 1317 (2001).
2. E. Katz, and I. Willner, Chem. Phys. Chem. **5**, 1085 (2004) and references herein.
3. P. G. Collins *et al.*, Science **287**, 1801 (2000); H. Hu *et al.*, J. Phys. Chem. B **109**, 4285 (2005).
4. F. Simon *et al.*, Chem. Phys. Lett. **383** 62 (2004); H. Kuzmany *et al.*, Synth. Metals **141** 113 (2004).
5. J. Chen *et al.*, Science **282**, 95 (1998).
6. R. J. Baierle *et al.* Phys. Rev. B **67**, 33405 (2003); S. B. Fagan *et al.*, Nano Lett. **4**, 1285 (2004).
7. D. Grimm, A. Latgé, R. B. Muniz, M. S. Ferreira, Phys. Rev. B **71**, 113408 (2005).
8. C. G. Rocha *et al.*, Phys. Rev. B **65**, 165431 (2002); A. Latgé *et al.*, Physica E **13**, 12624 (2002).
9. C. T. White and T. N. Todorov, Nature(London) **393**, 240 (1998).
10. C. Dekker, Phys. Today **52**, 22 (1999).
11. S. Latil, S. Roche, D. Mayou, and J.-C. Charlier, Phys. Rev. Lett. **92**, 256805 (2004).
12. M. S. Ferreira *et al.*, Phys. Rev. B **62**, 16040 (2000); *ibid.* **63**, 245111 (2001).
13. D. Grimm *et al.*, Phys. Rev. B **68**, 193407 (2003); *ibid.* J. of Molec. Catalysis A **228**, 125 (2005)
14. A. Latgé *et al.*, Phys. Rev. B **67**, 155413 (2003).
15. G. Treboux, P. Lapstun, Z. Wu, and K. Silverbrook, J. Phys. Chem. B **103**, 8671 (1999).

Fuctionalization of Single-Walled Carbon Nanotubes with Organo-Lithium Compounds: A Combined XPS, STM, and AFM study

Ralf Graupner*, Jürgen Abraham†, Andrea Vencelová*, Peter Lauffer*, Lothar Ley* and Andreas Hirsch†

*Universität Erlangen, Institut für Technische Physik, Erwin-Rommelstr. 1, D-91058 Erlangen
†Universität Erlangen, Institut für Organische Chemie, Henkestr. 42, D-91054 Erlangen

Abstract. A covalent sidewall functionalization of Single-Walled Carbon Nanotubes (SWCNTs) has been achieved by a reaction with butyllithium in benzene. The nucleophilic attack of the lithium organyl leads to a transfer of nagative charge to the SWCNTs which is balanced by the positive charge of the Li-ions. In our investigation we were able to verify the negative charge on the SWCNTs by a concomitant shift of the binding energy of the C 1s core levels of the SWCNTs in X-ray induced photoelectron spectroscopy (XPS). Moreover, scanning tunneling microscopy (STM) reveals a decoration of the tubewalls by features which exhibit the expected threefold symmetry of the butyl-groups. The charging of the SWCNTs in solution could also be verified by applying a homogeneous electric field using Au coated glass electrodes. The electrodes were chraracterized by XPS and atomic force microscopy (AFM). These measurements reveal the appearance of Li on one electrode, and the occurance of SWCNTs on the counter electrode.

Keywords: Single-Walled Carbon Nanotubes, Functionalization, XPS, STM
PACS: 79.60.Jv, 81.07.De, 82.30.Fi

INTRODUCTION

Chemical functionalization is a prerequisite for a widespread application of Single-Walled Carbon Nanotubes [1]. An attachment of functional groups enables the solubility of the SWCNTs in different solvents that is usually achieved by a non-covalent functionalization using surfactants. An increase in solubility may also be induced by a charging of the tubes. In this case, the Coulomb repulsion between the tubes may surpass the van der Waals interaction between the tubewalls, preventing them from re-bundeling in solution. Here, we study the reaction of SWCNTs with an organo-lithium compound, namely t-Butyllithium. The nucleophilic attack of the tubewalls results in a covalent sidewall functionalization with butyl groups, accompanied by a transfer of negative charge to the SWCNTs. Our aim is to verify the charge transfer and to detect the functional groups on the sidewalls of the tubes using scanning tunneling microscopy.

EXPERIMENTAL

The samples used for this study were purified SWCNTs, purchased from CNI. Func-tionalization was carried out under nitrogen atmosphere by ultrasonicating 10 mg of the starting material in benzene and then adding a few ml of t-Butyllithium. As a result, a

CP786, *Electronic Properties of Novel Nanostructures*, edited by H. Kuzmany, J. Fink, M. Mehring, and S. Roth
© 2005 American Institute of Physics 0-7354-0275-2/05/$22.50

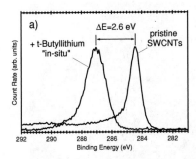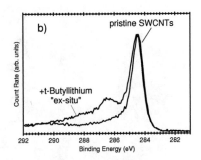

FIGURE 1. C 1s core level spectrum of the starting material compared to a) the C 1s core level of the functionalized SWCNTs measured "in-situ", i.e. without contact to air, and b) the C 1s core level spectra of the fuctionalized SWCNTs after "ex-situ" preparation.

black homogeneous solution forms which remains stable as long as it is kept under an inert atmosphere. For photoelectron spectroscopy the samples were usually filtered to obtain a solid "bucky-paper". As the charge transfer which results in the reaction with the organo-lithium compound is not stable if the sample is brought into contact with ambient air, a special "in-situ" preparation method was carried out to prevent a reaction of the functionalized SWCNTs. To do so, a small amount of the solution containing the functionalized SWCNTs was picked up with a syringe. The solution was then dropped through a rupper plug onto a Au coated sample holder in the load-lock chamber of the spectrometer which was vented using dry nitrogen under slight overpressure. After pump down of the solvent, a closed film of the functionalized SWCNTs is formed on the sample holder. XPS measurements were carried out using monochromized Al-K$_\alpha$ radiation ($h\nu = 1486.6$ eV).

The samples for the STM measurements were prepared by spraying a drop of the solution containing the functionalized SWCNTs onto a flame annealed Au substrate. The density of the SWCNTs on the sample was checked by AFM measurements before introduction into the STM chamber. STM measurements were carried out at liquid He temperature (4.7 K). The tips for STM measurements are chemically etched Pt-Ir tips. As the preparation is performed "ex-situ", no charge transfer is expected to be measured by STM.

RESULTS AND DISCUSSION

In Fig. 1, the C 1s core level spectra of pristine SWCNTs are compared to those measured on the functionalized samples. If the samples are prepared without contact to ambient air, shown in Fig. 1 a, the measured binding energy of the C 1s core level is 2.7 eV higher than that measured on the pristine, i.e. unfunctionalized SWCNTs. If, however, the sample is prepared as a regular bucky paper which was brought into contact with ambient air, the C 1s core level binding energy is virtually unaltered (Fig. 1 b) except for an additional component on the high binding energy side of the main C 1s core level.

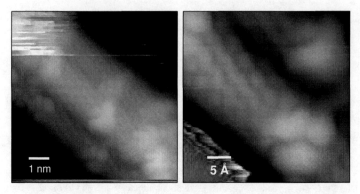

FIGURE 2. STM images of functionalized SWCNTs (U= −1 V, I= 15 pA), measured at 4.7 K. The right image is an expanded view of the central lower region of the left image with a linear background removed. Notice the three-lobed structure in the lower right corner which we attribute to a butyl group attached to a SWCNT.

Upon reaction with *t*-Butyllithium, negative charge is transfered to the SWCNTs. This charge is accomodated by previously unoccupied states of the carbon nanotubes. Therefore, the Fermi-level of the functionalized SWCNTs is located in the conduction band states of the SWCNTs. As in photoelectron spectroscopy binding energies are always measured with respect to the Fermi-level (=the zero energy of the binding energy scale), this shift of the Fermi-level is reflected in a binding energy shift of the C 1s core level towards higher binding energies [2]. Similar results are obtained from the stage one Li intercalation compound in graphite, LiC_6 [3], however, here the observed shift of the C 1s core level amounts only to 1.6 eV.

For the sample which was brought into contact with air, no shift of the main C 1s core level component is observed. As the functionalization of the sidewalls of the SWCNTs leads to the introduction of defects, the functionalized tubes are more sensitive towards oxidation than the untreated SWCNTs, which causes the chemically shifted component on the high binding energy side of main C 1s core line.

STM images, taken on the functionalized SWCNTs are shown in Fig. 2. They display an atomically resolved image of a small bundle of SWCNTs.

In the STM images, a decoration of the tubewalls is visible. We attribute these decorations to the butyl groups attached to the SWCNTs. In the expanded view, shown in the right image of Fig. 2, the expected threefold symmetry of the functional groups is evident. However, the STM studies revealed that not all tubes were found to be affected.

As the functionalized SWCNTs are charged in solution, an applied homogeneous electric field should result in a separation of functionalized SWCNTs. To do so, Au-coated glass plates were used as electrodes and mounted at a distance of 5 mm. These electrodes were put into solution and a voltage of 3 V was applied for 30 mins. The typical current which was measured amounted to 1 μA. Figure 3 shows AFM images of the cathode (left) and anode (right) after removal from the solution. The cathode shows round structures which by XPS measurements (not shown) are identified as oxidized Li. No Au is found on that electrode which proves that Li formed a closed layer. On

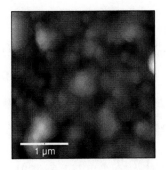 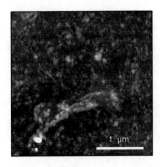

FIGURE 3. AFM images of the cathode (left) and the anode (right) after removal from the solution.

the anode, however, Au is still visible in the XPS measurements and no Li is found. In AFM images (Fig. 3, right), elongated structures are visible which presumabely are the functionalized SWCNTs. Here, the fact that no closed layer of the SWCNTs is formed is caused by the weak interaction between the SWCNTs and the Au electrode. After contact to the anode, the tubes release their charge and go back into solution.

CONCLUSION

Using an "in-situ" praparation technique for XPS measurements, we were able to confirm that the reaction of *t*-Butyllithium leads to negatively charged SWCNTs. The observed shift of the C 1s core level is found to be 2.6 eV. STM images reveal that the SWCNTs show a decoration of the sidewalls. These groups indeed show the expected threefold symmetry of the butyl groups. However, not all tubes were found to be affected. Applying a homogeneous electric field between Au contacts in solution results in the formation of a closed layer of Li on the cathode, whereas elongated structures are found on the anode which are due to functionalized SWCNTs.

ACKNOWLEDGMENTS

This work was supported by the Bayerische Forschungsstiftung under the auspices of the "Bayerische Forschungsverbund für Werkstoffe auf der Basis von Kohlenstoff, FORCARBON".

REFERENCES

1. A. Hirsch, *Angew. Chem. Int. Ed.*, **41**, 1853–1859 (2002).
2. R. Graupner, J. Abraham, A. Vencelová, T. Seyller, F. Hennrich, M. M. Kappes, A. Hirsch, and L. Ley, *Phys. Chem. Chem. Phys.*, **5**, 5472–5476 (2003).
3. G. K. Wertheim, P. M. T. M. V. Attekum, and S. Basu, *Solid State Commun.*, **33**, 1127–1130 (1980).

Functionalized Water-Soluble Multi-Walled Carbon Nanotubes: Synthesis, Purification and Length Separation by Flow Field-Flow Fractionation

Theodoros Felekis,[1] Nikos Tagmatarchis,[1]* Andrea Zattoni,[2] Pierluigi Reschiglian[2]* and Maurizio Prato[3]*

[1] National Hellenic Research Foundation, 48 Vass. Constantinou Ave., 116 35 Athens, Greece
[2] Dipartimento di Chimica 'G. Ciamician', Università di Bologna, Via Selmi 2, 40126 Bologna, Italy
[3] Dipartimento di Scienze Farmaceutiche, Università di Trieste, Piazzale Europa 1, 34127 Trieste, Italy

Abstract. Water-soluble, functionalized multi-walled carbon nanotubes (MWNTs) are length-separated and purified from amorphous material through direct flow field-flow fractionation. MWNTs subpopulations of relatively homogeneous, different length are obtained from collecting fractions of the raw, highly polydispersed (200-5000 nm) MWNT sample.

Keywords: carbon nanotubes, functionalization, azomethine ylides, water-soluble, purification, separation, field-flow fractionation.
PACS: 81.07.De

Carbon nanotubes (CNT) exhibit novel structural and electronic properties that bridge the bulk and molecular states and represent a flexible building block for the construction of new materials for diverse nanotechnological applications [1]. However, pristine carbon nanotubes in the forms of single-walled and multi-walled (SWNT and MWNT, respectively) are not a single compound but instead contain impurities such as metallic nanoparticles and amorphous carbon. These impurities and the lack of CNT monodispersity bring obstacles and hinder to a great extent the development and processibility of CNT for specialized applications. Recently, great efforts have been directed to resolve these issues and purify CNTs, including use of gel permeation chromatography (GPC) [2, 3], size exclusion chromatography (SEC) [4, 5] and capillary electrophoresis (CE) [6].

Field-flow fractionation (FFF) is a separation technique based on elution through a thin, empty channel. During elution separation is structured by physical interaction of sample components with an external field perpendicular to the fluid flow which drives the particles to different average positions across the thin channel. In flow FFF (FlFFF) the physical field is a hydrodynamic field generated by a second flow of

CP786, *Electronic Properties of Novel Nanostructures*, edited by H. Kuzmany, J. Fink, M. Mehring, and S. Roth
© 2005 American Institute of Physics 0-7354-0275-2/05/$22.50

carrier liquid that is applied to the longitudinal flow and separation occurs according to differences in the analyte diffusion coefficient (Fig. 1).

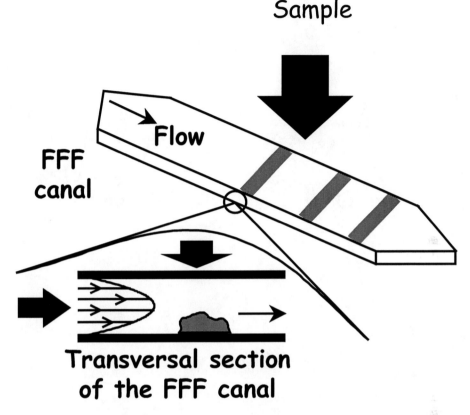

FIGURE 1. General concept of the FlFFF technique.

In the present work, very broadly dispersed, water-soluble functionalized MWNT purified and length-shorted utilizing the FlFFF technique. Briefly, water-soluble ammonium functionalized MWNT [7] obtained upon acidic deprotection of N-Boc functionalized MWNTs via 1,3-dipolar cycloaddition of azomethine ylides (Fig. 2).

The water-soluble modified MWNTs were subjected to FlFFF. The analyses were carried out with programmed field conditions to decrease the retention time of the longest MWNTs thus, reducing their interactions with the accumulation wall of the FlFFF channel. In the obtained fractogram, it is clear an intense peak corresponding to the void time and a broad band from 5 to 30 minutes (Fig. 3).

Six fractions were collected immediately after the void volume and each one was examined by transmission electron microscopy (TEM). Four of these fractions were found to contain MWNTs (fractions b, c, d and e), while the remaining two fractions (f and g) together with the fraction corresponding to the void peak (fraction a) were found to contain only amorphous material.

BocNH(CH₂CH₂O)₂CH₂CH₂NHCH₂COOH   +   HCHO

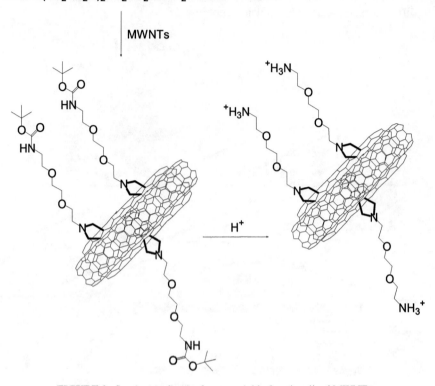

FIGURE 2. Synthetic scheme of water-soluble functionalized MWNTs.

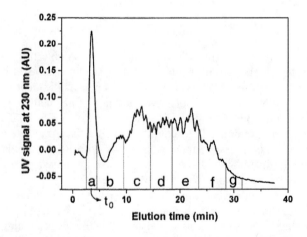

FIGURE 3. Fractogram of water-soluble functionalized MWNTs.

Figure 4 summarizes these findings in the form of TEM images. Evidently, short MWNTs with lengths of 200-600 nm appeared in fraction b, while MWNTs with lengths of 600-1500 nm were found in the following fraction c. Longer MWNTs were isolated in fraction d (1.5-5.0 μm), while very long MWNTs exceeding 5.0 μm length were isolated and visualized in fraction e. It should be noted that, in all cases, the reported micrographs were not randomly taken but they depict a statistically significant visualization of the actual fraction content.

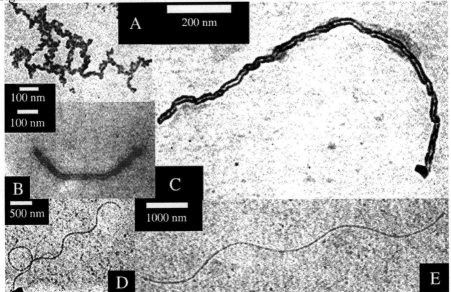

FIGURE 4. Representative TEM images of (A) first eluted FlFFF fraction (void) consisting of amorphous, nanosized matter; (B) to (E) following eluted fractions collected at increasing retention time, consisting of respectively short, moderate, and long water-soluble functionalized MWNTs.

MWNT of similar length were consistently observed within each separated fraction. It is important to focus that no retention foldback from nanosized to micronsized components was observed along the fractions. Water-solubility of functionalized MWNT allowed for their length-separation in homogeneous phase, which probably made the longest MWNTs able to wrap into a particle of spherical shape the size of which is much smaller than their sphere of rotation. In other words, the retention behavior of the micronsized MWNTs resulted similar to the retention behavior of nanosized spheres, which is known to depend on differences in the particle diffusion coefficient and, thus, to increase with increasing sphere size. This retention behavior was observed by other authors in FlFFF of water-insoluble SWNTs, and therein explained as due to the very high flexibility of the SWNTs [8]. It is thus proved that, through FlFFF, broadly dispersed water-soluble MWNTs can be sorted on a length basis and purified without sample pre-treatment to give different MWNT fractions of relatively uniform length.

In conclusion, it is shown that the FlFFF technique can be successfully applied to purify and sort by length water-soluble functionalized MWNTs. Although the resulting length-based MWNT sorting was performed on a micro-preparative scale, the isolation of purified and relatively uniform-length MWNTs is of fundamental importance for further characterization and applications requiring monodisperse MWNT material.

ACKNOWLEDGMENTS

NT acknowledges financial support by EUROHORCs/ESF through the European Young Investigator (EURYI-2004) Awards Scheme. This work was carried out with partial support from the EU (RTN programs "WONDERFULL" and "CASSIUSCLAYS") and MIUR (PRIN 2004, prot. 2004035502 and FIRB).

REFERENCES

1. Carbon Nanotubes, special issue: *Acc. Chem. Res.* **35**, 977-1113 (2002)
2. Chattopadhyay, D., Lastella, S., Kim, S., and Papadimitrakopoulos, F., *J. Am. Chem. Soc.* **124,** 728-729 (2002)
3. Zhao, B., Hu, H., Niyogi, S., Itkis, M. E., Hamon, M. A., Brownik, P., Meier, M. S., and Haddon, R. C., *J. Am. Chem. Soc.* **124**, 11673-11677 (2001)
4. Duesberg, G. S., Burghard, M., Muster, J., Phillip, G., and Roth, S., *Chem. Commun.* 435-436 (1998)
5. Holzinger, M., Hirsch, A., Bernier, P., Duesberg, G. S., and Burghard, M., *Appl. Phys. A* **70**, 599-602 (2000)
6. Doorn, S. K., Fields, R. E., Hu, H., Hamon, M. A., Haddon, R. C., Selegue, J.-P., and Majidi, V. *J. Am. Chem. Soc.* 124, 3169-3174 (2002)
7. Georgakilas, V., Tagmatarchis, N., Pantarotto, D., Bianco, A., Briand, J.-P., and Prato, M., *Chem Commun.* 3050-3051 (2002)
8. Chen, B., and Selegue, J. P., *Anal. Chem.* **74**, 4774-4780 (2002)

SWNTs with DNA in Aqueous Solution and Film

V.A. Karachevtsev[1], A.Yu.Glamazda[1], V.S. Leontiev[1], P.V. Mateichenko[2], and U. Dettlaff-Weglikowska[3].

[1]*Institute for Low Temperature Physics and Engineering, NASU, 61103, Kharkov, Ukraine,*
[2]*Institute for Single Crystals of NASU, Kharkov, 61103, Ukraine,*
[3]*Max-Planck Institute for Solid State Research, Heisenberg Str. 1, 70569 Stuttgart, Germany,*

Abstract. Raman and luminescence spectra of SWNTs in aqueous solutions with fragmented single-stranded DNA and films prepared from these solutions have been obtained. SEM film study shows that nanotubes aggregate in bundles after film drying. In Raman spectra it leads to the enhancement of the intensity of the low-frequency component of the tangential mode (G^-), to the shift and broadening of the high-frequency component of this mode (G^+). It was shown that the ratio of areas of the high-frequency component of the G mode to the low-frequency component (S_{G+}/S_{G-}) can serve as an indicator of nanotube aggregation into bundles for SWNTs suspended in solution. The film yields luminescence which indicates the presence of individual tubes or small bundles in the films. The luminescence bands of SWNTs in the film become wider and are attributed to the interaction of DNA with the nanotube surface in the solid state.

INTRODUCTION

The unique physical properties of carbon nanotubes in conjunction with the recognition capabilities of biomolecules attached to them could lead to biological electronics and chemical/biological sensors. The central direction in creating such devices is surface functionalization of nanotubes and selection/elaboration of interfaces between nanotubes and recognition biosystems. DNA is an ideal polymer for this purpose because it can be used for SWNTs biocompatibility and possesses itself a recognition function. In many cases the potential application of single-walled carbon nanotubes is associated with a design of arrangements based on individual nanotubes properties [1]. To regret, under usual conditions nanotubes aggregate in bundles separation of which into individual tubes is a difficult problem. Up to now separation of individual tubes from bulk material is possible only in aqueous solutions suspended with surfactants [1,2] or a polymer [3,4] after long-term sonication. Unfortunately, the solutions are not convenient for practical usage, in comparison with the solid state. Growing of individual nanotubes on a substrate is not a simple approach. Furthermore, the question whether individual tubes separated in solution by a surfactant or polymer will aggregate in the bundle or not at preparing a film from aqueous solution has not been clarified up to now.

In this report a comparative analysis of luminescence and Raman spectra of single-walled carbon nanotubes suspended in aqueous solutions with DNA and SDS as

CP786, *Electronic Properties of Novel Nanostructures*, edited by H. Kuzmany, J. Fink, M. Mehring, and S. Roth
© 2005 American Institute of Physics 0-7354-0275-2/05/$22.50

well as in films obtained from this suspension has been made. Nanotubes aggregation in films is controlled by scanning electron microscope (SEM).

EXPERIMENTAL

Aqueous suspensions of SWNTs (HiPCO) were prepared with DNA (extracted from chicken erythrocytes, purchased from Reanal, Budapest, Hungary) and the anionic surfactant – SDS (1%) (Aldrich, USA) by sonication with following ultracentrifugation (120000 g, 1 hour). Single-stranded DNA was prepared from native DNA by melting and quick cooling. The mean length of the DNA fragments estimated by the electrophoresis was in the range of 200-300 base pairs.

The nanotubes films for Raman and SEM experiments were obtained by dripping SWNTs suspension with DNA (or SDS) onto the Si substrate and drying in warm air. The Raman experiments were performed using the 632.8 nm excitation from a He-Ne laser (15 mW). Emission spectra were excited with a solid-state laser (λ_{exc}=532 nm, 20 mW) and measured using a double spectrometer with photomultiplier detection, operating in the single photon mode (400-1200 nm). SWNTs films were characterized using the JSM-820 scanning electron microscope (SEM).

RESULTS

As was shown earlier [5,6], the line shape and intensity of the low-frequency component of the G mode (G⁻) of SWNTs are very sensitive to an interaction among nanotubes in bundles. The asymmetric form of the G⁻ band is attributed to the metallic nanotubes and well described by Breit-Wigner-Fano (BWF) function [7]. The intensity of the BWF band was enhanced in thick bundles and decreased in thin ones. In aqueous solutions of SWNTs DNA-suspended the intensity of the G⁻ band is decreasing (Fig.1), in comparison with SWNTs in the film, and after ultracentrifugation becomes very weak in solution containing mainly individual nanotubes. The ratio of areas of the high-frequency component of the G mode to the low-frequency component (S_{G+}/S_{G-}) can serve as an indicator of nanotubes aggregation in bundles for aqueous solution as it was observed for the bundle in the solid state [5,6]. This ratio increases from 1,9 for tubes in solution to 8,6 for SWNTs in film. The width of the G⁺ (the high-frequency component of the G mode) band becomes lager (by about 10 cm⁻¹) in comparison with corresponding bands of SWNTs:DNA solution. In the SWNTs:DNA film the frequency of the G⁺ band is upshifted by 1.6 cm⁻¹ when compared to its position in solution spectrum. From this observation we suggest that the interaction between the wrapped DNA and nanotube is relatively stronger in film (in the absence of water molecules). For additional checking this supposition, we investigated Raman spectra of aqueous solution of SWNTs SDS-suspended and compared them with those of nanotubes film obtained from solution (Fig.1). As for SWNTs:DNA, we observed an enhancement of the intensity of the G⁻ band in film as well as an increase of the width of G⁺ band. The ratio S_{G+}/S_{G-}

decreased under nanotubes aggregation in film (1,4), in comparison with aqueous solution (Fig.1) before (3,7) and after (6,7) ultracentrifugation. In aqueous solution this ratio decreased also when SDS concentration was dropped (Fig.1). Enhancement of solution pH leads also to increasing the G$^-$ band intensity because micelles destruct, nanotubes aggregate and precipitate in some hours at pH10. As in the case of the SWNTs:DNA film, we observed an enhancement in the width of the G$^+$ band in the SWNT:SDS film.

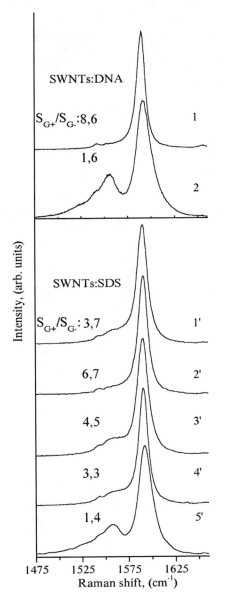

Nanotubes films on the Si substrate obtained from the aqueous solution of SWNTs with DNA were also studied by SEM. SEM images of the film formed after after evaporaton of SWNT:DNA solution and removing the excess of DNA by washing showed small bundles with a bundle diameter less than 50 nm (Fig. 2). This result suggests that SWNTs wrapped with DNA do aggregate into small bundles as the film begins to dry.

Dried SWNT:DNA films can be easily resuspend in water (3-5 min sonication) in contrast to the dried SWNT:SDS films, which need long sonication time and additional quantity of surfactant. This indicates that bundles in the film with wrapped DNA are different from usual bundles. In the film DNA wrapped around SWNT precludes the full nanotubes sticking along the whole tube length. It is obvious that the DNA layer between nanotubes in the bundle facilitates the process of their splitting upon sonication.

FIGURE 1. Raman spectra of SWNTs in range of G mode in aqueous solution (a.s.) and film with DNA (**1** – a.s. after ultracentrifugation; **2** - film obtained from **1**) or SDS (a.s.: before **1'** and after **2'** ultracentrifugation (pH7), **3'**- after water dilution of **2'** solution (SDS concentration was 0,1 %), **4'** – solution **2'** but with pH10; **5'** - films prepared from **2'** solution). Spectra were obtained using the 632.8 nm laser excitation

The luminescence spectra (the range 800-1100 nm) of semiconducting SWNTs in aqueous solution with DNA and of the SWNT:DNA film on the quartz substrate are shown in Fig.3. In this range 5 emission bands corresponding to 5 individual

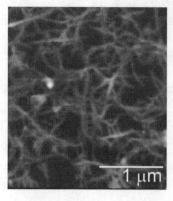

nanotubes of different diameters and chirality are observed for HiPCO SWNTs in aqueous solutions [2,8]. All emission bands can be assigned to the first electronic transition (E^{11}) of semiconducting nanotubes. In the Raman spectrum of aqueous solution of SWNTs with SDS obtained with the same laser excitation, five bands corresponding to RBM were observed. Among them only one band at the frequency $\nu=294$ cm^{-1} can be assigned to the semiconducting nanotube with (8,3) chirality.

FIGURE 2. SEM image of carbon nanotubes film prepared on Si substrate from aqueous solution containing DNA. (scale bar is 1 μm).

As can see from Fig. 3, the spectral position of all bands of SWNTs in solution with DNA is red shifted (by about 11-19 meV) relative to the spectrum of SDS-suspended nanotubes. This shift can be explained by different influences of DNA and SDS on the nanotubes electronic structure and also by the incomplete surface coverage of the polymer at the surface of the nanotube, that allows water molecules to reach the surface and interact with the SWNT [4]. Observation of nanotubes luminescence in the film indicates the presence of

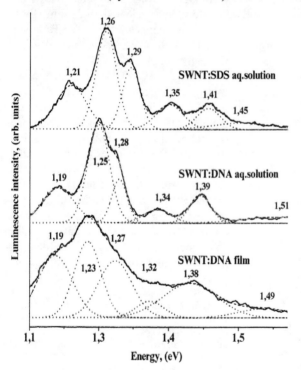

FIGURE 3. High-energy part of luminescence spectrum of HiPCO SWNTs aqueous solutions (suspended with SDS or with DNA) and nanotubes film on quarts substrate obtained from aqueous solution with DNA. Luminescence was excited by DPSS green laser at $\lambda_{ex}=532$ nm.

individual tubes in the film that do not aggregate in bundles and, possibly, the thin bundles consisted of only semiconducting SWNTs. DNA wrapped around SWNT precludes the nanotubes aggregation in the film. In the film the sharp luminescence spectrum of SWNTs in solution disappears, bands are broadened and shifted into the red field up to 16 meV.

The possible reasons of such behavior are: a) the interaction between DNA and the nanotube surface in the solid state can be stronger than in aqueous solution, b) nanotube interaction with substrates can increase also the emission bandwidth, c) intertubes interaction in thin bundles. It should be noted that the emission intensity of nanotubes in the film with DNA is very weak, essentially weaker than nanotubes luminescence in aqueous solution.

CONCLUSION

Under aggregation of nanotubes in solution the intensity of the G^- band increased and the G^+ band becomes wider. The ratio of areas of the high-frequency component of the G mode to the low-frequency component (S_{G+}/S_{G-}) can serve as an indicator of nanotubes aggregation in bundles for aqueous solution.

Nanotubes luminescence in the film indicates the presence of individual tubes in the film or bundles that are formed by tubes separated with the polymer. Although DNA does not hold individual nanotubes from sticking in bundles, but nevertheless, this polymer wrapped around tubes facilitates the following splitting of these bundles.

ACKNOWLEDGMENTS

This work was supported partly by Ministry of Education and Science of Ukraine, Program "Nanophysics and Nanoelectronics" (project No M228-2004).

REFERENCES

1. P.W. Barone, S. Baik, D. A. Heller and M.S. Strano, *Nature Mat.*, **4**, 86-92 (2005).
2. O'Connell, M. J.; Bachilo, S. M.; Huffman, C. B., Moore, V. C.; Strano, M. S.; Haroz, E. H.; Rialon, K. L.; Boul, P. J.; Noon, W. H., Kittrell, C.; Ma, J.; Hauge, R. H.; Weisman, R. B.; Smalley R. E., *Science* **279**, 593-596 (2002).
3. M. Zheng, A. Jagota, E. Semke, B. Diner, R. Mclean, S. Lustig, R. Richardson and N. Tassi, *Nature Materials* **2**, 338–342 (2003).
4. M. S. Strano, M. Zheng, A. Jagota, G. B. Onoa, D. A. Heller, P. W. Barone, M. L. Usrey, *Nano Letters* **4**, 543-550 (2004).
5. C. Jiang, K. Kempa, J. Zhao, U. Kolb, U. Schlech, T. Basche, M. Burghard, A. Mews, *Phys. Rev. B* **66**, 161404-4 (2002).
6. N. Bendiab, R. Almairac, M. Paillet, J.-L. Sauvajol, *Chem. Phys. Letters* **372**, 210-215 (2003).
7. A.M. Rao, P.C. Eklund, S. Bandow, A. Thess, R.E. Smalley, *Nature* **388**, 257-259 (1997).
8. A. Hartschuh, H.N. Pedrosa, L. Novotny, T.D. Krauss, *Science* **301**, 1354-1356 (2003).

Noncovalent Sidewall Functionalization of Carbon Nanotubes by Biomolecules: Single-stranded DNA and Hydrophobin

Sebastian Taeger*, Li Yi Xuang*, Katrin Günther* and Michael Mertig*

*Max-Bergmann-Zentrum für Biomaterialien, Technische Universität Dresden, 01069 Dresden, Germany

Abstract. Single-stranded DNA (ssDNA) is known to disperse individual carbon nanotubes (CNT) into aqueous suspensions. But other biomolecules are able to do so as well. We demonstrate a protein-assisted CNT dispersion by using hydrophobin. The yields of the suspensions are monitored by optical absorption spectroscopy (OAS). We perform atomic force microscopy (AFM) studies of DNA- and hydrophobin-functionalized CNT with a resolution that allows us to identify individual molecules attached to isolated CNT. We control the density of DNA on the nanotubes by the DNA:CNT ratio, and observe stable suspensions of CNT with surprisingly low surface coverages.

Keywords: carbon nanotubes, DNA, dispersion, functionalization
PACS: 81.07.DE, 87.14.Gg, 81.07.-b, 81.16.Fg

INTRODUCTION

In 2003 ssDNA was found to suspend and disperse CNT into water [1], a key step for many applications of, and experiments with CNT. To understand the principles of this process may help to find substances which perform even better than DNA. Continuing our previous work on ssDNA-CNT [2], we study the dependence of the DNA-assisted dispersion of CNT on several parameters, e. g. on the DNA:CNT ratio. Additionally, we started to investigate the interaction of CNT with hydrophobin. Hydrophobins are amphipathic proteins with hydrophobic domains [3], which are expected to interact strongly with the hydrophobic sidewalls of CNT.

EXPERIMENTAL

In our experiments we use laser-ablation CNT[1] after an acidic purification based on a method described by Rinzler [5]. For the results displayed in this work we suspended the CNT by synthetic ssDNA purchased from MWG-Biotech and VBC-Genomics: $(dTMP)_{30}$, a DNA oligomer consisting of 30 thymine bases (T30), and $(dTMP)_{60}$, consisting of 60 thymine bases (T60). The lengths of these molecules are 21 nm and 42 nm, respectively. For the hydrophobin-CNT suspensions, we use ABH1 of the fungus agaricus bisporus. The suspensions where obtained by ultrasonication on ice in an

---

[1] CNT produced by Oliver Jost, IfWW, TU-Dresden [4].

CP786, *Electronic Properties of Novel Nanostructures*, edited by H. Kuzmany, J. Fink, M. Mehring, and S. Roth
© 2005 American Institute of Physics 0-7354-0275-2/05/$22.50

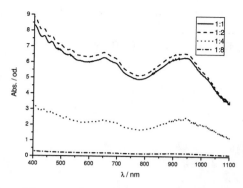

FIGURE 1. Dependence of the suspension efficiency on the DNA:CNT ratio for T30, monitored by OAS.

ultrasonic bath, followed by a centrifugation at 20,000 g for 90 minutes. OAS spectra are taken with an Cray 50 spectrometer. Samples for AFM are prepared on mica substrates and investigated by tapping mode AFM with a NanoScope IIIa (Digital Instruments, Santa Barbara). Image processing and analysis is being performed with WSxM [6]. Further details about sample preparation and investigation are reported elsewhere [2].

RESULTS AND DISCUSSION

We define the efficiency of a suspension process by the amount of suspended CNT material compared to the total amount of CNT inserted to the process. To our knowledge there is no reliable way, to determine the absolute concentration of CNT in a suspension from OAS. Therefore we use the optical density to qualitatively compare our suspensions among each other. The UV-Vis spectra (Fig. 1) show two distinguished peaks, the absorption of the second pair of van-Hove singularities of the semiconducting CNT at 950 nm and the absorption of the first pair of van-Hove singularities of the metallic CNT at 650 nm. At a mass ratio of 1:1 we achieve a high efficiency but observe a lot of free DNA molecules in these samples by AFM. The mass ratio can be lowered to 1:2 without much change in the concentration of suspended CNT (Fig. 1). A further reduction of DNA to 1:4 lowers the concentration of suspended CNT to approximately one half, and a reduction to 1:8 leads to a strong decline of absorption. For 1:2 or lower ratios, the amount of unbound DNA in the suspension is largely reduced.

The DuPont group calculated a relaxed structure of ssDNA-CNT [1]. According to this calculations, the DNA molecules should wrap around the CNT in a rather stretched manner with a wrapping angle to the tubes's axis of about 25 °. Based on this structure we expect an ideal mass ratio of DNA and CNT about 1:1.5, which corresponds to our observation of a saturation of the suspension yield between 1:2 and 1:1, although most of the CNT initially used for the suspension form a pellet during centrifugation.

By AFM, we find that most CNT have very similar heights around 1.5 nm. On top of these, DNA molecules with heights between 0.3 and 0.6 nm are observed (Fig. 2). We count the number of DNA molecules per CNT and calculate a surface coverage given in

TABLE 1. Dependence of the coverage of CNT with T30-DNA molecules on the DNA:CNT mass ratio.

DNA:CNT by mass	DNA molecules per 100 nm of CNT	Surface coverage by area
1:1	5.4	19.3%
1:2	4.9	17.5%
1:4	3.5	12.5%
1:8	2.9	10.4%

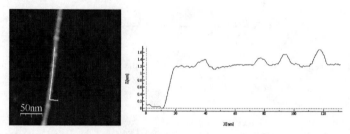

FIGURE 2. AFM image of an single nanotube wrapped by individual T30-DNA molecules.

Tab. 1, based on the assumption that the CNT are individualized. We observe that within one experiment the coverage varies from tube to tube. On average, like the suspension yield, also the coverage of CNT with DNA depends on the DNA:CNT ratio. We find that a surprisingly low coverage of 3 DNA molecules per 100 nm of a CNT is sufficient to keep it in a stable suspension. The maximum average value observed is 5.4 molecules per 100 nm of CNT, which is very close to 5.5, the value one derives using the wrapping angle of 25° from the calculated structure. But within one sample the density varies from 3 molecules per 100 nm for the least covered CNT to 8 molecules per 100 nm for the CNT with the highest coverage.

Unfortunately, due to tip-sample convolution, it is impossible to observe the wrapping angle of the DNA molecules by AFM directly. But it can be determined indirectly

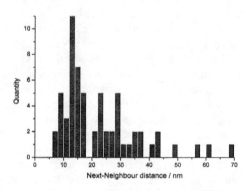

FIGURE 3. Statistics on the distances between neighboring DNA strands for T60-CNT.

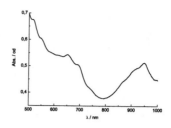

FIGURE 4. AFM image of a hydrophobin functionalized CNT on glass (z-scale is 15nm) and OAS spectrum of hydrophobin-CNT.

by measuring the distance between windings. The evaluation of the AFM images of T60-CNT shows a preferred distance between DNA strands of about 13 nm (Fig. 3) that clearly differs from the mean distance of 22 nm. A distance between windings of 13 nm corresponds to a wrapping angle of 20°. Thus we can conclude that in average wide wrapping is preferred for longer DNA strands. However, also for T60, we find irregularities, like varying coverage from tube to tube or significant gaps of almost 100 nm without any DNA. It is important to look for the reasons of the irregularities in the wrapping patterns, because they may be attributed to the properties of the wrapped CNT, like charged defects, the polarizability of the tube, or even it's chirality.

Furthermore, we demonstrate the successful dispersion of hydrophobin-CNT by AFM and OAS (Fig. 4). By now we can tell that hydrophobin suspends CNT almost as efficiently as random fragments of ssDNA. The results indicate that the interaction of hydrophobic domains of biomolecules with CNT is a general phenomenon, and not restricted to nucleic acids.

ACKNOWLEDGMENTS

This work is supported by the BMBF (13N8512) and the DFG (FOR335). The hydrophobin has been provided by BASF AG.

REFERENCES

1. M. Zheng, A. Jagota, E. D. Semke, B. A. Diner, R. S. McLean, S. R. Lustig, R. E. Richardson, and N. G. Tassi, *Nature Mater.*, **2**, 338–342 (2003).
2. S. Taeger, O. Jost, W. Pompe, and M. Mertig, "Purification and Dispersion of Carbon Nanotubes by Sidewall Functionalization with Single-Stranded DNA," in *ELECTRONIC PROPERTIES OF SYNTHETIC NANOSTRUCTURES: XVIII International Winterschool on Electronic Properties of Novel Materials*, edited by H. Kuzmany, S. Roth, M. Mehring, and J. Fink, AIP, 2004, pp. 185–189.
3. J. G. Wessels, *Adv. Microb. Physiol.*, **38**, 45–1 (1997).
4. O. Jost, A. Gorbunov, J. Möller, W. Pompe, X. Liu, P. Georgi, L. Dunsch, M. Golden, and J. Fink, *J. Phys. Chem. B*, **106**, 2875–2883 (2002).
5. A. Rinzler, J. Liu, H.Dai, P. Nikolaev, C. Huffman, F. Rodriguez-Macias, P. Boul, A. Lu, D.Heymann, D. Colbert, R. Lee, J. Fischer, A.M.Rao, P. Eklund, and R. Smalley, *Appl. Phys. A*, **67**, 29–37 (1998).
6. Nanotec Electronica S. L., WSxM SPM Software, http://www.nanotec.es (2005).

Dissolution Douce of Single Walled Carbon Nanotubes

Alain Pénicaud,[1] Philippe Poulin,[1] Eric Anglaret,[2] Pierre Petit,[3] Olivier Roubeau,[1] Shaïma Enouz,[4] Annick Loiseau[4]

[1]*Centre de Recherche Paul Pascal-CNRS, Université Bordeaux-I, Av. Schweitzer, 33600 Pessac, France,*

[2] *Laboratoire des Colloïdes, Verres et Nanomatériaux, UMR CNRS 5587, Université Montpellier-II, 34095 Montpellier cedex 5, France,*
[3] *Institut Charles Sadron, 6 rue Boussingault, 67000 Strasbourg, France*
[4] *LEM, CNRS-ONERA, 92322 Châtillon, France*

Abstract. This report deals with the "dissolution douce" or mild dissolution of single wall carbon nanotubes (SWNT), i.e a gentle way to obtain concentrated true solutions of SWNTs without functionalization, strongly acidic treatment and /or sonication thus keeping full integrity of the SWNT samples. Chemical reduction of SWNTs yields polyelectrolyte alkali metal salts of nanotubes, spontaneously soluble in polar organic solvent. The resulting solutions are indefinitely stable if kept under inert atmosphere.

Keywords: Single walled carbon nanotubes, polyelectrolyte, dissolution, exfoliation
PACS: 64.75.+g; 78.30.Na; 82.35.Rs

INTRODUCTION

In order to process carbon nanotubes (NT) and thus take advantage of their outstanding physical properties (aspect ratio, Young's modulus, thermal conductivity, ...), researchers have been trying to disperse the NT soot in various matrices and particularly liquids. Indeed chemical functionalization,[1] be it on the end or walls of the nanotubes, has been specially successful, albeit at the expense of the true chemical nature of the carbon nanotubes. Dispersing with surfactants,[2] polymers [3] or π-stacking molecules,[4] aided by sonication is another method of choice. More recently, dissolution of single walled carbon nanotubes (SWNTs) in superacids, has been reported.[5] Still, most authors recognize the need for a harmless and efficient dispersion method. Functionalization and sonication tend to cut the tubes at their defect sites, thus shortening them. Indeed it has been shown that sonication time and/or power can be related to deterioration of the nanotube performances.[6] An added difficulty is the fact that SWNTs can grow as bundles.[7] Debundling the nanotubes is a key issue if one expects to fully benefit from the physical (mechanical, electrical, optical) properties of carbon nanotubes, for using them either as individual objects or processed in composite materials.

CP786, *Electronic Properties of Novel Nanostructures*, edited by H. Kuzmany, J. Fink, M. Mehring, and S. Roth
© 2005 American Institute of Physics 0-7354-0275-2/05/$22.50

Polyelectrolyte, or charged polymers and macromolecules, dissolve spontaneously in polar solvents. An inspiring case for us was the spontaneous dissolution of an inorganic polyelectrolyte, $Mo_6Se_6^{2-}$ with Li^+ as a counter-ion reported some twenty years ago by Tarascon, Di Salvo and co-workers. [8] Indeed, $Li_2Mo_6Se_6$ is a soluble ceramic. We reasoned that by reducing carbon nanotubes with alkali metals, one should obtain a stiff polyelectrolyte, topologically not very different from $Mo_6Se_6^{2-}$, hence the physics of solubility should be similar and, indeed, they are.

Chimie douce [9] has been coined to describe room temperature sol-gel methods that yield ceramics and biomimetic materials without the high temperatures required by the classical hydrothermal methods. In a similar fashion, *dissolution douce* offers an acid-free, sonication-free, harmless method to dissolve SWNTs. [10]

RESULTS AND DISCUSSION

SWNT's can be reduced in THF (tetrahydrofuran) with alkali metals to yield a polyelectrolyte alkali metal salt of SWNT,[11] of formula $Na(THF)C_{10}$ in the case of sodium, hence one negative charge per 10 carbon atoms. The resulting salt is spontaneously soluble in polar organic solvents such as DMSO (dimethylsulfoxide) [10] without the need for sonication. (see video in supplementary materials of ref 10). For those reduced nanotubes, the higher electronic levels of the van Hove singularities are occupied, thereby canceling the well-known optical transitions in UV-visible spectroscopy. Raman spectroscopy of the solutions shows typical signature for reduced nanotubes, i.e. a broad, low intensity, upshifted TM band (tangential mode) and disappearance of the RBM band (radial breathing mode). [10]

Drying of the solution leaves a mat. When cut with a razor blade, this mat reveals its true nature i.e. entangled bundles of nanotubes (Figure 1).

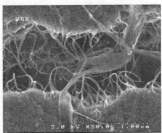

FIGURE 1. Scanning electron microscopy photograph of a dried extract of a polyelectrolyte NT solution.

An open question in any kind of SWNT dispersion or solution is the size of the bundles. In general, high power sonication, followed by ultracentrifugation at high velocity are required to unbundle the SWNT.[12,13] AFM pictures were obtained from the polyelectrolyte SWNT solutions to check the unbundling of the NT. Indeed, we find either fully exfoliated NT's (Figure 3 of ref 10) or small bundles, i.e. incomplete exfoliation.

One should bear in mind that no sonication has been used to dissolve the tubes. Why then should the nanotubes unbundle, even partially, when it requires so much energy in surfactant solutions ? A plausible hypothesis comes from the dissolution mechanism itself. Recent results indicate that the complex cations $Li^+(THF)$ can decorate the nanotube surfaces, including the triangular channels of the hexagonal lattice of the bundles, resulting in a dilatation of the bundle.[14] Dissolution in polar solvents requires the wetting of the cation, followed by swelling of the bundles and their exfoliation, as observed in the case of $Li_2Mo_6Se_6$[8]. Indeed, initial stage of the dissolution shows filament-like features (Figure 2).

FIGURE 2. $[Na(THF)]_nNT$ in DMSO. In the first seconds of the dissolution, filament-like features emerge from the solid. Scale of the picture is roughly 1 mm.

High resolution transmission electronic microscopy (HRTEM) images have also been obtained on the samples in the pristine state (Fig. 3a) and on dried extracts of the polyelectrolyte solutions (Fig.3b). The lattice fringes associated with the periodic packing of the tubes within the bundles are clearly visible in Fig 3a (arrows) and are no longer observed in dried extracts (Figure 3b). Instead of these fringes, one observes the contrast of isolated individual tubes and dilated bundles in which the tubes are packed in a disordered manner, i.e. without periodic packing.

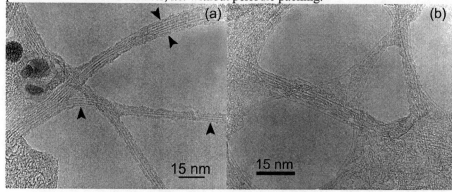

FIGURE 3. HRTEM images of pristine SWNTs bundles in a) and of dried extract of a SWNT polyelectrolyte solution in b).

True solutions of (even partially) unbundled nanotubes of significant concentration are extremely appealing for the fabrication of nanotube-based composites or other hybrid materials. Indeed, as a preliminary demonstration, composite PVA / SWNT films (PVA = polyvinylalcohol) have been obtained simply by mixing together [Na(THF)]$_n$NT solutions and PVA solutions in DMSO and drying the resulting solution at controlled temperature (60 °C). Initial results roughly show a doubling of all relevant mechanical parameters when compared with the pristine PVA film prepared in the same condition (Figure 4 and Table 1).

Eventually, specific applications or modifications of nanotubes are often sought through chemical, covalent, functionalization of the nanotubes. Such reactions can be realized in a controlled manner with the present salts, taking advantage of the controllable ratio Na/C. Quenching of [Na(THF)]$_n$NT solutions with electrophiles such as 6-methyl-2-chloropyridine and 1,n-dibromoalkanes (n=8, 10) in excess resulted in functionalized nanotubes. The amount of functionalization deduced by TGA experiments of the resulting soot or from the amount of salt formed matches with the initial Na/C ratio in the nanotube salt, thus showing the usefulness of the present polyelectrolyte nanotubes solutions for functionalization.

 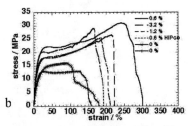

FIGURE 4. a: PVA / SWNT composite films. b: strain stress curves on the same films. Nanotubes used were from electric arc origin [7] unless otherwise indicated.

Table 1: Mechanical parameters for composite PVA/SWNT films				
SWNT / %wt	ε_r / %	σ_r / MPa	γ / GPa	E_r / J.g^{-1}
0	128	13	0.2	14
0	90	16	0.13	15
0.6	240	31	0.31	54
1.2	185	26	0.31	31
3.2	215	27	0.38	35
0.6 (HiPco)	163	29	0.33	31

ε_r : Maximum strain, σ_r : tensile strength, γ : Young's modulus, E_r : energy-to-break.

CONCLUSIONS AND PERSPECTIVES

Stable SWNT solutions of relatively high concentration (ca. 0.1 % in nanotubes) can be obtained under inert atmosphere through chemical reduction yielding a polyelectrolyte NT salt. Partial unbundling is observed. Efforts are now focused on fully debundling the SWNT's, exploiting the controlled chemical functionalization and the formation of composite films.

ACKNOWLEDGMENTS

We thank Muriel Alrivie for the SEM picture, Martin Cadek for advice and Jean-Louis Sauvajol et al. for sharing ref. 14 with us before publication. This work has been done in the framework of the GDRE n°2756 "Science and applications of nanotubes" - NANO-E

References
1. see reviews: Tasis, D.; Tagmatarchis, N.; Georgakilas, V.; Prato, M. Chem. Eur. J. 2003, 9, 4000-4008; S.S. Wong, S. Banerjee, Dekker encyclopedia of nanosciences and nanotechnology, Marcel Dekker, 2004, 1251-1268
2. J.M. Bonard et al., Advanced Materials 9, 827 (1997); Vigolo,B.; Pénicaud, A.; Coulon, C.; Sauder, C.; Pailler, R.; Journet, C.; Bernier, P.; Poulin, P., Science 2000, 290, 1331-1334; Islam, M.F.; Rojas, E.; Bergey, D.M.; Johnson, A.T.; Yodh, A.G. Nanoletters 2003, 3, 269-273; Moore, V.C.; Strano, M.S.; Haroz, E.H.; Hauge, R.H.; Smalley, R.E. Nanoletters 2003, 3, 1379-1382.
3. Dalton, A.B.; Stephan, C.; Coleman, J.N.; McCarthy, B.; Ajayan, P.M.; Lefrant, S.; Bernier, P.; Blau, W.J.; Byrne, H.J. J. Phys. Chem. B 2000, 104, 10012-10016; Star, A.; Fraser Stoddart J.; Steuerman, D.; Diehl, M.; Boukai, A.; Wong, E.W.; Yang, X.; Chung, S.W.; Choi, H.; Heath, J.R. Angew. Chem. Int. Ed. 2001, 40, 1721-1725; O'Connell, M.J.; Boul, P.; Ericson, L.M.; Huffman, C.; Wang, Y.; Haroz, E.; Kuper, C.; Tour, J.; Ausman, K.D.; Smalley, R.E. Chem. Phys. Lett. 2001, 342, 265-271; Chen, J.; Liu H.; Weimer, W.A.; Halls, M.D.; Waldeck, D.H.; Walker., G.C. J. Am. Chem. Soc. 2002, 124, 9034-9035.
4. Chen R.J. et al., J.Am. Chem. Soc., 123 (2001),3838-3839.
5. Ramesh, S.; Ericson, L.M.; Davis, V.A.; Saini, R.K.; Kittrell, C.; Pasquali, M.; Billups, W.E.; Adams, W.W.; Hauge, R.H.; Smalley, R.E. J. Phys. Chem. B 2004, 108, 8794-8798.
6. S. Badaire, P. Poulin, M. Maugey, C. Zakri, Langmuir 20 (2004), 10367-10370.
7 C. Journet, W.K. Maser, P. Bernier, M. Lamy de la Chapelle, S. Lefrant, P. Deniard, R. Lee, J.E. Fischer, Nature 388 (1997) 756
8 J. M. Tarascon, F.J. Di Salvo, C.H. Chen, P.J. Carroll, M. Walsh, L. Rupp, J. Solid State Chem. 58 (1985) 290-300.
9. J. Livage, New J. Chem. 25 (2001) 1.
10 A. Pénicaud, P. Poulin, A. Derré, E. Anglaret, P. Petit, J. Am. Chem. Soc. 127 (2005) 8-9.
11 P. Petit, C. Mathis, C. Journet, P. Bernier, Chem. Phys. Lett. 305 (1999) 370-374.
12 M.J. O'Connell, S.M. Bachilo, C.B. Huffman, V.C. Moore, M.S. Strano, E.H. Haroz, K.L. Rialon, P.J. Boul, W.H. Noon, C. Kitrell, J. Ma, R.H. Hauge, R.B. Weisman, R.E. Smalley, Science 297 (2002) 593-596.
13 N. Izard, D. Riehl, E. Anglaret, Phys. Rev. B, 2005, in the press.
14. J. Cambedouzou, S. Rols, N. Bendiab, R. Almairac, J.-L. Sauvajol, P. Petit, C. Mathis, I. Mirebeau, M. Johnson, to be published.

Separation and Assembly of DNA-dispersed Carbon Nanotubes by Dielectrophoresis

Daniel Sickert[*,†], Sebastian Taeger[**], Anita Neumann[‡], Oliver Jost[†],
Gerald Eckstein[*], Michael Mertig[**] and Wolfgang Pompe[†]

[*]Siemens AG, Corporate Technology, MM D2P, 81730 München, Germany
[†]Institut für Werkstoffwissenschaft, Technische Universität Dresden, 01062 Dresden, Germany
[**]Max-Bergmann-Zentrum, Technische Universität Dresden, 01069 Dresden, Germany
[‡]Infineon AG, Corporate Research, TL, 81730 München, Germany

Abstract. Current production methods for single wall carbon nanotubes (SWCNT) yield a mixture of metallic and semiconducting SWCNT, mostly bundled. Recent publications suggested the separation of metallic and semiconducting species by dielectrophoresis (DEP). We demonstrate the enrichment of metallic SWCNT in self-assembled wires deposited dielectrophoretically from a DNA dispersed suspension by applying resonant Raman spectroscopy. Our modification of the DEP separation process provides compatibility to bionanotechnology assembly approaches by avoiding tensides.

Keywords: carbon nanotubes, DNA, dielectrophoresis, resonant Raman spectroscopy, separation
PACS: 81.07.De, 81.16.-c, 82.45.-h, 81.20.Ym

INTRODUCTION

Single wall carbon nanotubes possess outstanding electronic properties that are applicable for nanoscale leads, transistors and sensors. Current production methods yield a mixture of metallic and semiconducting SWCNT, mostly appearing in form of bundles, which constitutes a major drawback for most applications. Separation of metallic and semiconducting SWCNT can be achieved by dielectrophoresis [1, 2]. This requires a suspension of individualized SWCNT that is obtained by using a tenside in combination with ultrasonication and ultracentrifugation. Alternatively SWCNT can be dispersed and singularized by DNA oligomers and ssDNA fragments as well [3, 4]. This hybrid structure provides a versatile building block for bionanotechnology and self assembly approaches [5]. Therefore we aim on the separation of DNA dispersed SWCNT by dielectrophoresis while avoiding tensides that are incompatible with biomolecules.

MATERIALS AND METHODS

SWCNT were produced by pulsed LASER evaporation [6] and purified by an enhanced acid treatment without ultrasonication. The purified SWCNT were dispersed in an aqueous suspension of T30 (a DNA oligomer with a sequence of 30 thymine bases) under ultrasonication (120 min) and finally centrifuged for 90 min at 20'000 g. The dispersion procedure yields a mean SWCNT length of 500 nm as judged from AFM images. The supernatant was diluted 1:1000 in deionized water for the dielectrophoresis experiment.

CP786, Electronic Properties of Novel Nanostructures, edited by H. Kuzmany, J. Fink, M. Mehring, and S. Roth
© 2005 American Institute of Physics 0-7354-0275-2/05/$22.50

Figure 1 shows the optical absorption spectrum (OAS) of the purified SWCNT after deposition on a quartz plate. The shape of each OAS peak can be described by a Gaussian function [7]. From this modeling a narrow diameter distribution can be derived with 85% of all SWCNT diameters between 1.1 and 1.35 nm.

The resonant Raman spectra of the same sample (Fig. 2) reveal a total range of diameters between 1.0 and 1.6 nm. This can be concluded from the relation $\omega_{RBM} = c_1/d + c_2$ with empirically derived values $c_1 = 223.5$ nm cm^{-1} and $c_2 = 12.5$ cm^{-1} [8]. The peaks can be attributed to semiconducting and metallic SWCNT by comparison of the exciting wavelength with empirically and theoretically determined transition energies [8, 9, 10]. The spectrum taken at 785 nm (1.58 eV) shows a strong metallic and a smaller semiconducting peak. At 514 nm (2.41 eV) excitation wavelength only semiconducting and no metallic tubes are found since SWCNT with a metallic transition (M11) around 2.4 eV would be smaller than 1 nm in diameter.

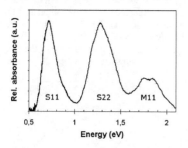

FIGURE 1. OAS peaks of semiconducting (S11, S22) and metallic (M11) band transitions after purification.

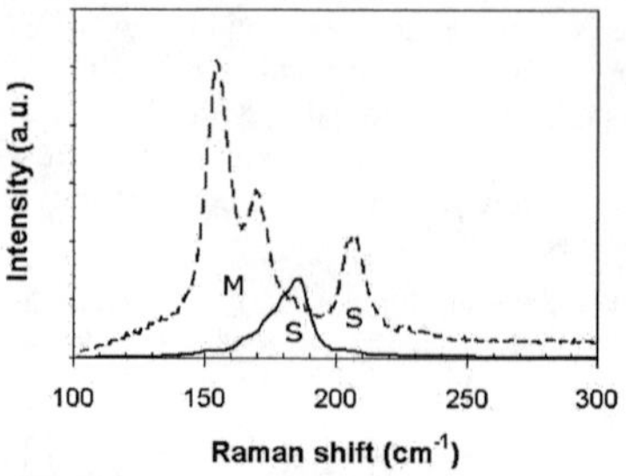

FIGURE 2. Raman spectra at two energies (dashed: 785 nm, solid: 514 nm) and peak assignment (M: metallic, S: semiconducting).

The schematic circuit of the DEP experiment can be seen in Fig. 3. Gold electrode patterns have been manufactured on glass substrates by a lift-off process. The five electrode pairs are 3 μm wide, 30 nm thick and separated by a 5 μm gap. They were contacted by a wafer prober and connected to a frequency generator and oscilloscope for voltage and frequency control. While the voltage was adjusted to 12 V$_{pp}$ for all experiments the frequency was varied between 300 kHz and 10 MHz. A 10 μl drop of the SWCNT suspension was pipetted on the electrode pairs for each single experiment. After 5 min deposition time the samples were blotted with nitrogen and rinsed in deionized water.

RESULTS

After each deposition, at least four of the five electrode gaps are visibly (interference contrast) bridged (Fig. 4). Obviously deposition did not stop after the formation of first conducting pathways and further SWCNT accumulated between the electrodes. The persistence of a sufficient field is presumably due to a high contact resistance between the electrodes and a high intertube resistance. The final resistance of the shown sample (five electrode pairs in parallel) is approx. 1.9 kΩ.

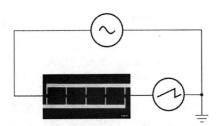

FIGURE 3. Schematic circuit of the DEP experiment. An ac voltage source and an oscilloscope are connected to the electrode pattern.

FIGURE 4. Oriented CNT films bridge the electrode pairs after deposition (10 MHz here). Micrograph is taken in interference contrast.

Figure 5 shows the Raman spectra of such SWCNT bridges that were deposited at different frequencies. All spectra are obtained at 514 nm excitation wavelength and normalized to the ω_{G-} peak. Note that Fig. 5a has got a magnified intensity scale. The shown curves were obtained by averaging three measurements on different spots.

The intensity of the radial breathing mode (RBM) peak is lowered with rising deposition frequency (Fig. 5a). Since this peak is associated to semiconducting SWCNT an enrichment of metallic SWCNT at higher deposition frequencies can be concluded. The high energy modes (HEM) (Fig. 5b) also indicate this preferential deposition. The broadening of the ω_{G-} peak and a decreasing intensity difference between ω_{G-} and ω_{G+} are typical features of samples with a higher concentration of metallic SWCNT [1, 9].

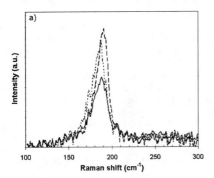

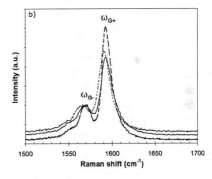

FIGURE 5. RBM (a) and HEM (b) region of Raman spectra measured on samples with different deposition frequency (dashed: 300 kHz, dotted: 3 MHz, solid: 10 MHz) that were obtained at 514 nm excitation wavelength. Spectra are normalized to ω_{G-} with different intensity scale in both plots.

DISCUSSION AND CONCLUSIONS

The dielectrophoretic force on particles in a suspension depends on their geometry and complex permittivity [11]. Differences in the permittivities can be utilized to separate particles by preferential attraction of one of the particle species towards higher electrical field strength. Above a critical value the frequency dependence of the permittivity results in the repulsion of particles with a permittivity lower than that of the surrounding medium. Krupke et al. demonstrated the utilization of this effect for the separation of metallic and semiconducting SWCNT in a sodium dodecyl benzene sulfonate (SDBS) suspension featuring a higher permittivity than semiconducting SWCNT [1, 2].

We substituted SDBS for DNA oligomers and achieved similar results, especially an enrichment of metallic SWCNT in the deposited material with rising frequency. However, the changes in our Raman spectra are less pronounced than the ones in the SDBS experiment. Possible explanations are an insufficient dispersion of the SWCNT, altered electronic properties due to attached DNA molecules or a different field distribution in our electrode pattern. Especially the first item is critical, since every bundle partially containing metallic SWCNT will be deposited and thus inhibit a complete separation. Future fluorescence measurements on the DNA suspension and Raman spectroscopy at other excitation wavelengths will enable us to disclose the reason for the incomplete separation.

ACKNOWLEDGMENTS

This work was supported by the DFG (FOR 335), the German federal ministry of research (BMBF: 13N8512) and the Siemens AG. D. Sickert is grateful to G.S. Düsberg (Infineon AG, Munich) for Raman measurements and discussion.

REFERENCES

1. R. Krupke, F. Hennrich, H. v. Löhneysen, and M. M. Kappes, *Science*, **301**, 344–347 (2003).
2. R. Krupke, F. Hennrich, M. M. Kappes, and H. v. Löhneysen, *Nano Lett.*, **4**, 1395–1399 (2004).
3. M. Zheng, A. Jagota, E. D. Semke, B. A. Diner, R. S. McLean, S. R. Lustig, R. E. Richardson, and N. G. Tassi, *Nature Mater.*, **2**, 338–342 (2003).
4. S. Taeger, O. Jost, W. Pompe, and M. Mertig, "Purification and dispersion of carbon nanotubes by sidewall functionalization with single-stranded DNA," in *Electronic Properties of Synthetic Nanostructures*, edited by H. Kuzmany, J. Fink, M. Mehring, and S. Roth, AIP Conference Proceedings 723, American Institute of Physics, New York, 2004, pp. 185–188.
5. M. Mertig, and W. Pompe, "Biomimetic fabrication of DNA-based metallic nanowires and networks," in *Nanobiotechnology - Concepts, Applications and Perspectives*, edited by C. M. Niemeyer, and C. A. Mirkin, WILEY-VCH, Weinheim, 2004, pp. 256–277.
6. O. Jost, A. A. Gorbunov, J. Möller, W. Pompe, X. Liu, P. Georgi, L. Dunsch, M. S. Golden, and J. Fink, *J. Phys. Chem. B*, **106**, 2875–2883 (2002).
7. X. Liu, T. Pichler, M. Knupfer, M. S. Golden, J. Fink, H. Kataura, and Y. Achiba, *Phys. Rev. B*, **66** (2002).
8. S. Bachilo, M. Strano, C. Kittrell, R. Hauge, R. Smalley, and R. Weisman, *Science*, **298**, 2361–2366 (2002).
9. M. Dresselhaus, G. Dresselhaus, A. Jorio, A. G. S. Filho, and R. Saito, *Carbon*, **40**, 2043–2061 (2002).
10. R. Saito, G. Dresselhaus, and M. S. Dresselhaus, *Phys. Rev. B*, **61**, 2981–2990 (2000).
11. M. Dimaki, and P. Bøggild, *Nanotechnology*, **15**, 1095–1102 (2004).

NMR Spectroscopy of Hydrogen Adsorption on Carbon Nanotubes

K. Shen*, S. Curran[†], J. Dewald[†] and T. Pietraß*

*Department of Chemistry, New Mexico Tech, Socorro, NM 87801, USA.
[†]Department of Physics, New Mexico State University, Las Cruces, NM 88001, USA.

Abstract. Hydrogen storage properties of both single- and multi-walled carbon nanotubes (CNTs) under pressures up to 15 MPa are studied by nuclear magnetic resonance (NMR) spectroscopy. The role of residual metallic catalyst in hydrogen adsorption is studied by comparing purified samples with parent materials. Structural features such as diameter and defect density are determined by resonance Raman spectroscopy.

Keywords: NMR spectroscopy, Raman spectroscopy, transmission electron microscopy, carbon nanotubes, hydrogen adsorption
PACS: 61.46.+w, 68.37.Lp, 81.07.De, 76.60.-k, 76.60.Cq

INTRODUCTION

The discovery of the potential of CNTs for efficient hydrogen storage sparked intense research activity.[1] In previous work,[2, 3] we investigated hydrogen adsorption in a large number of well-characterized carbon nanotube samples with magnetic resonance techniques. We concluded that at ambient temperature and hydrogen pressures up to 1.5 MPa, the storage capacity never exceeded 1 wt% (weight percent). Here, we report that exposure of the single- and multi-walled carbon nanotubes to pressures of 15 MPa may lead to a different type of hydrogen sorption.

EXPERIMENTAL

Single-walled (SWNT) and multi-walled (MWNT) carbon nanotube samples were obtained from Mer Corporation (Tucson, AZ) and manufactured by carbon arc discharge. The MWNT sample was produced without any catalyst. The samples were characterized using inductively coupled plasma analysis with mass spectrometry (ICP-MS) and transmission electron microscopy (TEM).[4] For purification, crude materials were refluxed in 3 M HNO_3 at 393 K for 16 h and the acid processed CNTs were then heated at 823 K for 30 min in stagnant air. This procedure halved the metal content for the SWNT sample. Prior to hydrogen loading, all samples were heated to 823 K for 2 h under vacuum ($\approx$ 1 Pa). For hydrogen pressures up to 1.5 MPa, the samples were exposed to hydrogen gas while in the NMR magnet. For hydrogen exposure at 15 MPa, the samples were contained in a high pressure apparatus and exposed to the gas for one or five days. Upon release of the excess hydrogen pressure, the samples were flame sealed in a nitrogen atmosphere.

CP786, *Electronic Properties of Novel Nanostructures*, edited by H. Kuzmany, J. Fink, M. Mehring, and S. Roth
© 2005 American Institute of Physics 0-7354-0275-2/05/$22.50

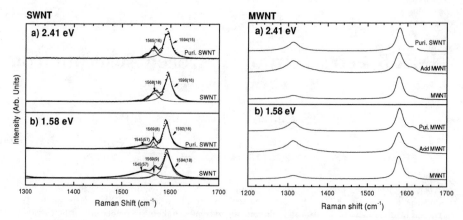

FIGURE 1. Raman spectra of SWNTs and MWNTs at different laser excitation energies.

NMR spectra were acquired on an Apollo (Tecmag)/MSL-400 (Bruker) and Discovery (Tecmag) NMR spectrometer at a magnetic field strength of 9.4 T. Raman spectra were recorded on a Renishaw inVia Raman microscope in a back scattering configuration with laser excitation wavelengths of 514 nm (2.41 eV) and 785 nm (1.58 eV) at 1 mW.

SINGLE-WALLED CARBON NANOTUBES

Transmission electron microscopy showed strong bundling and the presence of embedded metal particles.[5] Raman spectra for the SWNT sample for excitation at 1.58 eV and 2.41 eV are shown in Figure 1. According to Kataura,[6] the 1.58 eV line excites semiconducting tubes in the diameter range 0.9 - 1.1 nm, while 2.41 eV line excites tubes in the diameter range 1.2 - 1.4 nm (semi-conducting) or 0.9 - 1.1 nm (metallic). The shift in the radial breathing mode (RBM) to higher frequency after purification shows that smaller tubes are preferentially removed by the purification procedure. The ratio of the intensity of the D band to that of the G band, I_D/I_G, reveals the presence of defects. Moreover, after the final annealing step, the D band frequencies are the same as those of the parent materials. This is corroborated by the fact that I_D/I_G increases from 0.034 to 0.158 after acid treatment, and is then reduced again to 0.031.[5] Defects and functional groups introduced during the acid treatment are removed by the final annealing step.

Hydrogen adsorption was investigated previously with electron spin resonance (ESR) [3] and ^1H and ^2H NMR spectroscopy for hydrogen exposures up to 1.5 MPa.[2] The ESR signal was very weak for the SWNT sample and had a symmetric line shape indicative of isolated spins due to defect sites. Hydrogen adsorption at 136 kPa quenched the signal which suggests that the hydrogen preferentially adsorbs on defect sites. The low signal intensity is in agreement with the small I_D/I_G ratio found in Raman spectroscopy.

Knight shift contributions to the line position and line width give rise to a broad, unstructured resonance in the ^1H NMR at pressures up to 1.5 MPa. At this pressure,

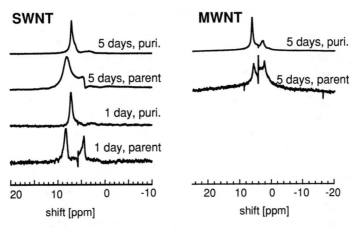

FIGURE 2. ^{1}H NMR spectra of SWNTs and MWNTs (parent and purified materials) after exposure to a H_2 pressure of 15 MPa. Note that the pressure was released before flame-sealing the NMR tubes.

the line width is about 16 kHz and the resonance is centered at 16 ppm.[2] Purification causes the line width to decrease to 2.4 kHz. Overall, much better resolution can be achieved when using ^{2}H NMR. Here, dipolar couplings are greatly reduced due to the smaller gyromagnetic ratio of [2]H .

^{1}H NMR spectra after exposure to a hydrogen pressure of 15 MPa are shown in Figure 2. First, it should be noted that the parent material gives rise to broader line widths than the purified material which is accompanied by an upfield shift (parent: 8 ppm; purified 7 ppm) when compared to a pressure of 1.5 MPa (parent: 16 ppm; purified 5.6 ppm). The difference in shift may be due to the fact that excess hydrogen at high pressure saturates defect sites that contribute to a Knight shift in the low pressure sample. After purification, the shift is closer to the one of the pure gas (7 ppm) due to stronger contributions from gas phase hydrogen. The component at 3 ppm is most likely due to probe background. ^{2}H NMR data will be recorded to resolve this issue.

It should be noted that at pressures up to 1.5 MPa the hydrogen adsorption was fast and reversible and had to be described as physisorption. Uptake capacities were estimated at 0.1 wt%.[2] The most interesting results of the 15 MPa pressure data are that 1) the hydrogen signal can be detected despite releasing the pressure; and 2) the signal is much stronger for for longer exposure times at a pressure of 15 MPa. ^{1}H NMR spin-lattice relaxation measurements immediately upon preparation of the sample yielded relaxation times on the order of 200 ms. Heating caused the appearance of a narrow gas resonance and a decrease in relaxation time to 87 ms, indicative of gas desorption and weaker interaction of the gas with the surface.

MULTI-WALLED CARBON NANOTUBES

Transmission electron microscopy revealed bundling, few defects, and a large number of graphitic particles and nano-onions. Purification does not damage the tube walls, and no change in average tube diameter upon purification is observed.[4] Figure 1 shows Raman spectra of the MWNTs at two different laser excitation energies. After acid purification, the D bands gain in intensity, suggesting the introduction of defects.

Hydrogen adsorption in MWNTs was studied by ESR and NMR spectroscopy. ESR revealed Dysonian line shapes, indicating the presence of conduction electrons. Hydrogen exposure caused a significant increase in ESR signal intensity.[3] Possible explanations may be an effect of pressure on the band gap, or 'dilution' of the conduction electrons through the diamagnetic species hydrogen.

For an exposure to 15 MPa of hydrogen, purification greatly improves ^{1}H NMR spectral quality (Figure 2). The component at 6 ppm strongly increases, while the component at 3 ppm is less affected. An analysis of the pulse width revealed that the component at 3 ppm is most likely due to probe background. Similar to the SWNTs, spin lattice relaxation times of the gas space reveal residual gas pressures in the samples of less than 100 kPa. It should be noted that determination of a 90° pulse width was not conclusive, indicative of significant conductivity as also shown by ESR. The absence of a metal catalyst in the manufacturing process implies that any electric conductivity must stem from the CNTs or residual graphitic particles.

Conclusions

We have shown previously that for a range of samples from commercial manufacturers at ambient temperature and in the pressure range from 0 to 1.5 MPa, hydrogen adsorption is fast and reversible and must be described as physisorption. However, exposure to much higher hydrogen pressures (15 MPa) with subsequent release of the pressure still allowed for the detection of adsorbed hydrogen. Moreover, exposure for five days led to stronger signal intensities than exposure for one day. These data suggest that the hydrogen sorption mechanism and capacity of carbon nanotubes may critically depend on pressure.

ACKNOWLEDGMENTS

This material is based upon work supported by the National Science Foundation under Grant No. 0107710.

REFERENCES

1. A. C. Dillon, K. M. Jones, T. A. Bekkedahl, C. H. Kiang, D. S. Bethune, and M. J. Heben, *Nature*, **386**, 377–379 (1997).
2. K. Shen and T. Pietraß, *J. Phys. Chem. B*, **108**, 9937–9942 (2004).
3. K. Shen, D. L. Tierney, and T. Pietraß, *Phys. Rev. B*, **68**, 165413 (2003).
4. K. Shen, H. Xu, Y. B. Jiang and T. Pietraß, *Carbon*, **42**, 2315–2322 (2004).
5. K. Shen, S. Curran, H. Xu, S. Rogelj, Y. Jiang, J. Dewald, and T. Pietraß, *J. Phys. Chem. B*, **109**, 4455–4463 (2005).
6. H. Kataura, Y. Kumazawa, Y. Maniwa, I. Umezu, S. Suzuki, Y. Ohtsuka, Y. Achiba, *Synth. Met.*, **103**, 2555–2558 (1999).

NANOTUBE FILLING AND
DOUBLE WALL CARBON NANOTUBES

A Photoemission Study of the Electronic Structure of Potassium-doped C_{60} Peapods

T. Pichler[1], H. Shiozawa[1], H. Rauf[1], M. Knupfer[1], J. Fink[1], B. Büchner[1], H. Kataura[2]

[1]*Leibniz Institute for Solid State and Materials Research Dresden, P.O. Box 270016, D-01171 Dresden, Germany*
[2] *Nanotechnology Research Institute, National Institute of Advanced Industrial Science and Technology (AIST) 1-1-1 Higashi, Tsukuba, Ibaraki 305-8562, Japan*

Abstract. We report on the doping dependence of the electronic structure of the valence band of mats of C_{60} filled single wall carbon nanotubes (C_{60} peapods) using high resolution photoemission spectroscopy as probe. For the pristine peapods about one third are metallic Tomonaga-Luttinger-Liquids (TLL), which is directly related to a power law scaling in the density of states. The changes in the valence band of these doped one-dimensional nanostructures explicitly show the effect of the dimensionality on the electronic structure. In addition to previous studies we observe a competitive charge transfer to the C_{60} peas and the SWCNT host. This yields to a more complex doping dependence and a crossover from a TLL behavior to a normal Fermi liquid at much higher doping levels as compared to intercalated SWCNT bundles.

INTRODUCTION

The charge transport properties of single wall carbon nanotubes (SWCNTs) and the controlled modification of their electronic properties by functionalization have been investigated intensively over the last years since these materials represent archetypes of a one-dimensional (1D) system [1]. For metallic 1D systems, conventional Fermi liquid theory fails since it becomes unstable due to long-range Coulomb interactions. Such a one dimensional paramagnetic metal is called a Tomonaga-Luttinger Liquid (TLL) and shows peculiar behavior such as spin charge separation and interaction dependent exponents in the density of states, correlation function and momentum distribution of the electrons [2]. One very efficient possibility to change the electronic properties by doping with electrons or holes is via intercalation. For SWCNTs [3-5] and C_{60} peapods [6,7] the intercalation physics has been analyzed in some detail. In contrast to graphite intercalation compounds no distinct intercalation stages have been observed as yet. For closed SWCNTs alkali metal intercalation takes place inside the channels of the triangular bundle lattice and leads to a shift of the Fermi energy, a loss of the optical transitions and an increase in conductivity by about a factor of thirty. A complete charge transfer between the donors and the SWCNTs and C_{60} peapods was observed up to doping saturation, which was achieved at a carbon to alkali metal ratio of about seven [4,6]. In addition, in the case of C_{60} peapods a competitive charge transfer to the C_{60} peas and the SWCNT host was reported [7]. Recently, first results on direct measurements of the low energy properties of the valence band as a function of doping have been reported for SWCNTs. Regarding intercalated SWCNTs, photoemission revealed a Fermi edge at high

CP786, *Electronic Properties of Novel Nanostructures*, edited by H. Kuzmany, J. Fink, M. Mehring, and S. Roth
© 2005 American Institute of Physics 0-7354-0275-2/05/$22.50

doping [8-10]. In addition, the density of states (DOS) of the valence band electrons of mats of SWCNTs was directly monitored by angle-integrated high-resolution photoemission experiments and explained regarding its low energy electronic properties within a TLL model [8-11]. Detailed studies of the change in the low energy electronic properties in mats of SWCNT as a function of potassium [9] and lithium [10] intercalation revealed a transition from a 1D to a 3D metallic behavior. In the case of lithium intercalation the charge transfer was not ionic and a finite hybridization between the intercalant and the SWCNT was observed. In this contribution we report on the doping dependence of potassium intercalated C_{60} peapods using high-resolution photoemission as a probe. The electronic properties of metallic peapods are expected to strongly depend on the type of tube and on the interaction between the SWCNTs and the fullerenes [12]. Hence, intercalation of C_{60} peapods are perfect test systems to study hybrid interactions and charge transfer.

EXPERIMENTAL

Mats of purified SWCNT which consist of a mixture of semiconducting and metallic SWCNT with a narrow diameter distribution which is peaked at 1.37 nm with a variance of about 0.05 nm [7] were produced by subsequent dropping of SWCNT suspended in acetone onto NaCl single crystals. The produced SWCNT film of about 500 nm thickness was floated off in distilled water and recaptured on sapphire plates. For the photoemission experiments the sample was mounted onto a copper sample holder, in-situ filled with C_{60} and cleaned in a preparation chamber under ultra high vacuum (UHV) conditions (base pressure 9×10^{-11} mbar) by vacuum sublimation and annealing up to 800 K, respectively. Electrical contact of the SWCNT film was established by contacting the surface to the sample holder via a Ta foil. Then the sample was cooled down to T=35 K and transferred under UHV conditions to the measuring chamber and its electronic properties were analyzed using a hemispherical high resolution Scienta SES 200 analyzer. For the angle integrated valence band photoemission spectra, using monochromatic HeIα (21.22 eV) excitation, the energy resolution was set to 10 meV. The core level photoemission measurements (XPS) were performed with an energy resolution of 400 meV using monochromatic Alkα excitation (1486.6 eV). The Fermi energy and overall resolution was measured using freshly cleaned gold. The intercalation was performed *in situ* after heating the sample to 450 K using commercial SAES potassium getter sources. After subsequent exposure to the dopant vapor an additional equilibration for about 30 min at 450 K was performed to increase the homogeneity of the sample.

RESULTS AND DISCUSSION

We first analyzed the sample stoichiometry and purity using core level photoemission (not shown). The sample purity was checked by an overview XPS spectrum. Only the signals of carbon could be observed with no contamination from oxygen or catalyst particles. The detailed analysis of the sample stoichiometry dependence was performed in the range of the C1s and K2p levels. The doping level was determined by the ratio of the C1s/K2p intensities taking into account the different photo-ionization cross section. We now turn to the detailed analysis of the low energy electronic properties as a function of doping using high-resolution valence band photoemission experiments at T=35K. The results are depicted in Fig. 1. In the left panel an overview of the valence band spectrum of the SWCNT mats and mats of C_{60} peapods are shown. For both samples two broad peaks are observed at about 3 eV and at about 8 eV binding energy. These features are similar to angle integrated photoemission of graphite and to previous results on SWCNT mats and correspond to the maxima in the DOS of the π and σ bands, respectively. In addition and in agreement with previous studies in the spectrum of the C_{60} peapods the molecular orbitals (MOs) of C_{60} are clearly observed and show the high filling of the *in situ* produced peapod samples [13].

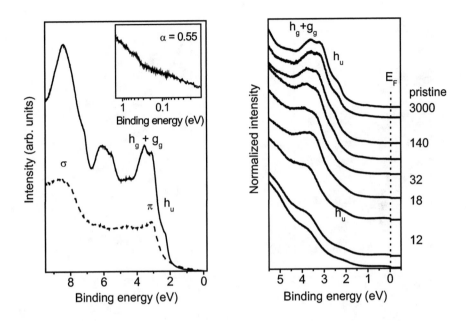

FIGURE 1: Left panel: Valence band photoemission of pristine SWCNT (dashed line) and C_{60} peapods (solid line) measured at T=35 K using Heiα excitation in the range of the π and σ electrons. The labels correspond to the SWCNT derived valence bands and the C_{60} derived MOs. The inset depicts the low energy region for the peapods on a double logarithmic scale in order to visualize the power law renormalization. Right panel: Doping dependence of the intercalated C_{60} peapods for five different doping levels as indicated by the C/K ratios.

In addition a detailed analysis of the low energy electronic properties of the C_{60} peapods was performed and is depicted as the inset in the left panel of figure 1 on a double logarithmic scale. The peaks corresponding to the first and second van Hove singularities (vHs) of the semiconducting SWCNT (S_1, S_2) and that of the first vHs of the metallic SWCNT (M_1) are observed at binding energies of 0.46, 0.78 and 1.04 eV with a full width half maximum of about 0.12 eV, respectively. A key manifestation of the TLL state is the renormalization of the DOS (n(E)) near the Fermi level which shows a power law dependence n(E) ~ E^{α} where α is depending on the size of the Coulomb interaction and can be expressed in terms of the TLL parameter g as α=(g+g^{-1}-2)/8 [2]. As can be seen in the inset we observe a power law scaling of α=0.55, which is within experimental error in good agreement to the previously reported values for pristine and intercalated SWCNT mats [8-11].

We now turn to an analysis of the doping dependence of the valence band of SWCNT as a function of potassium content. In the right panel of Fig. 1 a detailed analysis of the photoemission response of C_{60} peapods as a function of doping was performed. It is obvious that the position of the C_{60} MOs and of the SWCNT conduction bands are modified upon doping. In agreement with previous studies a complex doping dependence and a competitive charge transfer between the SWCNT hosts and the C_{60} peas is observed [7,14]. The results of the photoemission response in the vicinity of the Fermi level are depicted in Fig. 2. The evolution of the spectra at low doping (C/K > 50, charge transfer less than 0.018 e⁻/C, left panel), and at high intercalation (up to C/K=12, right panel) are also depicted in Figure 2.

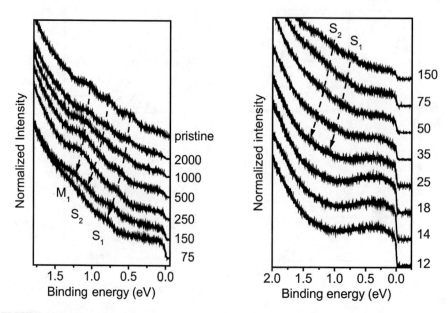

FIGURE 2: Doping dependence of the valence band photoemission spectra in the vicinity of the Fermi level (left panel: low doping, right panel: high doping levels). The numbers correspond to the C/K ratio derived from core level photoemission. The dotted lines are guidelines for the evolution of the S_1, S_2 and M_1 peaks with increasing doping.

With increasing doping, the peaks corresponding to the SWCNT vHs (S_1, S_2, and M_1) shift to higher binding energies due to a filling of the conduction band of the SWCNT with K 4s electrons. Interestingly, the conduction band of the semiconducting SWCNT is not filled at low doping (<0.002 e⁻/C, C/K= 500) since for this doping level the S_1 and S_2 peaks shift parallel up to 0.2 eV to higher binding energy which is not high enough to reach the first unoccupied vHs (S_1^*). In contrast to intercalation of SWCNTs for further increase of the doping up to <0.005 e⁻/C, C/K= 200 the unoccupied S_1^* vHs is still not filled and the unoccupied t_{1u}-derived MOs of the C_{60} peas are contributing. For this doping level the intercalation compound is still a 1D metal with a reduced value of the power law renormalization of $\alpha\sim0.25$. This reduction of α can be due to additional conduction channels from an metallic chain of C_{60} balls inside the SWCNTs as it has been previously observed for high doping levels [15], or due to a modified band structure in the peapod hybrid system. Such a decrease of α with an increasing number of conduction channels was also predicted and is consistent with our observation [16]. Only upon further doping the S_1 and S_2 vHs shift by an additional 0.2 eV to higher binding energies and lead to a occupation of the S_1^* vHs. Regarding the TTL scaling factor we observe a distinct change to a Fermi liquid at doping levels of C/K=50 [14]. This is exactly the amount of doping which is high enough to drive all peapods within a bundle of peapods metallic. This also means that up to this doping level the long range Coulomb interaction is essentially unaffected by the potential of the counter ions and C_{60} filling. Even higher doping leads to a shift of the S_1 peak beyond the position of the S_2 peak (see right panel of Fig. 2) concomitant with a smearing out and finally a vanishing of the peaks related to the SWCNT vHs. At very high doping levels the former unoccupied S_1^*, S_2^*, and M_1^* bands are filled with electrons and corresponding peaks should show up in photoemission below 1 eV binding energy. However, as can be seen in the figure there are no additional spectral features corresponding to now occupied t_{1u} derived MOs and now occupied SWCNT derived conduction bands. This fact can be explained by an increasing number of scattering centers (K^+ counter ions) and/or by an increasing intertube interaction within the SWCNT bundle in the intercalation compound [9,14]. The overall spectra of these highly doped samples are also very similar to equivalent GIC. The Fermi level shift can be extracted from the shift of the π band at about 3 eV. For the sample with C/K=12 (0.083 e$^-$/C) we observe a slightly lower value of ΔE_F=0.8 eV.

In summary, we have studied the doping dependence of the SWCNT valence band in bundles of C_{60} peapods using photoemission spectroscopy as probe. In agreement with previous studies on intercalated SWCNTs a change of the character of the electron liquid as a function of dopant concentration was observed. However, we observe a competitive charge transfer to the C_{60} peas and the SWCNT host which yields to a more complex doping dependence and a crossover from a TLL behavior to a normal Fermi liquid at much higher doping.

ACKNOWLEDGMENTS

This work was funded by the DFG PI440/1. We thank S. Leger, R. Hübel and R. Schönfelder for technical assistance. H.K. acknowledges support by an Industrial Technology Research Grant Program in '03 from the New Energy and Industrial Technology Development Organization (NEDO) of Japan. H.S. acknowledges funding by the Humboldt society.

REFERENCES

1. e.g. M.S. Dresselhaus, P.C. Eklund, Advances in physics **49**, 705 (2000).
2. R. Egger et al., *Phys. Rev. Lett.* **79**, 5082 (1997); C. Kane et al., *Phys. Rev. Lett.* **79**, 5086 (1997) ; J. Voit, "One-dimensional Fermi liquids" in *Rep. Prog. Phys.* **58**, 977 (1995)
3. S. Kazaoui et al. , *Phys. Rev. B* **60**, 13339 (1999).
4. T. Pichler et al., *Solid State Commun.* **109**, 721 (1999); X. Liu, et al., *Phys. Rev. B* **67**, 125403 (2003); T. Pichler, *New Journal of Physics* **5**, 23 (2003).
5. R.S. Lee et al., *Nature* **388**, 255 (1997).
6. X. Liu, et al., *Phys. Rev. B* **69**, 075417 (2004).
7. T. Pichler et al., *Phys. Rev. B* **67**, 125416 (2003).
8. S. Suzuki et al., *Phys. Rev. B* **67**, 115418 (2003).
9. H. Rauf et al., *Phys. Rev. Lett.* **87**, 267401 (2004).
10. A. Goldoni et al., *Phys. Rev. B* **71**, 115435 (2005).
11. H. Ishii et al., *Nature* **426**, 540 (2003).
12. S. Okada et al., Phys. Rev. B **67**, 205411 (2003).
13. H. Shiozawa et al. AIP Conf. Proc. **685**, 139 (2004).
14. H. Rauf et al. unpublished results.
15. T. Pichler et al., *Phys. Rev. Lett.* **87**, 267401 (2003).
16. R. Egger, *Phys. Rev. Lett.* **83**, 5547 (1999).

Heteronuclear carbon nanotubes

Ferenc Simon[1], Rudolf Pfeiffer[1], Christian Kramberger[1], Hans Kuzmany[1], Viktor Zólyomi[2], Jenő Kürti[2], Philip M. Singer[3], and Henri Alloul[3]

[1] Institute für Materialphysik, Universität Wien, Strudlhofgasse 4, 1090, Wien, Austria
[2] Department of Biological Physics, Eötvös University, Budapest, Hungary
[3] Laboratoire de Physique des Solides, UMR 8502, Université Paris-Sud, Bât. 510, 91405 Orsay, France

Abstract. The physical properties of double-wall carbon nanotubes (DWCNT) with highly ^{13}C enriched inner walls were studied with Raman spectroscopy and nuclear magnetic resonance (NMR). An inhomogeneous broadening of the vibrational modes is explained by the random distribution of ^{12}C and ^{13}C nuclei based on *ab-initio* calculations. The growth of DWCNTs from natural and ^{13}C enriched fullerene mixtures indicates that carbon does not diffuse freely along the tube axis during the inner tube growth. The high curvature of the small diameter inner tubes manifests in an increased distribution of the NMR chemical shift tensor components.

INTRODUCTION

Isotope engineering (IE) of materials provides an important degree of freedom for both fundamental studies and applications. The change in phonon energies upon isotope substitution, while leaving the electronic properties unaffected, has been used repeatedly to identify vibrational modes[1] and gave insight into underlying fundamental mechanisms, such as phonon-mediated superconductivity[2]. The SWCNT specific enrichment of SWCNTs was reported recently[3,4]. In brief, highly ^{13}C enriched fullerenes are encapsulated inside SWCNTs. After a high temperature heat treatment, the inner fullerenes are transformed to an inner tube that is also highly ^{13}C enriched. It was also shown that there is no carbon exchange between the inner and outer shells during the synthesis. The advantage of this method is that other carbonaceous phases, such as amorphous carbon or graphite are not expected to be ^{13}C enriched. This would be particularly advantageous for NMR studies, where a high contrast between the highly ^{13}C enriched inner tubes and other carbon phases, thus a highly SWCNT specific signal is expected.

Here, we study further this heteronuclear carbon nanotube system. An inhomogeneous, isotope distribution related broadening is observed that is explained by *ab-initio* calculations. It is shown that inner tube growth can be studied with the help of carbon isotopes. The NMR studies on the material indicate that other carbon phases are not ^{13}C enriched and the NMR signal is indeed specific to the inner tubes. A curvature related anomalous distribution of chemical shift tensor components is observed.

CP786, *Electronic Properties of Novel Nanostructures*, edited by H. Kuzmany, J. Fink, M. Mehring, and S. Roth
© 2005 American Institute of Physics 0-7354-0275-2/05/$22.50

EXPERIMENTAL

The sample preparation was described previously[3,4]. In brief ^{13}C isotope enriched fullerenes and commercial SWCNTT samples were used to prepare fullerene peapods C_{60},C_{70}@SWCNT. We used two grades of ^{13}C enriched fullerene mixtures: 25 and 89 %. Fullerene filling and DWCNT transformation was performed following Ref. 5 and 6, respectively. DWCNTs with different ^{13}C enrichment grades are denoted as ^{Nat}C-, ^{13}C$_{0.25}$- and ^{13}C$_{0.89}$-DWCNT. Vibrational analysis was performed on a Dilor xy triple Raman spectrometer. First principles calculations were performed with the Vienna ab initio Simulation Package (VASP)[7]. Magic angle spinning (MAS) and static ^{13}C-NMR spectra were measured at ambient conditions using a Chemagnets (Varian Inc.) MAS probe at 7.5 Tesla. The ^{13}C-NMR spectra were obtained by a Fourier transformation of the free induction decay following the excitation pulse.

RESULTS AND DISCUSSION

The Raman modes, including the radial breathing (RBM) and the G modes of DWCNTs with highly ^{13}C enriched inner walls were discussed previously[3,4]. In brief, an overall downshift of all the modes was observed for the enriched inner tubes. In the simplest continuum approximation, the shift originates from the increased mass of the inner tube walls. This gives $(\upsilon_0 - \upsilon)/\upsilon_0 = 1 - \sqrt{\dfrac{12 + c_0}{12 + c}}$, where υ_0 and υ are the original and downshifted phonon energies, respectively, c is the ^{13}C enrichment of the inner walls, c_0=0.011 is the natural abundance of ^{13}C in carbon.

The validity of the continuum approximation for the RBM was verified by performing first principles calculations on the (n,m)=(5,5) tube as an example. In the calculation, the Hessian was determined by DFT. A large number of random ^{13}C distributions were generated and the RBM frequencies were determined from the diagonalization of the dynamical matrix for each individual distribution. We observed that the distribution of the RBM frequencies can be approximated by a Gaussian whose center and variance determine the isotope shifted RBM frequency and the spread in these frequencies. The difference between the shift determined from the continuum model and from the *ab-initio* calculations is below 1 %.

The broadening for the ^{13}C enriched inner tubes was observed previously[3,4] and is explained by calculations herein.

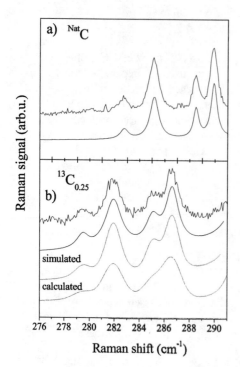

a) ^{Nat}C

b) $^{13}C_{0.25}$

simulated

calculated

Raman signal (arb.u.)

276 278 280 282 284 286 288 290

Raman shift (cm^{-1})

Figure 1. RBMs of some inner tubes at 676 nm laser excitation at 90 K with 0.5 cm^{-1} spectral resolution. a) ^{Nat}C, b) $^{13}C_{0.25}$. Smooth solid curves are the line-shapes after deconvolution by the spectrometer response. The dashed curve is a simulated line-shape with an extra Gaussian broadening to the intrinsic lines of the ^{Nat}C material. The dotted curve (lowest) is calculated line-shape (see text).

In Fig. 1a-b. we show the RBMs of some inner tubes for the ^{Nat}C and $^{13}C_{0.25}$ samples. Smooth solid curves are the line-shapes after deconvolution with the Gaussian response of our spectrometer. In Fig. 1a, this is a Lorentzian, but in Fig. 1b, the line-shape still contains a Gaussian component, as discussed below. The FWHMs of the resulting line-shapes are 0.76(4), 0.76(4), 0.44(4), 0.54(4) and 1.28(6), 1.30(6), 1.12(6), 1.16(6) for the inner tube RBMs shown in Fig. 1 of the ^{Nat}C and $^{13}C_{0.25}$ materials, respectively. The origin of the extra broadening is due to the random distribution of ^{12}C and ^{13}C nuclei. We found that the ratio between the half width of extra broadening and the shift, $\Delta v/(v_0-v)$, is approximately 0.19 for a 30 % ^{13}C enriched sample. The corresponding broadened line-shapes are shown by dotted curves in Fig. 1b. When the magnitude of the Gaussian randomness related broadening was fit (shown as dashed curve in Fig1b), we found that $\Delta v/(v_0-v)=0.15$. Similar broadening was observed for the 89 % sample which can also be reproduced by the calculation.

The freedom to manipulate the carbon enrichment on the inner walls allows to study the mechanism of inner tube growth. In Fig. 2. we show the RBMs for ^{Nat}C-, $^{13}C_{0.25}$-DWCNTs, together with a ^{Mix}C-DWCNT sample that was produced from a peapod containing mixed ^{Nat}C-C_{60} and $^{13}C_{0.25}$-C_{60}. The downshift of the inner RBMs for the ^{Mix}C-DWCNT sample lies half-way between the corresponding modes for the ^{Nat}C- and $^{13}C_{0.25}$-DWCNT samples reflecting the ~12 % overall enrichment of the inner walls that is a result of the 1:1 mixing of natural and enriched fullerenes.

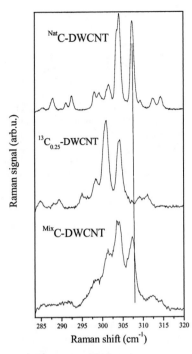

Figure 2. Inner tube RBMs at 676 nm laser excitation with 0.5 cm⁻¹ spectral resolution for NatC-, $^{13}C_{0.25}$-, and MixC-DWCNT. The vertical solid line is a guide to the eye to show the downshift of a mode for the MixC-DWCNT sample.

However, the widths of the inner tube modes are larger than expected from the homogeneously random distribution of ^{12}C and ^{13}C nuclei. This suggests that the two types of nuclei cannot form a homogeneously random mixture for such inner tubes, and instead ^{13}C rich and poor region alternates along the tube axis that gives rise to an increased inhomogeneity in the isotope distribution. This is supported by cluster-model simulations where the ^{13}C rich and poor regions with 60 randomly distributed nuclei vary randomly along the tube axis[8]. This proves that carbon cannot diffuse along the tube axis during the growth of the inner tube from fullerenes, and is stuck to its original position. This supports the growth model of inner tubes from pre-formed dimers with Stone-Wales transformations[9] rather than e.g. their growth through a complete evaporization into C_2 gas units.

The ^{13}C isotope enrichment of the inner wall of the DWCNTs allows NMR measurements with unprecedented contrast with respect to non-enriched carbon phases[3]. In Fig. 3, we show the static and MAS spectra of ^{13}C enriched DWCNTs, and the static spectrum for the SWCNT starting material. The typical chemical shift anisotropy (CSA) powder pattern is observed for the SWCNT sample in agreement with previous reports[10, 11]. However, the static DWCNT spectrum cannot be explained with a simple CSA powder pattern even though the spectrum is dominated by the inner tube signal. The structure of the spectrum suggests that the chemical shift tensor parameters are distributed for the inner tubes. It is the result of the higher curvature of inner tubes as compared to the outer ones: the variance of the diameter distribution is

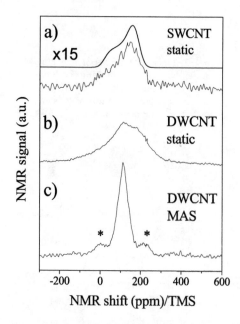

Figure 3. NMR spectra normalized by the total sample mass, taken with respect to the tetramethylsilane shift. (a) Static spectrum for non-enriched SWNT enlarged by a factor 15. Smooth solid line is a CSA powder pattern simulation with parameters published in the literature. (b) Static and (c) MAS spectra of $^{13}C_{0.89}$-DWCNT, respectively. Asterisks show the sidebands at the 8 kHz spinning frequency.

the same for the inner and outer tubes but the corresponding bonding angles show a larger variation. In addition, the residual line-width in the MAS experiment, which is a measure of the sample inhomogeneity, is 60(3) ppm, i.e. about twice as large as the ~35 ppm found previously for SWCNT samples[10, 11].

ACKNOWLEDGMENTS

This work was supported by the Austrian Science Funds (FWF) project Nr. 17345, by Hungarian State Grant (OTKA) Nr. T038014, and by the EU projects BIN2-2001-00580 and MEIF-CT-2003-501099 grants.

REFERENCES

1. M. C. Martin, J. Fabian, J. Godard, P. Bernier, J. M. Lambert, and L. Mihály, Phys. Rev. B **51**, 2844 (1995).
2. J. Bardeen, L. N. Cooper, and J. R. Schrieffer, Phys. Rev. **108**, 1175 (1957).
3. F. Simon et al., cond-mat/0406343, Phys. Rev. Lett. submitted.
4. F. Simon et al., in the Proc. of the XVIIIth IWEPNM, eds. H. Kuzmany et al., AIP Publishing 2004, p. 268.
5. H. Kataura et al., Synth. Met. **121**, 1195 (2001).
6. S. Bandow et al., Chem. Phys. Lett. **337**, 48 (2001).
7. G. Kresse and D. Joubert, Phys. Rev. B 59, 1758 (1999).
8. V. Zólyomi et al. unpublished.
9. Y. Zhao, B. I. Yakobson, R. E. Smalley, Phys. Rev. Lett. 88, 185501 (2002).
10. X.-P. Tang et al., Science 288, 492 (2000).
11. C. Goze-Bac et al., Carbon 40, 1825 (2002).

Carbon Nanotubes Investigated by N@C$_{60}$ and N@C$_{70}$ Spin Probes

B. Corzilius*, A. Gembus*, K.-P. Dinse*, F. Simon[#], and H. Kuzmany[#]

*) Chem. Dept., Darmstadt University of Technology, Petersenstr. 20, D-64287 Darmstadt, Germany
#) Institut für Materialphysik, Universität Wien, Strudlhofgasse 4, A-1090 Wien, Austria

Abstract. Nitrogen atoms encapsulated in C$_{60}$ can be used to detect small deviations from spherical symmetry via deformation-induced non-vanishing Zero-Field-Splitting (ZFS). In this context, experiments were performed by using these electronic quartet spin probes to investigate single wall carbon nanotubes. Time-fluctuating ZFS interaction would be indicative for rotational and/or translational degrees of freedom. Using pulsed EPR techniques, spin relaxation rates of N@C$_{60}$ and also of N@C$_{70}$ molecules with inherent static ZFS were measured. The analysis of their temperature dependence gave information about the dynamics of N@C$_{60}$ and N@C$_{70}$ molecules confined to the inside of the tubes.

INTRODUCTION

In contrast to the large group of metallo-endofullerenes (MEF) which can be classified as internal salts with strong Coulomb interactions, group 15-derived endofullerenes like N@C$_{60}$ and N@C$_{70}$ can be classified as atoms, which are freely suspended in a "chemical trap", thus forming a compound that is localized in space with negligible interaction with its confinement [1]. For this reason, these compounds are ideal sensors because spin or charge transfer is practically absent, and the localized paramagnetic atom in its quartet spin ground state probes the electric multipole moments at the site of the encased atom. In particular, deviations from spherical symmetry, as expected for N@C$_{60}$ embedded in single wall carbon nanotubes (SWCNT), could in principle be detected. Furthermore, thermally activated processes encountered by the incorporated spin probe, like reorientation within the multidimensional potential surface, should lead to temperature dependent spin relaxation processes.

EXPERIMENTAL

Peapods generated by inserting N@C$_{60}$ or N@C$_{70}$ in SWCNT were prepared using an optimized purification and insertion protocol as described in the contribution of Corzilius *et al.* in this proceedings [2]. Samples of N@C$_{60}$ and N@C$_{70}$ of various spin concentration were prepared as described elsewhere [3]. EPR spectra were measured using either a continuous wave (c. w.) EPR spectrometer (Bruker ESP 300E) or a pulse spectrometer (Bruker ElexSys 680), both equipped with Helium cryostats.

CP786, *Electronic Properties of Novel Nanostructures*, edited by H. Kuzmany, J. Fink, M. Mehring, and S. Roth
© 2005 American Institute of Physics 0-7354-0275-2/05/$22.50

RESULTS AND DISCUSSION

N@C$_{60}$ based peapods

The most striking result noted in the first study of N@C$_{60}$-based peapods by Simon *et al.* [4], was the observation that incorporation into nanotubes increased the EPR line width by more than an order of magnitude. This contrasts with the previous observation of record setting narrow lines of N@C$_{60}$ of less than 10 µT even in condensed matrices, which was taken as proof for the preserved near spherical symmetry at the site of the nitrogen atom. This observation was rationalized by assuming a "stiff" atomic cage, also allowing only negligible spin exchange. As seen in Fig. 1a, information about line width details were difficult to obtain in the first studies because of strong background signals, probably originating from defects in the SWCNT [4, 5]. Using the optimized preparation protocol [2], the early findings could be verified (Fig. 1b). For this comparison, N@C$_{60}$/C$_{60}$ samples of equal relative concentration (500 ppm) were used for peapod preparation.

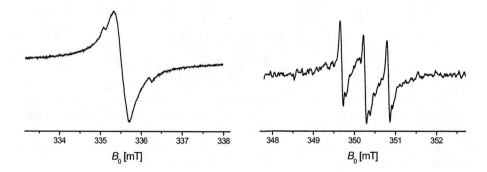

Fig. 1: Room temperature c. w. EPR spectra of N@C$_{60}$ peapods prepared as described previously (left) and using the improved protocol (right).

The interfering broad unstructured signal, which originates from empty SWCNT after preparation for insertion, can be suppressed by more than one order of magnitude, thus enabling spectral analysis of the nitrogen line triplet. In order to distinguish homogeneous and inhomogeneous line broadening, 2-pulse echo-induced EPR spectra were recorded. As shown in Fig. 2, N@C$_{60}$ as well as the broad line contribute to the echo-detected signal, i. e., both components are inhomogeneously broadened. Spectral decomposition is facilitated when measuring at 95 GHz, because first, the small g value difference of both signals leads to a more pronounced down field shift of the background signal, and, second, the width of the background signal scales nearly linearly with the field, thus increasing the relative N@C$_{60}$ signal intensity. This is in contrast to line width scaling behavior of N@C$_{60}$, for which only a factor of 3 increase in line width is observed.

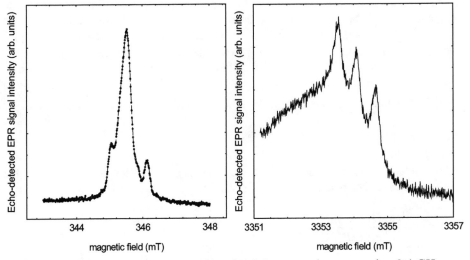

Fig. 2: Echo-induced EPR spectra of N@C_{60} peapods measured at 9.4 GHz (left) and 94 GHz (right) microwave frequencies.

For an identification of the dominant spin interaction, the temperature dependence of T_1 and T_2 relaxation times were studied. Using standard 2- and 3-pulse echo

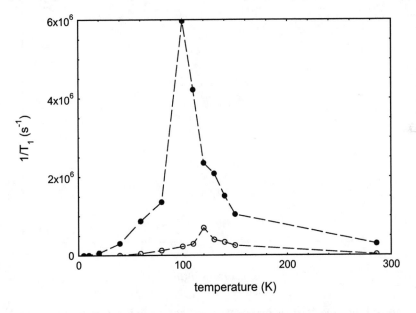

Fig. 3: Spin-lattice relaxation rates of N@C_{60}@SWCNT as function of temperature. Data are derived by fitting the echo signal intensity as function of T measured with a π-T-$\pi/2$-τ-π sequence. Both components of the bi-exponential fitting function are shown.

sequences for the determination of T_2 and T_1, respectively, the observed time dependence could be fitted by bi-exponential functions each. As shown in Fig. 3, the spin lattice relaxation rate T_1^{-1} shows a distinct rate maximum around 100 K. A spin-lattice relaxation rate maximum can be expected if the correlation time τ_c of the dynamic process is equal to the energy difference $\hbar\omega$ between electron spin levels, i. e., the condition $\omega\tau_c = 1$ is met. Assuming that spin relaxation is driven by a temperature-activated process, the temperature dependence of τ_c can be modeled as

$$\tau_c = \tau_0 \exp(E_a / k_B T) \tag{1}$$

with activation energy E_a and a fictitious high temperature lower bound τ_0 for the correlation time. The resulting T_1 rate is given by

$$T_1^{-1} \propto \frac{\exp(E_a / k_B T)}{1 + \omega^2 \tau_0^2 \exp(2 E_a / k_B T)} \tag{2}$$

Using this simple model, the experimental data set can be described with E_a and τ_0 as fit parameters. The result is depicted in Fig. 4.

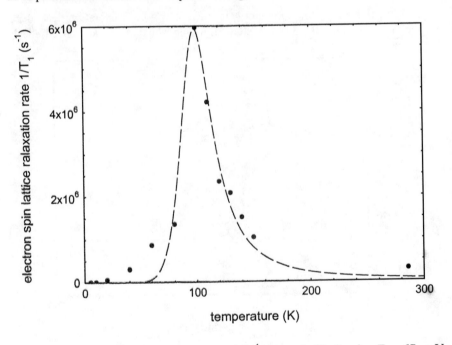

Fig. 4: Fit of the "fast" component of T_1^{-1} shown in Fig.3 using $E_a = 67$ meV and $\tau_0 = 6.3$ fs.

According to Eq. (2), the width of the relaxation peak is only determined by E_a, whereas the peak position is also dependent on τ_0. The value of E_a can be related to the binding potential of C_{60} to the inner wall of the SWCNT of a few eV. To find an activation energy of a few percent of the total binding energy is not unexpected, considering the various possible binding sites. What is more surprising is that the dispersion of tube diameters is not mapped into a corresponding variation of activation energies.

For verification of the observed effect the relaxation rates of $N@C_{70}@SWCNT$ were also determined. Here we found that two close lying rate maxima at 80 K and 110 K were observed. We tentatively correlate these maxima with different potential minima originating from C_{70} cages being stabilized parallel or perpendicular to the long axis of the SWCNT.

A further test for the validity of the presented relaxation model can be performed by increasing the Larmor frequency for instance by a factor of 10. Here from Eq.(2) it is predicted that first, the rate maximum is slightly shifted to higher temperatures, and second, the peak maximum is significantly depressed. This predicted reduction results from the decrease in spectral density because of a wider frequency spread of the total relaxation power. First experiments performed on the $N@C_{70}@SWCNT$ sample were in qualitative agreement with this prediction.

ACKNOWLEDGMENTS

Financial support by the Deutsche Forschungsgemeinschaft under various grants, the MEIF-CT-2003-501099 EU project and the Austrian Science Funds FWF 14893 are gratefully acknowledged.

REFERENCES

1. Akasaka, T. and Nagase, S. , Eds., *Endofullerenes, A New Family of Carbon Clusters*, Kluwer Academic Publishers, Dordrecht, (2002)
2. Corzilius, B., Gembus, A., Weiden, N., Dinse, K.-P., Simon, F., and Kuzmany, H. in: *Molecular Nanostructures*, Eds.: H. Kuzmany, J. Fink, M. Mehring, S. Roth; AIP Conference Proceedings, N. Y. (2005)
3. Jakes, P., Dinse, K.-P., Meyer, C., Harneit, W., and Weidinger, A., *Phys. Chem. Chem. Phys.* **5**, 4080-4083 (2003).
4. Simon, F., Kuzmany, H., Rauf, H., Pichler, T., Bernardi, J., Peterlik, H., Korecz, L., Fülöp, F., and Jánossy, A., *Chem. Phys. Lett.* **383**, 362-367 (2004)
5. Gembus, A., Simon, F., Jánossy, A., Kuzmany, H., and Dinse, K.-P. in *Electronic Properties of Synthetic Nanostructures*, Eds.: H. Kuzmany, J. Fink, M. Mehring, S. Roth; AIP Conference Proceedings 723, N. Y. (2004), p. 259-263.

Assembly of carbon tube-in-tube nanostructures

D. S. Su[a], Z.P. Zhu[a,b], X. Liu[a], G. Weinberg[a], N. Wang[c], R. Schlögl[a],

a: Department of Inorganic Chemistry, Fritz Haber Institute of the MPG, Faradayweg 4-6,
D-14195 Berlin, Germany
b: State Key Laboratory of Coal Conversion, Institute of Coal Chemistry, Chinese Academy of Sciences,
030001 Taiyuan, China
c: Physics Department, Hong Kong University of Science and Technology, Hong Kong, China

Abstract: Tube-in-tube carbon nanostructures were prepared by reorganization of graphitic impurity nanoparticles outside or inside of the pristine carbon nanotubes. Graphitic impurity nanoparticles were first disintegrated into small graphene fragments by a chemical oxidation with nitric acid, which also modifies the graphene fragments with carboxyl and hydroxyl groups at their edges. The functionalized graphene fragments were then reintegrated outside or inside of pristine carbon nanotubes to construct into tube-in-tube nanostructures. The combination of oxidatively functionalized graphene units, their solvate in a polar organic medium allowing for dispersive forces to effect supramolecular organization with carbon nanotubes acting as templates and their polycondensation by acid-catalysed esterification followed by pyrolysis of the oxygen functionalities lead to complex nanostructures inaccessible by direct synthesis.

INTRODUCTION

The surface and channel accessibility of carbon nanotubes (CNTs) bestow them with wide applications in catalysis,[1-4] gas sensing,[5-7] and template-based assembly of heterostructures.[8-11] However, the technical application of CNTs is handicapped by the high production costs and by the byproducts in all the high temperature methods developed so far. A rational, organic wet-chemical synthesis of CNTs becomes the dream of chemists and materials scientists during last 10 years. In the present work, as an intermediate approach between chemical vapor deposition and molecular organic synthesis concepts we report about the synthesis of carbon tube-in-tube[12,13] nanostructures by means of disassembly and assembly of graphitic fragments using ordinary carbon nanotubes as templates. We use non-covalent forces to manipulate graphene fragments to supramolecular tubular structures in a wet-chemical process rather different from all current high temperature methods for CNT synthesis(for instance, metal particle catalyzed chemical vapor decomposition of hydrocarbon or arc-discharging of graphite). On the one side, this experiment could provide a new way for the purification of carbon nanotube. It also sheds light to the new application of the carbon material because of the delicate structure and high surface area of carbon tube-in-tube nanomaterials that could be a promising material for hydrogen storage and support material for catalysts.

CP786, *Electronic Properties of Novel Nanostructures*, edited by H. Kuzmany, J. Fink, M. Mehring, and S. Roth
© 2005 American Institute of Physics 0-7354-0275-2/05/$22.50

EXPERIMENTAL

The procedure for re-assembling carbon tube-in-tube nanostructures consists of two steps: oxidation and esterification. Firstly 0.25 g of raw CNTs sample (Applied Sci. Ltd., OH) was combined with 25 ml concentrated HNO_3 with magnetic stirring and oxidized for 10 h under refluxing conditions. After filtration, the remaining solid was washed sequentially with deionized water and ethanol and dried at 110 °C in air for 12 h.

To synthesis the carbon tube-in-tube nanostructures, 50 mg of the oxidized sample was introduced in 50 ml tetrahydrofuran (THF) by ultrasonic dispersal for 10 min at room temperature. 0.2 ml of concentrated H_2SO_4 was added to the resulted suspension and refluxed for 16 h to allow a full re-integration of the carboxyl-hydroxyl-modified graphene fragments through an esterification linkage, in which H_2SO_4 served as the acid catalyst. After the reaction, the suspension was filtered, washed with ethanol, and dried at 110 °C in air for 12 h.

RESUTLS AND DISCUSSION

As shown in the TEM images (Figure1a, 1b), in raw material the carbon nanotubes are open-ended, have thin wall (about 20 nm) and big inside diameter (from 50nm to 200nm). There are many carbon nanoparticles inside and outside of carbon nanotubes as impurities. After the treatment mentioned above, carbon tube-in-tubes nanostructure were found, as revealed by TEM and SEM images (Figure 1c and 1d). The carbon nanotubes and carbon tube-in-tubes are mixed together (Figure 1c). The yield of carbon tube-in-tubes is more than 5%, which is estimated from the TEM micrographs. The range of diameters of inner and outer tubes is between 50 nm to 200 nm and the thickness of walls is about 20nm.The wall thickness of few inner tube is thinner than that of outside one. The carbon tube-in-tube nanostructure is well-constructed, the thickness is homogenous and the tube end is open. Comparing the electron micrographs before and after treatment, it is obvious that the carbon tube-in-tube nanostructures are formed through the oxidation and esterification processes.

The concentrated nitric acid used oxidizes the graphitic nanoparticles of carbon nanotubes. In this process, the advantages of acid treatment 1) introducing the NO_3^- into the interlayer spaces through defect-involved "doors" and exfoliating the large particles into small and thin graphene sheets; 2) modifying the edges of the graphene sheets with carboxyl and hydroxyl groups through oxidation reaction; 3) removing the amorphous carbon nanopartcles and iron catalyst particles. Although a part of the defected carbon nanotubes and nanoparticle is oxidized, the main carbon nanotubes are not influenced in morphology and structure, which is observed in TEM image (Figure 1c). After oxidation and esterification, there is significant change in morphology and composition, Carbonaceous impurity disappears and the new carbon nanotubes form along the pristine carbon nanotubes inside or outside. It is undoubted that the formation of carbon tube-in-tube is consequence of the assembly of the functionalized graphene sheets.

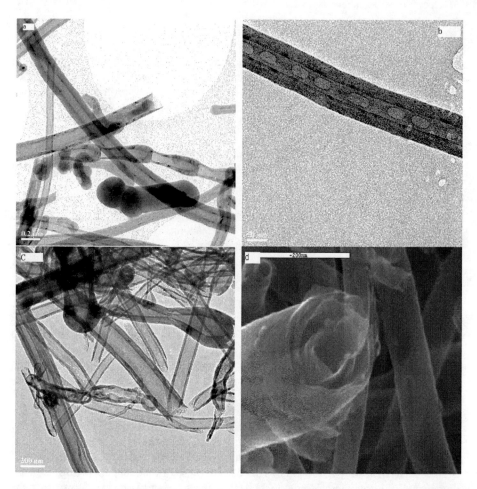

Figure1. a) TEM image of fresh carbon nanotubes with graphite nanoparticles (Applied Sci. Ltd., OH); b) TEM image of graphitic nanoparticles inside a fresh carbon nanotube; c) TEM image of carbon tube-in-tube nanostructures and carbon nanotubes; d) SEM image of carbon tube-in-tube structure.

Another impressive phenomenon is the bamboo-like carbon tube-in-tube nanostructure (Figure 2). A pristine bamboo-like carbon nanotube is shown in Figure 2a. After the treatment, a new bamboo-like tube is formed along with the pristine one , as it is shown in Figure 2b. Obviously the re-assembled carbon nanotube copies twists, nodes and other special shapes of carbon nanotube (inside or outside, see Figure 2b and 2c). This suggests that the pristine carbon nanotubes act as the structure-directing template. It is also clear that the graphene sheets re-assemble with the different mechanisms (outside, inside or both in the direction of pristine tubes). Evidently, the graphite particles oxidized and esterified form the carbon tube-in-tube, which morphology is strongly influenced by the matrix carbon nanotube.

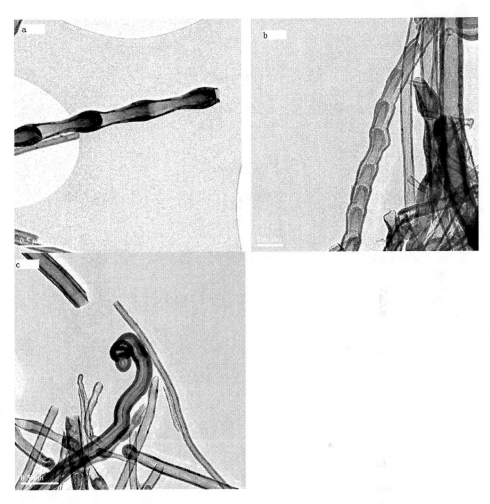

Figure2 a) TEM image of bamboo-like carbon nanotube in raw material; b) TEM image of bamboo-like carbon tube-in-tube; c) TEM image of carbon tube-in-tube, the new tube forms inside the carbon nanotube.

SUMMARY

The graphitic impurity among the carbon nanotubes is used as the carbon source for synthesis of new carbon nanotube. After oxidation and the esterification, the graphene sheets re-organize and assemble the carbon tube-in-tube. This strategy would give a light to the new synthesis method of carbon nanotube with low temperature and mild conditions, it also suggests that the new carbon material with the delicate morphology and high surface area would be the good material for hydrogen storage and support material for catalysts.

ACKNOWLEDGMENTS

The work is partially supported by NSFC (59872047) and by the Deutsche Forschungsgemeinschaft (Project SCHL 332), performed in frame of ELCASS.

REFERENCES

1. Kroto, H.W., Heath, J.R., O'Brien, S.C., Curl, R.F., Smalley, R.E., Nature 318, 162 (1985).
2. Iijima, S., Nature 354, 56 (1991).
3. Liu, J., Dai, H., Hafner, J. H., Colbert, D. T., Smalley, R. E. Nature 385, 780 (1997).
4. Sano, M., Kamino, A., Okamura, J., Shinkai, S. Science 293, 1299 (2001).
5. Martel, R., Shea, H. R., Avouris, Ph. Nature 398, 299 (1999).
6. Falvo, M. R., Clary, G. J., Taylor, R. M., Brooks, F. P., Washburn, Jr. S., Superfine, R. Nature, 389, 582 (1997).
7. Lu, J. P. Phys. Rev. Lett. 79, 1297(1997).
8. Chopra, N. G., Benedict, L. X., Crespi, V. H., Cohen, M. L., Louie, S. G., Zettl, A. Nature 377, 135 (1995).
9. Ugarte, D. Nature 359, 707 (1992).
10 Wang. Z.L., Yin, J.S., Chem. Phys. Lett. 289, 189(1998).
11. Rietmeijer, F.J.M.,, Schultz, P. H., Bunch, Th. E., Chem. Phys. Lett. 374, 464 (2003).
12. Zhu, Z.P., Su, D.S., Weinberg, G., Jentoft, R.E., Schlögl, R., Small 1, 1(2005).
13. Zhu, Z.P., Su, D.S., Weinberg, G., Schlögl, R., Nano Letter 4, 11 (2004).

A comparative study of field emission from single- and double-wall carbon nanotubes and carbon peapods

D.A. Lyashenko[1,2], A.N. Obraztsov[4], F. Simon[3], H. Kuzmany[3], E.D. Obraztsova[1], Yu.P. Svirko[2], K. Jefimovs[2]

[1]Natural Sciences Center of General Physics Inst., RAS, 38 Vavilov street, 119991, Moscow, Russia,
[2]Department of Physics, University of Joensuu, Yliopistokatu, 7, Joensuu 80101, Finland,
[3] Universitat Wien, Institut fur Materialphysik, Strudlhofgasse 4, A-1090, Wien, Austria,
[4]Physics Department of Moscow State University, 119992, Moscow, Russia

Abstract. In this work field electron emission was studied for a variety of single-wall carbon nanotube-based nanostructures (arc and HipCO nanotubes, C_{60}-and C_{70}-based carbon peapods, double-wall nanotubes) in the same conditions. A comparable surface roughness for all samples was confirmed by a scanning electronic microscopy, while distribution of emission centers was visualized with a phosphorescent anode screen. We developed an original approach that improves a reliability of the field emission threshold measurements. All materials showed the emission threshold values ranging from 0.5 to 2.5 V/μm. The lowest threshold was observed for HipCO and double-wall materials that contain thinner nanotubes than other studied materials.

INTRODUCTION

Carbon peapods and double-wall carbon nanotubes (DWNT) are novel nanostructures [1, 2] based on single-wall carbon nanotubes (SWNT). Since SWNT is an efficient field emission (FE) electron source [3-7], one may expect that both carbon peapods and DWNT should also show strong FE performance. However, since until now these materials have been obtained in rather small volumes and FE measurements have been performed in different geometries, the quantitative comparison has not been made yet. In this work we in the first time report on the comparative study of the FE in a variety of SWNT-based nanostructures (arc and HipCO SWNT, C_{60} -and C_{70}-based carbon peapods, DWNT) using the same experimental setup.

EXPERIMENTAL

Field emission measurements were performed in a vacuum diode configuration with the flat cathode and anode. The nanotube-based cathodes were placed on the holder, which position and tilt with respect to the anode can be changed. In our experiments, anode was made of either metal or glass with conductive ITO electrode covered by the phosphor layer. In the latter case we were able to visualize distribution

CP786, *Electronic Properties of Novel Nanostructures*, edited by H. Kuzmany, J. Fink, M. Mehring, and S. Roth
© 2005 American Institute of Physics 0-7354-0275-2/05/$22.50

of emission sites over the cathode surface. The current-voltage measurements were performed at room temperature, residual pressure of 10^{-5} Torr and bias up to 20kV. In order to observe the sample during the measurement process and to capture light emitted by the phosphor layer, we employed a measuring cell with a bell glass.

SWNT were synthesized by arc-discharge method with Ni/Co or Ni/Y_2O_3 catalyst. Commercially available raw and purified HipCO nanotubes were used. The peapods were formed from SWNT (arc-produced with Ni/Co) by filling them with C_{60} or C_{70} molecules at elevated temperature. The double-wall carbon nanotubes were synthesized by heating in vacuum of C_{60}-filled peapods.

In order to compare the FE properties of cathodes made from different nanocarbon materials, we prepared samples of the same shape and thickness and glued them onto Ni substrates. The morphology of each sample was examined by using Scanning Electron Microscope LEO-15500.

RESULTS AND DISCUSSION

Field emission measurements were performed in a vacuum diode configuration with the flat cathode [8]. In order to compare the FE efficiency in different materials we developed a new method that enabled to determine the threshold electric field (E_{th}) at cathode surface for all materials. This field can be found from $E_{th} = V_{th}/D$, where V_{th} is the "switch on" threshold voltage and D is an inter-electrode distance. However, in the experiment, we can directly measure only a relative displacement ΔD of the cathode rather than its distance from the anode, i.e. $D = D_0 + \Delta D$, where D_0 is *a priori* unknown. In order to overcome this difficulty, for each nano-carbon cathode we measured a current I as a function of V at several displacements ΔD (see *Figure 1 a*).

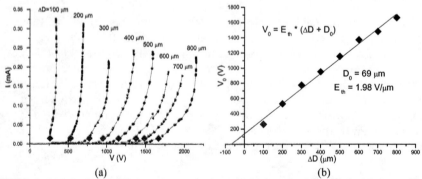

(a)   (b)

FIGURE 1. (a)A family of *I-V* dependences for the carbon peapods, fully filled with C_{60}, measured with different distances between electrodes D (mm). The emission threshold points are marked at the current level of 0.02 mA. (b) The emission threshold voltage dependence on the distance between electrodes. Points - experiment, line - its linear approximation.

From these measurements, we determined the voltage V_0 that corresponds to the current of $I_0 = 0.02$ mA. This voltage is slightly above the "switch on" threshold V_{th} for the particular sample at a given ΔD. Since the FE threshold $E_{th} = V_{th}/D$ is an intrinsic parameter of the cathode material that does not depend on applied voltage and distance between the cathode and anode, V_{th} is a linear function D. Correspondingly, the measured in the experiment V_0 should be a linear function of ΔD. Therefore, the offset of the measured dependence $V_0(\Delta D)$ in the horizontal axis (see Figure 1b) gives us the value of D_0 and the FE threshold E_{th} [8].

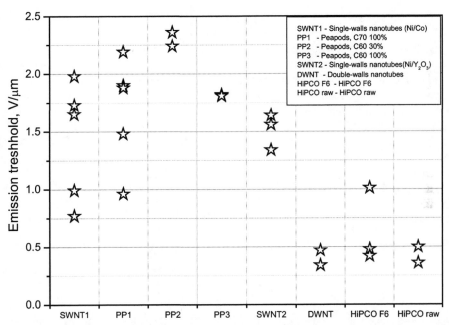

FIGURE 2. The emission threshold values measured for different carbon materials. The legend is in the insert.

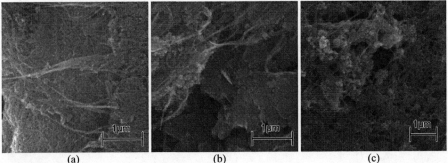

FIGURE 3. SEM images for SWNT synthesized by arc technique with Ni/Co catalyst (a), peapods filled with C_{70} (b) and SWNT synthesized with Ni/Y_2O_3 catalyst (c).

We employed the developed method to measure the FE threshold for all studied nanotube-based materials (see insert to Figure 2) and using several samples for each material. One can observe that the obtained threshold values are spreaded in the range

0.5-2.5 V/μm. The lowest threshold was observed for samples containing (both raw and purified) HipCO nanotubes and double-wall SWNT. The latter material was manufactured by heating of arc-SWNT based on C_{60} peapods. We suggest that such a reduction of the FE threshold is due to the presence of the nanotubes with the smallest diameter (0.6-1.0 nm) in HipCO nanotubes and double-wall SWNT.

CONCLUSION

We developed a novel approach that allows us to improve a reliability of the FE threshold measurements. This approach was demonstrated for nanocarbon cathodes made of different SWNT-based materials. The surface roughness of cathodes was investigated with a scanning electronic microscopy, while the phosphorescent anode screen was used to visualize a distribution of emission centers on the cathode surface. The nano-carbon films investigated showed the FE threshold values ranging from 0.5 to 2.5 V/μm. We suggest that the thinnest SWNT in HipCO- and DWNT-based cathodes are responsible for the relatively low FE threshold of 0.5 V/μm in these materials.

ACKNOWLEDGMENTS

The work is supported by RFBR-04-02-17618 and by RAS program "New materials".

REFERENCES

1. W. Smith, M. Montioux, et al., *Nature, 396 (1998) 323.*
2. R. Pfeiffer, H. Kuzmany, Ch. Kramberger, et al., *PRL 90 (2003) 225501.*
3. L.A. Chernozatonskii, Yu.V. Gulyaev, Z.Ya Kosakovskaya, et al., *Chem. Phys. Lett. 233 (1995) 6.*
4. W.A. de Heer, A. Chatelain, D. Ugarte, *Science 270 (1995) 1179.*
5. J.M. Bonard, J.P. Salvetat, W.A. de Heer, et al., *Appl. Phys. Lett. 73 (1998) 918.*
6. A.N. Obraztsov, I. Pavlovsky, A.P. Volkov et al., *J. of Vacuum Science & Tech. B 18 (2000) 1059.*
7. E.D. Obraztsova, A.S Pozharov, S.V. Terekhov et al., *Proc. of Material Research Society, Symp., 2002, vol. 706, p. Z6.7.1-Z6.7.6.*
8. A.N. Obraztsov et al., *Appl. Surf. Sci., 215 (2003) 214.*

A Raman Map of the DWCNT RBM Region

R. Pfeiffer*, F. Simon*, H. Kuzmany* and V. N. Popov†

*Institut für Materialphysik, Universität Wien, Vienna, Austria
†Faculty of Physics, University of Sofia, Sofia, Bulgaria

Abstract. Raman studies on double-wall carbon nanotubes (DWCNTs) obtained from peapods showed that the radial breathing mode (RBM) response of the inner tubes is split into more components than there are geometrically possible inner tubes. This was attributed to the possibility that one inner tube type may form in more than one outer tube type. Recent resonance Raman investigations revealed an even more complex and interesting behavior. The split components appear to be grouped into species belonging to the same inner tube chirality and the width of the splitting increases with decreasing tube diameter.

INTRODUCTION

Single-wall carbon nanotubes (SWCNTs) [1] have attracted a lot of scientific interest over the last decade due to their unique structural and electronic properties. By annealing fullerenes enclosed in SWCNTs (so-called peapods [2]) it is possible to transform the fullerenes into a secondary inner tube enclosed in the primary outer tube [3].

A Raman study of the radial breathing modes (RBMs) of the inner tubes revealed that these modes have intrinsic linewidths down to about $0.4 \, \mathrm{cm}^{-1}$ [4]. These small linewidths indicate long phonon lifetimes and therefore highly defect free inner tubes, which is a proof for a nano clean-room reactor on the inside of SWCNTs. A closer inspection of the RBM response of the inner tubes revealed that there are more Raman lines than geometrically allowed inner tubes. This was attributed to the possibility that one inner tube type may form in more than one outer tube type [5].

To study this splitting in more detail, we recorded a Raman map of the DWCNT RBM region between 1.54 and 2.54 eV (488 to 803 nm) with a spacing of about 15 meV (where possible). This experiment revealed that the large number of components of the RBMs are strongly grouped into species belonging to the same inner tube chirality. The grouping is retained during the growth process, i.e., members of the groups appear and grow collectively. Additionally, the groups seem to be unaffected of the carbon source filled into the primary tubes (C_{60}, C_{70}, or fullerenes mixed with toluene). Only the outer tubes diameter distribution can influence the intensity distribution within one series. However, the frequencies of the RBM components remain similar over several samples with different outer tubes diameters.

CP786, *Electronic Properties of Novel Nanostructures*, edited by H. Kuzmany, J. Fink, M. Mehring, and S. Roth
© 2005 American Institute of Physics 0-7354-0275-2/05/$22.50

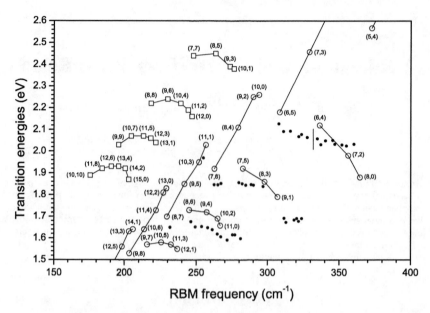

FIGURE 1. Experimental Kataura-plot for dispersed isolated HiPco tubes (open symbols, Refs. [6–9] and for DWCNTs (dots). The number of RBM components of the inner tubes in the DWCNTs was obtained by fitting the normal resolution spectra. Squares and circles denote the E_{11}^M and E_{22}^S transitions, respectively. Families were completed with calculations after Ref. [10].

EXPERIMENTAL

As starting material for our DWCNTs we used C_{60} peapods which were subsequently annealed at 1250 °C in a dynamic vacuum for 2 hours. The Raman spectra were measured with a Dilor xy triple spectrometer using various lines of an Ar/Kr laser, a dye laser with Rhodamine 6G and Rhodamine 101 as dyes and a Ti:sapphire laser. The spectra were recorded at ambient conditions in normal resolution mode. Selected spectra were recorded in high resolution mode at 90 K.

RESULTS AND DISCUSSION

Figure 1 shows the measured transition energies and RBM frequencies of the inner tubes of DWCNTs and compares it with the published values of dispersed isolated HiPco tubes. One can easily see that the number of RBM components is much larger than the number of geometrically allowed inner tubes. This is especially true for the (6, 5) and (6, 4) inner tube types. In normal resolution, each of these chiralities has to be assigned more than five components spread over 20cm^{-1} and more. In high resolution mode, the number of identifiable components increases to more than ten as can be seen from Fig. 2. For the larger diameter inner tubes the width of the splitting is reduced, although

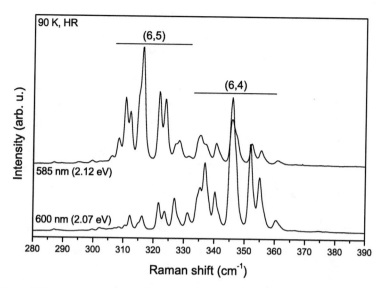

FIGURE 2. RBM response of the $(6,5)$ and $(6,4)$ inner tubes measured in high resolution mode at 90 K. A fit revealed 11 and 14 RBM components for these tubes, respectively.

overlappings of splitting series cannot be ruled out completely. Comparing the RBM frequencies with the HiPco data shows that the lowest frequency components in the groups correspond well to the HiPco frequencies.

The transition energies for all inner tubes are downshifted by about 50 meV as compared to HiPco. Additionally, the transition energies of the components of one inner tube type decrease slightly with increasing RBM frequency.

The transition energies E_{ii} for every inner tube RBM component were obtained by fitting the Stokes resonance Raman profiles with

$$I(E_L) = \frac{A}{\left|(E_L - E_{ii} - i\gamma)(E_L - (E_{ii} + E_{ph}) - i\gamma)\right|^2}, \tag{1}$$

where E_L is the laser excitation energy, E_{ph} is the phonon energy obtained from Raman, and γ is a damping constant. For the example of the 315.9 cm^{-1} RBM component of the $(6,5)$ inner tube (s. Fig. 3), one gets $E_{ph} = 0.0392$ eV, $E_{ii} = 2.0926(6)$ eV, and $\gamma = 0.043(1)$ eV. If one is only interested in the transition energy, one could also fit a Gaussian to the data points and subtract $E_{ph}/2$ from the central position (2.1122 eV for the example in Fig. 3). This procedure leads to the same transition energy as above. Usually, the damping constants scattered around 0.05 eV.

The origin of the splitting is still attributed to the interaction of one inner tube type with several outer tubes types of different diameter, but the possible combinations between inner and outer tunes is much larger than anticipated so far. Depending on the outer tube diameter, the inner tube experiences a different hydrostatic pressure which is responsible for the different frequency upshifts as well as for the slight transition energy

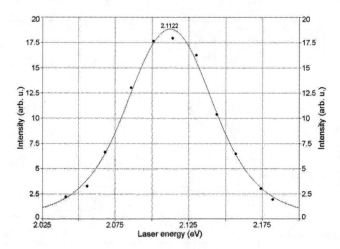

FIGURE 3. Resonance Raman Profile of the $315.9\,\mathrm{cm}^{-1}$ RBM component of the $(6,5)$ inner tube.

red-shifts.

ACKNOWLEDGMENTS

The authors acknowledge financial support from the FWF in Austria, project P17345, and from the EU, project PATONN Marie-Curie MEIF-CT-2003-501099. VNP was supported partly by Marie-Curie MEIF-CT-2003-501080 and partly by NATO CLG 980422. Valuable discussions with Prof. A. Jorio and Prof. A. Rubio are gratefully acknowledged.

REFERENCES

1. Iijima, S., and Ichihashi, T., *Nature*, **363**, 603–605 (1993).
2. Smith, B. W., Monthioux, M., and Luzzi, D. E., *Nature*, **396**, 323–324 (1998).
3. Bandow, S., Takizawa, M., Hirahara, K., Yudasaka, M., and Iijima, S., *Chem. Phys. Lett.*, **337**, 48–54 (2001).
4. Pfeiffer, R., Kuzmany, H., Kramberger, C., Schaman, C., Pichler, T., Kataura, H., Achiba, Y., Kürti, J., and Zólyomi, V., *Phys. Rev. Lett.*, **90**, 225501 (2003).
5. Pfeiffer, R., Kramberger, C., Simon, F., Kuzmany, H., Popov, V. N., and Kataura, H., *Eur. Phys. J. B*, **42**, 345–350 (2004).
6. Bachilo, S. M., Strano, M. S., Kittrell, C., Hauge, R. H., Smalley, R. E., and Weisman, R. B., *Science*, **298**, 2361–2366 (2002).
7. Fantini, C., Jorio, A., Souza, M., Strano, M. S., Dresselhaus, M. S., and Pimenta, M. A., *Phys. Rev. Lett.*, **93**, 147406 (2004).
8. Telg, H., Maultzsch, J., Reich, S., Hennrich, F., and Thomsen, C., *Phys. Rev. Lett.*, **93**, 177401 (2004).
9. Weisman, R. B., and Bachilo, S. M., *Nano Lett.*, **3**, 1235–1238 (2003).
10. Popov, V. N., *New J. Phys.*, **6**, 17 (2004).

Redox n-Doping of Fullerene C_{60} Peapods

Ladislav Kavan[1,2], Martin Kalbáč[1,2], Markéta Zukalová[1] and Lothar Dunsch[2]

[1]J. Heyrovský Institute of Physical Chemistry, Academy of Sciences of the Czech Republic,
Dolejškova 3, CZ-182 23 Prague 8
[2]Leibniz Institute of Solid State and Materials Research, Helmholtzstr. 20, D - 01069 Dresden

Abstract. Redox n-doping of C_{60}@SWCNT was carried out by liquid potassium amalgam and electrochemically in 1-butyl-3-methylimidazolium tetrafluoroborate (ionic liquid). The charge transfer induced effects were monitored by in-situ Raman spectroscopy. Doping with potassium amalgam causes reduction of intratubular C_{60} to C_{60}^{4-} and C_{60}^{5-}, but a complete re-oxidation to neutral fullerene occurs spontaneously upon contact to air. On the other hand, electrochemical n-doping does not generate any intratubular fulleride below C_{60}^{-} even at very negative potentials. The relative intensity of $A_g(2)$ mode of intratubular C_{60} does not seem to drop monotonously with decreasing potential. This recalls a similar intensity profile at chemical doping.

INTRODUCTION

Doping of C_{60}@SWCNT was previously studied by chemical reactions with vaporized potassium [1,2] and $FeCl_3$ [1] as well as by electrochemical methods [3,4]. Among the salient outputs were the specific "anodic Raman enhancement" of intratubular C_{60} modes in electrochemically p-doped peapods [3,4] and heuristic localization of potassium inside and outside the K-doped peapods [2]. The latter conclusion was recently confirmed by direct TEM observation [5]. However, critical review of doping of nanocarbons points at serious inconsistency of experiment versus theory already at the level of empty SWCNT [6]. Hence, we follow the suggestion of Iijima, Eklund et al. [6] that parallel investigation of Raman spectra of chemically and electrochemically doped species is important to learn more about charge-transfer induced effects. Here we present new data on the n-doping of C_{60}@SWCNT by potassium dissolved in mercury (potassium amalgam). Also novel results for electrochemical n-doping at extremely negative potentials (<-2.3 V vs. ferrocene reference electrode) are shown. The collected data allow comparison of electrochemical and chemical doping.

EXPERIMENTAL SECTION

Potassium amalgam (1.3 at% concentration) was prepared under vacuum by dissolving potassium metal in mercury. Buckypaper of C_{60}@SWCNT peapod (filling ratio 85%) [2] was contacted with the amalgam in a closed Raman optical cell. Pt-supported thin-film of peapods was fabricated by evaporation of the sonicated ethano-

CP786, *Electronic Properties of Novel Nanostructures*, edited by H. Kuzmany, J. Fink, M. Mehring, and S. Roth
© 2005 American Institute of Physics 0-7354-0275-2/05/$22.50

lic slurry and investigated by *in-situ* Raman spectroelectrochemistry. Electrochemical potentials were referred to the ferrocene reference electrode, Fc/Fc^+. The electrolyte was 1-butyl-3-methylimidazolium tetrafluoroborate (ionic liquid). Experimental details are given elsewhere [4]. Raman spectra were measured on a T-64000 spectrometer (Instruments, SA) interfaced to an Olympus BH2 microscope; the laser power impinging on the sample or cell window was between 2-5 mW. Spectra were excited by Ar^+ laser at 2.54 eV (Innova 305, Coherent). The Raman spectrometer was calibrated before each set of measurements by using for reference the F_{1g} line of Si at 520.2 cm^{-1}.

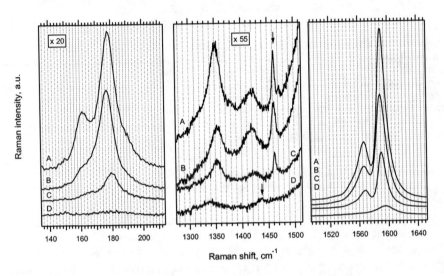

FIGURE 1. Raman spectra (excited at 2.54 eV) of C_{60}@SWCNT pristine [A] and doped with K-amalgam for 24 h at room temperature [B], for 72 h at room temperature [C] for 40 hours at 100 °C [D]. Spectra are offset for clarity. Intensities and frequencies are normalized against those of Si. Arrow points at the $A_g(2)$ mode.

RESULTS AND DISCUSSION

Fig. 1 shows the Raman spectra at various stages of n-doping with K-amalgam. Obviously, the reaction of K-amalgam with peapods is sluggish, and no significant changes are apparent after hours of contact at room temperature. When the sample is treated with K-amalgam at 100°C for 40 hours (spectrum C), the $A_g(2)$ peak softens by ca. 27 cm^{-1} and looses its high-frequency satellite. This softening is equivalent to a transfer of 4-5 electrons per C_{60} [2]. Compared to K-vapor doping, we do not produce the fully reduced C_{60}^{6-}[1,2]. The amalgam-doped sample spontaneously recovers to the undoped state upon short (minutes) contact to ambient air, when the characteristic $A_g(2)$ doublet of neutral peapod is restored. This is an explicit difference from K-vapor doped peapod, as the latter is not fully recoverable to the state of neutral C_{60} in air, water and even by anodic oxidation, when part of the intratubular fullerene still survives in the C_{60}^{4-} state [2]. The absence of hysteresis in re-oxidation of K-amalgam doped peapods indicates that the K^+ ions are located exclusively at the outer wall of peapod. They compensate the charge at SWCNT (pod) and at the intratubular anions, C_{60}^{4-} or C_{60}^{5-}. This system is sensitive to air exposure, which quickly restores the neutral C_{60} peas. On the other hand, if potassium penetrates into the peapod interior from the vapor phase, it makes, eventually, n air-stable intratubular fulleride K_4C_{60} [2]. The less efficient doping with K-amalgam is understandable, if we compare the standard redox potential of pure potassium (-2.934 V) with that of the K-amalgam (-1.975 V). After K-amalgam doping, the G-band is upshifted by 6 cm^{-1} and is single Lorentzian (Fig. 1, curve D). This situation is reminiscent of the final stage of Interval III doping of SWCNT with Cs-vapor [6], but the K-amalgam doping produces a final material exhibiting solely this anomalous G-band shift.

To get further insight into the n-doping chemistry of peapods, we present data for electrochemical charging of peapods in ionic liquid medium. Fig. 2 shows the doping feedback to Raman spectra. The potential of -2.3 V vs. Fc/Fc$^+$ (third curve from top, Fig. 2) represents similar value to that of K-amalgam.

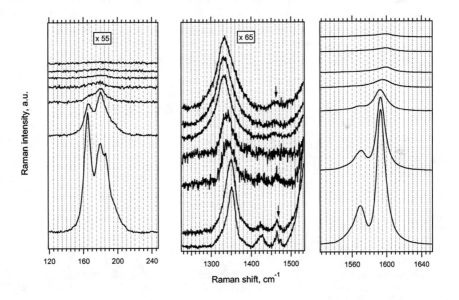

FIGURE 2. Raman spectra of C_{60}@SWCNT (excited at 2.54 eV) 1-butyl-3-methylimidazolium tetrafluoroborate, measured *in-situ*. The electrode potential (in V *vs*. Fc/Fc$^+$) was for curves from top to bottom: -3.0, -2.8, -2.3, -1.8, -1.3, -0.8, -0.3. Spectra are offset for clarity, but the intensity scale is identical for all spectra in the respective window. Arrow points at the $A_g(2)$ mode.

311

Due to comparable driving force for n-doping we can trace some similarities in the Raman spectra. Both RBM and G-bands are significantly attenuated. This is due to the quenching of E_{33}^S optical transition, which is responsible for resonance enhancement in pristine peapods at the laser energy used (2.54 eV). We note similar shape and upshift of the G-band in cathodically doped peapods (Fig. 2). The $A_g(2)$ mode is downshifted by ca. 5 cm^{-1} indicating the intratubular fulleride C_{60}^-, but the reduction of C_{60} does not progress further, if we apply still more negative potentials (Fig. 2). Of particular interest is the relative intensity of $A_g(2)$, which is first attenuated during cathodic potential sweep, achieving a minimum around -1.8 V, but enhances again at more negative potentials. This reminds the enhanced intensity of $A_g(2)$ in K-vapor doped peapod [2]. The non-monotonic doping response of $A_g(2)$ is traceable solely in ionic liquids, because classical electrolyte solutions do not allow that negative potentials to be applied.

Comparison of the data in ionic liquid (Fig. 2) with those in acetonitrile electrolyte solution [3] generates peculiar conclusions about the G-band. In acetonitrile medium, we can trace the "normal" monotonous downshift of the G-band during cathodic potential excursion within the fully accessible potential window [3]. This matches the behavior of empty SWCNT in the same electrolyte solution [7]. Eventually, non-monotonous shifts in SWCNTs were also traced for certain other electrolyte solutions, and we have to admit that this problem is far from being understood [6]. Qualitatively, we can conclude that the n-doping in ionic liquid is similar to the K-amalgam doping or to early stages of K- or Cs-vapor doping [1,6]. Presumably, the absence of solvent in ionic liquid mimics the conditions of chemical doping by alkali metal vapor or amalgam.

ACKNOWLEDGEMENTS

This work was supported by IFW Dresden, by the Academy of Sciences of the Czech Republic (contract No. A4040306) and by the Czech Ministry of Education (contract No. LC-510).

REFERENCES

1. Pichler T., Kukovecz A., Kuzmany H., Kataura H. and Achiba Y. *Phys.Rev.B* **67**, 125416-1254167 (2003).
2. Kalbac M., Kavan L., Zukalova M. and Dunsch L. *J.Phys.Chem.B* **108**, 6275-6280 (2004).
3. Kavan L., Dunsch L., Kataura H., Oshiyama A., Otani M. and Okada S. *J.Phys.Chem.B* **107**, 7666-7675 (2003).
4. Kavan L. and Dunsch L. *Chemphyschem* **4**, 944-950 (2003).
5. Guan L., Suenaga K., Shi Z., Gu Z. and Iijima S. *Phys.Rev.Lett.* **94**, 045502-0455024 (2005).
6. Chen G., Furtado C. A., Bandow S., Iijima S. and Eklund P. C. *Phys.Rev.B* **71**, 045408-0454086 (2005).
7. Kavan L., Rapta P., Dunsch L., Bronikowski M. J., Willis P. and Smalley R. E. *J.Phys.Chem.B* **105**, 10764-10771 (2001).

Reshaping of Peapods via Temperature and Laser Irradiation

C. Kramberger[1*], A. Waske[1], B. Büchner[1], T. Pichler[1], H. Kataura[2]

[1]*Leibniz Institute for Solid State and Materials Research Dresden, Helmholzstr. 20, 01069 Dresden, Germany*
[2]*Nanotechnology Research Institute, National Institute of Advanced Industrial Science and Technology (AIST) Central 4, Higashi 1-1-1, Tsukuba, Ibaraki 305-8562, Japan*

Abstract. We report a means to apply local and quick annealing using a focused laser beam on freestanding thin films of peapods. In situ characterization is performed by Fourier transform Raman spectroscopy. The findings are compared to traditional slow annealing in a furnace. With laser annealing we observe a fast transformation of the C_{60} "peas" to inner tubes and at even higher irradiation densities these new inner tubes are removed again. The removal of the inner tubes is not observed with conventional furnace annealing, however, carbon bicables, viz. two parallel nanotubes inside an oval shaped outer one, are obtained. These stark differences are a result of photo-induced processes in laser annealing.

Introduction

During the past decade Carbon nanotubes (CNT) have moved into the focus of research in physics, chemistry, and material sciences [1]. This holds for their synthesis, post synthesis processing, building nanoscale devices and various ways of characterization. Especially, the templated synthesis of novel 1D materials in the interior of SWCNT is an exciting opportunity for fundamental studies on low dimensional systems [2]. This contribution is focused on specialized low dimensional carbon nanostructures derived from peapods. We demonstrate the distinctive production of either double wall CNT or Carbon bicables by means of laser irradiation and furnace annealing. In addition, to distinctive production there is also the possibility of selective destruction in laser annealing.

Experimental

Thin (gray) films of peapods from the same batch were prepared on TEM grids and sapphire crystals. The sample morphology was checked by TEM micrographs (not shown), recorded with a high resolution analytical TEM Tecnai F30 FEI. An acceleration voltage of 100 kV was used to reduce specimen damage caused by the electrons [3]. FT Raman experiments were carried out with a Bruker FT Raman spectrometer equipped with an infrared microscope. The samples are mounted inside a Janis ST-500H microscope cryostat behind a quartz window. The 180° backscattered light is analyzed. The size of the laser spot on the sample is approximately 5 μm in

---

* Corresponding author email: C.Kramberger@ifw-dresden.de

CP786, *Electronic Properties of Novel Nanostructures*, edited by H. Kuzmany, J. Fink, M. Mehring, and S. Roth
© 2005 American Institute of Physics 0-7354-0275-2/05/$22.50

diameter. In addition to acting as a probe the 1064 nm laser of the FTR spectrometer can also be used to locally anneal the sample. Spectra were measured at 20 mW output power and for annealing as much as 500 mW could be applied. The output power is controlled electronically and is set within a few seconds. Since there is no need to touch the whole setup for (i) taking spectra, (ii) annealing the laser spot, (iii) waiting for cool down, (iv) and taking another spectrum, a whole sequence of different stages of annealing can be characterized in situ and on the same spot. Thicker (black) films of peapods on sapphire crystals were annealed in a furnace at various temperatures ranging from 1000 °C to 1550 °C and characterized by FT Raman spectroscopy. The samples were successively annealed at increasing temperatures. Before any annealing either by laser irradiation or in the furnace the samples were evacuated to a base pressure of $5 \cdot 10E^{-7}$ mbar in order to suppress uncontrolled oxidation.

Results

The FT Raman spectrum of the starting material is shown in Fig1 (pristine peapods). The total symmetric radial breathing mode (RBM) is observed in the low frequency region between 150 and 180 cm^{-1}. There is a well established linear relation between RBM frequency and inverse diameter $\omega_{RBM}=C_1/d+C_2$. Using values of 234 cm^{-1} and 13 cm^{-1} [4] this frequency region corresponds to diameters between 1.3 to 1.7 nm. Around 1590 cm^{-1} there is the response of the tangential modes wich are directly derived from in plane modes in graphite. The graphitic line (G-line) shows a slight dispersion with the diameter [5,6]. The G-line of very narrow nanotubes is downshifted by several wave numbers, while larger diameters are approximating the line position of regular graphite. At high output powers of the laser the material in the laser spot starts glowing immediately.

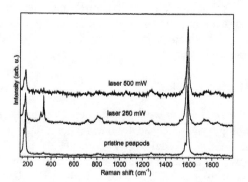

FIGURE 1. The shown FT Raman spectra are from bottom to top of (i) pristine peapods, (ii) material annealed at 260 mW for 60 sec, and (iii) further annealed material at 500 mW for 60 sec. All spectra are normalized to G line.

The peapods were successively annealed for 60 seconds at different laser powers. The first annealing step was performed at 100 mW and then the power was successively increased by 20 mW. The spectra taken after annealing at 260 and 500 mW are also shown in Fig 1. With the stepwise transformation of C_{60} to inner nanotubes (INT) there are significant changes in the observed Raman spectra [7]. The RBM of the INT arises between 300 and 345 cm^{-1}. This corresponds to INT diameters between 0.7 and 0.8 nm. At the same time the dip before the maximum of the G-line is covered by the downshifted response of the additional small diameter INT, resulting in one broadened peak with a downshifted center. If the laser power is further increased beyond the optimum transformation the response assigned to the INT is stepwise decreased and finally vanishes. The overall decrease in intensity is assigned to slight changes in the surface morphology affecting the alignment of the setup. With traditional annealing in the furnace, and consistent with previous results [8], a very different behavior is observed. When annealing up to 1300 °C there is a similar increase in the formation of INT as compared to laser annealing. Between 1300 °C and 1450 °C only very minor increases in INT are observed. Above 1450 °C a new peak arises on the high frequency side of the G-line. This is accompanied by a slight broadening of the outer tubes RBM. At its low frequency side the background is significantly raised. The RBM of the INT is preserved but significantly broadened as compared to laser annealing or furnace annealing at lower temperatures (e.g. furnace 1300 °C). Selected spectra obtained after furnace annealing are shown in Fig 2 (furnace 1300 °C and furnace 1550 °C).

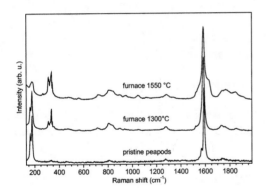

FIGURE 2. The shown FT Raman spectra are from bottom to top of (i) pristine peapods, (ii) material annealed 1300 °C for 1 h, and (iii) further annealed material at 1550 °C for 1h. All spectra are normalized to G line.

These changes in the spectra can be assigned to carbon bicables [9]. In this context the observed broadening of the RBM of the INT arises from symmetry breaking of the INTs local environment. While in DWCNTs every INT is in the center of its outer counterpart. In bicables another arbitrary INT is lying parallel and the individual INT are off center an oval shaped outer nanotube. This changed environment does not any longer fit to the cylinder symmetry of the RBM. Thus the environmental up-shift of the RBM frequency is no longer a uniform constant and the individual frequencies are smeared out. Both series of experiments show that the intermediate Raman response of CNT located in between the RBM and G-line is correlated to the relative amount of INT in the sample. First their intensities increase with a growing content of INT. In case of laser annealing their intensities are diminished during the removal of the INT while the intermediate Raman modes stay unaffected upon the formation of bicables. The observed behavior suggests a strongly increased Raman efficiency for more narrow CNT. The very fast transformation of encapsulated C_{60} to an INT in the first seconds is a remarkable finding in the case of laser annealing. Annealing series with just 10 seconds (not shown) irradiation times still exhibit the same overall behavior. This quick annealing is not possible in a furnace and is an interesting feature of laser annealing. In addition to this fast transformation there is also a fast decay of the INT at very high irradiation densities. While the formation of INT is also possible with furnace annealing their destruction is only observed in case of laser annealing. However, in furnace annealing the outer tubes are affected at very high temperatures forming the previously described bicables. These are not observed with laser annealing, even for long periods of time (2 hours). From this comparison it appears that temperature does not seem to be responsible for the selective destruction of INT. This selective removal is ascribed to destructive photo induced processes. Considering the strongly diameter dependent optical transitions responsible for absorption, this laser annealing effect might be a possible means to tailor the electronic properties of carbon nanotubes. In fact all small diameter INT that exhibited resonant Raman scattering were successfully removed while the outer shell withstood this treatment. In this context photo induced processes should also be taken into account in the accelerated formation of the INT. Another striking finding is the effect of very high temperature furnace annealing. The observed changes start at 1450 °C. Since purification is always in some way harmful to SWCNT a rather high defect concentration can be assumed in our opened and purified SWCNT. Then two adjacent DWCNT can form a bicable, as observed in Ref 9, too. Obviously the two outer shells can coalesce in this way but the INT stay unaffected. This is in agreement with their much lower concentration of defects as proposed in Ref 7.

Acknowledgments

The DFG project PI440/1 and the International Max Planck School "Dynamical Processes in Atoms, Molecules and Solids" is acknowledged for financial support. We also thank R. Schönfelder for technical support.

References

1. M.S. Dresselhaus, P.C. Eklund, Advances in Physics, 49, 705 (2000).
2. H. Kuzmany, R. Pfeiffer, C. Kramberger, T. Pichler, *Frontiers of Multifunctional Integrated Nanosystems*, pp. 171–184. Kluwer Academic Publishers, Dordrecht, 2004.
3. C. Kramberger, A. Waske, K. Biedermann, T. Pichler, T. Gemming, B. Büchner, H. Kataura, CHEM. PHYS. LETT, 407, (4-6): 254-259, 2005
4. C. Kramberger, R. Pfeiffer, H. Kuzmany, V. Zólyomi, J. Kürti, PHYS. REV. B, 68 235404, 2003
5. O. Dubay, G. Kresse, H. Kuzmany, PHYS. REV. LETT, 88 (23), 235506, 2002
6. A. Jorio, M. Pimenta, A. Souza Filho, G. Samsonidze, A. Swan, M. Ünlü, M. Goldberg, R. Saito, G. Dresselhaus, M. Dresselhaus, PHYS. REV. LETT, 90 (10), 107403 (1-4), 2003
7. R. Pfeiffer, H. Kuzmany, C. Kramberger, C. Schaman, T. Pichler, H. Kataura, Y. Achiba, J. Kürti, V. Zólyomi, PHYS. REV. LETT, 90(22), 225501, 2003
8. M. Holzweber, C. Kramberger, F. Simon, R. Pfeiffer, M. Mannsberger, F. Hasi, H. Kuzmany, AIP CONF. PROC. **723** , 234, 2004
9. M. Endo, T. Hayashi, H. Muramatsu, Y. Kim, H. Terrones, M. Terrones, M. Dresselhaus, NANO LETT, 4 (8): 1451-1454, 2004

Preparation and EPR characterization of N@C_{60} and N@C_{70} based peapods

B. Corzilius*, A. Gembus*, N. Weiden* and K.-P. Dinse*

*Eduard-Zintl-Institute for Inorganic and Physical Chemistry, Darmstadt University of
Technology, Petersenstr. 20, D-64291 Darmstadt, Germany*

Abstract. Using the quartet spin of encased nitrogen atoms as an electron paramagnetic resonance (EPR) probe, it is possible to examine the fullerene/nanotube interactions in a peapod. A purification method is developed which allows low temperature filling of nanotubes with endohedral fullerenes. The paramagnetic impurities of undoped single wall carbon nanotubes (SWNT) are characterized via EPR resulting in a broad superparamagnetic signal of the remaining catalyst particles and a rather narrow signal of carbonaceous material. Comparison of EPR spectra of several nitrogen endohedral doped peapods with their analogues obtained in a solid fullerene matrix shows a significant broadening of N@C_{60} and N@C_{70} EPR signals. This broadening is related to a non-vanishing zero-field splitting caused by deformation of the fullerene cage upon encapsulation.

INTRODUCTION

Since the discovery of peapods [1], many metallo-endohedral fullerene (MEF)-based peapods have been synthesized and characterized by X-ray techniques. No clear EPR signature of these compounds has been described, probably because of strong mixing of MEF and SWNT states. In contrast, endohedral N@C_{60} and N@C_{70} should be much less influenced by incorporation into SWNT, because the repulsive potential between fullerene cage and nitrogen atom prevents any significant charge and spin transfer to the cage. Furthermore, their insertion properties should be almost the same as found for the corresponding empty fullerenes.

With the recent development of several low temperature peapod preparation routines [2, 3, 4], it is now possible to synthesize peapods using N@C_{60} and N@C_{70}, which are thermally unstable above 450 K. The development of a reproducible insertion method is of great importance, because N@C_{60} and N@C_{70} can be provided only in small quantities. The weakly interacting nitrogen endohedral fullerenes can be used as probes to examine the interactions between the encapsulated fullerenes and also between the fullerene and the interior nanotube wall. It is also possible to resolve the kinetics of the fullerenes inside a nanotube.

EXPERIMENTAL

SWNT were obtained from Carbon Solutions Inc. (USA). They were produced using electric arc discharge with Ni/Y catalysts and had been received as produced. Additionally, some SWNT material from NanoCarbLabs (Russia) which had been purified by the manufacturer served as a reference. The mean diameter of all tubes is 1.4 nm with a

CP786, *Electronic Properties of Novel Nanostructures*, edited by H. Kuzmany, J. Fink, M. Mehring, and S. Roth
© 2005 American Institute of Physics 0-7354-0275-2/05/$22.50

narrow diameter distribution. Endohedral fullerenes $N@C_{60}$ and $N@C_{70}$ were prepared using the ion implantation technique described elsewhere [5]. Enrichment of the endohedral fullerenes was performed using high performance liquid chromatography [6], resulting in $N@C_{60}$ material with relative concentrations of 500 ppm as well as 0.5% and $N@C_{70}$ material with concentrations of 1000 ppm and 4%, respectively. EPR measurements were performed using a Bruker ESP 300E X-Band continuous wave (c. w.) spectrometer at a microwave (mw) frequency of 9.8 GHz. A rectangular cavity (Bruker 4102 ST) served as resonator. Temperature dependent measurements were done with a liquid helium flux cryostat (Oxford ESR 900) at a mw frequency of 9.4 GHz.

For a purification of the as-produced SWNT, several different protocols were tested including a HNO_3 treatment similar to the method of Rinzler et al. [7], ultrasonication in a H_2SO_4/HNO_3 mixture [8], oxidation in H_2O_2 [3], and oxidation in air. Unfortunately, none of these treatments led to SWNT material which could be filled at low temperature except for the combination of air-oxidation and HCl treatment, which is quite similar to the recently published purification method by Khlobystov et al. [4]. Best results were obtained as follows: As-produced SWNT were heated in air to 400 °C for 1 h. This was followed by a 1 h ultrasonication of the SWNT material in conc. HCl at 60 °C. The acid was removed via filtration. The solid was washed with de-ionized water and dried in high vacuum. After grinding, the black powder was heated again in air to 420 °C, followed by the same treatment in HCl as mentioned above. As a final step, the material was heated under high vacuum to 800 °C to remove adsorbates and functional groups which are likely introduced by the purification treatment. The overall weight loss is 72 %.

$N@C_{60}$ and $N@C_{70}$ peapods were prepared following the procedure by Yudasaka et al. [2]. Typically 5 mg of the endohedral fullerene were ultrasonically dispersed in 50 mL ethanol (abs.). 5 mg of pretreated SWNT material were added and the flask was sealed. After 24 h the supernatant ethanol was removed. The powder was dried in vacuum and suspended in toluene so that the outside fullerenes were removed by dissolution. The suspension was filtered, the solid washed thoroughly with toluene and dried in air.

RESULTS AND DISCUSSION

To quantize the removal of paramagnetic impurities in the SWNT material, c. w. EPR measurements were made before and after the purification. SWNT produced by the electric arc technique show two different EPR signals: a broad signal ranging from 0 to 600 mT at room temperature and a rather narrow signal at $g \approx 2$ as shown in figure 1. The latter signal probably originates from defect centers and is located near the minimum of the broad signal but with much lower intensity. During the purification procedure the intensity of the broad signal could be reduced by a factor of five, the intensity of the narrow signal drops by a factor of ten.

To further characterize the broad signal and to determine its origin, temperature dependent c. w. EPR experiments were performed. The plots in figure 2 show a significant broadening and resonance downshift of the signal with decreasing temperature. The correlation between the resonance shift δB_0 and the line width ΔB can be fitted according to the power law model of Nagata and Ishihara derived for anisotropic demagnetization

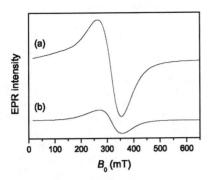
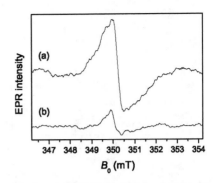

FIGURE 1. C. w. EPR spectra at 9.8 GHz (X-Band) of the broad (left) and narrow signal (right, after subtraction of the broad signal) of empty SWNT before (a) and after (b) purification.

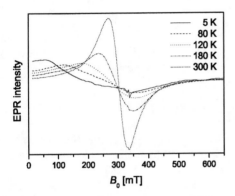
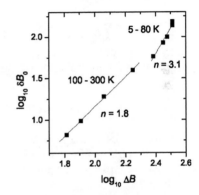

FIGURE 2. Temperature dependent c. w. EPR spectra of the broad signal of SWNT at 9.4 GHz (left) and the resulting correlation between the resonance shift and the peak to peak line width (right).

of randomly oriented superparamagnetic particles with the fitting equation

$$\delta B_0 \propto \Delta B^n \tag{1}$$

in which $n = 3$ applies for an ideal powder average [9]. At higher temperatures, n changes to smaller values, according to the observations of Nagata and Ishihara [9].

Encapsulation of endohedral fullerenes in nanotubes can be detected by the observation of a characteristic ^{14}N ($I = 1$) hyperfine triplet as shown by c. w. EPR measurements in figure 3. The hyperfine splitting constant of 15.85 MHz in the case of $^{14}N@C_{60}$ and 15.15 MHz for $^{14}N@C_{70}$ is not changed during encapsulation. The effect of line broadening is very large (up to a factor of 17) for small initial line widths and decreases with higher doping level of the endohedral fullerene. At a doping concentration of 4 % in case of $N@C_{70}@SWNT$, the line width is even less than observed for $N@C_{70}$ in a solid C_{70} matrix. The reduction of line width indicates a decrease of the dipolar interaction due to the one-dimensional order in comparison to the three-dimensional order in the fullerene solid state lattice. These findings imply a broadening by a mixture of

319

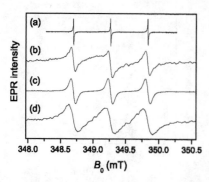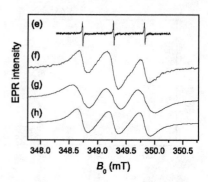

FIGURE 3. Room temperature c. w. EPR spectra of (a) N@C$_{60}$ (500 ppm), (b) N@C$_{60}$@SWNT (500 ppm), (c) N@C$_{60}$ (5000 ppm), (d) N@C$_{60}$@SWNT (5000 ppm), (e) N@C$_{70}$ (1000 ppm), (f) N@C$_{70}$@SWNT (1000 ppm), (g) N@C$_{70}$ (4 %), (h) N@C$_{70}$@SWNT (4 %) at 9.8 GHz.

inhomogeneous and homogenous effects, which is in agreement with the observation that standard 2 pulse electron spin echoes can be detected. For samples of low N@C$_{60}$ concentration a spin dephasing time of the order of 2 μs is measured.

ACKNOWLEDGMENTS

We thank P. Jakes for the preparation and enrichment of the N@C$_{60}$ and N@C$_{70}$ material and F. Simon for helpful discussions. Financial support by the Deutsche Forschungsgemeinschaft is gratefully acknowledged.

REFERENCES

1. Smith, B. W., Monthioux, M., and Luzzi, D. E., *Nature*, **396**, 323–324 (1998).
2. Yudasaka, M., Ajima, K., Suenaga, K., Ichihashi, T., Hashimoto, A., and Iijima, S., *Chem. Phys. Lett.*, **380**, 42–46 (2003).
3. Simon, F., Kuzmany, H., Rauf, H., Pichler, T., Bernardi, J., Peterlik, H., Korecz, L., Fülöp, F., and Jánossy, A., *Chem. Phys. Lett.*, **383**, 362–367 (2004).
4. Khlobystov, A. N., Britz, D. A., Wang, J., O'Neil, A., Poliakoff, M., and Briggs, G. A. D., *J. Mater. Chem.*, **14**, 2852–2857 (2004).
5. Pietzak, B., *Fullerenes as Chemical Atom Traps for Nitrogen and Phosphorus*, Ph.D. thesis, Technische Universität Berlin (1998).
6. Jakes, P., Dinse, K.-P., Meyer, C., Harneit, W., and Weidinger, A., *Phys. Chem. Chem. Phys.*, **5**, 4080–4083 (2003).
7. Rinzler, A. G., Liu, J., Dai, H., Nikolaev, P., Huffman, C. B., Rodríguez-Macías, F. J., Boul, P. J., Lu, A. H., Heyman, D., Colbert, D. T., Lee, R. S., Fischer, J. E., Rao, A. M., Eklund, P. C., and Smalley, R. E., *Appl. Phys. A*, **67**, 29–37 (1998).
8. Liu, J., Rinzler, A. G., Dai, H., Hafner, J. H., Bradley, R. K., Boul, P. J., Lu, A., Iverson, T., Shelimov, K., Huffman, C. B., Rodriguez-Macias, F., Shon, Y.-S., Lee, T. R., Colbert, D. T., and Smalley, R. E., *Science*, **280**, 1253–1256 (1998).
9. Nagata, K., and Ishihara, A., *J. Magn. Magn. Mater.*, **104-107**, 1571–1573 (1992).

Fuctionalization of single silicon nanocrystals by connecting with multiwalled carbon nanotubes

V. Švrček[1], T. Dintzer[1], F. Le Normand[2], O. Ersen[2], C. Pham-Huu[1], D. Begin[1], B. Louis[1], M-J. Ledoux[1]

[1]LMSPC-ECPM, 25, rue Becquerel, F67087 Strasbourg, France
[2]IPCMS, UMR 7504 CNRS, 23 rue du Loess, F-67037 Strasbourg, France

Abstract. Two approaches to connect single silicon nanocrystals (Si-nc) with multiwalled carbon nanotubes (MWCNTs) are studied. The first consists of direct growth of the MWCNTs in tapered element oscillating microbalance by chemical vapour deposition process on coated Si-nc with iron based catalyst. The second procedure, at room temperature connects and introduces dispersed Si-nc (2- 5 nm) in colloidal suspension of spin on glass by capillary forces.

Keywords: Silicon Nanocrystals, Carbon Nanotubes, Filling
PACS: 78.67.Bf , 66.20.+d, 68.37.Lp

INTRODUCTION

Among the nanoparticles that have attracted much interest in last decade are silicon nanocrystals (Si-nc). It has been demonstrated that efficient photoluminescence (PL) at room temperature can be observed from silicon in the form of quantum dots nanowires [1-3]. The quantum confinement together with surface states takes place leading in the opening of the bandgap and a considerable increase of radiative recombination probability [4]. One alternative technique of ex-situ preparation of free standing Si-nc by implementation into coloidal spin on glass (SOG) SiO_2 suspensions offers solutions to obtain structures with superior luminescence, good transport properties and an optical gain of Si-nc as an ensemble [5, 6].

However, the challenge is to adapt single Si-nc to real-world applications i.e. separate them, improve manipulation, connect them and most importantly establish good electrical contact to be able to incorporate them into electronic circuits. On the other hand, the carbon nanotubes themselves were subjected to increased attention as a unique 1D nano-material in last decade. Their relatively simple fabrication, unique characteristics and properties make them a promising material for the future. In particularly, the conducting and mechanical properties of carbon nanotubes due to their helicity would be very promising for several applications i.e. connecting nanowires of nanoparticles [7, 8].

Preparation of free standing Si-nc [5] would be a strategy to connect single Si-nc with conducting carbon nanotubes. Herein, we demonstrate two independent approaches to connect single Si-nc either by direct growing multiwalled carbon nanotubes (MWCNTs) on Si-nc coated with iron catalyst by chemical vapour

CP786, *Electronic Properties of Novel Nanostructures*, edited by H. Kuzmany, J. Fink, M. Mehring, and S. Roth
© 2005 American Institute of Physics 0-7354-0275-2/05/$22.50

deposition or a procedure at room temperature by capillary forces of Si-nc dispersed in organic spin on glass.

EXPERIMENTAL DETAILS AND DISCUSSION

The free floating Si-nc used in this work were prepared by electrochemical etching and the finest one harvested by sedimentation as described elsewhere (Cz silicon, p-type boron doped, <111>, 1 Ohm.cm) [5]. In order to grow directly MWCNTs, Si-nc were introduced into an aqueous solution of Fe (Fe(NO$_3$)$_3$·9 H$_2$O), with the iron concentration fixed at 20 wt. %. The deposited iron on Si-nc is the precursor of the catalyst. This solution was then kept in an ultrasonic bath for 30 min. Due to the difficulties with mixing with the hydrophobic Si-nc the impregnation process was repeated five times. The solution was oven-dried at 100 °C overnight and calcined in air at 350 °C for 2 h. To monitor the growing process of MWCNTs on Si-nc, Tapered Element Oscillating Microbalance (TEOM) has been used. To ensure repeatable results, the TEOM controls the automatic flow and the heating of the microreactor. Then 10 mg of the iron supported on the Si-nc catalyst was placed in the weighting pan of TEOM between quartz wool. The synthesis of the MWCNTs was performed by CVD from a mixture of ethane and hydrogen. The catalysts were reduced in situ under hydrogen at 400°C for 1 h prior to the CVD synthesis. After the reduction the reactor temperature was continually increased up to 750 °C. Then, the hydrogen flow was replaced by a mixture of ethane and hydrogen (C$_2$H$_6$:H$_2$ = 1:2 molar ratio) and the synthesis was conducted for 5 min. A helium purge gas flow passes around the tapered element. The tapered element frequency is determinate by optics and is inversely proportional to the change of the mass when ethane enters into weighting pan [9]. The system determinates the mass changes of the sample bed between times 0 and 1 using the following equation

$$\Delta m = const\left(\frac{1}{f_1^2} - \frac{1}{f_0^2} \right) \qquad (1)$$

where Δm is the mass change of the sample, f$_0$ the natural oscillating frequency at time inotial, and f$_1$ the natural oscillating frequency at a later time. When the mass of the sample bed increases, oscillating frequency tapered element decreases and vice versa.

For the room temperature filling, we have used carbon nanotubes commercially available (Pyrograf Products, Inc) with average inner diameters of 40 nm. As we have recently reported, before the filling and/or wiring appears, it is necessary to open the tips of MWCNTs and to declusterize the Si-nc [10]. The MWCNTs are opened by thermal annealing in oxygen atmosphere at 580 °C. They are then introduce into spin on glass matrix [3] previously diluted with methanol (methanol/SOG of 2/1) in order to decrease the overall viscosity of the solution (SOG is commercially available [5]). However, the Si-nc introduced into liquid SOG has a tendency to cluster each other, leading to the formation of large aggregates which are unfavorable for the mentioned objective [10]. Therefore, prior to the introduction of MWCNTs into the SOG-based solution (Si-nc/SOG/methanol) ammonia (30 %) (5 drops in 2 cm^3 solution -1 drop ~ 0.025 cm^3) was added. The MWCNTs were rapidly incorporated into the as-declusterized Si-nc solution. The Si-nc were well dispersed with particle size varying from 5 to 40 nm. After introduction of the MWCNTs, the solution was then kept in ultrasonic bath for 30 min, and then dried at room temperature for one day.

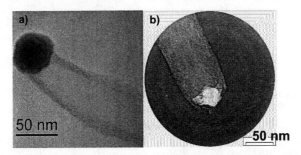

FIGURE 1. a) HR-TEM image of wired single silicon nanocrystal with size of 30 nm with multiwalled carbon nanotubes (MWCNTs) fabricated in TEOM. b) HR-TEM image of wired single silicon nanocrystal dispersed in spin on glass at room temperature with MWCNTs by capillary force.

After the CVD process, we were able to obtain well-separated and wired single Si-nc with conducting MWCNT grown directly on them. Figure 1a shows a HR-TEM micrograph of such wired single silicon nanocrystal with a size of 30 nm on MWCNT fabricated in TEOM. From interplane stripes (plane orientation along the <111> direction) and chemical analyses of the particle, the presence of Si-nc is witnessed. Apparently, the iron adhesion force is strong enough to avoid the loss of the Si-nc from the tube tip even during the sonication process before TEM observation. Surprisingly, after the CVD process, the Si-nc surface was not surrounded with iron [9] and it seems that MWCNTs has directly grown on Si-nc. The disappearance of the iron from the Si-nc surface during the MWCNTs growth process could be explained by the low interactions between the iron and the Si surface, leading to the removal of the iron by carbon segregation on the silicon surface.

On the other hand, the statistical HR-TEM observations revealed that the diameter of the MWCNTs fits with the diameter of the Si-nc. This corroborates with generally known observations from others suggesting that the smaller size of catalyst; the smaller diameter of carbon nanotube can be managed [8]. The great advantage of the controlled CVD process in TEOM allows evaluation of the growth rate on the Fe/Si-nc catalyst. In our case, the growth rate is 1,65 μg/sec [9]. A blank experiment on pure Si-nc was performed and no MWCNTs could grow directly on those nanocrystals without the presence of iron catalyst.

In Figure 1b is shown a Si-nc with an average diameter of about 30 nm capping the top of an open MWCNT with inner diameter 30 nm. It has been observed when Si-nc has size similar or larger, the inner diameter of MWCNTs stays stacked on top of the MWCNTs. The Si-nc embedded in SOG are very well attracted and stabilized on the top of the open MWCNT because of the relatively high capillarity forces of the tube channel. In the case of smaller Si-nc particles, it is also observed that the presence of Si-nc inside the carbon nanotube channel with larger inner diameter [11]. Such Si-nc seems to be transported via the SOG based solution which was sucked inside the tube channel by capillarity forces. To refer on the strong capillary effect, we have observed that Si-nc are localized far (500 nm) from the tip of the MWCNT [11].

It has to be stressed that in the case of the first approach, the wiring is achieved during the in situ growing process and the MWCNTs were directly connected to the Si-nc and generally the contact is much stronger compared to room temperature wiring. On the other hand, the room temperature approach of wiring offers advantages as low defects diffusion, low energetic budget, etc...

SUMMARY

To conclude, we have shown that free standing Si-nc formed by electrochemical etching can offer additional possibilities for application of single Si-nc. Two methods were successfully demonstrated in order to wire and/or connect single Si-nc. It has been shown that we are able to directly grow carbon nanotube on iron coated single Si-nc in tapered element oscillating microbalance at 750 °C. Then, the low temperature approach has shown the ability to wire (connect) single Si-nc embedded in colloidal suspension of spin on glass by capillary forces of MWCNTs. It was observed that when the inner diameter of MWCNTs is increased, the Si-nc have tendency to enter deeper into the channel of the MWCNTs. Both approaches evidenced the fact that we were able to obtain well-separated and wired single Si-nc. This, in principle, can help to solve very important problems as localization and handling of single Si-nc. Besides the direct application possibilities, these hybrid materials also allow to study peculiar physical phenomena at nanoscale level.

ACKNOWLEDGMENTS

This work was supported by the Centre National de la Recherche Scientifique and by the European NANOTEMP project (Contract No. HPRN-CT-2002-00192).

REFERENCES

1. Liu. A, Jones. R., Liao. L., Samara-Rubio. D., Rubin. D., Cohen. O., Nicolaescu. R., Paniccia. M., *Nature* **427**, 615-618 (2004)
2. Mathur. N, *Nature* **419**, 573-575 (2002)
3. NATO Proceedings: *Towards the First Silicon Laser*, editors: L. Pavesi, S.Gaponenko, L. D. Negro, NATO Science Serries, II. Mathematics, Physics and Chemistry – Vol. 93 (Kluwer Academic Publishers, 2003), p. 145, p. 197
4. Wolkin. M. V., Jorne. J., Fauchet P. M. , Allan. G., and Delerue. C. *Phys. Rev. Lett.* **82**, 197 (1999)
5. Švrček.V., Slaoui A., Muller. J-C. , *J. Appl. Phys.* **95** 3158 (2004)
6. Luterová, K., Dohnalová, K. Švrček, V., Pelant, I. Likforman, J.P. Cregut, O., Gilliot, P., Honerlage, B., *Appl. Phys. Lett.* **84**, 3280 (2004)
7. Iijima. S., *Nature* **56**, 354 (1991)
8. Monthioux, M., *Carbon* **40**, 1809 (2002)
9. Švrček V, et al submitted to Nanoletters.(2005)
10. Švrček,V., Slaoui, A., Rehspringer, J-L., Muller, J.-C,. *Semicond. Sci. Technol.* **20** 314-319 (2005)
11. Švrček V., et al submitted to J of Chem. Phys. (2005)

Effective valency of Dy ions in $Dy_3N@C_{80}$ metallofullerenes and peapods

H. Shiozawa[*1], H. Rauf[1], T. Pichler[1], M. Knupfer[1], M. Kalbac[1,2], S. Yang[1]
L. Dunsch[1], B. Büchner[1], D. Batchelor[3], and H. Kataura[4]

[1] IFW-Dresden, Institute for Solid State Research, Helmholtzstr. 20 01171 Dresden, German

[2] J. Heyrovský Institute of Physical Chemistry, ASCR, Dolejskova 3, CZ-18223 Prague 8, Czech Republic

[3] Universität Würzburg, BESSY II, Albert-Einstein-Straße 15 D-12489 Berlin, Germany

[4] Nanotechnology Research Institute, Advanced Industrial Science and Technology, Tsukuba 305-8562, Japan

Abstract. We report on the electronic properties of the trimetal-nitride fullerene $Dy_3N@C_{80}$ and its encapsulated form inside SWCNT, the metallofullerene peapod, using high-energy spectroscopic methods as probes. From a comparison of the Dy $4d$-$4f$ absorption edges and the valence-band photoemission spectra with atomic multiplet calculations the effective valency of the encapsulated rare-earth ions is evaluated. We observe that the Dy ions inside fullerene cage are essentially trivalent which is not affected by the peapod formation.

INTRODUCTION

After the first observation of metallofullerenes in 1985 [1], various metallofullerenes have been successfully synthesized and their endohedral nature has been reported by theoretical and experimental studies. Among them, the trimetal-nitride fullerenes are of very high stability similar to C_{60} and in strong contrast to the most of the other higher fullerenes or metallofullerenes. The most exciting feature of this endohedral family is that their carbon cage has high icosaheadral symmetry. A key to understand this exotic property is the charge transfer between the encapsulated metal atoms and carbon cage concomitant to inclusion of nitrogen atom inside the cage. The high-energy spectroscopy using synchrotron light source enables us to investigate such a molecule with strong element and site sensitivity. In the high-energy excitation valence-band photoemission spectrum, the carbon-derived valence band is smeared out and the characteristic $4f$ multiplet of the encaged metal cluster dominates because of the atomic photo-ionization cross-section which is much higher than those of carbon. This allows us to study the valency of the encaged metal ions in the fullerene [2-4]. In this contribution we present results of core-level absorption and high-energy excitation photoemission spectroscopy on newly synthesized trimetal-nitride fullerene $Dy_3N@C_{80}$ and their corresponding peapods consisting of metallofullerenes filled

CP786, *Electronic Properties of Novel Nanostructures*, edited by H. Kuzmany, J. Fink, M. Mehring, and S. Roth
© 2005 American Institute of Physics 0-7354-0275-2/05/$22.50

in single-wall carbon nanotubes (SWCNTs). The valency of Dy ions in the metallofullerene is evaluated by comparing the experimental response with atomic multiplet calculations.

EXPERIMENTAL

The trimetal-nitride fullerene $Dy_3N@C_{80}$ was produced using the modified Kratschmer-Huffman arc burning method and separated by the high performance liquid chromatography [5]. SWCNT material was synthesized by the laser ablation method and purified by H_2O_2 treatment [6]. SWCNT films were prepared by drop coating from acetone solution and annealed in ultrahigh vacuum (UHV). $Dy_3N@C_{80}$ was sublimated at 600°C onto the SWCNT film under UHV conditions using the evaporator specially designed for the purpose of subliming a tiny amount of molecules. The film was subsequently heated at 450°C overnight to encapsulate $Dy_3N@C_{80}$ into SWCNTs. Then the film was heated to 600°C in order to remove all $Dy_3N@C_{80}$ which did not enter the SWCNTs from the SWCNT film surface. The synthesized peapod film was in-situ transferred into the analysis chamber without breaking UHV. The film thickness was of the order of several hundreds of nanometer. The absorption spectrum was measured by measuring the drain current and the partial electron yield (PEY) at the beamline UE 52 PGM, Bessy II which has an energy resolving power of more than 10000 [7]. The photoemission experiment was performed using a hemispherical photoelectron energy analyzer SCIENTA SES 200 at an energy resolution set to 20 meV. The total experimental resolution and the Fermi energy (E_F) were determined from the Fermi edge of a clean gold film. The total experimental resolution was found to be 40 meV at the photon energy 125 eV and the temperature was 35 K. The base pressure in the experimental setup was kept below 3×10^{-10} mbar.

RESULTS AND DISCUSSION

Figure 1 shows the absorption spectra of $Dy_3N@C_{80}$ and of the peapods collected at 400 eV photon energy. The spectra, which are similar to each other, consist of several pre-edge structures and a giant peak around 161 eV and are very similar to the spectrum of Dy metal where Dy atoms are triply ionized [8]. The valence-band photoemission spectra of $Dy_3N@C_{80}$, SWCNT and peapods collected at 400 eV photon energy are depicted in Fig. 2. The photoemission spectrum of $Dy_3N@C_{80}$ has prominent peak structures that mainly correspond to the Dy $4f$ states. The peapod spectrum exhibits peak structures similar to those in the $Dy_3N@C_{80}$ spectrum. The subtraction of the SWCNT spectrum from the $Dy_3N@C_{80}$ spectrum yields the difference spectrum (Peapods – SWCNT) exhibiting the electronic structures of the $4f$ states of Dy ions in the peapod similar to those of $Dy_3N@C_{80}$ film. This indicates that the electronic structures of Dy ions inside the fullerene cage persist upon encapsulation of the metallofullerene into SWCNT. In addition, by comparing the intensity

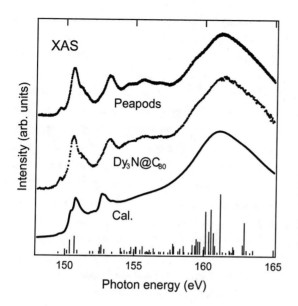

FIGURE 1. Absorption spectra of $Dy_3N@C_{80}$ and the peapods measured in the Dy $4d$-$4f$ excitation region. A calculated spectrum (solid line) is also shown with multiplets (vertical bars).

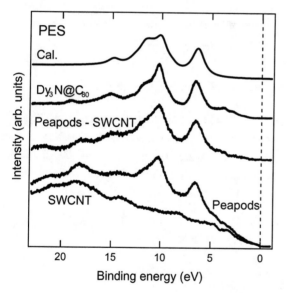

FIGURE 2. Photoemission spectra of $Dy_3N@C_{80}$, SWCNT and the peapods collected at 400 eV photon energy, together with the difference spectrum (Peapods - SWCNT) and calculated spectrum (upper line).

of the peapod spectrum with that of the SWCNT spectrum an effective bulk filling factor of about 60% was estimated [9]. A comparison of the absorption and valence-band photoemission spectra with atomic multiplets provides the information of the valency of the Dy atoms inside the carbon cage. The Dy $4d$-$4f$ absorption spectra were reproduced from the absorption multiplets (vertical bars) of the ground state configuration $^6H_{15/2}$ of trivalent Dy atom calculated with Cowan code using a Doniach-Sunjic line shape after subtraction of an offset linear background [10, 11]. The different width and asymmetry of the Doniach-Sunjic function were used for the multiplets in the energy region below and above 154 eV. The Slater integrals, F^k_{ff}, F^k_{df}, and G^k_{df}, are reduced to 70%, 90% and 65%, respectively. The calculated absorption spectrum (solid line) well reproduces the experimental data as seen in Fig. 1. This result indicates that the Dy ions inside the carbon cage are essentially trivalent. Further evidence for the trivalent Dy is obtained from the photoemission spectrum. The valence-band photoemission spectra measured at excitation energy 400 eV were again reproduced by atomic multiplets using a Doniach-Sunjic line shape plus step function with a proper choice of the parameters [11, 12]. The result using a Dy^{3+} multiplet is depicted as the solid line in Fig. 2. As can be seen in the figure the Dy^{3+} multiplets well reproduce the main features of the Dy $4f$ spectrum. Also there are slight changes in the relative peak intensity of the multiplet peak at the binding energy around 10 eV, which might be due to a small admixture from divalent Dy. The results further confirm that the Dy ions are predominantly trivalent.

In summary, we have investigated the electronic structure and the effective valency of the metal ions encaged in the trimetal-nitride fullerene $Dy_3N@C_{80}$ and their peapods using absorption and photoemission spectroscopy as probes. From comparison of the experimental spectra with atomic calculations, the valency of Dy atoms inside the fullerene was estimated to be essentially trivalent which is not affected by the peapod formation.

ACKNOWLEDGMENTS

This work was supported by the DFG PI440/1. We thank S. Leger, R. Hübel, R. Schönfelder, S. Döcke, K. Leger, F. Ziegs, and H. Zöller for technical assistance. H. K. acknowledges support by Industrial Technology Research Grant Program in 2003 from the New Energy and Industrial Technology Development Organization (NEDO) of Japan. H. S. acknowledges funding by the Alexander von Humboldt society.

REFERENCES

1. Heath, J. R. et al., *J. Am. Chem. Soc.* **107**, 779 (1985).
2. Pichler, T. et al., *Phys. Rev. Lett.* **79**, 3026 (1997).
3. Alvarez, L. et al., *Phys. Rev. B* **66**, 035107 (2002).
4. Liu, X. et al., unpublished result.
5. Dunsch, L. et al., *J. Phys. Chem. Solid* **65**, 309 (2004).
6. Kataura, H. et al., *Synthetic Metals* **103**, 2555 (1999).
7. Schöll, A. et al., *J. El. Spec. And Rel. Phen.* **129**, 1 (2003).
8. Muto, S. et al., *J. Phys. Soc. Jpn.* **63**, 1179 (1994).
9. Shiozawa, H. et al., in preparation.
10. Cowan, R. D., *The Theory of Atomic Structure and Spectra* (Univ. of Cal. Press, Berkley, 1981).
11. Doniach, S. and Sunjic, M., *J. Phys. C* **3**, 285 (1970).
12. Gerken, F. et al., *J. Phys. F* **13**, 703 (1983).

In-situ coalescence of aligned C_{60} molecules in Peapods

C. Arrondo[(1)], M. Monthioux[(1)], K. Kishita[(2)], M. Le Lay[(3)]

[1] *CEMES, UPR-8011 CNRS, BP 94347, F-31055 Toulouse cedex 4, France*
[2] *Toyota Motor Corp., Material Engineering Division 2, 1,Toyota-cho, Toyota, Aichi, 471-8572 Japan*
[3] *Toyota Motor Engineering and Manufacturing Europe, Tech. Cent., Hoge Wei 33B, Zaventern, Belgium*

Abstract. We report here the results of an extensive study of in-situ experiments combining both electron irradiation and low thermal treatment within a HRTEM. Parameters of the study were temperature and electron dose, flow, and energy. The range of parameter value for the actual DWNT formation at 490°C and 700°C was identified, apart which either the inner nanotube formation is not complete or the nanotubes are destroyed. The statistics of outer and inner tube diameters and inter-tube distance in DWNTs were systematically followed. The leading parameter was found to be the host (outer) nanotube diameter, whose the increase relates linearly to an increase of both the inter-tube distance and the inner tube diameter. This indicates that the coalescence mechanisms are driven by a compromise between two contradictory energetic requirements, i.e. tentatively enforcing the regular 0.34 nm van der Waals distance between the inner and outer graphenes, and tentatively maintaining the initial 0.7 nm diameter originating from the starting fullerenes. With respect to this overall mechanism, lowering the treatment temperature provides smaller tube diameters (thereby inducing larger inter-tube distances). On the other hand, none of the electron beam features were found to have a significant influence on the DWNT features, but possibly the electron dose. Beside, isolated peapods were found to behave differently than in bundles. Coalescence mechanisms are discussed.

INTRODUCTION

The principle of making double-wall carbon nanotubes (DWNTs) from "Peapods" (i.e., single-wall nanotubes – SWNTs - whose the cavity is filled with fullerene molecules [1]) was first proposed by Smith et al. [2] to explain the concomitant occurrence of such DWNTs (whose the inner tube had the diameter of C_{60}) and Peapods in purified then annealed SWNT-based samples from laser vaporisation. In this case, the formation of the inner tubes - thus forming DWNTs - was supposed to be due to the coalescence of the formerly encapsulated C_{60} molecules under the effect of the 1200°C annealing treatment. The latter assumption was then first experimentally confirmed by Smith and Luzzi [3]. On the other hand, C_{60} encapsulated in SWNTs have been shown able to coalesce into distorted elongated capsules upon electron irradiation within the microscope [1,2,4]. Hence, DWNTs cannot fully develop, due to the competition between damaging effect and coalescence effect.

However, some application of Peapod-derived DWNTs may require milder coalescence temperatures. It was then demonstrated that combining temperature and electron irradiation was successful in achieving the Peapods-to-DWNTs transformation at temperatures as low as 400°C [5]. Meanwhile, a preliminary study provided hints that the dimensional features of the DWNTs may possibly be somewhat controlled by adjusting properly the combined temperature and electron energy values [6]. The current paper summarizes the results and interpretations from a

CP786, *Electronic Properties of Novel Nanostructures*, edited by H. Kuzmany, J. Fink, M. Mehring, and S. Roth
© 2005 American Institute of Physics 0-7354-0275-2/05/$22.50

subsequent, more systematic study of the ability of combined temperature and electron irradiation treatments in promoting the full coalescence of fullerenes when encapsulated in SWNTs as Peapods.

EXPERIMENTAL

Peapods were prepared from purified, arc-prepared SWNTs supplied by NANOCARBLAB and 98% pure C_{60} molecules supplied by INTERCHIM. 20-30 mg of both materials were put in a sealed Pyrex vessel previously vacuumed (rotary pump) then filled with argon several times, then heat-treated at 500°C for 24 hours. The peapod material obtained was then deposited onto membrane-free copper microgrids (2000 mesh) from a droplet of a sonicated suspension in ethanol, and set into a PHILIPS CM30 transmission electron microscope (TEM) equipped with a LaB_6 electron source and a GATAN 652 heating sample holder. Peapods were then subjected to various in-situ treatments combining both electron irradiation and temperature (at 490 and 700°C). Electron beam parameters investigated were electron energy (120, 150, 200, or 300 keV), electron flow (from 10^4 to 7.10^5 e⁻/(nm².s)), and electron dose (from 10^6 to 10^8 e⁻/nm², which corresponds to the total amount of electrons received by the specimen at the end of a given experiment).

Measurement uncertainties were estimated considering the various sources of uncertainties and limitations including the resolution power of the microscope, the accuracy of the magnification value upon calibration, the variation related to that of focusing value, the accuracy of the measurement on the image (related to pixel size), and the error related to the dimension distortions induced in the image by the graphene curvature. As a resulting example, the apparent diameter for a 0.7 nm large C_{60} molecule was calculated to actually range from 0.58 to 0.73 nm in the images.

RESULTS AND DISCUSSION

Apart from any [heating + irradiation] experiments, statistics performed on the starting Peapods interestingly revealed that Peapods outer diameters appeared larger when bundled, as opposed to isolated Peapods. This was confirmed by a similar statistics performed on the starting (= empty) SWNTs, thereby suggesting that attraction between outer nanotubes or Peapods in bundles makes their overall diameter slightly expand. On the other hand, Peapod outer diameters appear smaller than the starting (empty) SWNTs, either bundled or isolated. This can be explained by the attractiveness exerted by the encapsulated fullerenes onto the containing nanotube, thereby inducing a slight diameter shrinking.

Then, the various treatment conditions used led to various results regarding the extension of encapsulated C_{60} molecule coalescence, whose Figure 1 provides few examples. Starting from Peapods (Fig. 1a), coalescence of encapsulated fullerenes can be partial (Fig. 1b), full (Fig. 1c), overstep (i.e., showing evidences for damages to the whole Peapod structure, Fig. 1d), or even not started yet (i.e., providing images similar to Fig. 1a). In the range 150-300 keV for electron energy and 490-700°C for heating temperature, the suitable window of electron beam conditions within which C_{60} full coalescence may occur – of course depending on the energy and temperature conditions - while preventing structure damaging corresponds to $10^7 <$ dose (e^- /nm^2)$< 10^8$ and $10^4 <$ flow (e^-/nm2.s)$< 2.10^5$.

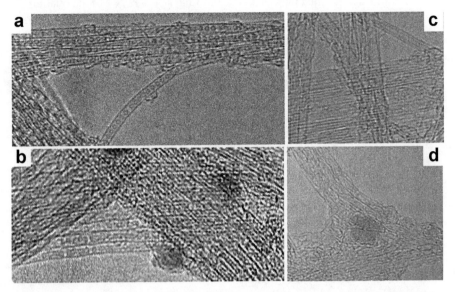

FIGURE 2 – Examples of HRTEM images of (a) as-prepared or unaltered Peapods; (b) partial coalescence of encapsulated C_{60}; (c) Full coalescence of encapsulated C_{60}, leading to DWNT formation; (d) damaged Peapod structure.

Generally speaking, within the coalescence window just defined, the parameters (dose, flow, energy) of the electron beam were surprisingly found to have nearly no influence on the dimension features of the resulting coalesced structures but driving the coalescence kinetics up to destruction. However, a slight increase in inner tube diameters seemed to follow the increase in electron dose (not shown). Correspondingly, the related plot of the intertube distance exhibited a slight decrease, while outer diameters were not affected.

On the other hand, obvious, linear relationships occurred between the geometric features of resulting coalesced structures themselves, whatever the irradiation and temperature conditions. Figure 2 below provides the example of the plot of the intertube distances versus outer diameters. A similar linear relation may be obtained when plotting inner coalesced structure diameters versus outer tube diameters, while there is no influence of the intertube distance onto the inner diameter value. This important finding indicates that the resulting diameter of the building-up inner tube is driven by the combined effect of a dual but contradictory energetics-based influence. One is to enforce the well-known van der Waals gap between graphenes, which could actually lead to a $[y = ax + b]$-type linear relation similar to that of Figure 2, but should meanwhile provide a flat relation between the inter-tube distance and the outer tube diameter (with a single inter-tube distance value equal to ~0.34 nm). The other is to maintain the inner tube diameter equal to that of the starting fullerene molecules whose the inner tube is built from, which could actually lead to a $[y = ax + b]$-type relation between the inter-tube distance and the outer tube diameter as observed, but should meanwhile provide a flat relation between the inner-tube distance and the outer tube diameter (with a single inner-tube distance value equal to ~0.7 nm). The combined influence resulted in both plots exhibiting a $[y = ax + b]$-type relation.

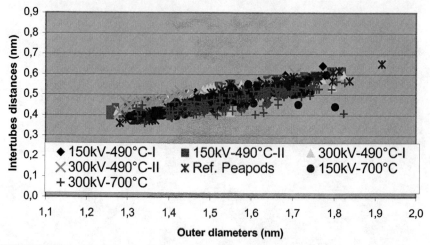

FIGURE 2 – Plot of the inter-tube distances versus outer diameters of ''DWNTs'' obtained for various combined heating + irradiation energy conditions while varying the electron dose and flow.

With respect to the linear relationships mentioned above, temperature appeared to be an important parameter to control the dimensional features of DWNTs. Typically, for a given electron energy, both resulting inner tube <u>and</u> outer tube diameters appeared smaller (and intertube distance larger) after 490°C than after 700°C heat-treatment, and both resulting inner tube <u>and</u> outer tube diameters were larger after 700°C heat treatment than for starting Peapods. Effects are more pronounced for (building-up) inner tubes than for (pre-existing) outer tubes. The latter set of results suggest distinct coalescence regimes according to the temperature range. At high temperature (700°C), thermal effects may prevail, resulting in larger (inner and outer) tube diameters somehow originating from a residual swollen state initially due to the thermal expansion of the structures, afterwards frozen for some reason yet to explain. On the other hand, at low temperature (490°C), irradiation effects may prevail, meaning that some atoms from the coalescing encapsulated C_{60} are knocked-off and hence no longer available to contribute to the inner tube structure, thereby inducing a relative shrinkage of the resulting inner tube diameter.

REFERENCES

1. Smith B.W., Monthioux, M., Luzzi D.E., Nature **396**, 323-324 (1998).
2. Smith B.W., Monthioux, M., Luzzi D.E., Chem. Phys. Lett. **315**, 31-36 (1999).
3. Smith B.W., Luzzi D.E., Chem. Phys. Lett. **321**, 169-174 (2000)
4. Luzzi D.E., Smith B.W., Carbon **38**, 1751-1756 (2000)
5. Sakurabayashi Y., Kondo T., Yamazawa Y., Suzuki Y., Monthioux M., Le Lay M., Europ. Pat. # 03023670.7, October 17 (2003)
6. Sakurabayashi Y., Monthioux M., Kishita K., Suzuki Y., Kondo T., Le Lay M. In : Molecular Nanostructures (eds. H. Kuzmany, J. Fink, M. Mehring, S. Roth), Amer. Inst. Phys. Conf. Proc. **685**, 302-305 (2003)

NON-CARBONACEOUS NANOTUBES

NO$_2$ gas adsorption on titania-based nanotubes

Denis Arčon*,†, Polona Umek†, Pavel Cevc†, Adolf Jesih†, Christopher
Paul Ewels** and Alexandre Gloter**,‡

*Faculty of Mathematics and Physics, University of Ljubljana, Jadranska 19, 1000 Ljubljana,
Slovenia.
†Institute Jožef Stefan, Jamova 39, 1000 Ljubljana, Slovenia
**Laboratoire de Physique des Solides, CNRS-UMR 8502, Université Paris sud, 91405 Orsay,
France
‡ICYS, National Institute for Materials Science, Namiki 1-1, Tsukuba 305-0044, Japan.

Abstract. TiO$_2$-based nanotubes were prepared using a hydrothermal technique and then investigated by high-resolution TEM. Nanotubes were found to be hollow scrolls with a typical outer diameter of about 10 nm, inner diameter 4-5 nm and length of several hundred nm. The surface of nanotubes is found to be largely hydrolised. The presence of surface OH groups also determines their adsorption properties for the NO$_2$ gas molecules. We demonstrate that TiO$_2$ based nanotubes strongly adsorb NO$_2$ molecules. Based on pulsed EPR experiments we suggest that the NO$_2$ molecules couple with the nanotube surface via the nonbonding p_y orbital of the oxygen atoms.

Keywords: nanotubes, adsorption, catalysis, electron paramagnetic resonance
PACS: 61.46.+w, 61.72.Ji, 68.35.-p

INTRODUCTION

Titania (TiO$_2$) has been proposed to be used for the photocatalytic degradation of the wastewater contaminants and air purification. However reduced quantum yield somehow limits its efficiency [1]. Therefore, to overcome these limitations different approaches were proposed, such as the incorporation of metallic particles to TiO$_2$ particles [2]. The increase in the photocatalytic oxidation rate is in this case a result of the improvement of the charge separation by the electron transfer to the metal [2, 3]. Another promising route is to increase the active surface area by synthesizing TiO$_2$ in the form of nanoparticles [4, 5, 6]. Improved photocatalytic activity of titania nanoparticles is in this case suggested to be due to the enhanced surface-to-volume ratio, which is then responsible for the favorable transfer of the photogenerated charge carriers to the adsorbed molecules [4].

In order to further increase the surface-to-volume ration one might try to prepare titania in the form of nanotubes. Kasuga et al. used a simple hydrothermal treatment of the crystalline titania to produce high quality, high purity titania-based nanotubes with a mean diameter of around 10 nm [7, 8]. Treatment of the crystalline TiO$_2$ with NaOH is suggested to lead to the delamination of the TiO$_2$ sheets, which under certain conditions tend to roll-up into the nanotubes with a rolling vector [001] [9].

In order to test the adsorption and catalytic properties of titania-based nanotubes we decided to expose TiO$_2$-based nanotubes to the NO$_2$ gas [10]. A strong adsorption of NO$_2$ has been observed. NO$_2$ gas adsorbs on the surface of the TiO$_2$ nanotube as clearly proved by the powder-like electron paramagnetic resonance (EPR) lineshape.

CP786, *Electronic Properties of Novel Nanostructures*, edited by H. Kuzmany, J. Fink, M. Mehring, and S. Roth
© 2005 American Institute of Physics 0-7354-0275-2/05/$22.50

Detailed low temperature EPR investigations revealed that the adsorbed NO_2 molecule performs small angle vibrations around the axis passing through the NO_2 oxygen atoms. Modeling of the EPR lineshape suggests that NO_2 molecule approaches titania surface with oxygen-atoms making a contact, i.e. NO_2 molecules couple with the nanotube surface via the nonbonding p_y orbital of the oxygen atoms.

EXPERIMENTAL DETAILS

TiO_2-based nanotubes were synthesized by using a similar synthesis procedure to that reported by Kasuga et al. [11]. TiO_2 powder in anatase form was dispersed in NaOH aqueous solution with a help of an ultrasonic bath. The reaction mixture was transferred into Teflon vessel and hydrothermally treated at 130 °C. The obtained white powder was in the next step washed with distilled water until the pH reached 6.

Finally the samples were dried under dynamic vacuum at temperatures not exceeding 200 °C in order to remove adsorbed water without altering the structure of the synthesized materials. Dried powders were then exposed at room temperature to NO_2 gas at pressures ranging from 50 to 450 torr. After exposure, the materials changed color from white to red-brown due to the NO_2 adsorption. The color remained, even after a gentle evacuation to replace NO_2 gas (unadsorbed atmosphere) with Ar.

A commercial Bruker E580 EPR spectrometer was used in pulsed EPR experiments. The EPR spectra were measured by recording the intensity of the echo signal as a function of the magnetic field. A Hahn-echo sequence ($\pi/2 - \tau - \pi - \tau - $ echo) with appropriate phase cycling has been used. Typically the length of the $\pi/2$ pulses was set to 24 ns and the delay time τ to 200 ns.

RESULTS AND DISCUSSION

High-resolution TEM images of prepared nanotubular samples clearly reveal that the titania-based nanotubes are scrolls. This is for instance revealed by the unequal number of walls on the two tube sides, and occasional end-on views within the images. The walls typically consist of 2 to 6 layers, with inter-layer separation of 0.756nm. The tubes are hollow with outer diameters of 8-11nm, inner diameters between 4-7 nm, and lengths of several hundred nanometers. They have open ends. Details of the structure of the titania-based nanotubes will be published elsewhere [12]. In as-prepared samples only a very weak EPR signal could be identified. Judging from the g-factor ($g = 2.0036$), this signal most probably reflects the small concentration of oxygen vacancies.

We now turn to the analysis of the EPR spectra measured in titania-based nanotubes exposed to NO_2 gas. Typical field-swept EPR spectra measured at T=150 K and at T=70 K are shown on Fig. (1). The low-temperature spectrum measured at T=70 K is composed of three broad lines. They all have a typical powder lineshape expected for the nitroxide paramagnetic center ($S = 1/2$) coupled to the nitrogen ^{14}N nuclear spin $I = 1$. The spectrum at T=70 K suggests that the physisorbed NO_2 molecules are static on the time scale of the pulsed-EPR experiment. It thus comes of no surprise that the EPR spectrum does not change any further on cooling below 70 K. We note, however

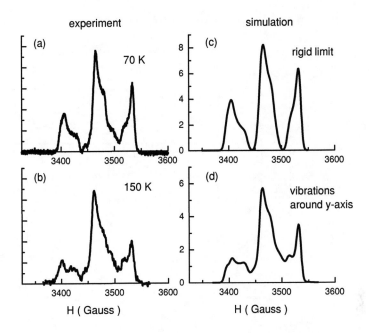

FIGURE 1. A comparison of the fi eld-swept echo detected EPR spectra of NO_2 adsorbed on titania-based nanotubes measured at (a) $T=70$ K and (b) $T=150$ K.

that in addition to these three resonances one may witness the presence of additional broad component centered just below the central line. A likely origin for this signal is the formation of NO molecules as a result of the catalytic activity of the titania-based nanotubes. Details of the analysis of this signal will be published elsewhere [10].

On warming the sample to $T=150$ K the basic characteristics of the EPR spectrum still remain, i.e. the spectrum is composed of three lines found at low temperatures. A close inspection, however, reveals the following change: although the width of the NO_2 signal does not change, the relative height of the shoulders of the triplet lines starts to change. For instance, if we focus for a moment on the low-field component, one may notice that the shoulder at $H=3426$ G has grown with respect to the singularity at $H=3400$ G. The same changes could be observed on the other side of the spectrum. Such changes in the EPR spectrum are a clear signature of some dynamics of the NO_2 molecule. Since the linewidth of the signal does not change, one can suspect that vibrations could be behind these changes.

Further evidence for the dynamics of the NO_2 molecules comes from the fact that the spin-spin relaxation time T_2 becomes field-dependent (Fig. 2). A clearly distinguishable minimum in T_2 (or maximum in the relaxation rate T_2^{-1}) is for instance seen on the low-field line. This means that the local magnetic-field fluctuations become angular dependent. Such a case could be realized only when one assumes small angle molecular vibrations.

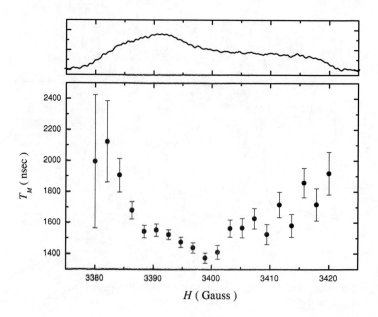

FIGURE 2. A field dependence of the spin-spin relaxation time T_2 measured at $T=70$ K. In the upper layer we show the EPR lineshape in this field region.

The EPR lineshape can be calculated from the following expression

$$F(H) = \sum_{m=-1}^{m=1} \int \exp[-2\tau/T_2] f(\omega - \omega_m(\vartheta, \varphi)) \sin \vartheta \, d\vartheta \, d\varphi . \qquad (1)$$

Here $\omega_m(\vartheta, \varphi)$ is the resonance frequency defined by the spin Hamiltonian. We assumed that the spin Hamiltonian has two contributions: a Zeeman term $\mathscr{H}_Z = \mu_B \vec{S} \cdot g \cdot \vec{H}$ and hyperfine coupling to the ^{14}N nuclear spin $(\vec{I})$ $\mathscr{H}_{hf} = \vec{S} \cdot A \cdot \vec{I}$. Small angle vibrations will not affect the resonance frequency ω_m, but they do influence the variation of the spin-spin relaxation time T_2 with the azimuthal and polar angles ϑ, φ. Full expressions for the T_2 are given in Ref. [13] Two different models for the NO_2 vibrations were tested. In the first model we assumed that the nitrogen atom will attach to the surface with oxygen atoms sticking out from the surface. In this case it is likely that NO_2 molecule will exhibit "torsional" oscillations where oxygen atoms will vibrate around their equilibrium positions. In the second model the oxygen atoms would get close to the nanotube surface and now the nitrogen atom is expected to vibrate around its equilibrium. A comparison with the experiment (Fig. (1)) demonstrates that only the second model is capable of describing the experiment.

At this point we should also stress that in electron spin echo envelope experiments (ESEEM) [14] performed on same samples we clearly noticed the hydrogen modulation. This suggests that the NO_2 molecules are physisorbed on the surface OH groups and that nanotubes have largely hydrolised surfaces. All these results suggest that NO_2 molecules couple with the OH groups on the nanotube surface via the nonbonding p_y orbital of the NO_2 oxygen atoms. Most probably even a weak hydrogen bond is formed during the physisorbtion process.

CONCLUSIONS

In conclusion, we have studied the adsorption properties of titania-based nanotubes. Physisorbed NO_2 molecules are proposed to bond to surface OH groups via weak hydrogen bond. The molecule vibrates around the molecular axis passing through the NO_2 oxygen atoms.

ACKNOWLEDGMENTS

CPE acknowledges the EU Marie Curie Individual Research fellowship scheme for funding.

REFERENCES

1. M.R. Hoffman, S.T. Martin, W. Choi, D.W. Bahnemann, *Chem. Rev.* **95**, 69 (1995).
2. X. Fu, W.A. Zeltner, M. Anderson, *Appl. Cat.* **B 6**, 209 (1995).
3. B. Krauetler, A.J. Bard, *J. Am. Chem. Soc.* **100**, 2239 (1998).
4. L. Cao, A. Huang, F.J. Spiess, S.L. Suib, *J. Catal.* **188**, 48 (1999).
5. A.J. Maira, K.L. Yeung, C.Y. Lee, P.L. Yue, C.K. Chan, *J. Catal.* **192**, 185 (2000).
6. H. Komiani, I.J. Kato, M. Kohno, Y. Kera, B. Ohtani, *Chem. Lett.* **12**, 1051 (1996).
7. T. Kasuga, M. Hiramatsu, A. Hoson, T. Sekino, and K. Niihara, *Langmuir* **14**, 3160 (1998).
8. T. Kasuga, M. Hiramatsu, A. Hoson, T. Sekino, and K. Niihara, *Adv. Mater.* **11**, 1307 (1999).
9. B.D. Yao, Y.F. Chan, X.Y. Zhang, W.F. Zhang, Z.Y. Yang, N. Wang, *Appl. Phys. Lett.* **82**, 281 (2003).
10. P. Umek, P. Cevc, A. Jesih, A. Gloter, C.P. Ewels, and D. Arčon, submitted to *J. Am. Chem. Soc.*.
11. T. Kasuga, M. Hiramatsu, A. Hoson, T. Sekino, K. Niiahara, *Adv. Mater.* **11**, 1307-1311 (1999).
12. A. Gloter, C. Ewels, P. Umek, D. Arčon and C. Colliex, submitted.
13. S.A. Dzuba, Yu. D. Tsetkov, A.G. Maryasov, *Chem Phys. Lett.* **188**, 217 (1992).
14. "Principles of pulse electron paramagnetic resonance", A. Schweiger and G. Jeschke, Oxford University Press, Oxford 2001.

Raman Spectroscopy Of Boron Nitride Nanotubes And Boron Nitride – Carbon Composites.

F. Hasi[1], F. Simon[1], H. Kuzmany[1], R. Arenal de la Concha[2] and A. Loiseau[2].

[1]Institut für Materialphysik, Universität Wien, Strudlhofgasse 4, A-1090, Wien, Austria
[2]LEM, Onera-Cnrs, 29 Avenue de la Division Leclerc, BP 72, 92322 Châtillon, France

Abstract. Boron nitride nanotubes (BN-NT) are topological analoges to single wall carbon nanotubes (SWCNT). We analysed and refined the filling process for SWCNTs and applied it to the BN-NTs. BN-NTs were first annealed in air to open the ends and to remove BN particles. A filling procedure with C_{60} fullerene via vapour phase was applied. Subsequently high temperature treatment was performed to transform the fullerenes. Some spectral features in the Raman spectra of the reaction products in the low frequency range may be assigned to small diameter carbon nanotubes inside the BN-NTs.

INTRODUCTION

The inside of carbon nanotubes [1] has been attracting the interest of chemists and physicists ever since the discovery of the fullerenes [2]. Smith et al. discovered the filling of single wall carbon nanotubes with C_{60} [3]. At temperatures of the order of 1400 K the fullerenes can be fused to a second tube, inside the master tube [4]. These inner tubes exhibit a strong Raman spectrum for the RBM [5]. BN-NTs are other tubular nanostructures which attract attention based on the theoretically predicted mechanical stability and electronic insulating behavior independent of chirality and number of tube walls [6]. Although BN-NTs and C-NTs are structurally very similar their electrical properties are different due to the polar lattice of boron nitride. The shielded inside of BN-NTs naturally presents the possibility to fill them with C_{60} which may enable the formation of boron nitride - carbon nanostructures.

EXPERIMENT

The BN-NTs were laboratory prepared by laser ablation of a hexagonal boron nitride (h-BN) target with a continuous CO_2 laser under a nitrogen flow at a pressure 1 bar [7]. From TEM analysis it was estimated that the yield of BN-NTs could be of the order of 25% of the ablated material and that about 80% of the nanotubes are single walled with diameter distribution centered at 1.4 nm (FWHM=0.6) [8]. SWCNTs were purchased from Med Chem Labs with purity of 50%. At first, in several experiments the dynamics of tube opening was investigated for carbon nanotubes. The SWCNTs

CP786, *Electronic Properties of Novel Nanostructures*, edited by H. Kuzmany, J. Fink, M. Mehring, and S. Roth

could be filled with C_{60} fullerenes as purchased by tempering them at 600 °C in a sealed and evacuated quartz ampoule for two hours together with excess amounts of C_{60}. This filling procedure was used throughout in the work presented here. The peapods were transformed to double wall carbon nanotubes by tempering in high vacuum at 1250 °C for two hours. This procedure of inner shell tube growth from the peapods was used throughout in this work. The transformed material was then checked by Raman spectroscopy in the range of the RBM for double wall carbon nanotubes. Closing of the tubes was performed at various temperatures between 800 °C and 1200 °C in an evacuated quartz tube at 10^{-6} Pa. Opening of the tubes was performed by exposure to air at various temperatures between 350 °C and 500 °C. BN-NTs were heat treated in air at 800 °C to open the tips and to remove boron particles. Filling and transformation was performed as for CNTs. Raman experiments were carried out with a Dilor xy triple spectrometer in the normal resolution mode. For the excitation various laser lines were used in the blue, red and yellow spectral range.

RESULTS FOR SWCNT FILLING

Figure 1a and 1b depicts some examples for the efficiency of C-NT filling after closing and reopening for various annealing conditions.

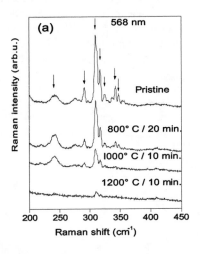
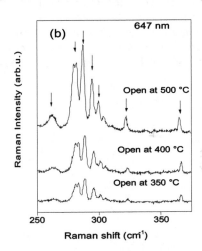

FIGURE 1. a) Raman response of the inner tube RBM of the untreated and standardized filled and transformed tube material (top) and after various annealing treatments as indicated before the filling and transformation processes; b) Raman response of the RBM of inner shell CNTs for standard filling and transformation procedures after the pristine tubes had been closed and reopened at temperatures indicated.

The top spectrum in Fig.1a represents the pristine tubes after standard filling and standard transformation. The three spectra below depict the response of the inner tube RBM for samples which were annealed under the conditions indicated in the figure before filling and transformation. As can be seen 800 °C annealing has almost no effect on the filling for the large diameter tubes but efficiently reduces filling of the smaller tubes with RBM frequencies larger than 300 cm^{-1}. After 20 minutes annealing, the signal at 340 cm^{-1} has reduced to 35% of the signal of the pristine sample. After 10 minutes annealing at 1200 °C, only a very small response from the inner shell tubes is left. Figure 1b depicts Raman spectra of the RBM recorded for 647 nm excitation for tubes which were closed at 1000 °C for two hours and then reopened by exposure to air at temperatures indicated. As a result, the opening process becomes less and less efficient with decreasing temperature. However, the reduction in efficiency is hardly depending on the tube diameter. For opening at 500 °C, 95% of the closed tubes could be reopened. The dependence of tube closing and opening on annealing temperature and time was studied in more detail and will be published elsewhere.

RESULTS FOR BN-NTS

The Raman experiments carried out on BN-NTs turned out to be more difficult due to the inhomogeneity of the sample and a strong luminescence. We investigated the samples with a 488 nm excitation wavelength laser at room temperature. Results are presented in Fig. 2. As can be seen from the spectra, (a), (b), (c), the sample appears to be very heterogeneous. Many of the Raman modes observed in the (a) spectra are not present in (b) and (c) spectra. In the high frequency range peaks at 1054, 1306 cm^{-1}, and 1368 cm^{-1} show the presence of cubic BN (c-BN) and hexagonal BN (h-BN) in the material. For comparison, Raman spectra of c-BN and h-BN are depicted in 2d and 2e. In the low frequency range, peaks at 129, 207, 498 and 881 cm^{-1} can be observed which are believed to originate from boron acids produced during the laser ablation process of the h-BN target. The absence of resonant Raman modes for BN-NTs in the spectra proves the theoretically predicted larger band gap than in C-NTs.

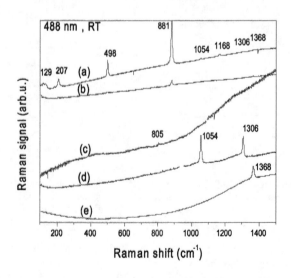

Figure 2. Raman spectra of BN- NT sample from three different spots (a), (b), (c). Raman spectra of cubic boron nitride and hexagonal boron nitride (d), (e) respectively.

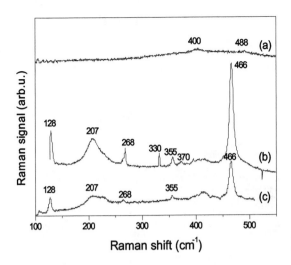

Figure 3. Raman response of the reaction product of BN-NTs with C_{60}. BN-NTs heated at 1250 °C for 2 h in dynamic vacuum, 568 nm, RT (a). Standardized transformation on BN-NTs, 568 nm, RT (b). Standardized transformation on BN-NTs, 647 nm, RT (c).

Similarly to the opening procedure for carbon nanotubes we applied oxidation in air at 800 °C to open the tips of the boron nitride nanotubes and to remove boron particles. Standardized reactions of the BN-NTs with C_{60} fullerene and transformation was performed as mentioned above. Figure 3 show the Raman response in the low frequency range of the reaction products of BN-NT material with C_{60} for 568 and 647 nm excitation wavelength. The Raman investigation of the reaction products shown in (b) and (c) for the radial breathing mode range for carbon nanotubes reveals new spectral features which are not present in the untreated BN-NT or in the BN-NT material heated at 1250 °C in dynamic vacuum without the presence of the fullerenes, spectrum (a). The structured peak at 268 cm^{-1} and the peaks at 330 and 355 cm^{-1} in the (b) and (c) for 568 and 647 nm excitation, respectively, may be assigned to reaction products of the carbon inside the BN-NTs. Also a sharp peak at 466 cm^{-1} is detected.

CONCLUSIONS

In conclusion, we have presented a Raman study of the reaction processes of fullerenes with SWCNTs and BN-NTs. A controlled closing and opening for SWCNTs was performed. The absence of the resonant Raman modes for BN-NTs in the spectra confirms the theoretically predicted larger band gap than in C-NTs. Critical temperatures for the opening process in BN-NTs must be determined. The investigation of the reaction product of BN-NTs with fullerenes revealed spectral features which may be assigned to phonon modes of inner shell carbon nanotubes inside boron nitride nanotubes.

ACKNOWLEDGMENTS

This work was supported by the Fonds zur Förderung der Wissenschaftlichen Forschung in Austria, project 14893 and by the EU project PATONN Marie Curie MEIF – CT- 2003 – 501099 grants.

REFERENCES

1. S. Iijima, Nature(London) 354, 56 (1991)
2. H. Kroto et al. Nature 318, 162, (1985)
3. B. W. Smith, M. Monthioux, D. E. Luzzi, Nature 396, 323 (1998)
4. S. Bandow, M. Takizaw, K. Hirahara, M. Yudasaka and S. Iijima. Chem. Phys. Lett. 337 48 (2001)
5. R. Pfeiffer, H. Kuzmany, C. Kramberger et al. Phys. Rev. Lett. 90, 225501 (2003)
6. A. Rubio et al, Phys. Rev. B 49, 5081 (1994)
7. R. S. Lee, J. Gavillet, A. Loiseau et al, Phys. Rew. B 64, 121405 (2001)
8. R. Arenal de la Concha et al., in Molecular Nanostructures edited by H. Kuzmany et al. AIP Conference Proceeings 685, 384-388 (2003)

Vibrational Spectroscopic Studies on the Formation of Ion-exchangeable Titania Nanotubes

Mária Hodos*, Henrik Haspel*, Endre Horváth*, Ákos Kukovecz*, Zoltán Kónya*, and Imre Kiricsi*

*Department of Applied and Environmental Chemistry, University of Szeged,H-6720 Szeged, Rerrich Béla tér 1., Hungary

Abstract. Ion-exchangeable titanium-oxide nanotubes have commanded considerable interest from the materials science community in the past five years. Synthesized under hydrothermal conditions from TiO_2, typical nanotubes are 150-200 nm long and 8-20 nm wide. High resolution TEM images revealed that unlike multiwall carbon nanotubes which are made of co-axial single-wall nanotubes, the titania tubes possess a spiral cross-section. An interesting feature of the titania tubes is their considerable ion-exchange capacity which could be utilized e.g. for enhancing their photocatalytic activity by doping the titania tubes with CdS nanoparticles. In this contribution we present a comprehensive TEM, FT-Raman and FT-farIR characterization study of the formation process.

Keywords: titanate nanotubes, Raman spectroscopy, synthesis conditions
PACS: 81.07.De, 81.16.Be

INTRODUCTION

The synthesis and characterization of titanium oxide nanostructures have commended considerable attention lately because of their numerous potential applications in solar cells, electronics, photocatalysis, sensorics and as catalyst supports [1]. Titania nanotubes are especially interesting as they combine high aspect ratio and high specific surface area with versatile chemistry, leading to e.g. larger Li-storage capacity than that of crystalline anatase [2].

In this paper we present a comprehensive experimental study on the effects of various synthesis parameters on hydrothermal nanotube formation. Factors varied were the type of the precursor titanium oxide material, synthesis length and temperature and synthesis mixture composition. The prepared samples were studied by transmission electron microscopy (TEM), powder X-ray diffraction (XRD), FT-Raman and far-IR spectroscopy. The optimal synthesis conditions were identified as well.

CP786, *Electronic Properties of Novel Nanostructures*, edited by H. Kuzmany, J. Fink, M. Mehring, and S. Roth
© 2005 American Institute of Physics 0-7354-0275-2/05/$22.50

EXPERIMENTAL

Analytical grade anatase TiO_2 and NaOH were purchased from Aldrich and used as received. The nanotubes were prepared by stirring 2 g of anatase TiO_2 in 140 ml c_1 (2.5–15.0) M NaOH aqueous solution until a white suspension was obtained, then aging the suspension in a closed, PTFE-lined autoclave at T_1 (90–170) °C for t_1 (3–72) h without shaking or stirring, and finally washing the product with deionized water to reach pH 8 at which point the slurry was filtered and the nanotubes were dried in air [3]. TEM observations were performed on a Philips CM10 instrument using copper mounted holey carbon grids. Stokes FT-Raman spectra were recorded on a Bio-Rad FT-Raman instrument operating with 1064 nm excitation wavelength at room temperature in air. Spectra were averaged from 256 scans performed at 4 cm^{-1} resolution. Far FT-IR spectra were measured on a Bio-Rad FTS-65v instrument (256 scans, 4 cm^{-1} resolution) in polyethylene (PE) pellets (3 mg sample/30 mg PE). All spectra presented in this paper are normalized to the intensity of the highest peak in the spectrum. The nitrogen adsorption isotherms were measured at 77 K with a QuantaChrome Nova 2000 surface area analyzer. Samples were outgassed at 393 K for 1 hour to remove adsorbed contaminants. The specific surface area (A_S) was calculated using the multipoint BET method on six points of the adsorption isotherm.

RESULTS AND DISCUSSION

It is seen in Fig. 1B(iv) that the Raman spectrum of the hydrothermally synthesized titanate nanotubes consists of a weak group of peaks centered at 180 cm^{-1} and four more intense and broader features at 278, 448, 660 and 905 cm^{-1}. The 448 cm^{-1} peak was assigned to a pure framework Ti-O-Ti vibration, the 278 and 660 cm^{-1} peaks to Ti-O-M vibrations involving the ion (M=Na$^+$ in our case) occupying an ion exchange position in the trititanate wall and the 905 cm^{-1} peak to four-coordinate Ti-O involving non-bridging oxygen atoms coordinated by the M ion [4].

Figure 1. shows that the system is not sensitive to temperature changes. Titanate nanotubes are readily formed in the 90–170 °C temperature range. At lower temperatures (90 and 110 °C) the reaction is slower; therefore, peaks assigned to anatase as well as those assigned to the nanotubes are present in the Raman spectrum. 130 °C and above are temperatures high enough to give full anatase conversion under the reaction conditions (10 M NaOH, 72 h). The specific surface area increases with temperature in the 90–150 °C range (70.9–163.1 m^2/g) and decreases to 139.6 m^2/g at 170 °C. Thus, we are unable to confirm the findings of Bavykin et al. [5] who observed a sharp drop in A_S at this point (244 m^2/g at 150 °C vs. 24 m^2/g at 170 °C).

The NaOH concentration dependence (Fig. 2) when fixing the reaction temperature and time at 130 °C and 72 h, respectively, is as follows. At 5 M NaOH a few scattered nanotubes appear, but the bulk of the system is anatase as evidenced by Raman, far-IR and XRD. 7.5 M NaOH corresponds to a mixed state in which anatase particles and nanotubes co-exist, whereas in 10 M NaOH the anatase conversion is complete and the system consists of trititanate nanotubes. The tubes remain stable in

346

12.5 M NaOH. In 15 M NaOH the spectral features broaden beyond recognition, indicating severe structural disorder: the trititanate structure is most likely destroyed.

Studying the time dependence of the reaction (in 10 M NaOH at 130 °C) is expected to reveal information about the nanotube formation mechanism. It is seen in Fig. 3 that some nanotubes are present already after 3 hours.

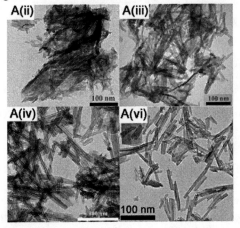

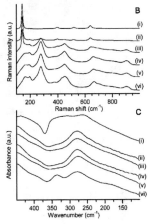

FIGURE 1. TEM micrographs (part A), Raman (part B) and far-IR spectra (part C) of trititanate nanotubes synthesized in 10 M NaOH for 72 h at various temperatures: (i) raw anatase, (ii)…(vi) nanotubes prepared at 90, 110, 130, 150 and 170 °C, respectively.

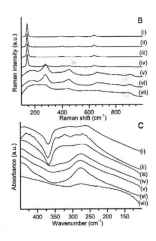

FIGURE 2. TEM micrographs (part A), Raman spectra (part B), far-IR spectra (part C) and XRD profiles (part D) of trititanate nanotubes synthesized at 130 °C for 72 h using various NaOH concentrations: (i) raw anatase, (ii)…(vii) nanotubes prepared in 2.5 M, 5 M, 7.5 M, 10 M, 12.5 M and 15 M NaOH, respectively.

However, the product of the 3 h and the 6 h reactions contains lots of nanoparticles besides the nanotubes, and the corresponding Raman, far-IR and XRD patterns

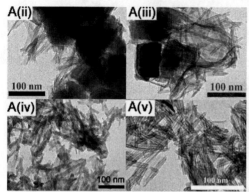
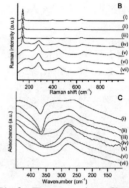

FIGURE 3.TEM micrographs (part A), Raman spectra (part B), far-IR spectra (part C) and XRD profiles (part D) of trititanate nanotubes synthesized at 130 °C in 10 M NaOH using various synthesis times: (i) raw anatase, (ii)...(vii) reaction times 3 h, 6 h, 12 h, 24 h, 48 h and 72 h, respectively.

evidence that the bulk of the material is still anatase. Nanoparticles and nanotubes are present in similar amounts at 12 h, whereas 24 h appears to be enough for the complete conversion of anatase into nanotubes. Indeed, the specific surface area of the synthesized nanotubes does not change any more after 24 hours. Thus, the optimal synthesis conditions leading to nanotubes with 100% raw anatase conversion are: 10 M NaOH, 130 °C and 24 hours in the autoclave.

ACKNOWLEDGMENTS

This work was supported by the Hungarian Ministry of Education through grants OTKA F046361 and F038249. A.K. and Z.K. acknowledge support from a Zoltan Magyary and a Janos Bolyai fellowship, respectively.

REFERENCES

1. Hodos, M., Horváth, E., Haspel, H., Kukovecz, A., Kónya, Z., and Kiricsi, I. *Chem. Phys. Letters* **399**, 512-516 (2004).
2. Kavan, L., Kalbac, M., Zukalova, M., Exnar, I., Lorenzen, V., Nesper, R., and Graetzel, M. *Chem. Mater.* **16**, 477-485 (2004).
3. Kasuga, M. H., Hosun, A., Sekino, T., and Niihara, K., Adv. Mater.**11**, 1307-1311 (1999).
4. Sun, X., and Li, Y., Chem. Eur. J. **9**, 2229-2238 (2003).
5. Bavykin, D. V., Parmon, V. N., Lapkin, A. A., and Walsh, F. C., *J. Mater. Chem.* **13**, 3370-3377 (2004).

Micro Raman Investigation of WS$_2$ Nanotubes

Konstantin Gartsman*, Ifat Kaplan-Ashiri*, Reshef Tenne*,
Peter Rafailov¶, and Christian Thomsen¶

*Weizmann Institute of Science, Rehovot, Israel
¶Technical University, Berlin, Germany

Abstract. Individual WS$_2$ multiwalled nanotubes, 2-3 micron in length, and with 15 - 25 nm diameter were mounted on AFM cantilevers tips. The nanotube orientation along the cantilever cone axis was confirmed by scanning electron microscopy (SEM). Micro-Raman spectra of these individual nanotubes showed similar vibrational frequencies as in the bulk material. The highly anisotropic shape of the nanotubes, however, leads to a strong antenna effect as it is known from single-walled carbon nanotubes. A qualitative assessment of the radiation field screening for polarization perpendicular to the nanotube axis is given.

Keywords: nanotubes, Raman spectra.

PACS: 61.46.+w; 62.25.+g; 63.22.+m

INTRODUCTION

Besides the carbon nanotubes, it was demonstrated in recent years that nanometer-sized tubules can be derived from a variety of materials with layered structure. The necessary conditions for the formation of such tubular structure are weak van der Waals forces between the distinct layers and strong intralayer bonds, as is the case, for instance, with the metal chalcogenide WS$_2$ and MoS$_2$ nanotubes. These nanotubes can be produced with high yield by sulfidization of a thin film of the respective transition metal oxide in a reducing atmosphere. Metal chalcogenide nanotubes are multi-walled nanotubes with typical outer diameters of 15 - 20 nm and 4 - 8 shells of S-M-S triple layers (M = Mo, W). In such triple layers each M atom is coordinated by six S atoms in a trigonal prismatic coordination, as in the corresponding basic material. In contrast to carbon nanotubes the vast majority of chalcogenide nanotubes are uncapped, i.e. they have open ends [1].

2H bulk WS$_2$ is an indirect gap semiconductor with two series of exciton absorption bands at the absorption edge [2]. At room temperature the indirect gap is 1.3 eV and the direct gap 2.05 eV, the A exciton energy lying at 1.95 eV [2]. Although the corresponding values for WS$_2$ nanotubes are not yet determined, optical absorption measurements show that they do not deviate significantly from those of the bulk material [2].

In this work we investigate individual WS$_2$ nanotubes with well defined orientation with Raman spectroscopy. We find the polarization behavior of their Raman active modes in apparent contradiction to the usual symmetry selection rules. The observed

CP786, *Electronic Properties of Novel Nanostructures*, edited by H. Kuzmany, J. Fink, M. Mehring, and S. Roth
© 2005 American Institute of Physics 0-7354-0275-2/05/$22.50

scattering intensity is only sensitive to the orientation of the exciting field vector relative to the nanotube axis. This is, however, a consequence of the antenna effect, i.e., the screening of the light electric field perpendicular to the nanotube axis due to its strongly anisotropic polarizability [3, 4, 5].

EXPERIMENT AND DISCUSSION

Individual WS_2 nanotubes, 2-3 micron in length and with 15-25 nm outer diameters were mounted on cone-shaped AFM cantilevers within the chamber of ESEM XL 30 microscope equipped with a micromanipulator. SEM images of thus prepared samples showed that the nanotube was attached to the cone top and oriented along the cone axis.

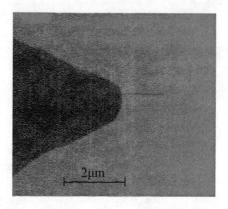

FIGURE 1. SEM image of the examined WS_2 nanotube mounted on cone-shaped AFM cantilever.

The Raman measurements were performed in backscattering geometry on a Dilor Labram spectrometer with a CCD-detector. A He-Ne-laser line at 632.8 nm was used as an excitation source, the laser beam being perpendicular to the axis of the AFM cantilever. The spectrometer slits were set to 4 cm^{-1} spectral width and absolute accuracy of 0.9 cm^{-1}. The laser beam was focused to a spot of diameter ~1μm using microscope optics.

In order to obtain a Raman signal from the nanotube we first focused on the very top of the AFM cantilever made of silicon and then performed several measurements at steps of ~1μm separation along the cantilever axis and perpendicular to it. A sample series of these measurements is shown in Fig. 2. The WS_2 nanotube signal could be maximized by focusing the laser just beyond the cantilever top in the virtually empty space. We see this from the increasing intensity ratio of the WS_2 signal to the signal of Si in Fig. 2. The T_{2g} mode of Si at 520 cm^{-1} was used also as an internal calibration source. Several vibrational features of WS_2 were identified in the spectra (See Fig. 2) but mainly the Raman signal of WS_2 is recognized in the two lines at 351 and 417 cm^{-1}. These lines completely resemble the A_{1g} and E_{2g} modes, respectively, as measured on the bulk material and various sizes and shapes of WS_2 nanoparticles [2].

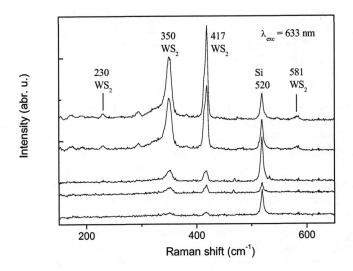

FIGURE 2. Series of Raman signal measurements from the WS$_2$ nanotube mounted on AFM cantilever.

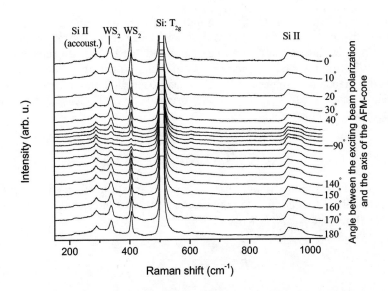

FIGURE 3. Raman signal measurements from the WS$_2$ nanotube varying the angle of the excitation polarization to the AFM cone axis from 0° to 180°.

The one-dimensionality of nanotubes is expected to give rise to a strongly anisotropic polarizability, i. e. the so-called antenna effect. This effect was predicted by Ajiki and Ando [3] who showed that a depolarizing field occurs for radiation polarized perpendicular to the nanotube axis and concluded that optical transitions from the valence to the conduction band are only observable for polarization parallel to the nanotube axis. The antenna effect can be explained also in the electrostatic limit by showing that the radial polarizability component undergoes a radius-dependent screening due to the cylindrical geometry while the axial component remains unaffected provided the nanotube is not capped [4].

To verify this we measured Raman spectra varying the angle of the excitation polarization to the AFM cone axis from $0°$ to $180°$. Due to the imperative necessity to preserve the same sample spot and to have a maximum nanotube signal we rotated the polarization of the incident and the scattered beam with a $\lambda/2$ plate.

The spectra at different angles are plotted in Fig. 3. We attribute the observed strong depolarization effect to the fact that at 1,96 eV the in-plane component of both the real and imaginary part of the dielectric tensor of a 2H-WS$_2$ layer is already several times larger than the out-of-plane component. For a nanotube consisting of such a layer this translates into a larger absorption-related dipole moment along the axis than perpendicular to it even in the absence of an antenna effect. The additional screening of the polarizabilities in X or Y direction through the antenna effect may then lead to the observed polar behavior of the Raman intensities.

SUMMARY

We measured Raman spectra from an individual WS$_2$ nanotube attached to the tip of an AFM cantilever and having a well defined orientation. By means of angular-dependent polarization measurements we established that a strong antenna effect takes place in WS$_2$ multiwalled nanotubes. We conclude that the antenna effect in cylindrically shaped nano-objects is not restricted to small radii, but can take place for significantly larger radii as well if the original layered material has strongly anisotropic polarizability.

ACKNOWLEDGMENTS

This work was supported by The Minerva Foundation (Munich); The G.M.J. Schmidt Minerva Center for Supramolecular Architectures, the German-Israeli Foundation (GIF) and NATO Science Foundation.

REFERENCES

1. G. Seifert, T. Köhler and R. Tenne, *J. Phys. Chem. B* **106**, 2497 (2002).
2. G. Frey, R. Tenne, M. J. Matthews, M. S. Dresselhaus and G. Dresselhaus, *J. Mater. Res.* **13**, 2412 (1998).
3. H. Ajiki and T. Ando, *Jpn. J. Appl. Phys., Suppl.* **34**, 107 (1995).
4. L. Benedict, S. G. Louie and M. L. Cohen, *Phys. Rev. B* **52**, 8541 (1995).
5. S. Reich, C. Thomsen and J. Maultzsch, *Carbon Nanotubes: Basic Concepts and Physical Properties*, edited by Wiley-VCH, Berlin, 2004.
6. D. Taverna, M. Kociak, V. Charbois and L. Henrard, *Phys. Rev. B* **66**, 235419 (2002).

Novel Elongated Molybdenum Sulfide Nanostructures

S. Gemming* and G. Seifert*

*Institut für Physikalische Chemie und Elektrochemie, Technische Universität, D-01062 Dresden, Germany

Abstract. Molybdenum sulfide nanostructures with a Mo:S stoichiometry of 3n : 3n+2 can be regarded as building blocks of the corresponding Chevrel phases. For alkali metals as counter ions the shortest ones of these nanostructures correspond to the negatively charged clusters which were observed recently by experiment. The transition from finite systems to the infinitely long wire was studied by calculations based on the density-functional theory. The relative stability of the shortest neutral structures exhibits an even-odd alternation, which decays towards n = 8. The infinitely long wire is metallic with an Mo-based conductance channel and more elongated bond lengths. For the intermediate neutral structures with n up to 8 the structural and electronic properties converge towards the values calculated for the infinitely long wire. An investigation of charged species indicated that the relative stability of the structures can also be tuned by the appropriate choice of counter ions.

INTRODUCTION

Clusters of the composition $Mo_{3n}S_{3n+2}$ can be regarded as one-dimensionally extended building blocks of the corresponding sulfur- and selenium-based Chevrel phases. Very recent experimental investigations also demonstrated the formation of three-dimensional networks of $[Mo_{3n}S_{3n+2}]^{d-}$ cluster anions [1] with K, Cs, and In as counter ions. These compounds are generally built from equal amounts of two $Mo_{3n}S_{3n+2}$ species with different values of n, thus the overall formula unit of the network is $Mo_{3n}S_{3n+4}$. For the alkali metals as counter ions the charge state of the single cluster building block is 1-, only for the less common case of M = Nb deviations from d = 1 occur. Thus, the clusters in these compounds occur in the same charge state as the free clusters observed by experimental studies [2].

Among the small, Mo-rich Mo_mS_n clusters, Mo_4S_6 was calculated to be the most stable small Mo_4S_n cluster with a large energy gap of 3 eV between the highest occupied (HOMO) and the lowest unoccupied (LUMO) cluster state [3]. Concomitant with these findings, mass spectra exhibit a pronounced signal at the corresponding mass [2], and the measured photoelectron spectrum agrees well with the calculated electronic structure. Higher signals in the mass spectra were observed, which correspond to the stoichiometries Mo_6S_8, Mo_9S_{11}, $Mo_{12}S_{14}$, and $Mo_{15}S_{17}$ [2].

The present density-functional calculations on these systems were carried out with the programs deMon [4] and ABINIT [5]. All structures were optimised at the local-density level, and the energies were refined employing the gradient-corrected PBE functional. Further details on the calculation are reported in ref. [6].

CP786, *Electronic Properties of Novel Nanostructures*, edited by H. Kuzmany, J. Fink, M. Mehring, and S. Roth
© 2005 American Institute of Physics 0-7354-0275-2/05/$22.50

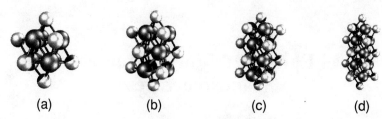

FIGURE 1. Schematic representation of the atom arrangement in the investigated clusters Mo_6S_8 (a), Mo_9S_{11} (b), $Mo_{12}S_{14}$ (c), and $Mo_{15}S_{17}$ (d) (from left to right, Mo in black, S in grey).

MODEL SYSTEMS

The four model systems studied in detail are depicted schematically in Fig. 1; furthermore the cluster series was investigated up to $n = 8$ by density-functional-based tight-binding calculations. As neutral species the smallest cluster, Mo_6S_8, undergoes a slight Jahn-Teller distortion. In the anionic species $Mo_6S_8^{4-}$ the orbitals responsible for the distortion are fully occupied and the cluster is composed of a perfect central Mo_6 octahedron, which is decorated by the 8 S atoms at the edge-bridging positions, i.e. in cubic arrangement. For creating the cluster sequence the Mo_6S_8 cluster is expanded by Mo_3S_3 units along one of the three-fold rotation axes of the central octahedron. Finally, the infinitely long wire consists of an AB-type stacking of regular sulfur-bridged Mo_3 triangles. The resulting Mo-Mo distances expand from a value of 2.68 Å in Mo_6S_8 to 2.73 Å in the more extended clusters. The Mo=S distances, on the other hand, remain rather constant around 2.49 Å.

The relative stability along the cluster sequence was evaluated from the total energy differences by the following formula:

$$\Delta_E = E(Mo_{3n}S_{3n+2}) + E(Mo_3S_3) - E(Mo_{3(n+1)}S_{3(n+1)+2}),$$

where $E(Mo_3S_3)$ is the total energy of a free, neutral Mo_3S_3 fragment; the other two values are the total energies of two successive clusters or cluster anions. For the neutral clusters Δ_E exhibits a weak even-odd-alternation, slightly in favour of the clusters with an odd value of n. The alternation reaches a maximum for $Mo_{12}S_{14}$, which is less stable than the preceding series member, fo_9S_{11}, by 0.18 eV, and it decays to 0.02 eV for longer clusters up to $Mo_{21}S_{23}$.

Furthermore, the infinitely long $[Mo_6S_6]_\infty$ wire was investigated by DFT band-structure calculations, employing periodic boundary conditions. The most stable structures are composed of Mo triangles, the edges of which are bridged by S atoms. The optimum arrangement of these triangular Mo_3S_3 units is the staggered stack with a length of the repeat unit of 8.25 bohr / 4.36 Å and bond lengths of d(Mo-Mo) = 2.74 Å and d(Mo-S) = 2.48 Å. Other periodic structures like the eclipsed arrangements Mo_3 triangles with S in bridging or in four-fold coordinated sites are less stable by rougly 5 eV per Mo_3S_3 unit.

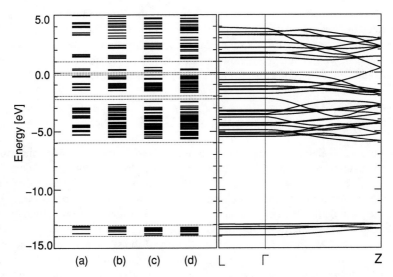

FIGURE 2. Electronic structure of the clusters Mo_6S_8 to $Mo_{15}S_{17}$ (left panel) and of the infinitely long wire (right panel). The Fermi level is set to zero in all graphs.

ELECTRONIC STRUCTURE

Both neutral and anionic species were investigated. All structures up to $Mo_{21}S_{23}$ exhibit a band gap of about 0.4 eV. For the three smallest clusters, Mo_6S_8, Mo_9S_{11}, and $Mo_{12}S_{14}$ larger band gaps of up to 1.4 eV can be obtained by adding negative charges, whereas the larger and more elongated clusters from $Mo_{15}S_{17}$ on would require a higher anionic charge for a similar stabilisation of the electronic structure. Especially for the four-fold negative cluster anions the next larger energy gap between the HOMO+2 and the HOMO+3 is reached for most of the structures. In that case the oscillation period of Δ_E is three units long, thus the relative stabilities of the different members of the cluster sequence is highly dependent on the charge state, i.e. on the number and type of counter ions in a solid phase.

Fig 2 depicts the electronic levels of the four shortest neutral clusters in more detail. The S(2s)-based levels at -14 to -13 eV are followed by a set of mixed Mo(4d)-S(2p) states between -6 and -2.5 eV, which account for the Mo-S bonding. The remaining Mo(4d) electrons from -2 eV to the Fermi level form Mo-Mo-bonds within each Mo_3 triangle and along the wire. Thus, the HOMO and LUMO levels are localised in the central Mo framework.

For the infinitely long wire the density of states, obtained by the band-structure calculations, exhibits a pseudogap close to the Fermi level. This pseudogap can be deepened by an elongation of the wire, thus, external stress can modify the electronic properties of the wire. The band structure along the $\Gamma = Z$ direction ($k_z = 0.0$ to $k_z = 0.5$) shows that two bands cross the band gap and provide metallic conductivity. A comparison with the subbands of the Mo-based and the S-based fragments shows, that the partially occupied bands are mainly located in the Mo core of the structure. The pronounced downward shift of the S-derived levels indicates the electron transfer from Mo to S and suggests the formation of sulfide ions surrounding the Mo-based central part of the nanowire. Thus, the nanowire can actually be regarded as a nanocable with an electronic conductor embedded in an insulating ionic coating.

CONCLUSIONS

The structural and electronic properties of elongated molybdenum sulfide nanostructures were investigated by density-functional calculations. Molybdenum-rich nanostructures with a Mo:S stoichiometry of 3n : 3n+2 can be regarded as building blocks of the corresponding Chevrel phases. For alkali metals as counter ions these nanostructures correspond to the negatively charged clusters which were observed recently by experiment. The change of the properties upon transition from finite systems to the infinitely long Mo_6S_6 wire was investigated. The relative stability of the shortest neutral structures (up to n = 8) exhibits a weak even-odd alternationi, which depends crucially on the charge state. The infinitely long wire is metallic with an Mo-based conductance channel. Dand Density-functional-based tight-binding calculations on the intermediate structures with n < 10 show that the structural and electronic properties converge towards the values of the infinitely long wire. A study of negatively charged $Mo_{3n}S_{3n+2}$ species indicates that the relative stability Garbled time appropriate choice of counter ions.

ACKNOWLEDGMENTS

The authors acknowledge financial support by the Deutsche Forschungsgemeinschaft and by the German-Israel Foundation (GIF).

REFERENCES

1. S. Picard, et al., Acta Cryst C60, i61 (2004); D. Salloum, et al. Sol. St. Chem. 177 1672 (2004).
2. N. Bertram, Y.D. Kim, G. Ganteför, Appl. Phys. A, submitted.
3. J. Tamuliene, S. Gemming, G. Seifert, TAMC04, (2005), in press.
4. A. M. Köster, R. Flores-Moreno, G. Geudtner, A. Goursot, T. Heine, A. Vela, S. Patchkovskii, D.R. Salahub, deMon 2004, NRC, Canada.
5. The ABINIT code is a common project of the Universite Catholique de Louvain, Corning Inc., and other contributors.
6. S. Gemming, J. Tamuliene, G. Seifert, n. Betram, Y.D. Kim, G. Ganteför, Appl. Phys. A, accepted (2005).

New inorganic nanotubes of dioxides
MO$_2$(M=Si, Ge, Sn)

L.A. Chernozatonskii[1], P.B. Sorokin[1,2], A.S. Fedorov[2]

(1)*Institute of Biochemical Physics, Russian Academy of Science, Moscow, 119991, Russia*
(2)*L.V.Kirensky Institute of Physics of Siberian Branch of the Russian Academy of Sciences, Krasnoyarsk, 660036, Russia*

Abstract. Structures, stability and electronic properties of inorganic dioxides MO$_2$ (M=Si, Ge, Sn) nanotubes with quasi-quadrone atomic lattice are studied by DFT method.

Keywords: nonorganic nanotubes, DFT calculations, structural and electronic properties
PACS: 61.46.+w; 68.65.–k

INTRODUCTION

The R. Tenne group discovery of (Mo, W)S$_2$ nanotubes in which atoms were arranged on three cylindrical surfaces promoted studies of noncarbon nanotubes [1]. Recently, Bromley group [2] has considered hypothetic hollow (SiO$_2$)$_N$ clusters (N= 6–12, 24). The structures of such a complex composition represent a new area for studying the unique physical and chemical properties of not only spheroidal but also tubular nanoclusters. In this communication, the author draws attention to the possibility of the existence of a new class of three-cylinder nanotubes composed of MO$_2$ oxides (M = Si, Ge, Sn), which, because of their dielectric and mechanical properties, can be used as protective sheaths of nanotubes and nanowires, and also as elements of nanophotonics.

It is known that the MO$_4$ tetrahedron with an M atom at the center and oxygen atoms at vertices is the basic element of all MO$_2$ structures (amorphous and crystalline). Such dioxides are dielectrics and semiconductors and can exist in the nature. As distinct from the dioxide structures studied previously, consider a three-atomic layer in which the tetrahedra are arranged in such a way that the oxygen and M-atoms are located only in the three planes of this layer. Calculations performed showed the stability of such MO$_2$ layers. The basic structural features of the MO$_2$ nanotubes are demonstrated by calculations of nanotubes rolled up of Si-dioxides (nanotubes composed of Ge- or Sn- dioxides demonstrate similar features).

CLASSIFICATION OF NANOTUBES

A single-layer (n,m) nanotube will be determined as a layer of a rectangular network of SiO$_2$ atoms rolled up into a cylinder (Fig. 1).

CP786, *Electronic Properties of Novel Nanostructures*, edited by H. Kuzmany, J. Fink, M. Mehring, and S. Roth
© 2005 American Institute of Physics 0-7354-0275-2/05/$22.50

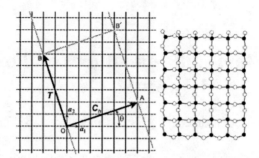

FIGURE 1. Nonrolled rectangular lattice of MO_2 nanotubes. A nanotube can be formed by connecting points O and A and also B and B'. Vectors OA and OB determine the chiral vector C_h and the translation vector T of the nanotube. The rectangle OAB' B determines the unit cell with the number of atoms N for the particular nanotube. The figure corresponds to C_h = (6, 2), T = (–2, 6), and N= 120. On the left: a fragment of the SiO_2 layer (5x6 cells), where the O(o)-atoms arranged along a_1 come out of the layer of Si atoms forward, and O(i)-atoms arranged along a_2 come out of the layer backward.

The M-atoms are located in its basic plane, and the oxygen atoms that come out of the plane forward (O(o)) or backward (O(i)) are located on lines parallel to the a_1 or a_2 unit vectors. In this ideal model, the tube diameter is tightly related to the lattice constant a, the distance between the nearest M-atoms, and to the numbers (n,m) [3].

The translation vector T determined as a unit vector of the nanotube under consideration is parallel to its axis and normal to the C_h vector in the non-rolled rectangular lattice in Fig. 1.The MO_2 nanotubes are formed from three cylinders: the O(o) atoms are located at the outer cylinder and the O(i) atoms are located at the inner cylinder with respect to the median cylinder with M(Si) atoms (see Figure 2).

(a)   (b)

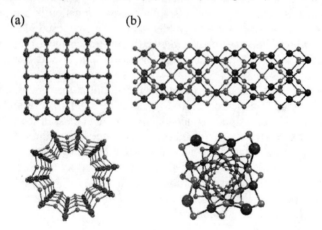

FIGURE 2. Models of (a) "linear" SiO_2-tube (10,0) and (b) "zigzag" SiO_2-tube (4,4).

The MO_2 nanotubes can be divided by the main symmetry differences into non-chiral $(n, 0)$ and $(0, n)$ tubes and chiral (n,m) tubes [3]. These tubes differ from the previously studied nanotubes composed of hexagonal layers: in their case, $(n, 0)$ tubes do not coincide with $(0,n)$ tubes, (n,m) tubes do not coincide with (m,n) tubes, and (n,n) tubes are chiral. For the SiO_2 and SnO_2 materials under consideration, layer deformation is significant, which leads to a dependence of the strain energy on the tube diameter differing from a $1/D^2$ behavior. Note that mixed nanotubes can also exist that can be obtained from layers with a changed arrangement of the O(i) and O(o) atoms by layers or even with a disordered arrangement of MO_4 tetrahedra in a layer [3].

METHOD

Primary geometry construction was performed by molecular mechanic MM+ and PM3 methods. The binding geometry and binding energy, and resulting electronic structure of SiO_2-tubes have been calculated by using the pseudopotential plane wave method within the local density approximation (VASP 4.4.4) [4]. The electronic structures were calculated with high k-point sampling of 33 irreducible k-points. A Gaussian smearing for the occupations was used with a smearing width of 0.1 eV. Geometry optimization was carried out until all the forces acting on all atoms decreased below 0.05 eV/Å.

GEOMETRY STRUCTURE AND ENERGY CHARACTERISTICS

The difference of radii between Si-cylinder and inner O-cylinder (outer O-cylinder and Si-cylinder) tend to distance between corresponding atomic planes in planar layer. Bond lengths along and across tube axis are distinguish. Bond lengths decrease and tend to bond lengths in plane structure (see Table 1).

TABLE 1. Main geometric characteristics for various SiO2-tubes

(n,m)	R (Å)			$\Delta(R_{Si\text{-}O(i)})$	$\Delta(R_{O(o)\text{-}Si})$	Bond (Å) Si-O		Bond (Å) Si-Si	
	O(i)	Si	O(o)			Along	Across	Along	Across
(6,0)	2.93	3.34	4.19	0.41	0.85	1.65	1.61	2.81	3.20
(8,0)	3.43	4.12	4.97	0.69	0.85	1.65	1.62	2.83	3.14
(10,0)	4.23	5.02	5.87	0.79	0.85	1.65	1.62	2.82	3.10
(15,0)	6.42	7.24	8.10	0.82	0.86	1.65	1.63	2.82	3.00
(3,3)	1.59	2.02	2.92	0.44	0.90	1.68	1.64	2.65	2.64
(4,4)	2.03	2.64	3.51	0.61	0.88	1.68	1.63	2.67	2.68
Plane	∞	∞	∞	0.87	0.87	1.64		2.78	

"Linear" (n,0) nanotubes are energetically advantageous than (0,n) and "zigzag" tubes. Lowest energy has "linear" (6,0) tube:

E (eV/atom)	-7.814	-7.813	-7.806	-7.789	-7.775	-7.427	-7.518	-7.193
(n,m)	(6,0)	(8,0)	(10,0)	(15,0)	(18,0)	(3,3)	(4,4)	(0,10)

Because Si-atom is situated in tetrahedral environment, more stable structure will be with closer parameters of configuration to ideal tetragon (see Table 2).
These nanotube can be prepared by using carbon nanotubes as a template [5].

TABLE 2. General angles for various SiO$_2$-tubes

(*n,m*)	Angles (degrees)		
	O$_1$-Si-O'$_1$	O$_2$-Si-O'$_2$	O$_1$-Si-O'$_2$
(6,0)	117.197	108.737	107.887
(8,0)	117.542	108.454	107.77
(10,0)	117.307	108.977	107.778
(15,0)	117.388	111.113	107.088
(3,3)	132.958	123.05	102.321
(4,4)	129.965	118.634	105.523
Plane	116.286	116.252	105.681

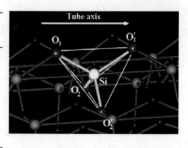

ELECTRONIC STRUCTURE

All tubes are dielectrics (Fig.3). Nanotube energy gap tends to the gap of plane structure with tube radius increasing. "Linear" tubes have indirect optical transition (Fig.3a) whereas "zigzag" tubes have direct optical transition given k=0 (Fig. 3b).

Because tubes haven't symmetry center they must have piezoelectric properties.

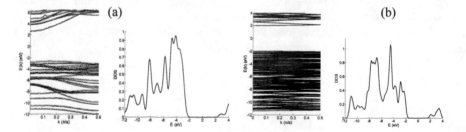

FIGURE 3 Band structure along k and density of states (DOS) for (a) linear (6,0) and (b) zigzag (4,4) tubes. The Fermi level is set as the zero point of the energy scale.

ACKNOWLEDGMENTS

This work was done within the RFBR grant N 05-02-17443 and DFG grant 2004. We appreciate the Institute of Computational Modeling SB RAS for opportunity to use the cluster computer for the calculations.

REFERENCES

1. *Carbon Nanotubes: Synthesis, Structure, Properties, and Applications*. Eds. M. S. Dresselhaus, G.Dresselhaus, Ph. Avouris, Topics in Applied Physics, **80**, Springer, Berlin, 2001.
2. Flikkema, E., Bromley, S. T., *J. Phys. Chem.* B **108**; 9638 (2004). S. T. Bromley *Nano Lett.* **4**, 1427 (2004).
3. Chernozatonskii, L. A., *JETP Lett.* **80**, 628 (2004).
4. Kresse, G., Hafner, J., *Phys.Rev.* B **47**, *558* (1993); *Phys.Rev.* B **49**, *14251* (1994); G. Kresse, J. Furthmüller, *Phys.Rev.* B **54**, *11169* (1996).
5. Colorado, Jr. R. and Barron A.R. Chem.Mat. **16** 2691 (2004)

Synthesis and structure of highly doped CN_x and CN_xB_y multiwalled nanotubes

S. Enouz[1,2], M. Castignolles[1,2], M. Glerup[3], O. Stephan[4], A. Loiseau[1]

[1] LEM, Onera-Cnrs, 29 Avenue de la Division Leclerc, BP 72, 92322 Châtillon, France
[2] LCVN, Université de MontpellierII, 34095 Montpellier, France
[3] Department of Chemistry, University of Oslo, 0315 Oslo, Norway
[4] Laboratoire de Physique Solides,CNRS UMR8502, Université Paris-Sud, 91405 Orsay, France

Abstract. Multiwalled nitrogen- doped carbon nanotubes (CN_x) with N ~ 20 at% in average and boron+nitrogen- doped carbon nanotubes (CN_xB_y) with N ~ 17 at% and B ~ 9 at% in average have been synthesized using a spray pyrolysis method. Degrees of purity, homogeneity and yield have been estimated from a systematic inspection of the samples by scanning electron microscopy. Combined analyses with Electron Energy Loss Spectroscopy (EELS) and High Resolution imaging modes of transmission electron microscopy, and spatially resolved EELS measurements in a STEM have been performed in order to correlate spectroscopic information with structural features at the nanometer scale.

INTRODUCTION

The electronic properties of pure carbon nanotubes vary between semi-conducting and metallic, with a band gap close to 1eV, depending on their chirality [1]. Boron nitride nanotubes possess the same hexagonal structure, with boron and nitrogen atoms on alternate lattice sites and theoretical investigations a predicted wide band gap close to 5.5eV, which is supposed to be independent on the tube diameter and chirality [2]. Up to now it is still difficult to tune the electronic properties by controlling the geometrical structure of the nanotubes. In this context, doping nanotubes with other chemical elements that can introduce donor or acceptor states near the Fermi level is supposed to be a promising approach for tailoring the electronic properties of nanotubes as a function of the composition. An efficient and well-controlled synthesis of such heteroatomics nanotubes is still a key-issue.

EXPERIMENT

Synthesis were carried out using the aerosol method as described elsewhere [3]. Its principle consists in spraying, in a hot furnace, liquid droplets of a mixture made of precursors of metallic catalysts and of carbon, nitrogen and boron sources. This method has the advantage to provide a population of metallic particles uniform in size, to favor this reaction with C, N and B reactants and to easily tune the proportion of doping elements. The synthesis conditions for the formation of CN_x tubes are given in Ref. [4, 5]. Carbon, nitrogen and boron sources were tetrahydrofuran, acetonitrile and

CP786, *Electronic Properties of Novel Nanostructures*, edited by H. Kuzmany, J. Fink, M. Mehring, and S. Roth
© 2005 American Institute of Physics 0-7354-0275-2/05/$22.50

boran pyridinic complex solvents and the metal catalyst precursor used was iron acetylacetonate. The flowing gas used for forming the aerosol spray is a well-defined mixture of hydrogen and argon. All reactions were carried out at 950°C. The nanotubes were ultrasonically dispersed in absolute ethanol before preparing the grids for electron microscopy studies. A high-resolution transmission electron microscopy (4000FX, 400/200kV) equipped with a Gatan model 666 parallel detection electron spectrometer was used for correlating the structure and the C, N and B ratio on individual nanotubes. Spatially resolved chemical and bonding analysis by EELS were carried out in a STEM VG-HB501 with a field emission gun operated a 100 kV. Such a STEM instrument delivers a 0.5 nm electron probe of high brightness and offers a 0.5 eV energy spread [6]. This provides appropriate conditions to record spectroscopic information at the nanometer scale on individual nanostructures in a mode that consists of the acquisition of one EEL spectrum for each position of the probe over a 2D region on the nanotube.

RESULTS AND DISCUSSION

Nitrogen doped multiwalled carbon nanotubes (CN$_x$) [7]

SEM and TEM studies revealed that the products obtained are carpets of several micrometers of aligned multiwalled nanotubes with a high degree of purity. The products mainly consist of two types of nanotubes (labeled I and II) as shown in figure 1a,b. The most prevalent tubes (type I) in the sample have a very frequent and regular compartmentalized morphology with a diameter of 20-60 nm. The other type of tubes (type II) observed is straighter with an unusual large inner diameter. They also have a bamboo-like structure, but with much longer and uneven compartments. EEL spectra measured on many individual nanotubes revealed well defined π^* and σ^* fine structure features which are evidences of sp^2-hybridisation in a graphitic structure and revealed a strong correlation between the nitrogen concentration and the nanotube morphology. Indeed, according to the EEL spectra (figure. 1c), type I tubes have an averaged nitrogen concentration ranging from 1 to 10at% whereas the type II exhibits a nitrogen concentration typically ranging from 10 to 25 at% (figure 1c) and nitrogen concentration can exceed 30 %at locally. The possibility of reaching very high nitrogen doping level in MWNT was confirmed by C. Tang et al. [8] who nearly reached the same nitrogen concentration by an aerosol assisted CVD synthesis method. One can assume that the pre-existing C-N bonds in the precursor mixture might be responsible for the high N concentration reached.

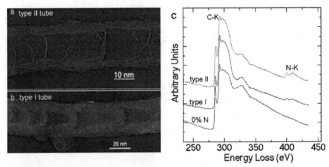

FIGURE 1 a, b HRTEM images of the two types of CN_x tubes. c) The corresponding EEL core electron K-shell spectra for the energy regions of carbon (280 eV - 340 eV) and nitrogen (399 eV - 425 eV), included is an EEL spectrum of an undoped MWNT.

Independently of doping concentration and morphology of the tubes, spatially resolved spectroscopy demonstrated that nitrogen is mainly located in the inner layers of the tubes and layers closing the compartments whereas outer layers are nearly nitrogen free (figure 2). This inhomogeneous distribution of nitrogen in CN_x tubes has earlier been observed [9, 10]. Furthermore, close inspection of fine structure of N and C absorption edges has revealed a dependence of the nature of C-N bonding to the local concentration in nitrogen. For lowest concentrations, bonding remains graphitic whereas for highest ones, it shows spectroscopic signatures close to pyrrolic or to intermediate situations corresponding to pyridinic environments.

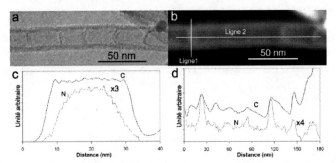

FIGURE 2 a, Bright field and b) dark field images of a type II tube. c),d) Concentration profiles for nitrogen and carbon each constructed from 128 EEL spectra.. The white lines indicate where the two measurement series have been carried out. c) measured across a nanotube on the bamboo walls (line 1), d) measured along the nanotube (line 2).

Boron and nitrogen doped multiwalled carbon nanotubes (CN_xB_y)

Structural characterizations by HRTEM and chemical investigations by EELS show multiwalled and compartmentalized boron and nitrogen doped carbon nanotubes as illustrated in figure 3a. The nanotubes exhibit heterogeneous and bamboo-like structures with diameters varying between 20-70 nm. Analysis of the EEL spectra was carried out for quantifying the content and the spatial distribution of the relevant chemical elements (carbon, boron and nitrogen). The concentration was determined

363

from the area of the characteristic signals after background substraction. The involved cross-sections were calculated applying the usual hydrogenic model for the C-K, B-K, and N-K edges. The nanotubes have an average N concentration ranging from 1 to 17 at% and an average B concentration ranging from 1 to 9 at%. The overall near edge fine structures confirm that the network is graphitic with the sp^2-type configuration. Scanning mode was used to get a series of spectra when scanning the incident electron probe across the CN_xB_y nanotube with a subnanometer step as illustrated in figure 3b. Each spectrum covers the three K edges with 0.5 eV energy dispersion (figure 3c). Figure 3d shows elemental profiles of C, B and N across the nanotubes extracted from these spectra. These profiles as well as the spectra reveal that outer layers are BN-rich whereas inner layers and compartments are N-rich as in CN_x nanotubes.

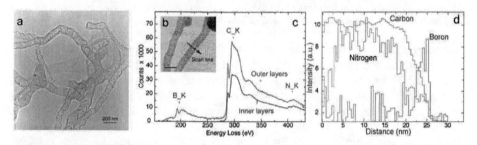

FIGURE 3 a, TEM image of CN_xB_y nanotubes b) Bright field image of a doped tube. The black arrow indicates where the measurements have been carried out. c) Corresponding EEL spectra of the inner and outer layers of the doped tube. d) Concentration profiles for C, B and N each constructed from 64 EEL spectra taken along the black arrow (b).

From the results discussed above, we can conclude that the incorporation of nitrogen and boron within the graphitic network takes place during the growth and not by atomic substitution. This must strongly relate to the solubility/diffusivity difference of carbon, nitrogen and boron in the used iron catalyst particle, which assists the nucleation of the tube.

CONCLUSION

Nitrogen- and nitrogen+boron- doped multiwalled carbon nanotubes can be produced by spray pyrolysis method with unprecedented high concentration. Furthermore, inhomogeneous incorporation of N and B elements into the graphitic structure has been for the first time achieved and will be studied in more details in order to have a better understanding of the influence of boron and nitrogen on CN_x and CB_yN_x nanotube structures and to better identify the chemical processes involved during the growth mechanisms.

ACKNOWLEDGMENTS

We are grateful to Dr. Gilles Hug and Dr. Lionel Bresson (LEM-ONERA, Châtillon) for fruitful discussions. This work has been done in the framework of the GdRE n°2756 "Science and applications of nanotubes", NANO-E. ONERA and ARKEMA (TOTAL) group are acknowledge for financial support.

REFERENCES

1   N. Hamada, S. Sawada, A. Oshiyama, *Phys. Rev. Lett.* **68**, 1579 (1992)
2   X. Blase, A. Rubio, S.G. Louie, M.L. Cohen *Europhys. Lett.* **28**, 335 (1994).
3   M. Glerup, H. Kanzo w, R. Almairac, M. Castignolles, P. Bernier, *Chem. Phys. Lett.* **377**, 293-298 (2003)
4   M.Castignolles A.Loiseau S.Enouz P.Bernier et M.Glerup, *Mat. Res. Soc. Symp. Proc.* **772**, (2003)
5   M. Glerup, M. Castignolles, M. Holzinger, G. Hug, A. Loiseau, P. Bernier, *Chem. Commun.*, 2542–2543 (2003)
6   O. Stephan, P.M. Ajayan, C. Colliex, P. Redlich, J.M. Lambert, P. Bernier, P. Lefin, *Science* **266**, 1683 (1994)
7   M. Castignolles, O. Stephan, M. Holzinger, A. Loiseau, P. Bernier M. Glerup, *To be published.*
8   C. Tang, Y. Bando, D. Golberg, F. Xu, *Carbon* **42**, 2625-2633 (2004)
9   W-Q Han, P. Kohler-Redlich, T. Seeger, F. Ernst, M. Rühle, *Appl. Phys. Lett.* **77**, 1807 (2000)
10 S. Trasobares, O. Stephan, C. Colliex, W.K. Hsu, W.W. Kroto, D.R.M. Walton, *J. of Chem. Phys.* **116**, 8966 (2002)
11 H. Sjöström, S. Stafström, M. Boman, J.-E. Sundgren, *Phys. Rev. Lett.* **75**, 1336 (1995)

Towards of Vanadium Pentoxide Nanotubes and Thiols using Gold Nanoparticles

V. Lavayen[2], G. Gonzalez[1], G. Cardenas[3], C. M. Sotomayor Torres[2]

[1]Departamento de Química. Facultad de Ciencias, Universidad de Chile, Po.Box. 653, Santiago, Chile.
[2] University College Cork, Tyndall National Institute, Lee Maltings, Prospect Row, Cork, Ireland.
[3] Departamento de Polímeros, Facultad de Ciencias Químicas, Universidad de Concepción, Po.Box. 160-C, Concepción, Chile.

Abstract. The template-directed synthesis is a promising route to realise 1-D nanostructures, an example of which is the formation of vanadium pentoxide nanotubes. In this work we report the interchange of long alkyl amines with alkyl thiols, this reaction was followed using gold nanoparticles prepared by the Chemical Liquid Deposition (CLD) method. The diameter of the gold clusters was 9 Å with a stability of about 85 days. SEM, TEM, EDAX and electron diffraction was the techniques used for the characterization of the reactions.

Keywords: Gold nanoparticles, vanadium pentoxide nanotubes, surfactants.
PACS: 78.67.Ch, 78.67.Bf

INTRODUCTION

The template-directed synthesis is a promising route to realize 1-D nanostructures, because it offers the possibility to design and build nanotubes and nanowires with specific electronics properties [1]. This is demonstrated by the increasing body of work describing the formation and morphology of nanotubes prepared with this approach [2-4].

In this work, we report the interchange of primaries amines in vanadium pentoxide nanotubes (VOx-NTs) with long alkyl thiols. This reaction was followed with the incorporation of gold nanoparticles in the nanotubes, which was prepared by Chemical Liquid Deposition (CLD) method [5]. The products are characterized by SEM, TEM, EDAX and electron diffraction. Indeed this work demonstrates that VOx-NTs can be functionalized with thiols and gold nanoparticles.

EXPERIMENTAL PART

Synthesis of vanadium pentoxide nanotubes (VOx-NTs): The method employed for the synthesis of vanadium pentoxide nanotubes was essentially the one described in references [3,7]. The preparation of Au-acetone colloid, was described elsewhere [8].

CP786, *Electronic Properties of Novel Nanostructures*, edited by H. Kuzmany, J. Fink, M. Mehring, and S. Roth
© 2005 American Institute of Physics 0-7354-0275-2/05/$22.50

Incorporation of gold Nanoparticles in VOx-NTs: 6 mg of the $V_2O_5(HDA)_{0.34}$ (VOx-Nts) were refluxed with 3 ml to *n*-dodecanothiol in 15 ml of ethanol anhydrous, (fraction distillation in ethanol 98% in $CaCl_2$) for 2 h. After that, we added the colloidal solution with gold (Au-nanoparticles) in acetone (average of the size of particles 9 Å). The system was then agitated for 12 h. In this case, the solvent was added in excess to ensure the solvatation of the metal atoms. During the warm-up to room temperature, clustering of metal atoms developed slowly [8].

Characterization of Samples:

The morphology of the vanadium nanotubes was measured using Scanning Electron Microscopy (SEM) with a Phillips XL-30 model equipped with EDAX detector, LV-SEM JSM-5900LV equipped with a Noran Voyager EDS detector and the chemical compositions of the samples were determined by elemental chemical analysis (SISONS model EA-1108). Transmission electron micrographs were obtained using equipment JEOL JEM 1200 EXII with electron diffraction

RESULTS AND DISCUSSION

The products of the hydrothermal treatment consist mainly of nanotubes; we found differences in the yield respect to report in the Ref [2,3], because our procedure of synthesis was different as the one reported there. The dimensions of the tubes were in outer diameter (60-150 nm) and the length of the tubes was up 2 μm [7].

We found that the concentration of the purple colloidal obtained was 4×10^{-3} M. The average of the particles size (15 Å) also was found using (TEM) see Figure 1. High grade of crystallinity was found using electron diffraction (ED). The stability of the colloidal suspension was 85 days, in agreement with the reported synthesis of metallic colloids reported by G. Cardenas *et al* [5,8].

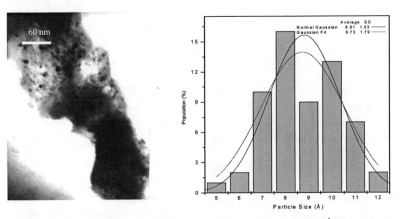

Figure 1. TEM image of gold nanoparticles in acetone and histogram (Au: 4×10^{-3} Mol/L)

We chose the acetone as the solvent in the gold nanoparticles because these present a higher solvatation (high stability in function of the time), compared with other kind of metals (all obtained using the same method and solvent) like Mn [8], Ni [6]. We found that the particle sizes in this colloid are smaller than those reported in [6,8]. Also the metallic colloid Au-nanoparticles/acetone exhibits a zeta potential of ζ= -40 mV solid.

We have mixed the thiol (dodecanothiol) and VOx-NTs in stoichiometrics relations (thiol:amine 4:1) in a reflux reaction. Using electron diffraction (ED) in the product, we found several interlaminar distances, these were measured and compared with references [2,3], e.g. the distance of the plane 003 is 10 Å. We can observe also the presence of one centre symmetric phase (vanadium oxide), see Table 1 and Figure 2.

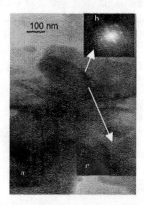 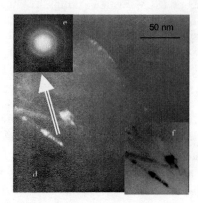

Figure 2. TEM image of VOx-NTs after the reflux reaction with thiols (**a**); magnification of some area of this image in the below right part (**c**); ED pattern insert in the top right part (**b**). TEM image in dark field of VOx-NTs with thiols and gold nanoparticles (**d**); ED pattern insert in the top left part (**e**) and bright field image in the bottom right of the same sample (**f**).

TABLE 1. Phase Indexation in the Compound $V_2O_5(HDA)_{0.3}$ with n-dodecanothiol.

Phase	d_{hkl} [Å]	Diffraction plane
VO_x-NTs	10.72	003
VO_x-NTs	2.69	210
VO_x-NTs	1.88	310
VO_x-NTs	1.30	420

We observe using TEM in the bright-field mode the sample contained VOx-NTs with thiol and gold nanoparticles (see Figure 2), in this mode we can not appreciate very well the presence of the particles, but using the dark-field mode we can observe in some regions brilliant points (metallic particles).

These regions correspond to the area where gold nanoparticles are present; for instance, using electron diffraction we can observe the presence of two phases in the system, one contained VOx-NTs-thiols (centre symmetric group) and the other gold nanoparticles, see Figure 2, and Table 2. With all this information we know that the gold nanoparticles are present in the system (on the surface or insert in the tubes), but due to the high aurophilicity of the thiols, we can suppose that the areas where the nanoparticles are present, the thiols molecules are present too.

TABLE 2. Phase Indexation of the Compound $V_2O_5(HDA)_{0.3}$ with *n*-dodecanothiol/gold particles.

Phase	d_{hkl} [Å]	Diffraction plane
Au	3.14	111
Au	2.21	200
Au	1.25	311
VO_x-NTs	25.14	---

The question now is in which zone of the tubes the nanoparticles are present, to clarify that, we analyzed some areas using Energy Dispersive Spectroscopy (EDS) in the surface mode. We observe that only a small fraction of gold and sulphur are in the surface of the tubes (spectrum not shown).

We conclude that most of the gold nanoparticles with thiols molecules are inside of the tubes and not on the surface of the tubes, also the spots to ED show the same kind of phase (Tables 1, 2) in the VOx-NTs before and after the interchange reactions (amine/thiol), that means, the tubes have not collapsed because the phases will swap positions, e.g. the indexed planes and the centre symmetric groups will change.

ACKNOWLEDGMENTS

This work was supported by FONDECYT (Project 1010891), the Postgraduate scholarship of the Chilean University and the Science Foundation Ireland. VLJ is grateful for the Deutscher Akademischer Austauschdienst doctoral scholarship. We are grateful to Antonio Zarate for the measurements in the electronic microscopy in the Brazilian Synchrotron Light Laboratory (LNLS).

REFERENCES

1. Special Issue *Adv. Mater.* **15**, 5 (2003).
2. F. Krumeich, H.-J. Muhr, M. Niederberger, F. Bieri, B. Schnyder, R. Nesper. *J. Am. Chem. Soc.* **121**, 8324 (1999).
3. H.-J. Muhr, F. Krumeich, U. P. Schönholzer, F. Bieri, M. Niederberger, L. J. Gauckler, R. Nesper. *Adv. Mater.* **12**, 231, (2000)
4. F. Bieri, F. Krumeich, H-J. Muhr, R. Nesper, *Helvetica Chimica Acta*, **84**, 3015 (2001).
5. G. Cárdenas. T., C. Retamal. C., K.J. Klabunde. *J. Appl. Polym. Sci. Polym.* Symp. 49 15 (1999).
6. G. Cárdenas, R. Segura. *Mat. Res. Bull.* **35**, 1369 (2000).
7. V. Lavayen, M. A. Santa Ana, J. Seekamp, C. Sotomayor Torres, E. Benavente, G. González. *Mol. Cryst. and Cryst Liq.* **416** 49 (2004).
8. G. Cárdenas, R. C. Oliva. *Mat. Res. Bull.* **35**, 2227 (2000).

The influence of annealing on transport properties of MoSIx nanowires

M. Uplaznik[1], B. Bercic[1], J. Strle[2], M. Ploscaru[1], M. Rangus[2], A. Mrzel[1], P. Panjan[1], D. Vengust[3], B. Podobnik,[4] D. Mihailovic[1,3]

[1]Jozef Stefan Institute, Jamova 39, 1000 Ljubljana
[2]Faculty of Mathematics and Physics, Jadranska 19, 1000 Ljubljana
[3]Mo6 d.o.o., Teslova 30, Tehnoloski park, 1000 Ljubljana
[4]LPKF Laser & Elektronika d.o.o., Planina 3, 4000 Kranj

Abstract. Results of a systematic set of conductivity measurements on inorganic MoSIx nanowires are reported. Different contact methods, including AC electrophoretic trapping, were compared. The influence of annealing on contact properties between bundles and metal is investigated and the conductivity is found to differ by several orders of magnitude. Annealing was performed also on bulk samples for comparison with single bundle measurements. We also compared x-ray spectra of the samples before and after annealing treatment to investigate possible phase transformations of the materials. Results of conductivity measurements using Pt/Ir coated AFM tips as one contact and lithographically manufactured electrode as another are also presented.

Keywords: inorganic nanowires, transport measurement, contact conductance, percolation conductance, SPM tip conductance measurement
PACS: 73.63.Fg, 73.63.Rt

INTRODUCTION

Inorganic nanowires present an important functional alternative to their carbon relatives. Both are seen as important milestones in the evolution of nanotechnology for they are one of the first materials that could be used in applications. To find the field of their application along with the proper treatment procedures and protocols it is necessary to determine their basic physical and chemical properties. The extraordinary intrinsic solubility of inorganic MoSIx nanowires in different solvents including water and alcohol lead us to the production of self-assembled devices using dielectrophoretical forces.

The aim of our work was to systematically investigate fundamental electrical properties of devices that included single bundles of inorganic MoSIx nanowires. We were particularly interested in interface properties between the metal and the bundle in the circuit. To elucidate the quality of the contacts we compared two different methods of introducing the contacts on the bundles and also tested the conductance as a function of annealing temperature.

CP786, *Electronic Properties of Novel Nanostructures*, edited by H. Kuzmany, J. Fink, M. Mehring, and S. Roth
© 2005 American Institute of Physics 0-7354-0275-2/05/$22.50

EXPERIMENTAL

a) Lithographic Two point Measurement Devices

For electronic conductivity measurements of single bundles we manufactured microdevices via e-beam lithography on a silicon wafer with a 600 nm oxide layer on top. The titanium contacts were 10 nm thick, 2 microns wide and 1 micron apart.

Our samples were bundles with diameters between 10 nm and 300 nm and were up to 5 microns long. Single bundles were attached to the contacts via dielectrophoretic force induced by a 10 MHz electric field applied to the two contacts. We obtained the conductivity with measurement of IV curves using a Keithley 2000 electrometer.

b) Conductive AFM Measurements

As an alternative to measurement in device configuration we developed a different method for making metal contacts on a single bundle. We produced metal contacts on a silicon wafer covered randomly with sample bundles using the same lithographic technique as for measurement circuits discussed above. Due to their high density, the possibility of producing bundles partly embedded under the contact metal was high so that they could be located with an AFM. Selected bundles were then pinned with the conductive AFM tip, creating the second contact of the measurement circuit (Fig.1b and the cross section in Fig.3a). The data was gathered with an AFM DI3100 microscope in conductive AFM mode.

c) Bulk sample measurements

Bulk measurements were performed on a pressed pellet with 3 mm diameter. Golden wires with 33 microns diameter were attached with silver paste on the surface of the sample creating closed circuit with the measurement instrument Keithley 2000.

Annealing treatment was performed in vacuum for single bundles and in argon atmosphere for bulk samples at 700 ^{0}C for one hour.

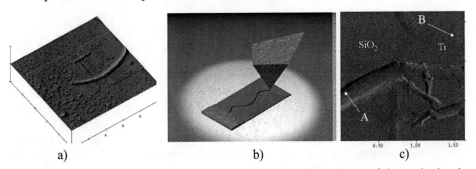

FIGURE 1: (a) A Picture of a bundle on titanium contacts. (b) A scheme of the conductive tip measurement setup. (c) An AFM picture of the bundle half imbedded under the titanium contact; spots A and B denote two different measurement points of the AFM tip.

RESULTS AND DISCUSSION

The measurements of single bundles with microdevices were performed on three types of materials with different stoichiometry: $Mo_6S_3I_6$, $Mo_6S_4I_4$ and $Mo_6S_2I_8$.

In order to make a comparison between different sized bundles we used the diameter and the length of the bundles extracted from AFM pictures to plot current density as a function of electric field, giving the conductance of the material.

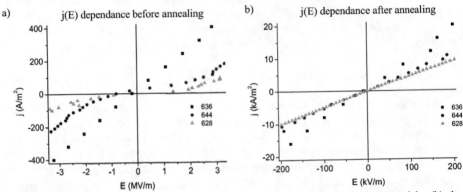

FIGURE 2: (a) The diagram shows current/voltage characteristic of as grown materials; (b) the diagram shows same dependence after annealing. The conductance increases by ~3 orders of magnitude after annealing.

TABLE 1. Single bundle conductance - dielectrophiretical two point contacts

Material	Before annealing [Sm^{-1}]	After annealing [Sm^{-1}]
$Mo_6S_3I_6$	$1.3\ 10^{-4}$	0.135
$Mo_6S_4I_4$	$3.7\ 10^{-5}$	0.057
$Mo_6S_2I_8$	$2.3\ 10^{-5}$	0.048

The results in Table 1 imply similar electronic properties for nanowires with different stoichiometry. This, together with linear characteristic of the curves on Fig2., suggests semimetallic behavior, where transport is limited by scattering.

More intriguing is the comparison of conductances before and after annealing which reveals up to three orders of magnitude higher conductance after the temperature treatment. To determine the influence of contact properties on that phenomena we have to compare this results with the measurements gathered with conductive AFM tip before annealing. In this experiment the tip was pressed on the half imbedded bundle, so that the current could flow from the tip, through the bundle towards the metal contact (Fig.3a). Measurements were performed as a comparison between the conductivity of the bundle (point A on Fig.1c) and of metal contacts (point b on Fig.1c). From the linear fit slopes of the results on Fig3b we find the conductivity of the bundle to be by a factor 50 lower then of the direct titanium contact. Using the size and the thickness of the bundle determined from the AFM image in Fig.1c, we calculated the specific conductance of our material. It was found to be 0.37 Sm^{-1}± 0.02. This value is closer to the results gathered with two point measurement devices (Table) for the samples after annealing, suggesting that the interface properties of the metal and the bundle changed significantly.

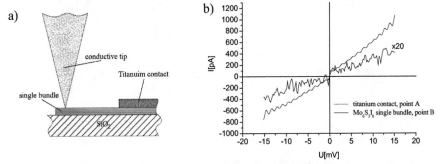

FIGURE 3: a) A schematic cross section of the CAFM experiment. b) A diagram of the conductance measurements of the titanium contact and the selected bundle from Fig1c.

Another cause for the large difference in conductance before and after temperature treatment is that a phase transformation occurs in the material. To test this we performed conductance measurements on bulk samples before and after annealing. The conductance differed up to two orders of magnitude. That could confirm our assumption, but for more definite answer x-rays were used to examine if any chemical changes occur upon annealing. A comparison in both spectra reveals two new peaks of small intensity, implying chemical change in the bulk material.

CONCLUSION

The presented measurements reveal interesting conductive properties of single bundle devices that depend strongly on annealing. This suggests that the interface properties of the bundles and the metal contacts contribute significantly on electric properties of the devices. We can conclude, that for any integration of inorganic MoSIx into electrical devices annealing treatment improoves the conductive properties of the devices.

ACKNOWLEDGMENTS

We would like to acknowledge Ministry of Education, Science and Sport of Slovenia for financial support.

REFERENCES

1. Vilfan, I., Mihailović, D.: *Atomic and Electronic Structure of Subnanometer Diameter Mo6S9- xIx Nanowires,* Internet adress: http://optlab.ijs.si/SpLABFiles/Nanotubes/MoS2.htm

2. Heim, T.: *Conductivity of DNA probed by conducting-atomic force microscopy: effects of contact electrode, DNA structure, surface interactions,* J. Appl. Phys. **96,** 2927-2936 (2004)

Self-assembly of gold particles to MoS$_x$I$_y$ nanowires ends

M. Ploscaru[1, 2], M. Uplaznik[1], A. Mrzel[1], M. Remskar[1], Sasa Jenko Kokalj[1], D. Turk[1], D. Vengust[1, 3] and D. Mihailovic[1]

[1]Josef Stefan Institute, Jamova 39, 1000 ljubljana, Slovenia,e-mail: mihaela.ploscaru@ijs.si
[2]National Institute for Laser, Plasma and Radiation Physics, P.O. Box MG-36, R-76900, Bucharest, Romania
[3]Mo6 d.o.o., Teslova 30, 1000 Ljubljana, Slovenia, www.Mo6.si

Abstract. We report here the first step towards functional derivates formed by self attachment of golden particles on MoS$_x$I$_y$ nanowires. The golden particles suspended in water solution reproducibly attach at the ends of nanowires. AFM, TEM measurements were performed in order to demonstrate the Au functionalized MoS$_x$I$_y$ nanowires. Attachment of thiol-containing proteins is also demonstrated.

Keywords: functionalization, attachment, nanowires
PACS: 73.63.Fg, 61.46. +w,

INTRODUCTION

The discovery of free-standing microscopic one-dimensional molecular structures, such as nanotubes of carbon nanotubes has attracted a great deal of attention because of their exceptional electronic and mechanical properties[1]. Despite all these qualities, CNT's have numerous disadvantages, such as bad dispersion characteristics in organic solvents or problematic separation of metallic from semiconducting CNT's.

Because of these inconveniences, the attention of researchers is now moving to other one-dimensional inorganic materials, like the family of nanowires made up of molybdenum, sulphur and iodine[2] (MoS$_x$I$_y$). The synthesis of these materials is a straightforward process in which the material is mixed in the desired stoichiometries. The resulting nanowires have monodispersed diameters, and can be easily dispersed in organic solvents into individual nanowires (Fig.1). The big advantage of MoS$_x$I$_y$ nanowires over CNT's is that are easily dispersed in organic solvents like isopropanol, dimethyl-formamide or chloroform[3].

Of recently interest is the functionalisation experiments of these novel materials, with the aim of exploiting new chemical interference between nanostructures and organic materials especially peptides via the bridges.

CP786, *Electronic Properties of Novel Nanostructures*, edited by H. Kuzmany, J. Fink, M. Mehring, and S. Roth
© 2005 American Institute of Physics 0-7354-0275-2/05/$22.50

1a.   1b.

FIGURE 1. SEM image of MoS_xI_y nanowires (1a) and HRTEM image of MoS_xI_y nanowires (1b)

Sample preparation

As synthesized MoS_xI_y nanowires (Fig. 1) were dispersed in isopropanol for 1 hour in a low power sonification bath. The concentration of the sample was 0.5 mg/ml. After sonification, the solution was left to settle down in order that the big bundles sediment out. The dispersion from the top of the solution was removed and analyzed with an atomic force microscope. The golden particles suspended in water were also analyzed with AFM, in order to find the right concentration which can be used and accurately measure the diameter.

In the end, the dispersed MoS_xI_y nanowires were mixed with the gold colloids in the proportion of 1:3. The mixture was left to react for 3 days and then studied with the AFM.

Experimental results

The solution was put on a mica substrate and then analyzed with DI Dimension 3100 AFM. Individual wires with golden particles at the ends were clearly visible (Fig. 2a and fig. 2c). The cross section of both the wire and the golden particle was measured. We find out that the wire has ~1±0.05nm diameter (Fig. 2b) and the golden particle has ~5±0.2 nm diameter (Fig. 2d).

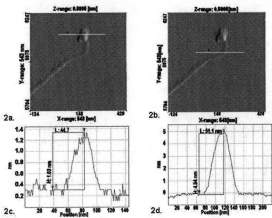

FIGURE 2   AFM images of MoS_xI_y nanowires functionalized with golden particle (2a and 2c) and cross section measurements of the individual wire (2b) and cross section measurements of the gold particle (2d)

375

These measurements give us the confirmation that the gold particle is attached at the end and not on the side of the nanowire. Further, TEM measurements confirmed the functionalisation of the wires with gold particle (Fig. 3a). From the structure of the MoS_xI_y nanowires we know that the individual wires are composed of Mo octahedra, separated by bridging anions (I or possibly S) shown in Fig. 3b. Moreover, it is well known from literature that Au is easily functionalized with the thiol group. The presence of S in the structure at the end of the wires contributes to the fact that the attachment occurs only at the end of the wires and not on the sides.

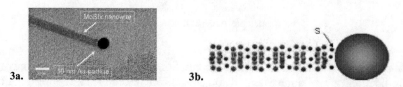

3a.   3b.

FIGURE 3   TEM image of gold particle attached at the end of the MoS_xI_y nanowires (3a.); Schematic image of an individual wire functionalized with 5 nm golden particle (3b)

This affinity for easily functionalisation with the gold particles is not shown by all the nanowires. We tried different kind of stoichiometries of MoS_xI_y nanowires and some of them were more successful than others (depends very much on the element that is present in the linking planes). So far $Mo_6S_3I_6$ and MoS_xI_y nanowires showed successful functionalisation.

The success of gold attachment to the MoS_xI_y nanowires straightens our attention to the field of biological applications. The next step was to try and functionalize proteins with the gold particle already attached on the MoS_xI_y nanowires. We used thyroglobulin, which is a dimeric iodinated and sulfonated protein, with a high molecular mass (660kDa). It is the major soluble protein of the thyroid gland. Thyroglobulin is stored in the lumen of the thyroid follicles as soluble dimmers and tetramers and insoluble multimers, formed through formation of disulfide and dityrosine bonds. This protein can be easily functionalized with Au particles due to its thiol groups. From the AFM measurements we observed that in fact the protein is attached direct on the MoS_xI_y nanowire and only after we have the functionalization with the gold particle (Fig. 4a and 4b).

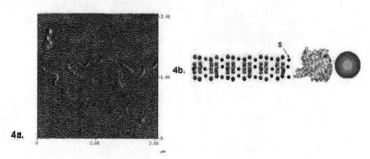

4a.

FIGURE 4 AFM image of the system MoS_xI_y nanowires-protein-gold particle (4a.); schematic image of MoS_xI_y functionalized with thyroglobulin which is protein attached with gold particles (4b.)

The cross section analyses (Fig. 5a and 5b) confirm that the protein which has a diameter ~2 nm it is attached on the 5 nm gold particle. Further experiments need to be done in order to understand better the statistical occurrence of these kinds of attachment.

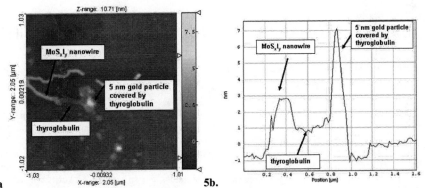

FIGURE 5 AFM image of the thyroglobulin protein attached on golden-MoS$_x$I$_y$ nanowire system (5a.); Cross section of the protein (5b)

Conclusions

We succeed to attach gold particle on MoS$_x$I$_y$ nanowires. The presence of the S in the linking planes in the structure facilitates the functionalisation of the gold particle at the end of the nanowires. We also succeed to functionalize the gold-nanowire system with proteins, results that can be used in biotechnology. The functionalisation research was made on several types of MoS$_x$I$_y$ nanowires but not all of them show this property.

Acknowledgments

We would like to acknowledge Ministry of Education, Science and Sport of Slovenia for financial support and also to NANOTEMP network. Acknowledges the help of Jonathan N. Coleman and V. Nicolosi for discussions made on solubility of Mo$_6$S$_x$I$_{9-x}$ nanowires.

References

1. R.H. Baughman, A.A. Zakhidov, W.A. de Heer, Science 297 (2002) 787
2. D. Vrbanic, M. Remskar, A. Mrzel, A. Jesih, P. Umek, M. Ponikvar, B. Jancar, A. Meden, B. Novosel, P. Venturini, J.N. Coleman, D. Mihailovic, Nanotechnology 15 (2004) 635
3. Valeria Nicolosi, D. Vrbanic, A. Mrzel, Joe McCauley, Sean O'Flaherty, D. Mihailovic, Werner J. Blau, Jonathan N. Coleman, Chemical Physics Letters 401 (2005) 13-18
4. M. Remskar, A. Mrzel, Zora Skraba, Adolf Jesih, Miran Ceh, Jure Demsar, Pierre Stadelman, Francis Levy and D. Mihailovic, Science 292, 479 (2001)
5. U. Berndorfer, H. Wilms, V. Herzog, J. Clin. Endocrinol. Metab. 81 (1996), 1918-26.
6. N. Baudry, P.J. Lejeune, F. Delom , L. Vinet, P. Carayon , B. Mallet , Biochem Biophys Res Commun. 242 (1998), 292-6.

Magnetic Properties of Nanometer-Sized Particles of the Superconductor $Mo_6S_6I_2$

M. Rangus[1], A. Omerzu[2], A. Mrzel[2], D. Vengust[2] and D. D. Mihailovic[1,2]

[1] Faculty of Mathematics and Physics, University of Ljubljana, Jadranska 19, 1000 Ljubljana, Slovenia
[2] Complex matter department, Jozef Stefan Institute, Jamova 39, 1000 Ljubljana, Slovenia

Abstract. We have conducted a study of magnetic field penetration in nanometer-sized particles of the superconductor $Mo_6S_6I_2$ (bulk $T_C = 13K$). Our aim was to obtain samples with dimensions above and below the superconductor's characteristic lengths, i.e. London penetration depth λ, correlation length ξ and Ginzburg-Landau parameter κ, which are typically *270 nm, 3 nm* and *79*, respectively in this class of materials. We find that the hysteresis in the magnetization curve *M(H)* widens with decreasing particle size, and secondly, the differences between zero-field and field cooled magnetization diminish as the particle size approaches the London penetration depth.

Keywords: nano-size superconductor, magnetic field penetration, critical fields.
PACS: 74.25.-q, 74.25.Ha, 74.78.Na

INTRODUCTION

$Mo_6S_6I_2$ is a superconductor which grows in the form of cubic single crystals. By grinding different size d particles can be obtained which allow systematic investigations of magnetic properties as a function of particle size d ranging from $d \approx \xi_0$ to $d > \lambda$. We find that the observed effects of H_{C1} and H_{C2} can not be understood in terms of standard Ginzburg-Landau theory.

EXPERIMENTAL

Samples of submicron sizes and well defined diameters ranging from *5 nm* to *1 μm* were obtained by grinding and were separated by size by sedimentation in a water dispersion. For the sedimentation we used a very long column and the process itself was very slow so we could obtain very narrow diameter distribution and consequently our particle size is very well defined. All samples were examined with AFM and SEM prior to magnetic measurements to determine particle size and diameter distribution (Figure 1). To avoid particle adhering, $Mo_6S_6I_2$ powder was mixed with GE-varnish and modeled into pellets. All magnetic measurements were made with Quantum Design MPMS magnetometer and for each sample we measured the magnetization dependence on the magnetic field at different temperatures (Figure 2).

CP786, *Electronic Properties of Novel Nanostructures*, edited by H. Kuzmany, J. Fink, M. Mehring, and S. Roth
© 2005 American Institute of Physics 0-7354-0275-2/05/$22.50

 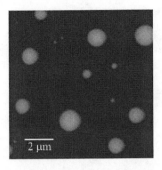

FIGURE 1. $Mo_6S_6I_2$ crystals as grown with diameters approximately *10 nm* (left) and sample with particle size *130 nm* (right). It can be seen that the particles were rounded by grinding and that their size distribution is quite narrow.

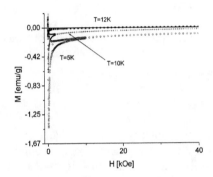 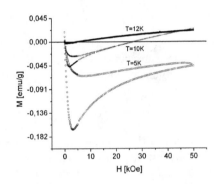

FIGURE 2. SQUID measurements of magnetization dependence on magnetic field at different temperatures. Magnetization curves look very different for smaller particle samples (for the *5 nm* sample in the right) than in the bulk (left).

RESULTS AND DISCUSSION

Lower Critical Field H_{C1}

The lower critical field H_{C1} was found to be highly dependent on particle size. Since particle diameters in our samples are in the same order of magnitude or smaller than the London penetration depth, H_{C1} is expected to change with particle size (Figure 3) according to [1,2]:

$$H_{C1} = \frac{\phi_0}{4\pi d^2} \ln \kappa .$$ (1)

Instead, rather unexpectedly, we observed that H_{C1} changes approximately as $d^{1/2}$ as shown by the dotted line in Figure 3.

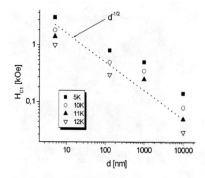 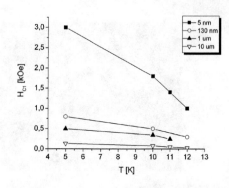

FIGURE 3. Lower critical field H_{Cl} changes with particle size as $d^{1/2}$ when the particles are smaller than the London penetration depth.

FIGURE 4. T-dependence of H_{Cl} for different particle size.

Upper Critical Field H_{C2}

When calculating the coherence length from the upper critical field H_{C2} obtained from our measurements, we took into account that it should change according to Ginzburg-Landau relations [1,3]:

$$H_{C2} = \frac{\phi_0}{2\pi\xi^2} \quad , \quad \xi = \xi_0 \left(1 - \frac{T}{T_C}\right)^{-1/2} . \tag{2}$$

The upper critical field stays almost constant with changing particle diameter and just slightly increases as we approach smaller diameters (Figure 5). This implies, according to Eq. (2) that the *coherence length* (which for the bulk is $\xi_0 = 3.4$ *nm*) *increases* with decreasing particle diameter and reaches the particle diameter in the *5 nm* particle sample (Figure 6). (*130 nm* and *1μm* samples had fewer measurements and this results in a larger error in the calculated coherence length.)

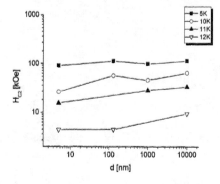 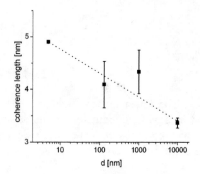

FIGURE 5. Upper critical field H_{C2} is roughly constant for different particle sizes and it varies only with temperature.

FIGURE 6. The coherence length ξ_0 calculated from Equations (2) was found to be dependent on the particle size and with *5 nm* particles it approaches their diameter.

Magnetization as a Function of Temperature

Magnetization as a function of temperature in zero-field cooled and field cooled samples has also been measured. We observed large differences for zero-field cooled and field cooled curves in samples with particles much larger than the London penetration depth ($\lambda=270$ nm). As the particle diameter becomes smaller than λ, the difference between ZFC and FC curves diminish (Figure 7).

FIGURE 7. The difference between ZFC and FC magnetization curves diminishes with particle size and are almost the same for particles with dimensions that are in the same order of magnitude than λ or smaller.

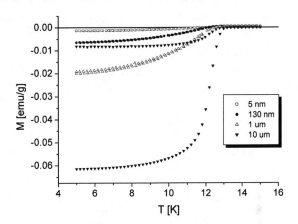

CONCLUSION

With the method described above we were able to obtain samples containing particles with well defined diameters. In particles smaller than characteristic lengths for $Mo_6S_6I_2$ superconductor our measurements show quite different mechanisms of magnetic field penetration. When the particle size is less than London penetration depth λ, the lower critical field H_{C1} scales as $1/d^{1/2}$ rather than $1/d^2$ as predicted. Unexpectedly [4], the coherence length deduced from H_{C2} is also highly dependent on the particle size and in 5 nm particle samples it approaches the particle diameter, which is not easily understood.

We have shown that nanoscale confined superconducting structures can have fundamentally different behavior than bulk material. Understanding the behavior of nanoscale superconducting particles is of fundamental interest as well as of great importance for the construction of nanoscale superconducting circuits and devices such as Josephson circuits and qubits.

M. Rangus would like to thank Jozef Stefan Institute for sponsoring our work and Center for Magnetic Measurements (*CMag*) for use of their magnetometer and Dr. V. V. Kabanov for useful discussions.

REFERENCES

1. Poole, C. P. Jr., Farach, H. A., and Creswick, R. J., *Superconductivity*, London: Academic Press Limited, 1995, pp. 265-342.
2. Bonca, J., and Kabanov, V. V., *Phys. Rev. B* **65**, 012509 (2001).
3. Buisson, O. et al., *Phys. Lett. A* **150**, 36-42 (1990).
4. Niu, H. J., and Hampshire, D. P., *Phys. Rev. Lett.* **91**, 027002 (2003).

Observation of Extremely Low Percolation Threshold in $Mo_6S_{4.5}I_{4.5}$ nanowire/polymer composites

Jonathan N. Coleman[*,a], Robert Murphy,[a] Valeria Nicolosi,[a] Yenny Hernandez,[a] Denis McCarthy,[a] David Rickard,[a] Daniel Vrbanic,[b] Ales Mrzel,[b] Dragan Mihailovic,[b] and Werner J. Blau[a]

[a] *Department of Physics, University of Dublin, Trinity College Dublin, Dublin 2, Ireland*
[b] *Jozef Stefan Institute, Jamova 39, 1000 Ljubljana, Slovenia*

Abstract. Homogenous composite films were fabricated from a polymer doped with $Mo_6S_{4.5}I_{4.5}$ nanowires. AC and DC electrical testing revealed percolative behavior characterized by an extremely low percolation threshold, $p_c=1.3\times10^{-5}$ and a maximum conductivity of 4×10^{-3} S/m. The low value of the percolation threshold can be explained by debundling of the nanowires at low concentration during sample preparation. This results in a homogenous dispersion of individual nanowires at low loading level.

INTRODUCTION

The enhancement in conductivity of insulating materials such as polymers and epoxies by the addition of a conductive filler is of critical importance for applications such as electromagnetic-interference shielding and electrostatic dissipation. This process, known as percolation, has been achieved with fillers such as conductive polymers, carbon black and graphitic fibers. Recently the filler-particles of choice have been carbon-nanotubes (CNTs) due to their high conductivity and aspect-ratio [1-3]. However their lack of solubility and propensity to aggregate are non-trivial problems.

Recently, a new form of inorganic nanowire made from molybdenum, sulphur and iodine, with stoichiometry $Mo_6S_{4.5}I_{4.5}$ has been reported [4-6]. Unlike CNTs, these nanowires exist as one electronic type (all metallic), are soluble in common solvents and tend to debundle in solution. This makes them ideal candidates as conductive fillers.

In this work we have measured AC and DC conductivity as a function of volume fraction for composites of $Mo_6S_{4.5}I_{4.5}$ nanowires in PMMA. We find an ultra-low percolation threshold of 1.3×10^{-5} and a conductivity of $\sim10^{-3}$ S/m by a volume fraction of 10^{-3}. These results compare well with the best results reported for CNT-based composites.

CP786, *Electronic Properties of Novel Nanostructures*, edited by H. Kuzmany, J. Fink, M. Mehring, and S. Roth
© 2005 American Institute of Physics 0-7354-0275-2/05/$22.50

EXPERIMENTAL METHODS

$Mo_6S_{4.5}I_{4.5}$ nanowires were produced by direct synthesis from elemental material mixed in the appropriate stoichiometries, as described in Ref. 4. Purified [5,6] nanowires were initially dispersed in dimethylformamide (DMF) at a concentration of 1mg/ml by high-power sonication. TEM studies showed that these nanowires were in the form of bundles with diameter and length distributions peaking at 40nm and 1200 nm respectively. In addition both distributions displayed a long tail to higher values.

Various volume fractions from 2.5×10^{-6} to 0.027 were prepared by blending this initial dispersion with PMMA (2.25mg/ml) solutions. For each volume fraction, the composite dispersion was drop-cast onto an 80nm pre-deposited gold electrode, on a glass substrate, forming a uniform film. These were then dried at 60°C to remove any residual solvent and six top gold contacts of thickness 80nm were evaporated onto each film. Reported conductivity measurements are averaged over the six electrodes. Film thicknesses were in the range of 3-4 μm as determined by Dektak surface profilometry. AC and DC measurements were performed using a Zahner IM6e impedance analyzer over a frequency range 1Hz-1MHz and a Keithley Model 2400 sourcemeter respectively.

RESULTS AND DISCUSSION

From the impedance spectroscopy measurements, the complex impedance ($Z*$) was obtained as a function of frequency for each sample. This allows the calculation of the electric modulus as a function of frequency, $\left|\sigma*(\omega)\right|$, which is defined as:

$$\left|\sigma*(\omega)\right| = \frac{1}{\left|Z*(\omega)\right|} \frac{d}{A}$$

(1)

where d and A are the film thickness and cross-sectional area respectively. Figure 1 shows the electric modulus versus frequency for a range of different nanowire loadings. The electric modulus for the PMMA sample is frequency dependent over the complete frequency range studied and has a slope of unity on a log-log scale, as expected for a pure dielectric. For the sample with highest nanowire volume fraction, the composite behaves as a pure conductor with no frequency dependence in electric modulus whatsoever. In between these extremes, the composites display both dielectric and conductive behavior with the latter dominating at low frequency. The AC conductivity, σ_{AC}, is taken from the low frequency part of this curve. The DC conductivity, σ_{DC}, is calculated from the slope of the current-voltage characteristics, which were Ohmic at all voltages.

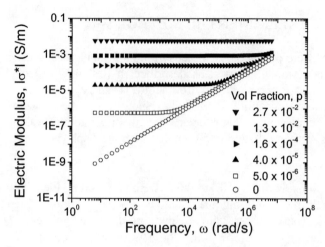

Figure 1. Frequency dependence of the electric modulus of composites with various nanowire volume fractions.

The AC and DC conductivities of the composites are shown together in Figure 2(a) as a function of nanowire volume fraction (p) and these values are in good agreement for each sample. A sharp increase in the conductivity, by over 8 orders of magnitude from the polymer itself, is observed over a very low volume fraction range. The maximum conductivity observed was 4×10^{-3} S/m at p=0.027. The sharp increase in conductivity indicates the formation of a percolating network [7] and is described above the percolation threshold, p_c, by:

$$\sigma = \sigma_0 (p\text{-}p_c)^t \qquad (2)$$

Figure 1 displays that the electric modulus appears to diverge from purely dielectric behavior by $p=5\times10^{-6}$ so an initial estimate for the percolation threshold was assumed to occur below this volume fraction. However, it was found by incrementally adjusting p_c and applying a linear fit to Eq. (2) (on a log-log scale) that a slightly higher value of $p_c=1.3\times10^{-5}$ provided the best fit using least squares regression as shown in Figure 2(b). This is an extremely low percolation threshold and compares favourably with the lowest threshold observed previously [1] of $p_c=2.5\times10^{-5}$ for a carbon nanotube doped system. A value of t=1.41±0.08 was obtained for the conductivity exponent which is in the range previously quoted for similar systems [1,2]. From the fit, σ_0=(81±58) S/m was obtained. It should be pointed out that a good fit is only obtained for $p\le1\times10^{-3}$. Above this value the experimental data deviates sharply from the fit.

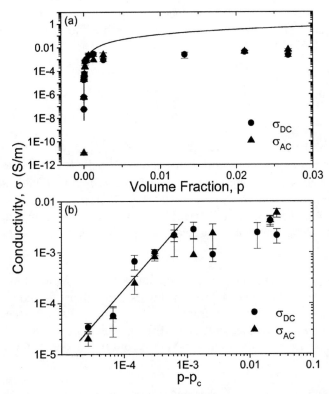

Figure 2. Log of AC and DC conductivities of (a) all composites studied in this work as a function of nanowire volume fraction. The solid line is a fit to Eq. (2); (b) those samples with volume fraction p above the percolation threshold p_c, plotted as a function of $\log(p-p_c)$, along with the percolation scaling law [Eq. (2)] for $p<1\times10^{-3}$.

These results raise questions about the nature of this low percolation threshold and the sharp divergence of the experimental data from the theoretical curve. For rod-like fillers, the percolation threshold can be approximated by the inverse aspect-ratio of the filler particles [8]. Applying this to the measured dimensions of the nanowire bundles used to prepare these samples, a value of $p_c\sim2.5\times10^{-2}$ would be expected. Clearly this conflicts with the measured value of $p_c=1.3\times10^{-5}$. However it has previously been observed [5,6] that dramatic debundling occurs when $Mo_6S_{4.5}I_{4.5}$ nanowires are dispersed at concentrations below $C_{NW}=1\times10^{-2}$ mg/mL. This concentration corresponds to a volume fraction of $p\sim1\times10^{-3}$ in this study (solution preparation phase). This debundling is manifested by a reduction in bundle diameter as the concentration is decreased [6]. Thus below $p\sim1\times10^{-3}$ debundling becomes important, resulting in a dispersion of individual nanowires of diameter 0.94nm and lengths ranging up to many microns at lower concentrations. Hence, this leads to a significant increase in aspect-ratio at lower concentrations which implies a much smaller p_c than predicted for large diameter bundles such as those found at higher concentrations.

Furthermore, according to Balberg [9], the presence of a distribution of lengths of rod-like fillers can reduce the percolation threshold compared to the case of identical filler particles. In this case

$$p_c = 0.7 \frac{\langle L \rangle^3}{\langle L^3 \rangle} \frac{D}{\langle L \rangle}$$

(3)

where L and D represent the length and diameter of the rods and $\langle L \rangle$ represents the mean of the length distribution. For any real distribution $\langle L \rangle^3 / \langle L^3 \rangle < 1$. For isolated nanowires D=0.94nm and we can conservatively estimate $\langle L \rangle \sim 1 \mu m$ giving an aspect ratio of approximately 1000. With $p_c = 1.3 \times 10^{-5}$, Eq. (3) then gives $\langle L \rangle^3 / \langle L^3 \rangle$ = 0.02. This very low value suggests that the electrical properties of the composite are dominated by very long nanowires associated with the tail of the length distribution.

That the average nanowire aspect ratio is approximately 1000 is supported by a recent computational study by Foygel et al [10]. This work showed that for rod like filler particles the critical exponent, t, falls from its 3-d value of 2.0 as the aspect ratio is increased. We have measured a value of 1.41 which Foygel's study associates with an aspect ratio in the range 500-2000 confirming our nanowire length estimate of 1000nm.

The presence of bundles only at higher concentrations also explains the breakdown in the percolation fit in Figure 2(b) for $p > 1 \times 10^{-3}$. Above this loading level the percolative network consists of nanowire bundles rather than individual nanowires. Such a network will be a less efficient conductor as it will contain fewer conductive paths for a given volume fraction, resulting in conductivities lower than those predicted by the extrapolated percolation fit.

One should also notice that the conductivities of the samples with p close to p_c are quite substantial. This could be due to possible doping of the polymer by iodine that has remained intercalated on the nanowires from production. Oxidative doping can occur whereby iodine can extract electrons from the polymer resulting in semi-mobile holes [11]. This additional contribution to overall conductivity also explains why the AC data for the $p = 5 \times 10^{-6}$ sample deviates from dielectric behaviour.

While the conductivities displayed by these composites, even at high volume fractions, are modest, they are in the technologically relevant range. Conductivities of 10^{-4} S/m are sufficient for anti-static coating applications. In this work we observe conductivities of 10^{-4} S/m at volume fractions as low as 10^{-4}. This means that extremely low quantities of filler are required resulting in the minimum perturbation of other properties of the host matrix.

CONCLUSIONS

In summary, we have shown that $Mo_6S_{4.5}I_{4.5}$ nanowires are highly conducting and form a percolation network with very low threshold of $p_c=1.3\times10^{-5}$. This appears to be the lowest threshold measured to date for any 1-d conductor/insulator composite system. This result can be explained by the debundling of the nanowires during sample fabrication at low nanowire concentrations. In addition the data shows the importance of the longer nanowires in the sample. The highest measured conductivity was 4×10^{-3} S/m which is high enough for anti-static applications.

ACKNOWLEDGEMENTS

The authors wish to acknowledge the support of the European Union in the form of the European Community's Human Potential Program under contract HPRN-CT-2002-00192, [NANOTEMP]. JNC and WJB thank the Science Foundation Ireland.

REFERENCES

1 J. K. W. Sandler, J. E. Kirk, I. A. Kinloch et al., Polymer **44**, 5893 (2003).
2 B. E. Kilbride, J. N. Coleman, J. Fraysse et al., J. Appl. Phys. **92**, 4024 (2002).
3 P.Pötschke, M. Abdel-Goad, I. Alig et al., Polymer **45**, 8863 (2004).
4 D. Vrbanic, M. Remskar, A. Jesih et al., Nanotechnology **15**, 635 (2004).
5 V. Nicolosi, D. Vrbanic, A. Mrzel et al., Chem. Phys. Lett. **401**, 13 (2005).
6 V. Nicolosi, D. Vrbanic, A. Mrzel et al., J. Phys. Chem. B, **109**, 7124 (2005)
7 D. Stauffer and A. Aharony, *Introduction to Percolation Theory*, Rev. 2nd ed. (Taylor & Francis, London, 1994).
8 I. Balberg, Phys. Rev. B **33**, 3618 (1986).
9 I. Balberg, Phys. Rev. B **31**, 4053 (1985).
10 M. Foygel, R. D. Morris, D. Anez et al., Phys. Rev. B **71**, 104201 (2005).
11 S. Roth and D. Carroll, *One-Dimensional Metals*, 2nd ed. (Wiley-VCH, Weinheim, 2004).

THEORY OF NANOSTRUCTURES

Optical Absorption of hexagonal Boron Nitride and BN nanotubes

Ludger Wirtz*, Andrea Marini[†] and Angel Rubio**

*Institute for Electronics, Microelectronics, and Nanotechnology (IEMN CNRS-UMR 8520),
B.P. 60069, 59652 Villeneuve d'Ascq Cedex, France
[†]Istituto Nazionale per la Fisica della Materia e Dipartimento di Fisica dell'Universitá di Roma
"Tor Vergata", Via della Ricerca Scientifica, I-00133 Roma, Italy
**Department of Material Physics, University of the Basque Country, Centro Mixto CSIC-UPV,
and Donostia International Physics Center (DIPC), Po. Manuel de Lardizabal 4, 20018
Donostia-San Sebastián, Spain

Abstract. We present calculations for the optical absorption spectra of hexagonal boron nitride (hBN) and BN nanotubes, using many-body perturbation theory. Solution of the Bethe-Salpeter equation for hBN leads to optical absorption and loss spectra where the positions and shapes of the peaks are strongly dominated by excitonic effects. The binding energy of the first exciton is about 0.71 eV. Comparison of the calculations with recently measured optical absorption and EELS demonstrates that DFT underestimates the "true" band gap of BN by 2.25 eV. This band gap difference can be partially (bot not completely) reproduced by a calculation of the quasi-particle band-structure on the level of the GW-approximation. We show, how the lower dimensionality of BN nanotubes leads to a much stronger excitonic binding energy and at the same time to a larger quasi-particle gap. This leaves the position of the first absorption peak almost unchanged. However, the difference in the series of excitonic peaks allows the spectroscopic distinction between BN nanotubes and bulk BN.

Keywords: BN nanotubes, excitonic effects, Bethe-Salpeter equation, GW-approximation
PACS: 78.67.Ch, 78.20.Bh, 71.35.Cc, 71.45.Gm

INTRODUCTION

Hexagonal boron nitride (hBN) is isoelectronic to graphite and has the same layered structure (only with different stacking). Graphite is a semi-metal with a linear crossing of the π and π^* bands at the K-point of the Brillouin zone. In hBN, however, the different electronegativities of boron and nitrogen lift the degeneracy at the K-point and lead to a considerable gap of 4.5 eV (in the LDA). The optical properties of hBN have been investigated experimentally (e.g. [1, 2]) and theoretically (e.g., [3, 4]) on the level of the Random Phase Approximation (RPA). Recently, hBN has regained interest due to the discovery of BN-nanotubes [5, 6] which can be considered as cylinders formed by rolling a single sheet of hBN onto itself. In the RPA, optical absorption is modeled as independent electron transitions from the valence to the conduction band. Since the electronic structure of the tubes can be constructed by rolling the sheet, the RPA optical absorption spectra for the 3-dimensional bulk-BN, for 2-dimensional single sheet and for the (quasi-)one dimensional tubes are very similar ([7, 8, 9] and see Fig. 2), for light polarization parallel to the layer-plane and tube axis and strong broadening of the spectra. For large radius tubes, the absorption spectrum must be the same as for the

CP786, Electronic Properties of Novel Nanostructures, edited by H. Kuzmany, J. Fink, M. Mehring, and S. Roth
© 2005 American Institute of Physics 0-7354-0275-2/05/$22.50

sheet, since the electronic structure converges towards the one of the sheet, and even for tubes with small radius, the presence of 1-dimensional van-Hove singularities in the density of states gives only small additional fine-structure in a spectrum that essentially resembles the one of the plane. Going beyond the simplistic picture of independent particle transitions and using the techniques of many-body perturbation theory, we will show that the optical properties of hBN are strongly dominated by excitonic effects. The difference in excitonic binding energies between hBN and the tubes will thereby lead to measurable differences in the optical spectra. The influence of excitonic effects in carbon nanotubes was recently shown [10, 11].

METHODS

Our calculations of optical absorption spectra, which are proportional to the imaginary part of the dielectric function, $\varepsilon_2(E)$, proceed in three steps:

I.) First we perform a self-consistent calculation of the (geometry-optimized) ground state density using density functional theory (DFT) in the local-density approximation (LDA). Then we calculate the Kohn-Sham wavefunctions ψ_{nk} of the occupied and a large number of unoccupied states (with a band-index n and a sufficiently fine discrete k-point sampling of the first Brillouin zone in reciprocal space). The calculations are performed with the code ABINIT [12]. Core electrons are substituted by pseudo-potentials and valence electrons are expanded in plane waves. Frequently, the LDA wavefunctions and energies are used directly to calculate the matrix elements for optical transitions. This corresponds to the random-phase approximation (RPA). For many materials, however, the thus-obtained spectra are in qualitative and quantitative disagreement with measured absorption spectra [13] and further calculations using the techniques of many-body perturbation theory are necessary. This is also the case for hBN and BN nanotubes.

II.) Within the GW-approximation [13], we calculate the quasi-particle energies ("true" single-particle excitation energies), E_{nk}, by solving the Dyson equation

$$\left[-\frac{\nabla^2}{2} + V_{ext} + V_{Hartree} + \Sigma(E_{nk}) \right] \psi_{nk} = E_{nk} \psi_{nk}. \tag{1}$$

The self-energy $\Sigma = iGW$ is approximated as the product of the one-particle Green's function G and the screened Coulomb interaction W. The GW-approximation increases considerably the LDA-band gap.

III.) An electron that is excited from the valence band to the conduction band can interact with the remaining "hole" through a screened Coulomb potential. This can lead to bound, excitonic, states within the band-gap. Quantitatively, the excitonic levels are calculated by the Bethe-Salpeter equation

$$(E_{ck} - E_{vk})A_{vck}^S + \Sigma_{k'v'c'} \langle vck|K_{eh}|v'c'k' \rangle A_{v'c'k'}^S = \Omega^S A_{vck}^S. \tag{2}$$

The electron-hole interaction kernel K_{eh} "mixes" different electron transitions from valence band states v, v' to conduction band states c, c'. This leads to transition energies Ω^s that can be quite different from the quasiparticle-energy differences $E_{ck} - E_{vk}$. The presence of bound excitons can lead to a considerable red-shift of oscillator strength. The calculations of step II and III are performed with the code SELF [16].

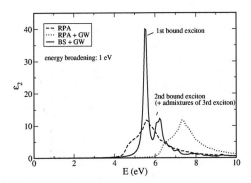

FIGURE 1. Optical absorption of bulk hBN calculated on three levels of approximation (see text).

RESULTS

In Fig. 1, we compare the results of the three levels of calculation for hBN [17]. The light-polarization is chosen parallel to the plane. The spectra are calculated with a Lorentzian broadening of 0.1 eV, corresponding approximately to experimental values for broadening. The dashed line shows the result of the RPA calculation (compare [3] and [4]). The broad peak with a maximum at 5.6 eV is entirely due to the continuum of inter-band transitions between the π and π^* bands. (Transitions between π and σ bands are forbidden due to selection rules). The GW-calculation leads to an almost uniform downward shift of the π band and upward shift of the π^* band (compare [14]). This results in a blue-shift with respect to the RPA-absorption spectrum by 1.7 eV while the shape remains approximately the same (dotted line). An important change takes place when we include excitonic effects through the Bether-Salpeter (BS) equation. This leads to a the double-peak spectrum (solid line in Fig. 1) with the high peak at almost the same position as the maximum in the RPA-spectrum. The blue-shift due to the GW corrections to the band-structure and the red-shift due to excitonic effects thus almost cancel each other. However, the shape of the spectrum has changed considerably. Contrary to the RPA spectrum which is due to a *continuum of inter-band transitions*, the BS spectrum is dominated by two *discrete bound excitons*. This can be seen more clearly in Fig. 2 A), where the same spectrum, but with a broadening of 0.025 eV, is plotted. There is a sequence of bound excitonic peaks below the onset of the continuum (which is determined by the GW-band gap). Most of the oscillator strength is collected by the first peak while at the onset of the continuum almost no absorption takes place. The binding energy of the first exciton is 0.71 eV. We note that the shape of the broadened BS spectrum of Fig. 1 is in very good agreement with the experimental data of Tarrio and Schnatterly (Ref. [1], Fig. 2, which was obtained by a Kramers-Kroning transform of the electron-energy loss function). It also matches well the optical absorption spectrum of hBN of Ref. [2]. Only the position of the BS+GW spectrum comes out too low by 0.55 eV in comparison with experimental data. We suppose that this is due to the limited accuracy of the GW-approximation and that the quasiparticle correction to the LDA band-gap should be rather 2.25 eV than the calculated shift of 1.7 eV mentioned above.

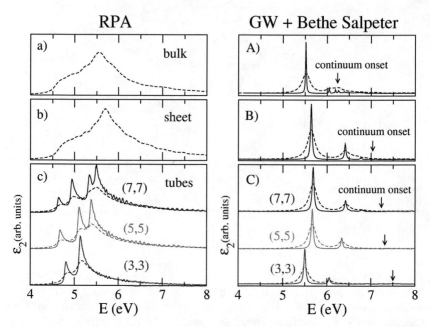

FIGURE 2. Optical absorption of hBN (a), BN-sheet (b) and three different BN tubes (c). We compare the results of the GW + Bethe-Salpeter approach (right hand side) with the random phase approximation (left hand side). Solid lines are calculated with a Lorentzian broadening of 0.025 eV, dashed lines with a broadening of 0.1 eV (for comparison with experimental data). The light polarization is parallel to the plane/ tube axis, respectively.

Keeping this limitation in mind, we conclude that the GW+BS method is well suited to explain the shape of the optical absorption (and EELS) spectra of hBN. We apply it consequently for the calculation of the spectra of the (quasi) two-dimensional BN sheet and the (quasi) 1-dimensional BN nanotubes. Our theoretical data is also in very good agreement with a recent EELS measurement of multiwalled BN-nanotubes [15].

On the left-hand side of Fig. 2 we present the RPA spectra of bulk hBN, of the single-sheet of hBN (calculated in a periodic supercell with an inter-sheet distance 20 a.u.), and of three different BN nanotubes (supercell with inter-tube distance of 20 a.u.) [18]. The RPA spectra of the bulk and the single sheet are almost indistinguishable. For the tubes, the RPA spectra display some peak structure below 5.5 eV which is due to the van-Hove singularities in the one-dimensional density of states. For tubes with larger radii, the density of the fine-structure peaks increases and the RPA spectrum approaches that of the sheet. Note that the fine structure is only visible in the calculation with small broadening (solid lines). In the calculation with 0.1 eV broadening, the RPA spectrum of the tubes is almost indistinguishable from the sheet and the bulk (compare [8] and [9]). As was already discussed for bulk BN, the BS+GW spectra are entirely different in shape, because they are dominated by excitonic peaks. The reduced dimensionality leads to a stronger binding energy of the first exciton. In the two-dimensional sheet we obtain a binding energy of 1.4 eV as compared with the bulk value of 0.71 eV. For the small diameter, quasi-one dimensional, BN(3,3) tube the binding energy even reaches 2 eV. With increasing tube diameter, the binding energy converges to the binding energy of the sheet (1.62 eV for the (5,5) tube and 1.58 eV for the (7,7) tube. At the same time, the reduced dimensionality leads to an increase of the GW gap [19]. Therefore, the position of the dominant absorption peak (the first excitonic peak) remains almost unchanged. The distance between the first and second excitonic peak increases with increasing tube diameter and rapidly converges towards the distance of the excitonic peaks in the sheet. This is an indication that the exciton is confined in a small region (a few atomic distances) and feels only the "local environment" Only in the strongly curved (3,3) tube, the exciton can be delocalized over the whole circumference of the tube. In the larger tube, the exciton only "sees" the locally flat environment as in a sheet.

In conclusion, the optical absorption spectra of hBN and BN nanotubes are dominated by excitonic effects. Most oscillator strength is collected by the first bound exciton. The binding energy of this peak increases strongly as the dimensionality is reduced from the 3-D bulk over the 2-D sheet to the 1-D tubes. At the same time the quasi-particle band gap increases with reduced dimensionality. This cancellation leaves the absolute position of the dominant absorption peak almost constant. Differences can be seen, however, in the energy differences between the excitonic peaks. We acknowledge support by the EU Network of Excellence NANOQUANTA (NMP4-CT-2004-500198). Calculations were performed at IDRIS (Project No. 51827) and CEPBA.

REFERENCES

1. C. Tarrio and S. E. Schnatterly, Phys. Rev. B **40**, 7852 (1989).
2. J. S. Lauret et al., Phys. Rev. Lett. **94**, 037405 (2005).
3. Y. Xu and W. Y. Ching, Phys. Rev. B **44**, 7787 (1991).
4. G. Cappellini, G. Satta, M. Palummo, and G. Onida, Phys. Rev. B **64**, 035104 (2001).
5. A. Rubio, J. L. Corkill, and M. L. Cohen, Phys. Rev. B **49**, R5081 (1994); X. Blase, A. Rubio, S. G. Louie, and M. L. Cohen, Europhys. Lett **28**, 335 (1994).
6. N. G. Chopra et al., Science **269**, 966 (1995).
7. L. Wirtz, V. Olevano, A. G. Marinopoulos, L. Reining, and A. Rubio, AIP Conf. Proc. **685**, 406 (2003).
8. A. G. Marinopoulos et al., and L. Reining, Appl. Phys. A **78**, 1157 (2004).
9. G.Y. Guo and J.C. Lin, Phys. Rev. B **71**, 165402 (2005).
10. C. D. Spataru et al., Phys. Rev. Lett. **92**, 077402 (2004); Appl. Phys. A **78**, 1129 (2004).
11. E. Chang, G. Bussi, A. Ruini, and E. Molinari, Phys. Rev. Lett. **92**, 196401 (2004).
12. X. Gonze et al., Comp. Mat. Sci. **25**, 478 (2002).
13. See, e.g., G. Onida, L. Reining, and A. Rubio, Rev. Mod. Phys. **74**, 601 (2002).
14. X. Blase, A. Rubio, S.G. Louie and M. L. Cohen, Phys. Rev. B **51**, 6868 (1995)
15. G. G. Fuentes et al., Phys. Rev. B **67**, 035429 (2003).
16. SELF , written by A. Marini (http://people.roma2.infn.it/ marini/self/)
17. Energy minimization leads to a bond-length of 1.441 Å and an inter-sheet distance of 3.251 Å. Energies and wave-functions are evaluated with an energy cutoff at 25 Hartree, using a 40x40x12 Monkhorst-Pack grid in reciprocal space. In the GW and BS calculations, we used a 16x16x6 Monkhorst-Pack sampling which is sufficient to converge the dominant discrete excitonic peaks. Unoccupied states with energies up to 40 eV above the band-gap are included.
18. The tubes are geometry-optimized. For the GW+BS calculations, the 1-D Brillouin zone is sampled by 20 k-points (corresponding to 11 irreducible k-points). Convergence of the spectra with respect to this number was checked numerically.
19. C. Delerue, G. Allan, and M. Lannoo, Phys. Rev. Lett. **90**, 076803 (2003)

Mechanical quantum resonators

A. N. Cleland* and M. R. Geller†

*Department of Physics, University of California, Santa Barbara CA 93106 USA
†Department of Physics and Astronomy, University of Georgia, Athens, Georgia 30602 USA

Abstract. We describe the design for a solid-state quantum computational architecture based on the integration of GHz-frequency mechanical resonators with Josephson phase qubits, which have the potential for demonstrating a variety of single- and multi-qubit operations critical to quantum computation. The computational qubits are eigenstates of large-area, current-biased Josephson junctions. Two or more qubits are capacitively coupled to a piezoelectric nanoelectromechanical disk resonator, which enables coherent coupling of the qubits. The integrated system is analogous to one or more few-level atoms (the Josephson junction qubits) in an electromagnetic cavity (the nanomechanical resonator). However, here we can individually tune the level spacing of the "atoms" and control their "electromagnetic" interaction strength. We show that quantum states prepared in a Josephson junction can be passed to the nanomechanical resonator and stored there, and then can be passed back to the original junction or transferred to another with high fidelity. The resonator can also be used to produce maximally entangled Bell states between a pair of Josephson junctions.

Keywords: Josephson junction; qubit; quantum computation
PACS: 03.67.Lx, 85.25.Cp, 85.85.+j

The lack of easily fabricated physical qubit elements, having both sufficiently long quantum-coherence lifetimes and the means for producing and controlling their entanglement, remains the principal roadblock to building a large-scale quantum computer. Superconducting devices have been understood for several years to be natural candidates for quantum computation, given that they exhibit robust macroscopic quantum behavior [1]. Demonstrations of long-lived Rabi oscillations in current-biased Josephson tunnel junctions [2, 3], and of both Rabi oscillations and Ramsey fringes in a Cooper-pair box [5], have generated significant new interest in the potential for superconductor–based quantum computation [6]. Coherence times τ_φ up to $5\,\mu$s have been reported in the current-biased devices [2], with corresponding quantum-coherent quality factors $Q_\varphi \equiv \tau_\varphi \Delta E/\hbar$ of the order of 10^5, indicating that these systems should be able to perform many logical operations during the available coherence lifetime (here ΔE is the qubit energy-level separation).

Here we describe a proposal that GHz-frequency nanoelectromechanical resonators can be used to coherently couple two or more current-biased Josephson junction devices together to make a flexible and scalable solid-state quantum-information-processing architecture [7]. The computational qubits are the energy eigenstates of the junctions. These superconducting phase qubits are capacitively coupled to a piezoelectric dilatational disk resonator, cooled on a dilution refrigerator to the quantum limit. We show that the integrated system is analogous to one or more few-level atoms in an electromagnetic cavity (the resonator). We can tune *in situ* the energy level spacing of each "atom", and control the "electromagnetic" interaction strength. This analogy makes it clear that our design is sufficiently flexible to be able to carry out any operation that can be done

CP786, *Electronic Properties of Novel Nanostructures*, edited by H. Kuzmany, J. Fink, M. Mehring, and S. Roth
© 2005 American Institute of Physics 0-7354-0275-2/05/$22.50

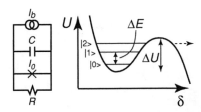

FIGURE 1. *Left:* Equivalent-circuit model for a current-biased Josephson junction. A capacitance C and resistance R in parallel with an ideal Josephson element with critical current I_0, all sharing a bias current I_b. *Right:* Potential in the cubic $s \to 1^-$ limit.

using other architectures, provided that there is enough coherence.

Several investigators have proposed the use of LC resonators [8], superconducting cavities [9], or other types of oscillators, to couple junctions together. Resonator-based coupling schemes, such as the one proposed here, have additional functionality resulting from the ability to tune the qubits relative to the resonator frequency, as well as to each other. By tuning the junctions in and out of resonance with the nanomechanical resonator, qubit states prepared in a junction can be passed to the resonator and stored there, and can later be passed back to the original junction or transferred to another junction with high fidelity. The resonator can also be used to produce controlled entangled states between a pair of junctions. The use of mechanical resonators to mediate multi-qubit operations in junction-based quantum information processors has not (to the best of our knowledge) been considered previously, but our proposal builds on the interesting recent work by Armour *et al.*[10] and Irish *et al.*[11].

Our implementation uses large-area current-biased Josephson junctions, with capacitance C and critical current I_0, as shown in Fig. 1. The largest relevant energy scale is the Josephson energy $E_J \equiv \hbar I_0/2e$, with charging energy $E_c \equiv (2e)^2/2C \ll E_J$. The dynamics of the phase difference δ is that of a particle of mass $M = \hbar^2 C/4e^2$ moving in an effective potential $U(\delta) \equiv -E_J(\cos\delta + s\,\delta)$, where $s \equiv I_b/I_0$ is the dimensionless bias current [12]. When $0 < s < 1$, $U(\delta)$ has metastable minima, separated from the continuum by a barrier of height ΔU, also shown in Fig. 1. The small-oscillation plasma frequency is $\omega_p = \omega_{p0}(1-s^2)^{1/4}$, with $\omega_{p0} = \sqrt{2E_J E_c}/\hbar$. The Hamiltonian for an isolated junction is $H_J = -E_c d^2/d\delta^2 + U(\delta)$, with quasi-bound states in the minima with energies ε_m. The lowest energy quasi-bound states $|0\rangle$ and $|1\rangle$ define the phase qubit, with $\Delta E \equiv \varepsilon_1 - \varepsilon_0$ the level spacing. We focus here on a single resonator coupled to one and two junctions; extensions to larger systems will be addressed in future work. The basic two-junction circuit is shown in Fig. 2. The disk-shaped element is the nanomechanical resonator, consisting of a piezoelectric crystal sandwiched between split metal electrodes, and the junctions are the crossed boxes.

The nanomechanical resonator is designed to have a fundamental thickness-resonance frequency $\omega_0/2\pi$ of a few GHz, and a high quality factor Q. Piezoelectric dilatational resonators with frequencies in this range, and with room-temperature quality factors around 10^3, have been fabricated from sputtered AlN [13]. We have performed RF network measurements down to 4.2 K for a similar piezoelectric 1.8 GHz resonator. The

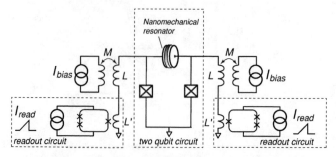

FIGURE 2. Two phase qubits coupled to a piezoelectric resonator.

observed low-temperature Q of 3500 corresponds to an energy lifetime τ of more than 300 ns, sufficient for the operations described below. Upon cooling to 20 mK, the 1.8 GHz dilatational mode will be in the quantum regime, with a probability of thermally occupying the first excited (one-phonon) state of about 10^{-2}. Using dilatational-phonon creation and annihilation operators, the resonator Hamiltonian is $H_{\text{res}} = \hbar\omega_0 a^\dagger a$.

An elastic strain in the resonator produces, through the piezoelectric effect, a charge q on the capacitor enclosing it, corresponding to a current $\dot{q}$. A model for a disk resonator of radius R and thickness b leads to $q = C_{\text{res}}(V - e_{33}bU_{zz}/\varepsilon_{33})$, where $C_{\text{res}} = \varepsilon_{33}\pi R^2/b$ is the resonator capacitance, V the voltage across it, e_{33} and ε_{33} the relevant elements of the piezoelectric modulus and dielectric tensors, and U_{zz} the spatially averaged strain. Strain induces an electric field $E_z = e_{33}U_{zz}/\varepsilon_{33}$ in the piezoelectric, and a charge of magnitude $\tilde{C}_{\text{res}}E_z b$ on the electrodes, where $\tilde{C}_{\text{res}}$ is a piezoelectrically-enhanced capacitance. The resonator adds the capacitance $\tilde{C}_{\text{res}}$ in parallel with the junction capacitance, reducing the charging energy E_c to $2e^2/(C + \tilde{C}_{\text{res}})$.

Quantizing the vibrational modes of the resonator in the presence of the appropriate mechanical and electrodynamic boundary conditions leads to a Hamiltonian for a single junction coupled to a resonator $H = H_J + H_{\text{res}} + \delta H$, where $\delta H = -ig(a - a^\dagger)\delta$, and g is a real-valued coupling constant with dimensions of energy. The eigenstates of $H_J + H_{\text{res}}$ are $|mn\rangle \equiv |m\rangle_J \otimes |n\rangle_{\text{res}}$, with energies $E_{mn} = \varepsilon_m + \hbar\omega_0 n$ (n is the resonator phonon occupation number), and an arbitrary state can be expanded as $|\psi(t)\rangle = \sum_{mn} c_{mn}(t) e^{-iE_{mn}t/\hbar} |mn\rangle$.

We first show that we can pass a qubit state from a junction to the resonator and store it there, using the adiabatic approximation combined with the rotating-wave approximation (RWA) [16]. We assume that s changes slowly on the time scale $\hbar/\Delta E$, and work at zero temperature. From the RWA, neglecting population and phase relaxation, we obtain from the equations of motion for the coefficients $c_{mn}(t)$.

At time $t = 0$ we prepare the junction in the state $\alpha|0\rangle_J + \beta|1\rangle_J$, leaving the resonator in the ground state $|0\rangle_{\text{res}}$. We then allow the junction and resonator to interact on resonance for a time interval $\Delta t = \pi/\Omega_0$, where Ω_0 is the vacuum Rabi frequency. We then bring the systems out of resonance and the resonator is found to be in the same pure state, apart from expected phase factors. The junction state has actually been swapped with that of the resonator.

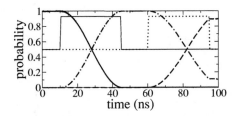

FIGURE 3. Qubit transfer between two junctions. Solid curve is $|c_{100}|^2$, dashed-dotted curve is $|c_{001}|^2$, and dashed curve is $|c_{010}|^2$. Thin solid and dotted curves show s_1 and s_2, respectively.

We have also solved the exact equations numerically, including all quasi-bound junction states present. The initial state is $|10\rangle$, corresponding to the case $\alpha = 0$, $\beta = 1$. After 10 ns, the bias current is adiabatically changed to bring the qubit in resonance with $\hbar\omega_0$. The junction is held in resonance for half a Rabi period, and then detuned. $s(t)$ has a trapezoidal shape with a crossover time of 0.5 ns. The storage operation is successful, and the magnitudes of the final probability amplitudes, are extremely close to the desired RWA values. The phases of the c_{mn} after storage, however, are not correctly given by the RWA unless $g/\Delta E$ is much smaller.

To transfer a qubit state $\alpha|0\rangle + \beta|1\rangle$ between two junctions, the state is loaded into the first junction and the bias current s_1 adjusted to bring that junction into resonance with the resonator for half a Rabi period. This stores the junction state in the resonator. After the first junction is taken out of resonance, the second one is brought into resonance for half a Rabi period, passing the state to the second junction. We have simulated this operation, assuming two identical junctions as in Fig. 2. Our results are shown in Fig. 3, where $c_{m_1 m_2 n}$ is the probability amplitude to find the system in the state $|m_1 m_2 n\rangle$, with m_1 and m_2 labelling the states of the two junctions and n the phonon occupation number of the resonator.

Finally, we can prepare an entangled state of two junctions connected to a common resonator by bringing the first junction into resonance with the resonator for one-quarter of a Rabi period, which produces the state $(|100\rangle + |001\rangle)/\sqrt{2}$. After bringing the second junction into resonance for half a Rabi period, the state of the resonator and second junction are swapped as $|001\rangle \rightarrow -|010\rangle$, leaving the system in the state $(|100\rangle - |010\rangle)/\sqrt{2}$, where the resonator is in the ground state and the junctions are maximally entangled. Our simulations of this operation, the results of which are presented in Fig. 4, demonstrate successful entanglement with a fidelity of 95%. The system parameters are the same as in Fig. 3.

The quantum-information-processing operations described here require a minimum coherence time of order 100 ns, a time already demonstrated in the phase qubit. More extensive operations could be performed with a coherence time of a few hundred nanoseconds, which have recently been achieved in the phase qubit [20]. The mechanical resonator must also achieve similar coherence times; using standard results for the coherence time of a particle coupled to a dissipative environment [21], we estimate the quantum coherence time of an n-phonon state to be the lesser of $\tau \approx \hbar Q/k_B T (n + 1/2)$ and the energy decay lifetime Q/ω_0. At 20 mK, the $|1\rangle$ state of our resonator is determined

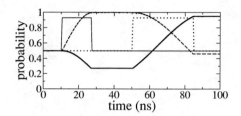

FIGURE 4. JJ entanglement. Dashed curve is the probability to be in the state (in the interaction representation) $(|100\rangle + |001\rangle)/\sqrt{2}$, and thick solid curve is for $(|100\rangle - |010\rangle)/\sqrt{2}$. Thin solid and dotted curves are s_1 and s_2.

by the decay lifetime, which for $Q \approx 3500$ is about 300 ns.

REFERENCES

1. Y. Makhlin, G. Schön, and A. Shnirman, *Rev. Mod. Phys.* **73**, 357 (2001).
2. Y. Yu, S. Han, X. Chu, S.-I. Chu, and Z. Wang, *Science* **296**, 889 (2002).
3. J. M. Martinis, S. Nam, J. Aumentado, and C. Urbina, *Phys. Rev. Lett.* **89**, 117901 (2002).
4. A. J. Berkley et al. *Science* **300**, 1548 (2003). F. W. Strauch et al., *Phys. Rev. Lett.* **91**, 167005 (2003).
5. Y. Nakamura, Yu. A. Pashkin, and J. S. Tsai, *Nature* **398**, 786 (1999). D. Vion et al., *Science* **296**, 886 (2002). T. Yamamoto, Yu. A. Pashkin, O. Astafiev, Y. Nakamura, and J. S. Tsai, *Nature* **425**, 941 (2003).
6. A. J. Leggett, *Science* **296**, 861 (2002).
7. A. N. Cleland and M . L. Geller, *Phys. Rev. Lett.* **93**, 70501 (2004).
8. A. Shnirman, G. Schön, and Z. Hermon, *Phys. Rev. Lett.* **79**, 2371 (1997). Y. Makhlin, G. Schön, and A. Shnirman, *Nature* **398**, 305 (1999). A. Blais, A. M. van den Brink, and A. M. Zagoskin, *Phys. Rev. Lett.* **90**, 127901 (2003). F. Plastina and G. Falci, *Phys. Rev. B* **67**, 224514 (2003).
9. O. Buisson and F. W. J. Hekking, in *Macroscopic Quantum Coherence and Quantum Computing*, edited by D. V. Averin, B. Ruggiero, and P. Silvestrini (Kluwer, New York, 2001), p. 137. F. Marquardt and C. Bruder, *Phys. Rev. B* **63**, 54514 (2001).
10. A. D. Armour, M. P. Blencowe, and K. C. Schwab, *Phys. Rev. Lett.* **88**, 148301 (2002).
11. E. K. Irish and K. Schwab, *Phys. Rev. B* **68**, 155311 (2003).
12. A. Barone and G. Paterno, *Physics and Applications of the Josephson Effect* (Wiley, New York, 1982).
13. R. Ruby, P. Bradley, J. Larson, Y. Oshmyansky, and D. Figueredo, *Tech. Digest 2001 IEEE Intl. Solid-State Circuits Conf.*, p. 120 (2001).
14. X. M. H. Huang, C. A. Zorman, M. Mehregany, and M. L. Roukes, *Nature* **421**, 496 (2003).
15. A. N. Cleland, M. Pophristic and I. Ferguson, *Appl. Phys. Lett.* **79**, 2070 (2001).
16. M. O. Scully and M. S. Zubairy, *Quantum Optics* (Cambridge University Press, Cambridge, 1997).
17. X. Maître, E. Hagley, G. Nogues, C. Wunderlich, P. Goy, M. Brune, J. M. Raimond, and S. Haroche, *Phys. Rev. Lett.* **79**, 769 (1997).
18. E. Hagley, X. Maître, G. Nogues, C. Wunderlich, M. Brune, J. M. Raimond, and S. Haroche, *Phys. Rev. Lett.* **79**, 1 (1997).
19. J. M. Martinis, private communication.
20. J. M. Martinis, S. Nam, J. Aumentado, and K. M. Lang, Phys. Rev. B 67, 94510 (2003).
21. E. Joos, in *Decoherence and the Appearance of a Classical World in Quantum Theory*, edited by D. Giulini *et al.* (Springer-Verlag, Berlin, 1996), p. 35.

Chiral-index assignment of carbon nanotubes by resonant Raman scattering

J. Maultzsch*, H. Telg*, S. Reich† and C. Thomsen*

*Institut für Festkörperphysik, Hardenbergstr. 36, 10623 Berlin, Germany
†Department of Engineering, University of Cambridge, Cambridge CB2 1PZ, United Kingdom

Abstract. We report a unique assignment of the chiral index (n_1, n_2) of carbon nanotubes to their optical transition energies and radial-breathing-mode frequencies. Our results provide a fast and easy method to assign nanotubes in an unknown sample by Raman spectroscopy using only one or a few excitation energies. Given the assignment, we can analyze how the optical and vibrational properties of carbon nanotubes depend on the chirality. We observe systematic dependences of the Raman scattering cross section on the chiral angle, indicating that the Raman and fluorescence intensity is not solely correlated to the chirality distribution.

The assignment of the chiral index (n_1, n_2) to the nanotubes observed in an experiment has been one of the major challenges in carbon nanotube research. For the fabrication of electronic devices, for example, it is important to know whether the tube is metallic or semiconducting. When optimizing growth methods, one wants to know the diameters of the tubes or whether a particular chirality is preferred in the growth. Direct imaging techniques like scanning tunneling microscopy are quite laborious and usually require a second experiment for cross-checking [1]. A more powerful method is optical spectroscopy since it can be applied to a large number of tubes at the same time.

In Raman spectroscopy, all assignment methods make use of the relationship between the radial breathing mode (RBM) and the tube diameter d, $\omega_{RBM} = c_1/d + c_2$. The first assignment by Raman scattering was reported by Jorio et al. [2], based on the "best" values for c_1 and c_2 that were consistent with other observations ($c_1 = 248\,\mathrm{cm}^{-1}\mathrm{nm}$, $c_2 = 0$). From photoluminescence experiments, Bachilo et al. [3] proposed an assignment to the first and second transition energies (E_{11}^S and E_{22}^S) of semiconducting tubes. Again, Raman spectroscopy was used to decide between several choices for the assignment ($c_1 = 223.5\,\mathrm{cm}^{-1}\,\mathrm{nm}$, $c_2 = 12.5\,\mathrm{cm}^{-1}$). Both methods therefore depend on the values of c_1 and c_2; in particular, the assignment of (n_1, n_2) to ω_{RBM} or E_{ii} changes if these values change. On the other hand, c_1 and c_2 predicted from theory or derived from experiments vary over a wide range in the literature [1]. Therefore, an assignment independent of the precise values of c_1 and c_2 is needed.

Here we present an independent chiral-index assignment based on the experimental information about the electronic transition energies in addition to the RBM frequency [4]. The electronic transition energies were determined from resonance Raman profiles and mapped vs. the inverse RBM frequency. Our assignment relies on pattern recognition between such an experimental Kataura plot and the theoretical Kataura plot. Based on the match of these patterns we derive experimental values for c_1 and c_2. The

CP786, *Electronic Properties of Novel Nanostructures*, edited by H. Kuzmany, J. Fink, M. Mehring, and S. Roth
© 2005 American Institute of Physics 0-7354-0275-2/05/$22.50

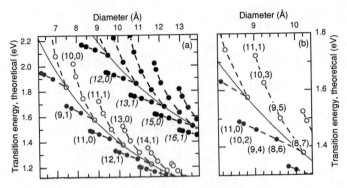

FIGURE 1. (a) Optical transition energies (E_{22}^S and E_{11}^M) as a function of tube diameter for light polarized parallel to the tube axis. For semicondcuting tubes, the full and open circles correspond to $\nu = -1$ and $\nu = +1$ tubes, respectively. (b) Same as in (a) with the chiralities of the (11,0) and the (11,1) branch indicated.

assignment remains the same even if c_1 and c_2 vary, say for different sample environments. Furthermore, we show how the RBM modes in a Raman spectrum can be easily assigned once the full resonance Raman experiment has been performed.

We recorded Raman spectra of HiPCO-produced carbon nanotubes in solution wrapped by SDS and SDBS [5]. We analyze the Raman intensity as a function of excitation wavelength. Whenever the excitation energy matches a real electronic transition of the particular nanotube, the Raman signal is strongly enhanced. From these resonance profiles we find the transition energies of ≈ 50 semiconducting and metallic tubes [4].

The systematics of the theoretical Kataura plot are shown in Fig. 1. The transition energies deviate from the $1/d$-dependence, forming "V"-shaped branches. Such a branch is found in the Raman spectra as a group of close-by RBM peaks, a so-called *la ola* [4]. For semiconducting tubes, the E_{22} branches on the lower (upper) side of the $1/d$ curve belong to the $\nu = -1$ ($\nu = +1$) tube families, where $\nu = (n_1 - n_2) \bmod 3$. For metallic tubes, the members of the upper and lower branches have the same chirality. The chiral angle varies systematically within a branch from (close to) zig-zag ($0°$) at the outermost positions to armchair ($30°$). In Fig. 1 (a), the first chiral index of each branch is indicated; the remaining chiral indices are subsequently derived by $(n_1', n_2') = (n_1 - 1, n_2 + 2)$, see Fig. 1 (b). We will use these patterns for our assignment.

When plotting E_{ii} vs. $1/\omega_{RBM}$, we obtain an experimental Kataura plot without using c_1 and c_2. We then shift and stretch the theoretical plot with respect to the experimental one to find the best match. Along the diameter ($1/\omega_{RBM}$) axis, this corresponds to varying the unknown c_1 and c_2. Shifting and stretching along the energy axis accounts for corrections of the calculations which neglect rehybridization and excitonic effects. In Fig. 2 we show the match between the experimental (colored, large symbols) and theoretical Kataura plot (gray, small symbols) [4]. The agreement along the diameter axis is excellent. Along the energy axis the experimental energies deviate systematically

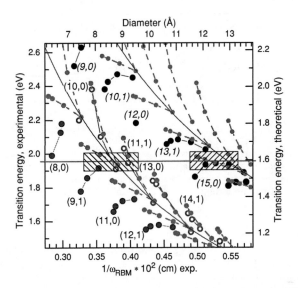

FIGURE 2. Experimental (colored, left and bottom axes) and theoretical (gray, right and top axes) Kataura plot. The first tubes of the $(n_1 - 1, n_2 + 2)$ branches are indicated. For semiconducting tubes, the full and open circles correspond to $\nu = -1$ and $\nu = +1$ tubes, respectively. The theoretical Kataura plot is calculated by the third-nearest neighbors tight-binding approximation [8]. The shaded boxes correspond to Fig. 3, see text.

from the theoretical data: the difference is small for armchair or near-armchair tubes (at the innermost positions of the "V"-shaped branches) and increases towards the zig-zag tubes (outer positions). This is consistent with the $\sigma - \pi$ rehybridization being stronger in zig-zag than in armchair tubes [6]. Similar experimental results were also obtained by Fantini *et al.* [7].

We can exclude other assignments of chiral index to RBM peaks for the following reasons: first, if the assignment changed slightly, we would lose the systematics of the branches. Second, if we shifted the assignment by an entire branch, we would observe more chiralities in some of the branches than actually exist. Therefore, the assignment shown in Fig. 2 is unambiguous. We can now fit the ω_{RBM}-diameter relationship to the experimental data and find $c_1 = 214 \pm 2\,\mathrm{nm\,cm^{-1}}$ and $c_2 = 19 \pm 2\,\mathrm{cm^{-1}}$. As we explain below, the knowledge of these coefficients, however, is not neccessary for finding (n_1, n_2) assignments by Raman scattering.

We now show how to assign the RBM frequencies observed in a single Raman spectrum to (n_1, n_2) by using the experimental Kataura plot in Fig. 2 [9]. In Fig. 3 we show an RBM spectrum with excitation energy $E_l = 1.96\,\mathrm{eV}$. To assign the RBM peaks, we look for both E_l and the inverse RBM frequency in the experimental Kataura plot. We can identify two groups of RBMs, one of them in the region of metallic tubes, the other belonging to semiconducting tubes. The width of the Raman resonance profiles is typically about 60 meV [4]; we can thus assume a window of $100 - 200\,\mathrm{meV}$ around

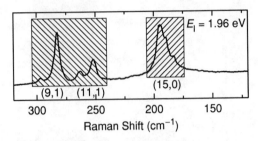

FIGURE 3. RBM spectrum of nanotubes with laser energy $E_l = 1.96\,\text{eV}$. For each group of peaks, the first member of the corresponding branch is indicated. The shaded boxes summarize the peaks originating from semiconducting and metallic tubes, respectively. The four largest peaks from semiconducting tubes are from the (8,3), (7,5), (7,6), and (10,3) tube. The metallic tubes giving the largest signal are (13,4), (12,6), and (11,8).

the laser energy containing nanotubes which might contribute to the spectra. The most important part is to identify the correct branches, only in a second step we determine the precise chirality of each RBM peak. For example, the group of metallic tubes giving rise to the RBMs at $170 - 200\,\text{cm}^{-1}$ are obviously all from the (15,0) branch, see Fig. 2. When comparing the RBM frequencies in detail with the experimental Kataura plot, we find for the three most prominent peaks (from larger to smaller RBM frequency): (13,4); (12,6); (11,8). The three remaining tubes of the (15,0) branch [(15,0), (14,2), and (10,10)] are present as weaker shoulders. Similarly, we determine the four largest peaks in the semiconducting region as (8,3); (7,5); (7,6); (10,3). Since we compare the Raman spectrum directly with the experimental Kataura plot, we do not need c_1 and c_2 for the identification of chiralities at all.

Given the (n_1, n_2) assignment, we can now analyze the chirality dependence of many properties in carbon nanotubes. For example, the characteristic lineshape of the high-energy Raman modes in metallic tubes can be verified independently [10]. Also the Raman scattering cross section depends strongly on the chiral angle. In Fig. 4 we show part of the experimental Kataura plot, where the size of the symbols indicates the Raman intensity. First, we observe a dependence on the family $v = \pm 1$; the tubes with $v = -1$ (lower branches) are in general stronger than tubes with $v = +1$. For example, the (7,5) tube and the (7,6) tube in Fig. 3 are both very close to the resonance; nevertheless, the signal of the (7,6) tube is much weaker. The metallic tubes exhibit the same behavior, even more, the upper branches could not be observed at all.

Within a single branch, the Raman intensity is at maximum for chiral angles between $10°$ and $15°$. The reason is that the Raman cross section contains the matrix elements for both the optical transitions and the electron-phonon coupling. To first approximation, the optical transitions become weaker for small chiral angle [3], depending on the index v and on E_{11} and E_{22} [11]. The electron-phonon coupling, on the other hand, increases for smaller chiral angles, as shown by *ab initio* calculations [12]. Thus our experimental results are in very good agreement with the theoretical predictions [12, 13].

In conclusion, we presented an (n_1, n_2) assignment for carbon nanotubes by Raman

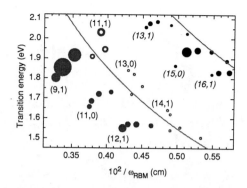

FIGURE 4. Section of the experimental Kataura plot where the area of the symbols corresponds to the Raman intensity. In general, the intensity is larger for the lower ($v = -1$) branches. Within a given branch, the maximum intensity occurs for chiral angles between $10°$ and $15°$.

spectroscopy independent of the values of c_1 and c_2. Since we assign (n_1, n_2) directly to the experimental RBM frequency combined with the transition energy E_{ii}, our assignment remains the same even if c_1 or c_2 should vary, *e.g.*, for environmental reasons. We showed how to identify the chiral indices in an experiment by one single Raman spectrum based on the experimental Kataura plot. The relevant information on the tubes is to which branch (and index v) they belong in the Kataura plot, because many properties are the same or very similar for tubes of the same branch.

We thank F. Hennrich for providing us with the sample. S.R. was supported by the Oppenheimer Fund and Newnham College.

REFERENCES

1. Reich, S., Thomsen, C., and Maultzsch, J., *Carbon Nanotubes: Basic Concepts and Physical Properties*, Wiley-VCH, Berlin, 2004.
2. Jorio, A., Saito, R., Hafner, J. H., Lieber, C. M., Hunter, M., McClure, T., Dresselhaus, G., and Dresselhaus, M. S., *Phys. Rev. Lett.*, **86**, 1118 (2001).
3. Bachilo, S. M., Strano, M. S., Kittrell, C., Hauge, R. H., Smalley, R. E., and Weisman, R. B., *Science*, **298**, 2361 (2002).
4. Telg, H., Maultzsch, J., Reich, S., Hennrich, F., and Thomsen, C., *Phys. Rev. Lett.*, **93**, 177401 (2004).
5. Lebedkin, S., Hennrich, F., Skipa, T., and Kappes, M. M., *J. Phys. Chem. B*, **107**, 1949 (2003).
6. Reich, S., Thomsen, C., and Ordejón, P., *Phys. Rev. B*, **65**, 155411 (2002).
7. Fantini, C., Jorio, A., Souza, M., Strano, M. S., Dresselhaus, M. S., and Pimenta, M. A., *Phys. Rev. Lett.*, **93**, 147406 (2004).
8. Reich, S., Maultzsch, J., Thomsen, C., and Ordejón, P., *Phys. Rev. B*, **66**, 035412 (2002).
9. Maultzsch, J., Telg, H., Reich, S., and Thomsen, C. (2005), to be published.
10. Telg, H., Maultzsch, J., Reich, S., and Thomsen, C. (2005), this volume.
11. Reich, S., Thomsen, C., and Robertson, J. (submitted 2004).
12. Machón, M., Reich, S., Telg, H., Maultzsch, J., Ordejón, P., and Thomsen, C., *Phys. Rev. B*, **71**, 035416 (2005).
13. Popov, V. N., and Lambin, L. H. P., *Nano Lett.*, **4**, 1795 (2004).

Spectroscopy of small diameter single-wall carbon nanotubes

A.Jorio[1], C.Fantini[1], L.G.Cançado[1], H.B.Ribeiro[1], A.P.Santos[2],
C.A.Furtado[2], M.S.Dresselhaus[3], G.Dresselhaus[3], Ge.G.Samsonidze[3],
S.G.Chou[3], A.Grueneis[4], J.Jiang[4], N.Kobayashi[4], R.Saito[4], M.A.Pimenta[1]

Departamento de Física, Universidade Federal de Minas Gerais, Belo Horizonte, MG, Brazil
Centro de Desenvolvimento de Tecnologia Nuclear, Belo Horizonte, MG, Brazil
Massachusetts Institute of Technology, Cambridge, MA, USA
Department of Physics, Tohoku University, Senday, Japan.

Abstract. Resonance Raman spectroscopy of small diameter (below 1.2 nm) single-wall carbon nanotubes (SWCNT) is presented. The diameter and chirality dependent many-body corrections to the tight binding based Kataura plot are discussed. The radial breathing modes also show small chirality dependence, giving evidence for the deviations of the small nanotube diameters from the ideal folded graphene structure. The use of spectroscopy for the characterization of small environmental effects and the (*n,m*) population on HiPco and CoMoCAT SWNT samples is pointed out. The richness of the intermediate frequency modes is highlighted.

Keywords: Carbon nanotubes, Raman spectroscopy, photoluminescence, many-body effects
PACS: 78.67.Ch, 78.30.Na, 73.22.-f, 78.55.-m, 71.35.-y.

OPTICAL TRANSITION ENERGIES AND RBM FREQUENCIES

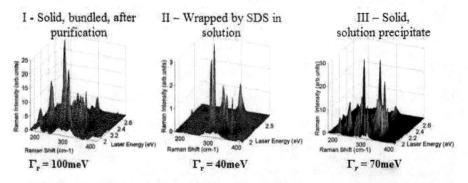

I - Solid, bundled, after purification $\Gamma_r = 100$meV

II – Wrapped by SDS in solution $\Gamma_r = 40$meV

III – Solid, solution precipitate $\Gamma_r = 70$meV

FIGURE 1. 3D plots for the RRS experiments on three different CoMoCAT samples. The average resonance window linewidth (FWHM) for the radial breathing mode peaks are given below each plot.

This work presents some highlights on resonance Raman spectroscopy (RRS) of single-wall carbon nanotubes (SWNT). The first highlight is about the development of theoretical models to describe the optical transition energies (E_{ii}) and the radial

CP786, *Electronic Properties of Novel Nanostructures*, edited by H. Kuzmany, J. Fink, M. Mehring, and S. Roth
© 2005 American Institute of Physics 0-7354-0275-2/05/$22.50

breathing mode (RBM) frequencies (ω_{RBM}) within experimental precision. Fig.1 shows three excitation laser (E_{laser}) dependent RRS spectra taken at the RBM spectral region for CoMoCAT samples just after purification (I), dispersed in solution with SDS (II) and for the solution precipitate (III) [1]. Each peak is associated with the RBM of an specific (n,m) SWNT, and the three experiments (I-III) can be used to study the E_{ii} and ω_{RBM} as a function of nanotube diameter and chiral angle [2], as well as the effect of environment and manipulation on SWNT electrons and vibrations [1]. The experimental precision is ± 10meV for E_{ii} and ± 1cm^{-1} for ω_{RBM} experimental determination [2].

The theoretical model describing the experimental E_{ii}'s is based on an extended tight binding (ETB) [3,4] that takes into account the distortion of the hexagonal lattice and hybridization of sigma and pi electrons, both effects due to the curvature of the graphene sheet in nanotubes. The E_{ii} values obtained with the ETB is corrected for electron-electron and electron-hole (excitons) many-body effects [5]. The procedures used for the many-body corrections are shown in Fig.2 and described in the caption.

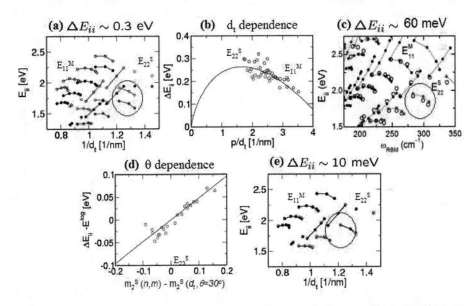

FIGURE 2. Routine for many-body corrections to the Kataura plot. (a) Comparison between E_{ii} obtained from the ETB method (filled circles) and from RRS measurements on HiPco samles (open circles). The deviations between experimental and theoretical ΔE_{ii} values are up to 0.3eV, and they are plot in (b); (b) correction for diameter dependent many-body effects. The open circles give ΔE_{ii} as a function of inverse diameter (p=1,2,3,4... for E_{11}^S, E_{22}^S, E_{11}^M, E_{33}^S...). The solid line gives the many-body correction using Kane and Mele logarithmic functional E^{log} [6]; (c) plot similar to (a), but after including the d_t dependent E^{log} many-body correction from (b). The remaining deviation between experiment and theory plot in (d) is up to 60meV; (d) the open circles give $\Delta E_{ii} - E^{log}$ as a function of the effective mass (m) deviation from the reference value of armchair nanotubes. The solid line gives de correlation between the θ dependent m's and the $\Delta E_{ii} - E^{log}$. (e) plot similar to (a) and (c), but after including both the d_t and θ dependent many-body correction given in (b) and (d).

The ω_{RBM} can be also described within experimental precision considering a functional that takes into account the chirality dependent distortion of the hexagonal lattice. The effect of the lattice distortion on ω_{RBM} was first shown by Kurti et al. [7] with ab initio calculations and confirmed experimentally [5]. The general functional is

$$\omega_{RBM} = 227/d_t - (1+\cos3\theta)/d_t^2 + B^S \qquad \text{(for semiconducting SWNTs)} \qquad (1)$$
$$\omega_{RBM} = 227/d_t - 3(1+\cos3\theta)/d_t^2 + B^M \qquad \text{(for metallic SWNTs)}$$

it describes ω_{RBM} for both HiPco and CoMoCAT SWNTs and it is in agreement with ab initio calculations [7]. The parameter $B^{S,M}$ measures the environmental effects and it is found to be within $10\pm3cm^{-1}$ for different samples. B^M was observed to be larger than B^S, indicating metallic tubes are slightly more affected by the environment. Furthermore, the environment is also observed to broaden asymmetrically and even double the RBM feature from one single (n,m) specie (see Fig.3). The observation of more than one RBM peak for a single (n,m) specie is even more drastic for the inner tubes of double-wall carbon nanotubes (DWNTs), where up to 10 peaks for one (n,m) inner tube has been observed [8]. To observe these effects, however, the use of a highly dispersive triplemonochromator is necessary.

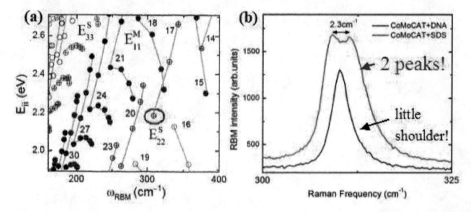

FIGURE 3. (a) Revised Kataura plot. The gray lines connect E_{ii} for SWNTs with same $(2n+m)$ values, and the values are indicated in the figure. The E_{22}^S for the $(6,5)$ SWNT is highlighted by a red ellipse. Radial breathing mode (RBM) spectra at E_{22}^S for the $(6,5)$, for CoMoCAT SWNTs wrapped by SDS (upper spectrum) and DNA (lower spectrum).

Environmental effects are also observed on the optical transition energies, and E_{ii} is observed to change by tens of meV when SWNTs are on different environments (wrapped by DNA, SDS, bundled, in suspension). Laser power also changes E_{ii} due to sample heating.

DETERMINATION OF SWNT POPULATION IN A SAMPLE

The relative RRS intensity for each RBM feature in a sample (I_{RBM}) should provide a good measure of the population of each specific (n,m) SWNT in the sample. However, this result is not directly obtained from the I_{RBM}. It should be corrected by the (n,m) dependent Raman cross section, since both electron-photon and electron-phonon interaction in SWNTs depend on diameter and chiral angle [9-11].

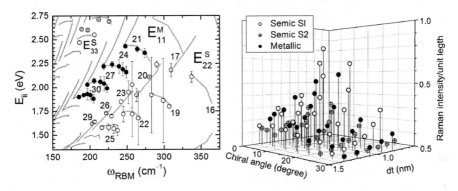

FIGURE 4. (a) Experimental Kataura plot for HiPco SWNT sample in SDS solution. The gray lines connect E_{ii} for SWNTs with same ($2n+m$) values, and the values are indicated in the figure. The open and filled circles are experimental results for semiconducting and metallic SWNTs, respectively. The error bars indicate the I_{RBM} of the respective RBM peak. (b) Calculated RRS intensity as a function of d_t and θ [9].

The circles in Fig.4(a) show the experimentally obtained E_{ii} [2], and the error bars give the relative intensities. Fig.4(b) shows the calculated I_{RBM} as a function of diameter and chiral angle. The calculations show that I_{RBM} should be larger for SWNTs with lower θ, and this result is indeed observed experimentally (Fig.4(a)). The ratio between the experimental (Fig.4(a)) and theoretical (Fig.4(b)) intensities give SWNT population [1,5].

THE RICHNESS OF THE IFM SPECTRAL REGION

Finally, we would like to call the attention to a spectral region that we call intermediate frequency modes (IFM), appearing between the well studied RBM and tangential modes (see Fig.5). The IFM spectral region is rich on features that exhibit a strong dependence on the excitation laser energy [12] as well as on sample differences. Fig.5(a) shows the IFMs for different SWNTs (see spectrum labels) [13]. The IFMs are very different for different samples. This result shows that the understanding of the IFM physics will probable largely advance the capability for spectroscopic characterization of SWNT samples. Although the IFM features are usually one or two orders of magnitude lower in intensity than the well studied RBM and G band features, they clearly exhibit resonant behavior, it can be connected to the

Kataura plot [12.13] and they are sometimes observed to be as intense as the RBM features, as shown in Fig.5(b).

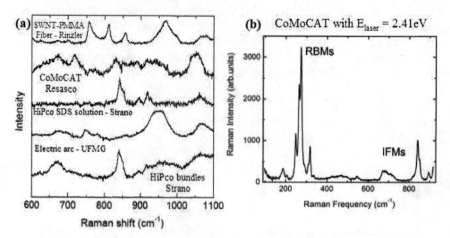

FIGURE 5. (a) The IFM spectra obtained with different SWNT samples. (b) The RBM and IFM spetrum obtained for CoMoCAT SWNT sample with the excitation laser energy $E_{laser} = 2.41eV$.

ACKNOWLEDGMENTS

The Brazilian authors acknowledge CNPq, FAPEMIG and PRPq-UFMG FOR financial support. T.U. acknowledges a Grant-in-Aid (No. 13440091) from the Ministry of Education, Japan. MIT authors acknowledge support under NSF Grants DMR 01-16042, and INT 00-00408.

REFERENCES

1. A.Jorio *et al.* (2005) Submitted.
2. C. Fantini *et al.* Phys. Rev. Lett. **93**, 147406 (2004).
3. Ge G. Samsonidge *et al.*, Appl. Phys. Lett. **85**, 5703 (2004).
4. V. Popov New J. Phys. **6**, 17 (2004).
5. A.Jorio *et al.*, Phys. Rev. B **71**, 075401 (2005).
6. C.L. Kane and E.J. Mele, Phys. Rev. Lett. **93**, 197402 (2004).
7. J. Kurti *et al.* New J. Phys. **5**, 125 (2003).
8. R. Pfeifer *et al.* IWEPNM 2005 – this proceedings.
9. J. Jiang *et al.*, Phys. Rev. B (2005).
10. V. Popov *et al.*, Nanotech. (2004).
11. M. Machon *et al.*, Phys. Rev. B **71**. 035416 (2005).
12. C. Fantini *et al.*, Phys. Rev. Lett. **93**, 087401 (2004).
13. C. Fantini *et al.*, (2005). In production.

Family Behavior of the Pressure and Temperature Dependences of the Band Gap of Semiconducting Carbon Nanotubes

Rodrigo B. Capaz[1,2,3], Catalin D. Spataru[1,2], Paul Tangney[4], Marvin L. Cohen[1,2], and Steven G. Louie[1,2]

[1]Department of Physics, University of California at Berkeley, Berkeley, CA 94720, USA
[2]Materials Science Division, Lawrence Berkeley National Laboratory, Berkeley, CA 94720, USA
[3]Instituto de Física, Universidade Federal do Rio de Janeiro, Caixa Postal 68528, Rio de Janeiro, RJ 21941-972, Brazil
[4]The Molecular Foundry, Lawrence Berkeley National Laboratory, Berkeley, CA 94720, USA

Abstract. Our recent work on pressure and temperature effects on the band gap of semiconducting carbon nanotubes is discussed. The family behavior in both phenomena is elucidated in terms of the structural dependent orientation of the bonding and antibonding characters of the band edge states.

Keywords: Carbon nanotubes, electronic properties, optical properties.
PACS: 61.46.+w, 62.25.+g, 62.50.+p, 73.22.–f, 63.22.+m, 71.38.–k

INTRODUCTION

The temperature and hydrostatic pressure dependences of the band gap (E_g) are two of the most fundamental properties of a semiconductor. Understanding of these shifts provides important information about the nature of the band-edge electronic states and their coupling with static and dynamic lattice distortions. In traditional semiconductors, direct band gaps typically increase with pressure and decrease with temperature [1].

Carbon nanotubes are relatively novel semiconductor materials [2] with a variety of potential applications [3]. We have recently calculated the shifts of the band gap with both hydrostatic pressure [4] and temperature [5] for isolated single-wall carbon nanotubes (SWNTs). In both cases, clear "family behaviors" (i.e., SWNTs with same value of ($n-m$) behaving similarly) have been observed and explained. In particular, we found that the pressure and temperature band gap shifts have different signs for different nanotube families, in contrast with traditional semiconductors. In this work, we review these results and explore the connections between the temperature and pressure dependences of the band gap. We show that both phenomena can be explained within a single framework, where the key concept is the anisotropy in the

CP786, *Electronic Properties of Novel Nanostructures*, edited by H. Kuzmany, J. Fink, M. Mehring, and S. Roth
© 2005 American Institute of Physics 0-7354-0275-2/05/$22.50

bonding and antibonding characters of the band-edge states with respect to the underlying geometric structure.

HYDROSTATIC PRESSURE EFFECTS

Figs. 1(a) and 1(b) summarize our results for the pressure coefficients of the band gap in the linear regime (dE_g/dP) for several SWNTs with various diameters and chiralities. Density-functional theory (DFT) has been used for obtaining the diameter-dependent elastic constants (and therefore the atomic relaxations under hydrostatic pressure conditions) and tight-binding (TB) electronic structure calculations have been used to calculate dE_g/dP. Details of the methodology are described in Ref. [4]. Fig. 1(a) shows that dE_g/dP displays family behavior and Fig. 1(b) shows that such apparently complicated dependence on diameter and chirality can be described by a simple phenomenological equation:

$$\frac{dE_g}{dP} = A + B(-1)^q d \cos(3\theta),\qquad (1)$$

where d and θ are the SWNTs diameter and chiral angle, respectively, $q = (n-m)\,\mathrm{MOD}\,3$, and A and B are constants. Analytical expressions for A and B can be obtained within a single-orbital TB model [4]. A physical interpretation of the second term in the right-hand-side of (1) has been offered, within a graphene band-folding scheme, in terms of the relative movement of the quantization lines with respect to the Fermi points [4].

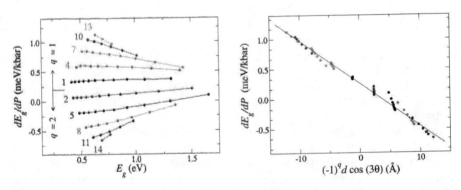

FIGURE 1. (a) Band-gap pressure coefficient as a function of the gap for a large number of semiconducting SWNTs. Tubes are grouped into according to their (n-m) family. The values of (n-m) for each family are also shown in the figure. (b) Collapse of dE_g/dP values to a single line when plotted against $(-1)^q d \cos(3\theta)$. Reproduced with permission from Ref. [4].

In this work, we propose an equivalent (although perhaps more intuitive) interpretation. Let us consider, for simplicity, the (10,0) and (11,0) tubes as prototypical examples of $q = 1$ and $q = 2$ SWNTs, respectively. Fig. 2 show $|\psi|^2$ for conduction (upper panels) and valence (lower panels) band edge states, for the (10,0) (left panels) and (11,0) (right panels) SWNTs, respectively. One notices that, for the (10,0) tube, the valence state has bonding character along the circumference and antibonding character along the tube axis, whereas the conduction state behaves in the opposite way. For the (11,0), the situation is reversed: The valence state is antibonding along the circumference and bonding along the axis, and the conduction state is again the opposite.

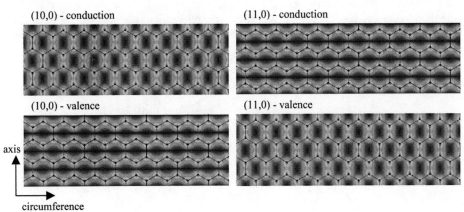

FIGURE 2. Electronic densities associated to the band edge states for the (10,0) and (11,0) SWNTs.

The two terms in expression (1) for dE_g/dP can now be explained in terms of the interplay of the anisotropic deformations introduced by hydrostatic pressure and the anisotropy of the band-edge states themselves. Carbon nanotubes have different elastic constants along the circumferential and axial directions: Upon hydrostatic pressure conditions, the radial strains are larger than axial ones [4]. An isotropic compression of the tubes would lead, in a band-folding scheme, to a constant and positive dE_g/dP for all tubes, due to the overall increase in the Fermi velocity of graphene. This is the meaning of the constant A, that we can now call "isotropic contribution" to dE_g/dP. However, since circumferential deformations are larger, gap shifts will also be more sensitive to the circumferential character (bonding or antibonding) of the band-edge states. For example, for the (10,0) SWNT, circumferential compression will lead to the valence (conduction) band state to shift down (up) in energy due to its bonding (antibonding) character along the circumference. Therefore, this effect will lead to gap increase in the (10,0) SWNT. The situation for the (11,0) SWNT will be precisely the opposite, leading to gap decrease. This is the physics behind the second term in Eq. (1), that we can now call the "anisotropic contribution" to dE_g/dP.

Pressure coefficients of the band gap have been measured in micelle-coated SWNTs in solution [6]. In contrast to our results for isolated tubes, only negative values of dE_g/dP have been found, regardless of the SWNT family. This discrepancy is still not clearly understood, but it is possible that interactions with the micelle coating could play a role.

413

TEMPERATURE EFFECTS

Fig. 3 illustrates the thermal shifts of the band gap for the (10,0) and (11,0) SWNTs. Once again, qualitatively different results are found depending on the value of q, as described in detail in Ref. [5]. In particular, at low temperatures there is a tendency for $dE_g/dT < 0$ for $q = 1$ and $dE_g/dT > 0$ for $q = 2$. Good semi-quantitative agreement is found in comparison with recent measurements in suspended tubes [7]. We have shown that this behavior can be traced to the contribution of a few low-frequency phonon branches ("shape-deformation modes") to the thermal shift of the band-edge states [5]. Here we show an interesting connection between this behavior and the pressure shifts of the band gap described in the previous section.

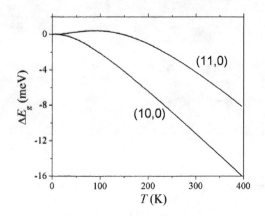

FIGURE 3. Thermal shifts of the band gap for the (10,0) and (11,0) SWNTs .

The harmonic contribution to the energy shift of a Bloch state $|\mathbf{k}\rangle$ given atomic displacements u_i $(i = 1,...,3N)$, can be written as a sum of two contributions [8]:

$$\Delta E_{\mathbf{k}} = \sum_{i,j}\left\{\sum_{\mathbf{k}'}\frac{\langle\mathbf{k}|\partial V/\partial u_i|\mathbf{k}'\rangle\langle\mathbf{k}'|\partial V/\partial u_j|\mathbf{k}\rangle}{E_{\mathbf{k}} - E_{\mathbf{k}'}} + \frac{1}{2}\langle\mathbf{k}|\frac{\partial^2 V}{\partial u_i\partial u_j}|\mathbf{k}\rangle\right\}u_i u_j , \qquad (2)$$

where V is the effective single-particle potential. The first ("self-energy") term is due to Fan [9] and the second ("Debye-Waller") is due to Antončik, Brooks and Yu [10]. The temperature dependence of the shift is then obtained by applying thermal averages over the atomic displacements. The self-energy term only contributes to decreasing the gap, therefore the q-dependent sign oscillation at low temperatures has to come from the Debye-Waller term. This term represents a weakening of pseudopotential form factors due to phonon displacements. In the language of tight-binding theory, this can be translated into a reduction of the magnitude of hopping matrix elements along certain bonds [11]. Since the shape-deformation modes involve only radial displacements, only bonds with circumferential components are affected. Therefore, as far as the gap response is concerned, this is *formally equivalent to a radial expansion of the nanotube*. Then, we can readily apply the results from the previous section and conclude that the signs of dE_g/dP and dE_g/dT (at low T) should be reversed, as found numerically.

CONCLUSIONS

The family behavior on the band gap shifts with pressure and temperature can be understood within a unified framework from the interplay between the anisotropy in the bonding and antibonding characters of valence and conduction band states and the anisotropy of the static (in the case of hydrostatic pressure) and dynamic (in the case of temperature) deformations.

ACKNOWLEDGMENTS

This work was partially supported by National Science Foundation Grant No. DMR04-39768 and by the Director, Office of Science, Office of Basic Energy Sciences, Division of Materials Sciences and Engineering, U.S. Department of Energy under Contract No. DE-AC03-76SF00098. RBC acknowledges financial support from the John Simon Guggenheim Memorial Foundation and Brazilian funding agencies CNPq, CAPES, FAPERJ, Instituto de Nanociências, FUJB-UFRJ and PRONEX-MCT. Computational resources were provided by NPACI and NERSC.

REFERENCES

1. Madelung. O., *Semiconductors : group IV elements and III-V compounds*, Springer-Verlag, Berlin, 1991.
2. Iijima, S., *Nature (London)* **354**, 56 (1991).
3. Baughman, R. H., Zakhidov, A. A., and de Heer, W. A., *Science* **297**, 787 (2002).
4. Capaz, R. B., Spataru, C. D., Tangney, P., Cohen, M. L., and Louie, S. G., Phys. Stat. Solidi (b) **241**, 3352 (2004).
5. Capaz, R. B., Spataru, C. D., Tangney, P., Cohen, M. L., and Louie, S. G., Phys. Rev. Lett. **94**, 036801 (2005).
6. Wu, J. *et al.*, Phys. Rev. Lett. **93**, 017404 (2004).
7. Lefebvre, J. *et al.*, Phys. Rev. B **70**, 045419 (2004).
8. Allen, P. B. and Cardona, M., Phys. Rev. B **23**, 1495 (1981).
9. Fan, H. Y., Phys. Rev. **82**, 900 (1951).
10. Antončik, E., Czech. J. Phys. **5**, 449 (1955); Brooks, H. and Yu, S. C., unpublished.
11. Biernacki, S., Scherz, U., and Meyer, B. K., Phys. Rev. B **49**, 4501 (1994).

The role of Van Hove singularities in disorder induced Raman scattering

Jenő Kürti*, János Koltai* and Viktor Zólyomi*

*Department of Biological Physics, Eötvös University Budapest, Pázmány Péter sétány 1/A,
H-1117 Budapest, Hungary

Abstract.
   The disorder induced D band in the Raman spectrum of graphite and carbon nanotubes has been shown to originate from a double resonance process, involving a scattering by a phonon, and a scattering by a defect. In the case of carbon nanotubes, the Van Hove singularities are expected to have a significant effect on the D band. We present a detailed study of this effect on the case of a simple one dimensional semiconducting model system, with quadratic electronic and phonon dispersion relations. The D band is calculated by exact integration of the perturbation formulas and the dependence of the position and the intensity on the laser excitation energy and the damping factor describing the finite lifetime is examined in detail. The intensity of the D band shows an extra enhancement when the energy of the incoming or the outgoing photon matches or nearly matches the energy of a Van Hove singularity of the nanotube.

Keywords: double resonance, D band, Van Hove singularity
PACS: 78.30.Ly, 78.67.Ch

INTRODUCTION

The disorder induced D band in the Raman spectrum of graphite and carbon nanotubes has been shown to originate from a double resonance scattering process, involving a scattering by a phonon, and a scattering by a defect [1]. The concept of double resonance, which is able to describe the shift of the D band as a function of the laser excitation energy (dispersion by $\approx$ 50-60 cm^{-1}/eV) [2, 3], as well as the fact that the anti-Stokes D band lies about 7-8 cm^{-1} higher than the Stokes D band [4, 5], has become widely accepted by now.

   Generally, the Raman cross section is the absolute square of the complex transition matrix element $K_{2f,10}$, which can be obtained with the well known formulas of perturbation theory as the sum of the complex amplitudes of many virtual processes [6]. In the usual, lowest order Raman only zone-centered phonons (q=0) are involved. The double resonance process, however, is a higher order Raman scattering, to which any phonon in the Brillouin zone can contribute, given that the phonon scattering is supplemented by another, phononless scattering due to a defect. This is necessary in order to fulfill the condition of the electron-hole recombination, for which the electron and the hole must have the same wave-vector. The small concentration of defects is compensated by the double resonance, where the real part of two of the energy denominators becomes zero at the same time, for a given term of the perturbation sum.

   In the case of graphite, zone boundary phonons with wave-vector near the K point are responsible for the D band. Because the dispersion relations of single wall carbon

CP786, *Electronic Properties of Novel Nanostructures*, edited by H. Kuzmany, J. Fink, M. Mehring, and S. Roth
© 2005 American Institute of Physics 0-7354-0275-2/05/$22.50

nanotubes (SWCNTs) are closely related to that of graphene, the D band of SWCNTs has the same origin as the D band of graphite. However, there is an extra effect in the case of SWCNTs which is not present for graphite, the Van Hove (VH) singularities.

There is still some confusion concerning the role of the VH singularities in the intensity of the D band of SWCNTs. In the works of Thomsen et al the VH singularities are completely neglected, the properties of the double resonance process in one dimension have always been demonstrated on the example of the linear electronic dispersion, where of course there is no VH singularity [1, 2, 7, 8]. On the other hand, it was shown by numerical integration of the Raman cross section that in addition to the double resonance, an extra enhancement occurs when the energy of either the incoming or the outgoing photon matches one of the VH singularities of the SWCNT [9]. However, care must be taken in order to get reliable result with numerical integration. The - multiple - integration should be carried out throughout the whole Brillouin zone using a high enough density of k-points. The problem with not enough k-points in the numerical integration is that it might lead to incorrect results due to not properly handling interference effects between the complex amplitudes. For example, as it was shown by Maultzsch et al, using analytical integration in the case of a linear gapless electronic dispersion, the contribution to the cross section of the D band by phonons from exactly the K point vanishes due to destructive interference [8].

We were motivated by the analytical calculation in Ref. [8]. However, as our aim was to investigate the effect of the VH singularity on the D band, a different one dimensional model dispersion was chosen where the electronic density of states diverges at some k-points, in contrast to the linear dispersion for which the density of states is constant. In order to exclude any deficiencies of numerical integrals, the perturbation formulas were evaluated fully analytically. The only approximation was the usual assumption that the matrix elements in the numerators were constant.

Clarifying this issue is important, because the correct interpretation of the experimental D band spectra inevitably requires the consideration of VH singularities due to the selective laser enhancement, both for bundles as well as single individual tubes [9, 10].

THE MODEL SYSTEM

The simplest one dimensional electronic dispersion which has VH singularities and which is able to simulate the D band of SWCNTs, is a quadratic one for both the electron and hole energies: $\varepsilon_{electron}(k) = -\varepsilon_{hole}(k) = \frac{\Delta}{2} + A(k \pm \frac{q_0}{2})^2$ where Δ is the gap between the symmetrically assumed electron and hole dispersions, and q_0 is the wavenumber connecting the minima of the two branches of the quadratic electronic dispersion curves. In order to simulate the behavior of a SWCNT, q_0 can be thought as of being the difference (or nearly the difference) between the K and K' (or equivalently between the Γ and K') points in the Brillouin zone of graphene, but actually the exact value of q_0 does not play any role in this model calculation. Similarly, a simple quadratic model dispersion of $\omega_{ph}(q) = \omega_0 + B(q - q_0)^2$ was chosen for the phonon frequencies. Realistic values of Δ=2 eV and ω_0=0.15 eV ($\approx$ 1200 cm^{-1}) were used in the calculations. The values for A and B were 4 eV/Å^2 and 0.05 eV/Å^2, respectively.

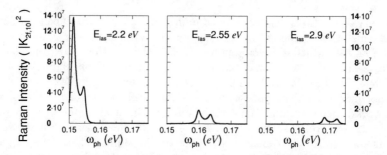

FIGURE 1. Calculated Raman D band with $\gamma=0.03$ eV damping parameter, for three different E_{las} laser excitation energies. The two peaks correspond to incoming and outgoing resonances.

RESULTS AND DISCUSSION

The $K_{2f,10}$ complex transition matrix element can be obtained as the sum of the contributions of different Feynman diagrams [9]. In our case: $K_{2f,10} = \Sigma_{\alpha,b,\psi} K_{\alpha b \psi}$, where the individual terms can be expressed, after straightforward calculations, as follows:

$$K_{\alpha b \psi} = i\pi \frac{(\kappa + \kappa')(\psi + \kappa)(\psi + \kappa')(\psi + \kappa + \kappa') - b^2[(1+\alpha)^2 \kappa \kappa' + \psi(\kappa + \kappa')]}{A^3 \psi \kappa \kappa'[(\psi + \kappa)^2 - b^2][(\psi + \kappa')^2 - b^2][(\kappa + \kappa')^2 - (1+\alpha)^2 b^2]}$$

where $\alpha = \pm 1$, $b = \frac{q_0 \pm q}{2}$, $\psi^2 + b^2 = \kappa^2$, κ'^2, $\kappa = \sqrt{\frac{E_{las} - \Delta - i\gamma}{2A}}$, $\kappa' = \sqrt{\frac{E_{las} - \Delta - \hbar\omega_{ph} - i\gamma}{2A}}$, and γ is the damping factor describing the finite lifetime of the electrons (and the phonons).

For $E_{las} < \Delta$ the intensity of the calculated Raman spectra is very low. There is an abrupt increase in the intensity at $q = q_0$ ($\omega_{ph} = \omega_0$), for $E_{las} = \Delta$ (incoming resonance) and a second one at $E_{las} = \Delta + \hbar\omega_0$ (outgoing resonance). Above this laser excitation energy, the Raman spectrum consists of two peaks, at least for low enough damping parameters. For large γ values (e.g. for 0.1 eV), the two peaks are smeared together. Increasing the laser energy results in an upward shift of the spectrum, this is the well known dispersion of the D band. At the same time the intensity of the D band decreases. Figure 1 illustrates all these for three different laser excitations, in the case of $\gamma = 0.03$eV.

In this work we concentrate ourselves on the intensity of the D band. Figure 2 shows the peak intensity as a function of the laser excitation energy, for three different damping parameters. For $\gamma=0.01$ eV and $\gamma=0.03$ eV the two peaks for the incoming and outgoing resonances can be clearly distinguished. The intensity of both peaks decreases dramatically after the maximum with increasing E_{las}. At $E_{las}=3$ eV the peak intensity is two orders of magnitude ($\gamma=0.01$eV) or a factor of 40 ($\gamma=0.03$eV) less than the maximum intensity which occurs when the incoming (or outgoing) photon matches the VH singularity at 2 eV. Even for $\gamma=0.1$eV one order of magnitude decrease can be observed when the laser energy is increased from 2.1 eV to 3 eV.

In conclusion: matching the VH singularity by the laser excitation causes a significant increase of the D band of SWCNTs, and has to be taken into account in the interpretation

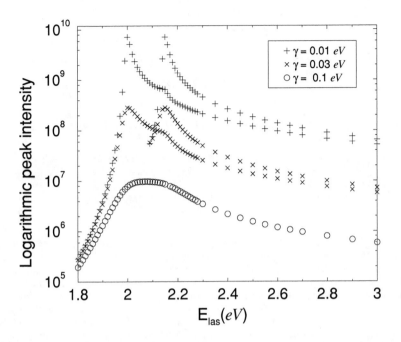

FIGURE 2. Peak intensity vs laser excitation energy (E_{las}) for three different damping parameters (γ). (Note the logarithmic scale.)

of experimental spectra.

ACKNOWLEDGMENTS

This work was supported by the OTKA T038014 and the OTKA T043685 grants in Hungary.

REFERENCES

1. C. Thomsen and S. Reich, Phys. Rev. Lett **85**, 5214 (2000).
2. S. Reich, C. Thomsen, and J. Maultzsch, *Carbon Nanotubes* (Wiley-VCH Verlag GmbH & Co. KGaA, Berlin, 2004).
3. I. Pócsik, M. Hundhausen, M. Koós, and L. Ley, J. Non-Cryst. Solids **227-230B**, 1083 (1998).
4. P. H. Tan, Y. M. Deng, and Q. Zhao, Phys. Rev. B **58**, 5435 (1998).
5. V. Zólyomi and J. Kürti, Phys. Rev. B **66**, 073418 (2002).
6. R. M. Martin and L. M. Falicov, in *Light Scattering in Solids*, edited by M. Cardona (Springer, Berlin, 1983), p. 79, topics in Applied Physics Vol. 8.
7. J. Maultzsch, S. Reich, and C. Thomsen, Phys. Rev. B **64**, 121407 (2001).
8. J. Maultzsch, S. Reich, and C. Thomsen, Phys. Rev. B **70**, 155403 (2004).
9. J. Kürti, V. Zólyomi, A. Grüneis, and H. Kuzmany, Phys. Rev. B **65**, 165433 (2002).
10. A. Jorio, R. Saito, J. H. Hafner, C. M. Lieber, M. Hunter, T. McClure, G. Dresselhaus, and M. S. Dresselhaus, Phys. Rev. Lett **86**, 1118 (2001).

Quantum Dots from Irradiated Carbon Nanotubes

Fabrizio Cleri* and Pawel Keblinski§

*Ente Nuove Tecnologie, Energia e Ambiente (ENEA), Unità Materiali e Nuove Tecnologie,
Centro Ricerche Casaccia, 00100 Roma A. D. (Italy)

§Department of Materials Science and Engineering, Rensselaer Polytechnic Institute,
Troy NY 12180-3590 (USA

Abstract. We modeled by finite-temperature tight-binding molecular dynamics the welding of crossed (5,5) metallic nanotubes, as demonstrated in recent electron-beam irradiation experiments. By studying the correlation between the atomic and electronic structure of the simulated systems, three classes of junctions among metallic nanotubes were individuated, as a function of the degree and type of disorder created by the irradiation in the junction region. The three classes of junction display, respectively, ohmic cunduction, hopping conduction, and quantum confinement, indicating the formation of a quantum dot contacted to metallic nanowires by tunnel barriers. Implications for molecular-electronics and quantum-computing devices based on arrays of such quantum dots are elaborated and discussed.

Keywords: Carbon nanotubes, quantum dots, nanostructures, molecular electronics.
PACS: 61.46.+w, 71.23.An, 73.21.La

INTRODUCTION

Carbon nanotubes (CNTs) have been proposed as a main component of future nanoscale electronics, either as fully active circuit elements in the form of linear, Y-shaped or T-shaped heterojunctions [1,2], or as interconnects in molecular-level devices [3]. Junctions between CNTs will be a key factor in the design of such nanoscale integrated devices. Notably, the properties of crossed nanotube junctions bonded by van der Waals interactions have been already investigated in some detail by both experimental and theoretical methods [4-6]. In such weakly-coupled junctions conduction may occur purely by tunneling, the conductance being dominated by the large contact resistance between the CNT.

An entirely different, and much less understood situation is encountered in the strongly-coupled, covalently-bonded junctions obtained by, e.g., electron-beam irradiation of crossed CNTs [7,8]. In this case a whole range of electrical conduction mechanisms, from purely ohmic, to phonon-assisted hopping, to quantum tunneling, is expected [9] depending on both the electrical properties of pristine nanotubes and the nature of the covalent bonds formed upon irradiation.

CP786, *Electronic Properties of Novel Nanostructures*, edited by H. Kuzmany, J. Fink, M. Mehring, and S. Roth
© 2005 American Institute of Physics 0-7354-0275-2/05/$22.50

MOLECULAR DYNAMICS AND ELECTRONIC STRUCTURE

We studied the electronic properties of covalently-bonded, crossed-CNT junctions by means of tight-binding molecular dynamics (TBMD) simulations, based on a well-established otrhogonal TB model for carbon [10]. We calculated the local density-of-states (LDOS) and localization coefficient ("participation ratio") for a number of different atomic structures, going from an "ideal" junction between two coplanar (5,5) CNTs with a small concentration of defects, to junctions between overlayed (5,5) CNTs with increasing density of covalent bonds, to a strongly disordered junction produced by a simulated electron beam irradiation process.

We have found [9] that the coplanar junction displays a substantially ohmic behavior. On the other hand, covalently-bonded junctions between overlayed CNTs at about 0.5 Å intertube distance display an increasing number of defect-induced electronic states around the Fermi energy E_F which perturb and eventually suppress the ohmic conduction, allowing for an increasing role of hopping conduction between strongly localized states. The most interesting behavior, however, is represented by the strongly disordered junctions, for which the spectrum around $E=E_F\pm0.5$ eV is composed of quantized levels weakly localized over the whole junction region, thus giving rise to a quantum dot (QD). Such an all-carbon QD structure appears very promising for several applications. It should be experimentally realizable by manipulating CNTs with the tip of an atomic-force microscope over properly arranged metallic contacts, and by subsequently irradiating the junction region as demonstrated in [7,8] to create a single QD, pairs of coupled QDs and, eventually, arrays of regularly spaced QDs by means of tailored CNT-growth techniques [11,12].

The atomic and electronic structure of a "realistic" junction is shown in Figure 1a. Two segments of (5,5) CNTs with 340 atoms each were overlayed at a distance of 0.5 Å along the z axis. A simulated electron-beam irradiation procedure was initially carried out by combined microcanonical and Langevin molecular dynamics with a Tersoff-Brenner empirical potential [9]. The resulting atomic configuration was subse quently relaxed by TBMD. The final junction appears highly disordered, with a sort of

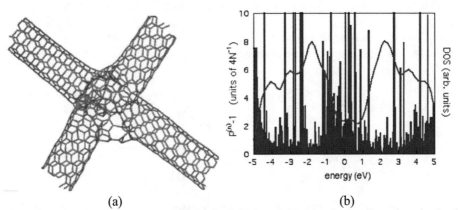

(a)   (b)

FIGURE 1. (a) atomic structure of the crossed (5,5) carbon nanotube junction after simulated irradiation. (b) Density of states (full curve) and localization factor (histogram) for this junction.

amorphous-like particle embedded within four CNT arms, including a large number of miscoordinated atoms, and several four-, five-, seven- and eightfold rings.

The electronic structure of this kind of junction, shown in fig. 1b (full curve) together with the participation ratio (histogram), is quite surprising. A whole band around the Fermi level is localized at the junction region. All the energy levels are quantized within the junction, although being finely spaced by $\Delta E=10\text{--}50$ meV. Such levels are defect states, weakly localized over the whole of the junction sites, and do not belong to the rest of the system. Indeed, the plot of the square modulus of the total wavefunction components for such eigenvalues (see Ref. [9], fig. 4) is substantially different from zero only in the junction region.

Such a combination of energy quantization and weak spatial localization is the typical behavior of a quantum dot (QD). From the local-projected DOS it can be seen that the first few carbon rings immediately adjacent to the QD show a forbidden band just above E_F, i.e., these rings behave as tunnel barriers between the junction and the adjoining perfect-CNT branches.

QUANTUM DEVICES FROM ALL-CARBON QUANTUM DOTS

The above results suggest that by combining nanomanipulation techniques and electron irradiation of CNTs, arrays of all-carbon-based QDs as small as 1 nm spaced by 5-10 nm could be obtained. Such QDs would be naturally connected to ballistic conductors (the CNT arms) by construction, separated by a tunnel-barrier gate allowing quantized charge injection into the QD. The junction region would be made chemically inert by saturation with hydrogen, which should not however alter its basic QD behavior. The application potential of such all-carbon quantum devices as building blocks of advanced quantum electronics and molecular-scale electronic circuits should be immediately evident. For example, reactive substitution of hydrogen with organic radicals should be feasible, to explore connections with the domain of electronically-active molecules.

The quantum-cellular automata (QCA) concept was proposed ten years ago as a transistorless alternative for nanoscale digital circuits [13-15]. A QCA cell consists of four QDs arranged in a 2x2 square, each QD pair containing one electron (Figure 2a). The QCA cell has two complementary polarization states, corresponding to the two electrons occupying antipodal sites in the QD-square because of electrostatic repulsion

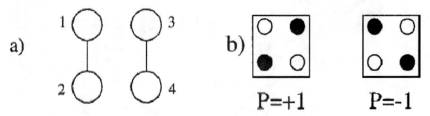

FIGURE 2. (a) schematic of a QCA cell: circles indicate quantum dots, lines indicate tunnel junctions. (b) equivalent polarization states of a QCA cell charged with one electron in each pair of dots.

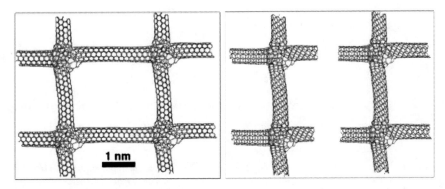

FIGURE 3. A QCA cell obtained by arranging four single-wall carbon nanotubes in a square pattern. (a) electron beam irradiation at the four junctions generates four quantum dots; (b) further irradiation at a higher dose-rate burns two of the connections and leaves two pairs of coupled quantum dots.

(fig. 2b). Notably, it was shown that arrays of such QCA cells can be combined into Boolean gates [15], and even perform all the basic logic operations of a quantum computer [16].

As we show schematically in Figure 3, the QCA basic cell could be obtained by arranging four CNTs in a square and forming QD junctions by electron irradiation (fig. 3a); further electron irradiation at a higher dose-rate [7] can be used to cut away the central branches, to achieve the proper QCA-cell configuration (fig. 3b). The fine spacing between quantized levels in the CNT-junction QDs might prevent charge quantization in the standard way. However, it has been recently demonstrated [17] that QDs can properly operate as QCAs even in the metallic limit [18], whenever the total number of electrons in a cell equals *4N+2* with *N* a possibly large integer.

REFERENCES

1. S. J. Tans, A. R. M. Verschueren, C. Dekker, *Nature* **393**, 49 (1998).
2. Z. Yao, H. W. Ch. Postma, L. Balents, C. Dekker, *Nature* **402**, 273 (1999).
3. P. W. Chiu, G. S. Duesberg, U. Dettlaff-Weglikowska, S. Roth, *Appl. Phys. Lett.* **80**, 3811 (2002).
4. M. S. Fuhrer et al., *Science* **288**, 494 (2000).
5. T. Rueckes et al., *Science* **289**, 94 (2000).
6. H. W. Ch. Postma, M. de Jonge, Z. Yao, C. Dekker, *Phys. Rev.* B **62**, 10653 (2000).
7. F. Banhart, *Nano Lett.* **1**, 329 (2001).
8. M. Terrones et al., *Phys. Rev. Lett.* **89**, 075505 (2002).
9. F. Cleri, P. Keblinski, I, Jang, S. B. Sinnott, *Phys. Rev.* B **69**, 121412 (2004).
10. C. H. Xu, C.Z. Wang, C.T. Chan and K.M. Ho, *J. Phys. Cond. Matter* **4** (1992) 6047.
11. B. Q. Wei et al., *Nature* **416**, 495 (2002).
12. S. Huang, B. Maynor, X. Cai, J. Liu, *Adv. Mater.* **15**, 1651 (2003).
13. C. S. Lent, P. D. Tougaw, W. Porod, *Appl. Phys. Lett.* **62**, 714 (1993).
14. P. D. Tougaw, C. S. Lent, *J. Appl. Phys.* **75**, 1818 (1994).
15. I. Amlani, A. O. Orlov, G. Toth, G. H. Bernstein, C. S. Lent, G. L. Snider, *Science* **284**, 289 (1997).
16. G. Tóth, C. S. Lent, *Phys. Rev.* A **63**, 052315 (2001).
17. C. S. Lent, P. D. Tougaw, *J. Appl. Phys.* **75**, 4077 (1994).
18. M. Girlanda, M. Governale, M. Macucci, G. Iannaccone, *Appl. Phys. Lett.* **75**, 3198 (1999).

Kohn Anomalies and Electron-Phonon Coupling in Carbon Nanotubes

S. Piscanec[1], A. C. Ferrari[1a], M. Lazzeri[2], F. Mauri[2], J. Robertson[1]

[1]Cambridge University, Engineering Department, Trumpington Street, Cambridge, CB2 1PZ, UK.
[2] Institut de Mineralogie et de Physique des Milieux Condenses, Paris, France.

Abstract. Kohn anomalies are distinct features of the phonon dispersion in metallic systems, associated to the presence of a Fermi surface. Graphene is a model system to understand the physical properties of carbon nanotubes. Two Kohn anomalies are present in the phonon dispersion of graphene and their slope is proportional to the square of the electron-phonon coupling. Kohn anomalies are enhanced in metallic nanotubes due to their reduced dimensionality and absent for semiconducting. The electron-phonon coupling for nanotubes of arbitrary chirality and diameter can be obtained from graphite.

INTRODUCTION

The understanding of the physical mechanisms ruling the phonon dispersions and the electron-phonon coupling (EPC) in graphite is a key step to derive the vibrational properties and the Raman intensities of carbon nanotubes. In graphite the interatomic force-constant matrix elements decay very slowly with the distance, and this long-range behavior strongly affects the phonon dispersion of the upper optical branches at **Γ** and **K**. Key to understand the phonons of graphite is the semi-metallic character of its electronic structure. In general, the atomic vibrations are partially screened by electronic states. In a metal this screening can change rapidly for vibrations associated to certain **q** points of the Brillouin Zone. The consequent anomalous behavior of the phonon dispersion is called Kohn Anomaly (KA) [1].

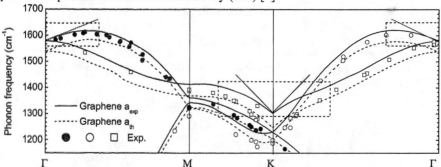

FIGURE 1. Graphene phonons calculated for the experimental (solid lines) and theoretical (dotted lines) lattice parameter. Points: experimental data [6]. The KA slopes are indicated by straight lines.

Graphite has two KA at the Γ-E_{2g} and K-A'1 modes [2]. They are revealed by two

---

[a] Email:acf26@eng.cam.ac.uk

CP786, *Electronic Properties of Novel Nanostructures*, edited by H. Kuzmany, J. Fink, M. Mehring, and S. Roth
© 2005 American Institute of Physics 0-7354-0275-2/05/$22.50

sharp kinks in the phonon dispersion, Fig. 1. The EPC of the Γ-E_{2g} and K-A'_1 modes is particularly large, whilst for the other modes at Γ and K is negligible [2]. These results have immediate implications for nanotubes, since they are either metals or semiconductors. Due to their reduced dimensionality, we expect metallic tubes to have much stronger KA than graphite. Thus, folded graphite cannot give the phonon dispersions of metallic tubes, no matter how accurate the graphite phonons. Semiconducting nanotubes cannot have KAs. In order to correctly describe the nanotube phonons one needs to explicitly account for the strong EPC. It is thus essential to devise an efficient approach to precisely calculate the nanotube phonons.

RESULTS AND DISCUSSION

We perform calculations using density functional perturbation theory (DFPT) within the general gradient approximation (GGA) [3]. We expand the valence electrons on a plane waves basis with a 90Ry cutoff and use Troullier-Martins pseudopotentials [4]. Bands occupancy is determined using a Methfessel-Paxton (graphite) or a Fermi Dirac (nanotubes) smearing function [5].

The calculated graphene phonon dispersions, Fig.1, are characterized by two sharp kinks in the highest optical branches at Γ and K, corresponding to E_{2g} and A'_1 phonons. The theoretical dispersions are compared with the experimental data of [6]. The difference between the experimental and calculated frequencies is less than 2%.

These kinks are clear examples of KAs. The occurrence of KAs is entirely determined by the geometry of the Fermi surface. In particular, they can happen only for phonons with a wave-vector $\mathbf{q}$, which connects 2 electronic states $\mathbf{k}_1$ and $\mathbf{k}_2=\mathbf{k}+\mathbf{q}$, both on the Fermi surface and such that the tangents to the Fermi surface at $\mathbf{k}_1$ and $\mathbf{k}_2$ are parallel [1]. The Fermi surface of graphene consists of the 2 non-equivalent $\mathbf{K}$ and $\mathbf{K'}$ points, with $\mathbf{K'}=2\mathbf{K}$. KAs can then occur for phonons connecting the $\mathbf{K}$ or $\mathbf{K'}$ points to themselves ($\mathbf{q}=\mathbf{K}-\mathbf{K}=\mathbf{0}=\Gamma$) or connecting $\mathbf{K}$ to $\mathbf{K'}$ ($\mathbf{q}=\mathbf{K'}-\mathbf{K}=2\mathbf{K}-\mathbf{K}=\mathbf{K}$). For a given $\mathbf{q}$ there are several phonon branches. In graphite, only the highest optical branches show KAs, due to their very high EPC [2]. It is possible to derive an exact analytical description of the slope of the two KA in graphite [2]:

$$\alpha_\Gamma^{LO} = \sqrt{3}\pi^2 \left\langle g_\Gamma^2 \right\rangle_F / \beta = 397 cm^{-1} \qquad \alpha_K = \sqrt{3}\pi^2 \left\langle g_K^2 \right\rangle_F / \beta = 973 cm^{-1} \qquad (1)$$

where $\langle g_\Gamma^2 \rangle$ is the average of the 4 EPCs corresponding to all the possible transition between the π and $\pi*$ bands at E_F; β=14.1 eV is the slope of the π and $\pi*$ bands near K. We get $\langle g_\Gamma^2 \rangle$=0.0405 eV^2 for the Γ-E_{2g} mode and zero for all the other branches. Similarly, at K the EPC is non negligible only for the A'_1 mode, $\langle g_K^2 \rangle$=0.0994eV^2. This proves that the only bands which are affected by the KA are those with non-zero EPCs. Our theoretical EPCs are validated by direct EPC measurement derived from the experimental optical phonon dispersions around Γ, from the Raman D peak dispersion and from the width of the Raman G peak [2,7].

KAs should then be present also in metallic nanotubes, since the geometrical condition for KAs is the same as graphene. Direct *ab-initio* calculations are necessary to properly describe nanotubes phonons, but this is practically unfeasible for nanotubes of arbitrary chirality, given their big unit cells. Nanotubes are identified by their chiral indexes (n,m), specifying the chiral vector $\mathbf{C_h}=n\mathbf{a_1}+m\mathbf{a_2}$, where $\mathbf{a_1}$ and $\mathbf{a_2}$

are the graphene lattice vectors. A nanotube is obtained by folding graphene so that that the two atoms connected by $\mathbf{C_h}$ coincide. In real space, a nanotube is periodic along its axis according to the translational vector $\mathbf{T}$. The electron-states correspond to the graphene states of periodicity $\mathbf{C_h}$, i.e. with $\mathbf{k} \cdot \mathbf{C_h} = 2\pi o$, o being an integer. The electron states are labeled by two-dimensional momentum vectors $\mathbf{k}$, which cut with a series of parallel lines the graphene Brillouin Zone (BZ), Fig.2(A). We refer to this set of lines as $ZF_{(n.m)}$. For nanotubes with diameters >0.8 nm the electron states are very well described by the folded graphene bands [8]. Then, the electronic structure of a (n,m) nanotube can be calculated on a graphene unit cell sampled with a set of k-points belonging to $ZF_{(n.m)}$. The EPCs of nanotubes with d>0.8nm can also be derived by folding the corresponding graphene EPCs [9]:

$$S \, | \, EPC_{tube} \, |^2 = \hat{S} \, | \, EPC_{graphene} \, |^2 \qquad (2)$$

where $\hat{S} = a_{th}^2 \sqrt{3} / 2$ is the graphene unit-cell surface, $a_{th} = 2.46$ Å is the graphene lattice spacing, and $S = |\mathbf{C_h} \mathbf{x} \mathbf{T}|$ is the nanotube unit-cell surface.

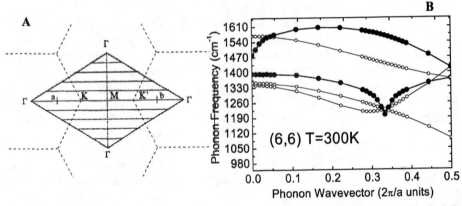

FIGURE 2. (A) graphite BZ. The dashed hexagons indicate hcp symmetry. The lines cutting the BZ are $ZF_{(5,5)}$. (B) Highest optical branches for a (6,6) nanotube calculated at 300K. Two KA are seen for phonons corresponding to the graphite Γ-E_{2g} and K-A'$_1$ modes.

Thus, the calculation of the phonon dispersions of nanotubes of arbitrary diameter and chirality is reduced to the calculation on a graphite unit cell sampled along $ZF_{(n,m)}$ [10]. This is by far less demanding than using the full unit cell, and can be easily performed within DFTP. Furthermore, in order to properly describe the shape of the KA, a equi-spaced k-point BZ sampling is very inefficient, requiring up to several thousands k-points for convergence. The most efficient way to numerically integrate the one-dimensional divergences arising from the KAs is to use a logarithmic sampling. Thus, we sample the BZ with logarithmic grids centered at **K** and **K'** [10]. This new adaptive refolding method allows us to efficiently and precisely calculate the complete temperature-dependent phonon-dispersions and the KA shape for any nanotube of any diameter and chirality.

A diameter dependence of phonon softening in metallic nanotubes was firstly reported in [11]. However, only the Γ modes were studied and the limited BZ sampling prevented to investigate the phonon softening as a function of temperature and EPCs. Our calculations show that metallic nanotubes have much stronger KA than graphite, for the phonons corresponding to the graphite Γ-E_{2g} and $\mathbf{K}$-A'_1 modes, Fig 2b. The mode softening depends on the electronic temperature and on tube diameter. The KAs are more intense for small diameters and at low temperatures. In one-dimensional metallic systems the EPC enhancement leads to a Peierls transition [12], which drives the system into an insulating phase by opening a gap. Thus, at 0 K all metallic tubes are not stable and undergo a Peierls distortion [10,12]. For diameters <0.4 nm, the KA is so strong to induce a Peierls distortion at room temperature [13]. We find that the Peierls distortion temperature decreases exponentially with the diameter [10]. Thus, for nanotubes with diameters >0.8 nm, this temperature is smaller than 10^{-8} K and, so, no Peierls distortion can be observed at room temperature.

Our findings also explain the differences in the Raman spectra of metallic and semiconducting nanotubes. For metallic nanotubes, the G⁻ peak is red-shifted, broader and more intense than in semiconductining nanotubes [14]. We interpret the G⁻ peak in metallic nanotubes as the signature of KA and of high EPC, and dismiss its assignment to a plasmon Fano resonance [15]. Indeed, the G⁻ peak experimental dependence on diameter is well reproduced by our calculations up to 2.5 nm diameter tubes [7,10].

CONCLUSIONS

Two KAs are present in the phonon dispersion of graphene and their slope is proportional to the EPC square. KAs are enhanced in metallic nanotubes, due to their reduced dimensionality. The EPC for nanotubes of arbitrary chirality and diameter can be obtained from graphite.

ACKNOWLEDGEMENTS

Calculations performed at HPCF (Cambridge) and IDRIS (Orsay) with PWscf [16]. SP acknowledges funding from EU project FAMOUS and Marie Curie Fellowship IHP-HPMT-CT-2000-00209. ACF acknowledges funding from the Royal Society.

REFERENCES

1. W. Kohn, Phys. Rev Lett. 2, 393 (1959)
2. S. Piscanec et al, Phys. Rev. Lett. 93, 185503 (2004)
3. S. Baroni et al., Rev. Mod. Phys. 73, 515, (2001)
4. N. Troullier and J. L. Martins, Phys. Rev. B 43, 1993 (1991)
5. M. Methfessel and A.T. Paxton, Phys. Rev. B 40, 3616 (1989)
6. J. Maultzsch et al, Phys. Rev. Lett 92, 075501 (2004)
7. M. Lazzeri et al. unpublished (2005)
8. V. Zolyomi , J. Kurti, Phys. Rev. B 70, 085403 (2004).
9. M. Lazzeri et al. cond-mat/0503278
10. S. Piscanec et al. unpublished (2005)
11. O. Dubay, G. Kresse, H. Kuzmany, Phys. Rev. Lett. 88, 235506 (2002)
12. R.E. Peierls, *Quantum Theory of Solids* (Oxford University Press, New York, 1955), p. 108.
13. K.P. Bohnen et al. Phys. Rev. Lett. 93, 245501 (2004); D. Connetable et al. ibid. 94, 015503 (2005).
14. A. Jorio et al. Phys. Rev. Lett. 90, 107403 (2003)
15. C. Jiang et al. Phys. Rev. B 66, 161404 (2002)
16. S. Baroni et al. http://www.pwscf.org

New Method for the Calculation of Hydrogen Adsorption at Nanotube Surfaces.

A.S.Fedorov, P.B.Sorokin

Kirensky Institute of Physics, Siberian Division,Russian Academy of Sciences, Akademgorodok, Krasnoyarsk

Abstract. We propose here new method for calculation of the thermodynamics of adsorbed particles. Wherewith the quantum and temperature effects are taken into account. The method is based on the solution of Schrödinger equation for adsorbate and using of Gibbs distribution for particle positions and occupation of energy levels. The method is applied for investigation of hydrogen adsorption at internal and external surfaces of CNT of (10,10) and (20,0)kinds. . It is shown the ρ (P,T) does not exceed ~3-4% (at both surfaces adsorption) at pressure up to 500 Bar and low temperatures.

Keywords: carbon nanotubes, adsorption, hydrogen, phase transitions
PACS: 61.46.+w, 68.35.Np, 68.43.De, 68.43.

INTRODUCTION

Development of new kinds of energy (hydrogen energy accumulators, etc.) demand the necessity of rigorous theoretical description of adsorption. One of interesting properties of carbon nanotubes (CNT) is the possibility to adsorb different atoms or molecules, including hydrogen. Ability to adsorb more than 3-5%(wt.) H_2 molecules inside CNT had been experimentally confirmed in some works, see e.g. [1]. But ambiguous experimental results of H2 adsorption and demand of the better understanding of adsorption phenomenon in nanostructures like CNT invoke rigorous theoretical investigations. Unfortunately, such theoretical investigations are typically restricted by lattice gas model, molecular dynamics or Monte-Carlo calculations. At that the adsorbate molecules are usually treated as classical particles without any quantum effects consideration. But quantum effects may be important when light molecule is adsorbed via weak van-der-Walls interactions [2]

In the proposed method, the hydrogen molecules are treated as quantum particles and Schrödinger equation for them is solved. At that the molecule interact with other hydrogen molecules via van-der-Walls interactions in the form of Silver-Goldman potential [3]. Similar potential is used for taking into account of interaction of hydrogen with carbon atoms of CNT surfaces. Schrödinger equation (1) is numerically solved inside unit cell of splay prism shape. The unit cell and its projection at the nanotube surface are shown at Fig.1. One can see 3 nearest hydrogen molecules at the right subfigure.

CP786, *Electronic Properties of Novel Nanostructures*, edited by H. Kuzmany, J. Fink, M. Mehring, and S. Roth
© 2005 American Institute of Physics 0-7354-0275-2/05/$22.50

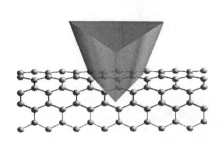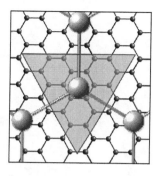

FIGURE 1. H_2 unit cell and its projection to the nanotubes surface

$$\left(-\frac{\Delta}{2m}+V(r)\right)\Psi_i(r)=\varepsilon_i\Psi_i(r)$$

$$V(r)=V_{C-H}(r)+V_{H-H}(r)$$

1)

After Schrödinger equation solving, at temperature T≠0 it is used Gibbs distribution (2,3) for occupation Pi of all energy levels ε_i. Nonzero excited level occupations Pi leads to the contribution in the entropy S1 (3).

$$\langle U\rangle=\frac{1}{Z}\sum_i\varepsilon_i\exp\left(-\frac{\varepsilon_i}{k_BT}\right);\quad Z=\sum_i\exp\left(-\frac{\varepsilon_i}{k_BT}\right)$$

2)

$$S_1=-k_B\sum_i P_i\ln(P_i);\quad P_i=\frac{1}{Z}\exp\left(-\frac{\varepsilon_i}{k_BT}\right)$$

3)

Gibbs distribution leads to the total energy E(P,T), entropy S (3), free energy G(P,T) (4) and thermodynamic Gibbs potential Φ(P,T) (5) dependence from pressure P and temperature T:

$$G=G_1+G_{phon};\quad G_1=\langle U\rangle-TS_1$$

4)

$$\Phi=\langle U\rangle-TS+PV=G+PV$$

5)

$$G_{phon}=\langle U\rangle_{phon}-TS_{phon}=\sum_i\left[\frac{\hbar\omega_i}{2}+k_BT\ln\left(1-\exp\left(-\frac{\hbar\omega_i}{k_BT}\right)\right)\right]$$

$$\omega_i\left(q_x,q_y,d,T\right)=\frac{\partial^2\langle U_{pot}(r,T)\rangle}{\partial r^2}*$$

$$\left(3\pm\sqrt{2\left[\cos\left(\{q_x-q_y\}d\right)+\cos(q_xd)+\cos\left(q_yd\right)\right]+3}\right)$$

6)

Correlation between positions of neighbour H_2 molecules is considered via phonon (with frequencies ω_i) contribution to G(P,T) (6). Minimum of Φ(P,T) leads to the dependence of hydrogen density ρ(P,T) from P,T using equation P= - $\partial G/\partial V$

Results:

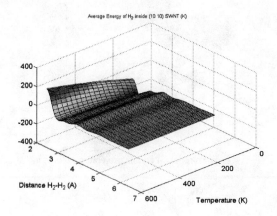

FIGURE 2. Free energy G(P,T) of H_2 molecule inside SWCNT (10,10)

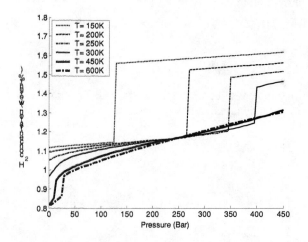

FIGURE 3. H_2 Adsorption rate (%wt.) inside SWCNT (10,10)

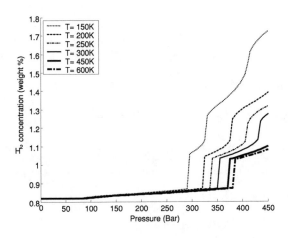

FIGURE 4. H_2 Adsorption rate (%wt.) outside SWCNT (10,10)

Conclusions:

1. It is calculated the dependence of adsorbed at CNT hydrogen density $\rho(P,T)$ on temperature T and pressure P
2. $\rho(P,T)$ pass through sequence of first-order phase transitions at P,T variation
3. H_2 adsorption rate is increased at SWCNT diameter decreased
4. Rate of adsorption inside nanotube is higher that that outside nanotubes
5. Rate of adsorption at SWNT both surfaces can reach 3-4%(wt.) at pressures $P \leq 500$ Bar.

REFERENCES

1. H. Zhu, A. Cao, X. Li et al., Applied Surface Science 178, p.50 (2001).
2. A.S. Fedorov, S. G. Ovchinnikov, "Density and Thermodynamics of Hydrogen Adsorbed Inside Narrow Carbon Nanotubes", Physics of the Solid State 46, No.3, pp. 584–589 (2004)
3. S. Silver, H. Goldman, Chem. Phys.Lett 320, p.3 (2000).

Curvature effects on vacancies in nanotubes

Johan M. Carlsson* and Matthias Scheffler*

*Fritz-Haber-Institut der Max-Planck-Gesellschaft, Faradayweg 4-6, D-14 195 Berlin, Germany

Abstract. Vacancies have a strong impact on the properties of nanotubes. We have therefore performed density-functional calculations for achiral single-wall nanotubes(CNTs) with single vacancies. Our calculations show that the curvature in the CNTs facilitates the relaxation leading to a local contraction. The vacancies prefer to align along the tube axis and the formation energy decrease with increasing curvature. The local magnetic moment at the vacancy disappears and the local charging decreases as the diameter of the nanotube gets smaller.

Keywords: Nanotubes, Vacancies, DFT-calculations
PACS: 73.22.Dj,61.72.Ji

Nanotubes have extraordinary properties, which are inspiring a number of applications such as nanotube transistors, field emission displays and sensors. However, most of these applications rely on the access to defect free nanotubes with a given chirality. It is still difficult to control the production of nanotubes and defects are formed either during growth or during the cleaning and separation process[1]. It is therefore vital to understand how these defects influence the properties of real, as grown, nanotubes. We have therefore studied how single vacancies influence the properties of metallic [(4,4), (6,6), and (8,8)] and semi conducting [(7,0), (10,0), and (14,0)] nanotubes. We focus on the curvature dependence of the formation energy and how the curvature and chirality influence the defect states.

We used density-functional theory (DFT) [2, 3] together with the generalized gradient approximation for the exchange-correlation functional [4] as implemented in the DMol^3-code [5]. The nanotubes were modeled in a supercell with periodic boundary conditions and the nanotubes were separated by 20 Å of vacuum to avoid tube-tube interactions. The supercell for the zigzag tubes contained 4 unit cells along the tube axis and 7 unit cells were used for armchair tubes giving a vacancy separation of 17 Å for both types of nanotubes. The perfect tubes contained up to 224 atoms and all atoms were fully relaxed. The Brillouin zone was sampled at 6 k-points along the tube axis. The formation energy of a vacancy is:

$$E_{\text{form}}(\mu_{(n,m)}) = \{E^{\text{V1}}_{(n,m)} + \mu_{(n,m)}\} - E_{(n,m)} \qquad (1)$$

where $E_{(n,m)}$ and $E^{\text{V1}}_{(n,m)}$ refer to the total energy of a (n,m) nanotube before and after the vacancy is created and $\mu_{(n,m)}$ denotes the chemical potential of an atom in this nanotube.

Our calculations show that the curvature in nanotubes has a major influence both on the energetics of vacancy formation and on the vacancy induced defect states. The curvature lowers the formation energy for vacancies in nanotubes as the diameter decreases as can be seen in Fig. 1a. This is in agreement with the TEM observation that more vacancies are formed in the smaller inner tube than in the outer tube of a double wall nanotube under electron irradiation [6]. Our results are also in agreement with

CP786, *Electronic Properties of Novel Nanostructures*, edited by H. Kuzmany, J. Fink, M. Mehring, and S. Roth
© 2005 American Institute of Physics 0-7354-0275-2/05/$22.50

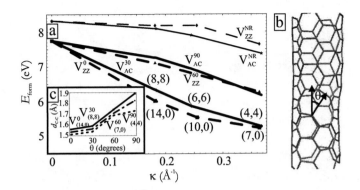

FIGURE 1. a) Formation energy E_{form} of single vacancies as function of nanotube curvature κ. Solid lines show E_{form} for armchair tubes V_{AC}^θ and dashed lines correspond to zigzag V_{ZZ}^θ tubes for the two inequivalent orientations of the vacancy with respect to the tube axis. The thin lines shows E_{form} for non-relaxed vacancies. b) The nanotube on the right side shows the definition of the rotation angle θ. The inset c) shows the bond length $d_{\text{C-C}}$ of the bond which completes the pentagon as function of θ. Solid line for (14,0) and (8,8), dashed-dotted line for (10,0) and (6,6) and dotted line for (7,0) and (4,4) tubes.

previous tight-binding [7] and DFT calculations [8]. The curvature has two geometric effects, which contribute to the decrease in formation energy. The curvature imposes strain, which weakens the bonds in nanotubes such that it is easier to break bonds as the diameter decreases. The relaxation of the atoms surrounding the vacancy is also facilitated, since the atoms can move inwards and thereby shorten the bond distance. The relaxation is particularly important for the rebonding between two undercoordinated C-atoms that changes an incomplete hexagon into an irregular pentagon. The relaxation therefore transforms the vacancy into a nine atom ring plus a pentagon by a Jahn-Teller distortion [7, 8, 9]. The nine atom ring contains one undercoordinated C-atom, which remains in a configuration similar to the zigzag edge of a graphene sheet and this zigzag atom points slightly outwards from the tube wall after relaxation. The Jahn-Teller distortion introduces orientation dependence for the relaxed vacancy, since the three possible rebonding directions are not equivalent in nanotubes as they are in graphene. However, they reduce to two inequivalent orientations separated by a 60° rotation for achiral tubes. These orientations can be characterized by the angle θ that the symmetry plane of the vacancy makes with the tube axis as sketched in Fig. 1b. Figure 1c shows that the largest relaxations occurs perpendicular to the tube axis, such that the bond that completes the pentagon has the shortest bond length at the V_{ZZ}^0 followed by V_{AC}^{30}. Comparing the formation energy and the bond length in Fig. 1 confirms that the vacancies oriented parallel to the tube axis (i.e. V_{ZZ}^0 in zigzag tubes and V_{AC}^{30} in armchair tubes) have the shortest bond length and the lowest formation energies. The presence of vacancies therefore leads to a contraction of the nanotube at the position of the vacancy due to the relaxation. The diameter of the (7,0)-tube decreases from 5.5 Å to 5.3 Å and the (4,4)-tube contracts to 5.2 Å at the vacancy. Figure 1a also shows that the relaxation energy is larger in nanotubes than in graphene, in particular for V_{ZZ}^0 and V_{AC}^{30}, indicating that the relaxation gives the largest contribution to the lowering of the formation energy in nanotubes.

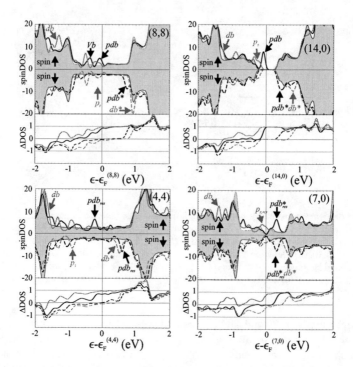

FIGURE 2. Spin polarized DOS for two armchair (left column) and two zigzag (right column) nanotubes with a vacancy. The grey region indicate the DOS in a perfect nanotube and the energy scale is set to zero at $\epsilon_F^{(n,m)}$ in the perfect (n,m) nanotube. The black and grey lines and arrows correspond to V_{AC}^{30} and V_{AC}^{90} respectively for armchair nanotubes and V_{ZZ}^{0} and V_{ZZ}^{60} respectively for zigzag nanotubes. The arrows marked *Vb*, *db*, p_z and *pdb* show the position of the valence band in the armchair tubes, the dangling bond, the p_z- and the mixed defect states. *res* indicate that the state is a resonance. Lower panel: The integrated difference density of states ΔDOS for the same vacancies where solid line is spin up and dashed line spin down.

The orientation of the vacancy has furthermore a noticeable effect on the electronic structure as can be observed in Fig 2. The defect states at vacancies making a large angle with the tube axis (V_{ZZ}^{60} and V_{AC}^{90}) have a similar character as in graphene i.e. a semi localized p_z-state and a localized dangling bond (*db*) state at the undercordinated zigzag atom which gives a local magnetic moment [8, 10]. The defect states at the low energy orientations (V_{ZZ}^{0} and V_{AC}^{30}) are on the other hand hybridized, such that they become a mixture of the p_z- and *db*-states and they do also mix with the π-band. This mixed defect state (indicated by the arrow marked *pdb* in Fig. 2) appears in the bandgap in (14,0) and (10,0) tubes, which means that its population and hence the magnetic moment can be manipulated by changing ϵ_F in the system. A similar defect state appears in the (8,8) and (6,6) tubes in the small band gap that opens up in the spin up channel due to the avoided band crossing of the π-band. The mixing of the defect states with the π-band get stronger in the smaller nanotubes, such that the wave function of the defect states get more delocalized and the defect state changes character into a resonance with the π-band. The splitting between the spin up and spin down components decreases successively with diminishing diameter. The effect is more pronounced in the zigzag tubes as can be seen by comparing the DOS for the (4,4) and (7,0)-tubes in Fig. 2. Figure 2 shows that the mixing is so strong in the (7,0)-tube that the magnetic moment is quenched at the V_{ZZ}^0 vacancy in agreement with previous DFT calculations [8].

We observed recently that vacancies in graphene can be negatively charged [10]. The integrated difference DOS, ΔDOS, in the lower panels of Fig. 2 indicates that the charging of vacancies in nanotubes is curvature and chirality dependent. The vacancies in the (8,8)-tube and (14,0)-tube both have a local charge similar to the vacancy in graphene, but the charging of the vacancies aligned along the axis of the tube decreases in the nanotubes with smaller diameter. Figure 2 indicates that the effect is stronger in zigzag tubes, such that the local charging disappears at the V_{ZZ}^0 vacancy in a (7,0)-tube in contrast to the V_{ZZ}^0 vacancy in (4,4)-tube. Scanning tunneling microscopy (STM) investigations of sputtered single and multiwall nanotubes [11] have shown hillocks similar to those found in sputtered graphite [12] indicating an increased number of states at the Fermi level. The diameters of the nanotubes in these experiments were even larger than the largest nanotubes studied here, which supports our results that vacancies in large nanotubes have properties similar to the corresponding defect in graphite. This suggests that curvature effects would only be visible for nanotubes with smaller diameter.

In summary, we have studied single vacancies in armchair and zigzag nanotubes of corresponding diameter. The curvature in the nanotubes facilitates relaxation of the atoms surrounding the vacancy, which leads to a local contraction of the nanotube. The relaxation furthermore lowers the formation energy for the vacancies as the diameter decreases and favours vacancies aligned along the tube axis. Vacancies oriented perpendicular to the tube axis maintain the character of vacancies in graphene, while the vacancies oriented parallel to the tube axis experience a stronger hybridization which increases as the diameter gets smaller. The curvature effects are stronger in zigzag tubes, such that the magnetic moment and local charging disappears at the V_{ZZ}^0 vacancy in a (7,0)-tube. This indicates that the properties of single vacancies in nanotubes are determined by the orientation of the vacancy together with the curvature and the chirality of the nanotube.

REFERENCES

1. T. W. Ebbesen and T. Takada, Carbon **33**, 973 (1995).
2. P. Hohenberg and W. Kohn, Phys. Rev. **136**, B 864 (1964).
3. W. Kohn and L. J. Sham, Phys. Rev. **140**, A 1133 (1965).
4. J. P. Perdew, K. Burke, and M. Ernzerhof, Phys. Rev. Lett. **77**, 3865 (1996).
5. DFT-code DMol^3, Academic version: B. Delley, J. Chem. Phys. **92**, 508 (1990).
6. K. Urita, K. Suenaga, T. Sugai, H. Shinohara, and S. Iijima, Phys. Rev. Lett. **94**, 155502 (2005).
7. A. J. Lu and B. C. Pan, Phys. Rev. Lett. **92**, 105504 (2004).
8. Y. Ma, P. O. Lehtinen, A. S. Foster, and R. M. Nieminen, New Journ. of Phys. **6**, 68 (2004).
9. P. M. Ajayan, V. Ravikumar, and J.-C. Charlier, Phys. Rev. Lett. **81**, 1437 (1998).
10. J. M. Carlsson and M. Scheffler, Submitted to Phys. Rev. Lett.
11. Z. Osváth, G. Vértesy, G. Petö, I. Szabó, J. Gyulai, W. Maser, and L. P. Biró, AIP Conference proceedings Vol. 723, pp 149 (2004).
12. J. R. Hahn, H. Kang, S. Song, and I. C. Jeon, Phys. Rev. B **53**, R1725 (1996).

Modelling of the Stabilization of the Complex of a Single Walled (5,5) Carbon Nanotube $C_{60}H_{20}$ with Cumulenic or Acetylenic Chain

T. C. Dinadayalane[a], Leonid Gorb[a], Helena Dodziuk[*,b], and Jerzy Leszczynski[*,a]

[a]Computational Center for Molecular Structure and Interactions, Department of Chemistry, Jackson State University, 1400 JR Lynch Street, P. O. Box 17910, Jackson, MS 39217, USA
[b]Institute of Physical Chemistry, Polish Academy of Sciences, 01-224 Warsaw, Kasprzaka 44, Poland

Abstract. Model calculations have been carried out for a (5,5) single walled carbon nanotube (SWNT) $C_{60}H_{20}$ and cumulene $C_{2n}H_4$ or acetylene $C_{2n}H_2$ (n = 1-4) chains and for their complexes obtained by insertion of a respective chain into the nanotube to check whether the theoretical simulation can (a) reproduce the stabilization of such supramolecular system and (b) propose which structure, *i. e.* acetylenic or cumulenic one, forms more stable complexes with SWNTs on the basis of quantum chemical calculations. In agreement with expectations, the calculations have revealed that the supramolecular systems are not stabilized at the DFT and HF level but, interestingly, reveal stabilization at the MP2/6-31G* level. Practically the same HOMO and LUMO energy gaps and very small charge transfer between the SWNT and chains have been found.
Keywords: SWNT, conjugated hydrocarbon chain, acetylene, cumulene, quantum chemical calculations, molecular mechanics, nonbonding interactions, HOMO-LUMO energy gap.
PACS: 03.67.Lx, 61.46.+w, 73.22-f

INTRODUCTION

Similarly to fullerene properties [1], the properties of carbon nanotubes, CNTs, undergo changes upon insertion of molecules or ions (e.g. filling CNTs with metals is known to influence their transport properties, magnetism and superconductivity [2]) allowing one to modify the tube properties for specific purposes. As evidenced by the volume "Computational Approaches in Supramolecular Chemistry" [3], until recently quantum chemical calculations have been only seldom applied to study supramolecular complexes in view of inability of the lower level theoretical approaches to correctly describe one of the main driving forces for the complex formation - nonbonding interactions. Therefore, a systematic study of the two types of systems consisting of cumulene $C_{2n}H_4$ or acetylene $C_{2n}H_2$ (n = 1-4) chains inserted as a guest along the principal axis of the host (5,5) nanotube (built of 60 carbon and 20 hydrogen atoms) has been undertaken to predict the stabilization energy using quantum chemical calculations. This system has been chosen in view of the observation of unsaturated carbon chains of unspecified (polyyne or cumulene) linear structure inside a carbon nanotube [4]. Another carbon chain in a multiwalled CNT

CP786, *Electronic Properties of Novel Nanostructures*, edited by H. Kuzmany, J. Fink, M. Mehring, and S. Roth
© 2005 American Institute of Physics 0-7354-0275-2/05/$22.50

has been recently reported [5]. Figure 1 shows the linear hydrocarbons and the nanotube considered in the study, and the complexes of nanotube with hydrocarbons are depicted in Fig. 2. In the nomenclature, **21** designates the hydrocarbon **2** inside the nanotube **1** along the principal axis, **31** specifies **3** inside **1**, and so on.

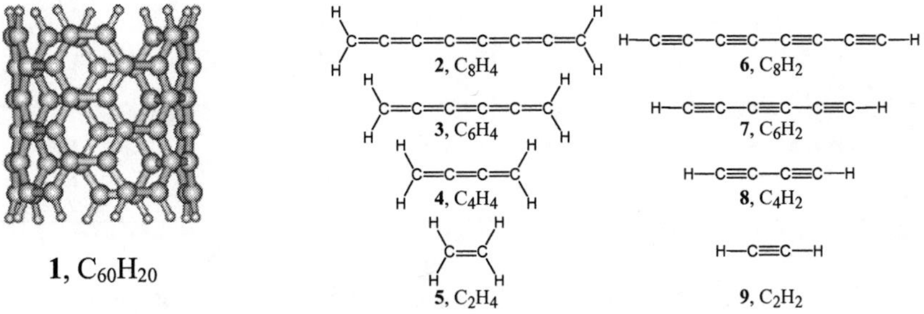

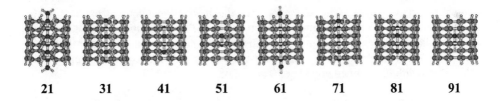

$$1, C_{60}H_{20}$$

2, C_8H_4
3, C_6H_4
4, C_4H_4
5, C_2H_4

6, C_8H_2
7, C_6H_2
8, C_4H_2
9, C_2H_2

FIGURE 1. Single walled carbon nanotube and the linear hydrocarbons considered in the study.

21   31   41   51   61   71   81   91

FIGURE 2. Complexes of linear hydrocarbons inserted into the nanotube.

Computational Methods

Interaction energies have been initially calculated using molecular mechanics, MM, with MM2 force field implemented in Chem3D pro 7.0. The molecular structures of the nanotube and the linear hydrocarbons have been optimized within the symmetry constraints at the MP2/6-31G* level. These optimized geometries of the nanotube and hydrocarbon have been kept rigid in the complexes in the subsequent calculations. Correction for Basis Set Superposition Error (BSSE) has been introduced using counterpoise option implemented in Gaussian 03 program package [6]. The calculations have been performed only for the finite tube length of the single walled carbon nanotube $C_{60}H_{20}$. The effect of tube length on the interaction energies has not been examined since the calculations for the nanotubes of longer tube length possessing more than 60 carbon atoms at the MP2 level could not be carried out within the available computational facility.

Results and Discussion

The interaction energies calculated for a series of the nanotube complexes under study are listed in Table 1. HF method predicts repulsion between the nanotube and the chain inserted into it for all analyzed complexes. In contrast, the results at the

MP2/6-31G* level show the attractive interaction indicating that electron correlation plays a crucial role in a proper modeling of weak nonbonding interactions.

TABLE 1. Interaction energies (in kcal/mol) of $C_{2n}H_4$ and $C_{2n}H_2$ with SWNT

Complex	MM2	HF/6-31G*[a]	MP2/6-31G*[a]
21	-18.9	32.0	-23.2
31	-15.6	32.5	-17.9
41	-10.9	22.8	-13.5
51	-6.7	22.3	-3.3
61	-16.2	33.7	-19.3
71	-14.3	28.3	-15.6
81	-11.4	26.3	-13.0
91	-8.2	14.5	-6.2

(a) Single point calculations performed using the MP2/6-31G* optimized geometries of the nanotube and hydrocarbons.

The MM values obtained using MM2 force field are in a close agreement with the results at MP2/6-31G* level. The present study indicates that the molecular mechanics with MM2 force field can be employed for modeling of the stability of these types of systems when the calculations at the MP2 level are prohibitively expensive or computationally not feasible. As expected, the interaction energy decreases as the hydrocarbon chain length becomes smaller. In case of hydrocarbons with eight and six carbons, the complexes involving nanotube and cumulene are more stable than those with the corresponding polyyne. In contrast, the complex with acetylene is more stable than that involving ethylene by about 3 kcal/mol.

TABLE 2. HOMO, LUMO energies and HOMO-LUMO energy gaps ($\Delta E_{H\text{-}L}$, in eV) obtained for the nanotubes and complexes along with the charge transfer values from hydrocarbon to nanotube at the HF/6-31G* level.[a]

Structure	HOMO	LUMO	$\Delta E_{H\text{-}L}$	Charge transfer[b]
1	-5.00	-0.01	4.99	-
21	-4.92	0.06	4.98	+0.005
31	-4.94	0.05	4.99	+0.025
41	-4.99	0.00	4.99	+0.014
51	-5.01	-0.02	4.99	-0.013
61	-4.89	0.09	4.98	-0.014
71	-4.90	0.09	4.99	+0.018
81	-4.94	0.03	4.97	+0.002
91	-4.99	0.01	5.00	0.000

(a) Single point calculations performed using the MP2/6-31G* optimized geometries of the nanotube and hydrocarbons.
(b) Positive values indicate the charge is transferred from nanotube to hydrocarbon and vice versa for negative values.

Table 2 gives the HOMO, LUMO energies and HOMO-LUMO energy gaps obtained at the HF/6-31G* level. The charge transfer values calculated using the Mulliken charges are also presented. Although the slight changes in HOMO and LUMO energies are observed for nanotube with hydrocarbons, practically the same values of the HOMO-LUMO energy gaps were obtained for the complexes and nanotube. The calculated HOMO-LUMO gap values should not be related to the values from the spectroscopic experiments (not reported for these systems), since they usually yield smaller band gap values, i.e., less than 2 eV for the nanotubes [7,8]. It is believed that theoretical HOMO-LUMO gaps provide an idea of variation of band gaps when other molecules interact with nanotube, and similar theoretical results to our observations have been reported [9]. A very small amount of charge transfer between nanotube and hydrocarbon is predicted in all the complexes except **91**.

CONCLUSIONS

The present study examines the interaction of linear hydrocarbons with (5,5) armchair single walled carbon nanotube and provides comprehensive investigations of the effect of chain length on interaction energies. Contrary to the experimental finding for a carbon chain in a nanotube, the HF method predicts repulsive interaction between the host nanotube and the guest chain, thus its destabilization. Interestingly, the calculations at MP2/6-31G* level show stabilization of the system upon the complex formation even without full optimization of the complexes. Interaction energies computed at the MM level with MM2 parameterization are in close agreement to those obtained at the MP2 level indicating that the empirical method can be used to analyze the stability of these types of complexes when the calculations at the post-HF level are computationally prohibitive. Minor variations in the HOMO and LUMO energies are observed in the complexes as compared to the nanotube, while the HOMO-LUMO energy gaps are practically the same. Slight charge transfer occurs between the nanotube and hydrocarbon.

ACKNOWLEDGMENTS

This research was supported by ONR grant # N00014-03-1-0116 and the contract from the US Army Engineer Research and Development Center. Mississippi Center for Supercomputing Research (MCSR) is acknowledged for the computational facilities.

REFERENCES

1. Dodziuk, H., Introduction to Supramolecular Chemistry, Dordrecht: Kluwer, 2002.
2. Dresselhaus, M. S., et al. Science of Fullerenes and Carbon Nanotubes, San Diego: Academic Press, 1996.
3. Computational Approaches in Supramolecular Chemistry, edited by G. Wipff, Dordrecht: Kluwer Academic Publisher, 1993.
4. Wang, Z., Ke, X., Zhu, Z., Zhang, F., Ruan, M., and Yang, J., *Phys. Rev. B* **61**, R2472-R2474 (2000).
5. Zhao, X. L., Ando, Y., Liu, Y., Jinno, M., Suzuki, T., *Phys. Rev. Lett.* **90**, 187401 (2003).
6. Gaussian 03, Revision C.02, Frish, M. J., Trucks, G. W., Schlegel, H. B., et al. Gaussian, Inc., Wallingford CT, 2004.
7. Wildoer, J. W. G., Venema, L. C., Rinzler, A. G., Smalley, R. E., Dekker, C., *Nature* **391**, 59-62 (1998).
8. Odom, T. W., Huang, J. -L., Kim, P., Lieber, C. M., *Nature* **391**, 62-64 (1998).
9. Bauschlicher, Jr. C. W., Ricca, A., *Phys. Rev. B* **70**, 115409-6 (2004).

Linear carbon chain in the interior of a single walled carbon nanotube

Viktor Zólyomi*, Ádám Rusznyák*, Jenő Kürti*, Shujiang Yang† and Miklos Kertesz†

*Department of Biological Physics, Eötvös University Budapest, Pázmány Péter sétány 1/A,
H-1117 Budapest, Hungary
†Department of Chemistry, Georgetown University, Washington DC 20057, USA

Abstract.
The physical properties of several kinds of single walled carbon nanotubes containing a single carbon chain in their interior are investigated with density functional theory, using the Vienna *ab initio* Simulation Package (VASP). The optimized geometry and the electronic band structure are both examined, and compared to the results of the isolated subsystems: the isolated carbon chain and the isolated nanotube. We find bondlength alternation in the optimized geometry of the isolated chain, as well as a gap in the band structure, clear signs of Peierls distortion. In the combined systems, hybridization and charge transfer are found between the tube and the chain, resulting in a partial or complete breakdown of the Peierls distortion of the carbon chain. The combined systems are always predicted to be metallic, even if both subsystems are semiconductors, and even if the chain still exhibits some bond length alternation.

Keywords: DFT, carbon nanotube, carbon chain, band structure, charge transfer, commensurability
PACS: 71.15.Mb, 73.22.-f

INTRODUCTION

The already rich science of single walled carbon nanotubes (SWCNTs) was further enriched by two fairly recent experimental observations of a multi-walled carbon nanotube which contains a single chain of carbon atoms inside the innermost tube [1, 2]. We were motivated by these two experimental works to perform theoretical investigations on the simplified version of these carbon nanostructures, a single walled carbon nanotube containing a single chain of carbon atoms. Our aim was to probe the interaction between the two subsystems, and determine how these interactions affect the physical properties of the combined system (which we will call chain@SWCNT, or more generally, CS). The electronic structure of such one-dimensional systems may exhibit a number of exotic properties (e.g. Peierls distortion or Luttinger liquid behavior).

In this work we present the results of our first principles calculations on the physical properties of several chain@SWCNT systems, and an isolated infinite carbon chain as reference. The calculations were performed with VASP 4.6 [3, 4, 5]. We used the local density approximation (LDA) [6] and the projector augmented wave (PAW) [7, 8] method. The geometries were optimized until all forces on all ions fell below the threshold value of 0.003 eV/Å. The plane wave cutoff energy was 400 eV, which equals to a large basis set. The number of irreducible k-points was different for each particular system examined, this is explicitly given in the corresponding text below.

CP786, *Electronic Properties of Novel Nanostructures*, edited by H. Kuzmany, J. Fink, M. Mehring, and S. Roth
© 2005 American Institute of Physics 0-7354-0275-2/05/$22.50

As expected, we find Peierls distortion for the isolated chain. If the chain is inside a SWCNT, charge transfer (CT) and hybridization between the tube and the chain leads to a partial or complete breakdown of the bond length alternation. Furthermore, the CT turns the combined systems into a metal, even if both the isolated chain and the isolated SWCNT are semiconductors.

ISOLATED INFINITE CARBON CHAIN

In order to serve as a reference to our calculations on the infinite carbon chain inside a nanotube, we have calculated the geometry of the isolated carbon chain. We used 31 irreducible k-points, and two atoms in the unit cell. Finite bond length alternation (BLA) was found, which is a consequence of Peierls instability, a trait inherent to one dimensional systems [9]. The optimal bond lengths were found to be 1.257 Å and 1.291 Å. The Peierls instability also leads to the opening of a gap of 0.3 eV at the Fermi level. The true value for the BLA is likely larger, as it is known that DFT methods underestimate the Peierls induced alternation in one dimensional systems [10]. Similarly, the band gap is also underestimated [10, 11, 12, 13, 14].

THE GEOMETRY OF THE COMBINED SYSTEMS

After determining the geometry of the isolated chain, we could turn our attention to the combined system (CS) of a carbon chain inside a SWCNT. In this work, we restrict ourselves to the case when the chain is positioned in the center of the nanotube, on the symmetry axis [15]. This means, that the radius of the tube cannot differ much from the carbon-carbon van der Waals distance of 3.4 Å. Therefore, the only SWCNTs that come into consideration, are the ones which fall into the diameter region of 6-8 Å. This leaves us with very limited choices for the nanotubes. We performed calculations on four SWCNTs, the (7,1), (5,5), (6,4), and (9,0) tubes, using 29, 29, 9, and 11 irreducible k-points, respectively. Given the differing unit cell sizes (c) these choices lead to similar accuracies for the k-space integrations. A commensurate model was used in all cases with appropriate supercells. A strictly commensurate model inevitably includes an artificial distortion of the chain geometry, which can however be separated with the help of an appropriate reference system: an isolated carbon chain, which has *exactly the same* unit cell length as the carbon chain inside the corresponding nanotube in the combined system's optimized geometry. The geometry of this "reference isolated chain" (RIC) must be optimized, but the unit cell length must be kept *fixed* at the value obtained from the optimization of the CS: only the bond length alternation is optimized. The commensurability requirement has a large effect on the alternation and plays a key role in the detailed analysis [16].

In all cases, we find a breakdown of the BLA. It is a partial breakdown in the case of the (7,1), where the BLA is decreased by 70% in the CS. In all other cases, we find a complete breakdown, the BLA vanishes completely. This breakdown of the BLA can be attributed partly to charge transfer, as discussed in the next Section.

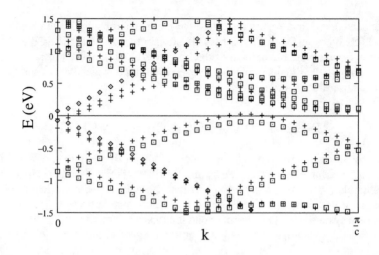

FIGURE 1. The band structure of the CS of a carbon chain inside the (7,1) nanotube, in comparison with the isolated subsystems. Crosses: combined system; diamonds: isolated carbon chain (degenerate); squares: isolated nanotube. The Fermi energy was shifted to 0 eV in all cases.

CHARGE TRANSFER AND HYBRIDIZATION

As we mentioned in the Introduction, charge transfer (CT) can be expected between the two subsystems. The CT can be quantitatively extracted from our calculations by means of comparing the band structures of the combined systems with the band structures of the corresponding isolated subsystems [10]. As an example, the band structure of the chain@(7,1) system is shown in Figure 1, depicted with diamonds. In all four cases, the band structure is metallic, and the CSs clearly possess a large density of states at the Fermi level, which makes them potential candidates for superconductivity, similar to doped multi-walled carbon nanotubes [17].

Beyond calculating the band structures of the four CSs, we also calculated the band structures of the isolated subsystems, these are plotted in Figure 1 (carbon chain: squares, SWCNT: crosses). In these calculations, the isolated subsystems were treated separately, both with the exactly same geometry, that resulted from the optimization of the CS.

Figure 1 shows, that the rigid band model can be used, but it must be applied with caution. There is some splitting of the degenerate bands of the subsystems, and, there is also some mixing of the bands of the separate subsystems. However, beyond this, the rigid band model works extremely well, as it is easy to see just which band of the CS originates from which band of one of the subsystems. Clearly, there is CT between the two subsystems, electrons are transferred from the tube to the chain, leading to metallicity in all cases.

The value of the CT between the carbon chain and its host SWCNT can be extracted from the band structures [16, 18]. One can study the effect of CT alone by reoptimizing the RIC with additional charges placed on it. We calculated the optimal BLA of the charged RICs with charges that correspond to the CT of the given CS. In the case of (7,1), we find that the BLA of the charged RIC is in fact larger than the BLA of the chain in the CS. While the BLA breakdown is 70 % in the CS, it is only 50 % in the charged RIC. This clearly means, that the full breakdown cannot be attributed to CT alone, but the hybridization of the bands of the subsystems also has a significant effect on the chain geometry.

SUMMARY

Using first principles methods, we have examined the optimized geometry and the electronic band structure of four SWCNTs containing an infinite carbon chain in their interior. We find that Peierls distortion induces bond length alternation (BLA) in the isolated carbon chain, and this BLA suffers a breakdown in the combined systems. We have shown that this breakdown is the result of two effects: hybridization between the bands of the subsystems, and charge transfer between the subsystems. In all cases, the combined system is metallic, with a large density of states at the Fermi level, making these systems a potential candidate for superconductivity.

ACKNOWLEDGMENTS

Support from OTKA (Grants No. T038014 and T043685) in Hungary and the National Science Foundation (Grant No. DMR-0331710) in the United States is gratefully acknowledged.

REFERENCES

1. Z. Wang, X. Ke, Z. Zhu, F. Zhang, M. Ruan, and J. Yang, Phys. Rev. B **61** (2000).
2. X. Zhao, Y. Ando, Y. Liu, M. Jinno, and T. Suzuki, Phys. Rev. Lett. **90**, 187401 (2003).
3. G. Kresse and J. Hafner, Phys. Rev. B **48**, 13115 (1993).
4. G. Kresse and J. Furthmüller, Comput. Mater. Sci. **6**, 15 (1996).
5. G. Kresse and J. Furthmüller, Phys. Rev. B **54**, 11169 (1996).
6. J. P. Perdew and A. Zunger, Phys. Rev. B **23**, 5048 (1981).
7. P. E. Blöchl, Phys. Rev. B **50**, 17953 (1994).
8. G. Kresse and D. Joubert, Phys. Rev. B **59**, 1758 (1999).
9. C. Kittel, *Introduction to Solid State Physics, 5th Edition* (John Wiley & Sons Inc., 1976).
10. G. Sun, J. Kürti, M. Kertesz, and R. Baughman, J. Chem. Phys. **117**, 7691 (2002).
11. R. W. Godby, M. Schlüter, and L. J. Sham, Phys. Rev. Lett. **56**, 2415 (1986).
12. J. W. Mintmire and C. T. White, Phys. Rev. B **35**, 4180 (1987).
13. R. W. Godby, M. Schlüter, and L. J. Sham, Phys. Rev. B **37**, 10159 (1988).
14. V. Zólyomi and J. Kürti, Phys. Rev. B **70**, 085403 (2004).
15. Y. Liu, R. O. Jones, X. Zhao, and Y. Ando, Phys. Rev. B **68**, 125413 (2003).
16. . Rusznyák, V. Zólyomi, J. Kürti, S. Yang, and M. Kertesz, to be published.
17. L. X. Benedict, V. H. Crespi, S. G. Louie, and M. L. Cohen, Phys. Rev. B **52**, 14935 (1995).
18. G. Sun, M. Kertesz, J. Kürti, and R. Baughman, Phys. Rev. B **68**, 125411 (2003).

Magnetism in Molecular Vanadium-Benzene Sandwiches

Y. Mokrousov, N. Atodiresei, G. Bihlmayer and S. Blügel

Institut für Festkörperforschung, Forschungszentrum Jülich, D-52425 Jülich, Germany,
y.mokrousov@fz-juelich.de, n.atodiresei@fz-juelich.de

Abstract. The magnetic moments of the vanadium-benzene sandwiches have been experimentally found to increase with the number of vanadium atoms in the molecule which suggests that the unpaired electrons on the metal atoms can couple ferromagnetically. We report on *ab initio* calculations of the equilibrium geometrical structures, electronic configurations and magnetic properties of complexes formed by vanadium and benzene, $V_n(C_6H_6)_{n+1}$ ($n < 5$). Employing density functional theory, our results confirm the increase of the magnetic moment in the considered sandwiches with the number of V atoms that was found at higher temperatures (300 K). At lower temperatures, however for certain clusters ($n = 3$) an antiferromagnetic coupling can appear, that gives rise to a lower average moment in the cluster.

Keywords: molecular magnetism, benzene sandwiches, organometallics, electronic structure
PACS: 31.15.Ew, 36.40.-c, 73.22.-f

INTRODUCTION

Recent developments in ultra-high density magnetic recording and the ability to reach very short access times rely on the miniaturization of the structures used in modern nanotechnology. Current technologies use nano-structured materials formed via layered inorganic solids [1]. The limit of nanostructuring and lithography at reach motivates the search for completely new approaches. For example, by using self-organization of magnetic nanoparticles in an organic matrix a recording density of 20 Tbit/in^2 can be achieved [2, 3]. Following this direction at the frontier of research, *atomic magnetic clusters* can be seen as the final point in a series of increasingly smaller magnetic units from bulk matter to atoms. Molecular magnets are still at an early stage of development but they seem to be the ideal candidates which exhibit novel properties for future applications in magnetic data storage, spin electronics and quantum computing. A novel class of high-spin organic molecular magnets are the one-dimensional (1D) multidecker organometallic sandwiches formed by a regular sequence of V and C_6H_6 ($V_n(C_6H_6)_{n+1}$). Reducing the dimensionality also increases the tendency of a system to become magnetic, thereby widening the spectrum of elements that can be used for magnetic applications. Moreover, the one-dimensionality of these nanomaterials opens the way for a whole range of phenomena that have no analogs in higher dimensional systems. In this context such organometallic compounds are studied with increasing interest as one can gain insight into the fundamental nature of magnetism.

CP786, *Electronic Properties of Novel Nanostructures*, edited by H. Kuzmany, J. Fink, M. Mehring, and S. Roth
© 2005 American Institute of Physics 0-7354-0275-2/05/$22.50

METHOD AND COMPUTATIONAL DETAILS

Non-spinpolarized and spin-polarized calculations were performed within the framework of the density functional theory (DFT) using the generalized gradient approximation (GGA-PBE) for the exchange-correlation potential. We employed a realization of the full-potential linearized augmented plane-wave method for one-dimensional systems (1D-FLAPW) [4], as implemented in the FLEUR code [5]. In contrast to conventional super-cell approaches, in this method the 1D-system is embedded in two dimensions in infinite vacuum. The requirement of periodicity along the third dimension (z-axis) calls for an efficient separation in z-direction between the single molecules, in order to avoid interactions coming from the neighboring molecules. Unit cell lengths of 10.6 Å ($n = 1$), 13.8 Å ($n = 2$), 17.0 Å ($n = 3$) and 20.1 Å ($n = 4$) used for the $V_n(C_6H_6)_{n+1}$ sandwiches proved to be sufficient to exclude unwanted side-effects on the electronic and magnetic structures due to interaction with the neighbors. The z-axis of the system cuts through the V atoms and the centers of gravity of the C_6H_6 molecules. Optimized geometries of the molecules were restricted to D_{6h} symmetry.

RESULTS AND DISCUSSION

Geometry. Our calculations show that stable structures of these molecules are sandwiches with the V atom between two C_6H_6 molecules, as was also confirmed experimentally [6]. In general, spin-polarization does not change the nonmagnetic equilibrium structural parameters by more than 1–2%. However, several total energy minima for a given molecule are possible, separated in energy up to 0.2 eV. For instance, this energy difference can characterize different rotational angles of the benzene rings relative to each other around the z-axis and is in agreement with [7], where a chirality of the complexes was also predicted. The specific equilibrium configurations preserve the magnetic properties of the molecules with the distances between V atoms and benzene molecules almost unchanged. For all sandwiches the planarity of the C_6H_6 is broken. The planes of C and H hexagons are parallel but slightly shifted relative to each other along the the z-axis. The C-H bond is practically the same as in the bare C_6H_6, but the C-C bond increases with decreasing the C-V distance. This is attributed to the bonding in the sandwich molecule by populating LUMOs type orbitals (with high number of nodes) of the benzene molecule.

Electronic structure. Insight into the mechanisms responsible for the bonding in the molecule $V_n(C_6H_6)_{n+1}$ can be gained on the basis of a schematic analysis of the orbitals of V and benzene which are classified in terms of the pseudoangular momenta around the z-axis. The five $3d$-orbitals of the metal atom can be divided according to their symmetries into one $d\sigma\,(d_{z^2})$, two $d\pi\,(d_{xz}, d_{yz})$, two $d\delta\,(d_{x^2-y^2}, d_{xy})$ and a $4s$ orbital classified as $s\sigma$ orbital. The six π-orbitals of C_6H_6 form one $L\sigma\,(\pi_1)$, two degenerate $L\pi\,(\pi_2, \pi_3)$ (HOMO), two degenerate $L\delta\,(\pi_4^*, \pi_5^*)$(LUMO) and one $L\phi\,(\pi_6^*)$ orbital. When the benzene molecules and V atoms are brought together, the HOMO and LUMO orbitals of C_6H_6 interact with the vanadium d orbitals of the same symmetry and hybridization occurs [8]. Yasuike and Yabushita [9] explain the electronic structure of the complexes by a schematic orbital interaction diagrams using extended Hückel or Hartree-Fock methods. In order to shed light on the validity of these schemes we have performed electronic structure calculations based on the DFT for the complexes $V_n(C_6H_6)_{n+1}$ with $n = 1, 2, 3, 4$ and an infinite wire $(V(C_6H_6))_\infty$. Based on the nonmagnetic local den-

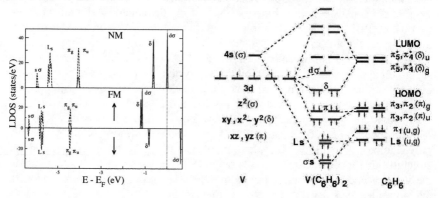

FIGURE 1. Left: Local densities of states (LDOS) for the $V(C_6H_6)_2$: LDOS at the carbon site (dashed line) and LDOS at the vanadium site (full line); the upper part shows the nonmagnetic (NM) densities of states and the bottom part shows the spin-polarized ferromagnetic (FM) DOS with spin-up and spin-down channels indicated by arrows. Right: schematic orbital interaction scheme for $V(C_6H_6)_2$.

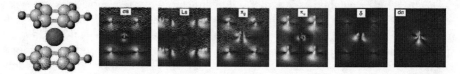

FIGURE 2. Charge density plots in a plane cutting the vanadium and four carbon and hydrogen atoms for the electronic states of $V(C_6H_6)_2$. The plots are labeled by the names of the states as indicated in Fig. 1 (left).

sities of states (LDOS) for the $V(C_6H_6)_2$ molecule (Figure 1, left) we briefly explain the interaction mechanisms responsible for the bonding in the sandwiches. When two C_6H_6 molecules are placed at the same distance from V, as in the $V(C_6H_6)_2$ complex, the $L(\sigma, \pi, \delta, \phi)$ orbitals form symmetry-adapted molecular orbitals indicated by an additional subscript (g, u) according to their symmetry. The atomic orbitals of V have g symmetry and interact only with $L(\sigma, \pi, \delta)_g$ orbitals of benzene [8]. The nonbonding $d\sigma$ orbital occupies the Fermi energy with one electron. Two degenerate δ-states that are lower in energy are created by hybridization of the V $d\delta$-orbitals and $(L\delta)_g$ orbitals of C_6H_6 and carry four V d-electrons (Fig. 1, right). The hybridization of $d\pi$ states of V and $L\pi_g$ orbitals of benzene produces a new π_g state which is in general slightly lower in energy than the nonbonding π_u located at the carbon sites. The charge density plots for the electronic states in the nonmagnetic LDOS of $V(C_6H_6)_2$ are presented in Figure 2. We conclude that in general, the nonmagnetic LDOS is consistent with the Hückel orbital scheme presented in the literature for this complex [9].

Magnetism. Spin-polarized LDOS for the $V(C_6H_6)_2$ complex (Figure 1, left) shows a large exchange splitting of the states near the Fermi level. This splitting is largest for the $d\sigma$ state of vanadium located at the Fermi level in the nonmagnetic case. A smaller split-

TABLE 1. Magnetic properties: M−NM (FM−AFM) represents the difference between the magnetic(ferromagnetic and antiferromagnetic) and non-magnetic solution. Negative numbers indicate a M (FM) ground state. In the last line the calculated magnetic moments of the molecules is given.

	$V(C_6H_6)_2$	$V_2(C_6H_6)_3$	$V_3(C_6H_6)_4$	$V_4(C_6H_6)_5$	$V(C_6H_6)_\infty$
M−NM (meV per V)	−459	−415	−300	−232	−106
FM−AFM (meV per V)	—	−3	+5	−31	−57
Magnetic moment μ_B	1	2	1	4	1_∞

ting can be seen for the δ states. Out of five vanadium electrons three have spin up (one at the $d\sigma$ level and two at the δ level) and two have spin down at the δ level. This results in a total magnetic moment of $1.00\mu_B$ for the $V(C_6H_6)_2$. Our calculations show that this magnetic moment is entirely located inside the muffin-tin sphere of V ($1.09\mu_B$). For all the sandwiches calculated the magnetic moments of H and C atoms are very small. The carbon magnetic moments (around $0.01\mu_B$) prefer to couple antiferromagnetically to the magnetic moments of the closest vanadium atoms. The magnetic moment of V is strongly localized for $n \leq 2$, while for $n > 2$ a delocalization of the $d\sigma$ orbitals along the z-axis occurs with the corresponding redistribution of the spin density among V atoms. We found that all the $V_n(C_6H_6)_{n+1}$, $n = 1, 2, 3, 4$ molecules are magnetic with a large gain in total energy compared to the nonmagnetic solution (see Table 1). For the molecule with $n = 1, 2, 4$ and in the infinite wire $V(C_6H_6)_\infty$, ferromagnetic ordering is preferred over the antiferromagnetic one. The molecule with $n = 3$ has an antiferromagnetic ground state with the total magnetic moment of $1\mu_B$, which is in agreement with the experimental value at 56 K [10]. The ferromagnetic solution for the latter sandwich is about 14 meV higher in energy, which corresponds to 167 K. Therefore, we conclude that at temperatures close to 300 K, the total magnetic moment of the sandwiches increases with the number of vanadium atoms in accord with the experimental measurements [11]. Our results indicate a possible explanation for the more pronounced temperature dependence of the magnetic moment that was found experimentally for the $V_3(C_6H_6)_4$ sandwich.

REFERENCES

1. S. Maekawa, and T. Shinjo, editors, *Spin dependent transport in magnetic Structures*, Taylor and Francis Group, London, 2002.
2. S. Sun, C. B. Murray, L. Folks, and A. Moser, *Science*, **287**, 1989–1992 (2000).
3. R. F. Service, *Science*, **271**, 920 (1996).
4. Y. Mokrousov and G. Bihlmayer and S. Blügel, *Phys. Rev. B*, accepted for publication, (2005)
5. http://www.flapw.de
6. D. Rayane, A.-R. Allouche, R. Antoine, M. Broyer, I. Compagnon, and P. Dugourd, *Chem. Phys. Lett.*, **375**, 506 (2003).
7. J. Wang, P. H. Acioli, and J. Jellinek, *J. Am. Chem. Soc.*, **127**, 2812 (2005).
8. I. Fleming, *Frontier Orbitals and Organic Chemical Reactions*, Wiley, New York, 1976.
9. T. Yasuike, and S. Yabushita, *J. Phys. Chem. A*, **103**, 4533 (1999).
10. K. Miyajima and M. B. Knickelbein and A. Nakajima, submitted to *Eur. Phys. Jour. D* (2005).
11. K. Miyajima, A. Nakajima, S. Yabushita, M. B. Knickelbein, and K. Kaya, *J. Am. Chem. Soc.*, **126**, 13203 (2004).

Wrapping carbon nanotubes with DNA: A theoretical study

N. Ranjan*,†, G. Seifert**, M. Mertig* and T. Heine**

*Technische Universität Dresden, Max-Bergmann-Zentrum für Biomaterialien
†Universität Regensburg, Fachbereich Physik
**Technische Universität Dresden, Institut für Physikalische Chemie

Abstract. A series of recent experimental studies indicate a considerable interaction of DNA with carbon nanotubes (CNTs). Experiments performed at DuPont [1] showed that CNT bundles can be effectively suspended by sonicating in a DNA solution. Motivated by these experiments, we have investigated the nature of these interactions theoretically. Our density functional (DFT) based calculations showed that there does not exists any covalent interaction between DNA and CNT, but there do exist long range van der Waal's interactions.

Keywords: Carbon nanotube, DNA, van der Waal's interaction, dispersion, electronic interaction
PACS: 81.07.De, 87.14.Gg, 71.15.Mb, 72.10.Bg

INTRODUCTION

The discovery of carbon nanotubes (CNTs) is one of the most important discovery of the last decade. The electronic structure of a single-walled nanotube [2] is either metallic or semi-conducting, depending on its diameter and chirality. After preparation, the CNTs form crystalline ropes and exists as bundles. DNA-assisted dispersion and separation of CNTs from the bundles was reported by the DuPont group [1]. Not much is known about the interaction forces which accomplish separation. We describe in this paper a theoretical model for the process of dispersion. We did sequence-dependent interaction energy calculations for DNA wrapped around the CNT. We considered two different types of DNA (poly(T) and poly(CT)) in our calculations. The files containing the coordinates of atoms were provided by *Ming Zheng et. al* [1]. van der Waal's interaction was calculated to see its effect on the dispersion of DNA coated CNTs. We also studied the effect of such interactions on the electronic properties of CNT.

VAN DER WAAL'S INTERACTION

The van der Waal's potential is weakly attractive as two uncharged molecules or atoms approach one another from a distance, but strongly repulsive when they get too close. This behavior is described empirically by the Lennard-Jones potential with the attractive r^{-6} and the repulsive r^{-12} term. The Lennard-Jones potential between two uncharged particles is given by

$$\phi_{LJ} = \varepsilon \left[(\frac{r_0}{r})^{12} - 2(\frac{r_0}{r})^6 \right]. \tag{1}$$

CP786, *Electronic Properties of Novel Nanostructures*, edited by H. Kuzmany, J. Fink, M. Mehring, and S. Roth
© 2005 American Institute of Physics 0-7354-0275-2/05/$22.50

Here, ε is the well-depth and r_0 is the van der Waal's bond length. We perfomed sequence dependent calculations with such model potential for poly(T) and poly(CT) DNA helically wrapped around the (10,0)CNT, as shown in Fig. 1. The calculated van der Waal's energies per unit length of CNT-DNA complex are given below:

- poly(CT) DNA = -1.03eV/nm,
- poly(T) DNA = -1.12eV/nm.

There also exist non-covalent interactions between nanotubes in bundles of nanotubes. These are also van der Waal's type. The interaction energy per unit length between two nanotubes of radius r (in Å) at optimal distance can be described by [3]

$$\phi(r) = -0.1135\sqrt{r} + 9.39 \times 10^{-3} eV/\text{Å}. \tag{2}$$

$\phi(r)$ represents the energy of interaction at the optimal separation of parallel tubes. With the (10,0) CNT of radius 3.96Å the interaction energy is -2.17eV/nm. Fig. 1 explains the energetics for the dispersion of CNT in DNA solution. The binding energy of poly(T) DNA by CNT is found to be -1.12eV/nm and for CNT-CNT as -1.08eV/nm per nanotube. Consequently, the binding of single stranded (ss) DNA onto CNT can compete with the known strong tendency of nanotubes to cling to each other to form bundles.

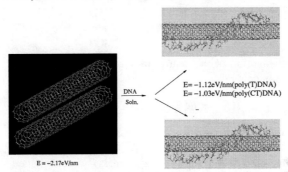

FIGURE 1. Dispersion effect of DNA solution onto the nanotube bundles. The net change of energy of the system in this process of dispersion is $\Delta E = -0.06eV/nm$ for poly(T) DNA and $\Delta E = 0.11eV/nm$ for poly(CT) DNA. From the calculation it seems that poly(CT) DNA is not effective for the process of dispersion. The contribution from the *polarization* and *solvation* energy has not been taken into account in the calculation. With their effect also taken into consideration, we expect the poly(CT) DNA solution to be also effective in dispersion.

ELECTRONIC INTERACTION

We adopted a different structural model to study the influence of CNT-DNA interaction on the electronic properties of the CNT. We created a simple GT sequence of oligonucleotide [4] as shown in Fig. 2. To study its effect on the conductance of CNT, we performed all calculations at the distance of 2.5Å from the surface of CNT, which represents a lower limit for the distance between a van der Waal's bound DNA-base to the CNT.

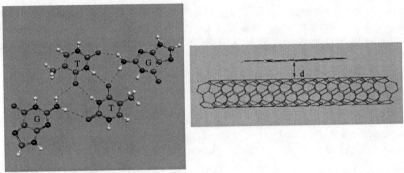

FIGURE 2. Left: Optimized structure of the oligonucleotide. Right: Vertical displacement of the oligonucleotide with respect to the (4,4)CNT.

We performed DFT based Green function calculations to obtain the density of states (DOS) and electronic transmission along the tube direction. The conductance $G(E)$ is related to the electronic transmission probability $T(E)$ according to the Landauer formula: $G = G_0 T(E)$. Here $G_0 (= 2e^2/h)$ is the quantum of conductance. $T(E)$ can be calculated using Green function techniques [5] via

$$T = Tr[\Gamma_L G_M \Gamma_R G_M^\dagger]. \tag{3}$$

G_M is the Green function of the central molecular region which can be calculated by means of the Dyson equation

$$G_M^{-1}(E) = [ES_M - H_M - \Sigma_L(E) - \Sigma_R(E)]. \tag{4}$$

$\Sigma_\alpha (\alpha = L, R)$ are the self-energies of the leads, while the spectral functions Γ_α are related to the self-energies by $i\Gamma_\alpha = (\Sigma_\alpha - \Sigma_\alpha^\dagger)$. We used non-orthogonal basis systems and the elements of H_M and S_M (overlap matrix) were obtained within the DFTB approach [6].

The results are shown in Fig. 3. There is no change in the DOS and transmission in and around the Fermi energy for the hybrid system. Any type of π electron/covalent interaction would have lead to a change in the density of states and conductivity of the system (CNT+DNA). There exist very little or no overlap between the electron clouds at such large distances. We did not take into account the existence of dipole-field arising due to the presence of charged and polar species in the DNA. The electronic properties of CNT may change in the presence of this external electric field.

CONCLUSION

We evaluated the short ranged electronic and the long ranged van der Waal's interaction between the CNT-DNA system. The van der Waal's energy of the hybrid system explains the energetics involved in the phenomena of dispersion. Our DFT and Green functional based calculations showed that there is no covalent interaction between the two systems.

FIGURE 3: Continuous lines: Transmission and total density of states (TDOS) of infinite (4,4)CNT. Broken lines: Transmission and TDOS for infinite (4,4)CNT with oligonucleotide. The orientation of oligonucleotide with respect to the CNT is shown in Fig. 2. The verticle line shows the Fermi energy for infinite (4,4)CNT.

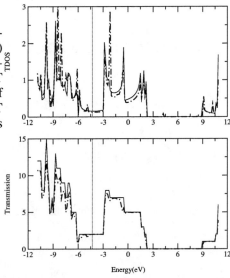

Energy(eV)

It also confirmed that the intrinsic electronic properties of CNTs are conserved when bound to DNA.

We work towards adding the effect of the solvation and polarization energies to our model which may explain other experimental results like the separation of metallic tubes from the semiconducting ones, diameter dependent separation, pitch of helix etc.

ACKNOWLEDGMENTS

We thank Wolfgang Pompe for valuable comments and suggestions. This work is supported by BMBF (13N8512).

REFERENCES

1. M. Zheng, A. Jagota, E. D. Semke, B. A. Diner, R. S. Mclean, S. R. Lustig, R. E. Richardson, and N. G. Tassi, *Nature Materials*, **2**, 338–342 (2003).
2. P. L. McEuen, M. S. Fuhrer, and H. Park, *IEEE Transactions on Nanotechnology*, **1**, 78–85 (2002).
3. L. A. Girifalco, M. Hodak, and R. S. Lee, *Phys. Rev. B*, **62**, 13104–13110 (2000).
4. M. Zheng *et al.*, *Science*, **302**, 1545–1548 (2003).
5. S. Datta, *Electronic transport in Mesoscopic Systems*, Cambridge University Press, Cambridge, 1995.
6. D. Porezag, T. Frauenheim, T. Köhler, G. Seifert, and R. Kaschner, *Phys. Rev. B*, **51**, 12947–12957 (1995).

Phonons and symmetry properties of (4,4) picotube crystals

M. Machón*, S. Reich†, J. Maultzsch*, R. Herges** and C. Thomsen*

*Institut für Festkörperphysik, Technische Universität Berlin, Hardenbergstr. 36, 10623 Berlin, Germany
†University of Cambridge, Department of Engineering, Trumpington Street, Cambridge CB2 1PZ, UK
**Institut für Organische Chemie, Universität Kiel, Otto-Hahn-Platz 4, 24098 Kiel, Germany

Abstract. The recently grown picotube crystals are the closest to a monochiral nanotube sample achieved up to now. We present an experimental and theoretical study of the vibrational properties of these crystals, including polarization dependent Raman spectra and *ab initio* calculations. We assign symmetries to the most intense peaks, A_1 in most cases. From *ab initio* calculations we obtain the underlying atomic displacements. We find, among others, modes related to the high-energy mode and the radial-breathing mode of carbon nanotubes.

Keywords: Molecular crystal, Raman, *ab initio*
PACS: 36.20.Ng,78.30.Na,31.15.Ar

Much of the existing knowledge about carbon nanotubes stems from the zone-folding of graphite properties[1]. In this way, many regularities of the nanotube properties can be explained, but the effect of the curvature of the nanotube walls is completely neglected. A complementary approach is to study the properties of very short molecule-like nanotubes: the system we present here is a realization of this idea. A picotube is a highly symmetric hydrocarbon which resembles a very short (4,4) nanotube. The synthetic procedure guarantees the existence of only one "chirality" and uniform geometry in the samples which facilitates the comparison with theory[2]. The achievement of growing picotube crystals offers currently the closest approximation to a monochiral nanotube crystal.

We present an experimental and theoretical study of the vibrations of picotube crystals. For the polarization-dependent micro-Raman measurements we used a single grating LABRAM spectrometer and an excitation laser wavelength of 633 nm in backscattering geometry. *Ab initio* calculations were performed with the SIESTA code using the local-density approximation[3, 4, 5]. The core electrons were replaced by nonlocal, norm-conserving pseudopotentials[6]. A double-ζ singly polarized (DZP) basis set of localized atomic orbitals was used for the valence electrons. The cut off radii were determined from an energy shift of 50 meV by localization. For the calculation of the picotube molecule we used a grid cutoff of $\approx$ 100 Ry in real space ($\approx$ 200 Ry for the phonon calculation).

In Fig. 1 we show the structure of a picotube as obtained from X-ray measurements of a picotube crystal[7]. It can be seen as a (4,4) hexagon ring with eight benzene-like wings. Due to the repulsion of these wings the inversion symmetry is broken: two of the upper wings bend towards the main axis of rotation (C_2), and the other two bend

CP786, *Electronic Properties of Novel Nanostructures*, edited by H. Kuzmany, J. Fink, M. Mehring, and S. Roth
© 2005 American Institute of Physics 0-7354-0275-2/05/$22.50

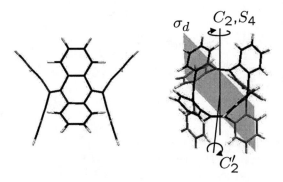

FIGURE 1. Structure of a picotube molecule. Side view: note the asymmetry of the upper and lower wings of each anthracene-like unit. Four of these wings bend towards the main rotational axis, and four away from it, forming an alternating structure. Top view: The black (gray) C atoms are in the image foreground (background). The H atoms are white. The symmetry axes and planes are indicated.

away from it, complementarily on the lower side. The molecule does not have D_{4h} but the lower D_{2d} symmetry. *Ab initio* relaxation of a single molecule yields a structure in excellent agreement with the measurements[8, 7].

In Fig. 2 we show Raman spectra for incident and outgoing light polarization parallel to one edge of a picotube crystal. It is dominated by a feature around 1600 cm^{-1}, typical for sp^2 carbon compounds. Taking a closer look we can differentiate two peaks at 1601 and 1592 cm^{-1}, plus a lower peak at 1569 cm^{-1} (see inset). The intermidiate frequency range is dominated by a peak at 1131 cm^{-1} and a group of three peaks at 1066, 1056, and 1042 cm^{-1}. At low frequencies we find a band around 480 cm^{-1} and a group of peaks around 270 cm^{-1}.

The dots in Fig. 3(a) show the intensity of the peak at 1601 cm^{-1} as function of the angle ψ between the polarization of the incident light and the x axis of the crystal. The trained eye will guess an A_1 symmetry for this mode. However, the intensity does not go to zero at $\psi = 45°$ for parallel incident and scattered light. Neither a single A_1 nor any other single Raman tensor can explain this behavior. Therefore, inspired by X-ray measurements we propose a unit cell with perpendicular molecules interacting weakly and scattering independently[7]. In this case the Raman intensity can be calculated as

$$I \propto (\mathbf{e}_i \cdot R_{PT} \cdot \mathbf{e}_s)^2 + (\mathbf{e}_i \cdot R_{PT}^{\perp} \cdot \mathbf{e}_s)^2 \tag{1}$$

where R_{PT} is the Raman tensor of a picotube molecule and $R_{PT}^{\perp}$ the same Raman tensor rotated by 90° with respect to the latter. $\mathbf{e}_{i(s)}$ denote the polarization of the incoming (scattered) light. We fitted the data for the 1601 cm^{-1} peak with the function obtained by applying Eq. (1) to the case of an A_1 Raman tensor. The resulting fit is excellent, as shown in Fig. 3. Therefore, we assign this peak to an A_1 symmetry mode from identical molecules with alternating orientations. Using this method we assigned the symmetries of the most intensive peaks, mostly A_1. The peak at 1592 $^{-1}$ is an exception. It can only be explained as the sum of different symmetries, including necessarily E or B_2. Infrared

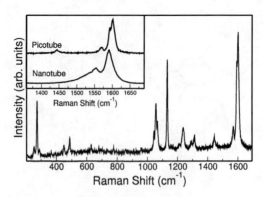

FIGURE 2. Raman spectrum of a picotube molecule along its x axis. Inset: Zoom of the high-energy region for the same picotube specrum and the spectrum of a nanotube bundle (diameter around 1 nm) .

Fourier Raman measurements show more than three peaks in this region, which confirms our observation[9].

We are now interested in the atomic displacements underlying the measured peaks. By *ab initio* calculations, we found three A_1 vibrational modes in the high-energy region shown in Fig. 3(b). The mode at 1621 cm^{-1} is equivalent to the high-energy mode of armchair carbon nanotubes, derived from the in-plane optical phonon of graphite. The high-energy mode of single-walled nanotubes has a double peak structure similar to the one found in the picotubes. However, in the case of armchair nanotubes it stems from only one phonon branch due to a double resonance process[10, 11, 1]. This is not the case for the picotubes, but the band shape is due to several modes. The lower symmetry of the molecule, for example the absence of inversion symmetry, allows more phonons to be Raman active. In addition, if we think of the picotube as a very short nanotube, we can see the picotube vibrations as the result of "folding" the Brillouin zone of the (4,4) nanotube. Therefore, for each Raman active branch of the nanotube, several modes will be found in the vibrational spectrum of the picotube. The number of active modes in the Raman spectrum of carbon nanotubes is further reduced by the antenna effect[1].

The fourth mode shown in Fig. 3(b), at a 262 cm^{-1} and with A_1 symmetry, is clearly related to the radial breathing mode of carbon nanotubes. The frequency is in good agreement with the measured peaks at 272 or 253 cm^{-1}. *Ab initio* calculations for the (4,4) nanotube yield a much higher frequency of 413 cm^{-1}. The frequency softening in the case of the picotube is related to its lower symmetry. All atoms in a nanotube are equivalent by symmetry. The radial-breathing mode is totally symmetric, therefore all atomic displacements must be equal. In the case of the picotube this condition is strongly relaxed. The wings are not equivalent and they are not connected by bonds. Part of the C-C bonds are not stretched since the mode is strongly mixed with a translation along the molecular axis, which lowers the frequency of the vibration.

Summarizing, we presented a study of the vibrational properties of picotube crystals. From polarization-dependent Raman measurements we inferred a low interaction between the molecules, supported by the excellent agreement between the theoretical structure of an isolated molecule and the measured structure in crystalline form. We

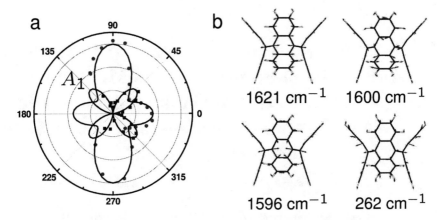

FIGURE 3. (a) Intensity of the peak at 1601 cm^{-1} as a function of the angle ψ between the polarization of the incident light and the x edge of the picotube crystal. The circles (squares) correspond to parallel (perpendicular) incident and scattered light. The lines are fits to a A_1 symmetry scattered by two perpendicular, non interacting picotube molecules. (b) Selected *ab initio* calculated vibrations of the picotube molecule.

were able to assign the symmetry of the most intensive peaks, mostly A_1. Comparing to *ab initio* calculations we found vibrations related to the in-plane optical phonon of graphite and a breathing-like mode, in strong analogy with carbon nanotubes.

ACKNOWLEDGMENTS

We thank P. Ordejón for his valuable help, U. Kuhlman for the symmetry analysis tool, and P. Rafailov and M. Tommasini for valuable discussions. S.R. was supported by the Oppenheimer Fund and Newnham College. We acknowledge the MCyT (Spain) and the DAAD (Germany) for a Spanish-German Research action.

REFERENCES

1. S. Reich, C. Thomsen, and J. Maultzsch, *Carbon Nanotubes, Basic Concepts and Physical Properties*, Wiley-VCH, Berlin, 2004.
2. S. Kammermeier, P. G. Jones, and R. Herges, *Angew. Chem. Int. Ed. Eng.*, **35**, 2669 (1996).
3. J. P. Perdew, and A. Zunger, *Phys. Rev. B*, **23**, 5048 (1981).
4. P. Ordejón, E. Artacho, and J. M. Soler, *Phys. Rev. B*, **53**, R10 441 (1996).
5. J. M. Soler, E. Artacho, J. D. Gale, A. García, J. Junquera, P. Ordejón, and D. Sánchez-Portal, *J. Phys. Condens. Mat.*, **14**, 2745 (2002).
6. N. Troullier, and J. Martins, *Phys. Rev. B*, **43**, 1993 (1991).
7. M. Machón, S. Reich, J. Maultzsch, R. Herges, and C. Thomsen (To be submitted).
8. R. Herges, M. Deichmann, J. Grunenberg, and G. Bucher, *Chem. Phys. Lett.*, **327**, 149 (2000).
9. C. Schaman (2005), private communication.
10. C. Thomsen, and S. Reich, *Phys. Rev. Lett.*, **85**, 5214 (2000).
11. J. Maultzsch, S. Reich, and C. Thomsen, *Phys. Rev. B*, **65**, 233402 (2002).

Conformation-dependent Molecular Orientation Deduced from First-principles Modeling of Oligo(ethylene glycol)-terminated and Amide Group Containing Alkanethiolates Self-assembled on Gold

L.Malysheva[*], A. Onipko[†], R. Valiokas[‡#], and B. Liedberg[‡]

[*]Bogolyubov Institute for Theoretical Phycis, Kiev, 03143 (Ukraine)
[†]Division of Physics, Luleå University of Technology, S-971 87 Luleå (Sweden)
[‡]Division of Sensor Science and Molecular Physics, Department of Physics and Measurement Technology, Linköping University, S-581 83 Linköping (Sweden)
[#]Present address: Molecular compounds physics laboratory, Institute of Physics, Savanoriu 231, LT-02300 Vilnius, Lithuania; E-mail: valiokas@ar.fi.lt

Abstract. We report orientation angles for the alkyl chain, amide group, and oligo(ethylene glycol) (OEG) portion within self-assembled monolayers (SAMs) of OEG-terminated and amide containing alkanethiolates which, depending on the OEG length and substrate temperature, display unique conformations - all-trans or helical. Optimized geometries of the molecular constituents, characteristic vibration frequencies and transition dipole moments are obtained by using DFT methods with gradient corrections. These *ab initio* data are subsequently used to simulate infrared reflection-absorption (RA) spectra associated with different conformations and orientations. The obtained results have generated a deeper knowledge of the internal SAM structure, which is crucial for understanding phase and folding characteristics, interaction with water and ultimately the protein repellent properties of OEG-containing SAMs.

Keywords: Self-assembled monolayers, infrared spectroscopy.
PACS: 81.16.Dn, 87.64.Je, 87.64.Aa.

INTRODUCTION

In the current communication, we address a specific class of OEG-terminated SAMs on Au(111) supporting surfaces, which has been thoroughly studied by means of temperature-programmed infrared RA spectroscopy [1]. These SAMs, formed by self-assembly of $HS(CH_2)_mCONH$-EG_n [where $EG_n = (CH_2CH_2O)_n$] exhibit unusual conformational properties depending on the length of the OEG tail and the substrate temperature. SAMs with $n \geq 6$ are known to undergo a reversible temperature-driven phase transitions from helical (at room temperature) to all-trans conformation (at $\sim 60°C$). In the case of a shorter OEG portion, the all-trans conformation is obtained at room temperature in SAMs on silver (for $m = 11$, $n = 3$ [2]) and on gold (for $m = 15$, $n \leq 4$ [3]). These findings are based on infrared RA spectra where the different molecular conformations leave markedly distinctive signatures. However, very little is known

CP786, *Electronic Properties of Novel Nanostructures*, edited by H. Kuzmany, J. Fink, M. Mehring, and S. Roth
© 2005 American Institute of Physics 0-7354-0275-2/05/$22.50

about molecular orientation within OEG-terminated SAMs and how it depends on the conformational state. Combining the results of density functional theory (DFT) calculations of the equilibrium molecular geometry, vibrational frequencies, and transition dipole moments with available experimental data on helical and all-trans OEG SAMs enables us to propose the optimized geometry of the $HS(CH_2)_{15}CONH-EG_6$ molecule and its orientation (including mutual orientation of the alkyl-, amide- and OEG portion) within the assembly on gold.

STRATEGY OF AB INITIO MODELING

In our modeling, the spatial orientation of alkyl chain, amide group, and OEG helix (h) or all-trans (t) chain is characterized by three sets of Euler angles defined for each of the molecular parts as shown in Fig. 1. Each set determines orientation with respect to the substrate surface coinciding with xy-plane, Figure 1.

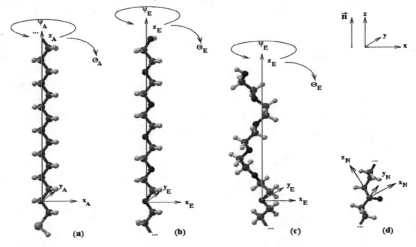

FIGURE 1. Definition of Euler angles and (x_A,y_A,z_A) axes for alkyl segment (a), (x_E,y_E,z_E) axes for OEG trans (b) and helix (c), and (x_N,y_N,z_N) for amide group (d). The axes are shown for the case when in laboratory frame of reference xyz, all the angles are zero. In Figs. 1a and 1b, x_Az_A- and x_Ez_E plane coincides with CCC- and all-trans COC plane respectively; for helical OEG conformation, axis x_E is directed along the bisector of the angle formed by the first COC group, axis z_E is the vector product of x_E and the bisector of the angle formed by the next COC group, and axis $y_E = z_E \times x_E$; for amide moiety CONH, z_N axis is parallel to C-N bond, axis y_N is perpendicular to O=C-N-H plane, and $x_N = y_N \times z_N$.

For any values of tilt angle θ_A, twisting angle ψ_A, and azimuth angle φ_A (the latter is defined as is commonly accepted [4]) specifying orientation of alkyl CCC plane, the two other sets of angles θ_N, ψ_N, φ_N and θ_E, ψ_E, φ_E are dictated by the equilibrium molecular geometry. That geometry was calculated by using BP86 exchange-correlation functional with 6-31G* basis set as provided by the Gaussian-03 suite of programs. The vibrational frequencies and transition dipole moments (TDMs) were obtained with the same accuracy. To deduce the orientation of the CCC plane, we have compared experimental and modeled RA spectra. The modeling procedure is described in detail elsewhere [5]. In brief, the modeling strategy can be outlined as follows.

The observed intensity of the symmetric and asymmetric CH_2-stretching peaks can be reproduced within a rather narrow range of angles θ_A and ψ_A. These peaks are very similar in their appearance in the RA spectra of both OEG-terminated alkanethiolates [1,2,6] and pure alkanethiolates [4]. From possible values of θ_A and ψ_A, giving a reasonable agreement with experimental RA spectra in the CH_2-stretching region (not shown here), the most likely molecular orientation is determined from the best fit of the modeled spectra in the fingerprint region, where one observes an intense amide II peak while the corresponding amide I vibration appears inactive [6]. The latter finding restricts the tilt and twisting angles θ_A and ψ_A. Finally, there are experimental indications of lateral hydrogen bonding in self-assemblies containing amide groups [1]. This has been used as a guide to obtain the azimuthal angle φ_A that has been deduced from the requirement of minimal O-H distance between oxygen and hydrogen atoms in nearest-neighbor amide groups.

MODEL DATA VERSUS EXPERIMENTAL DATA

Our findings regarding the molecular orientation within SAMs in focus are represented in Table 1. The upper two rows show the angles of the OEG tails and amide groups, as it would look like for the all trans (t) and helical (h) conformations, if the alkyl plane was oriented perpendicularly to the substrate as shown in Fig. 1a. As can be seen from these data, the respective OEG portions are tilted in opposite directions and the tilt is larger by $20°$ for the h conformation. In contrast, the orientation of amide portion appears to be less sensitive to the change in molecular conformation of the OEG tail.

The lower two rows in Table 1 reveals the actual orientation of the molecular parts adopting either the t or h conformation. A wealth of spectroscopic data supports the assumption that the alkyl angles θ_A and ψ_A change little under the conformational transition. This conclusion was recently confirmed by modeling the structure and RA spectra for SAMs with shorter alkyl and OEG portions [7]. Note that in both conformations, tilting and twisting angles are within the range reported for self-assemblies of alkanethiols [4]. As already mentioned, the values of θ_A and ψ_A, which are represented in Table 1, provides the best possible coincidence between the modeled and measured RA spectra. A few degrees deviation from the alkyl tilt and rotation angles specified in Table 1 results in a noticeable change in the spectral appearance and to a reduced reproducibility of the SAM properties (e.g. thickness). In this sense, the angles can be regarded as characteristic for a given type of SAMs. Once the orientation of the alkyl plane is found, the two other sets of Euler angles are calculated using appropriate transformations of coordinates [8].

One sound conclusion that can be drawn from these calculations is that the tilt of the z_E axis for the t conformation is substantially larger than for the h conformation. Moreover, the tilts occur in opposite directions with respect to the surface normal. We also found that the tilt of OEG portion within the SAMs increases from $22°$ to $33°$ during the h-t phase transition, Table 1. In terms of effective tilt (i.e., the angle between the normal and straight line connecting the sulfur atom and hydrogen atom of the OH end group) the corresponding increase is from $17°$ to $28°$. An immediate consequence of this result is that the difference in the SAM thicknesses for the *h* and *t* conformations should be considerably smaller than expected from a direct comparison of the molecular lengths, i.e. 38.6 and 43.8 Å for *h*- and *t* conformations, respectively. For the *h* conformation, the calculated SAM thickness 38.9 Å agrees well with the measured value 39.8 ± 1.7 Å [3]; the calculated thickness for the *t* conformation is 40.6 Å, but it remains to be verified experimentally. On the other hand, for SAMs with an OEG length of $n = 4$, for which the *t* conformation is stable up to at least 75 °C [1,3] the obtained SAM thickness equals 33.9 Å a value that is in good agreement with the experimental one, 33.3 ± 1.1 Å [3].

TABLE 1. Orientation of alkyl- (A), OEG- (E), and amide (N) part of molecule $HS(CH_2)_{15}CONH\text{-}EG_6$ in all-trans (t) and helical (h) conformations within SAM.

	θ_A	ψ_A	φ_A [a]	θ_E	ψ_E	φ_E [a]	θ_N	ψ_N	φ_N [a]
t	0	0	0	8°	− 67°	98°	44.5°	− 34°	− 63°
h	0	0	0	28°	87°	− 69°	41°	− 36°	− 77°
t	26°	− 65°	− 82°	33°	− 41°	− 74°	−34°	−4°	0°
h	26°	− 62°	−72°	22°	24.5°	− 146°	−27°	−4°	2°

[a] For the hexagonal lattice, the change of azimuth angle by $60° \cdot n$, n =0, 1, ... does not affect the SAM structure.

Our data also can be used to estimate the distance between oxygen and hydrogen atoms belonging to the nearest-neighbor amide groups in a hexagonal ($\sqrt{3} \times \sqrt{3}$)R30° lattice, where the sulfur atoms are separated from each other by ~5 Å. We obtained that this distance decreases from 2.5 Å for the *h* conformation to 2.2 Å for the *t* conformation implying that SAMs with an all-trans OEG terminus exhibit shorter and thereby stronger amide hydrogen bonds. This conclusion agrees with the observed shift of the amide II peak to higher frequencies under the *h* − t phase transition [3]. Such a shift is straightforwardly attributed to a strengthening of the NH···O=C bonds.

A meaningful test of the accuracy of the present approach is obtained by comparing the experimental and calculated RA spectra for SAMs with the OEG part in the *h*- and *t* conformation, Figure 2. The differences in conformation and orientation for the two phases are evident in a number of spectroscopic features [1,3,6]. For instance, peaks associated with rocking, wagging and bending vibrations practically disappear under the *h* to *t* transition. At the same time, the intensity of the amide II peak remains virtually the same for both conformations. These and other peculiarities are well reproduced by our calculations. It is also worthwhile mentioning that the main peak is blue-shifted in the *t* conformation by 29 cm[-1], which is almost the same shift as can be seen in the calculated spectra. Although most of observed spectral details find their counterparts in calculated spectra, there are some exceptions. One of them is the asymmetry of the main peak that is reproduced in the case of the helical conformation but not for all-trans conformation. We expect that further improvements (in which the first priority should be given to intermolecular interaction effects) might correct the results regarding the orientation angles of the molecular constituents within the SAMs.

In conclusion, our first principle modeling of the structure and orientation of $HS(CH_2)_{15}CONH\text{-}EG_6$ assemblies on Au(111) explains a number of experimental observations such as SAM thickness and tendency of hydrogen bond formation, as well as many specific features seen in the RA spectra. Our findings provide strong support to the hypothesis that the orientation of the alkyl portion remain unchanged during the phase transition, and resembles those observed for less complicated alkanethiolate SAMs. The CONH plane is not much affected by the OEG phase transition, whereas the OEG axis undergoes a substantial reorientation.

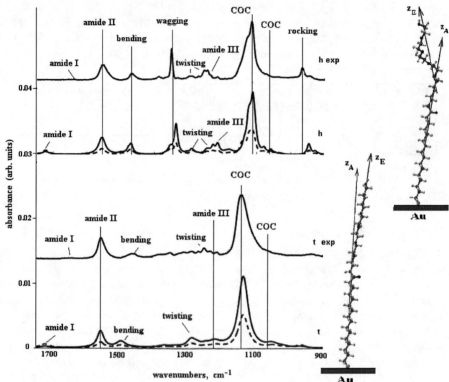

FIGURE 2. Measured [1] and model RA spectra in the fingerprint and amide regions for self-assemblies of HS(CH$_2$)$_{15}$CONH-EG$_6$ in all-trans (lower part) and helical (upper part) conformations. Molecules are shown in orientation specified in Table 1. In calculations, the spectrum is given by the sum of Lorentsian-shaped peaks with a certain half-width at half-maximum (HWHM); each peak is centred at the fundamental mode frequency and having the height proportional to the squared TDM's z component of the corresponding mode. HWHM = 12 cm^{-1} for amide peaks, and for all other bands HWHM = 12 cm^{-1} and 6 cm^{-1} for all-trans- and helical conformation, respectively. In model spectra, scaling factor equal to 1.037 is used only for amide II and amide III frequencies.

REFERENCES

1. Valiokas, R., Östblom, M., Svedhem, S., Svensson, S.C.T., Liedberg, B., *J. Phys. Chem. B* **104**, 7565 (2000).
2. Harder, P., Grunze, M., Dahint, R., Whitesides, G.M., and Laibinis, P.E *J. Phys. Chem. B* **102**, 426 (1998).
3. Valiokas, R., Östblom, M., Svedhem, S., Svensson, S.C.T., Liedberg, B., *J. Phys. Chem. B* **105**, 5459 (2001).
4. Parikh, A. N., and Allara, D. L. *J. Chem. Phys.* **96**, 927 (1992)
5. Malysheva, L., Klymenko, Yu., Onipko, A., Valiokas, R., and Liedberg, B., *Chem. Phys. Lett.* **37**, 451 (2003).
6. Valiokas, R., Svedhem, S., Svensson, S.C.T., and Liedberg, B., *Langmuir* **15**, 3390 (1999).
7. Malysheva, L, Onipko, A., Valiokas, R., and Liedberg, B., *in preparation.*.
8. Wilson, E.B. Jr, Decius, J.C., and Cross, P.C., *Molecular Vibrations*, McGraw-Hill: New York, 1955.

Dispersion relations of plasmons in carbon nanotubes

Ricardo Perez and William Que

Department of Physics, Ryerson University, 350 Victoria Street,
Toronto, Ontario, Canada M5B 2K3

Abstract. We propose a new theoretical interpretation of the electron energy-loss spectroscopy results of Pichler *et al.* on bulk carbon nanotube samples. The origin of the experimentally found nondispersive modes have been controversial, and at least three different interpretations have been offered in the literature. From our theoretical results of the loss functions for individual carbon nanotubes based on a tight-binding model, we find that the nondispersive modes could be due to collective electronic modes in chiral carbon nanotubes, while the observed dispersive mode should be due to collective electronic modes in armchair and zigzag carbon nanotubes, concluding that the nondispersive character is better related to the chirality of the nanotube than to the angular momentum of the collective excitations.

Keywords: Carbon nanotubes, plasmons, EELS, collective modes
PACS: 71.20.Tx, 78.67.Ch, 73.20.Mf, 71.10.Pm

Momentum-dependent electron energy loss spectroscopy (EELS) as carried out by Pichler *et al.* offers an excellent tool for studying plasmons in carbon nanotubes. Their experiment was performed first on bulk samples of single wall carbon nanotubes (SWNT) [1] and later on magnetically aligned bundles of SWNT [2]. In the low energy range of the spectrum, the experimental findings are: (i) a dispersive mode as function of momentum transfer in the $5-8$ eV range; (ii) several nondispersive modes at lower energies. The dispersive mode was attributed to the π-plasmon without controversy. As for the nondispersive modes Pichler *et al.* interpreted them, firstly, in terms of interband excitations between localized states polarized perpendicular to the nanotube axis [1]. Later this interpretation was modified in light of the new results coming from optical absorption measurements [2, 3, 4, 5].

The bulk sample used in the experiment of Pichler *et al.* had a mean diameter of 1.4 nm, and nondispersive modes were observed at 0.85, 1.45, 2.0, and 2.55 eV. Optical absorption measurements by Jost *et al.* [3] on carbon nanotube containing-soot revealed excitations at 0.72, 1.3, and 1.9 eV for the mean diameter of 1.29 nm. Since the gaps between Van Hove singularities in the electronic density of states is known to be inversely proportional to the diameter, single-particle excitation energies should be larger in smaller diameter carbon nanotubes. However, the observed excitations in the experiment of Jost *et al.* appear to be at smaller energies compared to those observed in the experiment of Pichler *et al.* To reconcile the two experiments, one has to assume that the nondispersive modes observed by Pichler *et al.* are collective rather than single-particle modes. Additionally recent experimental and theoretical results [4] on polarized optical absorption of aligned

CP786, *Electronic Properties of Novel Nanostructures*, edited by H. Kuzmany, J. Fink, M. Mehring, and S. Roth
© 2005 American Institute of Physics 0-7354-0275-2/05/$22.50

SWNT of 1.35 nm in average diameter show that when the light is polarized parallel to the tube axis, the absorption spectra have several peaks below 3 eV, but when the light is polarized perpendicular to the tube axis, the absorption spectra become essentially featureless. This result suggests that the nondispersive modes are not due to excitations polarized perpendicular to the tube axis.

A recent paper by Liu *et al.* [5] comparing optical absorption with EELS suggests that the nondispersive modes in the EELS are due to collective excitations of the optically allowed transitions. This could be a viable interpretation (barring the perpendicular polarization), making these modes analogous to the intersubband plasmons in quantum wires [6]. Although this interpretation seems to be more adequate, still no explanation was given about the nondispersive character.

Three years after the initial experiment, theorist Bose [7] gave his own explanation for the nondispersive character of the low-energy peaks. Based on a plasmon calculation using a model of free electron gas confined to a cylindrical surface, he suggested an alternative interpretation of the nondispersive modes in terms of optical plasmons carrying nonzero angular momenta. However, a close inspection of the calculated plasmon dispersion curves presented in an earlier paper by Longe and Bose [8] reveals difficulties with this interpretation regarding how to assign a specific angular momentum to the observed dispersive mode.

Today, the origin of the nondispersive modes remains a puzzle. In this paper, we present our theoretical results on the loss functions of individual SWNT, and shed some light on the origin of the nondispersive modes. In particular, we propose that the nondispersive modes are inter(sub)band plasmons from chiral carbon nanotubes which have small Brillouin zones. These collective modes generally are not polarized perpendicularly to the tube axis. Further experiments are suggested to decisively determine the validity of this interpretation.

It is well known that the electronic properties of carbon nanotubes are dependent on the chirality. Such important details are not captured by a free electron gas type model. On the other hand, a tight-binding model [9] is known to produce the electronic band structures of carbon nanotubes very well as long as the radius is not too small. We use such a tight-binding model for π band electrons to study the collective electronic excitations of individual carbon nanotubes, similar to the approach by Lin *et al.* [10, 11]. The theoretical framework is the well-used random phase approximation (RPA) theory, which has been applied successfully to many systems including quantum wires [6]. This theory have also proved its effectiveness when applied to Luttinger liquids [12, 13].

Due to the cylindrical symmetry of the individual nanotube, the electronic excitations have a well defined transfer momentum q_0 in the first Brillouin zone, angular momentum L and energy ω, these dependences will be transfered to the microscopic dielectric matrix $\epsilon_{G,G'}(q_0, L, \omega)$ where G, G' are lattice vectors in momentum space. The macroscopic dielectric function is given by $\epsilon_M(q_0, L, \omega) = 1/\epsilon_{0,0}^{-1}(q_0, L, \omega)$ and the loss function is calculated as $Im(-1/\epsilon_M(q_0, L, \omega))$ [14].

By scanning the loss functions to find peak positions at different wavevectors, we produce the dispersion curves of the collective electronic modes in Figs. 1 and 2, for wavevectors along the tube axis up to the Brillouin zone edge of the corresponding carbon nanotube. Assuming a carbon-carbon bond length of $a_{C-C} = 1.44$ Å, it

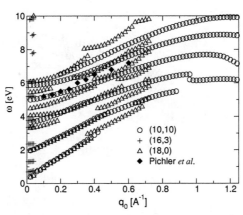

FIGURE 1. Dispersion curves for the collective electronic modes with angular momentum index $L = 0$. The solid diamonds are experimental results on the dispersive mode from Pichler *et al.*

can be shown that all armchair carbon nanotubes have the same Brillouin zone edge of $\pi/T = 1.26$ Å^{-1} (T is the length of the translational vector [9]), and all zigzag carbon nanotubes have the same Brillouin zone edge of $\pi/T = 0.73$ Å^{-1}, but different chiral carbon nanotubes have different Brillouin zone sizes. Those (n, m) chiral nanotubes for which the greatest common divisor among $2n + m$ and $2m + n$ is 1 have the smallest Brillouin zones, with $\pi/T = \pi/(3a_{C-C}\sqrt{n^2 + m^2 + nm})$. For the $(16, 3)$ chiral carbon nanotube, its Brillouin zone edge is at $\pi/T = 0.041$ Å^{-1}. Some of the curves for the $(10, 10)$ and $(18, 0)$ carbon nanotubes terminate before reaching the Brillouin zone edge due to vanishingly small peak amplitudes. Clearly, the $(10, 10)$ armchair tube and the $(18, 0)$ zigzag tube both have dispersive modes for all the computed L, and we find this to be generally true for armchair and zigzag tubes. On the other hand, the collective electronic modes of the $(16, 3)$ chiral tube have little dispersion, and so do many other chiral tubes. The reason for the lack of dispersion is the much smaller Brillouin zone.

If we compare the results in Figs. 1 and 2 with the results of Longe and Bose [8], a major difference is that in the latter, there is only one branch of collective mode for each angular momentum index L, while in our results, we find many branches for each angular momentum index. This is due to the band structures of carbon nanotubes with many occupied and many empty (sub)bands. Generally speaking, when L is increased, excitation energies increase, and dispersion is reduced. These qualitative features are already present in the free electron gas type model. Unlike Bose, we find there is no need for the nonzero angular momentum modes in order to explain the nondispersive modes. Since the experiment of Pichler *et al.* was performed on bulk samples (7 Å mean radius), the measured spectra contain contributions from many carbon nanotubes of different chirality. The nondispersive modes could be due to chiral carbon nanotubes, and the dispersive mode should be due to armchair and zigzag carbon nanotubes. Experimentally, only one dispersive mode was found, but since the peak of the dispersive mode was a couple eV broad,

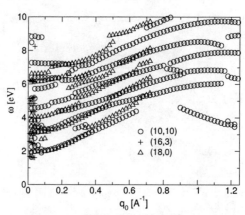

FIGURE 2. Dispersion curves for the collective electronic modes with angular momentum index $L = 1$, for the same three carbon nanotubes as in Fig. 1.

it is possible that several modes of large amplitude contributed to the broad peak.

Since intertube coupling shifts the energies of the collective electronic modes higher [15], it is not possible to match the calculated energies in this work for individual carbon nanotubes to experimental results on bulk samples where intertube coupling is present.

ACKNOWLEDGMENTS

This work was supported by the Natural Sciences and Engineering Research Council of Canada.

REFERENCES

1. T. Pichler *et al.*, Phys. Rev. Lett. **80**, 4729 (1998); M. Knupfer *et al.*, Carbon **37**, 733 (1999).
2. X. Liu *et al.*, Synth. Metals **121**, 1183 (2001).
3. O. Jost *et al.*, Appl. Phys. Lett. **75**, 2217 (1999).
4. M. F. Islam *et al.*, Phys. Rev. Lett. **93**, 037404 (2004).
5. X. Liu *et al.*, Phys. Rev. B **66**, 045411 (2002).
6. W. Que and G. Kirczenow, Phys. Rev. B **39**, 5998 (1989); W. Que, Phys. Rev. B **43**, 7127 (1991); Q. P. Li and S. Das Sarma, Phys. Rev. B **43**, 11768 (1991).
7. S. M. Bose, Phys. Lett. A **289**, 255 (2001).
8. P. Longe and S. M. Bose, Phys. Rev. B **48**, 18239 (1993).
9. R. Saito, G. Dresselhaus and M. S. Dresselhaus, *Physical properties of carbon nanotubes* (Imperial College Press, London, 1998).
10. M. F. Lin *et al.*, Phys. Rev. B **56**, 1430 (1997).
11. F. L. Shyu and M. F. Lin, Phys. Rev. B **62**, 8508, (2000).
12. Q. P. Li *et al.*, Phys. Rev. B **45**, 13713 (1992).
13. W. Que, Phys. Rev. B **66**, 193405 (2002); *ibid.* **67**, 129901 (2003).
14. Stephen L. Adler, Phys. Rev. **126**, 413 (1962).
15. W. Que, J. Phys.: Condens. Matter **14**, 5239 (2002).

Resonant Raman Intensity Of The Radial-Breathing Mode Of Single-Walled Carbon Nanotubes

Valentin N. Popov*, Luc Henrard, and Philippe Lambin

Laboratoire de Physique du Solide, Facultés Universitaires Notre-Dame de la Paix, 61 rue de Bruxelles, B-5000 Namur, Belgium
*E-mail: valentin.popov@fundp.ac.be; http://www.fundp.ac.be/~vpopov/

Abstract. The resonant Raman profile of the radial-breathing mode was calculated for all 300 single-walled carbon nanotubes in the radius range from 2 Å to 12 Å and for optical transition energies up to 3.5 eV within a symmetry-adapted non-orthogonal tight-binding model (V. N. Popov, New J. Phys. **6**, 17 (2004)). In the calculations, the electron-phonon and electron-photon interactions were accounted for explicitly. The simulated Raman spectrum for a nanotube sample is in fair agreement with the experimentally observed one. It is shown that the simplifying assumption for tube-independent matrix elements of these interactions can lead to erroneous predictions of the Raman spectra.

Keywords: carbon nanotube, resonant Raman scattering.
PACS: 63.22.+m, 73.63.Fg, 78.30.Na

INTRODUCTION

The carbon nanotubes have unusual vibrational, electronic, and optical properties due to their quasi-one-dimensionality [1]. Nowadays, it is possible to produce in large quantities nanotubes consisting of a single graphitic layer (single-walled nanotubes, SWNTs). The low-frequency vibrational mode of the nanotubes with uniform radial atomic displacements (the radial-breathing mode, RBM) gives rise to a high-intensity Raman line which can be used for the structural characterization of the samples by means of Raman spectroscopy [2,3]. The observation of these Raman lines is mostly possible under resonant conditions, i.e., when the laser excitation energy is close to the optical transitions of the tubes. The optical transitions of SWNTs, calculated within tight-binding [4] and first-principles [5] models, underestimate the observed absorption and emission energies. This disagreement is mainly due to the neglect of the self-energy and excitonic effects on the band energies [6]. However, major part of these effects cancel and the remaining part can be modeled by a logarithmic term [7].

The usual approach to Raman spectra assignment is based on the use of the RBM frequency and the optical transitions of the tubes. However, the assignment cannot be done unambiguously without the knowledge of the dependence of the Raman line intensity on the laser energy (the so-called resonance Raman profile, RRP) as well. The RRP is usually estimated within the πTB model with tube-independent

CP786, *Electronic Properties of Novel Nanostructures*, edited by H. Kuzmany, J. Fink, M. Mehring, and S. Roth
© 2005 American Institute of Physics 0-7354-0275-2/05/$22.50

momentum and electron-phonon coupling matrix elements [8]. In a few cases, the momentum matrix elements were calculated explicitly within the band structure picture [9,10]. To the best of our knowledge, no results for the RRP within the excitonic picture have been published so far. Recently, we have reported the results for the RRP of 50 narrow semiconducting nanotubes carried out within a well-tuned non-orthogonal tight-binding model [11] showing the existence of strong dependence of the maximum Raman intensity on the tube type. These RRPs were applied for the simulation of the Raman spectra of SWNT samples and a fair agreement was achieved with available Raman data on HiPco samples [2].

In this paper, we extend the study of Ref. [4] to encompass 300 SWNTs including metallic ones. We show that the assumption for tube-independent matrix elements of the electron-phonon and electron-photon interactions leads to unrealistic prediction of the Raman spectra of SWNT samples.

RESULTS AND DISCUSSION

The SA-NTB has already been implemented in the calculation of the band structure and dielectric function of many nanotubes [4]. The SA scheme allows one to use a two-atom unit cell instead of the translational unit cell of the nanotube and derive the band structure of the nanotube by solving the matrix eigenvalue equation

$$\sum_r \left(H_{klrr'} - E_{kl} S_{klrr'} \right) c_{klr'} = 0. \tag{1}$$

Here $H_{klrr'}$ and $S_{klrr'}$ are the matrix elements of the Hamiltonian and the overlap matrix elements, respectively, E_{kl} is the one-electron energy, and $c_{klr'}$ are the coefficients in the expansion of the one-electron wavefunction as a linear combination of the atomic orbitals of the two-atom unit cell. The index k is the one-dimensional wavevector and l is the integer quantum number. The index r labels the eight atomic orbitals of the two-atom unit cell. The solutions of Eq. 1 are E_{klm} and $c_{klmr'}$, where $m = 1, 2, ..., 8$.

The quantum-mechanical description of the Raman-scattering process can be done considering the system of electrons, photons and phonons, and their interactions. The most resonant Stokes process includes absorption of an incident photon with energy E_L with creation of an electron-hole pair, scattering of the electron (hole) by a phonon with frequency ω_{ph}, and annihilation of the electron-hole pair with emission of a photon with energy $E_S = E_L - \hbar\omega_{ph}$. We restrict ourselves to scattering configurations with parallel polarizations of the incident and scattered light along the tube axis when only electronic transitions $klm \rightarrow klm'$ are dipole-allowed. The Raman intensity of the process is given by

$$I(E_L, \omega_{ph}) \propto \left| \frac{1}{L} \sum_{cv} \frac{p_{cv} E'_{cv} p^*_{cv}}{\left(E_L - E_{cv} - i\gamma_{cv} \right)\left(E_L - E_{cv} - \hbar\omega_{ph} - i\gamma_{cv} \right)} \right|^2, \tag{2}$$

where E_{cv} is the vertical separation between two states in a valence band ($v = klm$) and a conduction band ($c = klm'$), and γ_{cv} is the excited state width. p_{cv} is the matrix element of the component of the momentum along the tube axis, E'_{cv} is the electron-phonon coupling matrix element, and L is the tube length. The summation in Eq. (2) is over the Brillouin zone of the given nanotube.

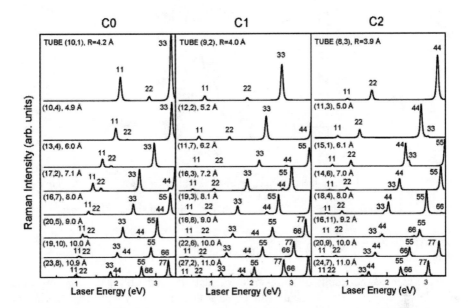

FIGURE 1. The calculated resonance Raman profiles for several metallic (C0) and semiconducting (C1 and C2) chiral SWNTs.

First, we calculated the RRPs of the RBM for the relaxed structure of all 300 SWNTs in the radius range 2 Å < R < 12 Å and laser excitation energies E_{ii} (ii = 11, 22, ...) up to 3.5 eV. The obtained RRPs for several chiral nanotubes are shown in Fig. 1. They consist of symmetric bell-like structures each of which is formed from two broadened resonances at E_{ii} and $E_{ii} + \hbar\omega_{ph}$. It is clear that peaks 11, 33, 55, ... in metallic C0 tubes have larger intensity that peaks 22, 44, 66, ... This fact corresponds to the experimental observation of higher intensity of the lower-energy component of the pair of optical transitions of metallic tubes originating from the trigonal-warping effect splitting [3]. Also, the intensity for transitions 22 in tubes C2 is much higher than for transitions 11 and 22 in tubes C1 and transitions 11 in C2 [3].

Secondly, we used the obtained RRPs to simulate the Raman spectra of a SWNT sample with a certain radius distribution of the tubes g and for a given laser excitation energy using the formula

$$I_{tot}(E_L, \omega) = \sum gI(E_L, \omega_{ph}) \frac{1}{\left(\omega - \omega_{ph}\right)^2 + \gamma_{ph}^2}, \qquad (3)$$

where the summation is over the tube types. The last factor in Eq. (3) describes the broadening of the RBM line due to the finite phonon lifetime; the linewidth γ_{ph} was chosen to be 3 cm^{-1}. The simulated Raman spectrum for a SWNT sample with a Gaussian diameter distribution is shown in Fig. 2. It is clear that the spectrum for assumed tube-independent electron-phonon and electron-photon matrix elements gives unrealistic prediction of the spectrum compared to that of the full calculation. The obtained RRPs can be used to determine the actual tube distribution g in a SWNT sample by fitting Eq. 3 to Raman spectra measured at different photon energies.

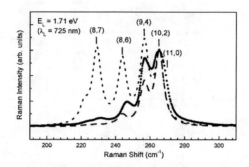

FIGURE 2. Simulated Raman spectrum in comparison with the experimentally measured one in Ref. [2] for a SWNT sample with an average radius $R_0 = 4.65$ Å and a standard deviation $\sigma = 0.4$ Å. Solid circles: experimental data, dashed line: full calculation, dotted line: tube-independent matrix elements.

CONCLUSIONS

We presented the results of large-scale calculations of the resonant Raman profiles of SWNTs. The comparison of the simulated Raman spectrum with the experimental one showed that the neglect of the tube-dependence of the electron-phonon and electron-photon interaction matrix elements can lead to erroneous predictions.

ACKNOWLEDGMENTS

V. N. P. was supported by the Marie-Curie Intra-European Fellowship MEIF-CT-2003-501080. L. H. was supported by the Belgian National Fund (FNRS). This work was partly funded by the Belgian Interuniversity Research Project on quantum size effects in nanostructured materials (PAI-IUAP P5/1). V. N. P., L. H., and Ph. L. were partly supported by the NATO CLG 980422.

REFERENCES

1. *Carbon nanotubes: Synthesis, Structure, Properties, and Applications*, edited by M. S. Dresselhaus, G. Dresselhaus, and Ph. Avouris, Berlin: Springer-Verlag, Berlin, 2001.
2. S. K. Doorn, D. A. Heller, P. W. Barone, M. L. Usrey, and M. S. Strano, *Appl. Phys.* **A 78**, 1147-1155 (2004).
3. H. Telg, J. Maultzsch, S. Reich, F. Hennrich, and C. Thomsen, Phys. Rev. Lett. **93**, 177401 (2004).
4. V. N. Popov, New J. Phys. **6**, 17 (2004).
5. S. Reich, C. Thomsen, and P. Ordejón, Phys. Rev. **B 65**, 155411 (2002).
6. C. D. Spataru, S. Ismail-Beigi, L. X. Benedict, and S. G. Louie, Phys. Rev. Lett. **92**, 077402 (2004).
7. C. L. Kane and E. J. Mele, Phys. Rev. Lett. **93**, 197402 (2004).
8. A. Jorio, A. G. Souza Filho, G. Dresselhaus, M. S. Dresselhaus, R. Saito, J. H. Hafner, C. M. Lieber, F. M. Matinaga, M. S. S. Dantas, and M. A. Pimenta, Phys. Rev. **B 63**, 245416 (2001).
9. E. Richter and K. R. Subbaswamy, Phys. Rev. Lett. **79**, 2738-2741(1997).
10. M. Canonico, G. B. Adams, C. Poweleit, J. Menendez, J. B. Page, G. Harris, H. P. van der Meulen, J. M. Calleja, and J. Rubio, Phys. Rev. **B 65**, 201402(R) (2002).
11. V. N. Popov, L. Henrard, and Ph. Lambin, Nano Letters **4**, 1795-1799 (2004).

Helicity dependent selection rules and correlated electron transport properties of double-walled carbon nanotubes

Shidong Wang, Milena Grifoni

Theoretische Physik, Universität Regensburg, 93040 Regensburg, Germany

Abstract. We analytically demonstrate helicity determined selection rules for intershell tunneling in double-walled nanotubes with commensurate (c-DWNTs) and incommensurate (i-DWNTs) shells. For i-DWNTs the coupling is negligible between lowest energy subbands, but it becomes important as the higher subbands become populated. In turn the elastic mean free path of i-DWNTs is reduced for increasing energy, with additional suppression at subband onsets. At low energies, a Luttinger liquid theory for DWNTs with metallic shells is derived. Interaction effects are more pronounced in i-DWNTs.

Since they are discovered in 1991 [1], carbon nanotubes have been under intensive investigations due to their unusual physical properties (See Ref. [2] for a review.). Carbon nanotubes can be single-walled (SWNT) or multi-walled (MWNT), depending on whether they consist of a single or of several graphene sheets wrapped onto coaxial cylinders, respectively. Electronic properties of SWNTs are mostly understood [2]. In particular, SWNTs are usually ballistic conductors [3], and whether a SWNT is metallic or semiconducting is solely determined by its so called chiral indices (n, m). Due to the one-dimensional character of the electronic bands at low energies, Luttinger liquid features at low energies have been predicted [4] and observed [5] The situation however, is much less clear for MWNTs. Except for few experiments [6], MWNTs are typically diffusive conductors [7, 8], with current being carried by the outermost shell at low bias [8, 9] and also by inner shells at high bias [10]. Intershell conductance measurements consistent with tunneling through orbitals of nearby shells have recently been reported [11]. Moreover, which kind of electron-electron correlation effects determine the observed zero-bias anomalies [12, 13] of MWNTs is still under debate. To better understand these features, i.e., the role of *inter*shell coupling on transport properties of MWNTs, some experimental [14] and theoretical [15, 16, 17, 18, 19] works focussed on the simplest MWNT's realization, namely on double-walled nanotubes (DWNTs). One main outcome is that a relation must exist between the intershell coupling, shell helicity and transport properties. Specifically, two shell are called commensurate (incommensurate), if the ratio between their respective unit cell lengths along the tube axis, is rational (irrational) [2]. For example, using tight-binding models, Saito *et al.* [15] numerically found energy gaps opened by the intershell coupling in a DWNT with two armchair (and hence commensurate) shells. Ab-initio calculations [16, 17] confirmed these results. In general, numerical evidence of a negligible intershell coupling in DWNTs with incommensurate shells (i-DWNT) at low energies is found [15, 17, 18].

CP786, *Electronic Properties of Novel Nanostructures*, edited by H. Kuzmany, J. Fink, M. Mehring, and S. Roth
© 2005 American Institute of Physics 0-7354-0275-2/05/$22.50

Here, we report on *helicity-dependent selection rules* for tunneling, which we recently derived [20]. For i-DWNTs the intershell coupling is negligible between the *lowest* subbands but it becomes important when *higher* subbands are involved. We show that this in turn yields an elastic mean free-path which decreases with energy and which shows a characteristic suppression at subbands onset. Then by including intra- and inter-shell Coulomb interactions, we show that metallic DWNTs can be described by Luttinger liquid theory at low energies. The tunneling density of states has a power-law behavior with different exponents for i-DWNTs and commensurate-shells DWNTs (c-DWNTs).

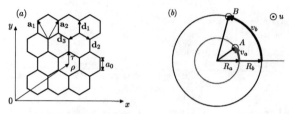

FIGURE 1. (a) Graphene lattice. The x and y axes are along the armchair and zigzag axes respectively. The distance between two nearest carbon atoms is $a_0 \sim 1.42$. The unit lattice vectors are $\mathbf{a}_1$ and $\mathbf{a}_2$. The vectors $\mathbf{d}_i$ connect three nearest neighbour atoms, while ρ and τ are two vectors required to specify the position of a carbon atom. (b) Cross section of a DWNT. Atoms A and B in two shells of radii R_a and R_b, respectively, are projected onto this cross section.

To derive the helicity-dependent selection rules, we use a tight-binding Hamiltonian for non interacting electrons with one π-orbital per carbon atom [2] and follow [21],

$$H_0 = \sum_\beta \sum_{\langle ij \rangle} \gamma_0 c_{\beta i}^\dagger c_{\beta j} + \sum_{ij} t_{\mathbf{r}_{ai},\mathbf{r}_{bj}} c_{ai}^\dagger c_{bj} + \text{H.c.} , \qquad (1)$$

where $\beta = a,b$ is the shell index, $\langle ij \rangle$ is a sum over nearest neighbors in a shell, $\gamma_0 \sim 2.7\,\text{eV}$ [2] is the intrashell nearest neighbour coupling. The intershell coupling is $t_{\mathbf{r}_{ai},\mathbf{r}_{bj}} = t_0 e^{-(d(\mathbf{r}_{ai},\mathbf{r}_{bj}) - \Delta)/a_t}$, where $t_0 \sim 1.1\,\text{eV}$, $\Delta \sim 0.34\,\text{nm}$, $d(\mathbf{r}_{ai},\mathbf{r}_{bj})$ is the distance between two atoms, and $a_t \sim 0.05\,\text{nm}$ [2]. We introduce the transformation $c_{\beta j} = \frac{1}{\sqrt{N_\beta}} \sum_{\mathbf{k}} e^{i\mathbf{k}\cdot\mathbf{r}_j} c_{\beta \eta(j)\mathbf{k}}$, where $\eta = \pm$ is the index for the two interpenetrating sublattices in a graphene sheet, and N_β is the number of graphene unit cells on a shell. The elements of the *inter*shell 2×2 coupling matrix are

$$\mathcal{T}_{\eta_a \eta_b}(\mathbf{k}_a, \mathbf{k}_b) = \sum_{\mathbf{G}_a \mathbf{G}_b} e^{i\mathbf{G}_a \cdot \mathbf{X}_a - i\mathbf{G}_b \cdot \mathbf{X}_b} t_{\mathbf{k}_a + \mathbf{G}_a, \mathbf{k}_b + \mathbf{G}_b} , \qquad (2)$$

with $\mathbf{X}_\beta = \rho_\beta + \eta_\beta \tau$, where $\rho_a - \rho_b$ describes the relative position of tow shells, cf. Fig. 1, and $\mathbf{G}$ is the graphene reciprocal lattice vector. For later purposes, cf. Eq. (8), we define the elements of the intershell tunneling matrix between two Bloch states in different shells $\tilde{\mathcal{T}}_{\zeta_a \zeta_b} = (U^\dagger \mathcal{T} U)_{\zeta_a \zeta_b}$, where the transformation U will diagonalize the Hamiltonian Eq. (1). Here, $\zeta = \mp$ is the index for bonding/ antibonding states corresponding to negative/positive energies $\varepsilon_{\beta,\zeta}(\mathbf{k})$ with $\beta = a,b$, respectively. Finally,

$$t_{\mathbf{q}_a, \mathbf{q}_b} = \frac{1}{A_{\text{cell}}^2 \sqrt{N_a N_b}} \int d\mathbf{r}_a d\mathbf{r}_b e^{i(\mathbf{q}_b \cdot \mathbf{r}_b - \mathbf{q}_a \cdot \mathbf{r}_a)} t_{\mathbf{r}_a, \mathbf{r}_b} \propto \delta(q_{va}R_a - q_{vb}R_b)\delta(q_{ua} - q_{ub}) ,$$

with A_{cell} the area of a graphene unit cell, and coordinates u and v along the tube axis and the circumference direction respectively, cf. Fig. 1(b). Due to the periodic boundary conditions, $\mathbf{k}_a \cdot \hat{\mathbf{v}} = k_{va}$ obeys $\ell_a = k_{va} R_a$, along the circumference. Likewise, $\ell_b = k_{vb} R_b$. Here the integers ℓ_a and ℓ_b characterize energy subbands. In contrast, $k_u = \mathbf{k} \cdot \hat{\mathbf{u}}$ is continuous, cf. Fig. 2(b). Therefore, according to the two δ-functions in $t_{\mathbf{q}_a, \mathbf{q}_b}$, the effective intershell couplings $\mathcal{T}_{\eta_a \eta_b}(\mathbf{k}_a, \mathbf{k}_b)$ between two shells (n_a, m_a) and (n_b, m_b) are nonzero if they satisfy the following *selection rules*,

$$\ell_a + (n_a l_{1a} + m_a l_{2a}) = \ell_b + (n_b l_{1b} + m_b l_{2b}) \,, \tag{3a}$$

$$k_{ua} + \mathcal{F}(n_a, m_a) = k_{ub} + \mathcal{F}(n_b, m_b) \,, \tag{3b}$$

with $\mathcal{F}(n,m) = \frac{2\pi}{3 a_0 \mathcal{L}(n,m)} \left((n+2m) l_1 - (2n+m) l_2 \right)$. Here $\sqrt{3} a_0 \mathcal{L}(n,m)$, with $\mathcal{L}(n,m) = \sqrt{n^2 + m^2 + nm}$, is the circumferential length of shell (n,m). At low energies only the lowest subband determined by $3\ell_\beta = 2n_\beta + m_\beta$ in each shell is important, which *fixes* the values ℓ_a and ℓ_b. For c-DWNTs, e.g. if the two shells are either both armchair or zig-zag, Eqs. (3) can always be satisfied. Moreover, the dominant contribution is for $k_{ua} = k_{ub}$. On the other hand, for i-DWNTs, e.g. a $(9,0)@(10,10)$, the selection rule Eq. (3b) can only be satisfied if the difference $k_{ua} - k_{ub}$ takes finite values. At low energies, this condition is never met. At higher energies higher subbands must be considered as well, and the selection rules can be satisfied. Notice that whenever the l.h.s. and r.h.s. of Eqs. (3) are not close to zero, the effective intershell coupling is exponentially suppressed [22].

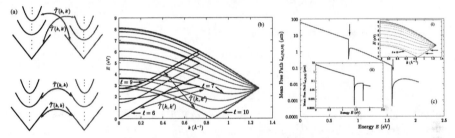

FIGURE 2. (a) Schematics of the effect of the selection rules for i-DWNT (upper) and c-DWNT (down). (b) Energy subbands of the i-DWNT $(9,0)@(10,10)$ in the absence of intershell coupling. The effect of the latter is to induce transitions between different subbands in different shells. The dominant coupling of the lowest armchair (zig-zag) subband is indicated. (c) Elastic mean free paths for electrons in the outer and inner (inset (i)) shells of the i-DWNT $(9,0)@(10,10)$. Notice the dips in correspondence of the first two subband onsets.

Then, we can evaluate the elastic mean free path by using a T-matrix approach [20]. As an example, the elastic mean free paths $l_{el,(10,10)}$ and $l_{el,(9,0)}$ for electrons in the outer and inner shell, respectively, of a $(9,0)@(10,10)$ DWNT are shown in Fig. 3. It is clearly shown that *before* the first subband onset, the motion is ballistic also for i-DWNTs of lengths up to $\simeq 5\mu m$. Therefore, for i-DWNTs, the increase of the inter-shell tunneling is at the origin of an elastic mean-free path l_{el} which decreases with increasing energy, and which shows a characteristic suppression at each subband onset. Our analytical results are in agreement with recent ab-initio calculations, showing a cross-over from ballistic to diffusive behavior in DWNT as the energy increases [18], as well as with the experimental observation that MWNT mostly exhibit diffusive behavior.

471

In the remaining of the paper we consider only metallic shells and include electron-electron correlation effects in the low energy regime where only the first subband of each shell is populated. In this regime transport is ballistic for c-DWNTs as well as for i-DWNTs. Due to the linearity of the dispersion relation, a multi-channel Luttinger liquid description can be used [23, 24].

At first, we consider an i-DWNT where the intershell coupling can be ignored. The unperturbed Hamiltonian can be written as

$$H_0 = -i\hbar v_F \sum_{r\alpha\sigma\beta} r \int du\, \psi^\dagger_{r\alpha\sigma\beta} \partial_u \psi_{r\alpha\sigma\beta} \,, \tag{4}$$

where $r = \pm$ is the index for right/left movers, $\alpha = \pm$ for the two independent Fermi points of a shell, and $\sigma = \pm$ for up and down spins. The electron density operator is $\rho_\beta(u) = \sum_{r\alpha\sigma\beta} \psi^\dagger_{r\alpha\sigma\beta}(u)\psi_{r\alpha\sigma\beta}(u)$. The two shells are only coupled by the Coulomb interaction, which gives rise to forward, backward, and Umklapp scattering processes. Experimentally, the Fermi points of nanotubes are usually shifted away from the half-filling due to doping or external gates. We assume that this is the case and hence neglect Umklapp processes. Since we are not interested in the extremely low temperature case, the backward scattering processes are also ignored here [23]. In the following, we only consider forward scattering processes described by the Hamiltonian

$$H_{FS} = \frac{1}{2}\sum_{\beta\beta'} \int du\,du'\, \rho_\beta(u) V_{\beta\beta'}(u - u')\rho_{\beta'}(u') \,, \tag{5}$$

with the effective one dimensional interaction [4]

$$V_{\beta\beta'}(u - u') = \int_0^{2\pi R_\beta} \int_0^{2\pi R_{\beta'}} \frac{dv\,dv'}{(2\pi)^2 R_\beta R_{\beta'}} U_{\beta\beta'}(\mathbf{r} - \mathbf{r}') \,,$$

where $U(\mathbf{r})$ is the Coulomb interaction. The Hamiltonian $H = H_0 + H_{FS}$ can be diagonalized by the bosonization procedure discussed in Ref. [24]. We introduce bosonic field operators for the total/relative ($\delta = \pm$) charge/spin ($j = c, s$) modes in shell β, as well the total/relative ($\xi = \pm$) modes with respect to the two shells obeying the commutation relation $[\Theta_{j\delta\xi}(u), \phi_{j'\delta'\xi'}(u')] = -(i/2)\delta_{jj'}\delta_{\delta\delta'}\delta_{\xi\xi'}\mathrm{sgn}(u - u')$. The Hamiltonian H can be then decoupled into 8 modes as

$$\sum_{j\delta\xi} \frac{\hbar v_{j\delta\xi}}{2} \int du \left(K_{j\delta\xi}\left(\partial_u \Theta_{j\delta\xi}(u)\right)^2 + \frac{1}{K_{j\delta\xi}}\left(\partial_u \phi_{j\delta\xi}(u)\right)^2 \right). \tag{6}$$

Only the two total charge modes are renormalized by the Coulomb interactions with velocities $v_{c+\pm} = v_F/K_{c+\pm}$ and interaction parameters

$$\frac{1}{K_{c+\pm}^2} = 1 + \frac{2}{\hbar\pi v_F}\left((\tilde{V}_{aa} + \tilde{V}_{bb}) \pm \sqrt{(\tilde{V}_{aa} - \tilde{V}_{bb})^2 + \tilde{V}_{ab}^2} \right), \tag{7}$$

where $\tilde{V}_{\beta\beta'} = \tilde{V}_{\beta\beta'}(2\pi/L)$ is the Fourier transform of the interaction potential at long-wave lengths, with L the nanotube length. The remaining modes are neutral with parameters $v_{j\delta\xi} = v_F$ and $K_{j\delta\xi} = 1$.

We consider now c-DWNTs where intershell tunneling is relevant. The intershell tunneling Hamiltonian is

$$H_t = \sum_{r\alpha\sigma} \tilde{\mathcal{T}}_0 \int du\, \psi^\dagger_{r\alpha\sigma a}(u)\psi_{r\alpha\sigma b}(u) + \text{H.c} \,, \tag{8}$$

where for simplicity the tunneling element $\tilde{\mathcal{T}}_{++}(\mathbf{k},\mathbf{k}')$ is evaluated at $\mathbf{k}=\mathbf{k}'=\mathbf{K}$ with the Fermi point $\mathbf{K}$ of graphene, and is the constant $\tilde{\mathcal{T}}_0$. As detailed in [23], the Hamiltonian $H_0 + H_t$ can be exactly diagonalized. One finds the same form as in Eq. (4) where now the index $\beta = 0, \pi$ stands for bonding and anti-bonding states, respectively. Moreover, the Fermi wave vectors of the two independent Fermi points are shifted as $k_F^\alpha \longrightarrow k_F^\alpha \pm (\tilde{\mathcal{T}}_0/\hbar v_F)$ where $\pm$ stand for π and 0, respectively. We retain again only (intraband and interband) forward scattering described by the Hamiltonian Eq. (5), where now the scattering potentials $\tilde{V}_{\beta\beta'}$ are $\tilde{V}_{00} = \tilde{V}_{\pi\pi} = \tilde{V}_{0\pi}/2 = (\tilde{V}_{aa} + \tilde{V}_{bb} + \tilde{V}_{ab})/4$, so that bosonization brings again the total Hamiltonian in the form Eq. (7) with 6 neutral modes and 2 renormalized total charge modes. The tunneling density of states (TDOS) of both shells, $\rho_{b/a}(\varepsilon)$, immediately follows [25]. For i-DWNT is $\rho_{b/a}(\varepsilon) \sim |\varepsilon|^{\alpha_{b/a}}$, with exponents $\alpha_{b/a}$ being different for electrons tunnelling into the middle or end of a nanotube:

$$\alpha_{\text{end},b/a} = \frac{1}{4} \sum_{\xi=\pm} A_{b/a}\left(\frac{1}{K_{c+\xi}} - 1\right), \quad \alpha_{\text{bulk},b/a} = \frac{1}{8} \sum_{\xi=\pm} A_{b/a}\left(K_{c+\xi} + \frac{1}{K_{c+\xi}} - 2\right). \tag{9}$$

Here the coefficients A_a, $A_b = 1 - A_a$ are related to the eigenvalue problem [23, 24]. For c-DWNT is

$$\rho_{b/a}(\varepsilon) \sim |\varepsilon|^{\alpha_0} + |\varepsilon|^{\alpha_\pi}\Theta(\varepsilon - 2\tilde{\mathcal{T}}_0) \,,$$

where $\Theta(x)$ is the Heaviside step function and $2\tilde{\mathcal{T}}_0$ is the gap between antibonding and bonding states. Because the intraband forward scattering potentials are equal, is $\alpha_0 = \alpha_\pi$. We find $\alpha_{\text{end/bulk}}$ given by Eq. (9) with $A_b = A_a = 1/2$. For illustration we calculate the tunneling exponents for the $(10,10)$ shell of a $(9,0)@(10,10)$ and of a $(5,5)@(10,10)$ with radii $R_a \approx 3.4$ and $R_b \approx 6.8$. We find $\alpha_{\text{end}} = 1.21, \alpha_{\text{bulk}} = 0.50$ for a $(9,0)@(10,10)$ DWNT and $\alpha_{\text{end}} = 0.80$, $\alpha_{\text{bulk}} = 0.34$ for a $(5,5)@(10,10)$ DWNT. For comparison, for a $(10,10)$ SWNT is $\alpha_{\text{end}} = 1.25$ and $\alpha_{\text{bulk}} = 0.52$. Hence, the exponents of DWNTs decrease due to the screening effect of the inner shell with respect to a SWNT. The intershell coupling reduces the exponents further. Notice that for Fermi liquids is $\alpha_{\text{end/bulk}} = 0$.

In summary, we derived selection rules according to which the intershell coupling is only negligible in i-DWNTs at low energies. An analytical expression in Born-approximation for the elastic mean free path was provided. Including the Coulomb interaction, we developed a low energy Luttinger liquid theory for metallic DWNTs according to which the intershell coupling strongly reduces the tunneling density of state exponents in c-DWNT with respect to those of i-DWNTs.

The authors would like to thank G. Cuniberti, J. Keller and C. Strunk for useful discussions.

REFERENCES

1. S. Ijima, Nature **354**, 56 (1991).
2. R. Saito, G. Dresselhaus, and M. S. Dresselhaus, *Physical Properties of Carbon Nanotubes* (Imperial College Press, London, 1998).
3. C. T. White and T. N. Todorov, Nature **393**, 240 (1998).
4. R. Egger and A. O. Gogolin, Phys. Rev. Lett. **79**, 5082 (1997); Eur. Phys. J. B **3**, 281 (1998); C. Kane, L. Balents, and M. P. A. Fisher, Phys. Rev. Lett. **79**, 5086 (1997).
5. Z. Yao *et al.*, Nature **402**, 273 (1999); M. Bockrath *et. al.*, Nature **397**, 598 (1999).
6. S. Frank *et al.*, Science **280**, 1744 (1998); A. Urbina *et al.*, Phys. Rev. Lett. **90**, 106603 (2003).
7. L. Langer *et al.*, Phys. Rev. Lett. **76**, 479 (1996).
8. A. Bachtold *et al.*, Nature **397**, 673 (1999).
9. A. Fujiwara, K. Tomiyama, and H. Suematsu, Phys. Rev. B **60**, 13492 (1999).
10. P. G. Collins *et. al*, Phys. Rev. Lett. **86**, 3128 (2001)
11. B. Bourlon *et al.*, Phys. Rev. Lett. **93**, 176806 (2004)
12. A. Bachtold *et al.*, Phys. Rev. Lett. **87**, 166801 (2001).
13. E. Graugnard *et al.*, Phys. Rev. B **64**, 125407 (2001).
14. M. Kociak *et al.*, Phys. Rev. Lett. **89**, 155501 (2002).
15. R. Saito, G. Dresselhaus, and M. S. Dresselhaus, J. Appl. Phys. **73**, 494 (1993).
16. Y.-K. Kwon and D. Tomanek, Phys. Rev. B **58**, R16001 (1998).
17. P. Lambin, V. Meunier, and A. Rubio, Phys. Rev. B **62**, 5129 (2000).
18. F. Triozon *et al*, Phys. Rev. B **69**, 121410(R) (2004); K.-H. Ahn *et al.*, Phys. Rev. Lett. **90**, 026601 (2003).
19. S. Uryu, Phys. Rev. B **69**, 075402 (2004); J. Chen and L. Yang, J. Phys. Cond. Matt. **17**, 957 (2005).
20. S. Wang and M. Grifoni, cond-mat/0505466, submitted for publication.
21. A. A. Maarouf, C. L. Kane, and E. J. Mele, Phys. Rev. B **61**, 11156 (2000).
22. This holds, e.g. for finite length DWNTs, where conservation law for the wave vectors along the tube axis is broken but the selection law for the wave vectors along the circumference still holds.
23. R. Egger, Phys. Rev. Lett. **83**, 5547 (1999).
24. K. A. Matveev and L. I. Glazman, Phys. Rev. Lett. **70**, 990 (1993).
25. When contacts are deposited on the outer shell, and in STM experiments, the tunneling current through a DWNT probes the TDOS of the *outer* shell.

474

ELECTRON TRANSPORT PROPERTIES

Electroluminescence in Carbon Nanotubes

Marcus Freitag, Jerry Tersoff, Jia Chen, Jim Tsang, and Phaedon Avouris

IBM Watson Research Center, 1101 Kitchawan Rd, Route 134, Yorktown Heights, NY 10598

Abstract. Electroluminescence in long-channel, ambipolar carbon nanotubes is highly localized in a small tube region where electron- and hole currents overlap. This region, reminiscent of a pn junction in a light-emitting diode, but produced without static dopants, can be moved along the entire carbon nanotube simply by adjusting the voltage at a uniform backgate. The movement of the spot, its size, the spectrum of the emitted light, and the efficiency of infrared generation allows a much better understanding of the electronics in ambipolar carbon nanotube field-effect transistors.

INTRODUCTION

Semiconducting single-wall carbon nanotubes, contacted with source and drain electrodes and gated by a nearby third electrode, show a strong field-effect switching behavior. [1,2] In recent years, the electronic characteristics of these tiny devices have improved to an extent [3-6] that they are seriously considered as candidates for the replacement of silicon MOSFETs. In an "ambipolar" carbon nanotube field-effect transistor (CNFET) the conduction is carried by a varying combination of electron and hole currents. [7] These devices are not appropriate for switching applications because their off currents become very large, especially under high bias. However, they turn out to be fascinating light-emitters [8-10] and photo-detectors. [11]

In these proceedings we focus on the electrically-induced light emission (electroluminescence) in ambipolar CNFETs with very long channel lengths on the order of 100μm. Ambipolar CNFETs emit infrared light [8] when electrons and holes, injected at opposite Schottky contacts, recombine inside the carbon nanotube. In long-channel carbon nanotube transistors, the infrared emission is not only localized laterally by the tube diameter, but also longitudinally, because the electron-hole recombination is faster than carrier transit times through the entire tube. [10] We measure and model the spatial distribution of the electroluminescence as a function of the applied voltages and thus gain unprecedented insight into conduction through carbon nanotube field-effect transistors.

EXPERIMENTAL DETAILS

Ambipolar CNFETs are fabricated from CVD tubes [12] grown directly on an oxidized silicon wafer (typically 100 nm SiO_2). Source and drain contacts are defined by e-beam lithography and evaporation of ~20 nm palladium with ~0.5 nm titanium adhesion layer. The degenerately-doped silicon substrate is used as a backgate. As-fabricated transistors show ambipolar conduction without further processing because the relatively large diameter tubes on the order of 2-3 nm have small bandgaps and therefore small Schottky barriers for electrons and holes. To reduce hysteresis due to surface traps on the SiO_2 the devices are usually covered with PMMA and annealed at 200C for 24h.

CP786, *Electronic Properties of Novel Nanostructures*, edited by H. Kuzmany, J. Fink, M. Mehring, and S. Roth
© 2005 American Institute of Physics 0-7354-0275-2/05/$22.50

[13] The electroluminescence in the infrared is measured with a 2-dimensional HgCdTe detector array and a set of 10 infrared band-pass filters in the range of 1.5μm to 2.4μm.

RESULTS AND DISCUSSION

Ambipolar carbon nanotube field-effect transistors emit infrared light when biased appropriately (gate voltage at roughly half the applied bias). [8] The light is generated when electrons that are injected at one contact recombine with holes that are injected at the opposite contact. As theoretically predicted for recombination involving two corresponding sub-bands (e.g. E_{11} or E_{22} transitions) the electroluminescence from nanotubes is linearly polarized along the axis of the carbon nanotube. [8] The spectrum of the emitted light has a low-energy cutoff at the nanotube bandgap and is quite broadband (around 100meV), suggesting considerable hot-carrier recombination. [9]

We have measured an electroluminescence efficiency of our devices on the order of 10^{-6} photons/electron-hole pair, [9] about two orders of magnitude below what is commonly observed in nanotube fluorescence. [14,15] It is possible that the presence of the SiO_2 surface and/or excess catalyst contributes to this diminished efficiency, analogous to the quenching that is observed in fluorescence for nanotubes in contact with a SiO_2 surface. [16] Non-radiative recombination through Auger processes [17] are also likely since current densities for electroluminescence are quite high ($10^9 A/cm^2$). In addition, multiple phonon emission might be important because the nanotube band gap is only 3 to 4 optical phonons wide.

An ambipolar CNFET does not have any chemically-defined pn-junction and it is not a-priori clear where light should be generated in such a device. The areas close to the contacts are obvious candidates whenever the current is dominated by one type of carrier and only a few minority carriers are injected at either source or drain. But what happens in the case of nearly equal electron and hole injection? To study this question, we have fabricated ambipolar CNFETs with extremely long channel lengths on the order of 100μm, where the infrared emission can be spatially resolved with an optical microscope and a sensitive detector array. We keep in mind that these long nanotube devices differ considerably from the usual sub-micrometer CNFETs because electronic conduction becomes diffusive and a voltage drop develops along the length of the tube, whereas conduction in very short nanotubes is quasi ballistic.

In our light-emission experiments with long nanotubes we find that the electron-hole recombination is so fast that an ambipolar region becomes localized on a micrometer-sized tube segment. This "ambipolar domain" can be positioned anywhere along the carbon nanotube simply by adjusting the backgate voltage. [10] In Fig. 1a a carbon nanotube is arranged vertically and biased with $V_d = -15V$ at the drain (top contact). For sufficiently negative gate voltage (Fig. 1c) light emission is observed solely at the drain contact (e.g. for $V_g = -4.1V$). In this case the current is carried predominately by holes that travel from the source (bottom) to the drain (top). The drain Schottky barrier allows injection of a small electron current and recombination with the majority carriers (holes) is observed close to the drain. When the gate voltage is increased, the electron current increases while the hole current decreases until a point is reached where both currents are roughly equal. Under these conditions the gate voltage not only controls the current, but also the position where recombination and therefore infrared emission takes place. The spot moves toward the source for increasing gate voltage until it reaches the source contact at $V_g = -1.2\text{V}$. Finally the roles of electron and hole currents are reversed with electrons being the majority carriers that solely carry the current from the drain contact to a spot close to the source.

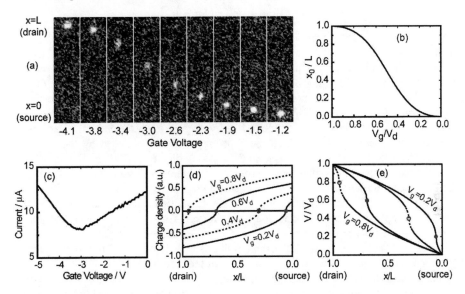

FIGURE 1. (a) Spatially resolved infrared emission along a carbon nanotube during a gate-voltage sweep. The size of the nanotube channel (from top-most to bottom-most spot) is 50μm. **(b)** Theoretical prediction for the spot movement with gate voltage. **(c)** Gate voltage characteristics during the scan in (a). **(d)** Predicted charge density in the tube for gate voltages between $0.2V_d$ and $0.8V_d$. **(e)** Potential drop in the device for the same range of gate voltages.

Figure 1b shows the result of a model to describe the spot movement. [18] In contrast to Guo and Alam [19] who performed numerical calculations on intermediate-length devices, we focus our attention on the long-channel limit, which is most relevant for these experiments, and where it is possible to derive an analytical expression for the spot movement. We assume that the current is diffusive and the back-gate couples capacitively to the tube: $V(x) = V_g + C^{-1}\rho(x)$, where V is the potential along the tube axis (x-direction), C is the nanotube-backgate capacitance and ρ is the charge density in the tube. The current is given by $I = -m \cdot \eta(x) \cdot dV(x)/dx$, where m is the mobility, assumed to be constant, η is the number density (the sum of electron and hole densities: $\eta(x) = \eta_e(x) + \eta_h(x)$), and $dV(x)/dx$ is the electric field along the tube. Charge and number densities are related by: $\rho(x) = \eta_h(x) - \eta_e(x)$. In staying with our long-channel limit we set the recombination length to zero so that at any position x along the tube there are either electrons or holes, but not both. Furthermore, the electric field from source and

479

drain is shielded by the back-gate and the contacts only enter through boundary conditions for the potential.

The equations for the potential and the current lead to a differential equation that can be easily solved using the above simplifications and we get:

$$\frac{x_0 - x_d}{L} = \frac{\left(1 - v_g^2\right)}{v_g^2 + \left(1 - v_g\right)^2}$$
(1)

Here $v_g = V_g/V_d$ is the scaled gate voltage, x_0 is the position of the infrared spot, x_d is the position of the drain contact, and L is the length of the nanotube. The good agreement between theory and experiment is evident in Figs. 1b and 1a. The S-shape movement of the spot with gate voltage is well reproduced. The gate-voltage range over which the spot moves is smaller however in the experiment (this model would predict that the spot moves from drain to source for gate voltages between V_d and 0V). The discrepancy can be resolved by including in our model a fixed voltage drop at the Schottky barriers. The predicted voltage range over which the spot moves is then reduced by the sum of the voltage drops at the contacts. A value of 6V for the voltage drop at one Schottky barrier, inferred from these light-emission experiments, compares well with threshold voltages in CNFETs with the thick (100nm) oxides as used in this study.

Figures 1d and 1e show the model's prediction for the charge density and potential drop in the carbon nanotube under different gating conditions. The potential drop is always steepest close to the ambipolar recombination region, which is in agreement with the hot spectral distribution that is usually observed in infrared emission. [9] The charge density (Fig. 1d) becomes zero at the recombination spot, which therefore assumes the potential of the gate. Because we required instantaneous e-h recombination in our model, the number density at the recombination spot is predicted to vanish as well. In reality, a finite recombination length will smear out the associated singularity in the electric field. It remains true however that the number density goes through a minimum due to the annihilation of carriers at the recombination spot. The remaining charges have to move faster than in other parts of the tube to preserve current continuity. This nicely explains the high field and fast potential drop at the recombination region.

In conclusion, we have imaged the recombination region in long-channel ambipolar carbon nanotube FETs for varying gate voltages and developed a model that reproduces the observed spot movement. We find that electron-hole recombination is much faster than carrier transit times, estimated from the saturation velocity $v_s = 5 \cdot 10^7 \text{cm/s}$ [20] around 100ps. The fact that a classical model can explain the spot movement to such a satisfying degree is evidence for the diffusive, rather than ballistic conduction along long-channel carbon nanotubes. The resulting potential along the tube is highly non-linear and a large part of the bias drops right at the recombination region. Additional voltage drops at the contacts are in agreement with prior work on the Schottky-barriers at carbon nanotube/metal contacts.

ACKNOWLEDGEMENTS

We acknowledge Qiang Fu and Jie Liu for the CVD growth of long carbon nanotubes and Bruce Ek for expert technical support.

REFERENCES

1. S. J. Tans, A. R. M. Verschueren, and C. Dekker, Nature **393**, 49 (1998).
2. R. Martel, T. Schmidt, H. R. Shea, et al., Appl. Phys. Lett. **73**, 2447 (1998).
3. S. J. Wind, J. Appenzeller, and Ph. Avouris, Phys. Rev. Lett. **91**, 058301 (2003).
4. J. Guo, A. Javey, H. Dai, S. Datta, and M. Lundstrom, cond-mat/0309039 (2003).
5. A. Javey, J. Guo, D. B. Farmer, et al., Nano Lett. **4**, 447 (2004).
6. Y. M. Lin, J. Appenzeller, and Ph. Avouris, IEEE DRC Digest **62**, 133 (2004).
7. R. Martel, V. Derycke, C. Lavoie, et al., Phys. Rev. Lett. **87**, 256805 (2001).
8. J. A. Misewich, R. Martel, Ph. Avouris, et al., Science **300**, 783 (2003).
9. M. Freitag, V. Perebeinos, J. Chen, et al., Nano Lett. **4**, 1063 (2004).
10. M. Freitag, J. Chen, J. Tersoff, et al., Phys. Rev. Lett. **93**, 076803 (2004).
11. M. Freitag, Y. Martin, J. A. Misewich, et al., Nano Lett. 3, 1067 (2003).
12. S. Huang, X. Cai, and J. Liu, J. Am. Chem. Soc. **125**, 5636 (2003).
13. W. Kim, A. Javey, O. Vermesh, et al., Nano Lett. **3**, 193 (2003).
14. S. M. Bachilo, M. S. Strano, C. Kittrell, et al., Science **298**, 2361 (2002).
15. S. Lebedkin, F. Hennrich, T. Skipa, and M. M. Kappes, J. Phys. Chem. B **107**, 1949 (2003).
16. J. Lefebvre, Y. Homma, and P. Finnie, Phys. Rev. Lett. **90**, 217401 (2003).
17. M. Y. Sfeir, F. Wang, L. Huang, et al., Science **306**, 1540 (2004).
18. J. Tersoff, M. Freitag, J. C. Tsang, and Ph. Avouris, submitted to APL.
19. J. Guo and M. A. Alam, Appl. Phys. Lett. **86**, 023105 (2005).
20. V. Perebeinos, J. Tersoff, and Ph. Avouris, Phys. Rev. Lett. **94**, 086802 (2005).

Tunable Orbital Pseudospin and Multi-level Kondo Effect in Carbon Nanotubes

Pablo Jarillo-Herrero, Jing Kong[†], Herre S.J. van der Zant, Cees Dekker,
Leo P. Kouwenhoven, Silvano De Franceschi[‡]

*Kavli Institute of Nanoscience, Delft University of Technology, PO Box 5046, 2600 GA Delft,
The Netherlands*
[†] *Present address: Department of Electrical Engineering, Massachusetts Institute of Technology,
Cambridge, Massachusetts 02139-4307, USA*
[‡] *Present address: Laboratorio Nazionale TASC-INFM, I-34012 Trieste, Italy*

Abstract. We report the observation of a multi-level Kondo effect resulting from orbital
degeneracy in carbon nanotube quantum dots. The orbital degree of freedom plays the role of a
pseudospin whose energy splitting is controlled by an applied magnetic field, and it is an order
of magnitude larger than the Zeeman splitting. At zero field and low temperature, spin and
orbital states form a 4-fold degenerate shell leading to a highly symmetric Kondo state and a
correspondingly high Kondo temperature. At finite magnetic field, a four-fold splitting of the
Kondo resonance is observed due to lifting both spin and orbital degeneracy.

Keywords: carbon nanotube, quantum dot, Kondo, orbital, spin.
PACS: 73.22.-f, 73.22.Dj, 73.23.Hk, 73.63.Fg

The Kondo effect, namely the interaction of a magnetic impurity with the
conduction electrons of a nonmagnetic host metal, is one of the basic topics in
condensed matter physics [1]. Nanotechnology provides new experimental tools to
nail down the deepest aspects of this fundamental phenomenon. Controllable single-
impurity Kondo systems have been realized in various nanostructures, including
quantum dots (QDs) formed in two-dimensional electron gases [2-4], carbon
nanotubes [5,6] and individual molecules [7,8]. Experiments performed until now
demonstrate that the electronic properties of such nanostructures can be strongly
affected by Kondo correlations.

The simplest Kondo system consists of a localized electron (spin S=1/2) coupled to
a Fermi sea via a Heisenberg-like exchange interaction, J S·σ, where σ is the spin of a
conduction electron interacting with the impurity site, and J the antiferromagnetic
coupling ($J > 0$) [9]. Below a characteristic temperature T_K, the so-called Kondo
temperature, a many-body singlet state forms between the impurity spin and the
surrounding conduction electrons. This bound state gives rise to a resonant level at the
Fermi energy. On average, a single conduction electron occupies this level and fully
screens the impurity spin (Fig. 1A). In the presence of both spin and orbital
degeneracy, the exchange interaction can be written as $J \sum_\alpha$ S$_\alpha$·σ$_\alpha$, where $\alpha = 1...4N^2-$
1, N being the number of degenerate orbitals [10]. The larger degeneracy effectively
yields a higher exchange coupling and a higher Kondo temperature ($T_K \sim \exp(-1/NJ)$).

CP786, *Electronic Properties of Novel Nanostructures*, edited by H. Kuzmany, J. Fink, M. Mehring, and S. Roth
© 2005 American Institute of Physics 0-7354-0275-2/05/$22.50

Also the symmetry is enhanced such that spin and charge degrees of freedom are fully entangled. The state of the impurity is represented by a $2N$-component "hyper-spin" and the Kondo Hamiltonian is invariant under special unitary transformations SU($2N$) in the "hyper-spin" space. In the case of double orbital degeneracy (N=2) the symmetry group is SU(4), a well-known symmetry in nuclear physics, where the spin and orbital roles are played by the nucleon spin and isospin. In the SU(4) Kondo effect, the screening of the local magnetic moment requires, on average, three conduction electrons (Fig. 1A). Despite the favourably higher T_K, the observation of SU(4) Kondo effect in QDs is challenging. The difficulty lies in the absence of orbital degeneracy in conventional semiconductor QDs, due to symmetry imperfections and level repulsion. Recently, elaborated device geometries [11-13] have been proposed to obtain orbital degeneracy and hence SU(4) Kondo. In this work we show that this condition can be met in carbon nanotubes owing to their unique band structure.

The electronic structure of carbon nanotubes (CNTs) can be derived from the two-dimensional band structure of graphene. The continuity of the electron wave function around the nanotube circumference imposes the quantization of the wave-vector component perpendicular to the nanotube axis, $k_\perp$. This leads to a set of one-dimensional subbands in the longitudinal direction [14]. Due to symmetry, for a given subband at $k_\perp = k_o$ there is a second degenerate subband at $k_\perp = -k_o$. Figure 1C shows in black solid lines the schematic 1D band structure of a gapped CNT near the energy band gap. Both valence and conduction bands have two degenerate subbands, labelled as "+" and "−". It was predicted [15] that the orbital degeneracy is lifted by a magnetic field, B, parallel to the nanotube axis (Fig.1C). Experimental evidence for this effect has been reported only recently [16-18]. In the case of finite-length nanotubes, a discrete energy spectrum is expected from size quantization. Assuming negligible inter-subband mixing [19], the level spectrum of a CNT QD can be described as two sets of spin-degenerate levels, $E_+^{(n)}$ and $E_-^{(n)}$ with $n = 1,2,3,...$ (see Fig. 1C). At B=0, there is a four-fold degenerate shell for every n (Fig. 1D). It is this degeneracy that gives rise to multi-level Kondo effects, as we will discuss below.

The four-fold shell filling emerges in a measurement of the linear conductance, G, versus gate voltage, V_G (Fig. 2A), for a QD device fabricated from a metallic nanotube with a small band gap [17,20] (Fig. 1B). The thick red trace, taken at T =8 K, shows Coulomb blockade oscillations corresponding to the filling of the "valence" band of the nanotube. Going from left to right, electrons are consecutively added to the last three electronic shells, n=3, 2 and 1, respectively. The shell structure emerges clearly from the V_G-spacing between the Coulomb peaks. The addition of an electron to a higher shell requires an extra energy cost corresponding to the energy spacing, $\Delta_{n,n+1}$, between shells (we define $\Delta_{n,n+1} \equiv E_+^{(n+1)} - E_+^{(n)} = E_-^{(n+1)} - E_-^{(n)}$). This translates into a larger width of the Coulomb valley associated with a full shell [21]. We estimate $\Delta_{1,2}$ ~3 meV, and $\Delta_{2,3}$ ~5 meV. The Coulomb charging energy for adding an electron to the same shell is U~5 meV [22]. The first group of four Coulomb peaks on the left-hand side of Fig. 2A (n=3) are strongly overlapped due to a large tunnel coupling to the leads. The coupling decreases with V_G and becomes very small near the band gap, which lies just beyond the right-hand side of the V_G-range shown. The Coulomb peaks associated with the last two electrons (n=1) are not visible.

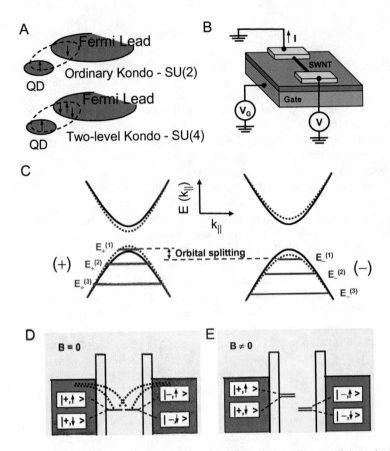

FIGURE 1. (A) Representation of the bound singlet state for an ordinary spin-1/2 impurity (SU(2) symmetry) and for a spin-1/2 impurity with both spin and (two-fold) orbital degeneracy (SU(4) symmetry). (B) Device scheme. Carbon nanotubes were grown by chemical vapor deposition on p-type Si substrates with a 250-nm-thick oxide. Individual nanotubes were located by atomic force microscopy and contacted with Ti/Au electrodes defined by e-beam lithography. The highly-doped Si substrate was used as a back-gate. A magnetic field was applied parallel to the substrate at an angle φ with respect to the nanotube axis. (C) Band structure of a CNT near its energy gap. Black lines represent the one-dimensional energy dispersion relation, $E(k_{||})$, at B=0 ($k_{||}$ is the wave vector along the nanotube axis). The valence (conduction) band has two degenerate maxima (minima). Size quantization in a finite-length nanotube results in a set of discrete levels with both spin and orbital degeneracy. The degeneracy is lifted by a magnetic field parallel to the nanotube. The 1D subbands (and the corresponding levels) at finite B are represented by red and blue dotted (solid) lines. Only the orbital splitting of the energy levels is shown in this figure. (D) Energy schematics of a CNT QD. At B=0 the ground state is four-fold degenerate. Spin and/or orbital states can "flip" by one-step cotunnelling processes (dotted lines). (E) At finite B, both orbital and spin degeneracy are simultaneously lifted, suppressing the Kondo effect.

The shell structure breaks up at finite magnetic field, B. The B-dependence of the orbital levels appears as a large shift in the position of the Coulomb peaks. This is shown in Fig. 2B, where G is plotted versus (V_G, B) for the same V_G-range as in Fig.

2A. The two groups of four Coulomb peaks exhibit similar evolution in B. The first (last) two Coulomb peaks in each group shift towards lower (higher) V_G, in full agreement with the situation depicted in the right inset to Fig. 2A. This is a clear demonstration of the B-dependence of the two-fold degenerate nanotube band structure, which was not observed before. For each shell, the orbital magnetic moment, μ_{orb}, can be extracted from the shift, $\Delta V_G(n)$, in the position of the corresponding Coulomb peaks. Neglecting the Zeeman splitting, we use the relation α $|\Delta V_G(n)| = |\mu_{orb}(n)\cos\varphi \, \Delta B|$, where ΔB is the change in B, φ is the angle between the nanotube and the B field, and α is a capacitance ratio extracted from non-linear measurements. The values obtained are 0.90, 0.80 and 0.88 meV/T, for $n=1$, 2 and 3, respectively. These values are an order of magnitude larger than the electron spin magnetic moment ($\frac{1}{2}g\mu_B=0.058$ meV/T for $g=2$), and in good agreement with an estimate of μ_{orb} based on the nanotube diameter [23].

Having acquired the necessary understanding of the energy level spectrum and its B-dependence, we now focus on the Kondo effect. At $B\sim0$, the conductance is strongly enhanced in certain regions of V_G (Fig. 2). The enhancement is due to Kondo correlations, as clearly demonstrated by the T-dependence (see below). The Kondo effect appears in the regions where the shells are partially filled, and it is absent in the case of full shells for which the ground state is a spin singlet. Starting from an empty shell (region 0, Fig. 2B), electrons are added by increasing V_G (1, 2, 3 and 4 electrons for regions I, II, III and IV, respectively). The first electron can enter any of the two degenerate orbital states and with either spin up or down. This four-fold degeneracy results in a multi-level Kondo effect [24], with an enhanced Kondo temperature associated to SU(4) symmetry, as we will substantiate below. An analogous situation occurs for three electrons since this is the same as one hole [25]. In both cases the total spin in the shell is $S=1/2$. For two electrons in the shell, there are two possible spin configurations. At $B=0$ the ground state is a spin triplet ($S=1$) due to orbital degeneracy [19] and Hund's rule. By lifting the orbital degeneracy with a magnetic field, the system undergoes a triplet-to-singlet transition when the orbital splitting overcomes the exchange energy, E_{exc}. We estimate $E_{exc}=0.5$ meV for $n=2$ [22]. At $B\sim0.35$ T, the triplet-to-singlet transition leads to a Kondo effect (Fig. 2B), in agreement with previous experiments [26]. In the $n=3$ shell, the Kondo temperature is much higher than the exchange energy, and there is Kondo effect over the entire region from 1 to 3 electrons. The shape of $G(V_G)$ is remarkably similar to predictions for a multi-level Kondo effect in quantum dots [27,28].

We now turn to the T-dependence of the linear conductance (Fig. 2A). Starting from $T=8$ K (thick red trace), G increases by lowering T in the regions corresponding to partially filled shells and decreases for full shells. In the centre of valleys I and III, G exhibits a characteristic logarithmic T-dependence with a saturation around $2e^2/h$ at low T, indicating a fully-developed Kondo effect (see Fig 2A, top inset, second from left). Similar T-dependences are observed for the $n=1$ and $n=3$ shell. From fits of $G(T)$, taken at the V_G values indicated by arrows in Fig. 2A, we find $T_K = 6.5$, 7.5, and 16 K, for $n=1,2$ and 3, respectively. These Kondo temperatures are an order of magnitude higher than those previously reported for nanotube QDs [5,6] and comparable to those reported for single-molecule devices [7,8]. Such high T_K values, and the fact that G exceeds $2e^2/h$ (the one-channel conductance limit), are signatures of a nonconventional Kondo effect. The bottom inset in Fig. 2A shows the normalized conductance, G/G_0, versus normalized temperature, T/T_K, for different shells and for both one and two electrons in the shell. The observed scaling reflects the universal character of the Kondo effect. The low-temperature behaviour is fully determined by a single energy scale, T_K, independent of the spin and orbital configuration responsible for the Kondo effect.

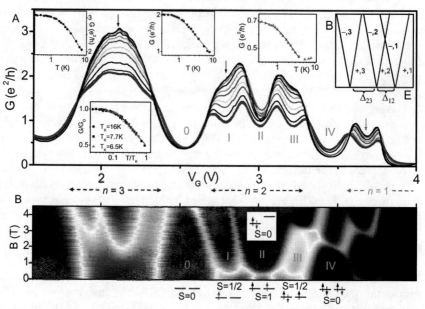

FIGURE 2. (A) Linear conductance, G, vs gate voltage, V_G, for different temperatures, T, between 0.34 K (black trace) and 8 K (red trace). The four-fold shell filling emerges more clearly at high T. At low T, the Kondo effect leads to an increase in conductance. Top-left insets: T-dependence for three values of V_G in shells 1, 2, and 3, respectively (see arrows). The solid lines in the insets are fits to the empirical function $G(T) \approx G_0/(1+(2^{1/s}-1)(T/T_K)^2)^s$ [31], where G_0, and T_K are fitting parameters, while s is fixed at 0.21. Bottom-left inset: scaling behaviour for the three T-dependences shown in the upper insets. Top-right inset: qualitative energy spectrum of the CNT QD as a function of magnetic field, B (Zeeman splitting and charging energy are neglected). (B) G vs B on a color scale at T=0.34 K (G increases from blue to red). The high-G regions in between Coulomb peaks are due to the Kondo effect. Within each shell, the first two Coulomb peaks move towards lower V_G with increasing B and the second two Coulomb peaks move towards higher V_G, reflecting the opposite magnetic moments of the two orbital states. The small diagrams indicate the spin and orbital filling of electrons for the n=2 shell.

In the non-linear regime the Kondo effect manifests itself as a resonance in the differential conductance, dI/dV, at source-drain bias $V\sim 0$. This is clearly seen in Fig. 3A, where $dI/dV(V,V_G)$ is shown for the second shell (n=2). Two large Coulomb diamonds separated by three small ones can be identified. These correspond to the regions 0 to IV in Fig. 2A. In regions I and III, there is a Kondo peak in dI/dV at zero bias. In the centre of the Kondo ridges, the full-width at half maximum (FWHM) of these peaks is 1.3 meV. This width gives $T_K\sim$8K (FWHM~$2k_B T_K$), in agreement with

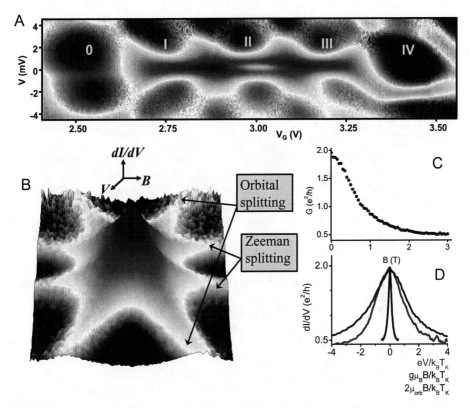

FIGURE 3. (A) Color-scale representation of the differential conductance, dI/dV, versus V and V_G for the second shell at $T\sim0.34$K and $B=0$ (dI/dV increases from blue to red). The regions labelled 0, I, II, III and IV, correspond to 0, 1, 2, 3 and 4 electrons, in the $n=2$ shell. (B) Three-dimensional plot of dI/dV versus V and B in the center of Coulomb valley I. V ranges from -2.5 to 2.5 mV, B ranges from -3 to 3T. The Kondo peak at $(V,B) = (0,0)$ splits in 4 peaks at finite B, due to the lifting of both orbital and spin degeneracies. (C) B-dependence of the zero-bias conductance. G decreases on a $\sim$1T scale, i.e. much faster than expected from Zeeman splitting. So the orbital splitting sets the scale for conductance suppression. (D) G is plotted vs normalized Zeeman energy, $g\mu_B B/k_B T_K$ (black trace), and vs normalized orbital splitting, $2\mu_{orb}B/k_B T_K$ (blue trace). $T_K = 7.5$ K as deduced from a fit of $G(T)$. (Note that $G(B)=0.5G(0)$ when $\mu_{orb}B/k_B T_K\sim1$). To compare the suppression due to magnetic field with that due to bias voltage, we also show a measurement of dI/dV vs normalized bias voltage, $eV/k_B T_K$ (red trace). The blue and red traces fall almost on top of each other.

the previous estimate based on the fits of G vs T. In region II, we observe two peaks at finite bias, reflecting the singlet-triplet splitting of the Kondo resonance [26].

At finite B, the Kondo resonance is expected to split. In an ordinary spin-1/2 Kondo system, this follows from lifting spin degeneracy. Two split peaks are observed in $dI/dV(V)$, separated by twice the Zeeman energy [2-4]. Here we show that a fundamentally different splitting is observed in the presence of orbital degeneracy. A four-fold peak splitting is found in region I as shown in Fig. 3B. The large zero-bias resonance opens up in four peaks that move linearly with B and become progressively smaller. The two inner peaks can be ascribed to Zeeman splitting and the two outer peaks to orbital splitting. The smaller splitting corresponds to inelastic cotunneling from $|+,\downarrow>$ to $|+,\uparrow>$, while the larger splitting reflects inelastic cotunneling from orbital "+" to orbital "−" (Fig. 1E). In the latter case, inter-orbital cotunneling processes can occur either with or without spin flip. The corresponding substructure, however, cannot be resolved due to the broadening of the outer peaks. The slope $|dV/dB|$ of a conductance peak directly yields the value of the magnetic moment associated with the splitting. We obtain a spin magnetic moment $\mu_{spin} = \frac{1}{2}|dV/dB|_{spin}$ =0.06 meV/T $\sim \mu_B$ from the inner peaks [29], and an orbital magnetic moment μ_{orb} $=\frac{1}{2}|dV/dB|_{orb}/\cos\varphi$ =0.8 meV/T $\sim$13 μ_B from the outer peaks.

The importance of the orbital degree of freedom appears in an explicit measurement of the B-dependence of G (Fig. 3C). If the Kondo effect was determined by spin only (this could be the case if one of the orbitals is weakly coupled), G should decrease due to Zeeman splitting on a field scale $B \sim k_B T_K/g\mu_B$ [30]. Instead we find that G is suppressed on a much smaller scale, $B \sim k_B T_K/\mu_{orb}$, being determined by the orbital splitting (see Fig. 3D). From the fact that the orbital splitting determines the Kondo resonance we come to the following important conclusion: the observed zero-bias peak largely originates from the presence of an orbital degeneracy. Degenerate orbital states, theoretically described by a pseudospin, can thus lead to a Kondo effect, analogous to degenerate spin states.

At B=0, however, we have both spin and pseudospin degeneracy. It is known from theory that in this case the Kondo ground state obeys SU(4) symmetry at low temperature ($T \ll T_K$) [11-13]. This SU(4) Kondo effect is substantially different from the Kondo effect for an impurity with S=3/2 (SU(2) symmetry), as well as from the Kondo effect for two independent parallel spin-1/2 systems (SU(2)×SU(2) symmetry). According to calculations reported in Ref. 13, the observation of a four-fold splitting constitutes direct experimental evidence of SU(4) symmetry. An SU(4) Kondo effect was recently reported also in vertical semiconductor quantum dots with no evidence, however, for such four-fold splitting [32].

We have shown that the spin and orbital degeneracy are simultaneously lifted by an applied magnetic field. Due to the large μ_{orb}, the orbital splitting can exceed the level spacing between shells for fields above a few Tesla. This offers the opportunity to recover orbital degeneracy at high fields, while having at the same time a large spin splitting. Eventually, this removal of spin degeneracy leads to a purely orbital Kondo effect [33]. We finally remark that the strong orbital Kondo effect discussed in this report has been measured in all devices (four devices obtained from three different fabrication runs) where Kondo effect and four-fold shell filling were observed.

ACKNOWLEDGMENTS

We thank G. Zaránd, and R. Aguado for helpful discussions. We acknowledge financial support from the Japanese Solution Oriented Research for Science and Technology and from the Dutch Organization for Fundamental Research, FOM.

REFERENCES

1. Brown, M. P., and Austin, K., *The New Physique*, Publisher City: Publisher Name, 1997, pp. 25-30.
2. Brown, M. P., and Austin, K., *Appl. Phys. Letters* **65**, 2503-2504 (1994).
3. Wang, R.T., "Title of Chapter," in *Classic Physiques*, edited by R. B. Hamil, Publisher City: Publisher Name, 1997, pp. 212-213.
4. Smith, C. D., and Jones, E. F., "Load-Cycling in Cubic Press" in *Shock Compression in Condensed Matter-1997*, edited by S. C. Schmidt et al., AIP Conference Proceedings 429, New York: American Institute of Physics, 1998, pp. 651-654.
1. Kondo, J., *Prog. Theor. Phys.* **32**, 37 (1964).
2. Goldhaber-Gordon, D., *et al.*, *Nature* **391**, 156 (1998).
3. Cronenwett, S. M., Oosterkamp, T. H., Kouwenhoven, L. P., *Science* **281**, 540 (1998).
4. Schmid, J., Weis, J., Eberl, K., von Klitzing, K., *Physica* B **256-258**, 182 (1998).
5. Nygård, J., Cobden, D. H., Lindelof, P. E., *Nature* **408**, 342 (2000).
6. Buitelaar, M. R., Bachtold, A., Nussbaumer, T., Iqbal, M., Schönenberger, C., *Phys. Rev. Lett.* **88**, 156801 (2002).
7. Park *et al.*, J., *Nature* **417**, 722 (2002).
8. Liang, W., Shores, M. P., Bockrath, M., Long, J. R., Park, H., *Nature* **417**, 725 (2002).
9. Hewson, A. C., *The Kondo Problem to Heavy Fermions*, Cambridge, Univ. Press Cambridge, 1993.
10. We assume that J is independent of spin and orbital quantum numbers. More generally, the exchange coupling can be written as $\Sigma_{\alpha\beta} J_{\alpha\beta} S_{\alpha\beta} \cdot \sigma_{\alpha\beta}$. In the strong coupling limit, which occurs at low T, this hamiltonian can be replaced by one where $J_{\alpha\beta}$=J, provided that the orbital splitting is smaller than T_K (11,12).
11. Borda, L., Zaránd, G., Hofstetter, W., Halperin, B.I., von Delft, J., *Phys. Rev. Lett.* **90**, 026602 (2003).
12. Zaránd, G., Brataas, A., Golhaber-Gordon, D., *Solid State Comm.* **126**, 463 (2003).
13. Choi, M.-S., *et al.*, e-Print available at http://xxx.lanl.gov/abs/cond-mat/0411665.
14. Dresselhaus, M. S., Dresselhaus, G. & Eklund, P. C. *Science of Fullerenes and Carbon Nanotubes*, San Diego, Academic Press, 1996.
15. Ajiki, H., Ando, T. J., *Phys. Soc. Jpn* **62**, 1255 (1993).
16. Minot, E., Yaish, Y., Sazonova, V., McEuen, P. L., *Nature* **428**, 536 (2004).
17. Zaric, S., *et al.*, *Science* **304**, 1129 (2004).
18. Coskun, U. C., Wei, T-C., Vishveshwara, S., Goldbart, P. M., Bezryadin, A., *Science* **304**, 1132 (2004).
19. A subband mixing may result in level repulsion and an orbital splitting. We observe no such splitting and conclude that this is much smaller than T_K. The condition for orbital degeneracy is thus fulfilled at the level required for the observation of SU(4) Kondo (11).
20. This band gap can be due to different perturbations, such as curvature or strain. The measured value of the band gap is ~30meV.
21. Liang, W., Bockrath, M., Park, H., *Phys. Rev. Lett.* **88**, 126801 (2002).
22. The value of U is extracted from the separation between the Coulomb peaks at high temperature. The accuracy in the values of U, $\Delta_{1,2}$ and $\Delta_{2,3}$ is determined by the accuracy in the α-factor, of order 20% due to large Γ broadening and the residual Kondo effect at T~8K. We estimate a 10% error bar in E_{exc} and μ_{orb}.
23. From the value for n=1, we extract a nanotube diameter D = $4\mu_{orb}/ev_F$ = 4.5 nm, in agreement with the measured diameter of 4.0±0.5 nm as determined by atomic force microscopy. For this device φ=37°. The decreasing value of μ_{orb} from n=1 to n=2 is in qualitative agreement with predictions (16). For n=3, the presence of Kondo effect almost throughout the entire B-range makes the determination less accurate.
24. Inoshita, T., Shimizu, A., Kuramoto, Y., Sakaki, H., *Phys. Rev. B* **48**, 14725 (1993).
25. In fact the situation for 1 and 3 electrons in n=2 is not completely symmetric, since $\Delta_{1,2} < \Delta_{2,3}$ and also the gate coupling varies with V_G. This may be related to the observed differences in *G* vs *B* in regions I and III (for both shells 2 and 3).
26. Sasaki, S. *et al.*, *Nature* **405**, 764 (2000).
27. Izumida, W., Sakai, O., Shimizu, Y., *J. Phys. Soc. Jpn* **67**, 2444 (1998).
28. Levy-Yeyati, A., Flores, F., Martín-Rodero, A., *Phys. Rev. Lett.* **83**, 600 (1999).
29. The linear fit yields the correct g-factor, but with an offset of ~300 μeV. See also Kogan, A., *et al.*, Phys. Rev. Lett. **93**, 166602 (2004).
30. Costi, T. A., *Phys. Rev. Lett.* **85**, 1504 (2000).
31. Golhaber-Gordon, D., *et al.*, *Phys. Rev. Lett.* **81**, 5225 (1998).
32. Sasaki, S., *et al.*, Phys. Rev. Lett. **93**, 017205 (2004).
33. Jarillo-Herrero, P., *et al.*, *Nature* **434**, 484 (2005).

Distance Dependence of the Electronic Contact of a Molecular Wire

L. Grill[1], F. Moresco[1], P. Jiang[2], S. Stojkovic[2],
A. Gourdon[2], C. Joachim[2], and K.-H. Rieder[1]

1) Freie Universität Berlin, Arnimallee 14, 14195 Berlin, Germany
2) CEMES-CNRS, 29 rue J.Marvig, P.O.Box 4347, 31055 Toulouse, France

Abstract. The central molecular wire of a so-called Reactive Lander molecule is brought in electronic contact with an atomic scale metallic nanostructure by manipulation with the STM tip. Several stable conformations are obtained in a controlled way, in accordance with calculations. An additional contribution to the tunneling current is observed at the end of the molecular board, reflecting the electronic interaction between the molecular wire and the nanostructure. The characteristic intensity of this electronic contact for different conformations is discussed by means of the vertical interatomic distance between the molecular wire and the metal atoms.

INTRODUCTION

It is of crucial importance for the use of devices in future molecular electronics to control precisely at the atomic scale the electronic contact between a single molecule and a metal electrode [1,2]. A well-defined electronic contact requires a planar molecular wire and an atomically clean metallic contact pad with suitable dimensions. The molecular wire must be decoupled from the metal substrate and must have chemical groups at the end of the wire, which provide a strong electronic interaction with the electrodes.

Low Temperature Scanning Tunneling Microscopy (LT-STM) is a very useful technique in this regard as it allows to follow experimentally the different steps of the creation of an electronic contact: imaging of the bare metal electrode with atomic precision, manipulation of the molecular wire for contacting and characterization of the electronic contact. Lander and Cu(II) phthalocyanine molecules have been manipulated by LT-STM to establish an electronic interaction between one end of the molecule and a metallic atomic scale structure or a mono-atomic step edge [3-6]. In the corresponding LT-STM images, an increased tunneling current, a so-called contact bump, was observed at the location of the molecule-metal junction. It was shown recently that it exhibits different intensities when the Lander molecule is brought in various conformations [6]. This contact bump at the location of a metal-molecular wire junction can be used as a measure of the metal-molecular wire-metal junction contact conductance [7]. In the present paper, the dependence of this electronic contact on the vertical distance between the molecular wire and the metallic pad is discussed.

We have studied so-called Reactive Lander (RL) molecules, which are part of the Lander "family" [8]. Lander molecules consist of a central molecular board, acting as a molecular wire, and four spacer legs (3,5-di-*tert*-butyl-phenyl groups) which are designed to lift the board up from the surface (Fig.1a). Accordingly, only the four legs

CP786, *Electronic Properties of Novel Nanostructures*, edited by H. Kuzmany, J. Fink, M. Mehring, and S. Roth
© 2005 American Institute of Physics 0-7354-0275-2/05/$22.50

are visible in STM images when these molecules are adsorbed on a copper surface [3,5,6,9,10].

All LT-STM experiments were done under ultra-high vacuum conditions (p = 10^{-10} mbar) at 7K. Molecules were deposited from a Knudsen cell onto the clean Cu(110) surface, kept at room temperature and subsequently annealed to 370 K [6]. STM images were taken in constant-current mode using bias voltages between 0.3 and 0.5 V (with respect to the tip) and tunneling currents between 0.3 and 0.6 nA.

RESULTS

Figure 1 shows the particular adsorption behavior of Reactive Lander molecules [6], which is equivalent to the similar Single Landers (SL) [9]. The molecules act as templates for the formation of a metallic nanostructure, connected to the step edge, underneath them. Their formation is a thermally driven, thus they can be observed only at low temperature after removal of the molecule (Fig.1c). As their dimensions (two close-packed rows of Cu atoms, seven atoms long) are defined by the molecule itself, they have a perfect shape to act as nanopads for contacting the central molecular wire of the RL molecules.

The most common conformation of the molecules on the nanostructure is the so-called parallel legs conformation (Fig.1d) where the entire molecular wire is located on the nanopad [6]. An arising question is whether there are other stable conformations of the molecule at the end of the nanostructure, suitable for an electronic contact. Figure 2 shows calculations of the potential energy of the molecule when moving it along the nanostructure, i.e. in [1-10] direction, for two orientations of the legs. These calculations are done for a SL molecule [5], but are presumably valid also for Reactive Landers as the molecules are very similar and differ only in the additional double bond at the end of the central board. Several minima are found in the curves, where B is not of interest in the regard of an electronic contact and C is the deepest minimum and corresponds to the initial conformation. In conformation D the molecular wire is tilted

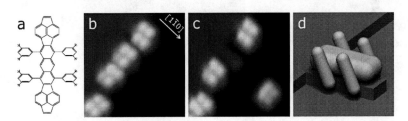

Figure 1. (a) Chemical structure of the Reactive Lander (RL) molecule. (b) RL molecules adsorbed at a Cu(110) step edge (80 × 80 Å^2). The indicated substrate orientation [1-10] is the same for all STM images. (c) After removal of a molecule by lateral manipulation, a nanostructure is visible at the adsorption site while the intact molecule is on the lower terrace (80 × 80 Å^2). (d) Schematic image of the adsorption geometry of the molecule on the Cu nanostructure (see text).

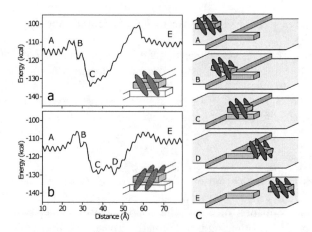

Figure 2. (a) and (b): Potential energy of the Single Lander molecule on the Cu nanostructure calculated as a function of the lateral position along the [1-10] direction [5]. The insets show schematically the molecular conformations of (a) (all legs are rotated towards the upper terrace) and (b) (legs are rotated towards the lower terrace). A-E indicate minima of interest in the curves. (c) shows the corresponding conformations A-E of the molecule (valid for both leg orientations (a) and (b)).

and has reached the end of the copper wire (all legs are rotated towards the lower terrace as there is no minimum D in Fig.2a). This conformation could therefore provide a suitable system to observe an electronic contact.

In the experiment, the molecule is moved with the STM tip laterally along the nanostructure by using suitable tunneling resistances (see ref. 5). Several stable molecular conformations are found in the STM images (Fig.3) and identified by

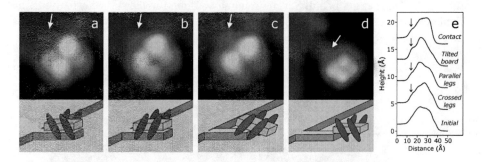

Figure 3. STM images and corresponding schemes (upper and lower panel respectively) of different conformations of the Reactive Lander molecule on the nanostructure, achieved by lateral manipulation with the STM: (a) crossed legs conformation, (b) parallel legs conformation, (c) tilted board conformation and (d) contact conformation. STM images are 35×35 Å^2 (a-c) and 45×45 Å^2 (d) in size. The white arrows mark the location of the contact bump, reflecting electronic interaction between the molecular wire and the nanostructure (see text). (e) Height profiles across the central molecular board along the substrate direction [1-10] for all conformations of the RL molecule (the upper terrace is on the left side in all height profiles). The arrows mark the contact bump with the apparent height Δh.

molecular mechanics - elastic scattering quantum chemistry (MM-ESQC) calculations [6,11]. The crossed and parallel legs conformations (Figs.3 a and b) reflect stable positions of the molecule upon a small shift of a few lattice constants of the copper substrate. The observation of the tilted board conformation (Fig.3c) confirms the calculations (minimum D). An interesting contact conformation is obtained when all four molecular legs have reached the lower terrace, but the central board is still in contact with the nanostructure (Fig.3d), i.e. a planar molecule-nanopad junction [6].

The arrows in the STM images in Fig.3 mark an additional contribution to the tunneling current, due to electronic interaction of the molecular wire with the metal atoms. This contact bump becomes clearly visible when taking height profiles from the STM images along the central molecular board (Fig.3e). It can be seen that an electronic contact between the molecular wire and the nanostructure is present in all conformations (in contrary to the SL case [5]) except the initial conformation (Fig.1d), where the molecular legs are rotated towards the unfavorable side and are therefore overshadowing this effect [6]. This means that, if the molecular wire is completely adsorbed on the nanostructure, at least one pair of legs needs to be rotated towards the lower terrace to have access to the end of the wire with the STM tip. In the tilted board case, the molecular board is only partially interacting with the nanopad and in the "ideal" contact conformation only its last end groups are in contact with the end of the copper double row.

While the lateral position of the contact bump in Fig.3e is the same for all cases, i.e. the location of the end group of the molecular wire, its intensity differs clearly. This indicates that the electronic interaction between the board and the nanopad depends on the molecular conformation. The reason for this effect seems to be the height of the board above the copper atoms as this is different for the presented conformations. In order to understand this behavior, the height Δh (a measure of the

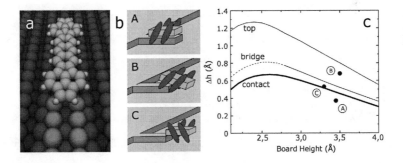

Figure 4. (a) Sphere model of the bare central board (oriented in the [1-10] direction) in the "contact" adsorption site. (b) Schemes of the three conformations parallel legs (A), tilted board (B) and contact conformation (C) of the RL molecule. (c) STM apparent height Δh of the contact bump as a function of the distance of the molecular board from the surface. Solid lines represent calculations whereas circles indicate experimental results for three different conformations A, B and C. At small board heights (below 2.7 Å), the bridge site adsorption curve is estimated (dashed line) as the contact bump from the double bond is overshadowed by contributions from other molecular orbitals.

493

electronic interaction) of the contact bump at the terminal double bond of the RL board, which is observed in an STM image, is calculated in Fig.4c as a function of the board height. Only the (planar) molecular board is considered to suppress the dominating role of the legs (Fig.4a). The resulting curves for three different adsorption sites are shown, where the contact curve is the one, which reflects the ideal geometry of the present system [6]. The maximum contribution is therefore predicted for an interatomic distance of about 2.7 Å, at larger distances the height of the contact bump decreases. While the board heights of the experimentally observed conformations are known from MM-ESQC calculations, the intensity of the contact bumps is determined from the height profiles [6]. These Δh values of the three qualitatively different cases A, B and C (Fig.4b) are 0.37 ± 0.03 Å (A), 0.68 ± 0.05 Å (B) and 0.53 ± 0.03 Å (C). Note that the differences between them are clearly larger than the error. As can be seen in Fig.4c, these experimental data points are in very good quantitative agreement with the theoretical curves. The relatively large value of the tilted board configuration (B) is probably due to the electronic interaction not only of the double bond end but also of the central part of the molecular wire with the end of the nanopad. However, the most important result is the qualitative accordance of the experimentally determined contact bump heights A and C with the calculations: The electronic interaction between the molecular wire and the metallic nanopad is smaller for the contact than for the parallel legs conformation, because the interatomic distance is decreased.

In conclusion, we have shown how the electronic interaction between a molecular wire and an atomic scale nanopad can be changed in a controlled way. Electronic contact is achieved by manipulation with the STM tip and observed as an additional contribution to the tunneling current, a so-called contact bump, in the STM images. The contact bump intensity depends on the molecular conformation and therefore the interatomic distance between the wire and the nanostructure, in good quantitative and qualitative agreement with calculations.

REFERENCES

1. R. H. M. Smit, Y. Noat, C. Untiedt, N.D. Lang, M. C. van Hemert, and J. M. van Ruitenbeek, Nature **419**, 906 (2002).
2. C. Joachim, J. K. Gimzewski, and A. Aviram, Nature **408**, 541 (2000).
3. F. Moresco, L. Gross, M. Alemani, K.-H. Rieder, H. Tang, A. Gourdon, and C. Joachim, Phys. Rev. Lett. **91**, 036601 (2003).
4. G. V. Nazin, X. H. Qiu, and W. Ho, Science **302**, 77 (2003).
5. L. Grill, F. Moresco, P. Jiang, C. Joachim, A. Gourdon, and K.-H. Rieder, Phys. Rev. B **69**, 035416 (2004).
6. L. Grill, K.-H. Rieder, F. Moresco, S. Stojkovic, A. Gourdon, and C. Joachim, Nano Letters **5**, 859 (2005).
7. S. Stojkovic, C. Joachim, L. Grill, and F. Moresco, Chem. Phys. Lett. **408**, 134 (2005).
8. A. Gourdon, Eur. J. Org. Chem. **1998**, 2797 (1998).
9. F. Rosei, M. Schunack, P. Jiang, A. Gourdon, E. Laegsgaard, I. Stensgaard, C. Joachim, and F. Besenbacher, Science **296**, 328 (2002).
10. L. Gross, F. Moresco, L. Savio, A. Gourdon, C. Joachim, and K.-H. Rieder, Phys. Rev. Lett. **93**, 056103 (2004).
11. P. Sautet and C. Joachim, Chem. Phys. Lett. **185**, 23 (1991).

SINGLE WALL CARBON NANOTUBES IN EPITAXIAL GROWN SEMICONDUCTOR HETEROSTRUCTURES

J.R. Hauptmann*, A. Jensen*, J. Nygård* and P.E. Lindelof*

*Niels Bohr Institute and Nano-Science Center, University of Copenhagen, Universitetsparken 5, DK-2100 Copenhagen, Denmark

Abstract. During a growth interrupt single wall carbon nanotubes (SWCNT) are incorporated in a Molecular Beam Epitaxy (MBE) grown semiconductor heterostructure. This is used to contact SWCNT with one and two ferromagnetic electrodes made of the diluted ferromagnetic semiconductor GaMnAs. In these devices we have observed strong hysteretic magnetoresistance below 30 K. The shape, sign and size of these magnetoresistance features are strongly dependent on parameters such as temperature, gate-voltage and bias voltage. The magnetoresistance is even present on devices with only one ferromagnetic electrode.

Keywords: carbon nanotubes, spintronics, spintransport, hybrid devices
PACS: 72.25.-b, 72.25.Hg, 73.63.-b, 73.63.Fg

Spintronics and Fullerene physics are both research areas that recently have received a lot of interest. The interest in spintronics is sparked by the belief that the abilities to implement and use the electron spin in macro -electronics will give an increased performance and functionality. To achieve this goal there is a need for finding materials with long spin scattering lengths. A promising material for this is carbon nanotubes. The combination of unique electrical properties, such as, long mean-free path and the absence of intrinsic magnetoresistance and negligible spin-orbit coupling makes it reasonable to believe that carbon nanotubes will have excellent properties for spintronics.

To gain knowledge of the spin-transport properties in carbon nanotubes, we studied two types of devices, namely single wall carbon nanotubes contacted by one and two ferromagnetic metal contacts. The devices were produced by incorporating carbon nanotubes into a Molecular Beam Epitaxy (MBE) grown heterostructure. Laser ablated carbon nanotubes, from Rice University, were deposited on a substrate, consisting of a cap layer of amorphous As placed on top of an undoped GaAs crystal separated from a heavily doped backgate by a AlAs/GaAs superlattice working as an electrical barrier. After depositing the tubes the substrate was overgrown with $Ga_{0.95}Mn_{0.05}As$. Electrodes and trenches were defined in the $Ga_{0.95}Mn_{0.05}As$ by standard electron beam-lithography, UV-lithography and wet chemical etch. A more detailed description of the device fabrication is found in [1]. All measurements on these devices from room temperature to 300 mK was DC-voltage controlled with a setup as sketched in Figure 1 (d). The magnetic field was applied in the plane of the sample, where the easy axis of GaMnAs is. Room temperature measurements on the tube devices were gate voltage independent i.e. the tubes were metallic. The magnetoresistance (MR) measurements were only performed on devices showing clear Coulomb-blockade behavior indicating coherent

CP786, *Electronic Properties of Novel Nanostructures*, edited by H. Kuzmany, J. Fink, M. Mehring, and S. Roth
© 2005 American Institute of Physics 0-7354-0275-2/05/$22.50

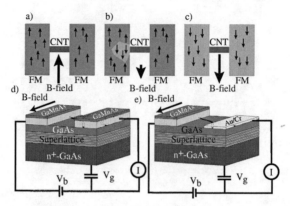

FIGURE 1. (a,b,c) Illustrates a magnetic domain reversal during a B-field sweep. (d,e) Schematic measurement setup for the two types of devices.

transport in the tube.

From studies of carbon nanotube quantum dots it is seen that even though the tube extends long under the electrode the contact between tube and electrode is happening at the edge of the electrode. This makes it reasonable that the tube is contacting a single domain, since the domain size of GaMnAs thin-films are micron sized [6]. So we imagine that the system under the influence of a magnetic field will behave as seen in figure 1 (a,b,c). This sudden change in magnetization direction could cause a resistance change. The resistance change in these kinds of systems have before been estimated by Julliere's model and described with a spin-valve model for single wall carbon nanotubes [2, 3] and multiwalled carbon nanotubes [4, 5]. Julliere's model might however not be applicable, since it describes a system of two magnets separated by an insulating barrier and not by a conducting channel, as in the case of two magnets contacting a metallic carbon nanotube.

A Curie temperature of 70 K in $Ga_{0.95}Mn_{0.05}As$ was found by MR measurements [1]. The in-plane coercive field was also determined from MR measurements to be around 0.05 T. The resistance change in the pure $Ga_{0.95}Mn_{0.05}As$ electrodes was a few ohm's which is much lower than the resistance in the tube devices and is therefore negligible in the MR measurements on the tube devices.

Figure 2 (a,b) shows measurements on a $Ga_{0.95}Mn_{0.05}As$-CNT-$Ga_{0.95}Mn_{0.05}As$ device i.e. a device with a single wall carbon nanotube connected to two ferromagnetic electrodes. The plots are chosen to show the diversity that we observe. The plots show the current as a function of magnetic field, each graph is an average over 5-10 runs and is fully reproducible. The measurements in (a) are taken at 310 mK and the three plots are measured at different biases. In (b) the measurement is made at 2 K with three different gate voltages. The graphs show a big diversity in size and sign of the MR. The sign of the MR can be changed by varying the bias or the gate voltage. The temperature has a similar influence although there was a general increase in the size of the MR feature as the sample was cooled down. At the lowest temperatures the conductance in the MR dip can be nearly completely quenched as seen in figure 3 (a) measured at 300 mK. The

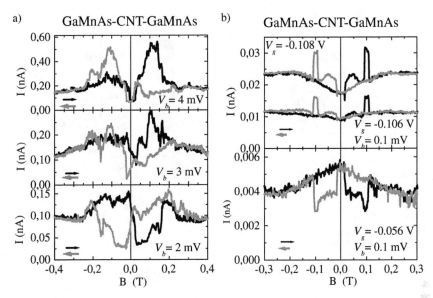

FIGURE 2. (a) Shows three plots of the current as a function of magnetic field taken at three different bias-voltages at $T = 310$ mK. The black curves describe a magnetic field sweep from - to + and the grey in the opposite direction. The magnetoresistance (MR) changes sign as a function of the bias-voltage. (b) Shows three plots of the current as a function of magnetic field, measured at 2 K at three different gate-voltages. The MR can be seen to change sign as a function of the gate voltage.

abrupt changes in the current are seen just as the magnetic field has passed zero also indicate that the tube is contacting a single ferromagnetic domain.

We also measured on reference devices consisting of a carbon nanotube contacted by only one ferromagnetic $Ga_{0.95}Mn_{0.05}As$ contact and one Cr/Au (5 nm/30 nm) non-magnetic contact. Measurements on such devices also showed MR and hysteresis as seen in figure 3 (b), where the current is plotted as a function of magnetic field. The three graphs are measured at different bias-voltages and at $T = 300$ mK.

Conclusion. We have measured the current as function of magnetic field on devices consisting of carbon nanotubes contacted by one or two ferromagnetic electrodes. In all these devices we observe a hysteretic MR. The MR change is always observed after the magnetic field has passed through zero field, indicating that it is a sudden change in magnetization direction of the single domain contacting the nanotube, that is the cause. The sign, shape and size of the MR change as a function of temperature, cooling cycle, gate and bias-voltage in a non-controllable manner. An increase in the size of the MR can be seen as the sample is cooled, in some cases resulting in a complete quenching of the current at 300 mK. The observation of MR in our reference devices with only one ferromagnetic electrode indicates that the MR behavior can not be explained as resulting from a simple spin-valve effect. The diversity of the results and the MR measured in our reference devices suggest that a better understanding of the precise contact mechanism between electrode and tube is needed to quantitatively understand the results.

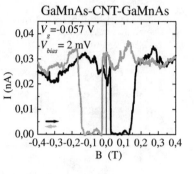
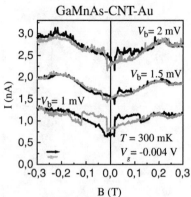

FIGURE 3. (a) Measurements of the current as a function of magnetic field made at $T = 300$ mK showing a quenching of the current right after the magnetic field has passed 0 T. (b) Three measurements made on a device with one ferromagnetic electrode and one non-magnetic Cr/Au electrode. The three measurements are made at three different bias-voltages. A clear MR can be seen in both devices.

ACKNOWLEDGMENTS

We acknowledge fruitful discussion with Pavel Strêda and Janusz Sadowski for MBE growth at III-V nanolab (NBIfAPG, University of Copenhagen) and at Max-Lab (Lund University). The work was supported by NEDO Spintronics and the Danish Research Councils (STVF, SNF).

REFERENCES

1. A. Jensen, J. R. Hauptmann, J. Nygård, J. Sadowski and P. E. Lindelof, *Nano Lett.*, **4**, 349 (2004).
2. K. Tsukagoshi and B. W. Alphenaar, *Nature*, **401**, 572 (1999).
3. A. Jensen, J. Nygård and J. Borggreen, *Proceedings of the International Symposium on Mesoscopic Superconductivity and Spintronics*, edited by K. Tsukagoshi and J. Nitta, World Scientific, Singapore, 2003 ,PP 33.
4. S. Sahoo, T. Kontos, C. Scönenberger and C. Sürgers, *Appl. Phys. Lett.* **86**, 112109 (2005).
5. D. Orgassa, G. J. Mankey and H. Fujiwara, *Nanotechnology*, **12**, 281 (2001).
6. U. Welp, V. K. Vlasko-Vlasov, X. Liu, J. K. Furdyna and T. Wojtowicz, *Phys. Rev. Lett.*, **90**, 167206, (2003).

Suppression of bias- and temperature-dependent conductance by gate-voltage in multi-walled carbon nanotube

T. Kanbara[a,b,*], K. Tsukagoshi[b,c] and Y. Aoyagi[b,d], and Y.Iwasa[a,e]

[a] *Institute for Materials Research, Tohoku University, Katahira 2-1-1, Aoba-ku, Sendai 980-8577, Japan.*

[b] *Riken, Hirosawa 2-1, Wako 351-0198, Japan*

[c] *PRESTO, Japan Science and Technology Agency, Kawaguchi 332-0012, Japan*

[d] *Tokyo Institute of Technology, 2-12-1 O-okayama, Meguro-ku, Tokyo 152-8551, Japan*

[e] *CREST, Japan Science and Technology Agency, Kawaguchi 332-0012, Japan*

Abstract. We observed an ambipolar behavior in multiwalled carbon nanotubes (MWNT) in a back-gate configuration. The four-terminal measurement indicates that the ambipolar behavior is originated from an intrinsic conductivity of MWNT. power-law behaviors, in temperature- and bias-dependent conductance, disappeared when a high gate voltage is applied, and conductance become temperature- and bias-independent. This result indicates an occurrence of gate-induced transformation from the unconventional to the normal metallic states in MWNT.

INTRODUCTION

The field effect transistor (FET) is known as a highly affective device for switching electron conduction in semiconductors, but researchers have also used this device to modulate electronic phase transition phenomena, such as ferromagnetism [1] and superconductivity [2,3]. This indicates that, although the effect of the gate field is generally weak, it is possible to tune electronic states of materials in FETs when devices are well prepared in combination with chemical tuning of material properties which will provide novel opportunities for FET devices.

In multiwalled carbon nanotubes (MWNTs), electrical conductance displays a power-law temperature and bias voltage dependence [4-7], which is ascribed to a Tomonaga-Luttinger liquid behavior driven by pure exchange Coulomb interaction [8] or to a kind of localized behavior induced by disorder combined with the Coulomb interaction [9,10]. In a previous paper, we reported that this temperature dependence is suppressed by application of gate voltage [11]. In this paper, we demonstrate that the conductance becomes bias-independent, being in fair agreement with the previous

CP786, *Electronic Properties of Novel Nanostructures*, edited by H. Kuzmany, J. Fink, M. Mehring, and S. Roth
© 2005 American Institute of Physics 0-7354-0275-2/05/$22.50

result. We also estimate contact resistance (R_{cont}) and confirmed R_{cont} can be ignored in measured region.

RESULTS AND DISCUSSION

MWNTs produced by an arc-discharge method were used as it is to avoid defects that may be introduced by chemical treatments. The radii of MWNTs, which were several µm in length, were estimated to be 2 - 5 nm from height profiles in atomic force microscope (AFM) images. MWNTs were sonicated in 1,2-dichloroethane, and deposited on SiO_2/Si substrates with Pt(5 nm)/Au(15 nm) markers. The positions of MWNTs were determined using a scanning electron microscope (SEM). The electrodes of Pt(5 nm)/Au(60 nm) were formed on a MWNT ('end-contact' configuration) by evaporation of metals through a resist mask made by an electron beam lithography system [12]. Figure 1 (a) shows a SEM image of a MWNT two-terminal device. The gate voltage V_G was applied from the back-gate through the SiO_2 insulator of thickness, t_{SiO2} = 200 nm. Figure 1 (b) shows a transfer characteristic at room temperature (RT). We observed an ambipolar behavior with a minimum of drain current in the MWNT device. The ambipolar behavior in MWNT has been observed by Kruger et al. in the electrochemical transistor configuration [13], since the electrochemical gate is much more effective than the back-gate. It is widely known that electronic conduction in MWNTs is dominated by the outermost shell [14],

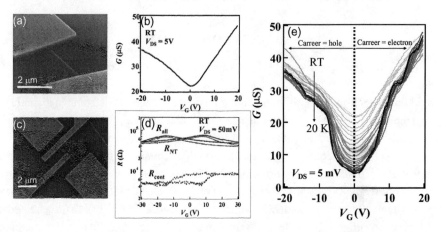

FIGURE. 1. (a) SEM image of the two-terminal MWNT-device, (b) transfer characteristics of the two-terminal device at RT, (c) SEM image of four-terminal device, (d) gate voltage dependency of estimated contact resistance from four-terminal measurement at RT, (e) transfer characteristics of the two terminal device from 20 K to RT.

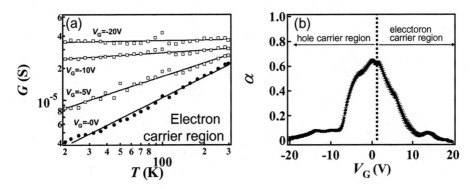

FIGURE 2. Observed and calculated difference spectra for V_G = +100 V. (a) Temperature dependence of G at V_{DS} = 50 mV with each V_G. (b) α vs V_G □ values are estimated from the slope of (a).

which is regarded to be a graphene sheet. Thus the minimum drain current should be attributed to the minimum density of states of MWNTs and the increased drain current with the negative and positive V_G is due to the accumulation of holes and electrons, respectively [13]. We also carried out a four-terminal experiment. Figure 1(c) displays a SEM image of a four terminal device. We estimated the R_{cont} ranging from V_G = -30V to 30 V. In this range, R_{cont} was found to be one-order lower than the resistance of MWNT itself (R_{NT}). Thus, the ambipolar behavior observed in Fig. 1(b) should be attributed to the intrinsic conductance of MWNT.

In order to clarify the transport properties near the current minimum, we measured detailed temperature T and V_{DS} - dependence of the transfer characteristics. Figure 1 (e) displays transfer characteristics of the two-terminal MWNT device at various temperatures. On-off ratio (G_{on} / G_{off}) become larger with decreasing temperature. In Fig. 2 (a), T dependency of G at some V_G is shown in both logarithmic scale. All the data are fitted well by a linear relation, indicating that G shows a power-law behavior proportional to T^α. The most important information gained from Fig. 2 (a) is that the slope of the log G – log T plot shows a systematic V_G dependence. The slope, which is the steepest at V_G = 0 V, becomes gentler with increasing $|V_G|$ in both negative and positive directions. In particular, at the highest gate voltages, the conductance becomes almost V_{DS}- independent irrespective of the sign of doped carriers. Since the relation $G \propto T^\alpha$ approximately holds at all gate voltages, we are able to define α values for all V_Gs. Figure 2 (b) summarizes the relationship between α and the applied V_G. When V_G is close to 0 V and the Fermi level is at the bottom of the density of states, α is

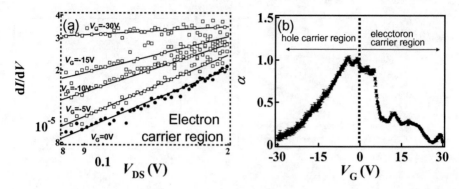

FIGURE 3. (a) V_{DS} dependence of differencial conductance at $T = 20$ K with different V_G. (b) Estimated α vs V_G. α values were obtained from the slope of (a).

approximately 0.64. The α value decreases with increasing $|V_G|$ and even becomes zero or increases slightly with decreasing temperature, indicating that MWNT behaves like a simple Fermi liquid. This phenomenon is interpreted as a crossover from an unconventional metallic state to a Fermi liquid in MWNTs. This behavior also observed in the bias dependent conductance. In Figure 3 (a), V_{DS} dependence of G is shown in a double logarithmic scale. The slope of the log G – log T plot shows a systematic decrease with increasing $|V_G|$. Figure 3 (b) displays a relationship between α which was estimated from slope of Fig. 3(a), and the applied V_G. The α value, being approximately 0.95 when V_G is close to 0 V, decreases with increasing $|V_G|$ and becomes almost zero at $|V_G| = 30$V. The nature of the unconventional metallic states in MWNTs is still in debate, since it is not yet established whether electron conduction in MWNT is in a ballistic or diffusive regime. There are two major models, Tomonaga-Luttinger (TLL) model with ballistic transport and nonconventional coulomb blockade (NCB) model with diffusive transport [8,9]. Because the results of the two models are qualitatively identical, we are not able to distinguish between these two models in the present experiment. According to these models, the exponent α decreases with increasing number of channels that participate in electron conduction, which corresponds to the number of subbands that cross the Fermi energy E_F [8-10,15]. If the Fermi level of MWNT reaches to subband, the number of conduction channel increase rapidly. This increase of channels seemed to cause the rapid decrease of α value.

CONCLUSION

We observed an ambipolar behavior in MWNTs in a back-gate configuration, which enabled us to make a systematic investigation of the transport properties of MWNTs. The result of the four-terminal measurement implies that this ambipolar-like behavior originated from intrinsic conductivity of MWNT. We observed a crossover behavior from an unconventional metallic state, where conductance shows a power-law dependence on T or V_{DS}, to a normal Fermi liquid state, where conductance is T and V_{DS} independent. The present result indicates that the electronic states of even nanoscale materials can be tuned by the gate electric field in the FET configuration by changing the carrier concentrations.

ACKNOWLEDGMENTS

This work was supported in part by Grant-in-Aid for JSPS Fellows.

REFERENCES

1 H. Ohno, D. Chiba, F. Matsukura, T. Omiya, E. Abe, T. Dietl, Y. Ohno, and K. Ohtani, Nature **408,** 944 (2000).

2 X. X. Xi, Q. Li, C. Doughty, C. Kwon, S. Bhattacharya, A. T. Findikoglu, and T. Venkatesan, Appl. Phys. Lett. **59,** 3470 (1991).

3 J. Mannhart, D. Schlom, J. Bednorz, and K. Muller, Phys. Rev. Lett. **67,** 2099 (1991).

4 C.Schonenberger, A. Bachtold, C. Strunk, J. -P. Salvetat, and L. Forro, Appl. Phys. A **69,** 283 (1999).

5 E. Graugnard, P. J. Pablo, B. Walsh, A. W. Ghosh, S. Datta, and R. Reifenberger, Phys. Rev. B **64,** 125407 (2001).

6 A. Kanda, K. Tsukagoshi, Y. Aoyagi, and Y. Ootuka, Phys. Rev. Lett. **92,** 036801 (2004).

7 E. Enomoto, K. Horiuchi, K. Miyamoto, Y. Matsunaga, N. Aoki, Y. Ochiai, Physica B **323,** 249 (2002).

8 R. Egger, Phys Rev. Lett. **83,** 5547 (1999).

9 R. Egger, & A. Gogolin, Chem. Phys. **281,** 447 (2002).

10 R. Egger and A. O. Gogolin, Phys. Rev. Lett. **87,** 066401 (2001).

11 T. Kanbara, T. Iwasa, K. Tsukagoshi, Y. Aoyagi, and Y. Iwasa, Appl, Phys. Lett. **85,** 6404 (2004).

12 K. Tsukagoshi, B. W. Alphenaar, and H. Ago, Nature **401,** 572 (1999).

13 M. Kruger, M. R. Buitelaar, T. Nussbaumer, C. Schonenberger, and L. Forro, Appl. Phys. Lett. **78,** 1291 (2001).

14 A. Bachtold, C. Strunk, J. -P Salvetat, T. -M. Bonard, L. Forro, T. Nussbaumer, and C. Schonenberger, Nature, **397,** 673 (1999).

15 A. Kawabata, and T. Brandes, J. Phys. Soc. Jpn 65, 3712 (1996).

Observation of Ballistic Thermal and Electrical Transport in Multi-wall Carbon Nanotubes

Elisabetta Brown, Ling Hao, John C. Gallop, Lesley F. Cohen* and John C. Macfarlane†

National Physical Laboratory, Hampton Road, Teddington, Middlesex, TW11 0LW, UK
*Imperial College, London, SW7 2BZ, UK
†University of Strathclyde, Glasgow, G4 0NG, UK

Abstract. We report thermal measurements on individual carbon nanotubes using a temperature sensing scanned microscope probe. An arc-grown bundle of multi-walled nanotubes (MWNTs) is mechanically attached to a thermal probe. The heat flow down individual MWNTs is recorded as a function of the temperature difference across them. Simultaneous measurements of thermal and electrical conductance are recorded. A study of the MWNTs size distribution using statistical methods on AFM images allowed us to model the available phonon channels. The size of the ballistic conductance steps observed at room temperature and the correlation between electrical and thermal conductance steps are discussed and we present the first evidence for ballistic transport of both phonons and electrons and in these tubes.
Keywords: Carbon nanotube, thermal conductivity, quantized conductance.
PACS: 65.80.+n

INTRODUCTION: BALLISTIC TRANSPORT

When the dimensions of a conductor become much less than the mean free path of its conduction carriers (electrons and/or phonons), transport is no longer diffusive but ballistic. In the ballistic regime the conductor behaves like a waveguide to the conduction carriers, and the conductance is unaffected by length but depends only on N, the number of 'channels' or 'modes' available to the conduction carriers.

For electrons, each mode contributes $1G_O$ to the electrical conductance [1]; while for phonons each mode will contribute a maximum of $1G_{TH}$ to the thermal conductance [2]. G_O and G_{TH} are the quantum of electrical and thermal conductance respectively. Their expressions are shown in equation (1); h: Planck's constant, k_B: Boltzmann constant, T: temperature, e: electron charge.

$$ G_O = \frac{2e^2}{h} = \frac{1}{12.9 k\Omega}, \; G_{TH} = \frac{\pi^2 k_B^2 T}{3h} = 9.456 \times 10^{-13} \frac{W}{K^2} \times T . \tag{1} $$

Several experiments have shown that carbon nanotubes (CNTs) display ballistic electron conductance at room temperature (RT). The expected contribution per conducting carbon shell is $2G_O$ [3] but experimental increments of $1G_O$ and $1/2G_O$ have also been reported [4].

CP786, *Electronic Properties of Novel Nanostructures*, edited by H. Kuzmany, J. Fink, M. Mehring, and S. Roth
2005 American Institute of Physics 0-7354-0275-2/05/$22.50

CNTs have high thermal conductivity (~6000 W/m K at RT) and high phonon velocities (~20 km/s – due to the strong sp^2 C-C bonding backbone) [5]. Their phonon mean free path has been measured to be several hundreds of nm at RT [6] so ballistic phonon transport should be possible at RT in sufficiently defect-free CNTs.

MEASUREMENTS AND RESULTS

In this work we used arc-grown MWNT bundles. These bundles are made of individual MWNTs of high quality; we estimated the average CNT diameter by imaging a dispersion of the nanotubes on graphite with a Nanoscope AFM. We found the mean diameter to be on average ~2nm.

A.

B.

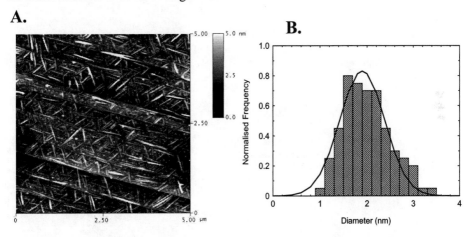

FIGURE 1. A. AFM image of the MWNT dispersed on graphite (HOPG). The tubes show a tendency to align with three-fold symmetry. This is understood as representing locking to the commensurate crystal structure of the underlying HOPG lattice [7]. **B.** The histogram is the result of the measurement of a sample of 100 nanotubes; the solid line is a fit to a normal distribution with mean of 1.9 ± 0.48 nm.

We attached an MWNT bundle to a Veeco thermal probe, mounted on an Explorer AFM. This arrangement allows for individual MWNTs to be studied whilst ensuring low CNT-probe interface scattering [4]. The piezo control of the AFM is used to move the bundle slowly towards a graphite substrate (HOPG – freshly cleaved) until contact is detected; a diagram of the experimental set-up is shown in Figure 2A. Simultaneous measurements of thermal and electrical conductance are performed when the MWNTs are in contact with the substrate. The force between tip and substrate is also measured. All measurements are done in high vacuum (~10^{-5} mbar).

The graph in Figure 3A shows an example of electrical, thermal and force measurements as a function of vertical displacement for an MWNT. The temperature difference, ΔT, across the MWNT was ~264K. The electrical conductance step is ~1G$_O$ and the corresponding thermal conductance step is ~260G$_{TH}$. The data in Figure 3A is a typical set and shows clear evidence for ballistic electron and phonon transport.

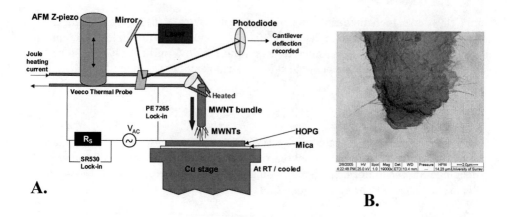

FIGURE 2. A. The experimental set-up. **B.** SEM image of the MWNTs; long, thin MWNTs bundles protrude from the bulk; most of them have only one MWNT at their extreme (scale bar is 2μm).

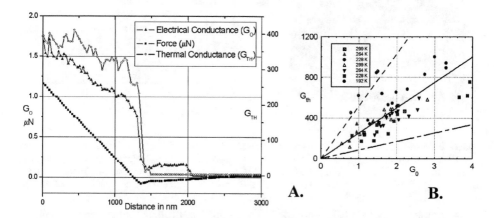

FIGURE 3. A. Simultaneous thermal, electrical and force measurements made as an MWNT is withdrawn from the HOPG substrate. **B.** Correlation between thermal and electrical conductance steps for MWNTs on HOPG for a range of temperature differences (see inset). The trend lines shown in the graph are the calculated ratios of G_{TH} to G_O for an armchair (long-short dash line) and a zig-zag tube (dash line). The armchair nanotube has n, m indices (7,7), diameter 0.96nm and 84 phonon channels. The zig-zag nanotube has n, m indices (42,0), diameter 3.33nm and 504 phonon channels. Thus, assuming each CNT contributes an amount of $1G_O$ to the electrical conductance, $G_{TH} - G_O$ slopes of 504 and 84 are plotted.

One piece of evidence is that both electrical and thermal conductance measurements show little change as the tube is withdrawn from the HOPG surface even though the length of the tube is expected to change by compression or bending in this process.

Further evidence for ballistic phonon conduction is that, if we assume the contrary, the estimated diffusive thermal conductivity for this tube (assuming a solid rod of diameter 1.9 nm and a minimum length estimate of 700 nm) is 2.5×10^4 W/m K! An unrealistically high value, a factor of 4 greater than the room temperature thermal conductivity of diamond.

Measurements were repeated for different values of ΔT across the MWNTs. In the graph in Figure 3B the thermal conductance step size is plotted as a function of the electrical step size. There is a clear evidence of a linear trend with mean slope of ~250 (solid line in the graph) which is the number of phonon modes per nanotube. This number is consistent with values expected for the range of diameters of nanotubes used (see Chapter 9 in Reference 3) thus providing further evidence for phonon ballistic transport in the nanotube.

CONCLUSION

We have measured simultaneously the electrical and thermal conductance of individual arc-grown MWNTs. Our measurements show strong evidence of ballistic transport for both electron and phonons.

ACKNOWLEDGMENTS

The authors would like to thank Mark Baxendale at Queen Mary University of London for supplying the MWNT bundles and David Cox at University of Surrey for the SEM images of the MWNT bundles.

REFERENCES

1. van Wees, B. J., et al, *Phys. Rev. Letters* **60**, 848 (1988).
2. Rego, L. G. C., et al, *Phys. Rev. Letters* **81**, 232 (1998); Schwab, K., et al, *Nature* **404**, 974 (2000).
3. Saito, R., Dresselhaus, G., and Dresselhaus M., "Chapter 8", in *Physical Properties of Carbon Nanotubes*, London: Imperial College Press, 1998, pp. 142-144.
4. Frank, S., et al, *Science* **280**, 1744 (1998).
5. Hone, J., et al, *Appl. Phys. A* **74**, 339 (2002).
6. Kim, P., et al, *Phys. Rev. Letters* **87**, 215502 (2001).
7. Falvo, M. R., et al, *Phys. Rev. B* **62**, R10665 (2000).

Organic Thin Film Transistor with Carbon Nanotube Electrodes

J. Y. Lee[1], S. Roth[2], Y. W. Park[1]

[1]School of Physics and NANO Systems Institute – National Core Research Center, Seoul National University, Seoul 151-747, Korea
[2]Max Planck Institute for Solid State Research, Heisenbergstrasse 1, D-70569 Stuttgart, Germany

Abstract. The contact resistance between organic semiconductors and metallic electrodes affects the performance of the organic thin film transistor (OTFT) negatively so that it may make the field effect mobility of charge carrier seem small. In order to reduce the contact resistance we used conducting Carbon Nanotube (CNT) films, which consist of the same element as the basic material of the organic semiconductors, as Source/ Drain electrodes. The measurements of transistor properties based on pentacene single crystals have been carried out by using both CNT film electrodes and metal electrodes.

INTRODUCTION

Finding good semiconducting organic materials is the first step for the high performance Organic Thin Film Transistor (OTFT), and some organic materials, like pentacene or rubrene, have already been found to be good candidates which could substitute the inorganic semiconductor[1, 2]. However, there are still difficulties to make application to the electronic device due to several factors, the incomplete understanding of the charge transport mechanism, the contact effects between the metallic electrodes and the organic active materials, the imperfection of the organic active layer - the insulating layer interface etc. The contact barrier sometimes limits the charge carrier transport of the OTFT or gating effect, which results in Non-linear I-V characteristics or the degradation of charge carrier's field effect mobility[4, 6]. Furthermore, it is unclear to distinguish between the intrinsic material property and the extrinsic deterioration caused by the contact regime. So, the technique to reduce the contact resistance is also an important part of the OTFT experiment in order to understand the essential nature of carriers in organic semiconductors, and to make applications of electronic devices.

We have recently fabricated TFT structure using anthracene single crystals in order to study the grain boundary-free properties of the charge carrier transport[5]. In that TFT structure the plate-shaped anthracene single crystal was put on the pre-patterned metal electrodes of the SiO_2/Si substrate, and then it was pressed down. The results showed that the high contact resistance could be reduced by chemical treatment for the metal electrodes. This means that charge injection was enhanced by the chemical treatment,

CP786, *Electronic Properties of Novel Nanostructures*, edited by H. Kuzmany, J. Fink, M. Mehring, and S. Roth
© 2005 American Institute of Physics 0-7354-0275-2/05/$22.50

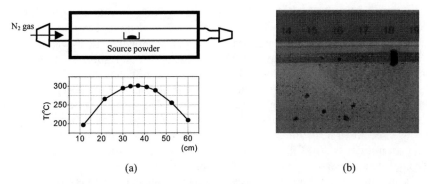

(a)   (b)

FIGURE 1. Schematic of crystal growth apparatus and temperature profiles of quartz tube furnace (a), and Pentacene single crystal grown by physical vapor deposition method [8] (b).

i.e. modifying the electrode-surface-materials. Graciela B. Blanchet et. al. also showed that non-metallic electrodes, namely printed organic conducting electrodes, could reduce the contact resistance[4]. In this paper we will present the properties of pentacene single crystal Thin Film Transistor with the conducting CNT film electrodes and compare with the result of metal electrode TFT.

EXPERIMENTS

Commercial pentacene powder purchased from Aldrich was used to grow a single crystal in a quartz tube furnace, 65cm long, which shows temperature gradient of ~100°C between the middle and the edge, as shown in Figure 1 (a). The crystals appeared at the 200~220°C temperature region after several hours keeping the temperature of the source zone at 300°C under slow nitrogen gas flow (20ml/min). Figure 1 (b) shows pentacene single crystals which are in the shape of plates with 2~5mm size. These palate-shaped crystals were put on CNT electrodes prepared as in Figure 2 (a).

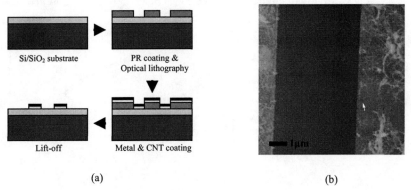

(a)   (b)

FIGURE 2. Procedure of the CNT electrodes fabrication on the Si/SiO$_2$ substrates (a), AFM image of the metal based CNT electrodes (b).

509

The heavily doped Si substrate with SiO_2 layer 200nm thick was prepared to be used as gate electrode and gate insulating layer. Optical lithography allows to make photo resist masks of 5, 30, 50 µm channel length, shadowing metal evaporation and CNT film coating. After the evaporations of Cr 3nm thick and AuPd 17nm thick, CNT film was deposited by air-brushing CNT suspensions in chloroform, DMF and SDS solutions respectively, in which the chloroform suspension gave the best results. When a dense CNT layer was deposited, it was difficult to get a clear electrode edge because the randomly entangled CNT have a tendency to prevent photo resist lift-off. Figure 2 (b) shows the AFM image of CNT electrodes based on metal of 5 µm channel length. As shown in Figure 2 (b) CNT passed cross the electrode edge, but it is small enough to be neglected compared with the whole channel length, 5 µm.

TFT structures were fabricated by combining pentacen single crystals and the CNT electrodes. Pressure was applied to enhance the contact between pentacene and CNT like in our previous works[5]. We measured field effect characteristics of 5, 30, 50 µm channel lengths CNT electrodes and metal electrodes, respectively, at room temperature after checking negligible the gate leakage current less than 10^{-11}.

RESULTS & DISCUSSIONS

For the most devices I-V curves show a linear region and a saturation region as shown in Figure 3 (a), but for 5 µm channel length devices the short channel effect[7] was also observed. The field effect parameters such as the field effect mobility (μ_{linear}, μ_{sat}), the subthreshold slop(S), the threshold voltage(V_{th}), and the on-off ratio (I_{on}/I_{off}), were calculated from the data of gate sweep graphs following the equations in reference [3] . All parameters are summarized in the Table 1. The field effect mobility of the best device amounts to 0.94 cm^2/Vs for linear region and 0.68 cm^2/Vs for saturation region, respectively and other parameters show mostly good numerical values. It is also interesting that the smaller ratio of channel width to channel length gives the better results. Comparing parameters of CNT electrodes to those of metallic electrodes, it is difficult to assert that CNT electrodes made the performance of TFT enhance because some parameters (S, I_{on}/I_{off}) become better but others(μ_{linear}, μ_{sat}) not. However, it is worth to notice that CNT electrodes made definitely improved some of

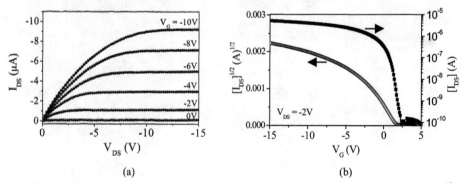

(a)   (b)

FIGURE 3. I-V characteristics at V_G = 0, -2, -4, -6, -8 and -10V (a) gate voltage dependence of $|I_{DS}^{1/2}|$ and $\log|I_{DS}|$ (b) for CNT electrode TFT with 30 µm channel length.

510

TABLE 1. Feld effect parameters where W/L is the ratio of channel width to channel length.

Electrode type	W/L	μ_{linear} (cm²/Vs)	μ_{sat} (cm²/Vs)	S (V/decade)	V_{th} (V)	I_{on}/I_{off}
AuPd 5μm	200	0.77	·	0.9	1.0	3.4×10^2
AuPd 30μm	30	0.65	0.43	1.0	1.2	4.1×10^2
AuPd 50μm	20	0.94	0.68	0.6	1.5	7.3×10^2
CNT 5μm	200	0.65	·	0.3	1.6	3.4×10^3
CNT 30μm	15	0.86	0.54	0.4	1.5	1.7×10^4
CNT 50μm	20	0.52	0.34	0.7	2.3	2.2×10^2

the TFT performances in spite of the fact that the surface of the electrodes had become quite rough, as shown in Figure 2 (b). It is expected that if we can make the surface of CNT film electrodes flat, the performance of TFT will improve much more. There is another problem to solve. Because pentacene single crystal might have anisotropic properties, simple comparison is meaningless. First of all we should find the direction which shows the best results for both CNT and metallic electrodes, and then the comparison will become meaningful.

In our future work, we will concentrate on uniform surfaces of CNT layers and on the anisotropy of organic single crystals.

ACKNOWLEDGMENTS

This work was supported by the NANO Systems Institute – National Core Research Center program of KOSEF. Partial support for JYL was provided by the International Research Internship Program of the Korea Research Foundation (KRF).

REFERENCES

1. S. F. Nelson, Y.-Y. Lin, D. J. Gundlach, and T. N. Jackson, *Appl. Phys. Lett.* **72**, 1854-1856 (1998).
2. V. C. Sundar, J. Zaumseil, V. Pozdrov, E. Menard, R. Willett, T. Someya, M. E. Gershenson, and J. A. Rogers, *Science* **303**, 1644-1646 (2004).
3. G. Horowitz, *Adv. Mater.* **10**, 365-377 (1998).
4. G. B. Blanchet, C. R. Fincher, M. Lefenfeld, and L. A. Rogers, *Appl. Phys. Lett.* **84**, 296-298 (2004).
5. A.N. Aleshin, J.Y. Lee, S.W. Chu, J.S. Kim, and Y.W. Park, *Appl. Phys. Lett.* **84**, 5383-5385 (2004).
6. A.N. Aleshin, S.W. Chu, V.I. Kozub , S.W. Lee, J.Y. Lee, S.H. Lee, D.W. Kim and Y.W. Park, *Curr. Appl. Phys.* **5**, 85-89 (2005)
7. M.D. Jacunski, M.S. Shur, A.A.A. Owusu, T. Ytterdal, M. Hack, and B. Iniguez, *IEEE Trans. Electron Dev.* **46**, 1146- 1158, (1999)
8. R.A. Aaudise, Ch. Kloc, P.G. Simpkins, and T. Siegrist, *J. Crys. Growth* **187**, 449-454 (1998).

Novel freestanding nanotube devices for combining TEM and electron diffraction with Raman and Transport

Jannik C. Meyer*, Matthieu Paillet†, Jean-Louis Sauvajol†, Dirk Obergfell*, Anita Neumann**, Georg Duesberg** and Siegmar Roth*

*Max-Planck Institute for solid state research, Stuttgart, Germany
†Laboratoire des Colloides, Verres et Nanomateriaux, Universite de Montpellier II, France
**Infineon Technologies CPR NP, Munich, Germany

Abstract. A versatile procedure for combining high-resolution transmission electron microscopy (TEM) and electron diffraction with Raman spectroscopy and transport measurements on the very same nanotube is presented. For this we prepare free-standing structures on the corner of a substrate by electron beam lithography and an etching process. Further, this procedure makes possible a TEM quality control of nanotubes grown directly on the substrate.

INTRODUCTION

After more than a decade of intensive research in carbon nanotubes it is still not possible to sort or grow a specific single nanotube structure, as defined by the tube indices (n,m). All production and sorting procedures end up with a mixture of different indices. For practical applications, it might be sufficient to sort metallic from semiconducting nanotubes. But for understanding the physics of a single molecule, it is desired to know the precise structure of the object under investigation. The analysis of individual single-walled nanotubes is carried out e.g. by Raman spectroscopy or electric transport measurements, providing information about the vibrational properties and electronic structure. We have developed a way to obtain an independent and unambiguous indentification of the nanotube indices of the same molecule that was present in the Raman or transport measurement: We etch away the substrate in such a way that the nanotube is accessible by transmission electron microscopy (TEM). Further, the process allows a quality control of nanotubes grown by chemical vapour deposition (CVD) directly on the substrate, which are usually also not accessible by TEM.

EXPERIMENTAL

Single-walled carbon nanotubes are grown by CVD [1, 2] on highly doped Si substrates with a 200nm oxide layer. A metal structure is created by electron beam lithography on top of the nanotubes. It consists of 3nm Cr and 110nm Au. The substrate is cleaved, so that the structure is on the edge of the substrate. The samples are etched in 30% KOH at

CP786, *Electronic Properties of Novel Nanostructures*, edited by H. Kuzmany, J. Fink, M. Mehring, and S. Roth

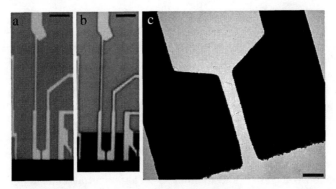

FIGURE 1. Sample for combining TEM and transport on carbon nanotubes: A contact structure is prepared close to a cleaved edge (a). By the etching process we obtain a free-standing structure that reaches out across the side edge of the substrate (b). (a) and (b) are optical microscope images. The free-standing part is now accessible for TEM (c) and electron diffraction investigations. Unfortunately, three tubes (instead of just one) bridge the contacts in this sample. Scale bar is 5μm in (a) and (b), and 500nm in (c).

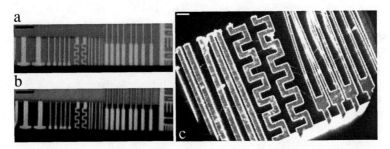

FIGURE 2. Sample for combining TEM and Raman spectroscopy on carbon nanotubes: A grid structure is prepared close to a cleaved edge (a). After etching (b) a free-standing structure is obtained which can be viewed by TEM. (a) and (b) are optical microscope images. A dark-field TEM image of a similar sample is shown in (c). It provides the location of the nanotubes with respect to the grid structure. Scale bar is 5μm in (a) and (b) and 1μm in (c).

60°C for 6 hours. The etch rate of the bulk silicon can be controlled by a bias potential between the substrate and the KOH solution, and the etching process is monitored with an optical microscope. The samples are then transferred into water, isopropanol and acetone, and dried by critical point drying. In this way arbitrary free-standing structures can be prepared. Depending on the geometry of the free-standing structure, the samples can be used for transport measurements or Raman spectroscopy.

RESULTS

A transport sample prepared in this way is shown in Figure 1. The free-standing structure is designed as a pair of contacts, connected to bond pads on the substrate. In this sample

513

there are three tubes bridging the gap, where a single tube is desired. This will be solved by producing a large number of contacts, with an appropriate density of nanotubes, and selecting only those contacts for transport measurements where a single tube is present. This is different from the approach published earlier [3], which required a marker system, location of individual nanotubes by AFM, and a total of three lithography steps. However, while the method shown here is more efficient for making many samples, the method shown in [3] provides a gated nanotube, facilitating a detailed analysis of the electronic structure of the nanotube.

Samples for Raman spectroscopy consist of a free-standing grid structure as shown in Figure 2. After sample preparation, TEM images are obtained in order to know the location of the nanotubes with respect to the metallic grid structure. This can be done at low acceleration voltages (60kV) in order to avoid structural damage to the nanotubes. As the metal structure can be seen in an optical microscope, and the location of the tubes is known, Raman spectroscopy can be carried out on sufficiently separated individual nanotubes. After the Raman measurement, the active tubes are investigated by high-resolution TEM and electron diffraction.

High-resolution TEM images of several freestanding nanotubes are shown in Figure 3. All these tubes were grown on a bulk substrate and turned into free-standing tubes by the described process. We see that CVD growth can provide individual SWNT with a very broad diameter distribution. However, CVD growth is very sensitive to the growth parameters. Figure 3 also gives some examples of tube-like objects that occurred, showing the need for quality control by TEM. In addition to the high-resolution images, it is possible to obtain electron diffraction patterns from sufficiently straight carbon nanotubes. Long, straight nanotube sections for diffraction experiments are easily found within these samples. Diffraction experiments are done at an acceleration voltage of 60kV, which is below the threshold for knock-on damage in carbon nanotubes [4].

CONCLUSIONS

We have developed a versatile approach for including TEM analysis into single-tube Raman and Transport investigations. Work is in progress to collect data on individual carbon nanotubes of which the structure was identified by electron diffraction. CVD growth on a substrate can produce a wide range of SWNT diameters, and other objects. Tubes grown on the substrate, which are normally not accessible by TEM, can be analyzed using our process. By combining the TEM analysis (including diffraction) with spectroscopy and transport, it is possible to obtain data for an exactly known structure. Thus, a direct and unambiguous comparison with the theoretical predictions and simulations will be possible.

ACKNOWLEDGEMENTS

We acknowledge financial support by the BMBF project INKONAMI (Contract No. 13N8402) and the EU project CANAPE.

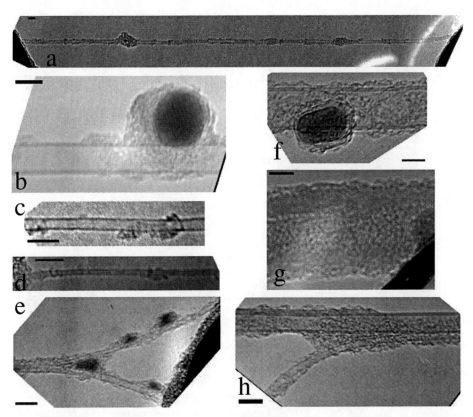

FIGURE 3. Freestanding nanotubes, prepared from nanotubes grown by CVD on Si substrates. A complete nanotube section between two metal contacts is shown in (a). A large SWNT with a diameter of 5.3nm is shown in (b). Tubes (a) and (b) are grown by process [1]. Tubes (c) (1.7nm) and the small-diameter tube (d) (0.9nm) are from [2]. Small bundles, which split near the ends, are common in dense samples (e). Under unfavourable growth conditions, there appear also DWNTs (f), cone-stacked objects (g), and tubes with different numbers of shells on each wall (h). All scale bars are 5nm.

REFERENCES

1. M. Paillet, V. Jourdain, P. Poncharal, J.-L. Sauvajol, A. Zahab, J. C. Meyer, S. Roth, N. Cordente, C. Amiens, and B. Chaudret. Versatile synthesis of individual single-walled carbon nanotubes from nickel nanoparticles for the study of their physical properties. *J. Phys. Chem. B*, 108:17112–17118, 2004.
2. R. Seidel, G. S. Duesberg, E. Unger, A. P. Graham, M. Liebau, and F. Kreupl. Chemical vapor deposition growth of single-walled carbon nanotubes at 600°c and simple growth model. *J. Phys. Chem. B*, 108:1888, 2004.
3. Jannik C. Meyer, Dirk Obergfell, Shihe Yang, Shangfeng Yang, and Siegmar Roth. Transmission electron microscopy and transistor characteristics of the same carbon nanotube. *Appl. Phys. Lett.*, 85:2911, 2004.
4. B. W. Smith and E. Luzzi. Electron irradiation effects in single wall carbon nanotubes. *J. Appl. Phys.*, 90:3509, 2001.

Nanotubes and Nanofibres: Evidence for Tunnelling Conduction and Luttinger Liquid?

S. K. Goh and A. B. Kaiser

*MacDiarmid Institute for Advanced Materials and Nanotechnology, SCPS,
Victoria University of Wellington, P O Box 600, Wellington, New Zealand*

Abstract. We analyze conduction data for vanadium pentoxide (V_2O_5) nanofibres and single-wall carbon nanotubes (SWNTs) for evidence of Luttinger liquid behaviour (power laws) or fluctuation-assisted electronic tunnelling. For V_2O_5 nanofibres, the current-voltage (*I-V*) characteristics and the temperature dependence of conductance agree well with the expressions for fluctuation-assisted tunnelling and activation; the temperature dependence appears too strong to be compatible with Luttinger liquid behaviour. For SWNTs, there is some evidence for nonohmic power laws in the *I-V* characteristics (but over a limited range of voltage) and in the temperature dependence of the conductance, with exponents in agreement with the Luttinger liquid relation. However, these data also appear consistent with fluctuation-assisted tunnelling.

INTRODUCTION

The discovery and development of nanoscale materials has led to increased interest in novel conduction mechanisms, for example ballistic conduction in near-perfect individual carbon nanotubes [1], and charged soliton pair creation by tunnelling of a segment of the conjugated-bond system along a polyacetylene chain [2]. At the nanoscale, quantum tunnelling often plays a key role.

In this paper, we investigate and compare conduction mechanisms in two different nanoscale conductors: vanadium pentoxide (V_2O_5) nanofibres as measured by Kim et al. [3]), and single-wall carbon nanotubes (SWNTs) as measured by Lee et al. [4]). While the SWNTs have a diameter of $1 - 2$ nm and Luttinger liquid behaviour might be expected [5], the V_2O_5 nanofibres have a flat cross-section (1.5×10 nm) and so are not obvious candidates for Luttinger liquid behaviour (although there are suggestions that a Luttinger liquid can also occur in a two-dimensional system [6]).

For tunnelling into a Luttinger liquid, the current I and the low-field conductance G_L are expected to follow the power laws [5,7]

$$I = bV^{\beta} \quad \text{(for } eV \gg k_BT) \tag{1}$$

$$G_L = aT^{\alpha} \text{ where } \alpha = \beta - 1 \quad \text{(for } eV \ll k_BT) \tag{2}$$

where a and b are constants, e is the electronic charge and k_B is Boltzmann's constant.

CP786, *Electronic Properties of Novel Nanostructures*, edited by H. Kuzmany, J. Fink, M. Mehring, and S. Roth
© 2005 American Institute of Physics 0-7354-0275-2/05/$22.50

For fluctuation-assisted electron tunnelling through and activation over barriers between metallic regions, the simplified expressions for the I-V characteristics and the low-field conductance G_L are [8,9]

$$I \approx G_L \, V \exp(V/V_0) \qquad (3)$$

$$G_L \approx G_H \exp[-T_b/(T+T_s)] \qquad (4)$$

where V_0, G_H, T_b and T_s are taken as constants.

V_2O_5 NANOFIBRES

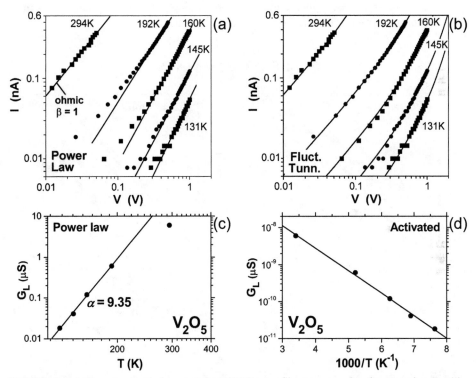

FIGURE 1. (a) Current-voltage characteristics of V_2O_5 nanofibres compared to the power law Eq. (1). (b) The same data fitted to the expression for fluctuation-assisted tunnelling and activation Eq. (3). (c) The temperature dependence of the conductance G_L as described by the power law Eq. (2). (d) G_L fitted to the expression for activated conduction (Eq. (4) with $T_s = 0$).

Kim et al. [3] deposited V_2O_5 nanofibres on an Si/SiO$_2$ substrate and then patterned Au/Pd electrodes separated by approximately 100 nm over the nanofibres. They measured the current-voltage (I-V) characteristics of seven individual V_2O_5 nanofibres between the electrodes. Their data are shown in Fig. 1(a) using logarithmic scales so that power laws are revealed by linearity of the data plots. At room temperature, approximately ohmic behaviour ($\beta \approx 1$) is observed, but as temperature is lowered the *I-V* characteristics develop nonlinear behaviour. The exponent of the power laws fitted to the data in Fig. 1(a) for larger applied voltages increases to $\beta \approx 2$ at 131 K.

The curvature of the data plots at smaller voltages in Fig. 1(a) indicates significant deviation from power law behaviour. Fig. 1(b) shows that these deviations agree well with the simplified generic behaviour Eq. (3) expected for fluctuation-assisted tunnelling and activation through or over barriers.

Fig. 1(c) shows that the low-field conductance does follow power-law behaviour below 200 K, but the very large value of 9.35 for the exponent α does not agree with the expectation Eq. (2) for a Luttinger liquid. Instead, we note from Fig. 1(d) that the temperature dependence of G_L over the whole range follows an activated behaviour, i.e. Eq. (4) with $T_s \to 0$ and activation energy $k_B T_b \sim 120$ meV). The fits in Figs 1(b) and 1(d) suggest that electronic thermal activation plays a key role in conduction in the V_2O_5 nanofibres rather than Luttinger liquid behaviour.

SUSPENDED SWNTS

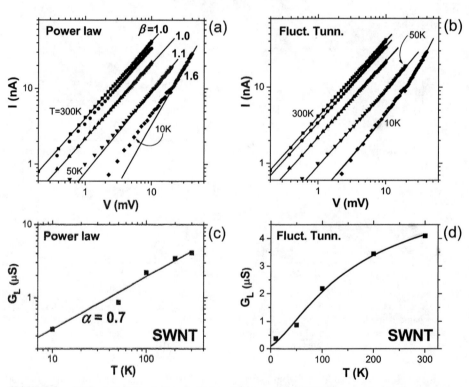

FIGURE 2. (a) Current-voltage characteristics of the suspended SWNTs compared to the power law Eq. (1). (b) The same data fitted to the expression for fluctuation-assisted tunnelling Eq. (3). (c) The temperature dependence of the low-field conductance G_L fitted to the power law Eq. (2). (d) G_L fitted to the expression for fluctuation-assisted tunnelling Eq.(4).

Lee et al. [4] measured the I-V characteristics of several SWNTs suspended in parallel between gold electrodes 300 nm apart). These data, shown in Fig. 2(a), also follow ohmic behaviour at high temperature and approximate power laws at temperatures of 50 K (exponent $\beta \approx 1.1$) and 10 K (exponent $\beta \approx 1.6$). At 10 K, the value of k_BT corresponds to 0.86 meV, so the deviation from the power law extends to voltages more than an order of magnitude greater.

Turning to the temperature variation of conductance G_L in Fig. 2(c), we find an approximate power law with exponent $\alpha \approx 0.7$, which if we use the value $\beta = 1.6$ at 10 K yields agreement with the Luttinger liquid relation between α and β in Eq. (2). Hence unlike the behaviour in the V_2O_5 nanofibres, that in the SWNTs is to this extent consistent with tunnelling into a Luttinger liquid.

However, as for V_2O_5 nanofibres, the generic nonlinearity shape for fluctuation-assisted tunnelling and activation Eq. (3) gives a good account of the nonlinearity in the I-V characteristics, as shown by the fits in Fig. 2(b). The G_L data are also consistent with fluctuation-assisted tunnelling, as shown in Fig. 2(d), with barrier energies of order $k_BT_b \sim 14$ meV.

CONCLUSION

For the SWNTs, although there is some evidence for power laws obeying the Luttinger liquid relation between exponents in Eq. (2), further data and analysis are required to distinguish between Luttinger liquid and electronic tunnelling behaviour. For the case of the V_2O_5 nanofibres, however, the temperature dependence of conductance follows activated behaviour and appears too strong to be consistent with Luttinger liquid behaviour.

ACKNOWLEDGEMENT

We thank Yung Woo Park and Gyu Tae Kim for sending original data files.

REFERENCES

1. S. J. Tans, M. H. Devoret, H. Dai, A. Thess, R. E. Smalley, L. J. Geerligs and C. Dekker, *Nature* 386, 474-477 (1997).
2. A. B. Kaiser and Y. W. Park, *Synth. Met.* 135, 245-247 (2003).
3. G. T. Kim, J. Muster, V. Krstic, J. G. Park, Y. W. Park, S. Roth and M. Burghard, *Appl. Phys. Lett.* 76, 1875-1877 (2000).
4. S. W. Lee, D. S. Lee, H. Y. Yu, E. E. B. Campbell and Y. W. Park, *Appl. Phys. A* 78, 283-286 (2004).
5. M. Bockrath, D. H. Cobden, J. Liu, A. G. Rinzler, R. E. Smalley, L. Balents and P. L. McEuen, *Nature* 397, 598-601 (1999).
6. P. W. Anderson, *Phys. Rev. Lett.* 64, 1839-1841 (1990).
7. S. V. Zaitsev-Zotov, Yu. A. Kumzerov, Yu. A. Firsov and P. Monceau, J. Phys.: Condens. Matt. 12, L303-L309 (2000).
8. A. B. Kaiser, S. A. Rogers and Y. W. Park, *Mol. Cryst. Liq. Cryst.* 415, 115-124 (2004).
9. P. Sheng, *Phys. Rev. B* 21, 2180-2195 (1980).

Electrical Characterization of Empty and Ferromagnetic Filled Multiwall Carbon Nanotubes

M. Milnera, H. Vinzelberg, I. Mönch

*Leibniz-Institute for Solid State and Materials Research Dresden,
Helmholtzstr. 20, D 01069 Dresden, Germany*

Abstract. Electron spin is not exploited in conventional semiconductor transistors. However, spin dependent devices have significant potential for data storage technology and future electronic devices. Due to their large spin-flip scattering length, carbon nanotubes (CNTs) are a promising candidate for spin dependent devices. High magneto resistivity effects were recently discovered on ferromagnetically contacted CNTs [1-2]. Ferromagnetically filled multiwall CNTs (MWCNTs) [3] represent an exciting compound material class, potentially providing excellent magnetic, electrical and mechanical properties. However, electric transport measurement results exist only on as grown two-dimensional arrays of aligned Fe-filled MWCNTs [4]. In order to realize future nanotube spintronic devices, it is necessary to understand the transport properties of both a nanotube (i.e. excluding contacts) and a transition zone between the tube and the electrode. Single tube devices were produced by using an AC-electrophoresis deposition on predefined Au- or Ti-finger structures. The measured magnetotransport data show a broad spectrum of behaviours: positive and negative magnetoresistance, oscillations and shoulders.
Keywords: Carbon nanotube, Dielectrophoresis, Magnetoresistance
PACS: 73.63.Fg

CHARACTERIZATION OF THE MWCNT RAW MATERIAL

Both the empty and filled MWCNTs were produced by CVD synthesis. The high resolution TEM image (Fig.1a) of a Fe-filled MWCNT shows a Fe core surrounded by a large number of graphene sheets [5]. The strength of the D-line in a Raman spectrum (Fig.1b) is a measure for disorder in the hexagonal structures of the graphene sheets. Due to an extreme strong D-line of the Fe-MWCNTs in comparision with other MWCNTs a transport regime for disordered systems is expected.

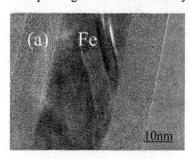
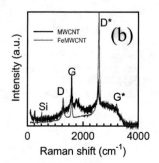

FIGURE 1. Fe-MWCNT (a). Overview Raman spectrum (b).

CP786, *Electronic Properties of Novel Nanostructures*, edited by H. Kuzmany, J. Fink, M. Mehring, and S. Roth
© 2005 American Institute of Physics 0-7354-0275-2/05/$22.50

PREPARATION OF DEVICES

The fast production of single tube devices suitable for low temperature measurements in a cryostat is a key point for the investigation of transport properties. A dielectrophoretic deposition technique (DDT) was developed to decrease the production time and to increase the amount of single tube devices arranged on a silicon-waver.

The result of a dielectrophoretic deposition is documented in Fig.2b (inversed optical dark field image). A 10V, 3 MHz sinusoidal voltage was applied to the electrodes. The tubes moved and aligned perpendicular to the electrode edges due to the force F and the angular moment M acting on the tube [6]:

$$F = p \cdot \nabla E = \frac{2\pi ab^2}{3} \frac{\varepsilon_2 - \varepsilon_1}{1 + \frac{\varepsilon_2 - \varepsilon_1}{\varepsilon_1} L_{\parallel}} \nabla E^2 \quad \text{and} \quad \vec{M} = p \times E, \tag{1}$$

where $L_{\parallel}$ is a geometrical factor depending on the aspect ratio (b/a) of the tube, ε_1 and ε_2 are the dielectric constants of the particle and the solvent (in all experiments ethanol), respectively. To understand this behavior, the field configuration of one electrode must be considered in plane (Fig.2c). The electric field is extremely enhanced at the edges leading to an effective alignment of the tubes. The contact resistances of devices with MWCNT on Au-electrodes are high and not stable, also after contamination lines written by an electron beam (Fig.2a). Low contact resistances were obtained with Ti-finger structures after a short annealing at 800°C.

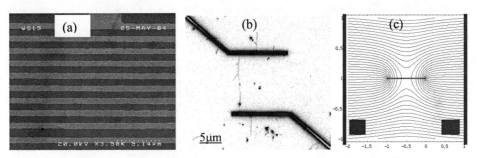

FIGURE 2. MWCNT on Au-contacts (a). Alignment in AC-fields (b). Electric field lines around an electrode cross section (black line in the middle) (c).

MAGNETOTRANSPORT OF MWCNT

A Co-filled MWCNT connected with low ohmic Ti-contact shows a smooth negative magnetoresistance (MR) (Fig.3a), which is well understood in the transport theory of disordered electron systems [7] and was fitted using

$$R = \frac{1}{1 / R(B = 0T) + \Delta G} \quad \text{with} \quad \Delta G = \frac{w}{L} \frac{e^2}{\pi h} \left(\Psi \left(\frac{1}{2} + \frac{1}{|x|} \right) + \ln|x| \right), \tag{2}$$

$$\text{where } x = 4L_\phi^2 / l_B^2, \text{ with } l_B^2 = h / eB. \tag{3}$$

w and L are the width and length of the conductor. Ψ is the digamma function and the result of the fit is the phase decoherence length L_ϕ (Fig.3b), which follows a power law: $L_\phi = cT^{p/2}$. The second fit gives $p = 0.67$ and $c = 180$.

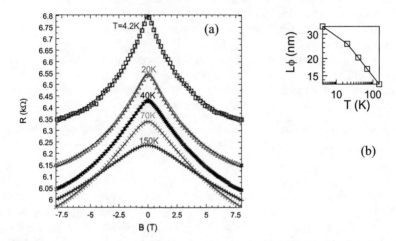

FIGURE 3. Resistance vs. magnetic flux density of a Co-filled MWCNT. The thin black line is the fit using weak localization theory and a nonlinear regression algorithm (a). Result of the fit is the temperature dependence of the phase coherence length (b).

The MR of low ohmic Ti-contacted empty MWCNTs was measured as a function of bias current and temperature (Fig.4).

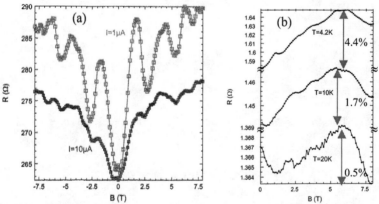

FIGURE 4. Resistance vs. magnetic flux density of an empty MWCNT for two different bias currents (device1, a). Fast disappearance of main fluctuation with increasing temperature (device2, b).

The disappearance of the fluctuation with increasing bias current (Fig.4a) and the fast disappearance of the major fluctuation with increasing temperature (Fig.4b) provides strong evidence for this effect to be interpreted as Universal Conductance Fluctuations.

The resistance of a MWCNT (e.g. of device1, Fig.5a) shows typical and characteristic logarithmic temperature dependence up to 100K as theoretically predicted for 2D electron systems:

$$\sigma_{2D} = \sigma_{2D,0} + \frac{p}{2}\frac{e^2}{\hbar\pi^2}\ln\left(\frac{T}{T_0}\right) \text{ (S)},\tag{4}$$

where p≈1. The differentiation of the R-values of device2 (Fig.5b) shows that also the fine structure (shoulders) is oscillating but not strictly periodic.

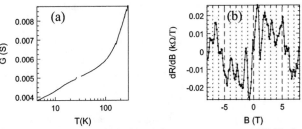

FIGURE 5. Conductance (device1, a) and Magnetoresistance spectrum at 4.2K (device2, b).

Summarizing the magnetotransport in devices with a ferromagnetically filled MWCNT, we have measured a smooth negative MR, which is also characteristic for some empty CVD MWCNT materials. The properties of the outer sheets dominate the transport. The MRs of the two investigated empty MWCNT devices show device specific Universal Conductance Fluctuations.

ACKNOWLEDGMENTS

The nanotubes raw materials were kindly provided by R. Kozhuharova and A. Leonhardt. We thank the technicians B. Eichler and S. Sieber for their contributions. The financial support by EU HPMD-CT-2000-00044 and the Austrian Research Association MOEL059 is gratefully acknowledged.

REFERENCES

1. Tsukagoshi, K., Alphenaar, B. W., Ago, H., *Nature* **401**, 572-574 (1999).
2. Zhao, B., Mönch, I., Vinzelberg, H., Mühl, T., Schneider, C. M., *Appl. Phys. Letters* **80**, 3144-3146 (2002).
3. Leonhardt, A., Ritschel, M., Kozhuharova, R., Graff, A., Mühl, T., Huhle, R., Mönch, I., Elefant, D., Schneider, C. M., *Diamond and Related Materials* **12**, 790-793 (2003).
4. Munoz-Sandoval, E., López-Urías, F., Díaz-Ortiz, A., Terrones, M., Reyes-Reyes, M., Morán-Lopez, J., *J. Magn. Magn. Mat.* **272**, 1255-1257 (2004).
5. Kozhuharova, R., Ritschel, M., Elefant, D., Graff, A., Leonhardt, A ., Mönch, I., Mühl, T., Schneider, C. M., *J. Mat. Sc.: Mat. Electronics* **14**, 798-791 (2003).
6. Jones, T. B., *Electromechanics of Particles*, Cambridge: Cambridge Univ. Press, 1995
7. Altshuler, B. L., Khmel'nitzkii, D., Larkin, A. I., Lee, P. A., *Phys. Rev. B*. **22**, 5142-5153(1980).

THE IMPORTANCE OF THE METAL-MOLECULE INTERFACE FOR CHARGE TRANSPORT: A THEORETICAL AND EXPERIMENTAL STUDY

Florian von Wrochem, Heinz-G. Nothofer, William E. Ford, Frank Scholz, Akio Yasuda, Jurina M. Wessels*

Sony Deutschland GmbH, Stuttgart Technology Center, Materials Science Laboratory, Hedelfinger Strasse 61, 70327 Stuttgart, Germany

Abstract. This contribution discusses the influence of the metal binding group on the charge transport and optical properties of percolated networks of nanoparticles interlinked with bis-dithiocarbamate or dithiol derivatives. DFT calculations show that the bis-dithiocarbamate derivatives form valence states that are delocalized on the metal and the molecule. This is a characteristic, which is classically not known from the thiol-gold interface. Conclusive experimental and theoretical evidence for the excellent electronic behavior of dithiocarbamate groups is provided.

Keywords: Charge transport, metal-molecule interface, nanoparticles, dithiocarbamate, density functional theory
PACS: 73.40.-c, 73.63.-b, 81.07.Nb, 68.43.-h, 31.15.Ew

INTRODUCTION

Substantial research work currently investigates nanoscale electronic devices based on organic molecules. However, one of the major tasks is to understand how the coupling of the molecules to the metal contacts and the molecular structure influences the current-voltage response of molecules within a contacting gap. Due to the intrinsic differences in the nature of metals and molecules, the metal-molecule interface generally resembles a barrier to charge transport and leads to a large potential drop across the interface. Most experiments, so far, have been carried out with molecules that are terminated by one or two thiol groups and are anchored to either Au or Hg. Theoretical and experimental results have shown that the Au-S bond, which has a low density of states, constitutes a pronounced barrier for charge transport.

In this contribution we provide conclusive experimental and theoretical evidence for the electronic behavior of dithiocarbamate groups, which provide an excellent electronic coupling to Au. The experimental results will be supported by ab-inio calculations at the DFT level.

CP786, *Electronic Properties of Novel Nanostructures*, edited by H. Kuzmany, J. Fink, M. Mehring, and S. Roth
© 2005 American Institute of Physics 0-7354-0275-2/05/$22.50

RESULTS AND DISCUSSION

Temperature dependent electrical and optical properties of self-assembled thin films of Au-clusters interlinked with the dithiols 1,4-bis(mercaptoacetamido)-cyclohexane (cHDMT), 1,4-bis(mercaptoacetamido)-cyclohexane (DMAAcH and 1,4-bis(mercaptoacetamido)-phenylenediamine (DMAAB), and with the bis-dithiocarbamates 1,4-cyclohexanebisdithiocarbamate (cHBDT) and disodium p-phenylenebisdithiocarbamate (PBDT) have been investigated (Table1). The three dimensional (3D) films were prepared using the layer by layer deposition process from solution onto functionalized glass substrates patterned with electrode structures. The final thickness of the as-deposited films was adjusted to an optical density of 0.3 in the plasmon band maximum resulting in a film thickness between 11 and 18 nm for the different linker molecules [1]. Temperature dependent conductivity measurements were performed in the range between 100K and 300 K at a bias of ± 1V using a HP4241 source-measure unit. The conductivity of the films was calculated from the electrode geometry, the resistance and the film thickness.

TABLE 1. Overview of molecules, with [a]S-S and [b]S⁻-S⁻ distance of the energy minimized structure.

Structure: details	Length (Å)	Structure: bis-dithiocarbamates	Length (Å)
(I) DMAAB	14.8^a	(III) PBDT	10.7^b
(II) DMAAcH	14.9^a	(IV) cHBDT	9.6^b

Theoretical calculations were performed using Density Functional Theory (DFT) at the generalized gradient approximation level with the BLYP functional as provided by the Dmol3 program [5]. A double numerical basis set extended with polarisation functions was used for all atoms except hydrogen and a real space cut-off of 4.5 Å was employed. The density functional semi-core pseudopotentials allow the explicit treatment of only the 19 valence electrons of Au, while the remaining 60 inner electrons are treated as a core. All structures including the Au_{13} cluster were first relaxed to the minimum energy structure before orbitals and energies were calculated.

Experimental Results

Pronounced differences in the optical and electrical properties of dithiol and bis-dithiocarbamate interlinked nanoparticle films were observed. The optical characteristics of the films are determined by the size of the nanoparticles (NP), their dielectric surrounding matrix and adsorbates. While the UV-VIS absorption spectrum of the di-mercaptoacetamides DMAAB and DAAcH films shows a clear plasmon resonance (554 nm and 573 nm), the peak position of which is influenced by the polarizability of the linker molecule , the films of Au-NPs interlinked with PBDT and cHBDT exhibit pronounced differences (Figure 1a). In case of PBDT, the plasmon band is nearly absent and the film shows a strong absorbance in the near infrared, similar to the absorption spectra of gold films. The plasmon resonance of the film interlinked with cHBDT is distinctively broadened and shows also a pronounced increase in the near infrared, which indicates that the phase coherence of the surface plasmon is strongly disturbed. According to the chemical interphase damping model [2], the plasmon band broadening could result from inelastic scattering processes at the cluster surfaces when excited electrons are transferred between electronic resonant states and the metal clusters. The prerequisite for the formation of electronic resonant states close to the Fermi level is the existence of an overlap between molecular orbital and metal wave functions. The strong electron withdrawing nature of the thiocarbonyl introduces a pronounced double bond characteristic to the C-N bond in the bis-dithiocarbamate moiety and inhibits the delocalization of the nitrogen electron pair into the benzene ring [3]. In addition, the electron withdrawing nature of the thiocarbonyl may also influence the electron distribution in the Au-S bond. It is anticipated that the formation of adsorbate states close to the Fermi level of Au leads to the pronounced changes in the adsorption characteristics and to a very good electrical coupling between dithiocarbamates and Au.

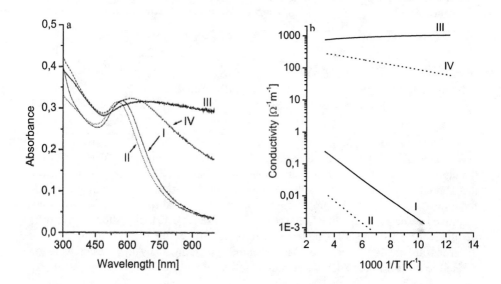

FIGURE 1. Absorption characteristics (a) and conductivity as a function of 1/T (b) of 3-D Au-NP films interlinked with the thiol derivatives DMAAB (I) and DMAAcH (II) and the dithiocarbamate derivatives PBDT (III) and cHBDT (IV).

Figure 1b shows a plot of the film conductivity as a function of T^{-1}. The pronounced differences apparent in the optical characteristics are also evident in the electrical properties of the films. Although there is only small difference in the length of the molecules, the conductivity of the films varies between the insulating limit and the metallic limit (PBDT). The films prepared from the bis-dithiocarbamates are 4 orders of magnitude more conductive than the films prepared with the dithiol derivatives. This is likely to result from the conjugated nature of the thiocarbamate group, which provides energy levels close to the Fermi level of the Au-cluster (*vide supra*). For both classes of binding groups (DMAAB and DMAAcH) and (PBDT and cHBDT), the conductivity of the film decreases by one order of magnitude upon substituting the benzene ring with a cyclohexane ring. We obtain a linear relationship when plotting the logarithm of the conductance as a function of the number of non-conjugated bonds (at the limit kT>>0) for the different molecules [1]. From the slope of the line we obtain a tunneling decay constant of $\beta=1$, which is comparable with values reported for alkane thiol SAMs [4]. This provides evidence that conjugated bonds do not contribute significantly to the tunneling decay constant, and suggests that a non-resonant tunneling process along non-conjugated parts of the molecule governs the electron tunneling decay constant β, while the contribution from the conjugated parts of the molecule are negligible corresponding to resonant tunneling through this part of the molecule.

Theoretical Results

The influence of the molecular structure and the metal-molecule interface on the conductivity of metal-cluster systems can be studied by electronic structure methods. The formation of a bond between molecule and metal leads to the hybridization of metal and molecular states, inducing a change in the electron density at the interface between metal and molecule. Valence states are expected to deliver the main contribution to the hybrid orbitals and are also mainly responsible for charge transport in metal-molecule-metal junctions. If small biases are applied, only the states closest to the Fermi energy provide a significant contribution to the net current, even if temperature dependent processes such as thermally activated electron transport can broaden the energy range of participating states.

Here, we exemplify the differences in the electronic coupling of thiols and thiocarbamates to Au, by comparing molecule-cluster systems consisting of DMAAB and PBDT attached with one end to an Au_{13} cluster. Au_{13} is the smallest "magic" cluster where the octahedral symmetry is roughly preserved upon coupling to the molecules and relaxation of the entire complex. The Na^+ counter-ion on PBDT is substituted with a H atom in order to avoid any influence of the electrostatic potential from the Na^+ ion on the electronic structure of PBDT-Au_{13}. This is a reasonable approach, since the Na^+ ion is replaced by Au during formation of the NP films.

DMAAB binds to Au as a thiolate (S^--Au) and after relaxation the most stable configuration is the coordination to Au_{13} at the threefold hollow site of the (111) plane. The S^--Au bond distance to the 3 Au atoms is 2.52 Å, 2.54 Å and 2.58 Å. respectively. The S-C bond distance is 1.89 Å, which is comparable with the S-C distance in DMAAB without Au_{13}-cluster (1.86 Å). The distance between the Au atoms participating in the Au-S bond is 3.70 Å, which is ~ 0.77 Å larger than in the unperturbed Au_{13}-cluster (2.93 Å). In contrast, the bond length between the central Au atom and the Au atoms in the plane connected to S is only slightly larger (3.07 Å) than the distance to atoms in the opposite plane (2.99 Å). This indicates that the cluster geometry is deformed mostly in the plane were the molecule is attached to the cluster.

The PBDT-Au$_{13}$ structure shows that both S atoms bind to Au. In one possible configuration (other possible minimum energy structures were not investigated here), one S atom binds at the bridge site between two Au atoms and one on top of an Au atom. The S-Au bond distances are 2.58 Å and 2.5 Å at the bridge site and 2.42 Å at the top site. The S-C distance is 1.75 Å. The distance between the Au atoms involved in the bond formation to PBDT is 3.52 Å, thus, the distortion of the Au$_{13}$ cluster is similar to the distortion obtained for the Au$_{13}$ cluster in the DMAAB-Au$_{13}$ system. The bond length between the central Au atom and the Au atoms in the plane connected to the S atoms (2.98 Å) is comparable with the distance to the opposite plane (3.01 Å).

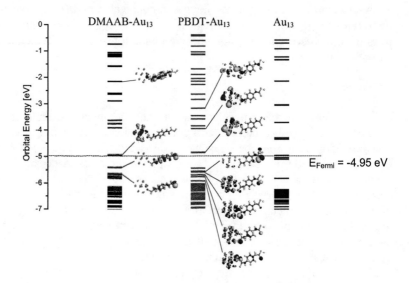

FIGURE 2. Orbital energy levels and isosurface plots for DMAAB-Au$_{13}$, PBDT-Au$_{13}$ and Au$_{13}$. The dashed line corresponds to the highest occupied orbital of the Au$_{13}$ cluster and is referred to as the Fermi level. The isosurface plots for DMAAB-Au$_{13}$ from top to bottom correspond to: LUMO+7, HOMO, HOMO-2, HOMO-3. The isosurface plots for PBDT-Au$_{13}$ from top to bottom correspond to: LUMO+4, LUMO, HOMO, HOMO-3, HOMO-4, HOMO-5, HOMO-6, HOMO-7. The isosurface plots of orbitals that are mainly localized on the Au$_{13}$ cluster are not shown.

In Fig. 2, the molecular orbital energy levels of DMAAB-Au$_{13}$, PBDT-Au$_{13}$ and Au$_{13}$ are presented. The Fermi level (-4.95eV) was defined as the energy level of the HOMO of the bare Au$_{13}$ cluster, even though such a cluster is only a rough approximation of a nanoparticle with a diameter of 5 nm, containing thousands of Au atoms. Molecular orbital isosurfaces that show a significant contribution on the molecular backbone were assigned to the orbital energy levels; the other levels are mainly cluster states. The isosurface plots of DMAAB-Au$_{13}$ show that the molecular orbitals and the cluster states do not undergo a significant mixing, indicating that the formation of DMAAB-Au$_{13}$ hybrid orbitals is suppressed. The DMAAB-Au$_{13}$ HOMO consists mainly of cluster states with only a small S 3p contribution from the thiol group. The HOMO-2 is solely localized on DMAAB, while the HOMO-3 shows a slight hybridization where a π orbital on DMAAB is combined with a small 5d contribution on Au. These results are in agreement with the literature and denote that the Au-S bond constitutes a barrier for charge transport.

In contrast, for most PBDT-Au$_{13}$ orbitals around the Fermi level the electron density is delocalized over the whole cluster-dithiocarbamate system (see isosurfaces in Fig. 2). The HOMO is a cluster state with S 3p contribution from the dithiocarbamate group. From HOMO-4 to HOMO-7 all orbitals show a strong mixing of the Au 5d states with molecular orbitals localized on the binding group of PBDT (in particular on N, C, and both S atoms). For HOMO-3 the mixing is weaker. HOMO-6 and HOMO-7 show evident contributions from the π orbitals of the benzene ring, the dithiocarbamate group and the cluster states. In conclusion, this suggests that the excellent electronic coupling of the PBDT derivative to Au is a result of the pronounced delocalization of the occupied orbitals.

It is further pertinent to note that, at small bias, the conductance for DMAAB and PBDT is probably governed by charge transport through occupied states, since the higher density of states and degree of conjugation occurs in the valence region below the Fermi level. For DMAAB-Au$_{13}$ the only unoccupied state with a contribution from a molecular orbital is the LUMO+7 at -2.15eV. This is 2.8eV above the HOMO of DMAAB-Au$_{13}$ and hence not available for conduction. PBDT-Au$_{13}$ shows unoccupied conjugated states which are closer to the Fermi level, however, still not as close as all the occupied orbitals from HOMO to HOMO-7. These orbitals are delocalized due to various contributions from Au$_{13}$, binding group and conjugated benzene ring. This, and the weak mixing of DMAAB molecular orbitals with Au13 cluster states, is probably the reason why Au-NP films interlinked with PBDT are significantly more conductive than Au-NP films interlinked with DMAAB *(vide infra)*.

In summary, these investigations show that the dithiocarbamate group provides an excellent electronic contact to Au, which is a result of the pronounced hybridization of dithiocarbamate and cluster states.

ACKNOWLEDGMENTS

This work was partially supported by the European Commission within the BIOAND project IST-1999-11974.

REFERENCES

1. Wessels, J. M., Nothofer, H.-G., Ford, W.E., von Wrochem, F., Scholz, F., Vossmeyer, T., Schroedter, A., Weller, H., Yasuda, A., *J. Am. Chem. Soc.*, **126**, 3349-3356 (2004)
2. Persson, N. J., *Surf. Sci.,* **172**, 153-162 (1993).
3. Humeres, E., Debacher, N.A., Franco, J.D., Lee, B., Martendal, A., *J. Org. Chem.*, **67**, 3662-3667 (2002).
4. Holmlin, R.E., Ismagilov, R.F., Haag, R., Mujica, V., Ratner, M.A., Rampi, M.A., Whitesides, G.M., *Angew. Chem. Int. Ed.*, **40**, 2316-2320 (2001).
5. DMol3, Density Functional Theory Electronic Structure Program, Accelrys Inc, 2004.

Highly nonlinear transport phenomena in fullerene based diodes

G. J. Matt*, T. Fromherz†, H. Neugebauer* and N. S. Sariciftci*

*LIOS, Physical Chemistry, Johannes Kepler University Linz, Altenbergerstraße 69, A-4040 Linz, Austria
†Institute for Semiconductor and Solid State Physics, Johannes Kepler University Linz, Altenbergerstraße 69, A-4040 Linz, Austria

Abstract. Voltage–current measurements of fullerene based diodes in the temperature range between $295 - 15\,K$ are presented. At temperatures below 95 K and at high current densities the diodes exhibit a voltage instability with a voltage hysteresis for opposite current sweep directions. This observation is interpreted with the formation of highly conductive current filaments in the fullerene film.

Keywords: Fullerenes, electronic transport, hopping transport, mobility edges
PACS: 71.20.Tx, 71.23.An, 71.23.Cq, 71.30.+h, 72.20.Ee, 73.61.Wp

INTRODUCTION

Fullerenes and derivatives of fullerenes exhibit a very special case of organic compounds which may possess semiconducting, metallic and even superconducting properties [1, 2, 3]. As in most organic compounds the mobility of the free charge carriers is rather low and is for spin cast fullerene films in the order of $10^{-3}\,cm^2/Vs$ [4]. The hindered transport can be described with capturing and thermal remission of the charge carries in and out of localized states (traps) [6]. According to this model the effective measured mobility is expected to decrease with decreasing temperature. In contrast to these considerations we report the conversion of fullerene films into a highly conductive state at low temperatures and high current densities in fullerene based diodes.

EXPERIMENTAL

Fullerene based diodes were prepared as follows: Indium–tin–oxide (ITO) covered glass substrates were cleaned in three steps with toluene, acetone and methanol in an ultra-sonic bath. An approximately 70 nm thick film of poly(3,4–ethylenedioxythiophene/poly(styrene sulfonate) (PEDOT:PSS) (Bayer AG) as metallic hole conductor was spin-coated on the ITO layer under ambient conditions. A methanofullerene [6,6]–phenyl C_{61}–butyric acid methyl ester (PCBM) film was then spin–coated from chlorobenzene solution (3% wt.) on top of the PEDOT:PPS covered substrate, resulting in a PCBM film of $\approx 150\,nm$ thickness. A top contact was evaporated on the PCBM film under dynamic vacuum ($p = 7 \cdot 10^{-6}\,mbar$) through a shadow mask and consists of a stack of lithium–fluoride (LiF) (0.7 nm), Al (70 nm) and Au (100 nm).

CP786, Electronic Properties of Novel Nanostructures, edited by H. Kuzmany, J. Fink, M. Mehring, and S. Roth
© 2005 American Institute of Physics 0-7354-0275-2/05/$22.50

Voltage–current $(V - I)$ measurements via applying an certain current density and measuring the resulting voltage drop between the ITO and the LiF/Al/Au electrodes have been performed in the temperature range from room temperature to 15 K.

RESULTS AND DISCUSSION

From the device structure and the energy–level diagram given in Fig. 1, a built–in voltage V_{bi} of ≈ 1 V is estimated from the energy difference between the lowest unoccupied molecular orbital (LUMO) of PCBM at -4.2 eV [5] and the Fermi level of PEDOT:PSS at -5.2 eV [5]. The low energy of the highest occupied molecular level (HOMO) of PCBM (-6.1 eV) [5] results in a large energy barrier of 0.9 eV between the Fermi level in the PEDOT:PSS contact and the HOMO of PCBM. This energy barrier effectively suppresses hole injection from PEDOT:PSS into PCBM. Therefore, for biasing the diode in forward direction (i.e. biasing the PEDOT/ITO contact positive with respect to the LiF/Al/Au contact) only an electron current is injected from the LiF/Al/Au contact into the LUMO of PCBM and no hole current has to be considered for this device.

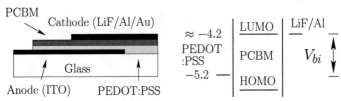

FIGURE 1. Geometric structure and energy-level diagram under flat band conditions for the PCBM based diode.

The current density for positive bias voltages features a power law dependency versus the effective applied bias voltage $V_{eff} = V - V_{bi}$ (see Fig. 2). Such an $I - V$ behavior is due to the formation of a space charge upon electron injection (space charge limited currents – SCLC) [6].

$$I \propto \theta \mu V_{eff}^2; \theta = n_f(E_f, 1/T)/n_t(E_f); E_f = f(V). \qquad (1)$$

θ is the fraction of free to trapped charge–carrier concentrations n_f/n_t, μ is the microscopic mobility, T the absolute temperature and E_f is the average Fermi level in the PCBM film. At room temperature θ is a voltage independent value and the current density is proportional to the square of V_{eff} (see Fig. 2a). Lowering the temperature θ becomes voltage dependent and the current density is proportional to V_{eff} with a higher power than 2 (e.g $I \propto V^{3.3}$ at $T = 182$ K in Fig. 2). Below 95 K the regime in the $V - I$ dependence becomes temperature independent at high current densities (indicated by a ($*$) in Fig. 2a and Fig. 3). Below 61 K the $V - I$ dependence features a voltage instability with a hysteresis for opposite current sweep directions (see Fig. 3). The hysteresis loops become more pronounced at lower temperatures and extend over larger area in the $V - I$ diagram (see inset in Fig. 3).

We interpret this observation as being due to a voltage dependent Fermi energy in PCBM and the formation of highly conducting current filaments between the diode contacts at low temperatures and high current densities [7]. The total number of trapped charges N_t, contributing to the space charge, is proportional to the applied bias voltage given by $N_t e = C \cdot V_{eff}$, where e is the elementary charge and C is directly related to the geometric capacitance. With applied bias voltage the trapped charges shift the Fermi level from E_{f0} (the Fermi–level when no bias voltage is applied) within the DOS of the traps (T–DOS) closer to E_c (the bottom edge of the conduction band; see Fig. 2b).

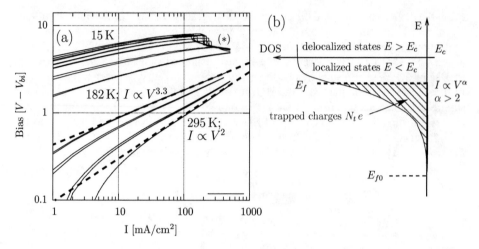

FIGURE 2. (a) Temperature dependent voltage–current density characteristics in double-logarithmic presentation. The dashed lines represent the $I - V$ power law dependency at two temperatures. The region with temperature independent $V - I$ characteristics is indicated by a ($*$). (b) Density of states and the respective average Fermi level in the PCBM film.

As the energy of the Fermi level is increasing with increasing bias voltage, θ becomes voltage dependent and the current density is proportional to V_{eff} with a higher power than 2. At low temperatures and high current densities the Fermi level reaches E_c, the free charge carrier concentration is sharply increasing and the fullerene film forms a highly conductive state. This transition causes a voltage instability in the $V - I$ diagram. As the value of E_f is directly related to the T–DOS and the T–DOS is nonuniform along the PCBM film, E_f and the free charge–carrier concentration is also nonuniform. These inhomogeneities lead to the formation of highly conductive filaments in the fullerene film. The process is meta-stable featuring a hysteresis for opposite current sweep directions.

SUMMARY

We have observed an anomalous voltage–current density behavior of fullerene based diodes at low temperatures. The process is interpreted in terms of space–charge limited currents which alters the Fermi level close to the conduction band of the fullerene film. At low temperatures and high current densities the Fermi level reaches the conduction band and the fullerene film forms highly conductive filaments. This filamented state,

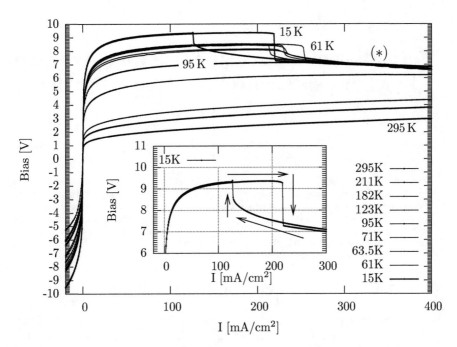

FIGURE 3. $V - I$ characteristics at various temperatures T from 295 to 15 K. The inset shows the hysteresis at 15 K in expanded scale. The region with temperature independent $V - I$ characteristics is indicated by a $(*)$.

which is metastable, shows a hysteresis for opposite current sweep directions and leads to a temperature independent $V - I$ behavior below 95 K and high current densities.

REFERENCES

1. M. S. Dresselhaus, G. Dresselhaus, and P. C. Eklund, *Science of fullerenes and carbon nanotubes*. Academic Press, Inc., (1995).
2. K. Tanigaki, I. Hirosawaand, T.W. Ebbesen, J. Mizuki, Y. Shimakawa, Y. Kubo, J. S. Tsai, and S. Kuroshima, *Nature* **356**, p. 419 (1992).
3. H. Sitter, A. Andreev, G. Matt, and N. S. Sariciftci, *Mol. Cryst. Liq. Cryst.* **385**. p. [171]/51 (2002).
4. G. J. Matt, T. Fromherz, and N. S. Sariciftci, *Appl. Phys. Lett.* **84**, 9, p. 1570 (2004).
5. C. Winder, D. Mühlbacher, H. Neugebauer, N. S. Sariciftci, C. Brabec, R. Janssen, and J. K. Hummelen, *Mol. Cryst. & Liq. Cryst.* **385**, p. 93 (2002).
6. A. Rose, *Phys. Rev.* **97**, p. 1538 (1955).
7. S. R. Ovshinsky, *Phys. Rev. Lett.* **21**, p. 1450 (1968).

Formation of Molecular-Scale Gold Nanogap Junctions via Controlled Electromigration

G. Esen, S.A. Getty, M.S. Fuhrer

Department of Physics and Center for Superconductivity Research, University of Maryland, College Park, MD 20742-4111, USA

Abstract. We have used a feedback method to control the electromigration of gold nanowires to form nanogap junctions, similar to recent reports [1, 2]. We have studied gold wires fabricated with conventional electron-beam and photolithography on thick SiO$_2$ and thin Al$_2$O$_3$ (over Al) substrates. We find that the role of feedback mechanism is to control the local temperature of the electromigrating nanowire, and controlled breaking of nanowires is possible only when the wire is in good thermal contact to either the substrate or the bulk metal leads.

INTRODUCTION

Electromigration (EM) is the electrical current-induced diffusion of atoms in a thin metal film, and is important as a serious mode of failure in integrated circuit interconnects in the semiconductor industry [3]. Failure of a narrow wire due to EM has been utilized extensively to prepare stable electrical contacts for single-molecule experiments [4] which rely on preparing a nanogap in the size of the molecule whose conductance properties are being measured. Although this technique is used extensively, the mechanism behind the formation of the nanogaps and how it depends on metal/substrate interaction is not well understood. When increasing current in a wire, two processes drive EM: the electric field increases linearly with the current, and the temperature increases due to Joule heating. The rate of electromigration depends on both the electric field and the temperature, depending on electric field as a power law, and temperature exponentially (since electromigration is electric field-biased motion of thermally-activated atoms). In order to properly design wires for electromigration to form nanogaps, it is important to understand whether the temperature or electric field dominates the EM process.

EXPERIMENTAL

For single-molecule experiments, it is advantageous to have a third gate electrode nearby the gap junction. This has been accomplished previously in two different configurations; in the first configuration the EM junctions are fabricated on top of thick (100-500 nm) SiO$_2$ gate dielectric over the degenerately-doped Si [4] gate, whereas in the second configuration the junctions are fabricated on a local gate consisting of a narrow Al wire with the thin (3-10 nm) native Al$_2$O$_3$ as gate dielectric [5].

CP786, *Electronic Properties of Novel Nanostructures*, edited by H. Kuzmany, J. Fink, M. Mehring, and S. Roth
© 2005 American Institute of Physics 0-7354-0275-2/05/$22.50

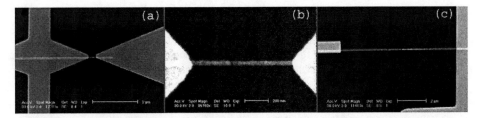

FIGURE 1. SEM microgrographs of the three different device geometries. Part (a) shows a short Au wire on SiO_2 substrate where the wire is 325 nm long, 80 nm wide and 30 nm thick. Part (b) shows an Au wire on Al_2O_3 (over Al) substrate where the wire is 900 nm long, 40 nm wide and 15 nm thick. Part (c) shows a long gold wire on SiO_2 substrate where the wire is 8 micrometers long, 100 nm wide and 30 nm thick. In both parts (a) and (c) the scale bar is 2 micrometers whereas in part (b) the scale bar is 200 nm.

We have fabricated and studied electromigration in three different device geometries using these two configurations, in order to vary the thermal coupling to the substrate and/or the bulk metal leads (see Figure 1): (a) Short Au wire on SiO_2 (over Si) substrate (good thermal coupling to the bulk metal leads), (b) Au wire on Al_2O_3 (over Al) substrate where due to thin Al_2O_3 oxide layer and high thermal conductivity of the Al this geometry has good thermal coupling to the substrate, and (c) long Au wire on SiO_2 (over Si) substrate (poor thermal coupling of the Au wire to the bulk metal leads and substrate). We expect that the thermal conductance of the center of the wire to the substrate is greatest in (a) and least in (c) (we roughly estimate the thermal conductance along the wire in (b) as comparable to that through the Au/Al_2O_3 interface; hence (a) has the greatest thermal conductance because of its geometry).

In fabrication of our devices we did not use a wetting layer for the gold deposition. To control the electromigration process we used a computer-controlled feedback scheme similar to ref. 2, consisting of the following steps: We first measure a reference conductance value, then we increase the voltage until the conductance drops by typically 2 to 5 percent from the reference conductance value. When we observe such a conductance drop we decrease the voltage and measure the new reference conductance value. We repeat this process until the desired conductance is reached. Experiments were performed in a gas flow ^{4}He cryostat at substrate temperatures from 1.3 K to room temperature (RT).

RESULTS AND DISCUSSION

A representative current vs. voltage (I-V_{bias}) curve taken through a feedback controlled electromigrating process for a short Au wire in SiO_2/Si [geometry (a)] is shown in Figure 2 below. The data labeled A show a smooth I vs. V_{bias} curve indicating that electromigration has not begun in the gold wire. Although there is a resistance increase with increasing bias in A, we found that if we stop the voltage bias before the wire begins to lose material due to electromigration, this resistance increase is reversible. Such a reversible resistance increase shows that the gold wire heats up before electromigration begins. The data labeled B show that after this initial heating the gold wire begins to lose material due to EM and its resistance increases irreversibly. The data labeled C show that one can stop and restart the voltage bias before the gold wire totally fails. I-V_{bias} curves of two bias processes perfectly match each other indicating that in the second biasing process the gold wire first heats up to the temperature where significant electromigration process takes place, and restarts to dissipate material.

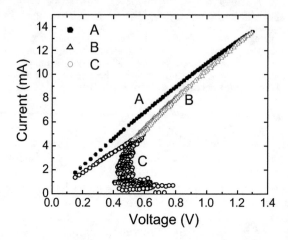

FIGURE 2. Current vs. bias Voltage curve of a feedback controlled electromigration process of an Au wire that has geometry (a) of Figure 1 at T=1.3 K. Part A is a smooth curve indicating than the EM has not begun whereas in part B the resistance of the line increases irreversibly due to EM. Both parts A and B are recorded in a single voltage biasing process, producing a final resistance of ~120 Ω. At this point the voltage was reduced to zero for some time. When the bias process was restarted in C, the wire resistance is the same, demonstrating that the EM process may be frozen by turning off the voltage.

We achieved this kind of feedback-controlled electromigration behavior only in device geometries (a) and (b), i.e. short Au wire on SiO_2/Si and Au wire on Al_2O_3/Al. We could not achieve same kind of controllability with geometry (c), a long Au wire over SiO_2/Si substrate. Although we observed that feedback control of electromigration requires good thermal heat-sinking of wires either to the substrate or to the bulk of the leads (Such as device geometries (a) and (b) in Figure 1), we did not observe any dependence on the temperature of the environment. Feedback control of electromigration was possible from 1.3 K to RT. Poorly heat-sunk wires failed abruptly, leading to large gaps (greater than 5 nm, as measured by SEM) which had immeasurably small conductance.

We calculated the resistance increase due to electromigration by modeling it as a serial resistor with the leads where the resistance due to electromigration, R_{EM}, is changing with applied bias as the wire electromigrates. Figure 3 shows the voltage drop at the junction, $V_{junction}$, after the voltage drop at the leads subtracted from the applied bias voltage vs. R_{EM} for two devices on which we performed feedback-controlled electromigration. The solid lines in both Figure 3a and 3b indicate slopes with R_{EM} proportional to $V_{junction}^2$. In our devices the increase of the resistance due to electromigration is proportional to $V_{junction}^n$ where $n = 2 \pm 0.5$ indicating roughly constant power dissipation during electromigration. It can also be seen that the power is greater in (a) than in (b), as expected considering the greater thermal conductance of the wire in (a). If the EM process were primarily bias-controlled, we would expect that $V_{junction}$ would be roughly constant, which is certainly not the case; the electromigration process occurs over nearly two orders of magnitude of $V_{junction}$.

In conclusion we were able to perform feedback controlled electromigration in gold wires with good thermal heat sinking over a temperature range from 1.3 K to RT in a gas flow ^{4}He cryostat. We observed that current-voltage relationship when EM occurs is roughly $I \sim V_{junction}^2$, indicating roughly constant temperature during EM and the EM process is temperature-controlled rather than bias-controlled.

This work was supported by the U.S. DOE under grant DOE-FG02-01ER45939, and the DCI Postdoctoral Fellowship Program.

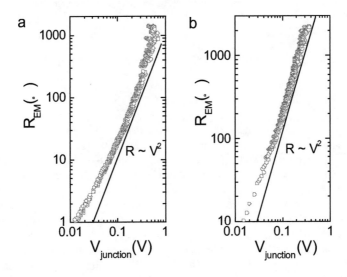

FIGURE 3. Voltage drop at the junction vs. the resistance increase of the junction due to electromigration, R_{EM}. (a) shows R_{EM} vs. $V_{junction}$ for geometry (a), short Au wire on SiO$_2$ substrate, whereas (b) shows R_{EM} vs. $V_{junction}$ for geometry (b), Au wire on Al$_2$O$_3$ substrate. Solid lines show the slope for R_{EM} proportional to $V_{junction}^2$.

REFERENCES

1. Houck *et.al.*, cond-mat/0410752 v1.
2. D. R. Strachan, *et al.*, *Appl. Phys. Lett.* **86**, 043109 (2005).
3 J. R. Lloyd, *Semicond. Sci. Technol.* **12**, 1177-1185 (1997), and the references therein.
4. H. Park *et al.*, *Appl. Phys.Letters* **75**, 301-303 (1999).
5. W. Liang *et al.*, *Nature* **417**, 725-729 (2002).

Organic Spin-Valves: Physics and Applications

Z. Valy Vardeny, Z. H. Xiong, Di Wu, Fujian Wang, and Jing Shi

Physics Department, University of Utah
Salt Lake City, Utah, 84112, USA

Abstract. Spin-valve devices of organic semiconductors in the vertical configuration using a variety of exotic and regular ferromagnetic electrodes were fabricated and studied as a function of applied magnetic field, temperature and applied bias voltage. These devices show that spin polarized carriers can be injected from ferromagnetic electrodes into organic semiconductors and diffuse without loss of spin polarization for distances of the order of 100 nm at low temperatures.

Keywords: Organic semiconductors, spin-valves, spin polarized carrier injection, spin-polarized transport.
PACS: 75.25.Dc; 72.25.Hg; 72.25.Rb; 72.80.Le.

INTRODUCTION

The discoveries of giant magnetoresistance (GMR), colossal magnetoresistance, and tunneling magnetoresistance have not only generated a great deal of excitement in condensed matter physics and materials sciences, but also found their applications in magnetic recording and memory technologies [1–3]. Spin-valve read heads and magnetic tunnel junction based random access memory devices are two examples of such applications. For both high-density recording and high-density non-volatile memory, incorporating semiconductor materials into the existing spintronic devices is highly desirable, since it would transform the usually passive devices into active devices. Semiconductor Spintronics can offer many other potential applications in information processing, transmission and storage [4, 5]. But to realize these potentials, efficient means of injecting spin-polarized charges from metallic or semimetallic electrodes into semiconductors must first be demonstrated. Spin injection from ferromagnetic (FM) metals into nonmagnetic metals has been well studied and documented [2, 3]. However spin injection by electrical means from FM into semiconductors remains a challenge.

Schmidt et al. [6] have pointed out that the basic obstacle for spin injection from a FM into a semiconductor originates from the conductance mismatch between the two materials. Recently Rashba [7], Smith and Silver [8] and Albrecht and Smith [9] have shown that the conductance mismatch problem could be circumvented if the injection occurs via tunneling. Since charge injection from metallic electrodes into organic semiconductors (OSEC) occurs mainly through tunneling [10], OSEC seem to be promising alternatives for semiconductor Spintronics [11]. In addition to efficient spin-polarized injection, a long spin relaxation time is also needed for spin transport applications in the transport layer. The building block atoms of OSEC are light (i.e. having low atomic number Z) with very small spin-orbit coupling. Moreover the π-electron wave function has zero amplitude on the nucleus sites, thereby minimizing the

CP786, *Electronic Properties of Novel Nanostructures*, edited by H. Kuzmany, J. Fink, M. Mehring, and S. Roth
© 2005 American Institute of Physics 0-7354-0275-2/05/$22.50

effect of hyperfine interaction. These unique properties show that OSEC may be rather effective for molecular Spintronics applications. In addition, OSEC have the potential to bring novel functionalities that do not exist in inorganic spintronic devices. One such functionality is the very efficient light emitting capability of OSEC. Here we review our recent work [12] where we demonstrated the first organic semiconductor spin-valve based on the small molecule tris-(8-hydroxyquinoline) aluminum (Alq$_3$).

DEVICE FABRICATION AND MEASUREMENTS

A vertical spin-valve device consists of three layers, two ferromagnetic electrodes (FM1 and FM2) and a non-magnetic spacer (Fig. 1a). By engineering the FM electrodes to have different coercive fields, the magnetizations in FM1 and FM2 can have either parallel or anti-parallel alignment in different magnetic field ranges. We have chosen Alq$_3$ as our OSEC spacer material in the spin-valves, since it can be easily evaporated and integrated with other electrode materials. The bottom ferromagnetic electrode (FM1) was a 100 nm-thick La$_{1/3}$Sr$_{2/3}$O$_3$ (LSMO) film, grown epitaxially on a LaAlO$_3$ substrate. LSMO is believed to be a half-metallic ferromagnet that possesses near 100% spin polarization [13] (For a recent review of colossal magnetoresistance see Ref. [14]). Unlike metallic films, the LSMO films are already stable against oxidation. In fact, our LSMO films were cleaned and reused multiple times without any degradation. The OSEC film (Alq$_3$) was then thermally evaporated onto the LSMO film, followed by Co film (FM2) evaporation in the same vacuum chamber, using a shadow mask. The active device size was about 2x3 mm^2. The OSEC film thickness ranged from 130 nm to 260 nm (Fig. 1b). The magneto-resistance (MR) of the obtained spin-valve device was measured in a close-cycled refrigerator from 11 to 300 K by sending a constant current through it via the two interfaces, while varying an external in-plane magnetic field, H (Fig. 1c). The hysteresis loops of the magnetization vs. H for the FM electrodes were measured by the magneto-optic Kerr effect (MOKE) over the same temperature range.

A schematic band diagram of a typical LSMO/Alq$_3$/Co device is shown in Fig. 1c. In the rigid band approximation, namely without taking into account the relaxation and polarization energy associated with charge injection, the highest occupied molecular orbital (HOMO) of Alq$_3$ lies about 0.9 eV below the Fermi levels, E$_F$ of the FM electrodes, whereas the lowest unoccupied molecular orbital (LUMO) lies about 2.00 eV above E$_F$. At low applied bias voltages, V, holes are injected from the anode into the HOMO level of the OSE mainly by tunnelling through the bottom potential barrier. In addition, the similar work function value, ϕ of the two electrodes (Fig. 1c) leads to a symmetric I-V response (Fig. 1d). For fabricated devices with d > 100 nm we found that the I-V characteristic was non-linear with a weak temperature dependence (Fig. 1d), indicative of carrier injection by tunnelling. Control devices with similar OSE thickness, in which ITO replaced the LSMO bottom electrode, showed electro-luminescence (EL) and also a conductivity detected magnetic resonance at g $\approx$ 2, indicating carrier injection into the OSE. In addition, at low bias voltages we measured a typical resistance of 10^4-10^5 Ω that depends on the deposition rate and thickness of the Co electrode; such resistance is also consistent with a dominant pinhole-free organic spacer. Devices with d < 100 nm, however showed a linear I-V response and lack of EL, leading us to believe that theses devices have an 'ill-defined' layer of up to 100 nm that may contain pinholes and Co inclusions. These findings suggest that the OSE spacers in the spin-valve devices fabricated with d >100 nm may be composed of two sub-layers: one sub-layer with a thickness $d_0 \sim 100$ nm thick immediately below the Co electrode that contains Co inclusions due to the inter-diffusion; and a second sub-layer of neatly deposited Alq_3 between this defected sub-layer and the LSMO film having a thickness $d-d_0$, in which carrier transport is dominated by carrier drift.

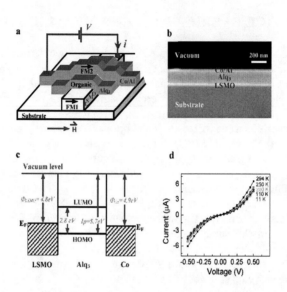

FIGURE 1. The structure and transport properties of the fabricated organic spin-valve devices. **a,** Schematic representation of a typical device that consists of two FM electrodes (FM_1 and FM_2) and an OSE spacer. Spin-polarized electrical current, I flows from FM_1 (LSMO), through the OSE spacer (Alq_3), to FM_2 (Co) when a positive bias, V is applied. **b,** Scanning electron micrograph of a functional organic spin valve consisting of 60-monolayer thick LSMO film, 160 nm Alq_3 spacer, 3.5 nm Co and 35 nm thick Al electrode. **c,** Schematic band diagram of the OSE device showing the Fermi levels and the work functions of the two FM electrodes, LSMO and Co, respectively, and the HOMO-LUMO levels of Alq_3. **d,** I-V response of the organic spin-valve device with d = 200 nm at several temperatures.

RESULTS AND DISCUSSION

Fig. 2a shows a typical magnetoresistance loop obtained in an LSMO/Alq_3/Co spin-valve device with d = 130 nm; the magnetoresistance curves of three other devices having larger d were also measured and their figure of merit $\Delta R/R$ is summarized in Fig. 2b. A sizeable $\Delta R/R$ of 40%, which is a giant magnetoresistance (GMR) response comparable to that obtained in metallic GMR spin-valves, is observed at 11 K. The GMR of the devices with larger d is progressively smaller, but still measurable up to d = 250 nm (Fig.

2b). MOKE measurements performed on the LSMO bottom electrode of the device in Fig. 1a indicate that the coercive field of the LSMO film is $H_{c1} \approx 30$ Oe at 11 K. Whereas the top Co electrode of this device is not accessible to MOKE due to the Al top contact, nevertheless the coercive field of a Co film of the same thickness deposited on Alq$_3$ under similar conditions was measured to be $H_{c2} \approx 150$ Oe at 11 K; much greater than that of the LSMO. Clearly then, the magnetization orientations in the two FM electrodes are anti-parallel to each other when the external field H is between H_{c1} and H_{c2}; in contrast, the magnetization orientations are parallel to each other when the field strength $H > H_{c2}$. Therefore, the observed GMR hysteresis is undoubtedly due to the spin-valve effect. We note that the resistance in the anti-parallel alignment is lower than that in the parallel alignment, which is opposite to the spin-valve effect usually obtained using two identical FM electrodes, or two different "d-band" metallic FM electrodes in some devices. This "inverse magnetoresistance" was also previously seen in LSMO/SrTiO$_3$/Co and LSMO/Ce$_{0.69}$La$_{0.31}$O$_{1.845}$/Co magnetic tunnel junction devices having extremely thin insulating spacers (~2 nm).

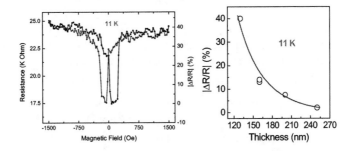

FIGURE 2. (a) The MR response of a LSMO/Alq$_3$/Co spin-valve at 11K showing a GMR response of ~ 40%. The blue (red) curve was measured in the forward (backward) magnetic field sweep direction. **(b)** The spin-valve figure of merit $\Delta R/R$ of spin-valves fabricated on the same LSMO film but with different Alq$_3$ OSEC layer thickness. The line through the data points is a theoretical fit as explained in the text.

We analyzed the obtained GMR effect and its dependence on d using a simple injection and diffusion model [12]. In the present organic spin-valves, the neatly deposited OSE sub-layer with thickness (d-d$_0$) is actually so thick (>30 nm) that simple quantum mechanical tunneling through it is not a viable possibility. Although the detailed physical picture is lacking at the moment, nevertheless we assume that there exists a potential barrier for spin injection at the Co/OSE interface (Fig. 1c), which may be self-adjusted [11]. Once carriers are injected through this interface they easily reach the neat sub-layer, where they drift under the influence of the electric field toward the other interface from which they can be extracted. As the injected carriers reach the end of the ill-defined sub-layer, the spin polarization is p$_1$; it further decays in the remaining neatly deposited sub-layer with a surviving probability exp[-(z-d$_0$)/λ_S], where z is the drift/diffusion distance along the normal direction to the interface, and λ_S is the spin diffusion length in the neatly deposited OSE sub-layer. The spin polarization p is defined

541

as the ability of the FM electrode to inject carriers with aligned spins. We express the thickness dependence of the GMR magnitude, $\Delta R/R$, which is the maximum relative change in electrical resistance, R within the spin-valve hysteresis loop, assuming no loss of spin memory at the interfaces due to the self-adjusting capability of the OSE [11] using Eq.(1) as given below [12]

$$\frac{\Delta R}{R} = \frac{R_{AP} - R_P}{R_{AP}} = \frac{2p_1p_2e^{-(d-d_0)/\lambda_s}}{1 + p_1p_2e^{-(d-d_0)/\lambda_s}}$$

where R_{AP} and R_P denote R in the anti-parallel and parallel magnetization configurations, respectively, and $(d-d_0)$ is the thickness of the neatly deposited OSE sub-layer. For inverse magnetoresistance, $R_{AP} < R_P$; therefore $\Delta R/R$ is negative according to Eq. (1). We used Eq. (1) with three adjustable parameters, namely p_1p_2, d_0 and λ_S to fit the data for $\Delta R/R$ in Fig. 2b. The good agreement shown in Fig. 2b was obtained with the following parameters: $p_1p_2 = -0.32$; $d_0 = 87$ nm, which is close to the lower limit of d below which the I-V response becomes linear; and $\lambda_S = 45$ nm, which is a reasonable value for the spin diffusion length in the neatly deposited OSE sub-layer. The obtained λ_S is smaller than what was extracted for T_6 (ref. 12), possibly due to the aluminium element in Alq$_3$ that may increase the spin-orbit coupling in this OSE. However, λ_S is similar to what was extracted from spin valve devices of single-walled carbon nanotubes.

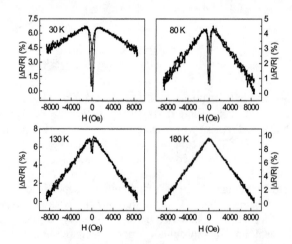

FIGURE 3. The magnetoresistance response of the device in Fig. 2a at different temperatures from 30 K to 180 K.

We measured the dependences of $\Delta R/R$ of the organic spin-valve devices on the applied bias voltage, V and temperature, T. We found that $\Delta R/R$ is in fact asymmetric with the bias V; the reason for that is not clear at the present time. $\Delta R/R$ also strongly depends on the temperature, T. A series of GMR hysteresis is shown in Fig. 3. The spin-valve effect diminishes with T and actually disappears at T = 180 K. The temperature dependence of

$\Delta R/R$ does not qualitatively follow that of the magnetization of the LSMO itself; $\Delta R/R$ vanishes at T < T_c of the LSMO (~ 300 K). We conclude that the spin-valve decrease is due to the decrease of λ_S with T, which has a stronger T dependence than that of the LSMO magnetization.

At elevated temperatures we found an additional interesting effect; an increase in high-field magnetoresistance (HFMR) accompanies the decrease of the low-field spin-valve related GMR (Fig. 3). We have also observed a similar but much smaller HFMR in the LSMO film itself. However the spin-valve device resistance is several orders of magnitude greater than that of the LSMO electrode. Therefore, the HFMR cannot be simply explained by the change in the serial resistance of the LSMO electrode. We found similar HFMR response in other two-terminal devices using LSMO electrodes [15] where the opposite electrode was not magnetic. We conclude therefore that the HFMR occurs at the LSMO/OSEC interface, and is probably due to the shift of the LSMO Fermi level with H. This model can also explain the recently measured room temperature resistance difference between two magnetic fields in planar LSMO/T_6/LSMO devices [16]. In that device it was claimed that spins injection and spin transport at room temperature was proven. This was a premature claim, since a simple MR at the LSMO/OSEC interface could explain the results. The spin diffusion within the OSEC, however diminishes at room temperature, as we concluded above.

In conclusion, the present study shows that spin-polarized carrier injection, transport and detection, which are the main ingredients of Spintronics, can be successfully achieved using π-conjugated OSE. This may initiate a variety of exciting new applications in organic Spintronics such as spin-OLEDs, enabled by the novel functionalities of the OSEC. This is therefore the debut of Organic Spintronics.

ACKNOWLEDGMENTS

This work was funded in part by the DOE grants # 04-ER-46109 and 04-ER-46161.

REFERENCES

1. Baibich, M. N. et al., *Phys. Rev. Lett.* 61, 2472-2475 (1988).
2. Wolf, S. A. et al., *Science* 294, 1488-1490 (2001).
3. Daughton, J., *J. Magn. Mat.* 192, 334-338 (1999).
4. Johnson, M., and Silsbee, R. H., *Phys. Rev. Lett.* 55, 1790-1793 (1985).
5. Merservey, M. and Tedrow, P. M., *Phys. Rep.* 238, 173 (1994).
6. Schmidt, G. et al., *Phys. Rev. B* 62, R4790-4793 (2000).
7. Rashba, E. I., *Phys. Rev. B* 62, R16267-16270 (2000).
8. Smith, D. L., and R. N. Silver, Phys. Rev. B 64, 045323 (2001).
9. Albrecht, J. D., and Smith, D. L., *Phys. Rev. B* 66, 113303 (2002).
10. Bassler, H., *Polym. Adv. Technol.* 9, 402 (1998).
11. Ruden, P. P., and Smith, D. L., *J. Appl. Phys.* 95, 4898-4903 (2004).
12. Xiong, Z. H., Wu, D., Vardeny, Z. V., and Shi, J. *Nature* 427, 821-824 (2004).
13. Hwang, H. Y. et al., *Phys. Rev. Lett.* 77, 2041-2044 (1996).
14. Ramirez, A. P., *J. Phys.: Cond. Mat.* 9, 8171 (1997).
15. Wu, D., Xiong, Z. H., Vardeny, Z. V., and Shi, J., *Phys. Rev. Lett.* (in press).
16. V. Dediu, V., Murgia, M., Matacotta, F. C., Taliani, C., and Barbanera, S., *Solid State Commun.* 122, 181-184 (2002).

Intershell Transport in Multiwall Carbon Nanotubes

B. Bourlon*, C. Mikó[†], L. Forró[†], D.C. Glattli*[‡], A. Bachtold*

*Laboratoire Pierre Aigrain, Ecole Normale Supérieure, 24 rue Lhomond, 75231 Paris 05, France.
[†]EPFL, CH-1015, Lausanne, Switzerland.
[‡]SPEC, CEA Saclay, F-91191 Gif-sur-Yvette, France.

Abstract. We have studied the intershell transport in multiwall carbon nanotubes (MWNT). To do this, nonlocal four-point measurements are used to study the current path through the different shells of a MWNT. Measurements agree with a model that describes MWNTs by resistive transmission lines. These measurements allow the first estimation of the intershell resistance, which is ≈ 10 kΩ for a length of 1µm.

INTRODUCTION

MWNTs that consist of several nested cylindrical graphene shells conduct charges exceptionally well in the direction parallel to the shells [1,2]. This high conductivity is attributed to the low level of diffusion experienced by conducting modes along the shells. These modes are delocalized along the shell surface. In MWNTs, neighboring shells are most often incommensurate. The resulting lack of periodicity is expected to inhibit the charge delocalization in the radial direction. The intershell conduction is thus thought to be governed by hopping. Notice that high conduction anisotropy is possible because MWNTs provide an elegant solution to the problem of current shortcuts through dangling bonds at the edge of finite parallel 2 dimensional sheets. Indeed, such edges are avoided by the topology that rolls the sheet into a cylinder. However, the intershell resistance remains to be experimentally quantified.

MEASUREMENTS

We select long MWNTs so that they can be electrically addressed by several electrodes. Figure 1 shows a MWNT contacted by 4 electrodes. The MWNT, synthesized by arc-discharge evaporation and carefully purified, was dispersed onto a 500 nm oxidized Si wafer from a dispersion in dichloroethane. Cr/Au electrodes were then patterned above the tube by electron beam lithography.

We now look at measurements of the nonlocal voltage $\Delta V_{nonlocal}$ (schematic of Fig. 2). $\Delta V_{nonlocal}$ is measured in the linear regime with eIR_{2P} below kT, I being the applied current and R_{2P} the two-point resistance between the current electrodes. $\Delta V_{nonlocal}$ is

CP786, Electronic Properties of Novel Nanostructures, edited by H. Kuzmany, J. Fink, M. Mehring, and S. Roth
© 2005 American Institute of Physics 0-7354-0275-2/05/$22.50

found to be linear with the applied current at such low biases. $\Delta V_{nonlocal}$ always has the same polarity for different MWNTs (see Fig. 2). Such a nonlocal voltage measured at room temperature is a signature of systems possessing complex current pathways. This has been reported previously in a Kirchberg proceeding [3] and attributed to multiple shells conduction with the intershell electron pathways occurring near defects or tube end caps. We show in this proceeding paper that the nonlocal voltage drop decreases exponentially with distance.

FIGURE 1. MWNT contacted by 4 electrodes

Figure 2 shows the nonlocal resistance $R_{nonlocal}$ at 300 K for different devices with different separation L between the current and the voltage electrodes. Interestingly, the non-local voltage significantly decreases with L for different MWNTs. Note that measurement points are quite scattered. This is attributed to the fact that separations between current electrodes or voltage electrodes are not the same for different devices. This might also originate from variations of the intrashell resistivity and the intershell conductivity.

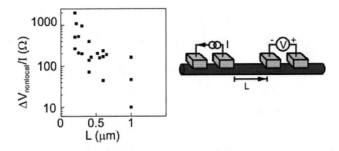

FIGURE 2. Nonlocal resistance for different MWNTs at 300 K. L is taken between the electrode borders. Separations between current electrodes and voltage electrodes are not the same for different MWNTs.

We now discuss measurements on a MWNT that is contacted by 11 electrodes. This allows us to probe the $R_{nonlocal}$ dependence on the length x for a single MWNT. Note that L in Fig. 2 is the separation between the electrode borders while x is the separation between the electrode centers. Figure 3 shows that the non-local voltage decreases exponentially with a decay length of 0.94 μm. We note that the current is zero far from the current electrodes such as at the tube end caps, showing that those play a negligible role in the current pathway.

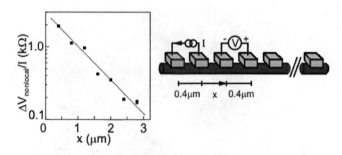

FIGURE 3. Nonlocal resistance for a single MWNT. The exponential decay length is 0.94 μm. x is taken between the electrode centers. The electrode width is 0.4 μm.

In ref. 4, we have shown that nonlocal and local measurements can be described by a current pathway that is schematized in Fig. 4. The 3 outermost shells are represented by lines. When current is externally applied, the current enters and exits the MWNT through the outermost shell. Importantly, part of I flows through to the second shell, so that the current pathway penetrates beside the region lying between the bias electrodes.

We discuss now the number of current carrying shells. The nonlocal voltage on the outermost shell is seen to decay exponentially on the scale of a micron. Thus, a fraction of the current leaves the outermost shell to penetrate into the inner shells over a typical length scale of $L_a \approx 1$ μm. The same phenomenon is expected to occur between the second shell and deeper shells. Due to the exponential dependence, only a small fraction of the current is thus expected to reach the third and the deeper shells when the separation between the current bias electrodes is lower than ≈ 1 μm.

We consider a model for a quantitative analysis of the results. The current transport is described by a resistive transmission line (Fig. 4). On a differential length of line, the intershell conductance is gδx and the intrashell resistances are $\rho_1 \delta x$ and $\rho_2 \delta x$. For an infinite long MWNT, it gives

$$R_{nonlocal}(x) \sim \exp(-x/L_a) \tag{1}$$

with $L_a = [g(\rho_1 + {}_2)]^{-1/2}$. Equation (1) predicts an exponential form of $R_{nonlocal}$ as a function of x, in excellent agreement with experiments.

These measurements allow the first estimation of the intershell resistance. The intrashell resistivity has been measured for similarly prepared MWNTs to be ≈ 10 kΩ per micron [5]. Using $L_a \approx 1\mu$m, this gives that $g \approx (10$ k$\Omega)^{-1}$ per micron. In ref. 4, we have shown that the intershell transport is tunnel-type with the transmission given by the overlap between pi-orbitals of nearby shells.

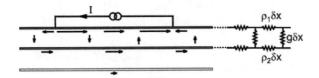

FIGURE 4. Current pathways and resistive transmission line, which models the two outermost shells.

In conclusion, the measure of the intershell conductance represents an important step in the characterisation of basic MWNT properties. We have found that the resistance between two shells is ≈ 10 kΩ for a section that is 1 μm long. These measurements are possible because the attenuation length is on the scale of a micron. Indeed, it is shorter than the longest existing MWNTs and it is longer than the minimum size accessible with nanofabrication techniques, allowing the measurement of a finite $\Delta V_{nonlocal}$.

ACKNOWLEDGMENTS

We thank B. Placais and C. Baroud for discussions and C. Delalande for support. LPA is CNRS-UMR8551 associated to Paris 6 and 7. The research has been supported by the DGA, ACN, sesame, the Swiss National Science Foundation and its NCCR "Nanoscale Science".

REFERENCES

1. Frank S., *et al.*, *Science 280*, 1744 (1998).
2. Bachtold A., *et al.*, *Nature 397*, 673 (1999).
3. Collins P.G., *et al.*, *Proc. 14th Int. Winter School on Electronic Properties of Novel Materials* (Am. Inst.Phys., New York, 2000).
4. Bourlon B., *et al.,*, Phys. Rev. Lett. **93**, 176806 (2004).
5. Bachtold A., *et al.,*, Phys. Rev. Lett. **84**, 6082 (2000).

Transport and TEM on the same individual carbon nanotubes and peapods

Dirk Obergfell[1], Jannik C. Meyer[1], Miroslav Haluska[1], Andrei Khlobystov[2], Shihe Yang[3], Louzhen Fan[3], Dongfang Liu[3], Siegmar Roth[1]

[1]*Max Planck Institute for Solid State Research, Stuttgart, Germany*
[2]*School of Chemistry, University of Nottingham, Nottingham, UK*
[3]*Department of Chemistry, The Hong Kong University of Science and Technology, Clear Water Bay, Kowloon, China*

Abstract. For the first time, full transistor characteristics – both output and transfer characteristics – in field-effect transistor configuration have been recorded and TEM investigations have been performed on the same individual nanotubes and metallofullerene peapods. Usual approaches for combining transport and TEM on the same nanotube only allow for measuring the current voltage characteristics, but no gate dependence can be acquired. Applying our new method of underetching a Si/SiO_2 substrate from the edge of a chip after the transport measurements, we can additionally get the transfer characteristics $I_{sd}(V_g)$, i.e. the gate response of the current, which provides crucial information about the electronic properties of the system investigated. After the transport measurements and the etching process the samples can be viewed in the TEM, which enables us to check, whether a contacted nanotube is really a single tube or a thin bundle and whether a tube is filled with fullerenes indeed. Gaining this additional structural information by TEM can significantly help to interpret electrical transport data correctly.

INTRODUCTION

Many publications on electrical transport in *single* single-walled carbon nanotubes (SWNTs) have appeared throughout the years since 1997, but only rarely authors mention the problem of proving that it is indeed a single tube located between the contacts. Even more important than showing that exclusively one SWNT is contacted is to prove that tubes are actually filled with fullerenes in the case of transport experiments on peapods and metallofullerene peapods. The common approach in the peapod case is to adsorb the peapod material onto a standard TEM grid and to carry out high-resolution transmission electron microscopy (HR-TEM) in order to prove the filling. If the tubes seen in the TEM images are filled, it often is claimed that *nearly all* SWNTs in the sample are filled, including those which subsequently are adsorbed onto chips and contacted by metal electrodes for the transport measurements. This approach is generally not correct. Judging the fraction of filled tubes based on TEM should rather be regarded as a crude estimation since TEM is a local probe only and often tubes deposited onto TEM grids lie on top and across each other which causes difficulties in judging the filling of individual tubes.

CP786, *Electronic Properties of Novel Nanostructures*, edited by H. Kuzmany, J. Fink, M. Mehring, and S. Roth
© 2005 American Institute of Physics 0-7354-0275-2/05/$22.50

Returning to the question how to prove that a contacted tube is single, one can state that AFM is not the method of choice in this respect: measuring the height of a nano-object by AFM usually results in an error of at least ±1 Å, additionally the diameter distribution of the SWNTs within a batch is not ultimately sharp. Even if the SWNTs adsorbed had a very narrow diameter distribution and one got the proper height value by AFM which one would assign to a single SWNT, a small bundle of SWNTs still could not be ruled out, since tubes lying next to each other could form a bundle and the AFM is not suitable for retrieving lateral dimensions. A suitable method how to decide whether a tube contacted is a single SWNT or a bundle and whether it is a tube filled with fullerenes is high resolution transmission electron microscopy (HR-TEM), which can only be applied for transparent substrates. That is why so far transport and TEM on the same individual carbon nanotube could only be realized either by utilizing transparent membranes as substrates [1], by putting the SWNTs across a slit [2] or by using clamping-mechanisms [3,4], whereat all these methods do not allow for the implementation of a well-defined gate electrode. Nevertheless having a gate electrode is indispensable for recording transfer characteristics $I_{sd}(V_g)$ being crucial for understanding the electronic properties of the device investigated.

Recently we introduced our new approach of combining electrical transport measurements (*including transfer characteristics*) and TEM on *the same* individual carbon nanotube [5], which solves the problems addressed above. In the following, we present the first results for transport plus TEM on the same metallofullerene peapods.

EXPERIMENTAL

Oxidised single-walled carbon nanotubes are filled with $Dy@C_{82}$ endohedral metallofullerenes in vacuum at 10^{-6} torr and 440 °C for 5 days yielding $(Dy@C_{82})_n@SWNT$ metallofullerene peapods. Both a HR-TEM micrograph of our sample material adsorbed onto a TEM grid and a schematic of a segment of a single metallofullerene peapod are depicted in Figure 1.

After production, $(Dy@C_{82})_n@SWNT$ metallofullerene peapods were dispersed in 1% aqueous SDS and adsorbed onto highly doped silicon substrates covered with 200 nm dry SiO_2 and provided with markers serving as coordinate systems. The positions of potentially single tubes with respect to the markers are determined by AFM. The chips were cleaved, tubes in vicinity to the cleaved edge were contacted with AuPd electrodes being approx. 30 nm high by standard e-beam lithography. We performed transport measurements at liquid helium temperature. The source drain voltage and the gate voltage were swept in loops to check for hysteresis. After the transport investigations massive support structures (5 nm Cr, 100 nm Au) were put on top of the electrode structures in a second e-beam step, leaving gaps at the positions of the contacted tubes. During the subsequent etching process and the critical point drying these support structures prevent the electrodes, which contact the nanotubes investigated, from being bent. The samples were etched in a 30% KOH solution for approx. 7 hours. The nanotubes now are freestanding and can be viewed in the TEM. For TEM imaging of the underetched samples a Philips CM200 microscope was used at 120 kV. Further details concerning the etching method can be extracted from [5].

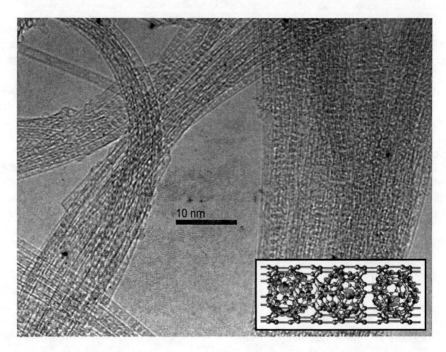

FIGURE 1. High-resolution TEM micrograph of the $(Dy@C_{82})_n@$ SWNT metallofullerene peapod material used within this study. In a crude estimation the fraction of the tubes filled with fullerenes is larger than 50%. **Inset:** schematic representation of a segment of a single $(Dy@C_{82})_n@SWNT$ metallofullerene peapod. The diameter of the nearly spherical $Dy@C_{82}$ metallofulleres is approx. 8 Å. Note: the scale bar holds for the TEM image only.

RESULTS AND DISCUSSION

At this early stage of the experiments combining transport and TEM as described above, results taken from two devices will be presented in the following.

An AFM image and HR-TEM micrograph of device A are depicted in Figure 2. Already by AFM it could be anticipated that a bundle of tubes was contacted, after HR-TEM it was evident that this bundle consisted of three tubes, whereat two are $(Dy@C_{82})_n@SWNT$ metallofullerene peapods and one is an empty SWNT. As for transport properties, device A behaves as a p-type semiconductor (Fig. 3). This is contrary to earlier publications on transport in metallofullerene peapods in which metallic and ambipolar transport behaviour for metallofullerene peapods at low temperatures was claimed respectively [6,7], but filling of the tubes could not be proved.

High-resolution TEM on device B (Fig. 4b) revealed that device B, the AFM image of which is depicted in Figure 4a, indeed is a single SWNT. Though being taken from the same peapod sample material as the bundle of device A, device B is not a peapod, but a single *empty* tube. Device B exhibits ambipolar transport behaviour (Fig. 5).

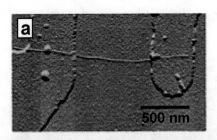

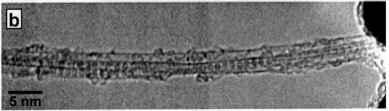

FIGURE 2. **(a)** Atomic force microscopy image of device A. According to AFM the channel of device A has a diameter of approx. 1.9 to 2.3 nm. **(b)** High-resolution TEM micrograph of device A revealing that sample A consists of two metallofullerene peapods and an empty SWNT within a bundle.

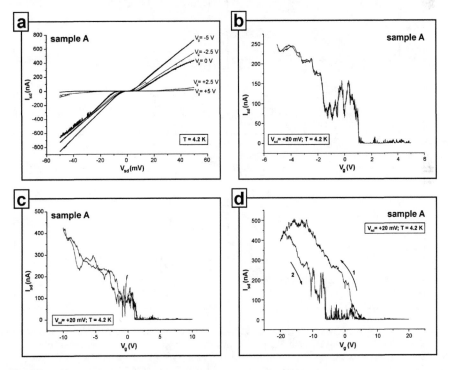

FIGURE 3. Transport data of device A at 4.2 K. **(a)** Output characteristics $I_{sd}(V_{sd})$ at different gate voltages V_g. **(b)-(d)** Transfer characteristics $I_{sd}(V_g)$ at $V_{sd} = +20$ mV showing that device A behaves as a p-type semiconductor. When staying at modest maximum gate voltages, the curve is highly reproducible, whereas an increasing maximum gate voltage $V_{g,max}$ leads to pronounced hysteresis.

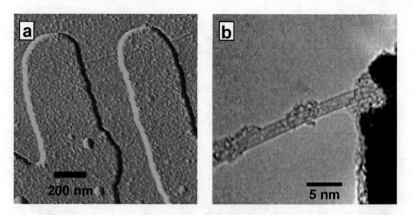

FIGURE 4. (a) Atomic force microscopy image of device B. According to AFM the channel of device B has a diameter of approx. 1.2 to 1.4 nm. (b) High-resolution TEM micrograph of sample B exposing that device B is a single empty SWNT with about 1.6 nm diameter.

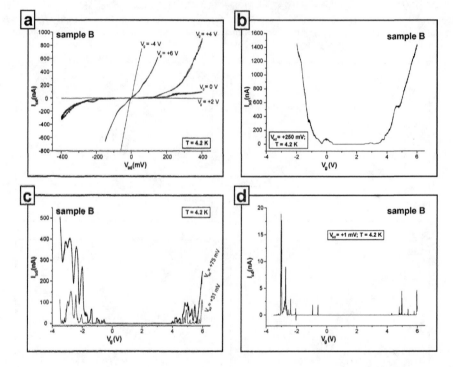

FIGURE 5. Transport data of device B at 4.2 K. (a) Output characteristics $I_{sd}(V_{sd})$ at different gate voltages V_g. (b) $I_{sd}(V_g)$ plot taken at $V_{sd} = +250$ mV displaying the ambipolar behaviour of sample B. (c) Decreasing the bias voltage V_{sd} results in lower $I_{sd}(V_g)$ curves which are provided with broad peaks both for conduction and valance band. (d) At $V_{sd} = +1$ mV Coulomb blockade oscillations are clearly visible, the nanotube behaves as a semiconducting quantum dot.

Lowering the source drain voltage V_{sd} to +1 mV, Coulomb blockade features occur, the single tube behaves as a semiconducting single-electron transistor at low temperatures and low V_{sd} values.

Like device B presented within this work, numerous *unfilled* carbon nanotubes investigated in several studies of other groups showed ambipolar transport behaviour. This is a clear indication that the mere occurance of ambipolar transport is no evidence for a nanotube being filled with metallofullerenes as proposed in [7]. Discussing ambipolar behaviour, the coupling strength between the nanotube and the gate electrode has to be accounted for, since it affects the transfer characteristics of a nanotube to a great extent.

CONCLUSIONS

The combination of transport measurements in transistor configuration and high-resolution TEM on the same individual nanotubes and peapods was accomplished, enabling us to acquire relevant structural information of the nanotubes contacted. This additional information significantly helps to interpret transport data properly, especially in the case of peapods. We now aim at transport plus TEM of the *same single* $(Dy@C_{82})_n@SWNT$ metallofullerene peapod.

ACKNOWLEDGEMENTS

The authors acknowledge financial support by the BMBF project INKONAMI and the Landesstiftung Baden-Wuerttemberg. Furthermore we thank xlith GmbH for producing the marker systems.

REFERENCES

1. Sagnes, M., Broto, J.-M., Raquet, B., Vieu, C., Conedera, V., Dubreuil, P., Ondarçuhu, T., Laurent, Ch., and Flahaut, E., *Microelectronic Engineering* **73-74**, 689-694 (2004).
2. Kasumov, A. Y., Deblock, R., Kociak, M., Reulet, B., Bouchiat, H., Khodos, I. I., Gorbatov, Y. B., Volkov, V. T., Journet, C., and Burghard, M., *Science* **284**, 1508-1511 (1999).
3. Kociak, M., Suenaga, K., Hirahara, K., Saito, Y., Nakahira, T., and Iijima, S., *Phys. Rev. Lett.* **89**, 155501 (2002).
4. Golberg, D., Mitome, M., Kurashima, K., Bando, Y., *Journal of Electron Microscopy* **52(2)**, 111-117 (2003).
5. Meyer, J. C., Obergfell, D., Roth, S., Yang, S., and Yang, S. F., *Appl. Phys. Lett.* **85**, 2911-2913 (2004).
6. Chiu, P. W., Gu, G., Kim, G. T., Philipp, G., Roth, S., Yang, S. F., and Yang, S., *Appl. Phys. Lett.* **79**, 3845-3847 (2001).
7. Shimada, T., Okazaki, T., Taniguchi, R., Sugai, T., Shinohara, H., Suenaga, K., Ohno, Y., Mizuno, S., Kishimoto, S., and Mizutani, T., *Appl. Phys. Lett.* **81**, 4067-4069 (2002).

Characterization of SWNT-Thin-Film Transistors

Masashi Shiraishi[1,4], Tomohiro Fukao[1], Shuichi Nakamura[1],
Taishi Takenobu[2,4], Y. Iwasa[2,4] and H. Kataura[3]

[1] Graduate School of Engineering Science, Osaka Univ., 560-8531 Machikaneyama-cho 1-3, Toyonaki, Osaka, Japan
[2] Institute of Materials Research, Tohoku Univ., 980-8531 Katahira 2-1, Aoba-ku, Sendai, Japan
[3] AIST, Nanotechnology Research Institite, Higashi1-1-1, Tsukuba 305-8562, Ibaraki, Japan
4 CREST-JST, Kawaguchi 332-0021, Japan

Abstract. Characteristics of network SWNT FETs (SWNT-TFTs) were examined. The SWNTs were dispersed in a solution of dimethylformamide in a narrow bundle structure to form non-aligned arrays, which became channels of FETs. The network-SWNT-FETs produced in this solution process was found to have a mobility of 10.9 cm^2/Vs, ≈ 100 times as high as those reported for other solution-processed organic thin-film FETs formed by solution processes, although the on/off ratio was 10^2. To improve the low on/off ratio, so-called electrical breakdown was introduced. By this procedure, 1.1 cm^2/Vs of mobility and 10^6 of the on/off ratio were simultaneously achieved.

INTRODUCTION

Single-walled carbon nanotubes (SWNTs) have quasi-one-dimensional structures with semiconducting or metallic properties, which promise various applications as molecular electronic devices with nano-sized structures. To analyze the semiconducting characters of SWNTs, field effect transistors (FETs) have been investigated intensively because of their unique transport properties, with special focus on their high mobility in operation and the origin of their polarity. Since the pioneer work by Tans et al. [1] on the p-type operation of SWNTs, a number of experimental and theoretical studies related to SWNT-FETs, multi-walled carbon nanotube (MWNT) FETs, and fullerene-encapsulated-SWNT-FETs (peapod-FETs) have been carried out [2,3]. Recent studies of ambipolar behavior of SWNT-FETs and their basic mechanisms [4] have utilized individual SWNTs as the channel of FETs. In other words, these studies have been directed to the use of single-molecule FETs.

As for organic thin film FETs (TFTs), such as pentacene or thiophene FETs, the channel is ordinarily formed with organic aggregates instead of a single molecule. The behavior of SWNT aggregates (network-SWNTs) seems to play an important role in FET characteristics because of its π-electron character for achieving high field-effect mobility (> 10 cm^2/Vs). As predicted by Dimitrakopoulos et al. [5], the mobility in

CP786, *Electronic Properties of Novel Nanostructures*, edited by H. Kuzmany, J. Fink, M. Mehring, and S. Roth
© 2005 American Institute of Physics 0-7354-0275-2/05/$22.50

organic aggregate FETs has an upper limit of 10 cm^2/Vs, whereas that of organic FETs yet achieved is only 0.1-1 cm^2/Vs; thus a novel material and design are strongly needed to reach and overcome this upper limit. The number of carrier hopping in the network channel can be decreased by a properly design a large aspect ratio in the shape of SWNTs. Thorough efforts have been made for production of network-SWNT-FETs [6,7] by growing SWNTs using chemical vapor deposition (CVD). High-quality SWNTs synthesized by laser ablation [8] are also expected for another type of network-SWNTs with enhanced π-electron character. The present manuscript reports on the network SWNT-FETs (SWNT-TFTs), in line with our effort for a large-scale and low-cost fabrication of high-quality FETs on substrates of low heat tolerance such as plastic materials.

EXPERIMENTS

The SWNTs were synthesized by laser ablation using Ni/Co as catalytic metals and then purified by a conventional method [8]. After purification, the SWNTs were ultrasonicated in dimethylformamide (DMF) solution (purity > 99.5 %, about 50 mg of SWNTs per 1 L of DMF solution) for 9 hr to dissolve the bundles. Then centrifugation at 5000 rpm for 20 min was carried out to select well-dispersed narrow bundle SWNTs. The centrifuged solution was dropped on to a back-gate FET substrate, and heat application at 353 K for 10-20 min was introduced to remove the DMF.

The mobility was calculated from the standard formula ($\mu = (dI_{sd}/dV_{g})/(\varepsilon V_{sd}W/L_{ox}L_{sd})$), where I_{sd} is the source-drain current, V_{g} the gate voltage, L_{ox} the thickness of SiO$_2$ (100 or 400 nm), L_{sd} the width between the source and drain electrodes (20 or 50 μm), ε the dielectric constant of SiO$_2$ (4.0), V_{sd} the source-drain voltage (0.1 V), and W the gate width (100 or 325 μm).

RESULTS AND DISCUSSION

Figure 1 and 2 show a typical device structure and the temperature dependence of a SWNT-FET device behavior after evacuation by pumping for one day, respectively. When the temperature was lowered, the FET still operated down to 4.6 K and the majority carrier (the hole) mobility was 2.4 cm^2/Vs, while the mobility at 300 K was about 11 cm^2/Vs. The On/Off ratio of this FET was 2.1×10^2 at 300 K and 4.6×10^2 at 4.6 K. This device showed an ambipolar behavior; thus the minority carrier (the electron) mobility was also calculated to be ≈ 0.70 cm^2/Vs at 300 K and ≈ 0.38 cm^2/Vs at 4.6 K. Such temperature dependence of the device characteristics was similar to that of an individual SWNT-FET. As for the hysteresis of the devices in the application of upward and downward sweeps of the gate voltage, the authors have observed similar behavior in an individual SWNT FET [9], which indicates that the trapped charges responsible for the present hysteresis are positive (hole), being similar to those proposed in Ref. [10]. The organic FETs fabricated by various solution processes have had much lower hole mobilities; 9.4×10^{-3} cm^2/Vs for polythiophene-FETs [11], $\geq 10^{-2}$ cm^2/Vs for organic blend FETs [12], and 0.1 cm^2/Vs for conjugated polymer FETs [13]. Hence, the hole mobility of the present solution-processed SWNT-FETs is much higher, by more than two orders of magnitude. For calculation of the mobility, the gate length, W, is assumed to be equal to the actual device width. However, since the SWNTs are dispersed on a smaller part of the electrodes, as judged from the SEM observation, the mobility calculated in this way sets the lower limit, and the mobility can be enhanced if the gate length is shortened.

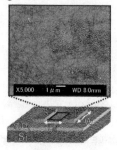

FIGURE 1. Device structure of SWNT-FET

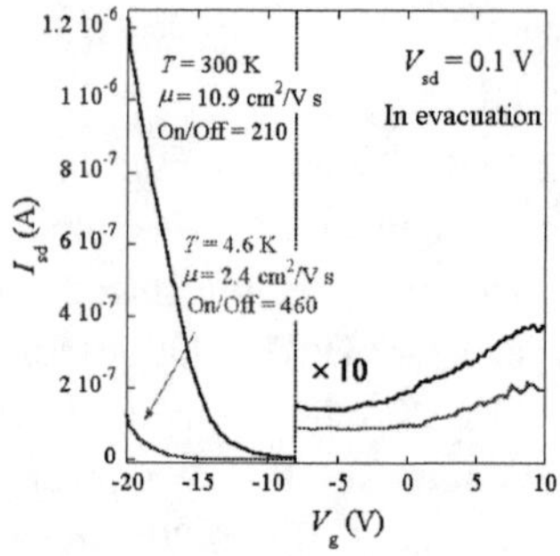

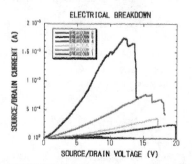

FIGURE 2. FET characteristics of the FFT.

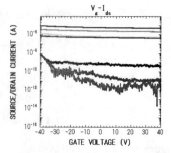

FIGURE 3a. I_{sd}-V_{sd} curves after the EB

FIGURE 3b. I_{sd}-V_g curves after the EB

The present study has shown a trade-off between the mobility and the On/Off ratio, as was also observed. This trade-off seems to be attributable to the continuous conduction paths of metallic SWNTs so that reduction of the fraction of metallic SWNTs contributes to the lowering of the On/Off ratio. To realize selective removing of the metallic SWNTs, we introduced a so-called electrical breakdown (EB) procedure which was proposed by Collins et al. [14]. By the procedure, the device characteristic was gradually improved as shown in Fig. 3, and after several times of removing the metallic SWNTs, comparably high FET performance ($\mu = 1.1$ cm^2/Vs, on/off $= 10^6$) was observed. Such simultaneous high performance indicates that this TFT device is potential enough compared the other organic TFTs. The barrier height for hole and electron injection were measured from a large number of I_{sd}-V_{sd} curves taken from 4.2 to 300 K., with $V_g = -40$ and $+40$ V, respectively. It was clarified that the conduction mechanism was governed by the thermionic current above at ~30 K, and the barrier height at zero field was estimated to be ~6.5 meV. In the electron injection, the barrier height was estimated to be ~26 meV, which was ~4 times larger than that for electrons. This finding corroborates the fact that the device operates as a p-type FET [15].

ACKNOWLEDGMENTS

This work was partially carried out by the financial aid of New Energy and Industrial Technology Development Organization (NEDO) and 21st COE program in Osaka Univ.

REFERENCES

1. S.J. Tans, A.R.M. Verschueren, C. Dekker, Nature (London) **393** (1998) 49.

2. S.J. Wind, J. Appenzeller, R. Martel, V. Derycke, Ph. Avouris, Appl. Phys. Lett. **80** (2002) 3817.

3. T. Shimada, T. Okazaki, R. Taniguchi, T. Sugai, H. Shinohara, K. Suenaga, Y. Ohno, S. Mizuno, S. Kishimoto, T. Mizutani, Appl. Phys. Lett. **81** (2002) 4067.

4. R. Martel, V. Derycke, C. Lavoie, J. Appenzeller, K.K. Chan, J. Tersoff, Ph. Avouris, Phys. Rev. Lett. **87** (2001) 256805.

5. C.D. Dimitrakopoulos, P.R.L. Malenfant, Adv. Matet. **14** (2002) 99.

6. E.S. Snow, J.P. Novak, P.M. Campbell, D. Park, Appl. Phys. Lett. **82** (2003) 2145.

7. K. Bradley, J-C.P. Gabriel, G. Gruener, Nano Lett. **3** (2003) 1353.

8. M. Shiraishi, T. Takenobu, A. Yamada, M. Ata, H. Kataura, Chem. Phys. Lett. **358** (2002) 213.

9. S. Heinze, J. Tersoff, R. Martel, V. Derycke, J. Appenzeller and Ph. Avouris, Phys. Rev. Lett. **89** (2003) 106801.

10. T. Duerkop, S.A. Getty, E. Cobas and M.S. Fuhrer, Nano Lett. **4** (2004) 35.

11. R.J. Kline, M.D. McGehee, E.N. Kadnikova, J. Liu, J.M.J. Frechet, Adv. Mater. **15** (2003) 1519.

12. E.J. Meijer, D.M. De Leeuw, S. Setayash, .E. Van Veenendaal, B.-H. Huisman, P.W.M. Blom, J.C. Hummelen, U. Scherf, T.M. Klapwijk, Nature Materials **2** (2003) 678.

13. H. Shirringhaus, N. Tessler, R.H. Friend, Science **280** (1998) 1741.

14. P.G. Collins et al., Science **292** (2001) 706.

15. M. Shiraishi, S. Nakamura, T. Fukao, T. Takenobu, Y. Iwasa and H. Kataura, Appl. Phys. Lett. in submission.

Coupling of Surface Acoustic Waves to Single Walled Carbon Nanotubes

V. Siegle*[1], D. Obergfell*, F.J. Ahlers,[†] and S. Roth*

*Max Plank Institute for Solid State Research, Heisenbergstr. 1, 70569 Stuttgart, Germany
[†]Physikalisch–Technische Bundesanstalt, Bundesallee 100, 38116 Braunschweig, Germany

Abstract.
In the presented paper we give an overview of existing approaches for single electron pumps. A single walled carbon nanotube (SWNT) based device, which is driven by surface acoustic waves (as proposed by Talyanskii et al., 2001), is discussed. The compatibility of the lithographic structures of the sample with the SWNT suspension and involved chemistry is studied. The results of this study are reported together with the progress in fabrication of such a device.

Keywords: surface acoustic waves, quantum wire, quantized charge transport, carbon nanotubes
PACS: 46.40.-f,78.20.Hp,71.10.Pm,73.21.Hb, 78.66.Tr

INTRODUCTION

The field of quantized charge transport in an adiabatically changing potential has atracted growing attention since first theoretical treatment by Thouless[1] (see also [2, 3, 4]). Various applications could be thought of for a device which is capable of delivering single electrons in a controlled manner. Such a device could be used as a platform for quantum computing[5] or for production of non–poissonian photons for quantum cryptography[6, 7]. One can also obtain a precise definition of electric current through $I = ef$, where f is the frequency at which the electrons are supplied and e is the electron charge. By taking advantage of Ohm's law and quantum effects for resistance and voltage it is also possible to precisely define a set of electrical units[8].

SINGLE–ELECTRON PUMPS

In the following we give a short overview of single–electron pumps described or proposed in the literature.

Turnstile device. One of the implementations of a device able to produce quantized current was realized by Geerligs et al.[9]. This so called *turnstile device* is based on a static quantum dot (QD). It was realized by coupling a metallic island capacitively to metal leads (see Fig. 1a).

---

[1] Corresponding author. E–mail: V.Siegle@fkf.mpg.de

CP786, *Electronic Properties of Novel Nanostructures*, edited by H. Kuzmany, J. Fink, M. Mehring, and S. Roth
© 2005 American Institute of Physics 0-7354-0275-2/05/$22.50

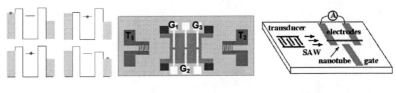

| (a) Principle of a turnstile device | (b) SAW based device with 3 split gates[13] | (c) SWNT based device suggested in [14] |

FIGURE 1. Different charge pumps

Due to the Coulomb blockade it was ensured that only a certain number of electrons was on the island. By applying two AC voltages to the leads, phase–shifted by $\pi/2$, electrons were single moved from the source to the drain electrode.

As a result, a very high precision of the charge current could be achieved. Devices based on the described principle reach a precision of 10^{-8} [10]. Unfortunately only very low currents (in the pA regime) can be achieved, so that precise and reliable measurements are difficult.

SAW based devices. To overcome this difficulty a device based on surface acoustic waves was proposed. Surface acoustic waves (SAW) are elastic waves confined to the surface within a distance of about a wave length. The movement of the atoms is restricted to the plane spanned by the wave vector and the normal of the surface (Rayleigh wave). On piezoelectrics as *GaAs* or *LiNbO₃* the mechanical wave is accompanied by an electromagnetic wave of the same velocity. SAW based devices are widely used in telecommunication[11] . The transformation of RF, which is supplied by a generator, into SAW occurs through interdigital transducers (IDT)[12]. Very high frequencies (in the GHz region) of SAW can be reached depending on the properties of the substrate and the finger spacing of the IDTs.

An experiment based on SAW was carried out by Shilton et al [15]. A typical device is depicted in Fig 1(b). In the middle of the 2DEC a voltage is applied by a split gate. The value of the voltage controls the width of a channel connecting the two parts of the 2DEC. With sufficiently high voltage the channel can be narrowed down to a quasi 1D conductor. By further voltage increase the potential barrier becomes too high for the electrons to pass – the channel is closed (*pinch–off*).

The electric potential of SAW is now used both to overcome the potential barrier and to confine a fixed number of electrons in a so called dynamic QD. The resulting electron transport leads to currents in the nanoampere range. This drawback of this method is the lower precision compared to the turnstile device.

SWNT Based Charge Pump. Talyanskii et al.[14] proposed to combine the SAW approach with a single walled carbon nanotube (SWNT) which is known to be a ballistic 1D–conductor. The experimental setup is outlined in Fig 1(c). A nanotube is placed between contacts on a piezocrystal, along the propagation of the SAW. A possible band gap in the nanotube together with the migrating electrical potential would confine

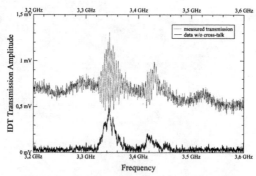

FIGURE 2. Transmission delay line characteristics

electrons in a quantum dot. Upon applying of SAW one would expect quantized charge transport in nanoampere range.

Some complications arise concerning the appropriate nanotube type. Metallic nanotubes have no band gap and do not suffice the necessary condition. The band gap in semiconducting nanotubes is in the order of some 100 meV and is too large to be modulated by the available electrical SAW potential.

Suitable nanotubes are so–called nearly metallic tubes. Despite of the fact that the electrical structure of a SWNT is to a large extend defined by its (n,m)–indices, small band gaps can open in an otherwise metallic (i.e. $(n-m) \mod 3 = 0$) nanotube due to perturbations such as high magnetic field, mechanical deformation or simply the curvature of the nanotube. Such a small band gap is in the order of some meV[16] which is accessible experimentally.

EXPERIMENTAL PROGRESS

To prepare a SWNT charge pump we decided to deposite nanotubes from an aqueous suspension. To study the compatibility of the IDT structures with the wet chemical handling, several samples were prepared and different types of SWNTs were deposited.

$LiNbO_3$ was chosen as substrate since its electromechanical coupling coefficient κ^2 is about two orders of magnitude higher than for $GaAs$ which results in higher electrical amplitude. IDTs were fabricated with electron beam lithography for a SAW wavelength of ca $1 \mu m$. IDTs were arranged pairwise in such an order that the SAW emitted by one IDT could be detected by another one — so called *transmission delay line* (TDL).

The samples were treated with an 0,1% aqueous solution of 3–Aminopropyl–trithoxysilan. This chemical surface modification allows the nanotubes to dock on the substrate. After placing a drop of SWNT solution, the substrate was rinsed with water and blow–dried.

In our tests one IDT was connected to a RF generator, at the second IDT the amplitude of the SAW was read out. This measurement was carried out before and after the deposition of SWNTs and the involved chemical treatment.

560

On several devices we were not able to pick up any signal after this procedure. For remaining devices the received amplitude war 3–5 times smaller compared to not treated samples.

On some samples IDTs were protected from influence of chemical treatment by a layer of photoresist which was removed after nanotube deposition. No improvement in transmission amplitude or failure rate could be detected for this case. Different SWNT suspensions had the same effect.

Fig 2 shows a typical response of a TDL. The measurement is very noisy due to electrical cross–talk and multiple scattering from neighbouring IDTs. The noise is filtered out through Fourier transformation of the measured signal, selecting in the time domain parts belonging to the actual SAW and back–transforming. The result is presented in the lower curve and shows a clear signal with distinct features.

CONCLUSION

In this work we showed that the IDTs are compatible with the processes necessary for SWNT deposition. The emerging losses have to be compensated through fine tuning of the processes and experimental setup. Having the IDT fabrication and nanotube deposition under control we now can proceed to building SWNT based devices.

ACKNOWLEDGMENTS

The help of Dr. Erol Sagol, Dr. Oliver Kieler, and Monika Riek for the sample preparation is gratefully acknowledged. The authors also acknowledge the financial support of the EU project INKONAMI.

REFERENCES

1. D. J. Thouless, *Physical Review B*, **27**, 6083–6087 (1983).
2. R. Citro, N. Andrei, and Q. Niu, *Physical Review B*, **68**, 165312 (2003).
3. S. Das, and S. Rao, *Physical Review B*, **71**, 165333 (2005).
4. Q. Niu, *Physical Review Letters*, **64**, 1812–1815 (1990).
5. C. H. W. Barnes, J. M. Shilton, and A. M. Robinson, *Physical Review B*, **62**, 8410–8419 (2000).
6. C. Wiele, et al., *Physical Review A*, **58**, R2680–R2683 (1998).
7. C. L. Foden, et al., *Physical Review A*, **62**, 011803 (2000).
8. F. Piquemal, et al., *metrologia*, **37**, 207 (2000).
9. L. Geerligs, et al., *Physical Review Letters*, **64**, 2692 (1990).
10. M. W. Keller, J. M. Martinis, and R. L. Kautz, *Physical Review Letters*, **80**, 4530–4533 (1998).
11. C. K. Campbell, *SAW Devices for Mobile and Wireless Communications*, Academic Press Inc, 1998.
12. R. White, and F. Voltmer, *Applied Physics Letters*, **7**, 314–316 (1965).
13. N. E. Fletcher, et al., *Physical Review B*, **68**, 245310 (2003).
14. V. Talyansii, et al., *Physical Review Letters*, **87**, 276802 (2001).
15. J. M. Shilton, et al., *Journal of Physics: Condensed Matter*, **8**, L531–L539 (1996).
16. M. S. Dresselhaus, *Science*, **292**, 650–651 (2001).

Electron Transport – from Buckypaper to Thin Single-Wall Nanotube Networks

V. Skákalová, A.B. Kaiser[*], M. Kaempgen and S. Roth

Max Planck Institute for Solid State Research, Heisenbergstr. 1, 70569 Stuttgart, Germany
[*]*MacDiarmid Institute for Advanced Materials and Nanotechnology, SCPS, Victoria University of Wellington, P O Box 600, Wellington, New Zealand*

Abstract. The mechanism of electronic transport in a buckypaper and in thinner networks of single-wall carbon nanotubes (SWNTs) of different densities is studied. A systematic trend in changes of the temperature dependences of the electrical conductance is observed as the thickness of the networks changes. Nevertheless, the behaviour typical for conductive networks is present even for very thin networks. The character of the temperature dependences of the electrical conductance of SWNT networks is different from that of an individual SWNT.
Various models of electronic transport developed for conductive networks were tested to fit the temperature dependences of the conductance for all samples. In the case of SWNT buckypaper, a model of metallic conduction interrupted by barriers accounts for the data, while for the less metallic thin SWNT networks, the Variable Range Hopping (VRH) model seemed to be more suitable.

INTRODUCTION

The discovery in 1976 of a new class of materials – conductive polymers, when polyacetylene was synthesized - brought with it new requirements. A number of different approaches were developed to explain the mechanism of electron transport in these conductive networks. The well-established models such as Variable Range Hopping (VRH) [1] and Fluctuation Assisted Tunneling (FAT) [2] have some difficulties to fit the experimental data of the electrical conductivity of conductive polymer films sufficiently in the full range of temperatures.

Although Endo discovered the carbon nanotubes already in 1976 [3], it is only after their rediscovery by Iijima in 1991 that a boom in research on these unique carbon structures took place. The size and shape of the new molecules allows investigating them individually. The effect of a Coulomb blockade together with a particular electronic structure of each nanotube can be the reason for the big variations in temperature dependences of electrical conductance measured for individual nanotubes [4].

On the other hand, self-standing films of single wall nanotubes (SWNTs) (buckypapers) behave similarly to those made of conducting polymers. In highly conductive samples, a maximum of conductivity in temperature dependences can appear. Also, the models that have been used to explain electron transport through conductive polymer networks more or less sufficiently fit the temperature

CP786, *Electronic Properties of Novel Nanostructures*, edited by H. Kuzmany, J. Fink, M. Mehring, and S. Roth

dependences of the electrical conductivity of SWNT networks as well. Nevertheless, models developed for the conductive networks might meet their limits when the density of the network is reduced significantly so that the effects observed in individual nanotubes might play a more visible role.

In the present work we study the temperature dependences of the electrical conductance of SWNT networks with different densities, from a buckypaper to very thin networks. The results are discussed in the framework of FAT and VRH models.

EXPERIMENTAL

Purified HiPCO SWNT were purchased from CNI (Texas). SWNT buckypaper was prepared from the suspension of SWNT in water with sodiumdodecylsulfur (SDS) by vacuum filtration through a membrane filter (0.4 μm pores). The thickness of the buckypaper was 35 μm.

SWNT thin networks of four different densities were air-brushed on surfaces of four glass and four Si/SiO$_2$ wafers so that one pair of glass and Si/SiO$_2$ wafers was covered with the same number (2, 4, 6 or 8) of layers of SWNT suspended in water with SDS. All the samples were then rinsed in de-ionized water to remove SDS from the wafer surfaces.

For optical transmission, a UV-VIS/NIR Perkin-Elmer Lambda Spectrometer was used in the range of wavelengths from 1100 to 200 nm.

The electrical conductance of SWNT buckypaper and of four thin SWNT networks on Si/SiO$_2$ wafers was measured by the four probe method, evaporating four chromium (20 nm) /gold (200 nm) strips of the same geometry for all the samples. A constant current of 10 mA was applied during the temperature dependent measurements from helium temperature up to room temperature.

RESULTS AND DISCUSSION

For characterization of the thin networks thickness, the correlation between the optical transmittance at 1100 nm and electrical conductance of SWNT thin networks of various densities, in semi-logarithmic scale, is used (Fig. 1).

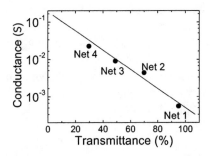

FIGURE 1. Correlation between electrical conductance and optical transmittance for transparent SWNT networks of various densities.

The set of curves in Fig. 2 shows the temperature dependence of the electrical conductivity of SWNT buckypaper and thin networks, with conductivity $\sigma(T)$ normalized to its value near 300 K. A stronger decrease of the curves as T decreases is observed, as the layers get thinner. As discussed earlier [5], the thick network shows clear signatures of metallic behaviour: the conductivity remains large in the zero-temperature limit (more than 40% of the room-temperature conductivity), and there is a change to metallic sign for the temperature dependence above 240 K (a weak decrease in conductivity $\sigma(T)$ as temperature increases above 240 K).

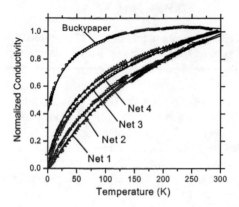

FIGURE 2. Temperature dependences of electrical conductance of SWNT buckypaper and thin networks normalized to the values at room temperature $\sigma(T)/\sigma(292\ K)$.

This behaviour is well explained by the fit shown to the expression [6] for quasi-1D metallic conduction interrupted by small barriers:

$$\frac{1}{\sigma(T)} = A\exp\left(-\frac{T_m}{T}\right) + B\exp\left(\frac{T_b}{T_s + T}\right) \qquad (1)$$

where the geometrical factors A and B can be taken as constant. The first term is for quasi-one-dimensional metallic conduction involving suppression of resistivity at lower temperatures where carriers are not easily backscattered (for scattering by lattice vibrations $k_B T_m$ would be the energy of the backscattering phonons [7]). The second term represents fluctuation-assisted tunneling [2] through barriers, with the order of magnitude of typical barrier energies indicated by the value of $k_B T_b$ and the extent of the decrease of conductivity at low temperatures is indicated by the ratio T_s / T_b. This expression still describes the data when the free carrier density is increased or decreased by different chemical treatments, although the metallic characteristics are strongly reduced for low carrier densities [5].

The situation is quite different, however, for the conductivity of the thin networks shown in Fig. 2: the metallic characteristics are essentially eliminated, with the conductivity extrapolating to almost zero in the zero-temperature limit, and with no sign of a changeover to metallic temperature dependence at higher temperatures. In fact, the conductivity continues to increase near room temperature in a rather linear fashion that does not agree well with the fluctuation-assisted tunneling expression (1). As shown by the fits in Fig. 2, these thin network data show behaviour closer to that for variable-range hopping (VRH) [1]:

$$\sigma(T) = \sigma_0 \exp\left[-\left(\frac{T_0}{T}\right)^{1/(1+d)}\right], \tag{2}$$

where σ_0 and T_0 are constants and d is the dimensionality of the hopping conduction. We note that there is a discrepancy at low temperatures where the conductivity does not decrease as fast as expected for VRH, an effect expected if some tunneling between metallic regions is still occurring as the hopping conduction is frozen out.

We ascribe the change to hopping–type conduction as the networks become thin to the greater difficulty of achieving good metallic percolation paths through the samples as the sample thickness is reduced, as indicated by the dramatically reduced conductance shown in Fig. 1, especially in the thinnest network (Net 1). This interpretation is also supported by our observation that Net 1 gives slightly better agreement with the VRH expression for dimensionality $d = 2$, whereas the others agree better with the $d = 3$ expression (the better agreement with $d = 2$ and 3 than with $d = 1$ suggests that in each case the greatest resistance is associated with hopping between tubes rather than along a single tube).

Summarizing, the mechanism of electron transport is sensitive to the proportion of the nanotube's intrinsic resistance to the inter-nanotube contact resistance participating in electron transport through a SWNT network. This depends on the network density.

ACKNOWLEDGMENTS

This work was supported by EU projects SPANK and CANAPE.

REFERENCES

1. Mott, N. F. and Davis, E. A., *Electronic Processes in Non-Crystalline Materials, 2nd ed.* (Clarendon, Oxford, 1979).
2. Sheng, P., *Phys. Rev. B* **21** (1980) 2180-2195.
3. Oberlin, A., Endo, M., Koyama, T.: *J. Crys. Growth* **32** (1976) 335-349.
4. Ebbesen, T. W., Lezec, H. J., Hiura, H., Bennett, J. W., Ghaemi, H. F. and Thio, T., *Nature (London)* **382**, (1996) 54.
5. Skákalová, V., Kaiser, A. B., Dettlaff-Weglikowska, U., Hrnčariková, K. and Roth, S., *J. Phys. Chem. B* **109** (2005) 7174-7181.
6. Kaiser, A. B. *Rep. Prog. Phys.* **64** (2001) 1-49.
7. Pietronero, L. *Synth. Met.* **8** (1983) 225-231.

Effect of Contact Improvement on the FET Characteristics of an Individual Single Walled Carbon Nanotube

Yunsung Woo and Siegmar Roth

Max-Planck Institute for Solid State Research, Heisenbergstra. 1, 70569 Stuttgart, Germany

Abstract We report that the enhancement in the contact resistance after pulse annealing using the electric-current-induced joule heating in Pd electrode make an effect on the field effect transistor (FET) characteristics consisted of an individual single walled carbon nanotube (SWNT). The SWNTs are deposited onto Pd electrode patterned on SiO_2/Si substrate, through which the electrical currents flow for microseconds to achieve a low contact resistance by meting the Pd electrode for an instant. Here, the transport measurement was performed to an identical SWNT before and after the pulse annealing at various temperatures from room temperature to 4K. From the results, a decrease of barrier height was observed after applying the pulse annealing, but much smaller than the value estimated from an increase of source/drain current. Therefore, we suggest that the enhancement in the contact resistance after the pulse annealing results from an increase of the effective contact length, causing a higher transmission probability, rather than a decrease of the barrier height between SWNT and metal electrode.
Keywords: Single walled carbon nanotube, contact resistance, field effect transistor

INRTODUCTION

The contact between SWNT and metal electrode has become a critical issue to explain the characteristics of carbon nanotube field effect transistor (CNTFET). On the other hand, it has been also reported that a substantial Schottky barrier at the contact between SWNT and source/drain metal electrode limits the transistor conductance and reduce the current delivery capabilities [1]. Therefore, various experimental ways have been attempted to reduce the contact resistance such as thermal annealing [2], metal deposition on nanotube [3], and soldering using electron beam [4]. Theoretical studies also explain the degree of coupling between CNT and metal depending on the type of metal, contact length, chirality, and disorder of CNT [5, 6].

Although the comparative study between the samples having high and low resistance have described the different transport behavior based on the barrier dominated transport, the resistance caused by the deformation and defects in CNT cannot be excluded entirely. Therefore, we studied the transport characteristics of an identical SWNT with decreasing the contact resistance using the pulse annealing. Moreover, it was also studied how the pulse annealing help in lowering the contact resistance between SWNT and metal electrode.

CP786, *Electronic Properties of Novel Nanostructures*, edited by H. Kuzmany, J. Fink, M. Mehring, and S. Roth
© 2005 American Institute of Physics 0-7354-0275-2/05/$22.50

EXPERIMENTAL

The SWNTs used in this experiment are synthesized by HiPco at CNI. The SWNTs are dissolved and sonicated in a 1% aqueous solution of sodium dodecyl sulfate, and then centrifuged to remove the metal particle. Then, they are dispersed on the Pd electrodes patterned on highly doped silicon substrate with a 200 nm thermally grown silicon oxide layer. The Pd electrode was deposited with a width of 1.2 μm, a length of 3.9 μm, and a thickness of 20 nm by electron beam lithography. In particular, the bridge-shaped electrodes were designed to make the pulse passing through the metal electrode, as shown in Fig. 1. After positioning an individual SWNT between two electrodes, microseconds-voltage pulses were applied to the Pd electrodes in sequence and produce Joule heat making the Pd melt and wet to SWNT for an instant.

Electrical transport measurements were performed by employing the conventional current-voltage measurements. The gate voltage dependence as well as the current-voltage characteristics of an individual SWNT are compared before and after pulse annealing at various temperature from room temperature to 4K.

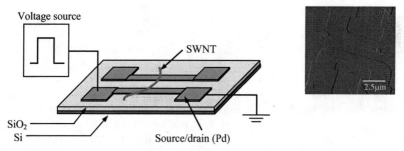

FIGURE 1. Schematic apparatus of the electric-current-induced joule heating method. The pulsed voltages of 6-7 V were applied for microseconds to Pd electrode with a length of 3.9 μm, a width of 1.2 μm, and a thickness of 20 nm.

RESULTS AND DISCUSSION

Figure 2 (a) shows the change of the current-voltage characteristics of an individual SWNT deposited onto Pd electrode before and after the pulse annealing at 293K (room temperature) and 4K, respectively. Here, the measured SWNT have a diameter of about 0.98 nm and a distance between source and drain electrode was 600 nm. The resistance at room temperature, calculated around zero bias in Fig. 2 (a), was decreased from 27.2 MΩ to 3.3 MΩ by applying a pulse of 7 V with 1 μs. As shown in Fig. 2 (b), the conductance around zero bias was almost zero at 14K and 4K for the both devices with and without pulse annealing. As the bias voltage increases, however, the conductance of the annealed device begins to increase slightly, but that of the device before annealing remains to be almost zero to the bias voltage of 0.4 V. In addition, the width between the two biases showing a dramatic increase in conductance, indicated in Fig. 2 (b) as ΔE_1 and ΔE_2, seems to be slightly smaller for the device before the pulse

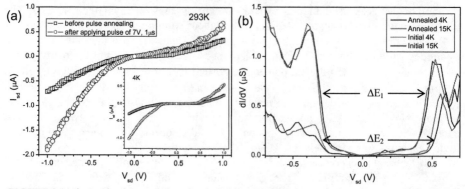

FIGURE 2 (a) the I_{sd}-V_{sd} characteristics of an individual SWNT before and after pulse annealing (the pulse with 7V of width and 1µs of duration time) and measured at 293K and 4K (inset). (b) the conductance derived from the I_{sd}-V_{sd} plot measured at 14K and 4K for the case with and without pulse annealing, respectively.

annealing than that after ($\Delta E_1 < \Delta E_2$). This can be explained by two reasons of the decrease of barrier height and the diminution of the interfacial layer between a SWNT and source/drain metal.

Figure 3 (a) shows the gate dependence of CNTFET before and after pulse annealing, respectively. As clearly seen in Fig. 3, the measured SWNT exhibits p-type characteristics as expected from the Fermi level of Pd close to the valence band of SWNT due to its large work function. However, it is worthy to note that the 'on' state current of the device before annealing is larger than that of the device after annealing. The I_{ON}/I_{OFF} ratio versus temperature was plotted in Fig. 3 (b), compared between the devices before and after annealing. The CNTFET before the pulse annealing showed the large I_{ON}/I_{OFF} ratio, which decrease somewhat as the temperature decrease. On the other hand, the I_{ON}/I_{OFF} ratio of CNTFET after the pulse annealing does not vary considerably over the whole temperature range. This result can be explained by the removal of adsorbed

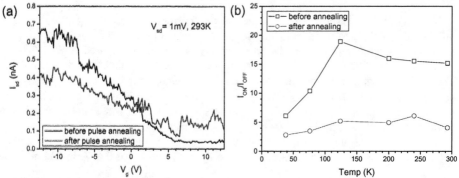

FIGURE 3 (a) the plot of source/drain current versus gate voltage measured at 1 mV of constant source/drain voltage at room temperature (b) the ratio of I_{ON}/I_{OFF} as a function of temperature. The CNTFET characteristics was compared before and after pulse annealing in both of (a) and (b).

oxygen on the surface by heating, and therefore the p-type characteristics are suppressed although the 'off' state current even increase after the pulse annealing.

The barrier height was estimated roughly from the slope of Arrehnius plot assuming the thermionic emission from metal to SWNT dominant at high temperature region. Surprisingly, the barrier height estimated at 100mV of source/drain voltage was found to increase with an amount of about 10 meV after the pulse annealing although the current increased with an order of magnitude from 2.80 nA to 27.8 nA after annealing. Hence, it is suggested that the pulse annealing make the Pd metal melt and wet to the SWNT, and cause the increase of the effective contact length between metal and SWNT resulting in the higher transmission probability of electron from metal to SWNT after all [6].

SUMMARY

We have presented that the enhancement in the contact resistance by the pulse annealing using an electric-current induced joule heating cause a change of transistor characteristics as well as a dramatic increase of source/drain current. The estimated barrier height between SWNT and metal electrode before and after pulse annealing imply that the enhanced current density after the pulse annealing is caused by the increase in the effective contact length not by the reduced barrier height.

ACKNOWLEDGMENT

This work is supported by INKONAMI project.

REFERENCES

1. 1. S. Heinze, J. Tersoff, R. Martel, V. Derycke, J. Appenzeller, and Ph. Avouris, Phys. Rev. Lett., **89**, 106801, **2002**
2. 2. J. O. Lee, C. Park, Ju-Jin Kim, Jinhee Kim, Jong Wan Park, and Kyoung-Hwa Yoo, J. Phys. D, **33**, 1953, **2000**
3. M. Liebau, E. Unger, G. S. Duesberg, A. P. Graham, R. Seidel, F. Kreupl, and W. Hoenlein, Appl. Phys. A, **77**, 731, **2003**
4. D. N. Madsen, K. Mølhave, R. Mateiu, A. M. Rasmussen, M. Brorson, C. J. H. Jacobsen, and P. Bøggild, Nano Letters, **3**, 47, **2003**
5. A. N. Andriotis, M. Menon, and G. E. Froudakis, Appl. Phsy. Lett., **76**, 3890, **2000**
6. M. P. Ananatram, S. Data, and Y. Xue, Phys. Rev. B, **61**, 14219, **2000**

Investigating Schottky Barriers Effects in Carbon Nanotube FETs

Alberto Ansaldo*, Davide Ricci*, Flavio Gatti[†], Ermanno Di Zitti* and Silvano Cincotti*

*Department of Biophysical and Electronic Engineering, University of Genoa,
Via All'Opera Pia 11a, 16145 Genoa, Italy
[†] Department of Physics, University of Genoa, V. Dodecaneso 33, 16146 Genoa, Italy

Abstract The paper addresses the issue of Schottky barrier effects in carbon Nanotube field effect transistors and outlines some solutions to tackle such problem. Microfabricated test patterns have been designed to allow the deposition in a controlled way of single wall nanotubes between different couples of source and drain electrodes in a backgate device architecture. Current-voltage characteristics depending on the back gate applied voltage have been recorded and briefly discussed. Through a proper exploitation of ad-hoc developed contact arrays for CNFET fabrication, a strategy for the direct measurement of metal-CN and CN-CN junction characteristics has been described.

INTRODUCTION

Carbon nanotubes are drawing an ever increasing attention for nanoelectronic device development thanks to their many attractive properties [1]-[2]. For instance, field effect transistors that use carbon nanotubes as channels [3] are claimed to show very high transconductance values with respect to their size and exceptional carrier mobility. Moreover, the use of single wall carbon nanotubes in field effect transistors opens the possibility to develop monolithically integrated nanoelectronic devices [4]-[5] and hybrid fabrication technologies.

However, Schottky barrier effects still need to be studied to fully understand the behavior of carbon nanotube FET (CNFET) devices [6]. While simple current-versus-voltage curves are useful to characterize the channel conductance and the transconductance of the device, they can't reveal the influence and the electrical behavior of the single barriers.

This paper presents a methodology to investigate such effects. Using ad-hoc developed contact arrays for CNFET fabrication through dielectrophoresis deposition of carbon nanotubes onto metal electrodes, we single out a four terminal device made by two crossed nanotubes contacted by four metal pads on a backgate structure. An appropriate measurement strategy allows us to exploit the second carbon nanotube both as a probe to characterize the single Schottky barriers of the device, and as part of a CN-CN junction.

CP786, *Electronic Properties of Novel Nanostructures*, edited by H. Kuzmany, J. Fink, M. Mehring, and S. Roth
© 2005 American Institute of Physics 0-7354-0275-2/05/$22.50

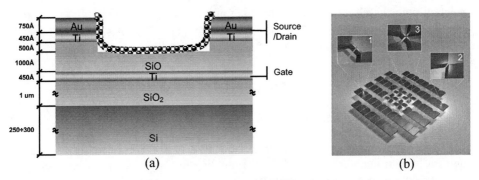

Figure 1. a) schematic cross section our device; b) a sketch of the matrix of contacts.

DEVICE FABRICATION

Standard UV lithography and subsequent evaporation of thin films has been used to build-up surface contact structures (gold on titanium) onto a p-doped silicon wafer with a 1 μm thick thermally grown oxide provided with a titanium back gate layer (figure 1).

The array of contacts were designed with the purpose of minimizing the number of nanotubes between surface electrodes, aiming to test both three-terminal carbon nanotube field effect transistors and five terminal devices (four pads and one gate) that allow the four-point measurements.

Finally we obtained 24 couples of 2 μm wide contacts spaced by a 3 μm gap (b1), 4 couples of triangular shape contacts (b2) and 4 cross-shaped four-contact structures (b3).

In order to directly place the nanotubes between the source and drain electrodes, we followed the electric field method described in [7]. The method has been optimized for our case, taking into account the electrode geometry: good results are obtained using a sonicated 0,5 mg/ml nanotube in ethanol solution and by casting a 1 pl drop of

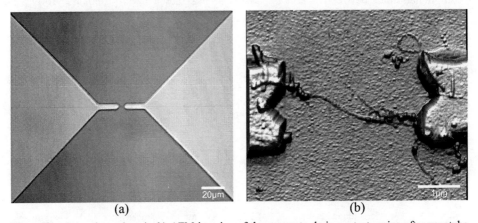

Figure 2. a) An electrode pair; b) AFM imaging of the source-to-drain contact region after nanotube deposition. A nanotube rope that joins metal electrodes is shown.

solution onto the electrodes while applying an ac voltage of 5 V peak-to-peak @ 1 MHz plus a dc bias of 2 V. Figure 2a shows a microphotograph of an electrode pair; a nanotube rope between the electrodes is shown in figure 2b.

RESULTS AND DISCUSSION

The measured characteristics of the devices are shown in figure 3. The devices exhibit a negative threshold voltage, as it can be derived from the mutual characteristics shown in figure 3b, thus showing a typical p-type behavior, in good agreement with literature results. It is worth noting that a strong influence of the gate potential on channel conductivity has been achieved, as shown in the output characteristics (figure 3a).

We also noticed a strong and unexplained asymmetry in the device characteristics. This behavior could be related to the application of a dc component during the dielectrophoresis deposition. An investigation of a possible effect of the offset voltage on the barrier characteristics will be undertaken to have greater insight into such phenomenon.

The shape of the output characteristics is rather different from a conventional FET, and it reminds a rectifying behavior. This effect can be related to a significant influence of the Schottky barriers between metal and nanotubes. In order to thoroughly investigate these effects, fabrication of palladium contacts and crossed nanotubes junctions has been initiated: the idea is to use a scheme based on crossed nanotubes to measure CN-CN junctions and to measure the single Schottky barriers of the CNFET.

The CNT dielectrophoresis deposition technique will be optimized through electric field modelling. Nanotubes are going to be placed crossing one another by two different depositions using the two contact couples separately.

Figure 4a shows the physical layout of the four-point experiment and its corresponding circuit model; a more detailed equivalent circuit is illustrated in figure 4b. If no current flows in R_{cnj} and in the upper Schottky diodes, the voltmeter V can measure the voltage drop between node ε and node γ. According to this scheme, the carbon nanotube between β and δ pads is acting like a nano-probe that allows to

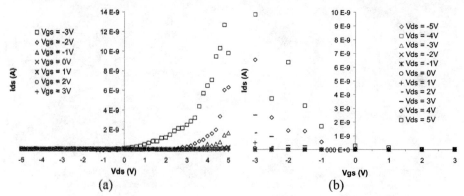

(a) (b)

Figure 3. Electrical characteristics of a CNFET: a) Output characteristics I_{ds}-V_{ds} varying V_{gs}; b) Mutual characteristic I_{ds}-V_{gs} varying V_{ds}. Devices show a typical p-MOS behaviour.

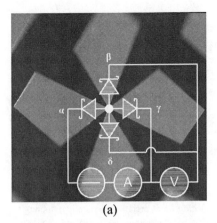
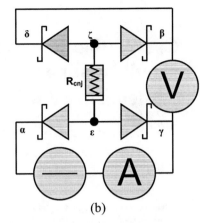

(a)   (b)

Figure 4. a) the pattern of the four-point experiment; b) a more detailed equivalent circuit

investigate the inner behavior of the CNFET between α and γ.

By switching the connections, the other nanotubes can be measured in a similar way, thus making it possible to characterize all four metal-CN junctions.

Moreover, if a current is imposed between nodes γ and β and a voltmeter is placed between node δ and γ, it is possible to measure the R_{cnj} (CN-CN junction) characteristic by a standard four-point measurement.

By changing the metals used to make the contacting pads, the outlined approach allows also to study the influence of electronegativity values within the Schottky barriers.

ACKNOWLEDGMENTS

We thank Matteo Ventrella for his valuable contribution. Work supported by the Italian Ministry of University and Scientific and Technological Research (National Research Program *"Carbon nanotubes for electronics: synthesis, characterization and manipulation"*) and *Fondazione CARIGE*.

REFERENCES

1. M.S. Dresselhaus, G. Dresselhaus and Ph. Avouris, *Carbon nanotubes: synthesis, structures, properties and applications*, Springer, Springler, 2001, pp. 1-420.
2. G.S. Duesberg, J. Muster, H.J. Byrne, S. Roth, and M. Burghard, "Towards processing of carbon nanotubes for technical applications", *Appl. Phys A* **69**, 269-274 (1999).
3. J. Appenzeller, J. Knoch, R. Martel, V. Derycke, S.J. Wind, and Ph. Avouris, "Carbon nanotube electronics", *IEEE Trans. Nanotechnology*, **1**, no. **4**, 184-189 (2002).
4. H.T. Soh, C.F. Quate, A.F. Morpurgo, C.M. Marcus, J. Kong, H.Dai, "Integrated nanotube circuits: controlled growth and ohmic contacting of single-walled carbon nanotubes", *Appl. Phys. Lett* **75**, 627-629 (1999).
5. Y. Tseng, P. Xuan, A. Javey, R. Malloy, Q. Wang, J. Bokor, and H. Dai, "Monolithic integration of carbon nanotube devices with silicon MOS technology", *Nanoletters*, **4**, no.1, 123-124 (2004).
6. P. Avouris, J. Appenzeller, R. Martel, and S.J. Wind, "Carbon Nanotube Electronics", Proc. IEEE, **91**, no. **11**, 1772-1784 (2003)
7. L.A. Naghara, I. Amlani, J. Lewenstein, and R.K. Tsui, "Directed placement of suspended carbon nanotubes for nanometer-scale assembly", *Appl. Phys Lett.* **80**, 3826-3828 (2002).

Correlation between Transport Measurements and Resonant Raman Spectroscopy on site-deposited Individual Carbon Nanotubes

Matti Oron[1,2], Frank Hennrich[2], Manfred M. Kappes[1,2] and Ralph Krupke[2]

[1]Universität Karlsruhe, Institut für Physikalische Chemie, D-76128 Karlsruhe, Germany,
[2]Forschungszentrum Karlsruhe, Institut für Nanotechnologie, D-76021 Karlsruhe, Germany.

Abstract. We performed a comprehensive study on 25 individual metallic and semiconducting single-walled carbon nanotubes, by electron transport measurements and resonant Raman spectroscopy. The tubes were deposited via low-frequency dielectrophoresis, a technique which allows the deposition of both individual metallic and semiconducting tubes from suspension. The tubes were first proven to be individual by atomic force microscopy, then subjected to electron transport measurement characterizations, and subsequently measured by resonant Raman spectroscopy. The prime result is a clear correlation between Raman spectra and electron transport measurements without exception.

INTRODUCTION

Resonant Raman spectroscopy is a powerful tool to characterize single-walled carbon nanotubes (SWNTs) [1]. The technique provides information on the structural and electronic properties due to resonances that occur when the energy of the incoming or outgoing photon matches an allowed optical transition. Various Raman studies were done on tubes where the assignment to either metallic or semiconducting character followed theoretical models [2]. In most of the studies, no additional electrical transport measurements have been performed to support the interpretation of the Raman data, except for very few tubes as in the work of Balasubramanian et al. and Cronin et al. [3-4]. Our aim is to investigate whether the metallic or the semiconducting character of a SWNT, which can be unambiguously determined from electron transport measurements, indeed correlates with a type specific Raman spectrum. For that purpose, we have characterized 25 individual metallic and semiconducting tubes by electron transport measurements and resonant Raman spectroscopy.

The experimental set up was already published in the proceedings of 2004 as well as an example of a wired individual tube [5]. We emphasize that the frequency used in our experiments is much smaller than the one reported for tube separation [6]. At low frequencies ($\omega \ll \omega_C$), with ω_C defined as the surfactant-concentration dependent crossover frequency, both types of tubes experience a positive dielectrophoretic force and are thus attracted towards the electrodes. The tubes were subjected to electrical transport measurements and resonant Raman spectroscopy. Transport measurements

CP786, *Electronic Properties of Novel Nanostructures*, edited by H. Kuzmany, J. Fink, M. Mehring, and S. Roth
© 2005 American Institute of Physics 0-7354-0275-2/05/$22.50

were done at a constant source drain voltage, where the source drain current, I_{SD}, was measured while changing the gate voltage. All the transport measurements were done at room temperature under vacuum (10^{-5} mbar). For the Raman measurements, we used confocal resonant Raman spectroscopy, excited with an Ar^+ ion laser at 2.41 eV or Ne/He laser at 1.96 eV. The laser spot had a diameter of ~500 nm on the sample with a power density of ~1.4 MW/cm^2. The Raman spectra were recorded at room temperature in air with the polarization of the incident light parallel to the long axis of the electrodes and with a spectral resolution of 3.75 cm^{-1}. According to figure 1, we excite at 632.8 nm primarily metallic SWNTs through the first pair of van Hove singularities (E_{11}^M) and at 514.5 nm primarily semiconducting SWNTs through the third pair for optical transition (E_{33}^S).

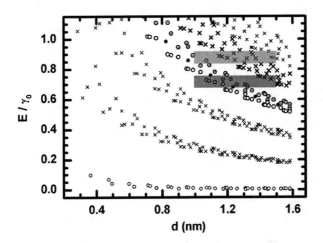

FIGURE 1. Calculation of normalized SWNTs excitation energies as a function of diameter, d, based on a tight-binding model of Ding *et al.* [7] with the nearest neighbor overlap integral $\gamma_0 = 2.75$ eV and the C-C bond length $a_{C-C} = 0.144$ nm. Excitation energies are related to semiconducting SWNTs (X), metallic armchair SWNTs with n = m (full circle) and metallic non-armchair tubes (α,) with split excitation energies [1]. Upper and lower horizontal rectangles contain semiconducting and metallic tubes, which might be in resonance with the corresponding laser wavelengths of 514.5 nm and 632.8 nm, respectively. Their vertical width accounts for the uncertainty of the calculated excitation energy and the horizontal width corresponds to the nanotube diameter range, 1-1.5 nm.

RESULTS AND DISCUSSION

We note that only individual tubes, which showed both transport measurements and a clear radial breathing mode (RBM) signal, were studied in this work i.e., the other contacts were either bridged with long non-resonant individual tubes, possessed too short tube segments or bridged by bundles. Figure 2 shows transport measurements and the corresponding Raman signals of two representative individual SWNTs. Figure

2a (left) represents a typical gate dependence as observed on 15 individual tubes, where the source drain current, I_{SD}, strongly depends on the applied gate voltage, V_G. Switching between the on and off states is accompanied by a rather strong hysteresis originating from the reorientation of dipolar surfactant or water molecules [8], or alternatively due to trapped charges in the gate oxide [9]. The result is typical for a *semiconducting* tube based Schottky-barrier type field effect transistor (SB-FET), which exhibits a p-type behavior due to oxygen exposure [10]. The corresponding Raman spectrum (2a, right) shows a RBM peak at 193.7 cm^{-1}, a disorder D mode peak at 1348 cm^{-1}, and a *narrow* G mode peak at around 1600 cm^{-1}. Figure 2b (left) represent another type of gate dependence as observed on an additional 10 individual tubes. The device can not be switched off at any gate voltage and exhibits a nearly constant I_{SD} for any V_G applied. This behavior is typical for a wired *metallic* tube, where the gate potential does not change the number of conduction channels. The corresponding Raman spectrum (2b, right) reveals a sharp RBM peak at 194.9 cm^{-1}, a D mode peak at 1310 cm^{-1} and a *wide* G mode peak ranging between 1500-1600 cm^{-1}.

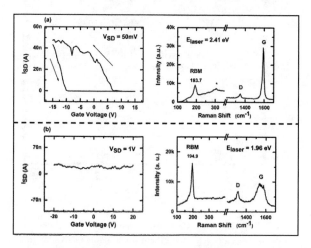

FIGURE 2. Comparison between the electrical transport measurements (left) and the corresponding Raman spectra (right) of an individual semiconducting SWNT (a) and an individual metallic SWNT (b). The type of tube is identified by the specific gate voltage dependence of the source drain current, I_{SD}, and the spectral shape of the G mode (see text). For all semiconducting SWNTs, a gate dependence with a hysteresis, as indicated, is observed in conjunction with a narrow G mode. In contrast, for all metallic SWNTs no gate dependence is observed together with a broad G mode. For most metallic / semiconducting tubes, the G mode appeared rather weak / strong compared to the RBM mode, respectively. Laser excitation energies, RBM frequencies and source drain voltage, V_{SD}, are indicated. The star in the Raman spectrum indicates a contribution from the Si substrate accompanied with a background from the metal electrodes.

Tentative (n, m) assignments were made only for tubes which showed a RBM signal since its appearance indicates that an optical transition is resonant with either an incident or a scattered photon [1]. Within both the single resonance and the double resonance theories, the shape of the G mode is only a characteristic feature of the tube type if the laser energy matches an optical transition.

The prime result of this study is that the shape of the G mode of an individual tube can indeed be used to determine the metallic or semiconducting character, if a clear RBM signal is observed.

In order to assign (n,m) values to our measured tubes, we compared the RBM frequencies of our individual metallic tubes with measured values from Telg et al. [11] and Fantini et al. [12], taking into account an excitation resonance window of $E_{laser} = 1.96 \pm 0.1$ eV. The data from our individual semiconducting tubes we compared with excitation energy calculations of Ding et al. [7], taking into account an excitation resonance window of $E_{laser} = 2.41 \pm 0.1$ eV, while the RBM frequencies for our semiconducting tubes were calculated by the formula given in reference 13. Our entire data, which will be published elsewhere, shows good agreement between experimental and calculated RBM frequencies, which allowed for many of our samples, the assignment of one particular (n, m) tube.

SUMMARY

In conclusion, we have shown, for the first time, the systematic correlation of transport measurements and resonant Raman spectroscopy for both metallic and semiconducting individual SWNTs. All nanotubes, which have been identified by transport measurements as metallic or semiconducting, do show the theoretically expected wide or narrow G mode, respectively. Our results show that the shape of the G mode of an individual tube can indeed be used to determine its metallic or semiconducting character, at least if a clear RBM signal is observed.

REFERENCES

1. Dresselhaus M. S.; Dresselhaus G.; Jorio A.; Souza Filho A. G.; Saito R. Carbon **40**, 2043-2061(2002).
2. Jorio A.; Souza Filho A. G.; Dresselhaus G.; Dresselhaus M. S.; Swan A. K.; Ünlü M. S.; Goldberg B. B.; Pimenta M. A.; Hafner J. H.; Lieber C. M.; Saito R. *Phys. Rev. B* **65**, 155412 (2002).
3. Balasubramanian K.; Fan Y.; Burghard M.; Kern K. *Appl. Phys. Lett.* **84**, 2400-2402 (2004).
4. Cronin S. B.; Barnett R.; Tinkham M.; Chou S. G.; Rabin O.; Dresselhaus M. S.; Swan A. K.; Ünlü M. S.; Goldberg B. B. *Appl. Phys. Lett.* **84**, 2052-2054 (2004).
5. Oron M.; Krupke R.; Hennrich F.; Weber H. B.; Beckmann D.; Löhneysen H. v.; Kappes M. M. Electronic properties of synthetic nanostructures, Kuzmany H., Fink J., Mehring M., Roth S., Eds.; AIP Conference proceedings, American Institute of Physics: New York, 2004.
6. Krupke R.; Hennrich F.; Löhneysen H. v.; Kappes M. M. *Science* **301**, 344-347 (2003).
7. Ding J. W.; Yan X. H.; Cao J. X. *Phys. Rev. B*, **66**, 73401-1 (2002).
8. Woong K.; Javey .A; Vermesh O.; Wang Q.; Li Y.; Dai H. *Nano Lett.* **3**, 193-198 (2003).
9. Fuhrer M. S.; Kim B. M.; Dürkop T.; Brintlinger T. *Nano Lett.* **2**, 755-759 (2002).
10. Xiaodong C.; Freitag M.; Martel R.; Brus L.; Avouris Ph. *Nano Lett.* **3**, 783-787 (2003).
11. Telg H; Maultzsch J.; Reich S.; Hennrich F.; Thomsen C. *Phys. Rev. Lett.* **93**, 177401-1 (2004).
12. Fantini C.; Jorio A.; Souza M.; Strano M.S.; Dresselhaus M.S.; Pimenta M.A. *Phys. Rev. Lett.* **93**, 147406-1 (2004).
13. Bachilo S. M.; Strano M. S.; Kittrell C.; Hauge R. H.; Smalley R. E.; Weisman R. B. *Science* **298**, 2361-2366 (2002).

APPLICATIONS

Fullerene Derivatives for Medical Applications

Andreas Hirsch

Institut für Organische Chemie
Universität Erlangen Nürnberg,
Henkestraße 42
91054 Erlangen, Germany

Abstract. Water-soluble fullerene derivatives have a potential for a variety of medical applications. This is due to the unique structural-, electronic and chemical properties of the fullerene core. The biological properties of suitably functionalized fullerenes range from enzyme inhibition/receptor binding, anticancer and antiviral activity, cell signalling, DNA- and genomic applications, photodynamic activation and most importantly antioxidant properties. This review focusses on the anti-HIV and antioxidant properties of a couple of water soluble fullerene derivatives.

Keywords: Fullerenes, Medical Applications, Antioxidants, Enzyme Inhibitors

HIV-INHIBITORS

The first example of biologically active fullerenes was reported by Friedman and Wudl.[1,2] They proposed that a C_{60} molecule could fit into the hydrophobic cavity of the HIV protease and may inhibit the enzyme. This assumption was soon corroborated by experimental investigations. A polar cylopropanated fullerene was shown to be a good inhibitor of HIV-1 protease with a K_i value of 5.3 μM.[2] For comparison, the best peptide-based protease inhibitors known to date are effective in the nanomolar range. By computer manipulation of the structure the binding of fullerene HIV-1 protease inhibitors was optimized and the the increase of in hydrophobic desolvation computationally predicted. The affinity and activity of fullerene-based HIV inhibitors were increased by evaluation of hydrophobic contacts of hypothetical fullerene derivatives docked into the active site of the enzyme.[3] New target molecules were identified by this method and synthesized, showing improved anti-HIV-1 activities with K_i values in the 100nM range.[4]

Water-soluble C_{60} malonate derivatives with a broad variety of groups R including functional units like e.g. dendrimeric carboxylic acids (e.g. **1**)[5] have been prepared by regioselective cyclopropanation of C_{60} and subsequent transformation of the malonate function (Fig. 1).

The monoaddition pattern retains the C_{60} properties. The dendritic fullerene octadecaacid **1** was highly soluble in water (34 mg/mL at pH 7.4) and no aggregation was observed at pH 7.[6] The substance was non-toxic and excreted *via* the kidneys. It revealed activity in primary human lymphocytes accutely infected with HIV-1LA1 with EC_{50} values of = 0.22 μM.[7] No cytotoxicity was observed, the K_i value against

CP786, *Electronic Properties of Novel Nanostructures*, edited by H. Kuzmany, J. Fink, M. Mehring, and S. Roth
© 2005 American Institute of Physics 0-7354-0275-2/05/$22.50

HIV-protease was determined with 0.10 ± 0.01 μM, activity against HIV-1 p66/51 reverse transcriptase I_{50} was 0.9 μM, in addition, the C_{60} derivative was active against mutant molecular infectious clones of HIV-1, resistant to AZT and/or 3TC reverse transcriptase inhibitors.[7]

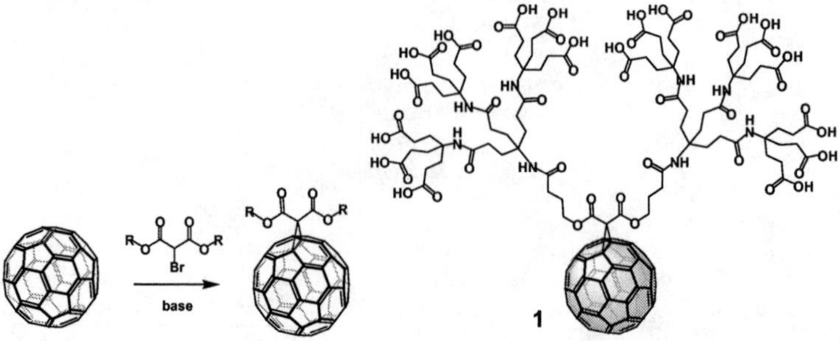

FIGURE 1. Bingel cyclopropanation of fullerene C_{60} with bromomalonate, R = H, alkyl, spacers, functional units, dendrimeric branches (e.g. C_{60} octadecaacid **1**).

Computer modelling based on X-ray crystal structure of the HIV-protease revealed no direct interaction of the fullerene acid **1** with the active aspartate sites. However, **1** fitted well into the hydrophobic pocket of the enzyme, obviously blocking the access to the active site by tight competitive binding.

NEUROPROTECTIVE FULLERENE DERIVATIVES

The free-radical-mediated oxidative injuries have been implicated in the pathogenesis of a number of human diseases. These injuries may occur during accute insults as well as in chronic neurodegenerative conditions. Reactive oxgen species, such as superoxide ($O_2^{\bullet -}$) and hydroxyl ($^{\bullet}OH$) radicals, and the closed shell molecule H_2O_2 may cause oxidative damage to cellular components like cellular membrane lipids, nucleic acids, transport proteins and mitochondria, resulting in glutamate-receptor-mediated excitotoxicity and apoptotic cell death.

Fullerenes can be recognized as effective intercellular antioxidants, since the low lying LUMO of C_{60} can easily take up an electron of the superoxide anion. This fullerene superoxide quenching process appears to be catalytic, i.e., the fullerene can react with many superoxides without being consumed. Also a number of other radicals involving energetically lower lying SOMO levels can be quenched *via* addition reactions. The fullerene antioxidants can be localized within the cell at the mitochondria and other cell compartment sites where free radical production occurs in disease states. Interesting biological activities of water-soluble polyhydroxylated fullerols as potent antioxidants and radical scavengers were reported.[8] Water-soluble fullerene malonate derivatives decreased the toxicity of active oxygen generators,[9] possibly quenching radicals or superoxide. Fullerene compounds revealed important effects on nitric oxide signalling pathways.[10] C_{60} malonic acid caused an inhibitory effect on nitric oxide-dependent relaxation of smooth muscle,[10] producing an augmentation of muscle tone and reducing maximum relaxation.

C_3 and D_3-isomeric and water-soluble C_{60} tris(malonic) acids have been prepared by cyclopropanation of C_{60} and the different regioisomers separated by HPLC. By a recently developed synthesis technique using trismalonate functionalized macrocycles as template addends, defined regioisomeric trisadducts of C_{60} could be obtained in very high (e.g. 96 % *e,e,e*) to complete (*trans-3,trans-3,trans-3*, D_3) regioselectivity.[12,13] Fig. 2 exemplarily shows the synthesis of the *e,e,e*-isomer using a C_8 spacer).

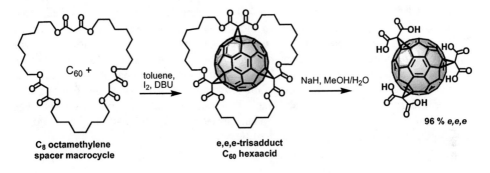

C$_8$ octamethylene spacer macrocycle   **e,e,e-trisadduct C$_{60}$ hexaacid**

FIGURE 2. Highly regioselective synthesis of fullerene C_{60} hexaacid trisadducts using cyclic malonates as addition templates.

The two isomeric C_{60} tris(malonic) acids revealed potent antioxidant properties and appeared as effective scavengers of $O_2^{\bullet-}$ and even more avid scavengers of $O_2^{\bullet-}$ than the spin-trapping agent DMPO, as it was demonstrated by different assays. The C_3 fullerene derivative was able to effectively out-compete DMPO for $O_2^{\bullet-}$, and also reacted as an effective radical scavenger in one-electron Fenton-type reactions.[14-16]

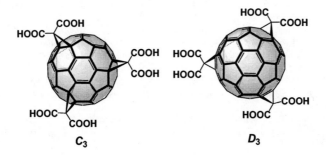

FIGURE 3. Regiomeric fullerene C_{60} trisadduct hexaacids with C_3 and D_3 symmetry.

The water-soluble fullerenes represented effective inhibitors of lipid peroxidation, showing robust neuroprotection against excitotoxic, apoptotic and metabolic insults in cortical neurons. Neuroprotective and cytoprotective properties of C_{60} malonic acids were tested with different cell cultures in coapplication experiments with NMDA and in head-to-head comparison of the C_3 isomer with several other benchmark antioxidants. The C_3 isomer showed complete blocking of NMDA receptor-mediated toxicity (in contrast to all other classes of scavengers) by exposure of the probe to high-dose oxidant.[17]

Reactive oxygen species contribute importantly to apoptotic cell death. Oxidative stress can trigger apoptosis, and free radical production is an early process in programmed cell death. Thus, antioxidants will act as neuroprotectives against programmed cell death/apoptosis. When C_3 was applied as antioxidant agent, neuronal cells failed to exhibit enhanced free radical production and were rescued from apoptotic death.[14] C_{60} derivatives also blocked apoptosis triggered by other insults. Very recently, neuroprotection was demonstrated by C_{60} malonic acid derivatives against toxin-induced death of dopaminergic neurons.[18] In additional ongoing studies *in vivo*, C_3 fullerene derivatives provided neuroprotection in animal models of nervous system disease and enhanced motor function recovery. Rats after impact trauma to their chord gained locomotoric activity.[17] In summary, water-soluble C_{60} derivatives were found to be highly effective neuroprotective agents, blocking excitotoxic, apoptotic, and toxin-induced cell death of neurons and glial cells.

ACKNOWLEDGMENTS

We thank the C Sixty Inc. (USA), the Deutsche Forschungsgemeinschaft DFG, the Fond der Chemischen Industrie, and the Friedrich-Alexander Universität Erlangen-Nürnberg for financial support.

REFERENCES

1. Friedman, S. H., DeCamp, D. L., Sijbesma, R. P., Srdanov, G., Wudl, F., and Kenyon, G. L., *J. Am. Chem. Soc.* **115**, 6506-6508 (1993).
2. Sijbesma, R. P., Srdanov, G., Wudl, F., Castoro, J. A., Wilkins, C., Friedman, S. H., DeCamp, D. L., and Kenyon, G. L., *J. Am. Chem. Soc.* **115**, 6510-6512 (1993).
3. Friedman, S. H., Ganpathi, P. S., Rubin, Y., and Kenyon, G. L., *J. Med. Chem.* **41**, 2424 (1998).
4. Wilson, S. R., "Biological Aspects of Fullerenes," in *Fullerenes, Chemistry, Physics and Technology,* edited by K. M. Kadish and R. S. Ruoff, Wiley Interscience, pp. 437-465, (2000).
5. Brettreich, M., and Hirsch, A., *Tetrahedron Letters* **39**, 2731 (1998).
6. Quaranta, A., McGarvey, D. J., Land, E. J., Brettreich, M., Burghardt, S., Schoenberger, H., Hirsch, A., Gharbi, N., Moussa, F., Leach, S., Goettinger, H., and Bensasson, R.V., *Phys. Chem. Chem. Phys.* **5**, 843 (2003).
7. Schuster, D. I., Wilson, S. R., Kirschner, A. N., Schinazi, R. F., Schlueter-Wirtz, S.,Tharnish, P., Barnett, T., Ermolieff, J., Tang, J.,Brettreich, M., Hirsch, A., *Proc. - Electrochem. Soc.,* **2000-11**, 267-270 (2000).

8. Chiang, L. Y., Wang, L.-Y., Swirczewski, J. W., Soled, S., and Cameron, S., *J. Org. Chem.* **59**, 3960 (1994).

9. Okuda, K., Hirobe, M., Mochizuki, M., and Mashino, T., *Proc. Electrochem. Soc.* **97**, 337 (1997).

10. Lincoln, J., Hoyle, H. V., and Burnstock, G., *Nitric Oxide in Health and Disease.* Cambridge University Press, Cambridge (1997).

11. Satoh, M., Matsuo, K., Kiriya, H., Mashino, T., Hirobe, M., and Takayanagi, I., *Gen. Pharmacol.* **29**, 345 (1997).

12. Lamparth, I., and Hirsch, A., *J. Chem. Soc., Chem. Commun.* **1994**, 1727.

13. Reuther, U., Brandmüller, T., Donaubauer, W., Hampel, F., and Hirsch, A., *Chem. Eur. J.* **8**, 2261 (2002).

14. Dugan, L. L., Turetsky, D. M., Du, C., Lobner, D., Wheeler, M., Almli, R., Shen, C. K. F., Luh, T. Y., Choi, D. W., and Lin, T. S., *Proc. Natl. Acad. Sci. USA* **94**, 9434 (1997).

15. Grayson, M., Lovett, E., Dugan, L., Wang, Y., and Gross, M., *Abs. Am. Soc. Mass Spectrosc.* **1999**, 929.

16. Anderson, D. K., Dugan, L. L., Means, E. D., and Horrocks, L. A., *Brain Res.* **637**, 119 (1994).

17. Dugan, L. L., Lovett, E., Cuddihy, S., Ma, B.-W., Lin, T.-S., and Choi, D. W., "Carboxyfullerenes as Neuroprotective Antioxidants," in *Fullerenes, Chemistry, Physics and Technology,* edited by K. M. Kadish and R. S. Ruoff, Wiley Interscience, pp. 467-479 (2000).

18. Lotharius, J., Dugan, L. L., and O'Malley, K. L., *J. Neurosci.* (1999).

Consideration Of The Toxicity of Manufactured Nanoparticles

Mary L. Haasch[1], Patricia McClellan-Green[2] and Eva Oberdörster[2,3]

[1]*The University of Mississippi, School of Pharmacy, National Center for Natural Products Research, Environmental Toxicology Research Program, University, MS 38677 USA*
[2]*Duke Marine Lab, Beaufort, NC 28516 USA*
[3]*Southern Methodist University, Department of Biological Sciences, Dallas, TX 75275 USA*

Abstract. Fullerene (C60 and single- and multi-wall carbon nanotubes, SWCNT and MWCNT, respectively) is engineered to be redox active and it is thought that the potential toxicity of fullerene exposure is related to the formation of reactive oxygen species. During manufacture, transport or during scientific investigation, there is a potential for human or environmental exposure to nanoparticles. Several studies regarding human exposure have indicated reasons for concern. There is a lack of studies addressing the toxicity of engineered nanoparticles in aquatic species but one study using the fish, largemouth bass, exposed to fullerene has shown increased (10-17-fold) lipid peroxidation (LPO) in the brain. It is likely that repair enzymes or anti-oxidants may have been induced in gill and liver tissues that had reduced LPO compared to control tissues (Oberdörster, 2004). In support of that hypothesis, suppressive subtractive hybridization was used with liver tissue and the biotransformation enzyme, cytochrome P450, specifically CYP2K4, and other oxidoreductases related to metabolism, along with repair enzymes, were increased while proteins related to normal physiological homeostasis were decreased in fullerene-exposed fish. In a new study involving the exposure of a toxicological model fish species, the fathead minnow (*Pimephales promelas*) to water-soluble fullerene (nC60), uptake and distribution indicated that nC60 elevated LPO in the brain and induced expression of CYP2 family isozymes in the liver. In an *in vitro* study, BSA-coated SWCNT interfered with biotransformation enzyme activity. These studies taken together provide support to the hypothesis that the toxicity of manufactured nanoparticles is related to oxidative stress and provide insight into possible mechanisms of toxicity as well as providing information for evaluating the risk to aquatic organisms exposed to manufactured nanoparticles.

Keywords: Fullerene, biotransformation, aquatic, fish, lipid peroxidation.
PACS: 61.48.+c, 61.46.+w, 82.39.Rt, 89.60.Ec, 06.60.Wa, 07.60.Rd

INTRODUCTION

Carbon-based nanomaterials such as fullerene and single- and multi-wall carbon nanotubes are intrinsically redox active. During the manufacture, transport or waste disposal, or during research efforts on these materials, there is a potential for human or environmental C60 or SW/MWCNT exposure. Humans and animals have been exposed to natural sources of nanoparticles throughout the existence of life on earth due to the release of nanomaterials by volcanoes and forest fires or through exposure to viruses. Unlike these natural exposures, anthropogenic activities throughout human history have caused incremental increases in nanoparticle exposure from metal fumes, internal combustion engines, jet engines and within the last 20 years the direct manufacture of nanomaterials. Commercial uses of nanomaterials of all types continue to increase and they are now found in common items such as fuel cells, solar

CP786, *Electronic Properties of Novel Nanostructures*, edited by H. Kuzmany, J. Fink, M. Mehring, and S. Roth
© 2005 American Institute of Physics 0-7354-0275-2/05/$22.50

cells, tires, brake linings, cosmetics and sunscreens. It is anticipated that fullerene will be manufactured by the thousands of tons by 2007. This recent increase, with expanding use in the future, has produced some concern over the possible human and environmental adverse effects due to nanoparticle exposure. Likely routes of exposure therefore include, inhalation/respiration, dermal, ingestion and in the case of biomedical applications, injection. Exposure to the biota can occur by similar routes with the exception of injection. Exposure may lead to translocation into the blood or lymphatic system followed by transportation throughout the organism. The human lung airway and alveoli represent a significant surface area for particle uptake and translocation to the blood stream. Nanoparticles entering the air passages can be eliminated by the mucociliary elevator or may be translocated via the olfactory bulb neurons to the brain, or via the trigeminal nerve bundle to the spinal cord [1]. Consequences of nanoparticle movement into the central nervous system are largely unknown but increased oxidative stress has been shown [1]. It has also been shown that fullerene localizes into plasma membranes [2] and in *in vitro* studies can cause membrane leakiness [3]. A recent study has indicated that the majority of inhaled nanoparticles do not reach the lung [1]. Nevertheless, the long-term consequences of a single bolus exposure or of intermittent or low-level chronic exposures are virtually unknown.

Environmental exposures of the biota can occur during manufacture, transport or waste disposal. The consequences of such exposures are just beginning to be explored. In a study of lethal toxicity using a standard Environmental Protection Agency (USEPA) toxicity test, the lethal concentration resulting in 50% organism mortality (LC_{50}) in *Daphnia magna* was ~800 ppb or moderately toxic when compared to other common chemical contaminants [4]. In additional exposures to a fish, the largemouth bass (LMB; *Micropterus salmoides*), to water-soluble C60 (nC60 solubilized using tetrahydrofuran followed by water) found increased lipid peroxidation in brain tissue indicating neural translocation [4] similar to mammalian studies [1].

For this project, the LMB study was expanded to include suppressive subtractive hybridization to investigate changes in gene expression and an additional study using a standard toxicological model, the fathead minnow (*Pimephales promelas*), was undertaken to include water-solubilized nC60 and an investigation of CYP2 protein expression. An *in vitro* study of the effect of bovine serum albumin (BSA)-coated SWCNT on biotransformation enzyme activity was also performed.

MATERIALS AND METHODS

Suppressive subtractive hybridization was performed as previously described [5] on samples derived from a previous study [4]. Differentially expressed mRNA were cloned and compared to known sequences (GenBank) for similarity.

Fathead minnow (FHM) studies were performed in reconstituted hard water (RHW) and 48 hr treatments included: control (RHW), and 0.5 ppm nC60 (stirred for 2 wks in MilliQ system water and isolated by centrifugation 2 x 10 min at 4,000xg). Fish were individually exposed in 1 L glass aquaria with a 50% water change at 24 hr. Lipid peroxidation (LPO) and immunodetection were performed as previously described [4, 6].

BSA-coated SWCNT were added to fish liver microsomal preparations prior to determination of ethoxycoumarin-*O*-deethylase activity, a well-known biomarker activity for several P450 isozymes in fish [7]. Microsomes were pre-incubated for 30 min with 0, 25, 50, 100 or 500 µg BSA-SWCNT and the ability to metabolize 7-ethoxycoumarin was determined [8].

587

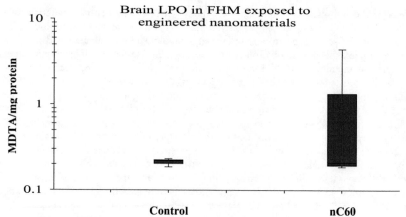

FIGURE 1. Brain lipid peroxidation in male FHM exposed to 0.5 ppm nC60 (n = 5).

RESULTS

Additional findings from the previous study in LMB [4] using suppressive subtractive hybridization, included possible increased repair mechanisms in the gill and liver and altered gene expression in these tissues including genes related to the immune response, inflammation and tissue repair. Interestingly, the expression of the mRNA for the biotransformation enzyme, CYP2K4, along with other oxidoreductases and repair enzymes was increased while the mRNAs for proteins related to normal homeostasis were decreased in nC60-exposed fish compared to control (data not shown).

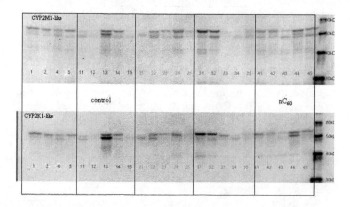

FIGURE 2. SDS-PAGE and immunodetection of CYP2M1/2K1-like proteins in FHM. Control: lanes 11-15; 0.5 ppm nC60: lanes 41-45 (as labeled). Other lanes are of samples not described in this manuscript.

FHM treated with nC60 exhibited elevated LPO in brain tissue (Fig. 1) and it was determined that the expression of CYP2 family proteins (CYP2M1/2K1-like) in liver was significantly induced by 0.5 ppm nC60 exposure (Fig. 2) after removal of lane 13 from the data analysis.

In an *in vitro* study, fish microsomes pre-incubated for 30 min with varying amounts of BSA-coated SWCNT exhibited significant inhibition of 7-ethoxycoumarin-*O*-deethylase activity (Fig. 3).

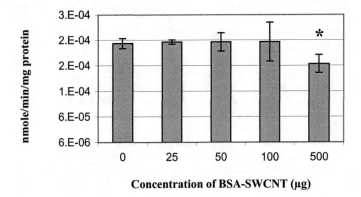

FIGURE 3. Effects of BSA-coated SWCNT on *in vitro* 7-ethoxycoumarin-*O*-deethylase activity in fish liver microsomes.

CONCLUSIONS

Evidence exists for reasonable concern regarding possible health effects in humans exposed to manufactured nanoparticles [1]. Environmental exposure of the biota is just beginning to be explored. Along with a previous study [4], these initial findings indicate changes in gene expression, lipid peroxidation (increased oxidative stress) and effects on metabolism (biotransformation enzyme activity) in nanoparticle exposed aquatic organisms.

ACKNOWLEDGEMENTS

The authors wish to thank Ya-Ping Sun, Ph.D., Clemson University, Clemson, SC for providing the iron-free BSA-SWCNT and Shiqian Zhu, The University of Mississippi, Environmental Toxicology Research Program, National Center for Natural Products Research, University, MS for performing the immunodetection of CYP2M1/2K1-like proteins in fathead minnow.

REFERENCES

1. Oberdörster, G., E. Oberdörster, and J. Oberdörster, *Nanotoxicology: An emerging discipline evolving from studies of ultrafine particles.* Environ Health Perspect, 2005. **doi:10.1289/ehp.7339.**
2. Foley, S., et al., *Cellular localisation of a water-soluble fullerene derivative.* Biochem Biophys Res Commun, 2002. **294**(1): p. 116-119.
3. Sayes, C., et al., *The differential cytotoxicity of water-soluble fullerenes.* Nano Letters, 2004. **4**(10): p. 1881-1887.
4. Oberdörster, E., *Manufactured Nanomaterials (Fullerenes, C60) Induce Oxidative Stress in Brain of Juvenile Largemouth Bass.* Environ Health Perspect, 2004. **112**(10): p. 1058-1062.
5. Larkin, P., et al., *Gene expression analysis of largemouth bass exposed to estradiol, nonylphenol, and p,p'-DDE.* Biochem Mol Biol, 2002. **133**: p. 543-557.
6. Haasch, M.L., *Effects of vehicle, diet and gender on the expression of PMP70- and CYP2K1/2M1-like proteins in the mummichog.* Mar Environ Res, 2002. **54**: p. 297-301.
7. Haasch, M.L., et al., *Use of 7-alkoxyphenoxazones, 7-alkoxycoumarins and 7-alkoxyquinolines as fluorescent substrates for rainbow trout hepatic microsomes after treatment with various inducers.* Biochem. Pharmacol., 1994. **47**(5): p. 893-903.
8. Lake, B.G., *Preparation and characterization of microsomal fractions for studies on xenobiotic metabolism*, in *Biochemical Toxicology: A practical approach*, K. Snell and B. Mullock, Editors. 1987, IRL Press: Washington, DC. p. 183-215.

Ultra microelectrodes from MWCNT Bundles

M. Kaempgen[a], S. Roth[a]

[a]Max-Planck-Institute for Solid State Research, Heisenbergstrasse 1, 70569 Stuttgart, Germany

Abstract. Bundles of Multi Wall Carbon Nanotubes (MWCNT) are used as ultra microelectrodes. They show featureless cyclic voltammograms in 0.1m KCl indicating the absence of surface functionalities. Typical behavior for radial diffusion was found for faradaic processes as expected for low dimensional electrodes. Since CNTs are hydrophobic by nature, strategies are explored to overcome the surface tension of aqueous electrolytes. Additionally, manipulations with insulating coatings are demonstrated. These more technological aspects are very interesting in terms of practical use for all applications of individual CNT electrodes.

INTRODUCTION

Carbon Nanotubes (CNTs) are widely used as electrodes in electrochemistry because they are chemically related to the commonly used carbon electrode structures. For example, CNT sheets, so called Bucky paper (BP), are promising candidates for electrochemical energy conversion in super capacitors [1,2] or lithium batteries [3,4], whereas free standing CNTs have been proposed as electrode arrays for electrochemical bio sensors [5,6] where the possibility of functionalization is a key property. In such arrays, the freestanding CNT are widely separated from each other and work as individual ultra microelectrodes in order to maximize the sensitivity for a desired reaction. Since the preparation of freestanding individual CNTs (single tubes or arrays) is more or less laborious, the use of single CNT bundles could be much easier for basic investigations of such electrodes. Although CNT bundles have a much larger diameter than individual CNTs the chemistry can be expected to be comparable because all devices are from the same material. Moreover, such bundles have a high aspect ratio and the dimensions are still small (μ-scale) which could be interesting in terms of practical use in small volumes or spacings (holes, ditches). Tip functionalized ultra microsensors for local measurement (pH, O_2 concentrations) or for locally controlled reactions could be a potential application of such ultra microelectrodes.

In this paper we investigated some basic electrochemical properties of MWCNT bundles. Additionally, we experimented with different parameters in order to assess the feasibility for use in practical applications.

EXPERIMENTAL

MWCNTs are grown in a typical CVD (chemical vapor deposition) process. For that, a 10 nm iron layer was evaporated on a silicon substrate as catalyst. The substrate was placed in a tube furnace at 800 C° and 1 mbar acetylene atmosphere. Under these

CP786, *Electronic Properties of Novel Nanostructures*, edited by H. Kuzmany, J. Fink, M. Mehring, and S. Roth
© 2005 American Institute of Physics 0-7354-0275-2/05/$22.50

conditions, we typically get a 'forest' of aligned MWCNTs. For the electrode preparation, a thin wire was covered at the tip with Ag-epoxy as a conductive adhesive. The prepared wire was then attached to a CNT bundle protruding from the forest of aligned MWCNT. This procedure was carried out with assistance of an optical stereo microscope (Wild Heerburg) and an xyz-translation stage (Leitz, Wetzlar, Germany). A commercial glass pipette was used as electrochemical cell where the open tip (diameter 0.8 mm) was used as access for the working electrode and the open end for the reference and counter electrode. The wire with the glued MWCNT bundle was fixed to the xyz-translation stage in order to control position and immersion depth of the CNT ultra microelectrode. For Bucky paper (BP) preparation, a CNT suspension in 1% Sodium dodecyl sulfate (SDS, Sigma Aldrich) was prepared with aid of ultrasound. The concentration was typically 1-2mg/ml. All electrochemical characterizations were carried out with a potentiostat (Jaissle 83 PC, Germany) vs. a saturated calomel electrode (SCE) as reference. For electrolyte preparation, deionized water and all other chemicals are used as delivered without any further purification. Fig. 1 shows a microscope image of our typical setup and SEM images of the MWCNT bundle attached to a Pt wire.

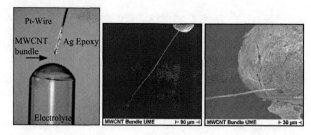

Figure 1. Optical microscope image of the set up and SEM images of a MWCNT bundle, attached to a Pt wire.

RESULTS AND DISCUSSION

Electrochemical Characterization

The first electrochemical characterization of the CNT ultra microelectrode is carried out in aqueous electrolyte without any redox couple in order to investigate oxygen functionalities. Such functional groups would cause some features around 0.2 V vs. SCE [7]. Fig. 2 shows a typical cyclic voltammogram (CV) of our MWCNT bundle. The potential window is limited by H_2 and O_2 evolution at around -0.5 V and 1.2 V, respectively. The absence of any features within this potential window indicates that the MWCNT bundle has no oxygen functionalities that are observable at this level of sensitivity. So one can start from the assumption that the surface properties of essentially pure carbon will be measured.

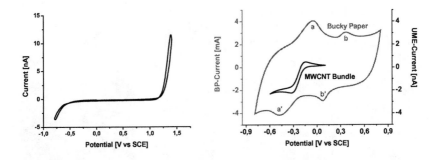

Figure 2. (left): CV of a single MWCNT bundle in 0.1M KCl, scan rate: 100mV/s
Figure 3. (right): CV of macroscopic BP (grey) and a μ-scale MWCNT bundle (black) in 5 mM Ru(NH₃)₆Cl₃ in 0.1M KCl. The scale on the right side belongs to the inner CV (MWCNT bundle) whereas the left scale to the BP. Scan rate was 0.1V/s.

However, faradaic processes within this potential window can be used to investigate ultra microelectrode behavior. In order to compare a MWCNT bundle electrode with macroscopic electrodes the use of a MWCNT bucky paper is ideal. Unfortunately, the adhesive forces between MWCNTs are much weaker than between Single Wall Carbon Nanotubes (SWCNTs), and we did not succeeded in making a BP from MWCNTs. Instead, we prepared a BP from SWCNT and used it for comparison purposes. Fig. 3 shows a CV of a MWCNT ultra microelectrode (inner curve) in comparison to a macroscopic BP (outer curve) for the Ru^{3+}/Ru^{2+} redox couple known for fast electron transfer at MWCNTs [8]. The CV of the BP is dominated by capacitive charging due to its high surface area. Nevertheless, two pairs of redox processes can be observed which can be assigned to the Ru^{3+}/Ru^{2+} redox couple (a/a': $Ru(NH_3)_6^{3+} \leftrightarrow Ru(NH_3)_6^{2+}$) and a protonation/deprotonation reaction of oxygen functionalities at the surface (b/b': =CO ↔ =C-OH) [7]. However, for MWCNT bundles well defined peaks are missing but a saturation in current can be observed for the Ru^{3+}/Ru^{2+} redox couple. This limiting current is a characteristic for radial diffusion and known for small electrode dimensions [8]. We did not extend the potential range towards more positive values (as done for the BP) because oxygen functionalities were not expected (Fig. 2). The reason for such oxygen functionalities in the SWCNT BP is likely due to damage resulting from necessary treatment of the material during preparation. Additionally we observed for all our samples a small scan-rate-independent hysteresis. This feature has also been observed in previous experiments, but the origin is still unclear [8].

Manipulation

For practical use of such MWCNT ultra microelectrodes it is essential to control several properties such as insulating or wetting behavior. The following describes a possible approach to these two parameters.

Wetting behavior: Since CNT are hydrophobic by nature the surface tension of the electrolyte must be overcome, otherwise the bundles bend and swim on the surface of the aqueous electrolyte, as demonstrated in Fig. 4a. It is possible that short and stiff bundles could be forced to be immersed, but this is not possible for our long bundles which can be over 1mm in length. In principle, one can change either the surface tension of the electrolyte or the hydrophobic surface of the CNT. In Fig. 4 we demonstrate both approaches:

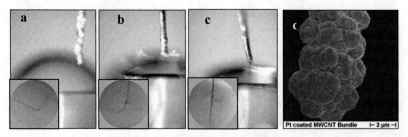

Figure 4. Wetting behavior of a MWCNT bundle dipped into a) water, b) 1% SDS, c) water after coating with platinum and d) SEM image of a platinum coated MWCNT bundle.

Whereas in an aqueous electrolyte the MWCNT bundle swims on the surface (Fig. 4a), an immersion into the electrolyte is achieved if some surfactant is added (1% SDS). An alternative approach would be to apply a metallic coating. For that we used platinum which can be easily electroplated out of a platinum salt solution (1mM K_2PtCl_6 in 0.1M KCl). Fig. 4c demonstrates that a Pt coated MWCNT bundle shows not only improved wetting behavior but also a stiffening of the whole bundle. Fig. 4d shows SEM images of such a platinum coated MWCNT bundle in more detail:

Insulating Coating: An insulating coating of the MWCNT ultra microelectrode is desirable for reactions that should only occur on the very tip. Other advantages of coating include possible stiffening and an improved wetting behavior. Polyphenol was used as insulating coating since it can be easily polymerized electrochemically on the CNT surface. In order to demonstrate the effectiveness of an insulating coating, we measured the limiting current at different immersions depths before and after coating. Fig. 6 a and 6b show clearly that the limiting current of an uncoated MWCNT bundle scales linearly with immersion depth. The intercept in Fig. 6b gives the current for the very tip and therefore the diameter of the bundle can be calculated using the Cottrell equation for disc electrodes [9]. For this sample we found a diameter of about 1.5 µm. In order to get a very small active electrode area we tried to remove the insulating coating from the very tip of the MWCNT bundle. Therefore we immersed just the very tip into the electrolyte and applied a negative potential of –1V. Such a negative potential leads to H_2 generation which blows away the coating. In principle the size of the skinned part is only limited by the resolution of the optical microscope and the xyz-translation stage which is always used for positioning. Although the insulating coating is permeable for protons, larger ions can be hindered effectively to reach the surface. For the insulated part, the limiting current does not depend linearly on the immersion depth any more, as shown in Fig. 8c. The nearly zero slope of the insulating part clearly proves the effectiveness of the insulating polyphenol coating.

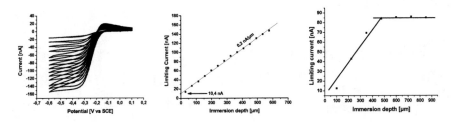

Figure 5a (left): CVs of uncoated MWCNT ultra micro electrode at different immersion depths in 5 mM Ru(NH₃)₆Cl₃ in 0.1M KCl.
Figure 5b (middle): Linear dependence of limiting current from immersion depth.
Figure 5c (rigth): Dependence of limiting current from immersion depth in 5 mM Ru(NH₃)₆Cl₃ in 0.1M KCl. The MWCNT bundle was first completely coated with polyphenol which is then removed from the very tip only.

CONCLUSIONS

Single MWCNT bundles were used as electrodes for electrochemical applications. We observed typical radial diffusion behavior of ultra microelectrodes and the absence of oxygen functionalities for our samples within the sensitivity limits of our set up. We showed that the hydrophobic nature of the MWCNT can be overcome either simply by a surfactant or metallic coatings. Effective insulating coatings are used in order to get very small electrode dimensions at the tip. Such modifications are of great interest for electrochemical applications with CNTs.

ACKNOWLEDGMENTS

This work was supported by the EU project CANAPE.

REFERENCES

1. J.H. Chen et al., Carbon 40, 1193, 2002
2. Ch. Emmenberger et al., J. Pow. Sour. 124, 321, 2003
3. G. Maurin et al., Chem. Phys. Lett., 312, 12, 1999
4. E. Frackowiak et al., Carbon 40, 1775, 2002
5. Y. Lue et al., Nano Lett. 4 (2) 191, 2004
6. J. Li et al., Nano Lett. 3 (5) 597, 2003
7. J. Barisci et al., J. Electroanal. Chem. 488, 92, 2000
8. J. K. Campbell, J. Am. Chem. Soc. 121, 3779, 1999
9. Bard Faulkner, J. Wiley & Sons, Inc., 2001

Purification and Percolation – Unexpected Phenomena in Nanotube Polymer Composites

P. Pötschke[1], B. Hornbostel[2], S. Roth[2], U. Vohrer[3], S.M. Dudkin[4], I. Alig[4]

[1]Leibniz Institute of Polymer Research Dresden, Hohe Str. 6, 01069 Dresden, Germany
[2]Max Planck Institute for Solid State Research, Heisenbergstr. 1, 70569 Stuttgart, Germany
[3]Fraunhofer Institute of Interfacial Engineering and Biotechnology, Nobelstr.12, 70569 Stuttgart, Germany
[4]Deutsches Kunststoff-Institut Darmstadt, Schlossgartenstr. 6, 64289 Darmstadt, Germany

Abstract. SWNT were synthesized by arc-discharge methods. The as produced material was treated by different procedures in order to get purified SWNT. The treatments led to higher conductivity values as measured on bucky papers. The SWNT were incorporated into a polymer matrix of polycarbonate by melt mixing using a small scale compounder. The as produced SWNT material led to electrical percolation between 2 and 3wt%. Unexpectedly, the composites with the treated SWNT showed higher percolation concentrations. This is contradictory to theoretical expectations. These results would suggest that increasing the purity of SWNT does not lead automatically to lower percolation contents when dispersing SWNT in a molten polymer by shear mixing. During the purification the interactions between SWNT bundles might be increased, thus leading to a more agglomerated SWNT material which is more difficult to wet, infiltrate and disperse during the melt mixing procedure.

INTRODUCTION

One of the application of carbon nanotubes (CNT) in polymeric matrices is to get conductive materials which is attained by forming a percolation structure of the conductive CNT in the insulating polymer matrix. This percolation concentration is quite low as compared to those achieved with carbon black structures because of the fibrous structure of the CNT. However, it is known that the percolation composition does not only depends on the aspect ratio but also on CNT structure and purity. In previous investigations on multi-walled carbon nanotubes (MWNT) we could show that smaller tube diameter and higher CNT purity can shift the percolation composition towards lower MWNT amounts [1,2]. Here, very thin MWNT (Nanocyl S.A. Belgium) showed electrical conductive composites at 3 wt% for the as produced material (purity >60%) but at 2 wt% after MWNT purification (purity >95%).

Thus, it was our intention to apply the incorporation of CNT material into a polymeric matrix and to measure the percolation threshold composition as an indirect method in order to characterize the purity of single-walled carbon nanotube (SWNT) materials. When enhancing the SWNT purity the amount of SWNT material needed to form the percolation pathway through the polymer matrix should be lower, thus reaching percolation at lower SWNT concentrations. Melt mixing was used as method of incorporation since it was already shown as a suitable method [3,4]. Also, in context with industrial applications of nanotube/polymer systems, melt mixing is the preferred method of composite preparation.

CP786, Electronic Properties of Novel Nanostructures, edited by H. Kuzmany, J. Fink, M. Mehring, and S. Roth
© 2005 American Institute of Physics 0-7354-0275-2/05/$22.50

Using the direct incorporation method, first a wetting of the nanotube surface with polymer has to be achieved as well as an infiltration of the polymer into existing nanotube agglomerates. Wetting depends mainly on the CNT surface characteristics, the interfacial tension between nanotube surface and polymer melt, and the polymer melt viscosity. For the infiltration the density of the CNT agglomeration structure seems to have an influence, as discussed in [5] for different MWNT structures having different network morphologies in epoxy. The next task is to achieve a good distribution and dispersion of the CNT into homogenously distributed single tubes.

EXPERIMENTAL

The SWNT material was synthesized by the arc-discharge method in a standard Krätschmer reactor [6]. The melting-off carbon anode was doped with Ni and Y (1.3 at%/0.2 at%). The as synthesized material was collected from defined regions in the reactor with high SWNT ratio (no wall- and no cathode-material). The SWNT materials from different runs were harvested and macroscopically homogenized by extensive stirring in an ethanol emulsion. After the evaporation of the ethanol the strongly agglomerated material was pounded in a mortar to a fine powder. A TEM micrograph of the as synthesized SWNT is shown in Figure 1 indicating single tube diameters of 1.0-1.3 nm and a bundled structure.

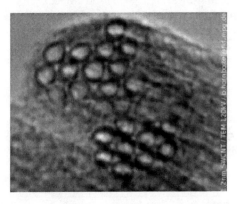

FIGURE 1. TEM micrographs of the as produced SWNT material indicating the bundled structure

The purification was done as following: Purification V1) dry purification in a furnace under air at 355°C for 400 min, weight loss during purification: 15 wt%; Purification V2) dry purification in a furnace under air at 370°C for 13 h, weight loss after purification: 26 wt%; Purification / functionalisation V3) as in V2, followed by storage in a flask with $SOCl_2$ at room temperature over night; Purification V4) as in V2, as treated SWNT were suspended in SDS using ultrasonic agitation, two runs of centrifugation at 5000 min^{-1} were applied using the lighter material for the next step, filtration and rinsing with distilled water as a final step.

For comparison, SWNT produced by the HiPco method from by Carbon Nanotechnology Inc, (Houston, USA) was used (according to CNI 5% impurities). Except using a furnace temperature of 300°C the material was treated accordingly to V4.

The conductivity of SWNT material was measured on bucky papers using the four-leads-method. A strip of bucky paper with a width of 5 mm and a press-on force of 0.6 N were used. The conductivity was calculated using the measured thickness of the bucky paper and the distance between the measuring electrodes (10 mm). Raman spectroscopy was performed on the up-side of the bucky papers using a Jobin Yvon LabRAM 010 at 632.817 nm. Powder X-ray diffraction was measured using a turning flat sample holder, Cu-Kα radiation and a semicircle detection plate (STOE IP PSD).

As polymer matrix polycarbonate (PC) Iupilon E2000 (Mitsubishi Engineering Plastics, Japan) with a zero shear viscosity at 260°C ($\eta_{0,260}$) of 6800 Pa s was used. Melt compounding was carried out using a DACA Micro Compounder (capacity 4.5 cm^3) operated at 280°C, 50 rpm, and 15 min. After drying of polymer and SWNT (100°C, 2 h vacuum) the powdery materials were premixed by shaking in a glass bottle until optical homogeneity and fed into the running compounder. Electrical volume resistivity was measured on compression molded sheets (diameter > 60 mm, thickness ~ 0.35 mm). A Keithley electrometer Model 6517A equipped with an 8009 Resistivity Test Fixture was used to measure high resistivity samples (full symbols). Lower resistivity composites were measured by a four point test fixture combined with a Keithley electrometer Model 2000 on strips (20 mm x 3 mm) cut from the sheets (open symbols). In addition, dielectric (AC) measurements were performed at room temperature in a frequency range from 10^{-3} to 10^7 Hz using a frequency response analysis system consisting of Solartron SI1260 Impedance/Gain Phase Analyzer and Novocontrol broadband dielectric converter with the BDC active sample cell [7]. Discs (d=20 mm) were cut from the sheets and gold layers were sputtered onto both sides as electrodes. Scanning electron microscopy (SEM) was performed using a LEO VP 1530 microscope on raw materials or fractured composite samples.

RESULTS & DISCUSSION

The purification of the as produced materials led to an increased conductivity of the bucky papers as shown in Table 1. The highest effect is visible after thermal treatment and centrifugation leading to 4 times higher value as compared to the starting SWNT.

TABLE 1: Conductivity of the bucky papers prepared from differently purified SWNT

		Conductivity in S/cm
V0	as produced SWNT material	55
V1	purified 2x200 min at 355°C	108
V2	purified 13h at 370°C	107
V3	purified 13h at 370°C and treated with SOCl₂ for 25h	-
V4	purified as V3 but additionally centrifuged	200

Figure 2 shows SEM micrographs of the bucky paper structures indicating clearly the reduction of amorphous carbon, graphite material and catalyst particles from the material after purification V4. Raman spectroscopy and X-ray diffraction (Fig. 3) also indicate the effect of purification. In the Raman spectra the D/G ratio is reduced upon purification indicating less amount of disordered structures within the material. X-ray diffraction spectra show that the amount of the nickel catalyst is reduced and the corresponding peaks are shifted towards nickeloxide bands. Based on these investigations, we consider the material as "purified".

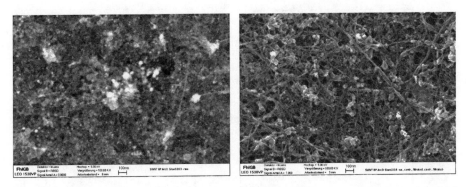

FIGURE 2. SEM micrographs of bucky paper of as produced SWNT (left), purified V4 material (right)

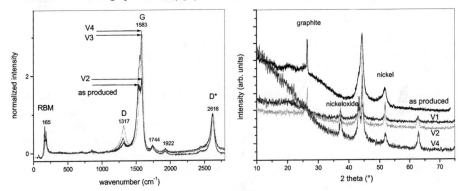

FIGURE 3. Raman spectroscopy (left) and X-ray diffractometry (right) on differently purified SWNT

The SWNT materials were incorporated into PC using melt mixing conditions as adopted in PC/MWNT composites [7,8]. The measured volume conductivity values are summarized in Figure 4. Whereas the as produced SWNT material shows percolation between 2 and 3 wt% the percolation thresholds of the purified SWNT are shifted towards higher values. In case of the purification/functionalisation V3 no percolation at all was observed up to 5 wt% SWNT. These results were also confirmed by dielectric spectroscopy as shown in Figure 5.

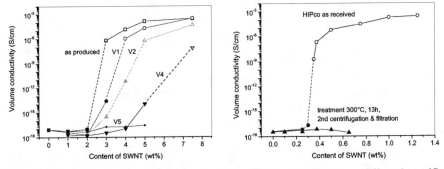

FIGURE 4. Electrical volume conductivity for PC composites prepared from differently purified SWNT: DC conductivity on arc-discharge material (left) and on HIPco SWNT treated like V4 (right)

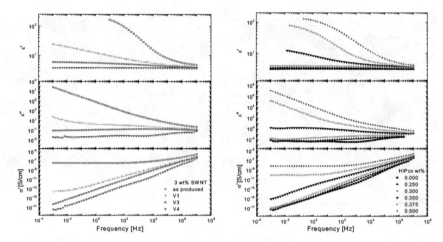

FIGURE 5. AC measurements with ε' real part, ε''imaginary part of complex permittivity, and σ' real part of complex conductivity for PC composites prepared from differently purified SWNT: arc-discharge material (left) and on HIPco SWNT treated like V4 (right)

These findings are contradictory to theoretical expectations and to the found conductivity values for the bucky papers. Therefore, we additionally applied a similar treatment like in V4 to HiPco material. Conductivity of the bucky paper increased from 155 S/cm to 573 S/cm after purification. The composite with the as-produced SWNT material percolates between 0.35 and 0.375 wt% SWNT content, whereas in the treated HiPco SWNT no percolation can be found up to 0.65 wt%.

These results imply that purified SWNT material in a polymer matrix does not automatically lead to lower percolation thresholds when being dispersed by shear mixing. One possible explanation might be that defects caused by the cleaning process weaken the lattice and, thus, the tubes break more easily in the intense shear mixing process. As a consequence, the aspect ratio would be reduced leading to higher percolation thresholds. However, we suppose another effect to be mainly responsible for increased percolation threshold. This effect could be addressed as "dispersability". By the removal of non-SWNT allotropes we achieve increased interactions between the SWNT bundles. This leads to dense agglomerated structures of the SWNT which are difficult to disperse in the melt mixing procedure. The initial impregnation step (wetting and infiltration of polymer melt into the SWNT structures) is hindered and the applied shear forces are not sufficient to untie these agglomerates efficiently.

Breton at al. [5] discussed a similar case comparing MWNT structures synthesized using different catalysts and therefore leading to different densities and network entanglement. The MWNT having higher entanglement were more difficult to be impregnated by the epoxy resin and consequently to disperse. Thus, we conclude that we have a comparable effect induced by the applied purification processes. Regarding the morphology of the composites by SEM this assumption might be applicable (Figure 6). In comparison some non-wetted agglomerates and also nicely distributed bundles of SWNT can be observed in the composite containing the as-produced SWNT material. However, in case of purified SWNT the amount of agglomerates increases. Figure 6 depicts a sample with a highly compacted SWNT agglomerate being not wetted at all by polymer.

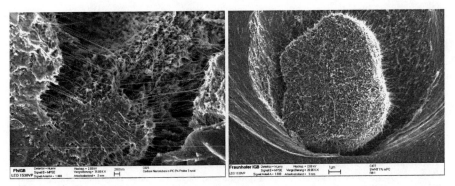

FIGURE 6. SEM micrographs of nanocomposite samples of SWNT in PC: 1 wt% SWNT as produced (left), 1 wt% SWNT purified V4 (right)

CONCLUSIONS

The microstructure of the CNT material as used for melt mixing with polymers plays an important role on the distribution and dispersion and, thus, on the property performance of polymer-CNT nanocomposites. We have shown that melt mixing can be a quite useful method in order to get nanocomposites with percolation threshold as low as 0.35 wt% HiPco as-produced SWNTs. However, when using purification methods the higher density and agglomeration structure of the CNTs becomes more difficult to be wetted, infiltrated and dispersed in the melt mixing procedure. There is a need to quantify the dispersability of CNT structures and to develop easy and applicable test methods. Further work has to be done to adapt the melt mixing procedure to CNT incorporation, e.g. adapting screw configuration and temperature regime to influence wetting, infiltration, distribution and dispersion.

ACKNWOLEDGMENTS

The authors thank Dr. Ursula Dettlaff, MPI-FKF, Stuttgart, and Martin Dubosc, CNRS-IMN, Nantes-France, formerly MPI-FKF for their helpful assistance. P. P. acknowledges financial support from the German Federation of Industrial Cooperative Research Associations "Otto von Guericke" (AIF) within the project 122Z.

REFERENCES

[1] P. Pötschke., A.R. Bhattacharyya, A. Janke, S. Pegel, A. Leonhardt, C. Täschner, M. Ritschel, S. Roth, B. Hornbostel, J. Cech, *Fullerenes, Nanotubes, and Carbon Nanosrt.*, **13**, 211-224 (2005).

[2] P. Pötschke, A.R. Bhattacharyya, A. Janke, H. Goering, *GAK*, **58**, 1, 45-51 (2005).

[3] D.W. Ferguson, E.W.S. Bryant, H.C. Fowler, *ANTEC'98*, 1998, pp. 1219-1222.

[4] P. Pötschke, T.D. Fornes, D.R. Paul *Polymer* **43**, 3247-3255 (2002).

[5] Y. Breton, G. Désarmot, J. P. Salvetat, S. Delpeux, C. Sinturel, F. Béguin, S. Bonnamy, *Carbon* **42**, 1027-1030 (2004)

[6] U. Detlaff-Weglikowska , J. Wang, J. Liang, B. Hornbostel and S. Roth, Purity Evaluation of Bulk Single Wall Carbon Nanotube Materials, Kirchberg Proceedings 2005

[7] P. Pötschke, S.M. Dudkin, I. Alig, *Polymer* **44**, 5023-5030 (2003).

[8] P. Pötschke, A.R. Bhattacharyya, I. Alig, S. M. Dudkin, A. Leonhardt, C. Täschner, M. Ritschel, S. Roth, B. Hornbostel, J. Cech, *AIP Conference Proceedings* 723, 2004, pp.478-484.

The Space Elevator

Bryan E. Laubscher

Los Alamos National Laboratory

Abstract. The Space Elevator is conceived to be a carbon nanotube ribbon stretching from an Earth station in the ocean on the equator to far beyond geosynchronous altitude. This elevator co-rotates with the Earth. Climbers ascend the ribbon using power beamed from Earth to launch spacecraft in orbit or to other worlds. The requirements of the ribbon material, challenges to the building of the space elevator, deployment and the promise of the space elevator are briefly discussed in this paper.

Keywords: space elevator, breakthrough technology, space access, carbon nanotubes
PACS: 95.40.+s, 81.07.De, 42.68.Wt, 89.30.Cc, 94.30.Ch, 52.75.Di

INTRODUCTION

This paper is a short description of the space elevator (SE) based upon the conceptual outline described in the book, *The Space Elevator*[1] by Edwards and Westling. This concept involves a meter wide carbon nanotube (CNT) ribbon thinner than a sheet of paper that extends from a point on the Earth's equator to 100,000 km above that point to a counterweight. To use the ribbon to access space, mechanical climbers ascend the ribbon laden with payloads. Power for these climbers is expected to be beamed from the ground in the form of infrared, laser energy.

To first order, the center of mass of the system resides near geosynchronous altitude: thus the system co-rotates with the Earth. The counterweight, in uniform circular motion around Earth, provides a restoring force. At geosynchronous altitude the ribbon is widest because the forces are greatest. Below geosynchronous Earth orbit (GEO), the net force on a ribbon segment is toward the earth and the tensile strength of the segment must support all below it. Above the ribbon the net force is away from Earth and a segment must hold all segments beyond itself. As a climber ascends the ribbon the tension in the ribbon provides a restoring force that continually increases the angular momentum of the climber.

The SE system is analogous to a railroad. Once the rails are laid, the cost to transport across the system is low. Such a system is subject to the economy of scale. In the case of the SE, it is expected that the cost to low Earth orbit (LEO) would fall by a factor of 10 to 100. Currently, the space shuttle costs approximately $23,000 per kilogram to LEO. The cost to middle Earth orbit (MEO) and GEO would decrease by a greater factor. This dramatic drop in launch costs will enable the exploitation of space to solve Earth's problems.

The SE is a new paradigm for accessing space and as such is an enabling technology. Commerce in space including power generation, tourism, manufacturing,

CP786, *Electronic Properties of Novel Nanostructures*, edited by H. Kuzmany, J. Fink, M. Mehring, and S. Roth
© 2005 American Institute of Physics 0-7354-0275-2/05/$22.50

mining and commercial/government exploration will be developed since transportation will have dropped to a small part of the cost of the enterprise.

HISTORY AND MATERIALS

Many individuals may have conceived of an SE since the concept is a part of human consciousness. The Tower of Babel, Jack and the Beanstalk and references to "stairways to heaven" have existed for a long time. More recently, the academician, Konstantin Tsiolokovsky, a Pole living in Russia around 1895 wrote of towers extending from Earth up into the cosmos while visiting Paris and seeing the Eiffel Tower. In 1945, Sir Arthur C. Clarke patented the geosynchronous satellite and pointed out its usefulness for communications. There is a report that an American John McCarthy studied a "Skyhook" in the early 1950s.

However, the first published discussion of an SE-like structure was by the Russian engineer Y.N. Artsutanov[2]. In 1960 he published a short article in *Pravda* for children. The concepts and numbers presented in this article make it clear that he understood the concept and had done calculations to back up his statements. The first journal article was by four scientists from Woods Hole Institute[3] in 1966. Presumably the long cables that oceanographers used to explore ocean trenches inspired the authors to apply this technology to space. The last person to discover the space elevator was Jerome Pearson in 1975. Pearson published the most detailed study[4] at that time and has continued to investigate the concept up to the present.

These discoverers understood the promise of the SE but all recognized that no material existed that was strong and lightweight enough to make the SE a reality. Therefore, the SE existed in many forms in science fiction for many years[1].

In 1991 Iijima[5] discovered carbon nanotubes (CNTs). CNTs have been reported theoretically to possess a tensile strength in the range of 300 GPa[6-7]. The Earth SE requires around 100 GPa of tensile strength minimum[1]. An actual ribbon would be built with strength contingency so consider 130 GPa or higher as the required strength. Thus these new materials offer, for the first time, the possibility of building an SE cable.

A macroscopic SE ribbon could be formed with CNTs by spinning individual fibers or creating a composite material. Spinning of nano-scale CNTs require long tubes (possibly millimeters or centimeters) so that the Van der Waals forces are effective over a length sufficient to provide strong bonding between tubes. Handling nano-scale fibers is a challenge so spinning CNTs into micron size threads is another option. Composites possess other challenges that include the chemical bonding of the CNTs to the matrix material, achieving uniform distribution of CNTs throughout the matrix and the alignment of the CNTs in the matrix. Whatever technique is adopted for the SE ribbon, the process must be adaptable to manufacturing so that 100,000 km of ribbon can be fabricated in a timely fashion. Finally, the cost of CNTs must drop dramatically before an 800 ton SE cable can be manufactured by any of the previous techniques.

In 1999 NASA held a conference on the SE. The scenario that emerged from the meeting was that an asteroid would be captured and placed into GEO around the earth.

The carbon on the asteroid would be used to build a massive, CNT tower from the asteroid down to Earth's surface. Four magnetically levitated trains would run up the tower at very high speeds, delivering people and cargo to the GEO station. The asteroid would end up slightly above GEO to preserve the center of mass residing at GEO and would act as a counterweight. One estimate of the time to realize this vision was 300 years!

At that point Bradley C. Edwards became interested in the SE. He proposed to NIAC and was awarded a Phase I and later a Phase II contract. The conceptual design that he developed is a simpler, bootstrapping method that is realizable in a much shorter time – possibly 15 years if a well funded, focused, concerted program were initiated. The rest of the paper will discuss the aspects of this conceptual design as it has evolved since 2000 when other scientists and engineers have begun to study and refine aspects of the plan. Such research has only just begun and many more questions (beyond the materials) need to be answered before construction can begin.

DEPLOYMENT SCENARIO

Deployment of the first space elevator will be a major effort. The basic plan is to launch the components, deployment spacecraft and pilot ribbon, into LEO. This will take many launches as even a narrow (about 20 cm) ribbon and its deployment spacecraft are massive. Modern rocket launch vehicles do not have the capacity to carry such massive payloads to LEO in a single launch. In LEO the system is assembled possibly with the help of astronauts. Then the system must be lofted to GEO above the desired point on Earth, probably the ocean on the equator about 300 km west of the Galapagos Islands.

Two possible scenarios are being considered to thrust the spacecraft to GEO. The first scenario is to launch all the fuel to LEO and use an efficient high specific impulse engine to rocket the system to GEO. This increases the required number of launches. The second scenario is to launch propellant for an engine (say an ion engine) and then beam the power up to the spacecraft from 3 or more power beaming station floating on the Earth's oceans. These stations will eventually converge on the ground point and be used to power climbers up the ribbon. The second technique will require fewer launches and will prove the power beaming systems capabilities.

Once the deployment spacecraft is at GEO over the ground station, the ribbon can be let out. The end may need a small propulsion system to get the ribbon started toward Earth and to handle the angular momentum change as it descends. A "homing" device on the end will facilitate intercepting the ribbon and affixing it to the ground station. During the spooling out of the ribbon, the spacecraft has risen above GEO so that the center of mass remains there. The spacecraft also must thrust to stay above the ground point because of the angular momentum change at different altitudes. Once the ribbon is completely deployed the spacecraft acts as the counterweight.

The pilot ribbon has a lifetime of only a few years because of impacts from small debris that cannot be actively avoided. Therefore, immediately, climbers must be sent up the ribbon to attach more ribbon to the SE. These climbers must be engineered for the small capacity of the pilot ribbon. As the ribbon capacity increases, the climbers will be larger and will carry more ribbon. After two years, a meter wide ribbon rated at 20 tons (extra tension) would be completed. Subsequent, higher capacity SEs will be built with an existing SE in much less time, possibly 5 months. Indeed, each elevator's first task may be to build it successor ribbon.

CHALLENGES

Even if the ribbon and deployment system was ready to go, there exist issues that must be dealt with before the SE can be deployed. These issues include the technologies required to ascend the ribbon, hazards to the ribbon and the problem of humans traveling on the SE above LEO.

Climbers that ascend the ribbon must operate over a wide range of environments, possess high reliability and climb the ribbon without damaging it. Climbers are assembled, loaded and launched in the troposphere. The regions encountered during ascent include the stratosphere, thermosphere, ionosphere, and magnetosphere. The atmospheric pressure, temperature and composition change dramatically as the altitude increases. The climbers must operate in all these environments including the radiation environment of the magnetosphere. Climbers must be reliable enough to make at least one 100,000 km trip if not a round trip. A stranded climber could compromise the use of the SE.

Climbers also must have a wide range of uses and so will differ in design. Most climbers will carry and launch payloads. Others will carry out diagnostic measurements, repair tasks, ribbon laying, science experiment and rescue of stranded climbers.

The power beaming system that will energize the climbers has three major system components; the infrared laser, the large (~12 meter) telescope and the adaptive optics system. Each of these components (or a close example) exists separately but has never been operated as a system. An adaptive optics system is required to focus the energy onto the climber photocells through the atmosphere. Current adaptive optics systems are not capable of phasing the entire aperture of large telescopes.

Hazards to the SE include winds, lightning and aircraft in the troposphere; atomic oxygen in the upper atmosphere; and radiation, solar storms, orbital debris, orbiting satellites and meteorites in the magnetosphere. Placing the ground station on the equator in a region with few storms and defending a no-fly zone around the SE mitigate the troposphere hazards. Coating the SE ribbon with a metal like nickel or aluminum would protect it from atomic oxygen. LEO space satellites and debris greater in size than 10cm can be avoided by moving the lower end of the elevator. Therefore, the ground station may be a ship or navigable floating platform. The width of the ribbon (1 meter) and its curved cross-section allow the SE to survive micrometeor and small debris collisions. The statistical rarity of large meteors renders this hazard a low probability. The CNT ribbon is hard to proton and electron radiation.

Another hazard is the dynamics of the ribbon. Oscillations can be induced by winds, moving the ground station, the gravitational attraction of the sun and moon, solar storms, solar wind pressure, magnetospheric electromagnetic interactions, thermal heating/cooling and climbers ascending the ribbon. The ribbon will need to be stabilized by active damping from the counterweight, base and possibly at GEO.

The climbers are envisioned to travel at about 200 km/hr. This means that the climbers spend a significant amount of time in the Van Allen radiation belts. The lower belt is mainly trapped energetic protons and the outer belt is primarily energetic electrons. Humans riding on the elevator spend so much time in the radiation belt that the radiation dose is beyond the safe level. Therefore, mitigation techniques must be developed or climbers must significantly increase in speed before humans ride the elevator beyond LEO.

PROMISE OF THE SPACE ELEVATOR

The low cost of access to space promised by the space elevator will enable the exploitation of space. Currently, commercial space business is only profitable in the case of communications. With lower cost to space, many types of commerce will be profitable. Solar power satellites that beam power to Earth will provide clean, inexpensive electrical power. Earth observation and scientific space missions will be expanded in number and capability. Human and robotic exploration as well as colonization will be possible at much lower cost. SEs could be thrown to the moon and Mars and deployed to enable supplying settlements and two way trade.

Space tourism, now being vitalized by the success of Space Ship One, will be stimulated by the SE as well. Humans will ride to LEO at first, returning either back down the elevator or by dropping off the elevator and re-entering the atmosphere.

REFERENCES

1. Edwards, B.C. and Westling, E.A., *The Space Elevator*, Houston, BC Edwards, 2002.
2. Artsutanov, Y., V Kosmos na Elektrovoze, *Komsomolskaya Pravda*, 1960.
   (contents described in Lvov, *Science*, 158:946)
3. Isaacs, J.D., Vine, A.C., Bradner, H., Bachus, G.E., Elongation into a true 'Skyhook', *Science*, 151:682, 1966.
4. Pearson, J, The Orbital tower: a spacecraft launcher using the Earth's rotational energy, *Acta Astronautica*, 2:785, 1975.
5. Iijima, S., *Science*, 254:56, 1991
6. Yu, M.F., Lourie, O., Dyer, M.J., Molini, K., Kelley, T.F., and Ruoff, R.S., *Science*, 287:637, 2000
7. Yu, M.F., Files, B.S., Arepalli, S., and Ruoff, R.S., *Physical Review Letters, 84, no. 24:5552, 2000*

Tunable Nanoresonator

K. Jensen, Ç. Girit, W. Mickelson, and A. Zettl

Department of Physics, University of California at Berkeley
Center of Integrated Nanomechanical Systems, University of California at Berkeley
The Molecular Foundry, Lawrence Berkeley National Laboratory, and
Materials Sciences Division, Lawrence Berkeley National Laboratory,
Berkeley, CA 94720 U.S.A.

Abstract. We have created a tunable mechanical nanoscale resonator with potential applications in precise mass, force, position, and frequency measurement. The device consists of a specially prepared multiwalled carbon nanotube (MWNT) suspended between a metal electrode and a mobile, piezo-controlled contact. By exploiting the unique telescoping ability of MWNTs, we controllably slide an inner nanotube core from its outer nanotube casing, effectively changing its length and tuning its flexural resonance frequency.

INTRODUCTION

Due to their low mass, low force constants, and high resonance frequencies, nanoscale resonators are currently being used to weigh single bacteria, detect single spins in magnetic resonance systems, and probe quantum mechanics in macroscopic systems [1]. These resonators are typically micromachined from silicon; however, because of their low density, high Young's modulus, and atomically perfect structure, carbon nanotubes provide an alternate, nearly ideal building material. Some progress has been made in constructing nanotube-based resonators [2]. However, these resonators have a narrow frequency range and obey a complicated physical model. We propose a fundamentally different nanotube resonator, which takes advantage of one of carbon nanotubes' most interesting properties. MWNTs, which consist of multiple, concentric nanotubes precisely nested within one another, exhibit a striking telescoping property whereby an inner nanotube core may slide within the atomically smooth casing of an outer nanotube shell [3]. Already this property has been exploited to build a rotational nanomotor [4] and a nanorheostat [5]. Future nanomachines such as a gigahertz mechanical oscillator are envisioned [6]. By harnessing this versatile telescoping property in a new fashion, we have created a tunable nanoscale resonator.

CP786, *Electronic Properties of Novel Nanostructures*, edited by H. Kuzmany, J. Fink, M. Mehring, and S. Roth
© 2005 American Institute of Physics 0-7354-0275-2/05/$22.50

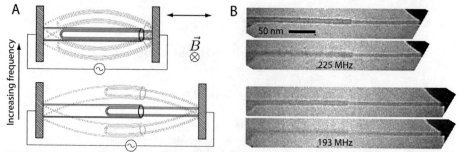

FIGURE 1. (A) Schematic of tunable nanoresonator. A specially prepared MWNT is suspended between a metal electrode and a mobile contact. Telescoping the nanotube in or out changes the effective length of the beam and tunes its resonance frequency. (B) Device in action. The top two images show the nanoresonator at one extension before resonance (sharp) and during resonance (blurred). The bottom two images show the nanoresonator with a lower resonance frequency after the inner nanotube has been telescoped out.

DEVICE OPERATION

Figure 1(A) is a schematic drawing of our tunable nanoresonator. A MWNT is suspended between a metal electrode and a mobile, piezo-controlled contact. By peeling the outer shell of the MWNT [7] and exposing the inner core, we exploit its unique telescoping ability. Like a trombone player shifting notes, we controllably slide the inner nanotube from its casing using the mobile contact, effectively changing the length of the MWNT and tuning its resonance frequency. In the top image, the resonator is fully retracted and has a relatively high resonance frequency. In the bottom image, the resonator is extended and consequently has a lower resonance frequency. By operating the device in an external magnetic field and applying alternating current through the nanotube, we can excite the mechanical vibrations of the nanotube via the Lorentz force. With a transmission electron microscope (TEM) it is possible to detect these vibrations through the physical displacement of the beam.

TEM micrographs in Fig. 1(B) show our tunable nanoresonator in action. The first two images show the nanotube beam at one extension before resonance (sharp) and during resonance at 225 MHz (blurred). The final two images show the nanotube beam after the inner nanotube has been telescoped out 50 nm. The resonance frequency has shifted downward to 193 MHz.

The resonance frequency of our tunable nanoresonator obeys the Euler-Bernoulli beam equation, which makes explicit the dependence of the resonance frequency on the total length of the tube. Specifically, the frequency f_n of the nth mode is given by

$$f_n = \frac{\beta_n^2}{8\pi} \frac{\sqrt{r_{outer}^2 + r_{inner}^2}}{L^2} \sqrt{\frac{E}{\rho}}; \qquad \beta_n \approx 4.74, 7.85, 11.00, \dots \qquad (1)$$

where r_{outer} is the effective radius of the outer wall of the nanotube system, r_{inner} is the effective radius of the inner core of the nanotube system, L is the length of the nanotube system, E is the Young's modulus, and ρ is the density. Effective radii are used to account for the fact that the actual radii are not constant across the length of the nanotube and change during telescoping. Notably, the tension of the beam does

608

not appear in the equation because it is held at a fixed, almost negligible value by the van der Waals attraction between the nanotubes.

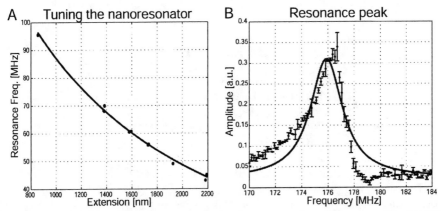

FIGURE 2. (A) Plotting frequency versus extension demonstrates the tuning process. The data follow the $1/L^2$ dependence expected from the Euler-Bernoulli equation. (B) Typical resonance peak with a Lorentzian fit.

RESULTS AND DISCUSSION

As a demonstration of our ability to tune the resonator, we plot the resonance frequency versus extension length for a typical nanoresonator in Fig. 2(A). The resonance frequency follows the predicted $1/L^2$ dependence. Moreover, from the fit we calculate the Young's modulus of a MWNT to be approximately 1.2 TPa in agreement with accepted values [8]. Also apparent in the graph is the extreme sensitivity of the resonance frequency to the extension length; 100 nm corresponds to a 5 MHz shift. Some of our other tunable resonator devices have shown sensitivities as great as 1 nm per 1 MHz shift. The extreme sensitivity of the resonance frequency to the length of the beam suggests possible application as a nanoscale positioning device or an extremely sensitive strain gauge.

Figure 2(B) shows a typical resonance peak that was observed by sweeping frequencies while maintaining the same extension. Our current detection techniques require that the resonator be driven at large amplitudes, likely in the non-linear regime. This could explain the relatively low quality factor (Q=244) and the odd shape of the resonance peak. Magnetomotive detection [9] should allow for smaller amplitudes, which may increase the quality factor.

By harnessing the almost frictionless sliding in telescoping MWNTs, we have created a fundamentally new breed of tunable nanoscale resonator. Our resonator is unique in that it is the only nanoresonator tunable via an effective length change, which suggests various nanoscale-positioning or strain measurement applications. Also because the frequency follows a $1/L^2$ dependence, our resonator has a relatively wide frequency range compared to other tunable resonators. These advantages make our tunable nanoresonator an interesting candidate for further study.

ACKNOWLEDGMENTS

This research was supported in part by the U.S. National Science Foundation and the U.S. Department of Energy.

REFERENCES

1. H. G. Craighead, *Science* **290**, 1532 (2000).
2. V. Sazonova, Y. Yaish, H. Ustunel et al., *Nature* **431**, 284 (2004).
3. J. Cumings and A. Zettl, *Science* **289**, 602 (2000).
4. A. M. Fennimore, T. D. Yuzvinsky, W. Q. Han et al., *Nature* **424**, 408 (2003).
5. J. Cumings and A. Zettl, *Phys. Rev. Lett.* **93** (2004).
6. Q. S. Zheng and Q. Jiang, *Phys. Rev. Lett.* **88** (2002).
7. J. Cumings, P. G. Collins, and A. Zettl, *Nature* **406**, 586 (2000).
8. P. Poncharal, Z. L. Wang, D. Ugarte et al., *Science* **283**, 1513 (1999).
9. A. N. Cleland and M. L. Roukes, *Appl. Phys. Lett.* **69**, 2653 (1996).

Single-Wall Carbon Nanotube Suspension as a Passive Q-Switch for Self Mode-Locking of Solid State Lasers

N.N. Il'ichev[1], S.V. Garnov[1], E.D. Obraztsova[2]

[1] *A.M.Prokhorov General Physics Institute, RAS, 38 Vavilov Street, 119991, Moscow, Russia*
[2] *Natural Sciences Center of A.M.Prokhorov General Physics Institute, RAS, 38 Vavilov St., 119991, Moscow, Russia*

Abstract. Single-wall carbon nanotube suspension in D_2O was used as a passive Q-switch for self mode-locking in flash-lamp pumped solid state lasers with different active media providing the output radiation with different wavelengths: YAG:Nd (λ=1064 nm), YAP:Nd (λ=1340 nm) and Nd-doped glass (λ=1055 nm).

Keywords: self-mode locking, nanotubes.
PACS: 42.55.Rz; 78.67.Ch

INTRODUCTION

Single wall carbon nanotubes (SWNT) are under intensive investigation now. The studies are aimed to improve the nanotube synthesis [1], to investigate the tube electronic structure [2], which manifests itself in optical absorption, luminescence and resonance Raman spectra [1,3-6]. A majority of synthesis techniques [7-9] gives bundles of nanotubes with different geometry and conductivity. A substantial progress in obtaining of individual SWNT has been achieved after development of a new purification procedure including an intensive ultrasonication of SWNT together with a surfactant in water, followed by ultracentrifugation, resulting in a mass-separation of the dispersed material fragments. An upper fraction of the final suspension contained a considerable percentage of individual nanotubes rather than ropes [1]. The rope disintegration is confirmed by an appearance of well-resolved peaks, corresponding to individual tubes, in optical absorption and photoluminescence spectra [1,4,6] and by a loss of the Fano-type contour of the tangential mode [10] in the Raman spectrum.

Pump-probe measurements have revealed short relaxation times (from 50 fs to 50 ps) of electronic excitations in SWNT [11-13]. This observation, together with a wide spectral range covered by SWNT absorption, is very promising for development of SWNT-based fast passive Q-switches.

Until recently only one example of SWNT-based passive Q-switch has been demonstrated. SWNT films, deposited [14] or grown [15] onto the fiber cross-section, were used in Er-doped continuous wave (*cw*) diode- pumped *fiber* laser (λ=1540 nm).

A different approach is explored in our experiments. A *liquid SWNT suspension in D_2O* is considered as an efficient and thermo-stable nonlinear optical medium. A

CP786, *Electronic Properties of Novel Nanostructures*, edited by H. Kuzmany, J. Fink, M. Mehring, and S. Roth
© 2005 American Institute of Physics 0-7354-0275-2/05/$22.50

nonlinear transmission of SWNT suspension has been measured at wavelength 1540 nm [16]. In this case the saturation intensity for a probing pulse with duration 250 ns has been estimated as 1 MW/cm^2. The suspension demonstrated significant residual losses. Under the probing pulse peak intensity of 40 MW/cm^2 a decrease of the absorption coefficient was 3.6 cm^{-1}, while the background coefficient value was 17 cm^{-1}.

The nonlinear characteristics of SWNT suspension in D_2O allow to consider it as an efficient passive Q-switch for a number of solid state lasers. Recently we have demonstrated that such suspension serves as an efficient and thermo-stable passive Q-switch in 2 solid state lasers with different active media: Er^{3+}-doped glass (λ=1540 nm) [16] and LiF crystals with F_2^- color centers (λ=1150 nm). [17]. To the best of our knowledge for LiF: F_2^- - laser the passive self-mode locking was obtained in the first time. *Cw* diode-pumped Yb-doped fiber laser was used as a pump source. The pump radiation was chopped and 0.5 ms pulses were used for pumping the laser. The chopping was necessary to prevent the passive Q-switch from boiling under pump radiation. In Fig. 1 the oscilloscope trace of this laser output intensity is shown.

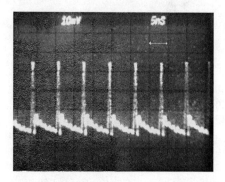

FIGURE 1. Oscilloscope trace (5 ns/div) of LiF: F_2^- - laser with a passive Q-switch based on SWNT.

In this work we report about obtaining a self-mode locking regime with the passive Q-switch based on suspension of SWNT in D_2O in YAG:Nd, YAP:Nd and Nd-doped glass flash-lamp pumped solid state lasers. Optical characteristics of SWNT suspension in D_2O (optical absorption and Raman spectra) are presented.

EXPERIMENT

Raw HiPCO [18] material was used for preparation of SWNT suspension in D_2O. For bundle disintegration the suspension was exposed to ultrasound during 1 hour and subjected to ultracenrifugation (170000 g, 1 hour). The preparation procedure and optical characteristics of this suspension are described in detail elsewhere [6]. In Fig. 2 the Raman spectra of our HipCO SWNT material before the rope disintegration *(powder)* and after this procedure *(aqueous suspension)* are presented. A loss of the Fano-type contour of the tangential Raman mode (at 1592 cm^{-1}) [10] after treatment may be considered as an indicator of the rope disintegration.

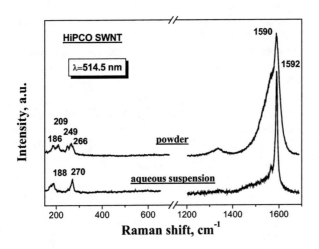

FIGURE 2. The Raman spectra of HipCO SWNT before the rope desintegration *(powder)* and after this procedure *(aqueous suspension)*

The laser resonator was linear. The cell with SWNT suspension was placed nearby the resonator mirror. The cell thickness was 0.1 mm. An optical transmission spectrum of the cell in the range 1000 – 1700 nm is shown in Fig. 3. The cell and the active element were installed in resonator at some angle with respect to the resonator axis to prevent a frequency discrimination due to reflection of the laser radiation by the optical faces of the resonator elements.

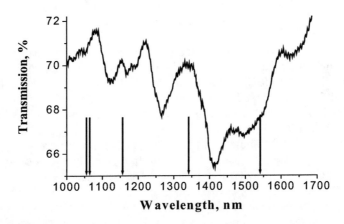

FIGURE 3. An optical transmission spectrum for the cell containing SWNT suspension in D_2O. The Q-switch based on this suspension has been used in our work for realization of a passive self-mode locking regime in several lasers. Arrows indicate the working wavelengths of these lasers.

An oscilloscope TDS-3052 and a photodiode LFD-2a with a total resolution time about 1 ns were used for registration of the self-mode locking traces.

In Fig. 4 the oscilloscope traces for YAP:Nd laser (λ=1340 nm), locked by SWNT passive Q-switch, are shown. A time-dependent laser radiation output is a train of short (about 1 ns) pulses. The time interval between two successive pulses (see Fig. 4b) is equal to the photon round trip through the laser resonator.

A self-mode locking has been obtained also for YAG:Nd (λ=1064 nm), and Nd-doped glass (λ=1055 nm) solid state lasers, pumped with a flash lamp. For these two lasers the time-dependent outputs are very similar to those shown in Fig 4. Due to this we have omitted their images in this paper.

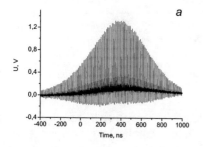 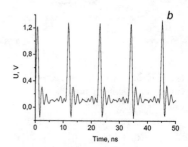

FIGURE 4. The oscilloscope traces of YAP:Nd laser with a passive Q-switch based on SWNT; a) 200 ns/div, b) 20 ns/div.

CONCLUSION

We have realized a self mode-locking regime with SWNT suspension -based passive Q-switch in several solid state lasers with different active media and different wavelengths: Er-doped glass (λ=1540 nm) [16], LiF crystals with F_2^- color centers (λ=1150 nm) [17], Nd-doped glass (λ==1055 nm), YAG:Nd (λ=1064 nm) and YAP:Nd (1340 nm). In Fig. 3 the arrows denote the laser wavelengths where a self-mode locking has been obtained.

The wavelengths of the above mentioned lasers cover a wide spectral range 1055-1540 nm. This is unique for a passive Q-switch. In case of dyes application for a fast passive Q-switching a particular dye is needed for each laser because an absorption spectral width of these dyes is usually less than 100 nm. On the base of our experiments we could conclude that the SWNT liquid suspension is a new thermo-stable and wide-range material for passive Q-switching in solid state lasers.

ACKNOWLEDGMENTS

Authors are sincerely grateful to Ursula Dettlaff-Weglikowska for offering the raw HiPCO material, A.L.Mulyukin for ultracenrifugation of suspension, G.A.Bufetova for measurement of the optical transmission spectra of SWNT, S.E.Mosalva for a help during experiments and Russian Foundation for Basic Research for financial support in frames of projects # 03-02-17316, # 04-02-17618, and # 04-02-08017 ofi-a.

REFERENCES

1.   M.J. O'Connell, S.M.Bachilo, C.B.Huffman et al., *Science* **297**, 593 (2002).
2.   Avouris P. Acc. *Chem. Res.*, **35**, 1026 (2002).
3.   S.M.Bachilo, M.S.Strano, C.Kittrell et al., *Science*, **298**, 2361 (2002).
4.   Hagen A., Hertel T., *Nano letters*, **3**, 383 (2003).
5.   M.S.Strano, C.A.Duke, M.L.Usrey, P.W.Barone, M.J.Allen, H.Shan, C.Kittrell, R.H.Hauge, J.M.Tour, R.E.Smalley, *Science*, **301**, 1519 (2003).
6.   E.D.Obraztsova, M.Fujii, S.Hayashi, A.S.Lobach, I.I.Vlasov, A.V.Khomich, V.Yu.Timoshenko, W.Wenseleers, E.Goovaerts, *Proc. of NATO ASI "Nanoengineered Nanofibrios Materials"*, Turkey, September 2003, Kluwer (2004) pp. 389-398.
7.   M.S.Dresselhaus, G.Dresselhaus, P.S.Eklund, "Science of Fullerenes and Carbon Nanotubes", Academic Press, San Diego, CA, 1996.
8.   M.J.Bronikowski, P.A.Willis, D.T.Crobert et al., *J. Vac. Sci, Technol.* **A19**, 1800 (2001).
9.   R.Saito, M.S.Dresselhaus, G.Dresselhaus, "Physical Properties of Carbon Nanotubes", Imperial College Press, London, 1998.
10. R.Kempa, *Phys.Rev.* **B 66**, 195406 (2002).
11. Hagen A., Moos G., Talalaev et al., *Appl. Phys.* **A 78**, 1137 (2004).
12. G.N. Ostojic, S. Zaric, J. Kono et al., *Phys. Rev. Letts.*, **92**, 117402 (2004).
13. J.-P. Yang, M.M. Kappes, H. Hippler et al., *Phys. Chem. Chem. Phys.*, **7**, 512 (2005).
14. S.Y.Set, H..Yaguchi, Y. Tanaka et al., *Book of Abstracts of OFC'03*, Atlanta, USA, 2003, # PDP44.
15. S. Yamashita, Y. Inoue, S. Maruyama et al., *Opt. Lett.*, **29**, 1581 (2004).
16. N.N.Il'ichev, E.D.Obraztsova, S.V.Garnov, S.E.Mosaleva., *Quantum Electronics*, **34**, 572 (2004).
17. N.N.Il'ichev, E.D.Obraztsova, P.P.Pashinin, V.I.Konov, S.V.Garnov., *Quantum Electronics*, **34**, 785 (2004).
18. P.Nikolaev, M.J.Bronikovski, R.K.Bradley et al., *Chem. Phys. Lett.*, **313**, 91 (1999).

Single Walled Carbon Nanotubes as Active Elements in Nanotransducers

Christoph Stampfer*, Alain Jungen* and Christofer Hierold*

*Micro and Nanosystems, Department of Mechanical and Process Engineering, ETH Zurich,
Tannenstrasse 3, CH-8092 Zurich, Switzerland.

Abstract. We present the fabrication and characterization of a nanometer-scale electromechanical transducer based on individual single walled carbon nanotubes (SWNTs). Each nano electromechanical device incorporates one electrically connected suspended SWNT, which is placed underneath a freestanding metal bridge. By introducing a mechanical load to the freestanding bridge (e.g. by pushing with an AFM tip) the suspended SWNT is mechanically deformed (i.e. local deformation at the edges and global strain in the branches) leading to a remarkable change in resistance of the SWNT. We present mechanical and electromechanical measurements and show that the resistance changes significantly under mechanical load.

Keywords: Carbon Nanotubes, Nanosystems, Nanosensors, NEMS, Nanomechanics.
PACS: 61.46.+w, 62.25.+g, 73.63.Fg

INTRODUCTION

Single walled carbon nanotubes (SWNTs) are, due to their extraordinary electrical, mechanical and electromechanical properties [1], very promising candidates for highly sensitive elements in future nanosensors and nanosystems. SWNTs withstand very high current densities (10^9 A/cm^2 [2]), are highly elastic (the Youngs modulus is in the range of 1TPa [2]) and show strong electromechanical effects: Several experiments [3, 4, 5] have shown recently that mechanical deformations (bending, stretching and kinking) of a suspended SWNTs leads to a remarkable change of conductance (effective piezoresistive gauge factors of up to 1000 [5]). Moreover, first carbon nanotube based nanosystems (e.g. switches and sensors) have been presented recently [6, 7] showing that novel device concepts at nanoscale become feasible.

In this paper we focus on a simple configuration acting as a possible SWNT based force sensor, which allows the study of the electromechanical response of SWNTs integrated in a nano electromechanical system (NEMS). The presented nano transducer consists of an electrically connected suspended SWNT, which is placed underneath a freestanding bridge (Fig. 1b) or a cantilever (Fig. 1c). Deflecting the bridge or the cantilever by an external force leads to mechanical deformation of the underlying suspended SWNT. This finally leads to a significant change in conductance of the SWNT, which can be measured electrically. The presented functional building blocks are actuated by using an atomic force microscope (AFM)-tip to introduce the required external out-of-plane force (see Fig. 1c). However, the AFM tip is not required in a real nanosensor configuration, as shown in Fig. 1a. A suspended SWNT is mounted underneath a freestanding support spring, which is connected to a suspended disc with a seismic mass m.

CP786, *Electronic Properties of Novel Nanostructures*, edited by H. Kuzmany, J. Fink, M. Mehring, and S. Roth
© 2005 American Institute of Physics 0-7354-0275-2/05/$22.50

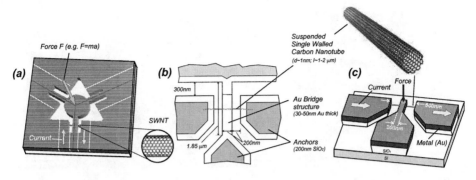

FIGURE 1. (a) Concept study of a SWNT based accelerometer. (b,c) Schematic illustration of the fabricated SWNT based nano-electromechanical system: (b) bridge and (c) cantilever based configuration.

Introducing an external (out-of-plane) force F onto the disc (e.g. due to an acceleration a; thus, $F = ma$) leads to a deflection of the support springs and finally to a mechanical deformation of the SWNT itself, which can be detected electrically. Note that this is still a purely conceptual image of a SWNT based accelerometer and has not been fabricated and tested experimentally yet. In the next section we show by investigating a SWNT bridge based nanotransducer that the basic working principle of these devices can be confirmed.

AU BRIDGE BASED NANOTRANSDUCER

We fabricated suspended Au bridges with and without SWNTs (Fig. 1b). The fabrication technique involves the adsorption of SWNTs (dispersed HiPco-SWNTs in SDS) onto SiO$_2$ with pre-patterned reference alignment markers. AFM images are recorded to determine the relative location and orientation of the SWNTs, e-beam lithography is subsequently used to define the nanotransducer and HF etching followed by critical point drying releases the structures. For a more detailed fabrication description please refer to Ref. [8]. In Fig. 2 we show the final bridge based device with a 200nm wide and 1.85μm long suspended Au bridge with a clamped SWNT. The suspended SWNT can clearly be seen in Fig. 2a, whereas Fig. 2b, which shows the same device under an angle of 15 degree, highlights the underetch (see marked etch edge) of the released Au bridge.

Characterization

For mechanical and electromechanical characterization of the presented device an AFM (Asylum MFP 3D) has been used to introduce a mechanical load onto the suspended Au bridges. Pure mechanical investigations have been performed by AFM based nano beam deflection [2] experiments in the small deflection regime. A series of force versus deflection (FvD) measurements [2] have been performed as a function of the contact point x (see inset in Fig. 3a; the origin of the coordinate x is set in the center of the

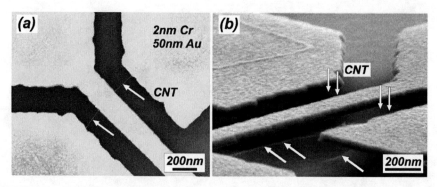

FIGURE 2. SEM image of the fabricated single walled carbon nanotube based NEMS. (a) Top view of the structure: the SWNT is indicated by the with arrows. (b) SEM image under an angle of 15 degree where the under etching can be seen.

bridge), which is defined as the point where the AFM tip deflects the bridge. In Fig. 3a we show the measured deflection Δz of the Au bridge based systems as a function of the contact point x at the constant pushing force $F_p = 30$ nN. The maximum deflection Δz of the pure Au bridge (without SWNT; gray triangles) is in the range of 15 nm; whereas the maximum deflection of the system with integrated SWNT (circles) is found to be in the range of 5nm [9]. This shows that the SWNT is stiffening the mechanical system significantly. Since we are in the small deflection regime, we can assume that the simple elastic beam theory (Euler-Bernoulli Theory of beams [10]) can be applied for the bending problem. In this framework the bridge (without SWNT) deflection Δz can be expressed as: $\Delta z(\tilde{x} = x + l/2) = F/(EI_x)\left[1/(2l^2) - (l + 2\tilde{x})/(6l^3)\right](l - \tilde{x})^2\tilde{x}^3$, where E is the Youngs modulus, $I_x = wt^3/12$ the polar moment of inertia and l is the length of the bridge. By evaluating this equation at $l = 2.1\mu m$ (underetch has been considered), $w = 200$ nm, $t = 41$ nm and $E_{Au} = 78$ GPa we find good agreement with the experimental results (see solid line in Fig. 3a). In Figs. 3b-d we show electromechanical measurements of the suspended SWNT placed underneath the Au bridge as a function of maximum deflection Δz_{Max} in the center of the bridge. The base resistance of the SWNT ($d \approx 1nm$) between electrode A and B (see inset in Fig. 3a) is measured to $R \approx 400k\Omega$ and the IV-characteristic in the range of $V_{sd} = [-60, +60mV]$ showed linear behavior. Note, that in all three cases (Figs. 3b-d) the resistance change is reversible, making the system suitable for force or displacement sensing purposes. Moreover we see in Figs. 3b-d that the maximum change in resistance increases significantly as a function of the max. deflection, which agrees with Ref. [3]. However, an analytic description of the correlation between bridge deflection, SWNT deformation and their electrical response is still missing.

CONCLUSION

We presented a brief description of the fabrication process for SWNT based nano electromechanical structures. The feasibility of the SWNT Au bridge based transducer

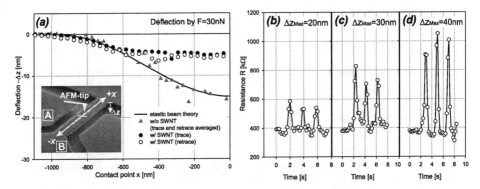

FIGURE 3. (a) AFM based deflection (Δz) versus contact point x (i.e. force vs. displacement) measurement performed on the Au bridge (inset) with (circles) and without (triangles) integrated SWNT. The solid line plots the result of the elastic beam theory (see text) for the system without SWNT. (b-d) Electromechanical measurements: Resistance of SWNT as a function of time (deflection); each peak corresponds to continuous (2sec for up and down) deflection of the AFM cantilever at a fixed contact point $x = 0$ for three different max. bridge deflections Δz_{Max}.

is assessed from a mechanical and electrical point of view and a new concept of deformation based nanoscaled sensors has been shown. The force sensing mechanism is due to the reversible deformation of a suspended SWNT. We showed SEM images of the fabricated device and electromechanical measurements prove the basic functionality.

ACKNOWLEDGMENTS

The authors wish to thank D. Obergfell, R. Leturcq, T. Helbling, F. Robin, D. Junker and S. Vuillmers for helpful discussions. Support by the ETH FIRST Lab and financial support by the ETH Zurich (TH-18/03-1) are gratefully acknowledged.

REFERENCES

1. S. Reich, C. Thomsen, J. Maultzsch, "Carbon Nanotubes", Wiley-VCH, 2003.
2. B. Bhushan, "Handbook of Nanotechnology", Springer, pp. 66, 2004.
3. T.W. Tombler, et al., "Reversible electromechanical characteristics of carbon nanotubes under local-probe manipulation", Nature, vol. 405, pp. 769-772, 2000.
4. E. D. Minot, Y. Yaish, V. Sazonova, J.-Y. Park, M. Brink, and P. L. McEuen, "Tuning carbon nanotube band gaps with strain", Phys. Rev. Lett., vol. 90, pp. 156401-156404, 2003.
5. J. Cao, Q. Wang and H. Dai, "Electromechanical properties of metallic, quasimetallic, and semiconducting carbon nanotubes under stretching", Phys. Rev. Lett., vol. 90, pp. 157601-157604, 2003.
6. S.N.Cha et al., "Fabrication of a nanoelectromechanical switch using suspended carbon nanotube", Applied Physics Letters, Vol. 86, pp. 083105 1-3, 2005.
7. A.M. Fennimore, et al., " Rotational actuators based on CNTs", Nature, vol. 424, pp. 408-410, 2003.
8. C. Stampfer, A. Jungen and C. Hierold, " Fabrication of discrete carbon nanotube based nano-scaled force sensors", IEEE Sensors 2004, T4L-F.5, pp. 1056-1059, 2004.
9. The contact point of the SWNT to the Au brigde has been measured to $x_{SWNT} \approx +70nm$.
10. A. N. Cleland, "Foundations of Nanomechanics", Springer, 2003.

Field Emission from Multi-Walled Carbon Nanotubes – Its applications in NEMS

M. Sveningsson, S. W. Lee[#], Y. W. Park[#], E. E. B. Campbell

Department of Physics, Göteborg University, SE-412 96 Göteborg Sweden.
[#]School of Physics and Nano Systems Institute-National Core Research Center, Seoul National University, Seoul 151-747, Korea.

Abstract. Nanoelectromechanical systems (NEMS) are a rapidly growing area of research. We discuss here the fabrication of two different three terminal nanorelay structures based on multi-walled carbon nanotubes (MWNT). Using a gate voltage the whole MWNT can be deflected and the tip of the carbon nanotube can be moved closer to a drain electrode making physical contact. In another device structure the carbon nanotubes are shorter so they never can get real physical contact with the drain electrode, instead this device is based on the process of electron field emission from the MWNT.

INTRODUCTION

The basic idea underlying NEMS is the strong electromechanical coupling in devices on the nanometer scale in which the Coulomb forces associated with device operation are comparable with the chemical binding forces. Carbon nanotubes [1] are almost ideal materials in order to build future NEMS, they have low mass, high Young's modulus and can be considered to be 1-dimensional objects due to their high aspect ratio. NEMS based on carbon nanotubes have already experimentally been demonstrated, and some examples are: nanotweezers [2, 3], sensors [4, 5], nano-rotors [6], two terminal switches [7] and lately also three terminal structures by Sazonova et al. [8] and by our group [9]. The structure we are looking into is theoretically described by Kinaret et al. [10] and Jonsson et al. [11] and has been shown to potentially have internal operating frequencies in the GHz regime, making it suitable for various applications.

The basic system, illustrated in Fig. 1d, consists of a conducting MWNT placed on a patterned Si substrate. A source electrode is connected to one end of the tube while the other end is free to move over a drain electrode. A gate is also positioned beneath the MWNT. The electrostatic force due to the gate capacitance can control the position of the tube and in the ideal case be used to switch between an ON state (closed circuit) and OFF state (open circuit). So, by applying a small source-drain voltage to this structure no current will flow in the device unless a gate voltage is applied and starts bending the tube so that the tip comes into contact with the drain electrode.

CP786, *Electronic Properties of Novel Nanostructures*, edited by H. Kuzmany, J. Fink, M. Mehring, and S. Roth
© 2005 American Institute of Physics 0-7354-0275-2/05/$22.50

EXPERIMENTAL

The method used for fabricating the three terminal relay structures is similar to a method reported by Lee et al. [12], where suspended SWNTs were produced. The electrode material is gold and, to support the deposited nanotubes, PMMA was spin-coated on top of the gate and drain electrodes and oxygen plasma ashed to be the same height as the source electrode. The MWNTs were aligned and positioned with the ac-electrophoresis technique [13]. A scheme of how the samples were fabricated is seen in Fig. 1.

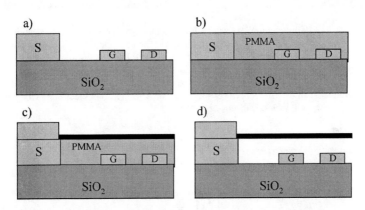

FIGURE 1. The fabrication of the nanorelays is a multiple electron beam lithography process. a) First the source (S), gate (G) and drain (D) electrodes are fabricated on SiO_2. Later PMMA is spin-coated (b) and tubes are aligned and clamped (c) with another top layer on the source. Finally the PMMA is carefully removed leaving a suspended MWNT [9].

RESULTS AND DISCUSSION

A scanning electron microscope (SEM) image of a successful fabrication of the relay is given in Fig. 2a. A more careful examination of structures like these is presented by Lee et al. [9]. These devices were investigated by measuring the source-drain (S-D) current as a function of the gate voltage while applying a S-D bias of 0.5 V. In [9] we report about mainly two different outcomes of these measurements. The first example shows a non-linear current as the gate voltage is increased with a small hysteresis effect as the voltage is decreased to zero again. This behavior was reproduced several times and for several samples. A second sample showed the same behavior for low gate voltages but at around 5 V the current rapidly increased by a factor of 200, see Fig. 3. At this point the MWNT has made a good physical contact with the drain and in this device the tube was permanently stuck to the drain electrode which meant that the S-D current could no longer be controlled by changing the gate voltage.

FIGURE 2. left: 2a) Carbon nanotube based relay structures. A MWNT is clamped with Au electrodes at one end (source) while the other end is suspended over the gate and drain electrode. Here the MWNT is very long and extends beyond the drain electrode to the left. In this SEM picture one can also observe a shorter tube that does not extend all the way to the drain electrode. A single structure like this is shown to the right 2b). This tube does not reach the drain electrode and this will solve the problem that the carbon nanotubes, in the first structure, sometime stick to the drain electrode. This device is based on electron field emission, which is discussed at the end of this paper.

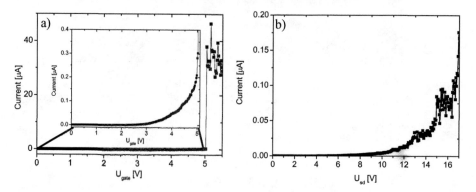

FIGURE 3. a) Source drain current as a function of gate voltage for a carbon nanotube relay [9]. At 5 V there is a high increase in the current which is a result of physical contact with the drain electrode. The carbon nanotube was permanently stuck to the drain after this and the relay was non-operable. b) Source-drain current as a function of s-d voltage for the field-emission configuration of the nanotube relay.

The second type of structure is a three-terminal device that operates in the field emission regime. Field emission is a process by which electrons are emitted from a surface (i.e. tip of MWNT) due to an applied electric field resulting in a high local field at the emitter tip due the tubes high aspect ratio. A prototype structure of this device is displayed in Fig. 2b. This field emission based relay is basically the same as the previous geometry but with a shorter suspended carbon nanotube. The first consequence of having a horizontal separation between the drain and the tip (see Fig. 2b) of the tube is a smaller capacitive coupling between the tube and drain contact. A second consequence is that the nanotube never gets into physical contact with the drain electrode. The operation conditions are slightly different with this configuration. A larger source-drain voltage is needed since we are working in a field emission regime. A high voltage will, on the other hand, give a large capacitive force between the drain and nanotube, which might result in worse control of the vertical position of the tube by changing the gate voltage. Therefore it is important to apply a small S-D bias in this configuration, but still high enough to get electron field emission from the carbon nanotube. First results with this configuration do indeed show the expected relay behavior, Fig. 3b.

ACKNOWLEDGEMENTS

This work was financial supported by the Swedish Foundation for Strategic Research (SSF) within the "CMOS integrated carbon-based nanoelectromechanical systems" and EC FP6 funding (contract no. FP6-2004-IST-003673, CANEL). This publication reflects the views of the authors and not necessarily those of the EC. The Community is not liable for any use that may be made of the information contained herein. The Korean co-authors were supported by a STINT collaboration project between Göteborg University and Seoul National University and the NSI-NCRC program of KOSEF and BK21 of MOE, Korea. The authors acknowledge fruitful discussions with J. M. Kinaret and L. M. Jonsson.

REFERENCES

1. R. Saito;G. Dresselhaus, M. S. Dresselhaus, Physical Properties of Carbon Nanotubes; *Imperial College Press:* London, (1998).
2. P. Kim, C. M. Lieber, *Science*, **286**, 2148. (1999).
3. S. Akita, Y. Nakayama, S. Mizooka, Y. Takano; T. Okawa, Y. Miyatake, S. Yamanaka, M. Tsuji; T. Nosaka, *Appl. Phys. Lett.*, **79**, 1691. (2001).
4. P. G. Collins, K. B. Bradley, M. Ishigami, A. Zettl, *Science*, **287**, 1801, (2000).
5. C. K. W. Adu, G. U. Sumanesekera, B. K. Pradhan, H. E. Romero, P. C. Eklund, *Chem. Phys. Lett.*, **337**, 31. (2001)
6. A. M. Fennimore, T. D. Yuzvinsky, W.-Q. Han, M. S. Fuhrer, J. Cumings, A. Zettl, *Nature*, **424**, 6947, 408-410, (2003)
7. Dequesnes, M.; Rotkin, S. V.; Aluru, N. R, *Nanotechnology*, **13**, 120. (2002)
8. V. Sazonova Y. Yaish H. Uestuenel D. Roundy T. A. Arias P. L. McEuen, *Nature*, **431**, 7006, 284-287 (2004).
9. S. W. Lee, D. S. Lee, R. E. Morjan, S. H. Jhang, M. Sveningsson, O. A. Nerushev, Y. W. Park, E. E. B. Campbell, *Nano Lett.* 4(10), 2027-2030 (2004)
10. Kinaret, J. M.; Nord, T. Viefers, S. *Appl. Phys. Lett.*, **82**, 1287 (2003).
11. Jonsson, L. M.; Nord, T.; Viefers, S.; Kinaret, J. M. *J. Appl. Phys.*, **96**, 629 (2004).
12. Lee, S. W.; Lee, D. S.; Yu, H. Y.; Campbell, E. E. B.; Park, Y. W. *Appl. Phys. A*, **78**, 283 (2004).
13. Chen, X. Q.; Saito, T.; Yamada, H.; Matsushige, K. *Appl. Phys. Lett.*, **78**, 3714 (2001).

WORKSHOP ON "APPLICATION OF MOLECULAR NANOSTRUCTURES"

Workshop on "Application of Molecular Nanostructures"
Tuesday, March 15, 2005

In order to stress the importance of molecular nanostructures for application a special workshop was organized on Tuesday, March 15, 2005 as part of the Winterschool. It consisted of invited oral contributions, reports from companies, discussions, and contributing posters.

The workshop demonstrated the high application potential of carbon nanophases. A substantial part of the reported contributions in the mini-workshop came from start up companies which documents the importance of the field in high technology industrial development.

Titles of oral contributions and reports of companies together with authors and affiliation are listed below.

Invited oral lectures:

J. Robertson, Cambridge, GB: *Application of carbon nanophases (tutorial)*
M. Endo, Shinshu, J: *Large scale synthesis of carbon nanotubes by catalytic CVD*
A. Hirsch, Erlangen, D: *Fullerene derivatives for medical applications*
Strano, Illinois, US: *Solution phase, near-infrared optical sensors based on single walled Carbon Nanotubes*
Freitag, Yorktown Hights, US: *Electro-optics with carbon nanotubes*
M. Haasch, Missisipi, US: *The environmental toxicology of manufactured nanoparticles in aquatic species*
P. Pötschke, Dresden, D: *Purification and percolation – unexpected phenomena in nanotube polymer composites*
R. Baughman, Dallas, US
    Drawn-twist-spun multiwalled nanotube yarns for artificial muscle and other multifunctional applications

Miniworkshop on Scientific supplies and industrial needs:

How can we make money from carbon nanophases?
Discussion

C-sixty, Huston, US;
represented by A. Hirsch: *Fullerenes as neuroprotectants*
FutureCarbon Inc.: Bayreuth, D;
 represented by W. Schütz: *Production of carbon nanomaterials for high technology application*
UK-Ionbond, London, GB;
represented by P. Hatto: *International actions and standardization in nanotechnology*
Mitsui-Bussan Inc., Tsukuba, J;
represented by Y. M. Endo: *Applications with multi wall and single wall carbon nanotubes*
Nanocyl, Namur, B;
represented by A. Fonseca: *Production and Marketing of Carbon Nanotubes*
Nanoledge, Montpelliere, F;
represented by K. Schierholz: *Business opportunities for carbon nanotubes: from nanoscience to nanotechnology*
Samsung, Seoul, COR;
represented by I.T. Han: *Carbon nanotubes for display technology*